GILLIAN MORRIS

Third Edition

# Genetics

## Ursula Goodenough
Washington University

**SAUNDERS COLLEGE PUBLISHING**

Philadelphia   New York   Chicago
San Francisco   Montreal   Toronto
London   Sydney   Tokyo   Mexico City
Rio de Janeiro   Madrid

**HOLT-SAUNDERS JAPAN**

*Address orders to:*
383 Madison Avenue
New York, NY 10017

*Address editorial correspondence to:*
West Washington Square
Philadelphia, PA 19105

Text Typeface: 10/12 Palatino
Compositor: York Graphic Services
Acquisitions Editor: Michael Brown
Project Editor: Carol Field, Diane Ramanauskas
Copyeditor: Bonnie Boehme
Managing Editor & Art Director: Richard L. Moore
Art/Design Assistant: Virginia A. Bollard
Text Design: Caliber Design Planning
Text Artwork: Vantage Art, Inc.
Production Manager: Tim Frelick
Assistant Production Manager: Maureen Iannuzzi

**Library of Congress Cataloging in Publication Data**

Goodenough, Ursula.
   Genetics.

   (Saunders golden sunburst series)
   Bibliography: p.
   Includes indexes.

   1. Genetics. I. Title. [DNLM: 1. Genetics.
QH 430 G649g]
QH430.G66 1983     575.1     83-4438

ISBN 0-03-058212-1

GENETICS

ISBN 0-03-058212-1 (US College Edition)
ISBN 4-8337-0183-9 (Holt-Saunders International Edition)

**Library of Congress catalog card number 83-4438.**

Printed in Japan, 1984

3456  16  98765432

**CBS COLLEGE PUBLISHING**
Saunders College Publishing
Holt, Rinehart and Winston
The Dryden Press

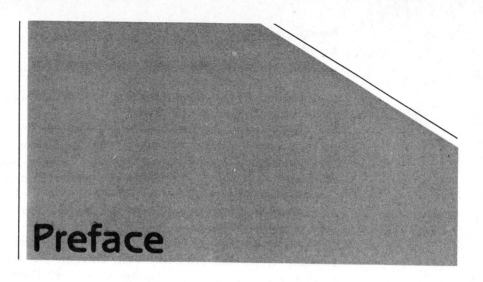

# Preface

Do DNA sequences belong in a genetics textbook? This was the most important question I faced as I undertook writing a third edition of this text in the midst of the cloning/sequencing "revolution." The DNA sequences prevailed, as a glance through the text will indicate. My reasoning can be summarized as follows.

Genetics seems to have been embroiled in an identity crisis since 1953, when genes began to be actively analyzed as molecular units as well as heritable units of function. Scientists who studied the molecular properties of genes came to call themselves molecular biologists, whereas those who continued to study genes as heritable units of function called themselves geneticists. Pedagogical distinctions soon followed: molecular biology textbooks were devoid of mapping functions and epistasis, whereas genetics textbooks rarely mentioned histone phosphorylation or $C_0t$ plots. A dichotomy also developed in teaching styles: whereas undergraduate courses in molecular biology were full of the "latest news," with relatively little attention paid to the development of the field, courses in genetics were usually taught in an historical fashion, with much emphasis placed on the intellectual approaches taken during and since Mendel's original experiments.

I have written this edition with the conviction that these distinctions are already obsolete in practice and should be made obsolete in the classroom. Molecular biologists, now able to clone genes at will and sequence them in weeks, have come knocking on the doors of geneticists to obtain strains with interesting genetic properties. Geneticists, meanwhile, have come to realize that gene fine structure will never again be productively tackled by the heroic genetic approaches of Benzer, Sherman, or Judd, and they are increasingly receptive to the knocks on the door from their molecular colleagues. Population geneticists are using restriction enzymes to detect polymorphisms; evolutionists are discovering that the genomes of modern organisms contain a rich fossil record of their genetic ancestry;

molecular biologists attend conferences on hybrid dysgenesis in *Drosophila* and the dissociator-activator system in maize. With the old distinctions rapidly blurring in the laboratory, why should they continue to be perpetrated on undergraduates?

I usually hear three answers to this question. One is the notion that the molecular material is "too hard," that students won't understand it. This is certainly true for non-science majors, for whom this text is definitely not written, but biology majors take required chemistry courses that provide ample background for understanding the principles of molecular genetics when they are carefully presented. Indeed, it is my experience that most students are able to understand the implications of a DNA sequence far more readily than a coefficient of coincidence, but perhaps other instructors have a different impression. Then there is the intellectual argument. Once students grasp the construction of a brilliant genetic proof, are they not better trained in science? I strongly agree that they are, and the text includes detailed presentations of many "strictly genetic" experiments. But if these experiments are presented in a molecular vacuum, then the training is incomplete: the student emerges steeped in genetic logic but has not been allowed to combine that logic with a knowledge of how genes are constructed and how they recombine, mutate, and express themselves.

Finally, it is argued that students will learn "that stuff" in another course. I agree that such subjects as DNA replication, renaturation kinetics, and sequencing are rightly considered in considerable depth in biochemistry and molecular biology courses. They are presented quite superficially in this text. It hardly follows, however, that genes should not be the focal subject of a genetics course.

In a field moving as rapidly as genetics, it proves impossible to simply "update" a previous edition. Therefore, this is once again an almost entirely new book. Major changes from the second edition include the following:

1. New chapters have been added on somatic cell genetics, transposable elements, and immunogenetics, three of the most rapidly advancing fields in the past five years.
2. Such topics as mutation, recombination, chromosome structure, and gene structure are each presented in two chapters, the first giving a general overview and the second a more detailed molecular view (*see* Table of Contents). This should allow instructors more flexibility in determining the "molecular depth" of their assignments.
3. Population and quantitative genetics have been expanded considerably.
4. By popular request, Mendel's experiments are included *per se*.

I have been fortunate to obtain excellent critiques on the previous and current editions. Particular thanks go to Drs. Julian Adams (University of Michigan), Fred Allendorf (University of Montana), Alan Atherly (Iowa

State), C. William Birky (Ohio State), Barbara Brownstein (Temple University), Darrel Falk (Syracuse University), Peter Kuempel (University of Colorado), Gustavo Maroni (University of North Carolina), H. James Price (Texas A & M University), W. Stuart Riggsby (University of Tennessee), Raymond Rodriquez (University of California, Davis), Carol Sibley (University of Washington), Richard Siegal (U.C.L.A.), Edward Simon (Purdue University), Millard Susman (University of Wisconsin), Eric Weinberg (University of Pennsylvania), and Herbert Wiesmeyer (Vanderbilt University) for their overall reviews and to Drs. Douglas Berg (Washington University), Richard Borowsky (New York University), Nam-Hai Chua (Rockefeller University), Jeffrey Davidson (Eleanor Roosevelt Institute, Denver), W. J. Dickinson (University of Utah), John Drake (National Institute of Environmental Health Sciences), Sarah Elgin (Washington University), Nancy Martin (University of Texas, Dallas), Janice Pero and Vicki Sato (Harvard University), and Christopher Woodcock (University of Massachusetts, Amherst) for their individual chapter reviews. Michael Brown, Biology Editor at Saunders, provided important help at many stages in the production process.

This edition was written almost entirely in the serenity of Chilmark, Massachusetts, in the loving presence of my four children—Jason, Mathea, Jessica, and Thomas—and my husband, John Heuser. The book is the product of the peace of mind, and hence the clarity of thought, that they all bring to me.

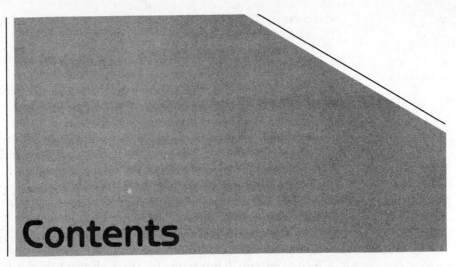

# Contents

The 20 chapters displayed in black present the "core material" of present-day genetics; the 8 chapters displayed in gray offer detailed presentations of topics in molecular genetics.

Preface    iii

## 1 DNA (and RNA) as the Genetic Material in Chromosomes    1

Introduction    1
The Requirements to Be Met by Genetic Material    2
The Structure of DNA and RNA    4
Relating DNA Structure to Its Genetic Requirements    14
Experiments Indicating DNA and RNA as the Genetic Material    17
*Questions and Problems*    23

## 2 Cell Cycles, Chromosome Duplication, and Mitosis    25

Introduction    25
The Bacterial Cell Cycle and Chromosome Replication    26
The Eukaryotic Cell and Cell Cycle    34
The Karyotype    48
Atypical Eukaryotic Cell Cycles    57
*Questions and Problems*    60

# 3 Molecular Organization of Chromosomes        62

Introduction        62
Protein Structure        63
Chromosomal Proteins        73
The 10 nm Nucleosome Filament of Chromatin        74
Higher Orders of Chromatin Organization        81
*Questions and Problems*        92

# 4 Molecular Analysis of Chromosomal DNA and Genetic Engineering        93

Introduction        93
Overall Composition of Genomic DNA        94
Reassociation Kinetics ($C_0T$ Plots) of Genomic DNA        98
The Kinetic Classes of Eukaryotic DNA        102
Restricting, Sequencing, and Cloning DNA        108
*Questions and Problems*        122

# 5 The Meiotic Transmission of Chromosomes        124

Introduction        124
Meiosis        125
Life Cycles of Sexually Reproducing Organisms: Mitosis-Meiosis
        Alternations        137
Meiotic Errors        145
*Questions and Problems*        153

# 6 Mendelian Inheritance of Genes Carried by Autosomes and Sex Chromosomes        153

Introduction        154
Mendel's Experiments        154
Segregation of Alleles        159
Independent Assortment        170
Sex-Linked Inheritance        182
*Questions and Problems*        196

## 7 Mutation: Induction and Detection of Mutant Organisms and Chromosomes    201

Introduction    201
Screening Procedures    204
Characterizing Mutant Karyotypes    212
Mutagens, Clastogens, and Carcinogens    220
*Questions and Problems*    227

## 8 DNA Replication and Repair Mechanisms and Their Contribution to Mutagenesis    230

Introduction    230
DNA Replication Mechanisms    231
Direct Mutagenesis Mechanisms    237
Repair and Misrepair Mechanisms    245
*Questions and Problems*    255

## 9 Genes and Gene Transcripts: General Features    258

Introduction    258
General Features of Transcription    259
Anatomy of Structural Genes    267
"Split Genes" in Eukaryotes    272
Visualizing Structural Gene Transcription    277
*Questions and Problems*    282

## 10 Genes and Gene Transcripts: Specific Genes    285

Introduction    285
Transfer RNA Genes    286
Ribosomal RNA Genes    294
Structural Eukaryotic Genes    300
*Questions and Problems*    306

# 11 Structural Gene Expression and the Genetic Code    308

Introduction    308
Protein Synthesis    309
"Cracking" the Genetic Code    317
Nonsense Mutations and Chain Termination    330
Suppressor Mutations    332
*Questions and Problems*    *336*

# 12 Mapping Viral Chromosomes    340

Introduction    341
Viral Infection Cycles    341
Complementation Analysis    348
Recombination-Frequency Mapping    352
Deletion Mapping    367
Mapping without Recombination    375
Approaches to Solving Mapping and Complementation
    Problems    385
*Questions and Problems*    *389*

# 13 Mapping Bacterial Chromosomes and Plasmids    399

Introduction    400
Molecular Overview of Bacterial Conjugation    400
Mapping by Bacterial Conjugation    405
Bacterial Transformation    413
Generalized Transduction    417
Specialized Transduction    421
Plasmids    425
Approaches to Solving Mapping Problems    432
*Questions and Problems*    *433*

# 14 Mapping Eukaryotic Chromosomes in Sexual Crosses    437

Introduction    438
Classical Studies on Linkage and Recombination    438
Mapping *Drosophila* in Sexual Crosses    444
Cytological Mapping    450
Linkage Groups and Chromosomes    456

Mapping by Tetrad Analysis     459
Approaches to Solving Mapping Problems in Eukaryotes     472
*Questions and Problems     475*


## 15  Somatic Cell Genetics     482

Introduction     482
The Parasexual Cycle of *Aspergillus*     483
Genetic Analysis of Cultured Somatic Cells     490
Gene Transfer or Eukaryotic Transformation     498
The Human Chromosome Map     503
*Questions and Problems     507*


## 16  Extranuclear Genetic Systems     511

Introduction     512
Molecular Studies of Mitochondrial Genetic Functions     512
Genetic Analysis of the Yeast Mitochondrial Genome     515
Mitochondrial Genomes of Higher Eukaryotes     523
Chloroplast Genomes     527
Endosymbiosis and the Origins of Organelle Genetic Systems     531
Inheritance of Preformed Structures     533
*Questions and Problems     535*


## 17  General Recombination Mechanisms     538

Introduction     539
Models of General Recombination     539
Enzymes that Mediate General Recombination     547
Formation and Segregations of Physical and Genetic DNA Hybrids
    during General Recombination     550
Mismatch Repair of Heteroduplex DNA during General
    Recombination     554
*Questions and Problems     565*


## 18  Transposition and Mutagenesis by Temperate Viruses and Transposable Elements     569

Introduction     569
Integration and Excision by Temperate Bacteriophages     570
Integration and Excision by Transposable Elements     577
*Questions and Problems     589*

# 19  Related Genes: Alleles, Isoloci, and Gene Families   591

Introduction   591
Traits Controlled by a Single Gene Locus   592
Isozymes Specified by Isoloci   604
Gene Families   606
Distinguishing Alleles, Isoloci, and Gene Families in Genetic Crosses   613
*Questions and Problems*   615


# 20  Immunogenetics   618

Introduction   619
Properties of the Immunoglobins   619
Construction and Expression of Light-Chain Genes   625
Construction and Expression of Heavy-Chain Genes   634
Somatic Mutation of Antibody Genes, and a Summary of Antibody
    Diversity Mechanisms   637
*Questions and Problems*   640


# 21  Genes that Cooperate to Produce Complex Phenotypes and Quantitative Traits   642

Introduction   642
Clustered Genes Specifying One Trait   643
Dispersed Genes Specifying One Trait   652
Biochemical Genetics   654
Polygenes and Continuous Variation   664
*Questions and Problems*   675


# 22  Control of Gene Expression in Bacteria   681

Introduction   682
General Features of Gene Regulation   682
Regulation of Lactose Utilization   687
Regulation of Tryptophan Biosynthesis   697
Translational Control   704
*Questions and Problems*   705

## 23 Control of Gene Expression in Bacteriophages and Eukaryotic Viruses   709

Introduction   709
Regulation of Gene Expression by Lytic Bacteriophages   710
Regulation of Gene Expression during Phage λ Infection   713
Gene Regulation during SV40 Infection   723
*Questions and Problems*   726

## 24 Control of Gene Expression in Eukaryotes: Short-Term Regulation   728

Introduction   728
Short-Term Regulation in Fungi   729
Short-Term Regulation in Higher Eukaryotes   733
Mechanisms of Short-Term Regulation in Eukaryotes   739
*Questions and Problems*   742

## 25 Control of Gene Expression in Eukaryotes: Long-Term Regulation   745

Introduction   746
General Features of Long-Term Differentiation   746
The Differentiation of the Egg and Maternal Influences on
   Development   753
Development Genetics of *Drosphila*   760
Developmental Genetics of Vertebrates   774
Differential Expression of Hemoglobin Genes   783
*Questions and Problems*   788

## 26 Population Genetics I: General Principles and Mendelian Populations   792

Introduction   792
General Principles of Population Genetics   793
Genetic Variability in Populations   797
Mendelian Populations   802
*Questions and Problems*   814

## 27 Population Genetics II: Evolutionary Agents    819

Introduction    820
Fitness    820
Selection    823
Migration    842
Random Drift in Small Populations    846
The Contributions of Selection and Drift to Polymorphism    851
*Questions and Problems*    855

## 28 Populations Genetics III: Speciation and Molecular Evolution    858

Introduction    858
Speciation    859
Molecular Evolution    864
*Questions and Problems*    875

## Boxed Reference Material

CsCl Gradient Centrifugation    30
Autoradiography    32
Antibodies and Antibody Testing    45
Staining and Banding Chromosomes    53
Chromotography    67
Gel Electrophoresis    69
Properties of Enzymes Relevant to Molecular Genetics    79
DNA Sequencing    114
Sucrose Gradient Centrifugation    273
Southern Blot Hybridization    381

## Index    879

# 1 | DNA (and RNA) as the Genetic Material in Chromosomes

A. **Introduction**

B. **The Requirements to Be Met By Genetic Material**

C. **The Structure of DNA and RNA**
1.1 The Polynucleotide
1.2 The Double Helix

D. **Relating DNA Structure to Its Genetic Requirements**
1.3 DNA as a Coded Molecule

1.4 DNA Replication
1.5 DNA Expression
1.6 DNA Variation

E. **Experiments Indicating DNA and RNA as the Genetic Material**
1.7 Transformation Experiments
1.8 Hershey-Chase Experiments
1.9 Experiments with RNA Viruses

F. **Questions and Problems**

## Introduction

Today, when the terms **gene** or **genetics** are mentioned, most biologists immediately think of **DNA**. DNA, or **deoxyribonucleic acid,** is well known as the chemical bearer of genetic information; **RNA (ribonucleic acid)** serves this function in certain viruses.

In the history of genetics as a science, DNA became the center of attention only relatively recently. Focus first centered on **heredity,** on the patterns of inheritance of a given trait (blue eyes, red flower color, short tail) from parent to offspring. It was postulated that these inherited traits were somehow dictated by genes and that genes were linearly arranged along the chromosomes of higher animals and plants. "Maps" of gene order on chromosomes were constructed, and many of the details of gene transmission from generation to generation were worked out well before much was known about what a gene is and how it acts.

As the science of genetics developed, increased attention was given to how genes function, and more experimental use was made of microorganisms, notably bacteria and bacterial viruses. During this period it was proposed, with good evidence, that **the function of most genes is to specify the formation of proteins.** When it was eventually established that most genes are borne within molecules of DNA, primary attention was given to the chemical nature of the gene itself.

In beginning our text with DNA and RNA and in developing a molecular picture of genes and gene function at the same time as we establish patterns of heredity, we are, in one sense, violating the sequence set by scientific history. In another sense, of course, we are more closely following evolutionary history, since genes almost certainly developed their fundamental properties well before the hereditary patterns exhibited by modern organisms were established.

# The Requirements to be Met by Genetic Material

Certain requirements must be met by any molecules if they are to qualify as the substances that transmit genetic information. These requirements extend directly from what is known about the continuity of species and the process of evolutionary change.

1. **Genetic material must contain biologically useful information that is maintained in a stable form.**
2. **Genetic information must be reproduced and transmitted faithfully** from cell to cell or from generation to generation.
3. **Genetic material must be able to express itself** so that other biological molecules, and ultimately cells and organisms, will be produced and maintained. Implicit in this requirement is that some mechanism be available for decoding, or translating, the information contained in the genetic material into its "productive" form. A narrow, but important, distinction is thus made between a molecule that can generate only its own kind and a molecule that can also generate new kinds of molecules. A salt crystal can "seed" a salt solution so that new salt crystals are formed, but this is the extent of its influence over its surroundings.
4. **Genetic material must be capable of variation.** This requirement is somewhat contradictory to the first requirement, which demanded stability of the genetic material. There is, in fact, no *a priori* reason why genetic material should have built-in provisions for change; one could certainly design a hypothetical genetic system in which information would be rigidly conserved from one generation to another. The dominant theme in the history of life is, however, organic evolution, and

this demands that genetic material be capable of change, if only infrequently.

Two sources of change have been recognized in present-day genetic systems: **mutations** and **recombination.** A mutation changes the nature of the information transmitted from parent to offspring and thus represents a relatively drastic way of bringing about variation. If the change is deleterious (and it usually is), the offspring may be greatly handicapped and may die soon after conception, or else it may introduce a deleterious gene into the population. Recombination is a more moderate way of producing variation. It occurs during the course of some sort of sexual process, and it involves the precise shuffling of parental genetic information such that new combinations of genes are produced. These are then inherited by the offspring.

With these four requirements in mind, we can study the physical and chemical properties of DNA and RNA, putting the molecular facts into a genetic context. Table 1-1 reviews some key definitions from organic chemistry that are relevant to the next few sections of this chapter.

**TABLE 1-1   Definitions from Organic Chemistry**

| | |
|---|---|
| 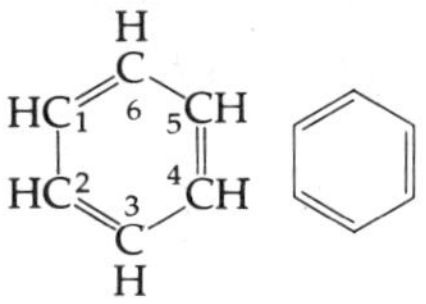 | Benzene ring, with = indicating double bonds where two carbon atoms share four electrons between them. The common abbreviated version of a benzene ring is shown at right. |
| $-CH_3$ | Methyl group |
| $-OH$ | Hydroxyl group |
| $=O$ | Keto group |
| | Aldehyde group |
| | Carboxyl group, characteristically acidic ($-COO^-$) |
| $-NH_2$ | Amino group, characteristically basic ($-NH_3^+$) |
| Covalent bond | A strong bond formed when two atoms share a pair of electrons between them. |
| Hydrogen bond | A weak attractive force between an electronegative (electron-seeking) atom (usually N or O) and a hydrogen atom covalently linked to a second electronegative atom (usually O—H or N—H). |
| Hydrolysis | Breaking a large molecule into two or more smaller molecules by adding water. |

# The Structure of DNA and RNA

## 1.1 The Polynucleotide

DNA and RNA are composed of two different classes of nitrogen-containing bases, the **purines** and the **pyrimidines**. The two most commonly occurring purines in DNA are **adenine** and **guanine,** and the common pyrimidines are **cytosine** and **thymine**. Their structures are shown in Figure 1-1. Thymine is not found in most species of RNA; instead, one finds the pyrimidine **uracil,** which is also shown in Figure 1-1. Modified forms of these bases (5-methyl cytosine, for example) are occasionally found as well, particularly in certain specialized forms of RNA described in Chapter 10.

The purines and the pyrimidines can be seen in Figure 1-1 to contain several conjugated double bonds. Molecules containing such bonds are potentially able to exist in a number of different chemical forms, for their hydrogen atoms have a certain freedom. A hydrogen atom can, for example, move away from an amino group (—NH$_2$), leaving an imino group (—NH) and a net negative charge that is absorbed by the conjugated ring system of the molecule. Such chemical fluctuations are called **tautomeric shifts,** and the different molecular structures that result are called **tautomers**. It turns out that under physiological conditions, the purines and pyrimidines exist almost invariably in the forms that have been drawn in Figure 1-1; the other tautomeric forms of these bases rarely occur. In other

**FIGURE 1-1 Purine and pyrimidine bases.** Molecular weights (mw) are given in dalton units (see Table 1-2).

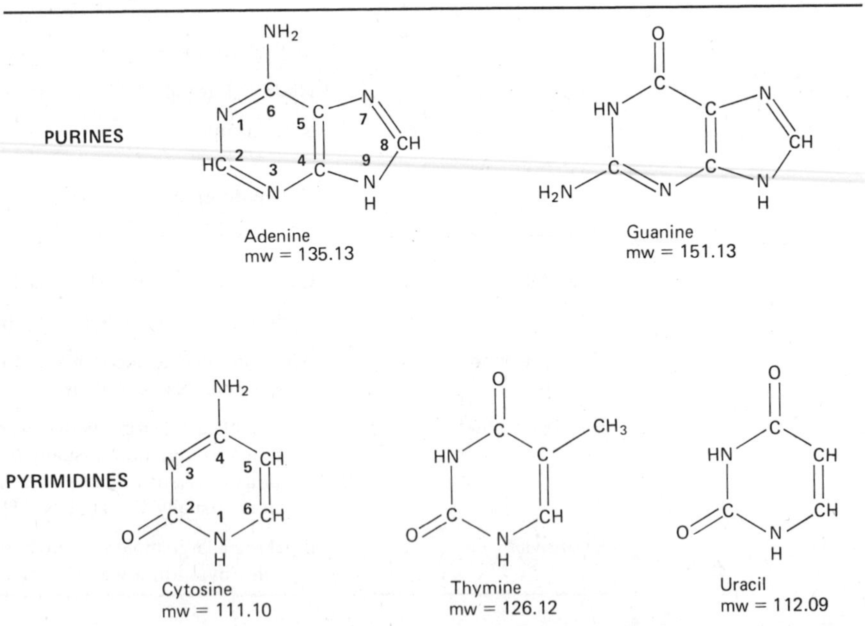

PURINES

Adenine
mw = 135.13

Guanine
mw = 151.13

PYRIMIDINES

Cytosine
mw = 111.10

Thymine
mw = 126.12

Uracil
mw = 112.09

FIGURE 1-2 Two
nucleosides.
Adenosine is an
RNA precursor; 2-
deoxyadenosine is
a DNA precursor.

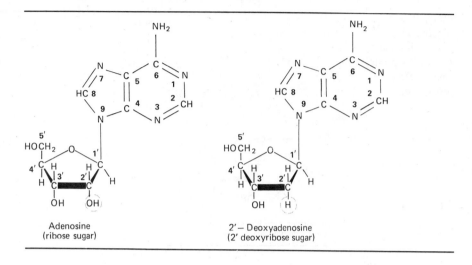

FIGURE 1-2 Two nucleosides. Adenosine is an RNA precursor; 2-deoxyadenosine is a DNA precursor.

words, even though the bases possess potentially unstable bonds, they remain chemically stable, in one tautomeric form, most of the time. This stability is, of course, an important genetic attribute.

Purines and pyrimidines can form chemical linkages with pentose (5-carbon) sugars. The carbon atoms on the sugars are designated 1′, 2′, 3′, 4′, and 5′, and it is the 1′ carbon of the sugar that forms a **glycosylic bond** to the nitrogen atom in position 1 of a pyrimidine or the nitrogen atom in position 9 of a purine. The resulting molecules are known as **nucleosides** (Figure 1-2), and they can serve as elementary precursors for DNA or RNA synthesis. DNA precursors contain the pentose **deoxyribose;** in RNA the sugar is **ribose** and not deoxyribose, the difference being that ribose contains an additional oxygen atom at position 2′ (Figure 1-2).

Before a nucleoside can become part of a DNA or RNA molecule it must become complexed with a phosphate group to form a **nucleotide** (Figure 1-3) or, more specifically, a deoxyribonucleotide or ribonucleotide. Nucleotides with a single phosphate group are known as nucleoside mon-ophosphates, an example being adenine ribonucleoside monophosphate, or AMP. Nucleotides can also possess two or three phosphate groups (for example, ADP or ATP). It is the nucleoside triphosphates that serve directly as precursors for DNA and RNA synthesis, as we shall see in Chapter 8.

DNA and RNA are simply polymers of nucleotides: When two nucle-otides are joined together, the resultant molecule is called a **dinucleo-tide;** three form a **trinucleotide;** several form an oligonucleotide; and many form a **polynucleotide.** Only one phosphate group of each precursor tri-phosphate is included in the polymer. This phosphate group, which is bound to the 5′-carbon of the pentose sugar on one nucleotide (Figure 1-3), also becomes chemically bound to the 3′-carbon of the sugar of a second

**FIGURE 1-3
Nucleotide,**
deoxyadenosine 5′-
monophosphate,
also a DNA
precursor.

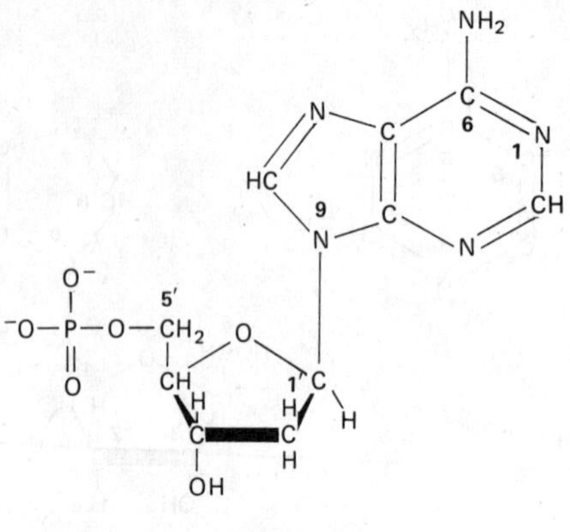

Deoxyadenosine 5′—monophosphate
(deoxyadenylic acid; dAMP)

**FIGURE 1-4
Trinucleotide of
DNA.**

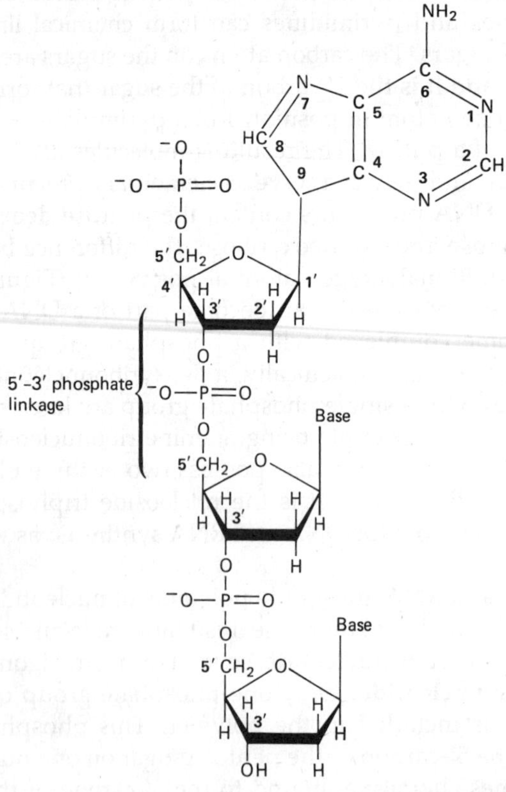

**TABLE 1-2   Units of Length and Mass**

**Length**
1 angstrom (Å) = $10^{-8}$ cm ≃ diameter of a hydrogen nucleus
1 nanometer (nm) = 10 Å
1 micron (μm) = 1000 nm
1 millimeter (mm) = 1000 μm
1 meter (m) = 1000 mm

**Mass**
1 dalton = mass of 1 hydrogen atom = $3.32 \times 10^{-24}$ g

nucleotide, so that a series of **5'-3' phosphate linkages** holds the nucleotides together along the length of the polymer, as illustrated in Figure 1-4. The phosphate bonds are known as covalent ester bonds, or **phosphodiester bonds,** and are extremely strong. The phosphate residues ($PO_4^-$) along the chain are acidic, leading to the term **nucleic acid:** they are commonly neutralized, however, so that DNA and RNA are more appropriately considered as salts.

Once a sugar-phosphate "backbone" has been formed, the position of the purine and pyrimidine bases in a nucleic acid is quite rigidly fixed. Each base will be stacked on top of the next, like a stack of pennies, with one base 3.4 Å away from its neighbor (see Table 1-2 for a list of physical units). Since the purines and pyrimidines are essentially flat, two-dimensional structures (Figure 1-1), there is no stereochemical interference between the bases in a stack. Indeed, weak bonds form between the stacked bases and add to the stability of the molecule.

The structure of a nucleotide chain can be abbreviated as in Figure 1-5, where adjacent base-sugar complexes are represented by the parallel

**FIGURE 1-5
Structural polarity** of a DNA hexanucleotide. The 5'-P end is at the top and the 3'-OH end is at the bottom, giving the same polarity as the more detailed rendition shown in Figure 1-4.

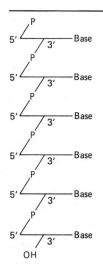

lines that are oriented at right angles to the long axis of the molecule, and where the covalent 5'-3' phosphate linkages are represented by the sloping lines that extend from median (3') positions to terminal (5') positions on adjacent sugar molecules. This schematic rendering emphasizes an important fact: **polynucleotides are structurally polarized,** meaning that a **3'-hydroxyl (3'-OH) end** and a **5'-phosphate (5'-P) end** can be identified for any chain.

Single-stranded polynucleotide chains have been adopted as the genetic material in certain viruses. Five kinds of virus are currently known to contain single-stranded DNA chromosomes, and most RNA-containing viruses, including the familiar polio and influenza viruses, possess single-stranded RNA. Single-stranded forms of DNA and RNA should not be confused with the more familiar double helix forms of DNA (and RNA), which are described in the next section.

## 1.2 The Double Helix

In 1953 J. D. Watson and F. H. C. Crick proposed that most DNA is found in a double helix having very specific properties. Their hypothesis was based on a number of important findings that had been contributed by the work of others.

1. The primary structure of a single nucleotide chain, diagrammed in Figure 1-4, was known at that time. It was also suspected that most naturally occurring DNAs did not exist as single chains; instead, two or more chains appeared to interact with one another in some way. The resulting macromolecules were known to be long, thin, and rigid, such that they formed highly viscous solutions in water.

2. E. Chargaff had established that when DNA from any particular species is subjected to chemical hydrolysis in order to release its component purines and pyrimidines, the total amount of adenine released is always equal to the total amount of thymine (A = T), as seen in the third and fourth columns of Table 1-3. Similarly, the total amount of guanine always equals that of cytosine (G = C), as shown in the fifth and sixth columns of Table 1-3. **Chargaff's Rule,** in other words, states that **in natural DNAs the base ratio A/T is always close to unity and the G/C ratio is always close to unity.**

3. R. Franklin, working at the same time as Watson and Crick, was studying the X ray diffraction patterns produced by isolated DNA fibers. These patterns revealed that the fibers contained highly ordered, helical molecules. Knowing that DNA was constructed from polynucleotide chains, she realized that two or more of these chains must be coiled around one another, in a helical fashion, to produce the secondary structure of a DNA macromolecule.

Watson and Crick, in attempting to visualize such a macromolecule, started with the simplest assumption, namely, that two polynucleotide

**TABLE 1-3    DNA Base Compositions** (as Moles of Nitrogenous Constituents per 100 g-Atoms Phosphate in Hydrolysate)

| Organism | Tissue | Adenine | Thymine | Guanine | Cytosine | $\dfrac{A + T}{G + C}$ |
|---|---|---|---|---|---|---|
| *Escherichia coli* (K-12) | – | 26.0 | 23.9 | 24.9 | 25.2 | 1.00 |
| *Diplococcus pneumoniae* | – | 29.8 | 31.6 | 20.5 | 18.0 | 1.59 |
| *Mycobacterium tuberculosis* | – | 15.1 | 14.6 | 34.9 | 35.4 | 0.42 |
| Yeast | – | 31.3 | 32.9 | 18.7 | 17.1 | 1.79 |
| *Paracentrotus lividus* (sea urchin) | Sperm | 32.8 | 32.1 | 17.7 | 18.4 | 1.85 |
| Herring | Sperm | 27.8 | 27.5 | 22.2 | 22.6 | 1.23 |
| Rat | Bone marrow | 28.6 | 28.4 | 21.4 | 21.5 | 1.33 |
| Human | Thymus | 30.9 | 29.4 | 19.9 | 19.8 | 1.52 |
| Human | Liver | 30.3 | 30.3 | 19.5 | 19.9 | 1.53 |
| Human | Sperm | 30.7 | 31.2 | 19.3 | 18.8 | 1.62 |

From E. Chargaff and J. Davidson, Eds., *The Nucleic Acids*. New York: Academic, 1955, pp. 356–359.

chains form a **double helix.** Working with "ball and stick" type models, they experimented with the idea of placing the sugar-phosphate backbones on the outside and pointing the bases toward the inside of the helix. By placing the two polynucleotide chains in phase with each other—an arrangement suggested by the regular X ray diffraction patterns—opposing bases on the two adjacent chains tended to line up with one another. Here Watson and Crick made an exciting observation.

They realized that **a stable helical structure could form** *only* **if adenine lined up opposite thymine or if guanine lined up opposite cytosine.** In the former case the molecular structures of the adenine and thymine were such that two hydrogen bonds could readily form between them (Figure 1-6). In the latter case they found that cytosine could form three hydrogen bonds with guanine (Figure 1-6). These bonds, it was reasoned, would greatly stabilize a large helical molecule and would help to explain the physical properties of naturally occurring DNA.

By assuming a double-helical or **duplex** molecule containing only A—T and G—C base pairs, Watson and Crick further found that they could construct a model with a uniform diameter corresponding to about 20 Å. Purine-purine base pairs, they discovered, were too large to fit within such a helix, whereas pyrimidine-pyrimidine pairs were so far apart that hydrogen bonds could not form between them at all. A distortion of the helical structure also occurred when an attempt was made to pair A with C or G with T.

In short, when all possible base-pair combinations were considered, it

FIGURE 1-6
Hydrogen bonding
between bases.

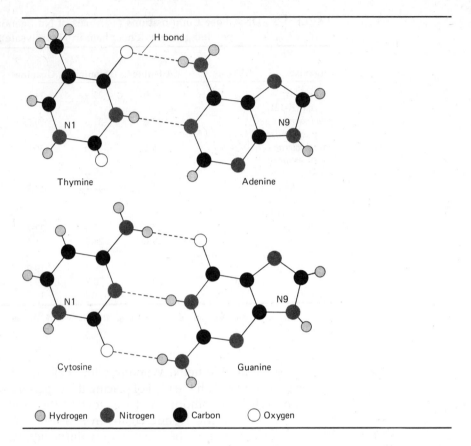

became clear that A—T and G—C pairs were the only pairs that could form stable, hydrogen-bonded entities having the correct molecular dimensions to fit inside a polynucleotide double helix with a regular diameter of 20 Å. For this reason A and T were said to be **complementary** to each other, and G and C were said to be complementary to each other.

This conclusion related immediately to Chargaff's Rule, which, as summarized earlier, states that in the DNA of any species, the A/T and G/C ratios are invariably close to unity (Table 1-3). No such consistent relationship had been found for any other sets of bases. Thus the Watson-Crick model was supported at its inception by some careful experimental evidence.

The full structure for the DNA molecule proposed by Watson and Crick is shown in Figure 1-7, and a more detailed molecular model appears in Figure 1-8. Having such a model in hand, it is possible to count **10 bases in each full turn** of a helical polynucleotide strand. It will be recalled that in a polynucleotide chain the distance between bases in the penny-like stack was 3.4 Å, and it follows that in the double helix **a strand makes one complete turn every 34 Å.** The actual numbers are less important to the

FIGURE 1-7
**Watson-Crick**
**model** of a DNA
double helix. (After
J. D. Watson and F.
Crick, *Cold Spring*
*Harbor Symposia on*
*Quantitative Biology*
**18:**123, 1953.)

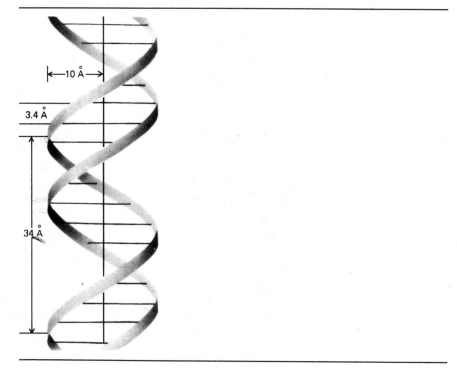

geneticist than the concept that **the DNA duplex is an extremely ordered**
**structure composed of relatively stable organic molecules that pair in a**
**precise manner.**

It is important to look at the complementary base pairs in Figure 1-6
and note that they are drawn in a specific way, with the relevant hydrogen-bonding groups precisely aligned. In the DNA duplex such an alignment of the bases can occur *only* when the two polynucleotide strands are
oriented in opposite directions. As pointed out earlier and stressed in Figure 1-5, a polynucleotide possesses a 3'-OH and a 5'-P end. **The Watson-**
**Crick model requires that one strand be oriented in a 3' $\longrightarrow$ 5' fashion**
**in the helix and the other in a 5' $\longrightarrow$ 3' fashion,** as shown schematically
in Figure 1-9. When an attempt is made to construct a double helix with
two similarly directed strands, the bases do not align with each other to
allow the formation of the requisite hydrogen bonds.

The Watson-Crick hypothesis was advanced for DNA, but double-
helical RNA molecules have since been described for certain RNA-containing viruses. Moreover, most single-stranded RNA molecules probably contain at least some regions in which the strand folds back on itself to form
short double-helical segments. When RNA exists as a double helix and
serves as the genetic material of viruses, most of the statements we have

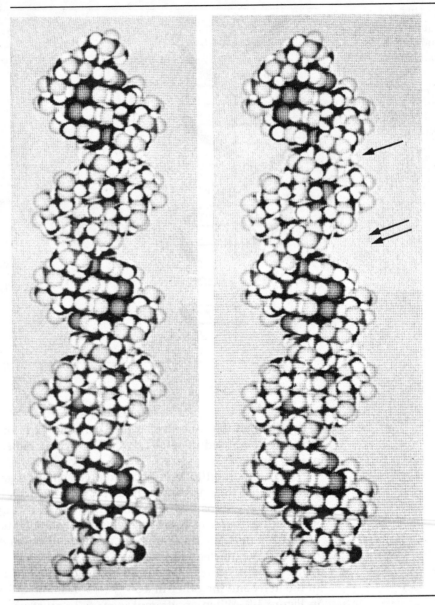

FIGURE 1-8 Molecular model of DNA double helix, shown as a stereo pair. The single arrow points to the **minor groove** in the helix; the double arrows point to the **major groove.** The sugar-phosphate backbone spirals around the exterior, while the bases are sequestered in the interior. Note, however, that many of the atoms in the bases are exposed in the major groove. (To fuse the two halves of the stereo pair, it is necessary to diverge your eyes and yet focus on the two images. Some people find this easy to do and some difficult. If you have difficulty, hold the book up in front of you at arm's length, then look just above it at something about 20 feet away (i.e., across the room); this will cause your eyes to diverge. Now slowly raise the book up into your field of gaze, and you should see the DNA in 3-D.) (Courtesy of Dr. Richard Feldman, NIH.)

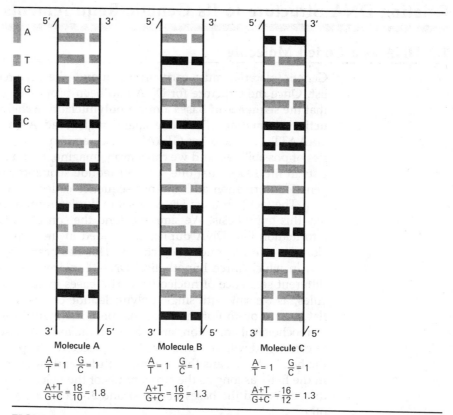

**FIGURE 1-9 "Ladder" scheme showing three DNA helices.** The sides of the ladder represent the sugar-phosphate "backbone," and the rungs are formed by the base pairs. Arrowheads indicate a 5'-phosphate. All three molecules carry distinct nucleotide sequences, but molecules B and C happen to have the same overall nucleotide composition.

made about the secondary structure of DNA are applicable except that adenine forms hydrogen bonds with uracil and not with thymine.

We now leave the physical and chemical properties of helical nucleic acids and go on to consider some of their genetic properties, focusing particularly on DNA. In so doing, however, it seems appropriate to quote from the concluding paragraph of a review article by J. Josse and J. Eigner.

*What has perhaps emerged most clearly from all the work which has been reported here is the enduring validity of the double-helical model of DNA structure proposed by Watson and Crick. Down to almost all of its finest details, this model has been convincingly confirmed from virtually every physical approach with which DNA can be examined.*

# Relating DNA Structure to Its Genetic Requirements

## 1.3 DNA as a Coded Molecule

Genetic material must carry information—the first requirement on our list. Once the structure for DNA had been proposed, it became apparent that **the sequence of bases along a polynucleotide chain could contain genetic information.** Thus one chain might read AAAATT . . . , another GAAATT . . . , another GTAATT . . . , and so on through an endless series of possibilities, and we can simply imagine, as did a number of geneticists in the 1950s, that one such nucleotide sequence specifies one sort of genetic information and another sequence codes for another sort.

The fact that most DNA exists in duplex form does not at all contradict this hypothesis. We simply extend the concept and propose that the information in a DNA duplex is encoded in the sequence of nucleotides along its **two** polynucleotide chains. This has been diagrammed in Figure 1-9, in which three hypothetical DNA duplexes are shown, each with a different sequence of nucleotides. The bases in such double-helical molecules, as we saw with single polynucleotide chains, are stacked, with one flat surface on top of the next, like pennies (Figure 1-8). Accordingly, no stereochemical limitations are placed on the **order** of bases within the helix, as contrasted with the strict limitation placed on the **pairing** of bases within the helix. This means that **any nucleotide sequence is chemically possible in the helix** as long as the two strands of the helix are oriented in opposite directions and the bases in one strand are complementary to those in the other.

If the genetic code is indeed inherent in the nucleotide sequence of a DNA molecule, we might predict that different species of organisms, having different genetic information, will possess DNAs with different nucleotide sequences, whereas all sources of DNA from the same species will be identical. Biochemical techniques in the 1950s did not permit determination of the base sequences in DNA, but the prediction was certainly supported by Chargaff's experiments (Table 1-3). We have already noted that Chargaff's data established the universality of the A = T and G = C rule. In addition, his data reveal that there is not necessarily any similarity between the **total nucleotide composition** of DNA isolated from one type of organism and the DNA isolated from another type of organism. Expressing total nucleotide composition as the ratio A + T/G + C, we find (Table 1-3, last column) a ratio of 1.00 for the colon bacterium *Escherichia coli*, 1.59 for the bacterium *Diplococcus pneumoniae*, and so on, a result that is readily explained by assuming that the various species possess DNA with different nucleotide sequences (Figure 1-9). When, on the other hand, we compare different cell types from the *same* species of organism—from the thymus, liver, and sperm cells of humans, for example—all are found to have nearly identical A + T/G + C ratios (Table 1-3), as if all possessed the same genetic information.

Nucleotide sequences and nucleotide compositions are two quite distinct properties of a DNA molecule, as we demonstrate by example in Figure 1-9. We therefore stress that the Chargaff data do not *prove* that DNA contains genetic information. They served, however, to provide important support for the concept of DNA as the genetic material at the time this concept was being formulated. Proof that DNA is a coded molecule is forthcoming in succeeding chapters of this book.

## 1.4  DNA Replication

The double-helical model for the structure of DNA immediately suggested a mechanism for faithful DNA replication, the second on our list of requirements for the genetic material. Watson and Crick saw that once a particular nucleotide sequence was established on one strand of the helix, the sequence on the other strand had to be complementary and could be predicted. For example, knowing that one strand reads 5' . . . ACGAT . . . 3', the other strand, by the rules of base pairing, must read 3' . . . TGCTA . . . 5'.

Watson and Crick therefore proposed that if two strands in a given helix were to separate by an unwinding process and so become exposed to a solution containing nucleotides, **each strand might serve as a template for the laying down of new polynucleotide strands.** The sequences of bases in the new daughter strands would, by the rules of base pairing, complement the sequences in the parental strands, as shown in Figure 1-10. At the end of this round of replication, two new helices would be formed, each containing one parent and one daughter strand, and both new helices would be exact replicas of the original helix.

This proposal is a particularly ingenious feature of the Watson-Crick model. Moreover, it proved to be correct, as we describe in Chapter 8.

## 1.5  DNA Expression

The third requirement we imposed on the genetic material was that its encoded information be somehow "expressible," that is, translatable into such processes as growth and development. At the time the Watson-Crick model was proposed, it was not clear how a code composed of a sequence of bases could be biologically effective, but the concept soon evolved that **the sequence of nucleotides in a DNA molecule might be translated into the sequence of amino acids in a polypeptide** and that the group of nucleotides that specified such a single polypeptide could be considered a **gene.** It also became clear that genes did not ordinarily participate directly in the synthesis of polypeptides: instead their nucleotide sequences were first **transcribed** into complementary nucleotide sequences in the RNA molecules known as **messenger RNA.** These messenger RNA molecules then went on to engage in polypeptide synthesis. The transcription and translation of genes are discussed in detail in Chapters 9 and 11.

FIGURE 1-10
Replication of
DNA according to
Watson and Crick.
Purines are drawn
as long rectangles
and pyrimidines as
short rectangles;
the new strands are
incomplete. (After
J. D. Watson and
F. H. Crick, *Nature*
171:737–738, 1953.)

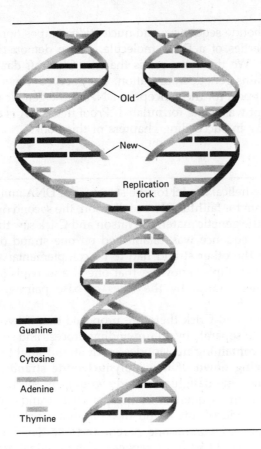

Old

New

Replication
fork

Guanine

Cytosine

Adenine

Thymine

## 1.6 DNA Variation

Our fourth requirement of the genetic material was that it be subject to occasional variation by mutation and recombination. The process of mutation could be readily visualized with the Watson-Crick model, for any alteration in the sequence of its nucleotides, produced either physically or chemically, could bring about a change in the information contained in the DNA molecule and would thereby change, in some aspect, the construction of the cell. The physical and chemical basis for mutation, the subject of Chapter 8, could in fact be meaningfully explored only after a structure for the genetic material had been proposed.

On the other hand, the Watson-Crick model did not, in itself, immediately suggest a mechanism for recombination. As we shall see in Chapter 17, the physical basis for recombination has only recently begun to be understood at a molecular level.

# Experiments Indicating DNA and RNA as the Genetic Material

Watson and Crick did not choose DNA as the subject for their model-building simply out of hand. A paper describing the nucleic acids was published as early as 1874 by F. Miescher, and by 1953 it was clear that proteins, long favored as the genetic material, were poor candidates for a number of reasons. R. Shoenheimer had noted, in 1938, that the DNA of a cell was unusually stable, in contrast to the rapidly turning-over proteins, and A. Mirsky and H. Ris had found, in 1949, that whereas all cells of an organism contained similar amounts of DNA, different cell types contained quite different amounts and kinds of protein. Its stability and constancy therefore favored DNA as the genetic material.

A number of experiments, moreover, had indicated that DNA alone could transmit genetic information from one generation to the next. Several of these experiments are described in the following paragraphs, including a few that were performed after 1953. In the course of this presentation, certain organisms and laboratory techniques are discussed rather sketchily and only as they relate to the experimental results. In later chapters the organisms and the techniques are covered in much greater detail. At this point the experiments serve as an introduction to the kinds of approaches used by the molecular geneticist.

## 1.7 Transformation Experiments

One series of experiments indicating DNA as the genetic material utilized the bacterium *Diplococcus pneumoniae*, or pneumococcus. To understand the experiments, several facts are pertinent.

1. Pneumococcus can exist in two forms: Smooth (*S*) cells are surrounded by a polysaccharide capsule and are **virulent,** or infectious, whereas rough (*R*) cells lack the capsule and are **nonvirulent.**
2. Several *S* strains have been isolated (Types II*S* and III*S*, for example), each having a polysaccharide coat with a slightly different chemical composition. The various *S* and the *R* coat types are inherited from generation to generation and are thus said to be genetically determined.
3. A change from *S* $\longrightarrow$ *R* will occur by mutation in about one *S* cell in 10 million. The *R* colony that grows from this mutant will, in turn, sometimes give rise to an *S* cell by a second reverse mutation, *R* $\longrightarrow$ *S*. The resultant *S* colony is found to possess the same capsule type as the *S* strain from which the *R* mutant was originally derived; for example, Type II*S* $\longrightarrow$ *R* $\longrightarrow$ Type II*S*, not III*S*.

An early experiment with *S* and *R* pneumococcus, performed by F. Griffith, is diagrammed in Figure 1-11. Mice were injected with a small

FIGURE 1-11
Pneumococcus
transformation
experiment with
mice.

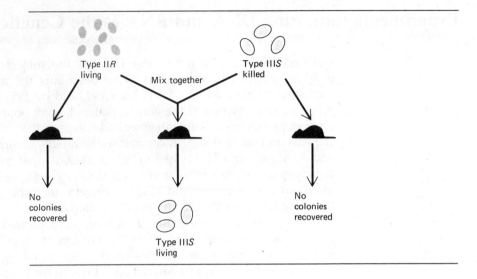

number of living (but nonvirulent) R pneumococci that had originally arisen by mutation from the Type IIS strain. At the same time the mice were injected with a large number of heat-killed (and therefore no longer virulent) Type IIIS bacteria. Surprisingly, many of the mice developed pneumonia, and blood samples of the diseased mice revealed the presence of large numbers of living, virulent, smooth pneumococci. Moreover, these were all of Type IIIS, meaning that they could not have arisen by reverse mutation of the living injected R cells that had been co-injected. The conclusion, therefore, was that some of the dead Type IIIS bacteria had **transformed** the living R bacteria into smooth, virulent Type IIIS cells during their coexistence within the mouse (Figure 1-11). It could be further shown that the newly formed Type IIIS cells reproduced true to type for many generations, indicating that the transformation had directly affected their genetic material.

As other investigators repeated these experiments, it was found that the mouse could be eliminated: as diagrammed in Figure 1-12, heat-killed S cells and live R cells could be grown together in a test tube, and virulent S cells with the same capsule type as the heat-killed S cells were recovered (Figure 1-12). In a further experiment it was found that even intact, heat-killed S cells were not required: **an extract of heat-killed S cells could transform the R cells in vitro.** The remaining task, therefore, was to determine which chemical substance in the extract was responsible for this transformation process.

The substance, or **transforming principle,** as it was then called, proved to be DNA. In 1944 O. T. Avery, C. M. MacLeod, and M. McCarty incubated R cells in the presence of a highly purified DNA fraction obtained from Type IIIS bacteria, and some virulent Type IIIS bacteria were recovered from the test tube. If the Type IIIS DNA fraction was first treated

**FIGURE 1-12** *In vitro* **pneumococcus transformation experiment.**

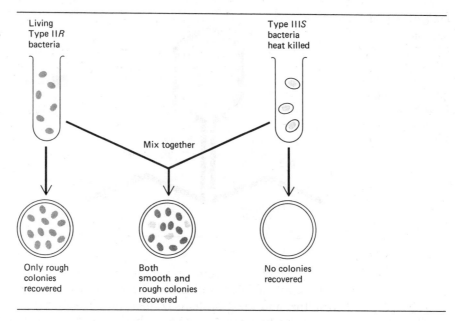

with **deoxyribonuclease,** an enzyme that destroys DNA, no transformation of *R* cells occurred in the subsequent experiment.

At the time these experiments were published, they were interpreted by some to mean that the transforming principle, or DNA, was capable of inducing a type-specific, "directed" mutation in the genetic material of the *R* cells such that *S* cells were produced. The mechanism by which such a directed mutation might be induced was obscure, but at that time it was certainly no more obscure than the correct interpretation of the experiments, namely, that the purified, *S*-type DNA is capable of entering the *R* cells and recombining with the *R*-cell DNA to produce a small number of living, smooth (IIIS) recombinants. The details of this bacterial transformation process as it is now understood are discussed in Chapters 13 and 17.

## 1.8 Hershey-Chase Experiments

In 1952 A. D. Hershey and M. Chase published experiments that indicated DNA as the genetic material in a more direct manner. Their experiments were concerned with the bacteriophage **T2,** a virus that replicates in the bacterium *Escherichia coli.* The structure of T2 is diagrammed in Figure 1-13; it has a hexagonal head containing DNA, plus a tail and tail fibers. The first step in the T2 infection of *E. coli* was already known at that time to be the attachment, or **adsorption,** of a phage, via its tail fibers, to the outside of the host cell. Phage material then entered the bacterium in some way and multiplied at the expense of the bacterium until the bacterium finally burst open, or **lysed,** liberating perhaps a hundred new phage progeny.

FIGURE 1-13 T2
bacteriophage.
(Redrawn from
Sutton, *Genes,
Enzymes, and
Inherited Diseases*.
New York: Holt,
Rinehart, and
Winston, 1961.)

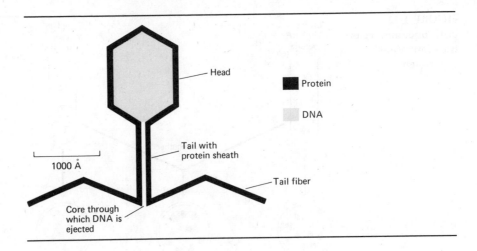

T2 phages were known to be composed of approximately equal amounts of DNA and protein. Since DNA contains phosphorus but no sulfur, and most proteins contain no phosphorus but (usually) some sulfur, the two can be distinguished by the use of radioactive isotopes of phosphorus and sulfur. Hershey and Chase therefore grew *E. coli* in a medium containing either the radioactive isotope of phosphorus ($^{32}$P) or that of sulfur ($^{35}$S) and then allowed T2 phages to infect and multiply within these labeled host cells (Figure 1-14a). The progeny phages that appeared after lysis of the bacteria were collected and found to be similarly labeled (Figure 1-14b). In this way Hershey and Chase acquired two stocks of radioactive T2, one containing $^{32}$P-labeled DNA and the other, $^{35}$S-labeled protein.

Labeled phages were then used to infect unlabeled *E. coli* cells (Figure 1-14c), and a significant result was obtained (Figure 1-14d). When the infection was brought about by the $^{32}$P-labeled phages, almost all the radioactivity was subsequently found *within* the host bacteria; moreover, following bacterial lysis, some of the label appeared in the progeny phages. On the other hand, when $^{35}$S-labeled phages were used, very little label appeared within the host bacteria or in the progeny phages; almost all of it remained on the *outside* of the bacterium, adsorbed to the bacterial cell wall. Thus it was demonstrated that the phage DNA and protein become separated during the infection process: the DNA enters the host cell, the site of phage replication, whereas the protein coat appears to function primarily in the external adsorption process.

The Hershey-Chase experiment does not, in fact, prove that DNA is the phage genetic material, for about 20 percent of the $^{35}$S label enters the host along with the DNA, and it could certainly be argued that this small amount of protein carries the genetic information. The Watson-Crick model was published in the following year, however, and the era of DNA-oriented research was launched.

**FIGURE 1-14
Hershey-Chase
experiments** with
isotopically labeled
bacteriophages.

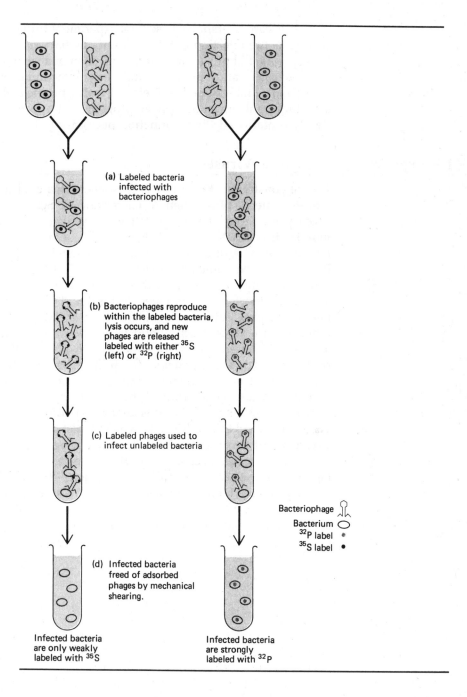

(a) Labeled bacteria
infected with
bacteriophages

(b) Bacteriophages reproduce
within the labeled bacteria,
lysis occurs, and new
phages are released
labeled with either $^{35}$S
(left) or $^{32}$P (right)

(c) Labeled phages used to
infect unlabeled bacteria

Bacteriophage
Bacterium
$^{32}$P label
$^{35}$S label

(d) Infected bacteria
freed of adsorbed
phages by mechanical
shearing.

Infected bacteria
are only weakly
labeled with $^{35}$S

Infected bacteria
are strongly
labeled with $^{32}$P

A "clean" Hershey-Chase type experiment proved not to be possible so long as the infection of intact bacteria by bacteriophages was a part of the experimental procedure, for the injection of a small amount of protein is an obligate feature of the natural phage infection process. If, however, bacteria are first stripped of their cell walls, purified phage DNA can be introduced and infectious phage progeny will appear. Thus **DNA alone clearly contains all the information necessary to construct T2 phages.**

## 1.9 Experiments with RNA Viruses

The demonstration that RNA is the genetic material in RNA-containing viruses came in 1956, when A. Gierer and G. Schramm showed that to-bacco plants could be inoculated with purified RNA from the **tobacco mosaic virus (TMV)** and TMV-like lesions could later be identified on the tobacco leaves. A different approach was taken by H. Fraenkel-Conrat and B. Singer in experiments published in 1957. They first found that they could experimentally separate the RNA from the protein of TMV viruses. They then developed techniques for forming "reconstituted" viruses con-taining the protein from one mutant strain of TMV and the RNA from an-other, or vice versa. Such hybrid viruses were allowed to infect tobacco leaves, and the progeny were examined. In all cases **the progeny were of the parental RNA type and *not* the parental protein type.** Such an outcome is reminiscent of the pneumococcus transformation results in which the parental DNA type determined the progeny type.

In a final refinement of this kind of experiment, N. Pace and S. Spie-gelman in 1966 purified RNA of two quite distinct base compositions from two different mutant strains of the RNA phage **Qβ**. The isolated RNAs were then incubated separately in the presence of *E. coli* cell extracts con-taining enzymes capable of RNA replication. The new RNA synthesized was in each case identical in base composition to the particular phage RNA introduced into the *in vitro* system, thus indicating that the phage RNA can serve as a template for self-replication.

(a) Rosalind Franklin performed early X-ray diffraction studies of DNA. (b) Erwin Chargaff (Columbia University) performed early biochemical studies on the base composition of DNA.

B. Singer (c) and (d) H. Fraenkel-Conrat (University of California, Berkeley) have studied the genetic properties of viral RNA chromosomes.

# Questions and Problems

1. Distinguish between a purine, pyrimidine, nucleoside, nucleotide, polynucleotide, and double helix. How do these structures differ in DNA and RNA?
2. Diagram the pairing between adenine and uracil. Is there any difference between the way in which adenine pairs with uracil or thymine?
3. If one strand of a helix reads 5'-AGCAGCA-3', what would be the structure of the complementary strand in a DNA helix? An RNA helix?

4. Diagram two DNA duplexes, each having 15 nucleotide pairs and an A + T/ G + C ratio of 2.0, but each having different nucleotide sequences.

5. Diagram a hypothetical duplex with five base pairs to represent a portion of the gene specifying the Type IIS pneumococcus capsule. Now make an arbitrary change in the sequence to represent the following mutational event: Type IIS $\longrightarrow$ R. What change would be required for the event R $\longrightarrow$ Type IIS? Would you expect both events to occur with equal frequency?

6. Why does the Watson-Crick model require that a DNA helix must unwind before it is replicated?

7. Outline experiments using radioactive isotopes in which you could show that transformation occurs when S-type pneumococcal DNA, and not protein, enters R cells.

8. At a certain time after the addition of $^3$H-thymidine to a growing culture of E. coli, all the DNA is found to possess one radioactive strand and one nonradioactive strand. Is this result demanded by the Watson-Crick hypothesis for the replication of DNA? Explain.

9. A DNA duplex possesses major and minor "grooves" (Figure 1-8). Are any of the atoms in the purines and pyrimidines exposed in these grooves? Are the hydrogen bond–forming atoms exposed? (This question can most readily be answered if a space-filling model of a double helix is available.)

10. Which of the following base-composition ratios are compatible with the Watson-Crick DNA model? Which contradict it? Explain for each.

(a) $\dfrac{G + C}{A + T} = 1$     (c) $\dfrac{\text{pyrimidines}}{\text{purines}} = 1$     (e) $\dfrac{GA}{C} = T$

(b) $\dfrac{G + A}{C + T} = 1$     (d) $\dfrac{G + C + A}{T} = 1$

11. Using the convention shown in Figure 1-5 for the structure of a DNA strand, draw a duplex hexanucleotide, giving the strands the correct polarity with respect to one another.

# 2

# Cell Cycles, Chromosome Duplication, and Mitosis

**A. Introduction**

**B. The Bacterial Cell Cycle and Chromosome Replication**
- 2.1 Bacterial Cell and Chromosome Structure
- 2.2 The Bacterial Cell Cycle
- 2.3 Semiconservative DNA Replication in Bacteria: The Meselson-Stahl Experiment
- 2.4 Visualizing Bacterial Chromosome Replication

**C. The Eukaryotic Cell and Cell Cycle**
- 2.5 Eukaryotic Cell Structure
- 2.6 The Typical Eukaryotic Cell Cycle
- 2.7 DNA and Chromosome Replication in Eukaryotes
- 2.8 Chromosome Replication in the Cell Cycle
- 2.9 Mitosis

**D. The Karyotype**
- 2.10 General Features of Karyotypes
- 2.11 Karyotypes Using Banded Chromosomes
- 2.12 Chromosome Purification

**E. Atypical Eukaryotic Cell Cycles**
- 2.13 Cell Cycle of Ciliated Protozoa
- 2.14 Cell Cycle of Dipteran Larva Cells

**F. Questions and Problems**

## Introduction

This and the next few chapters consider the structure, organization, and duplication of **chromosomes,** the physical bearers of genetic information. Bacterial cells possess a single **main chromosome** plus an optional endowment of small, circular DNA molecules called **plasmids.** If we think of all the essential genetic information possessed by an individual as the **genome,** then the genome of a bacterium is encoded within its single large chromosome. For nuclear organisms, by contrast, the genome is subdivided into a **collection of chromosomes,** each carrying a distinct subset of genes and, therefore, distinct nucleotide sequences.

It is clearly essential to the survival of a species that when one cell gives rise to two daughter cells, each daughter receives a complete copy of the genome. During the **cell cycle** of a bacterium, therefore, the main chromosome first undergoes **replication** to form two duplicate copies. Each daughter cell then receives a copy at the time of cell division. For nucleate organisms, the cell cycle entails not only chromosome duplication but also the equitable allotment of a complete chromosome collection to each daughter, a process known as **mitosis.** The detailed enzymology of DNA replication during these cell cycles will be considered in Chapter 8; in this chapter, attention will focus on overall patterns of chromosome duplication and distribution. We conclude by examining the morphology of mitotic chromosomes in cytological preparations known as **karyotypes.** These are widely used by cytogeneticists to analyze the number and integrity of chromosomes in particular cell types.

# The Bacterial Cell Cycle and Chromosome Replication

## 2.1 Bacterial Cell and Chromosome Structure

The bacteria *Escherichia coli, Bacillus subtilis, Salmonella typhimurium,* and pneumococcus represent the prokaryotic species most widely used in genetic research. Although each bacterial species has its own distinctive properties, all are small cells (perhaps 1 to 2 μm long and 0.5 μm wide) surrounded by one or more membranes and walls (Figure 2-1). The **cytoplasm** of the cell is filled with particulate **ribosomes** (Figure 2-1), structures

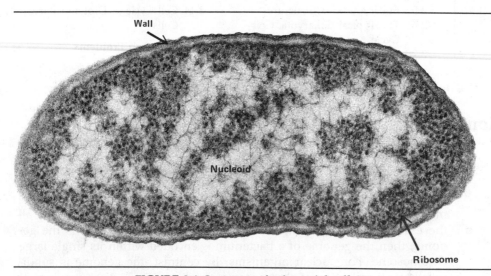

FIGURE 2-1 Structure of a bacterial cell.

TABLE 2-1   Rule-of-Thumb Conversion Factors for Duplex DNA

Average molecular weight of 1 nucleotide pair $\simeq$ 660 daltons.
1 pg ($10^{-12}$ g) duplex DNA $\simeq 6 \times 10^{11}$ daltons $\simeq 10^9$ nucleotide pairs.
1 $\mu$m duplex DNA $\simeq 2 \times 10^6$ daltons $\simeq 3 \times 10^3$ nucleotide pairs = 3 kb (kilobases).
1 average structural gene (coding portion) $\simeq 10^3$ nucleotide pairs = 2 kb.

involved in protein synthesis and described in Chapter 11, plus a wide variety of soluble enzymes and metabolic products. Distinct DNA-containing regions known as **nucleoids** occupy the rest of the cell (Figure 2-1). The DNA is so densely packed into a nucleoid that most cytoplasmic particles are excluded, but the nucleoid is not surrounded by a membrane. For this reason, bacteria are said to be **prokaryotes:** they lack the membrane-enveloped nucleus possessed by **eukaryotes** (a **karyon** is nucleus in Greek, and the prefixes *pro-* and *eu-* denote "before" (in time) and "true.")

Bacterial DNA is largely in the form of a single **main chromosome,** which, in *E. coli.,* contains about 1300 $\mu$m of DNA or $4 \times 10^6$ base pairs and has a molecular weight of about $2.6 \times 10^9$ daltons (see Table 2-1 for some conversion factors). This DNA takes the form of a giant circle: the duplex chromosome comes back on itself and seals covalently so that if one were to scan along the DNA molecule, it would be impossible to locate its "beginning" or "end." In addition to the major chromosome, bacterial cells often possess one or more minor chromosomes, each called a **plasmid,** which constitute from 0.5 to 2 percent of the DNA of the cell. A plasmid may contain 4 $\mu$m to 35 $\mu$m of duplex DNA, and this, too, circles about itself to form a small ring. (Plasmids are considered in detail in Chapters 13 and 18.) Figure 2-2 shows DNA spilling out of a lysed *E. coli* cell. In addition to the large main chromosome, a circular plasmid is present.

## 2.2 The Bacterial Cell Cycle

The cell cycle of *E. coli* at slow growth rates (>60 minutes per generation) consists of three periods: preparation for initiation of chromosome replication, chromosome replication, and a period between termination of chromosome replication and cell division. At rapid growth rates, a second round of DNA replication may be initiated before the first has been completed; therefore, the three periods become blurred and the synthesis of DNA can be said to be essentially continuous.

At the time of cell division, the two daughter chromosomes must **segregate** from one another and come to reside in the two daughter cells. Since cell division is accomplished by an ingrowing shelf of bacterial membrane called the **mesosome,** the simplest model of bacterial chromosome segregation states that a bacterial chromosome attaches to the membrane at the start of replication via an **attachment point,** that a "daughter attachment point" is synthesized during the course of replication, and that the ingrowing mesosome separates these two attachment points spatially,

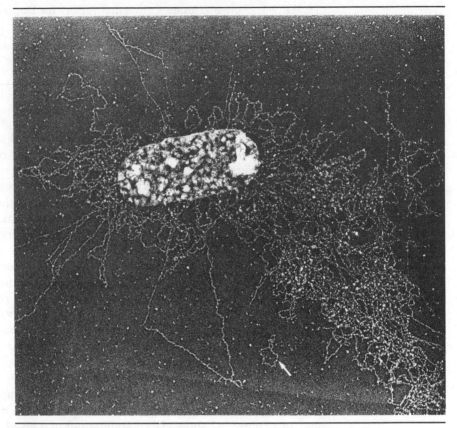

**FIGURE 2-2 Bacterial DNA.** An *E. coli* cell has been disrupted by mild detergent treatment, releasing its enormous chromosome onto the surface of a plastic-coated electron microscope grid. Arrow denotes a circular plasmid that is also released. (Courtesy of Dr. Jack Griffith.)

thereby segregating the daughter chromosomes into daughter cells. Consistent with this model are reports showing that *E. coli* DNA indeed binds to specific membrane proteins at the time of its replication, but details of the mechanism are still unclear.

An *E. coli* cell will divide as rapidly as once every 20 minutes (yeast, one of the fastest growing eukaryotes, divides every 1 1/4 hours under comparable growth conditions). This rapid **generation time** is, of course, an attractive feature of *E. coli* for genetic research, since it means that large numbers of genetically identical organisms will arise from a single cell in a very short period of time. If a very dilute suspension of bacteria is spread on the surface of nutrient medium solidified by the addition of agar, each cell will give rise to a discrete mound of such identical cells, which is known as a **bacterial colony;** the sister cells within the colony are frequently referred to as a **clone.**

## 2.3 Semiconservative DNA Replication in Bacteria: The Meselson-Stahl Experiment

We have already considered, in Chapter 1, the basic features of the Watson-Crick model for DNA replication. As diagrammed in Figure 1-10, the model suggests that once chromosome replication is initiated, the two orig-

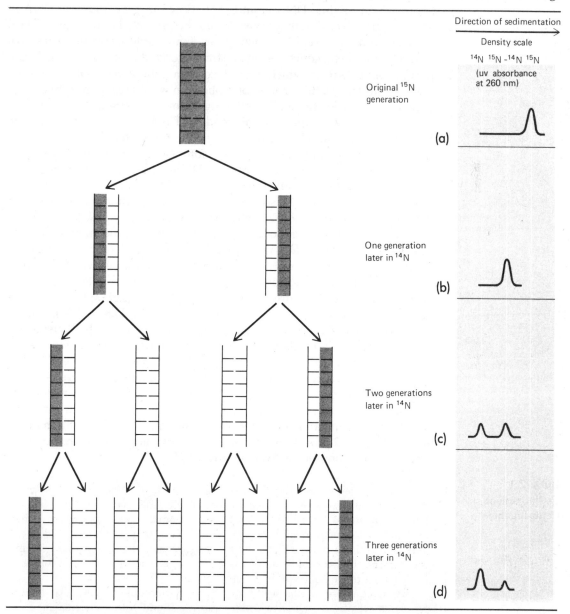

FIGURE 2-3 **The semiconservative replication of DNA** and its demonstration in the Meselson-Stahl experiment. Gray color indicates "old" ($^{15}N$-labeled) DNA; black color indicates "new" ($^{14}N$-labeled) DNA.

inal polynucleotide strands of the helix will unwind, at least locally, so that each can serve as a template for a new strand. An immediate prediction follows from this proposal: **both duplexes that result from replication should be hybrid in nature,** each containing an *old* strand derived from the original molecule and a *new* strand that has been formed during the replication process. This prediction is diagrammed in the left portion of Figure 2-3, in which original DNA strands are shown in gray and new DNA in black. Figure 2-3 also outlines what would be predicted if these two hybrid duplexes went on to replicate themselves. Four duplexes would result, two of which would still contain a single strand derived from the original chromosome and two of which would contain totally new DNA in both strands. Similarly, a third round of replication would result in eight duplex chromosomes, with two still containing an original strand.

Experimental support for this mode of DNA replication came in 1958 from M. Meselson and F. Stahl. They utilized the fact that if cells are grown in a medium containing $^{15}N$, the heavy isotope of nitrogen, the DNA of the cells will become heavier than ordinary DNA, and this density difference can be distinguished by centrifugation in a CsCl gradient (see Box 2.1). By growing *E. coli* in the $^{15}N$ isotope for several generations, Meselson and Stahl obtained a population of cells that contained fully $^{15}N$-labeled molecules. The DNA from these cells was shown to exhibit the buoyant density expected of $^{15}N$-labeled DNA (Figure 2-3a). The cells were then transferred to a medium containing only $^{14}N$ as a nitrogen source. It was found that after one round of replication the DNA in the daughter cells had neither the original $^{15}N$ nor the pure $^{14}N$ density. Instead, a single species with the intermediate density of a $^{15}N$-$^{14}N$ hybrid was found (Figure 2-3b), exactly as one would expect from the mode of replication proposed by Watson and Crick. If the daughter cells were then allowed to divide again, two species of DNA appeared in the gradient: roughly half of the DNA showed $^{15}N$-$^{14}N$ hybrid density, whereas the rest showed pure $^{14}N$ density (Figure 2-3c). As the cells continued to divide in the $^{14}N$ medium, a greater and greater proportion of the DNA in the population showed $^{14}N$ density, but for many generations a faint band was detectable at the $^{15}N$-$^{14}N$ DNA position in the gradient (Figure 2-3d). Again, these results are compatible with the mode of replication proposed by Watson and Crick.

**BOX 2.1**
CsCl Gradient
Centrifugation

If a concentrated solution of CsCl is centrifuged at high speeds for many hours, centrifugal forces will induce the CsCl to sediment toward the bottom of the tube, a process that will be counteracted by back-diffusion of the CsCl toward the top of the tube. Eventually, at equilibrium, the opposing processes will balance, and a stable CsCl concentration gradient will be established in which there is progessively more and more CsCl as one scans from the top to the bottom of the tube. The density of the solution at each position in the gradient is, in turn, proportional to its CsCl concentration.

If pieces of DNA are mixed with the CsCl and both are centrifuged to equilibrium, each piece will tend to move and become concentrated into that region of the gradient in which the density of the solution matches the **buoyant density** (symbolized by ρ and expressed as g/cm³) of the DNA (see accompanying illustration). Thus if all the pieces in a sample have the same net buoyant density, a single band of DNA will form in the gradient, whereas if the pieces are markedly different in density, two or more bands may form. To visualize the position and width of the various bands, advantage is taken of the fact that nucleic acids strongly absorb ultraviolet light at 260 nm and can be detected by optical devices. The ρ values of each class of fragment can then be calculated.

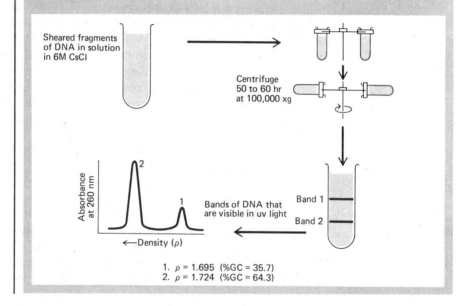

1. ρ = 1.695  (%GC = 35.7)
2. ρ = 1.724  (%GC = 64.3)

The Meselson-Stahl experiment is particularly valuable in that it not only establishes the mode of replication of DNA but also eliminates the possibility of other modes. For example, it might be postulated that replication is **conservative**: the original duplex serves as a template for a new duplex, but it also remains intact such that at the end of one round of replication an old and a new double helix are observed. Hybrid $^{15}$N-$^{14}$N DNA should not form in this case. The experimental result, however, disproves this. A **dispersive** mode of DNA replication is also ruled out by the experiment: this mode predicts that no pattern of transmission exists and that parental DNA strands break at random during the replication process in such a way that the DNA duplexes in daughter cells contain varying amounts of both old and new DNA. In this case a wide spectrum of DNA densities, rather than a single $^{15}$N-$^{14}$N species, should be detected after one round of duplication.

The mode by which DNA replicates has been termed **semiconservative** to indicate that the parental strands of DNA *are* conserved (as opposed to dispersed) but that as replication proceeds, they wind up, as it were, in two different helices.

## 2.4 Visualizing Bacterial Chromosome Replication

Replicating *E. coli* chromosomes have been visualized by autoradiography, a technique described in Box 2.2. The first such study was made by J. Cairns, who grew cells in the presence of $^3$H-thymidine for about two generations, carefully isolated their DNA, and made autoradiographs from the isolated preparations. The most favorable autoradiograph, shown in Figure 2-4, revealed an intact circular structure with two diverging segments. Cairns counted the number of exposed silver grains along the various lengths of the molecule and found that certain segments were roughly twice as radioactive as others. From these observations, he argued convincingly that the chromosome is replicated as an intact circle and that its DNA replication is semiconservative.

The Cairns autoradiogram was interpreted to indicate that the *E. coli* chromosome is replicated in a unidirectional fashion, but subsequent studies have shown that the replication of circular bacterial chromosomes is usually **bidirectional**: each chromosome possesses a unique **origin** at which replication is initiated; two growing points then travel in opposite directions around the circular chromosome, each copying 50 percent of the genome; and both meet at the **terminus** on the opposite side (Figure 2-5). Within each growing point, the original strands separate to create a **replication fork,** and each prong of the fork is copied semiconservatively. Since the entire *E. coli* chromosome is typically copied *in vivo* in 40 minutes, and since the chromosome is 1300 μm long, 15 μm of chromosome must be copied per minute within each fork. This corresponds to 750 base pairs added to each growing point per second. The elaborate enzymatic apparatus that has evolved to accomplish this feat is described in Chapter 8.

**BOX 2.2**
Autoradiography

The subatomic particles (for example, β-particles) emitted from a radioisotope will expose the silver halide grains in a photographic emulsion. This fact can be used to localize the position of molecules carrying such isotopes as $^{14}$C, $^3$H, or $^{125}$I. The sample to be studied may be overlaid by a photographic film, or a slurry of photographic emulsion can be layered over the sample and allowed to dry. In either case the sample is stored in the dark after preparation to allow for radioactive decay; the film or the emulsion-covered sample is then developed in much the same way as a photographic film. The developed silver grains in the emulsion, which appear as black dots, localize the position of the isotope with great precision.

**FIGURE 2-4
Chromosome
replication in *E.
coli.*** The original
autoradiograph is
shown; the inset
diagrams the same
structure and
divides it into three
sections (A, B, and
C) that arise at the
two forks (X and
Y). Double-solid
lines indicate DNA
with both strands
labeled; solid-
dashed regions
have only one
strand labeled.
(From J. Cairns,
*Cold Spring Harbor
Symposia on
Quantitative Biology*
**28**:44, 1963.)

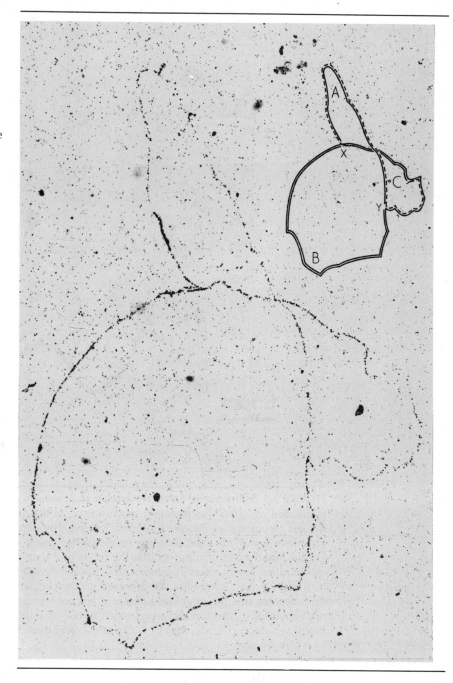

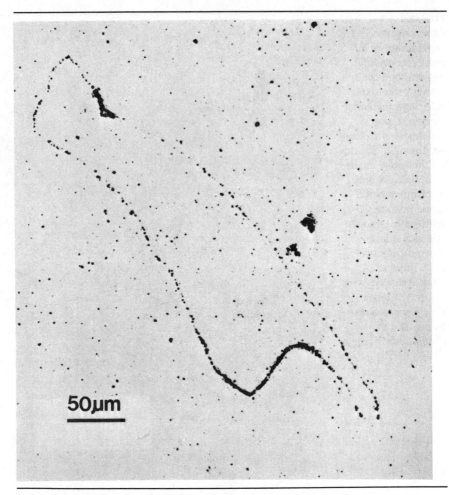

**FIGURE 2-5 Bidirectional DNA replication in *E. coli.*** A synchronous
population of *E. coli* was provided low levels of $^3$H-thymine so that their
circular chromosomes became uniformly labeled. High levels of $^3$H-thymine
were then provided at the time one round of replication was terminating and a
second round was initiating, so that the origin and terminus of each
chromosome became heavily labeled. The autoradiograph shows two regions of
high grain density symmetrically disposed on the circular chromosome. This
demonstrates that the origin and terminus are not contiguous and that
replication must therefore have proceeded in two directions. (From R. L.
Rodriguez, M. S. Dalbey, and C. I. Davern, *J. Mol. Biol.* **74:**599–604, 1973.)

# The Eukaryotic Cell and Cell Cycle

As noted in Section 2.1, eukaryotes are defined as those organisms whose
chromosomes are ordinarily confined within a membrane-limited region

called the **nucleus.** Of the vast array of single-celled eukaryotic organisms in existence, those most extensively used at present in genetic research include the yeast *Saccharomyces cerevisiae*, the cellular slime mold *Dictyostelium discoideum*, the alga *Chlamydomonas reinhardi*, and the ciliate *Paramecium aurelia*. Certain multicellular fungi such as *Neurospora crassa* and *Aspergillus nidulans* also enjoy the attention of geneticists. All such creatures are often referred to as "lower" eukaryotes to distinguish them from the "higher" eukaryotes; of the latter, the fruit fly (*Drosophila melanogaster*), corn (*Zea mays*), the mouse (*Mus musculus*), and the human being (*Homo sapiens*) are perhaps the most extensively studied from a genetic point of view.

In subsequent sections, the structure of a eukaryotic cell is first examined, with emphasis on its nucleus. There follows a description of the eukaryotic cell cycle and its culminating event, the mitotic apportionment of chromosomes from parent to daughter cells.

## 2.5  Eukaryotic Cell Structure

A eukaryotic nucleus as seen with the electron microscope is illustrated in Figure 2-6. This nucleus is not engaging in mitosis and is thus called an **interphase** nucleus. The most prominent landmark of an interphase nucleus is its **nucleolus** (Figure 2-6)—a large, deeply staining spherical body that contains DNA, RNA, and protein and represents the site of synthesis and storage of the cell's cytoplasmic ribosomes.

The chromosomes of an interphase nucleus cannot be distinguished as individual entities. Instead, they appear as an amorphous network to which early light microscopists gave the name **chromatin** (substance colored with stain), a term that has since been adopted to describe the material seen with the electron microscope (Figure 2-6).

The chromatin in an interphase nucleus is surrounded by a membrane that folds back on itself to form an **envelope.** The envelope appears in Figure 2-6 as two dense lines—the inner and outer membranes—that enclose a narrow channel or cisterna. At intervals the envelope is perforated by **pores,** each perhaps 40 nm in diameter, which are usually covered by a septum of a moderately dense and still unidentified material (Figure 2-6). At other intervals the envelope extends out into the cytoplasm as a network of channels and large cisternae known as the **endoplasmic reticulum (ER).** Some of the cell's ribosomes are often bound to the outer surface of the ER, forming what is known as **"rough"** ER (Figure 2-6).

The cytoplasm of a eukaryotic cell contains soluble enzymes, free ribosomes, and additional systems of membranes that divide the cell into a number of compartments called **organelles.** One of these organelles is the endoplasmic reticulum, and this is often associated with the Golgi apparatus (Figure 2-6), a relationship important for the synthesis of many proteins that are secreted from the cell. Other organelles include mitochondria and chloroplasts (Figure 2-6). The former, found in all eukaryotic cells, are de-

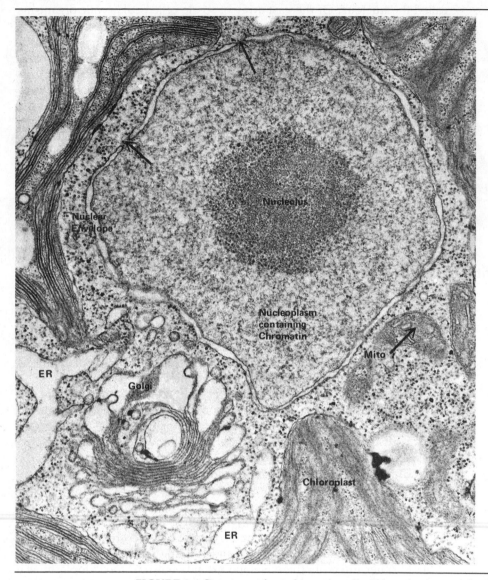

**FIGURE 2-6 Structure of a eukaryotic cell,** *Chlamydomonas reinhardi.* Arrows point to nuclear pores.

voted primarily to respiration; the latter, found only in eukaryotic plant cells, are devoted primarily to photosynthesis. Chloroplasts and mitochondria possess chromosomes of their own, and the genetics of these chromosomes has become of major interest in recent years, as will become apparent in Chapter 16.

## 2.6 The Typical Eukaryotic Cell Cycle

In simple eukaryotes such as yeast, the DNA synthesis leading to chromosome replication takes place throughout the interphase and ceases only during the brief period of mitotic nuclear and cell division. In higher plants and animals, however, DNA replication occurs only during a discrete interval of the interphase known as the **S period** (S standing for synthesis). Before and after the S period, cells engage in growth and metabolic activity but not in chromosome replication and are said to be in the **$G_1$** or **$G_2$** (G standing for gap) period of interphase. The $G_2$ period is ordinarily followed by mitosis **(M),** and the sequence $G_1 \longrightarrow S \longrightarrow G_2 \longrightarrow M$, followed by another $G_1$, is known as the **cell cycle.** When nutrients become scarce, the cells shift into a **stationary,** or **$G_0$,** phase in which cellular metabolism essentially shifts into a holding pattern until nutrients are replenished.

As shown in Figure 2-7, different eukaryotic cells vary in the length of time they take to complete an entire cell cycle; they also differ in the relative proportions of time allotted to each of the four stages of the cycle. In general, the length of an S phase is a constant property of any one cell type, but the length of its $G_1$ phase may vary considerably with environmental fluctuations, such as nutrient supply. It is also apparent in Figure

**FIGURE 2-7 The cell cycle** in various cell types. (From B. Kihlman, T. Eriksson, and G. Odmark, *Hereditas* **55**:386, 1966.)

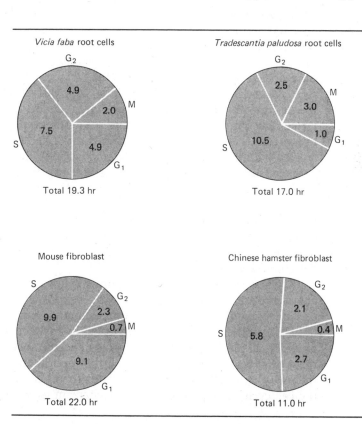

2-7 that the amount of time allotted to the actual mitotic segregation of daughter chromosomes is usually relatively brief. Even here, however, exceptions are known. In the grasshopper neuroblast, for example, mitosis may take as long as 8 hours after an interphase of only 30 minutes.

From a geneticist's point of view, the two key events in the eukaryotic cell cycle are the DNA/chromosome replication, which occurs during the S phase, and the precise allotment of these replicated chromosomes to daughter cells, which occurs at mitosis.

## 2.7 DNA and Chromosome Replication in Eukaryotes

Each chromosome in a eukaryotic cell is believed to contain a single, enormous DNA helix that is associated with a variety of proteins, a statement amplified in the next chapter. During the S phase, this helix must be copied so that it will yield two daughter helices, and a variety of studies similiar in design to the Meselson-Stahl experiment (Figure 2-4) have demonstrated that this replication is semiconservative.

In human cells, DNA is replicated at an average rate of about 0.5 μm per minute. An average human chromosome contains about 50,000 μm of DNA. Therefore, if replication were to initiate at one end of a chromosome and proceed sequentially to the other end, replication would take 1000 hours, a prediction clearly in conflict with the typical S phase duration of 6 to 8 hours. J. Huberman and A. Riggs discovered a solution to this paradox when they exposed cultured human cells in S phase to thymidine labeled with the radioactive isotope tritium ($^3$H) (recall that thymine is a DNA-specific base) for brief periods; they then isolated large DNA fragments from these cells, spread the fragments on a flat surface, and subjected the fragment preparation to autoradiography (*see Box* 2.2). They found that a given DNA fiber exhibits multiple sections that are heavily labeled with tritium (Figure 2-8), demonstrating that **each chromosome is "worked on" by replication enzymes at many different places at once.** In other words, each eukaryotic chromosome contains a number of **replication units, or replicons.** A typical replicon is about 30 μm long, and the labeling patterns indicate that DNA synthesis within a replicon is bidirectional, much as we saw for *E. coli* (*Section* 2.4). The short DNA segments that are formed in each unit eventually link up to form long, continuous daughter strands.

In many respects, replication units behave as if they were independent of one another: in Figure 2-8, a short exposure to $^3$H-thymidine labels certain segments, whereas adjacent segments remain unlabeled. Since all the DNA in the nucleus must be copied in advance of mitosis if daughter nuclei are to receive complete sets of genetic information, some mechanism must exist that recognizes when all the replication units have completed their synthetic activities and triggers the onset of the $G_2$ and M phases of the cell cycle. The nature of this mechanism is unknown.

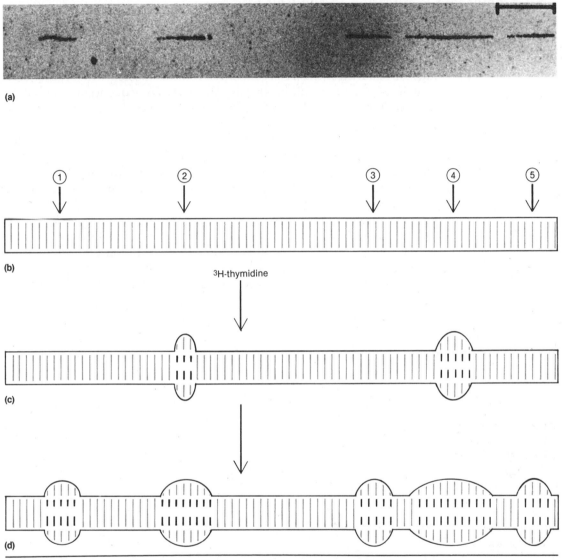

**FIGURE 2-8 Replicons.** (a) Autoradiogram of eukaryotic DNA that has been pulse-labeled with ³H-thymidine during replication. Scale indicates 50 μm. (b) to (d) Diagram of the labeling experiment. (b) Unlabeled, unreplicating DNA showing five origins of replication. (c) During a very brief exposure to ³H-thymidine, only replicons ② and ④ have initiated replication (radiolabel in black). (d) In the continued presence of ³H-thymidine, the remaining three replicons initiate replication, while the two "early" replicons expand the lengths of DNA that are copied, producing the pattern shown in (a). (Used with permission from J. A. Huberman and A. Tsai, *J. Mol. Biol.* **75:**5–12, 1973. Copyright by Academic Press, Inc. [London] Ltd.)

## 2.8  Chromosome Replication in the Cell Cycle

As we noted earlier, the genome of a eukaryotic cell is typically found not in one chromosome but in several chromosomes. As an example we can cite the cellular slime mold *Dictyostelium discoideum*. In the $G_1$ phase, each nucleus of a *Dictyostelium* cell contains seven chromosomes that collectively contain all the information (that is, all the DNA sequences) required to make a *Dictyostelium* cell. During the S phase, all of the chromosomes are replicated, so that each $G_2$ nucleus has twice the number of chromosomes; for the $G_2$ nucleus of *Dictyostelium*, this number is 14. Finally, during mitosis, one copy of each chromosome goes to each of the two daughter cells so that each daughter nucleus enters its own $G_1$ phase with the fundamental chromosome number of 7 again. These relationships are diagrammed in Figure 2-9a.

If we focus next on one particular chromosome of the seven in *Dictyostelium* (Figure 2-9b), the net result of semiconservative DNA replication is that two identical copies of its information are present in the nucleus, where only one copy was present before. To avoid confusion, these identical copies are best called **sister chromatids,** with the term chromosome

**FIGURE 2-9 The mitotic replication and segregation of chromosomes.** The replication and allotment of the entire chromosome set of *Dictyostelium discoideum* is shown in (a); a single chromosome is shown in (b). The centromere is represented by an oval structure in (b).

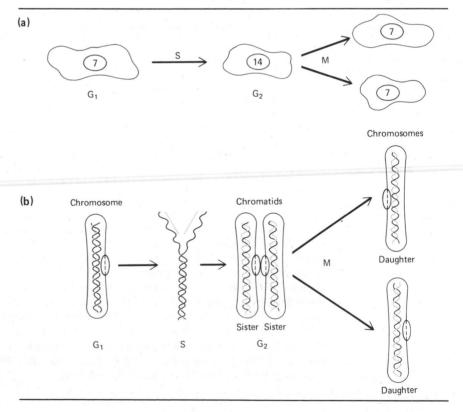

being reserved to specify an unreplicated structure. (In practice, "chromosome" is used as a catchall term to mean both replicated and unreplicated forms of genetic material; in this text we distinguish between chromatids and chromosomes only when to do so increases the clarity of an explanation.) Sister chromatids, like unreplicated chromosomes, cannot be recognized as such in an interphase (that is, in the $G_1$, S, or $G_2$ phase) nucleus. It is only when the nucleus is preparing to divide, either by mitosis or by meiosis (Chapter 5), that individual chromatids undergo a dramatic condensation process and become distinguishable through the microscope as small, deeply staining bodies (the original structures for which the term chromosome was coined). Each sister chromatid pair is seen at this time to be connected at a region known as a **centromere** (Figure 2-9b). During mitosis, the centromere region will effectively split in two, with one sister chromatid and its centromere-half going to one daughter nucleus and the other chromatid with its centromere-half to the second daughter nucleus. The separated chromatids are now called **daughter chromosomes,** as diagrammed in Figure 2-9b.

## 2.9 Mitosis

Strictly speaking, mitosis is unrelated to DNA and chromosome replication, since the latter events occur during the S period of interphase (Figure 2-9b). Mitosis is, however, an essential adjunct to chromosome duplication, since it ensures that each daughter cell gets one sister of each chromatid pair and therefore a complete set of chromosomes. The four mitotic stages—**prophase, metaphase, anaphase,** and **telophase**—are in fact arbitrarily defined, since mitosis takes place as a continuous process. Figure 2-10 diagrams the sequence of events during mitosis as it appears under the light microscope, and Figure 2-11 is a series of photographs of mitosis in the peony.

**Prophase** Mitotic prophase is heralded by the **onset of chromosome coiling** (Figure 2-10b), a phenomenon considered in the next chapter. As coiling and condensation progress (Figure 2-10c and d), the two sister chromatids of each chromosome can often be distinguished. The other notable morphological feature of prophase is that the nucleolus becomes undetectable with the light microscope. With the electron microscope it is apparent that the component particles of the nucleolus disperse throughout the nucleus during mitotic prophase, but the reason for this dispersal is not known.

**Metaphase** The onset of metaphase is marked in most cells by the **breakdown of the nuclear envelope,** releasing the chromosomes and nucleoplasm to fill the greater part of the cell (Figure 2-10e). In many protozoa and fungi, however, the nuclear envelope and nuclear integrity character-

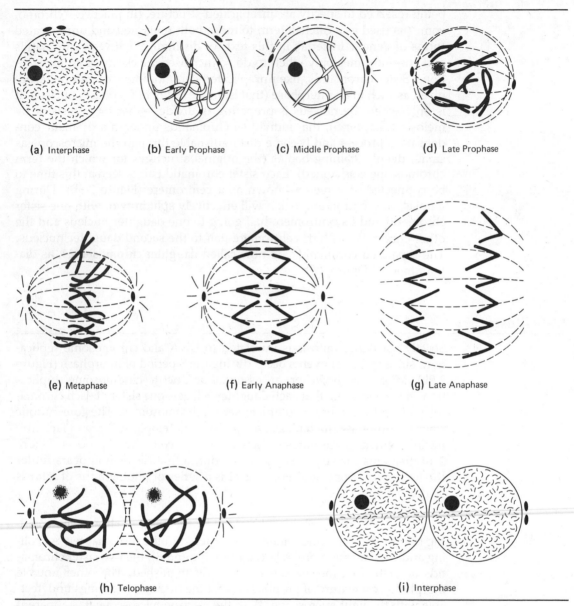

(a) Interphase    (b) Early Prophase    (c) Middle Prophase    (d) Late Prophase

(e) Metaphase    (f) Early Anaphase    (g) Late Anaphase

(h) Telophase    (i) Interphase

FIGURE 2-10 Mitosis in an animal cell.

FIGURE 2-11 Mitosis in the peony. (a) Early prophase, (b) late prophase, (c) metaphase, (d) early anaphase, (e) late anaphase, and (f) telophase. (Courtesy of M. Walters and S. Brown.)

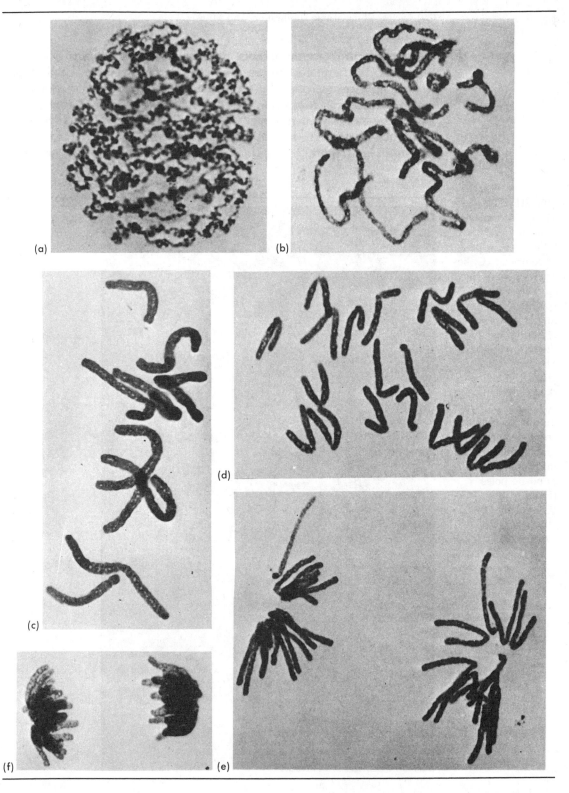

istically remain intact throughout mitosis. A more diagnostic feature of metaphase is, therefore, the presence of a complete **spindle.** Under a polarizing light microscope, the spindle appears as a refractile zone from which cytoplasmic components of the cell are excluded (Figure 2-13a). Under the electron microscope, each spindle is seen to be composed of numerous long fibers, each approximately 25 nm in diameter, known as **microtubules** (Figure 2-12, M). Antibodies (see Box 2.3) can be raised

**FIGURE 2-12**
**Mitotic metaphase** in a cat kidney tubule cell. A pair of centrioles (C) lies at one pole of the spindle, with spindle microtubules (M) radiating toward the chromosomes. The two kinetochore faces (K) of a sister-chromatid pair are indicated, and chromosomal microtubules insert into each face from opposite poles. (From P. Jokelainen, *J. Ultrastruct. Res.* **19**:19, 1967.)

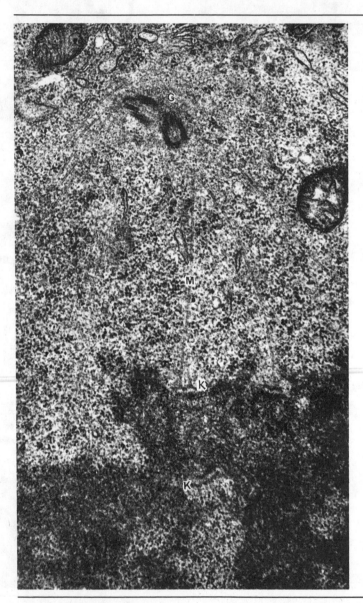

against tubulin, the major protein of microtubules, and allowed to penetrate into mitotic cells. The distribution of antitubulin is then probed using a fluorescent "second antibody" raised against the first (Box 2.3). Figure 2-13b shows the results: the full dimensions of the spindle are vividly displayed.

In most animal cells, the spindle exhibits two **poles** located on opposite sides of the cell, each pole being marked by the presence of one or two small, complex structures known as **centrioles** (Figure 2-12, C). Plant cells do not ordinarily possess centrioles, and their spindles, unlike those of animal cells, are not tapered at the poles; otherwise, the plant and animal spindle structures are similar.

Spindle microtubules originate (or terminate, depending on one's point of view) at the poles. In animal cells they typically appear to insert into dense material of unknown composition that surrounds the centriole. Some spindle microtubules, known as continuous microtubules, extend from one pole towards the other. Others, called chromosomal microtubules, extend only to the chromosomes, where they insert into the centromere regions of the sister chromatid pairs.

The site of microtubule insertion in the centromere region is called the **kinetochore.** The kinetochore typically appears as a platelike, often banded, structure (Figure 2-12, K). In favorable material a kinetochore at metaphase can be seen to exhibit two faces, one pointing toward one pole and the other toward the other pole, and comparable numbers of chromosomal microtubules attach themselves to either face (Figure 2-12).

With the insertion of the chromosomal microtubules into the kinetochores at metaphase, the sister chromatid pairs become aligned in a plane that usually passes approximately through the cell midline (the **metaphase plate**). Since the sister chromatid pairs are held together only at their common centromere regions, the arms of each pair are free to dangle (Figure 2-10e).

**Anaphase** Mitotic anaphase begins with the **division of the centromeres** or, more precisely, of the kinetochores. The sister chromatids of a given pair now separate, and each moves, kinetochore first, toward the spindle pole nearest to it (Figure 2-10f). Once the chromatids have separated, they are considered **daughter chromosomes** (*Figure 2-9b*). The difference between sister chromatids and daughter chromosomes is thus an operational one; physically, they are equivalent structures except that sister chromatids share a centromere, whereas daughter chromosomes do not. Obviously, a daughter chromosome is synonymous with a chromosome, and the term is encountered only when mitotic events are being discussed.

**BOX 2.3**
Antibodies and
Antibody Testing

A highly specific attribute of a protein is its **antigenic** properties. When a protein is injected into a foreign vertebrate animal—for example, a chicken protein into a rabbit—the rabbit will respond by producing an **antibody** to that protein. The blood serum from the rabbit (the **antiserum**) is then collected. If this is mixed with a small amount of the chicken antigen, a precipitate known as the **precipitin** is formed. Precipitin is formed because each antibody has two binding sites for the antigen (as described in Chapter 20) so that an antibody-antigen network is created.

A version of this test known as **double immunodiffusion** requires only small amounts of material. Antiserum and antigen are placed in separate wells in a gel matrix and allowed to diffuse toward each other. A solid band of precipitin forms where the antibody and antigen meet. No such band appears when the antiserum instead meets a protein against which the rabbit has not been immunized.

The location of a particular antigen in a cell is commonly determined by **indirect immunofluorescence.** Thus if the chicken protein in our example is tubulin, and the rabbit has

produced an antitubulin antiserum, the experiment is performed as follows: chicken cells are fixed, permeabilized with a membrane solvent, and incubated with the rabbit antiserum; any excess unbound rabbit antiserum is washed away, and the cell is then incubated in a second antiserum; this antiserum, which has been raised in a third vertebrate type (e.g., a goat) against injected rabbit antibody, has been tagged with a fluorescent dye. This goat-antirabbit antiserum reacts with the antigen-antibody complexes in the cell and renders them fluorescent.

**Indirect immunoautoradiography** can also be performed at the light microscope or electron microscope level. In this case, the second antibody (the goat antiserum) is tagged with a radioactive label, and the location of antibody-antigen complexes is detected by autoradiography (*Box 2.2*).

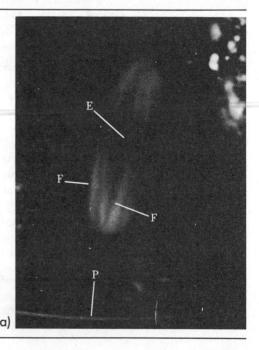

(a)

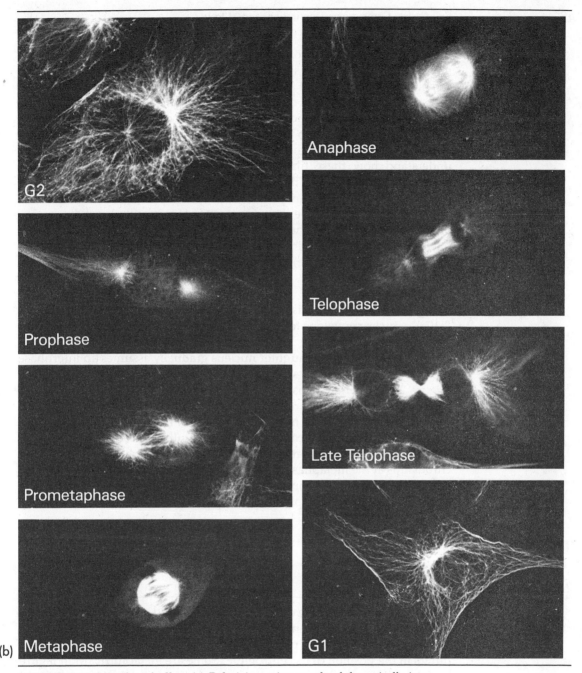

**FIGURE 2-13 Mitotic spindles.** (a) Polarizing micrograph of the spindle in a living egg from the worm *Chaetopterus*. Chromosomes (not visible) lie along the equator (E). F, fibrous microtubules; P, plasma membrane of cell. (Courtesy of Shinya Inoué, from *Chromosoma* **5**:491, 1952). (b) Cultured mouse cells stained with antibody against microtubules, showing the dramatic shifts in microtubule distribution during the mitotic process. Interphase cells (G1 and G2) display a delicate meshwork of microtubules that are thought to provide the cell with structural support. (Courtesy of Bill Brinkley)

The alignment of chromosomes along the metaphase plate and their migration to the poles at anaphase constitute one of the most dramatic activities exhibited by eukaryotic cells, and much research is currently focused on how such movements are generated. There is general agreement that the spindle microtubules are involved, since the drug **colchicine** has been shown to interfere specifically with spindle microtubule formation and to block chromosomal movements. The challenge is to understand how these filamentous macromolecules interact with one another to generate both the medial movements of metaphase and the poleward movements of anaphase. Current research indicates that some mitotic movements may be the consequence of microtubules sliding past one another, whereas other movements may result from the fact that microtubules are able to increase their length (polymerize) or decrease their length (depolymerize) in response to local changes in the ionic environment of the cell.

**Telophase and Cytokinesis**   Once poleward migration is complete (Figure 2-10g), a nuclear envelope forms around each set of daughter chromosomes, a nucleolus reforms, the spindle microtubules disappear, the chromosomes uncoil, and each daughter nucleus gradually assumes an interphase morphology. These events characterize telophase, the final mitotic stage (Figure 2-10h). Under most circumstances, telophase is followed by cell division, or **cytokinesis,** and the nuclei return to interphase (Figure 2-10i).

At the conclusion of mitosis, therefore, both daughter nuclei contain an identical chromosome complement, the result of the equal separation of the identical sister chromatids at anaphase. The orderly alignment of the chromatids at the metaphase plate and the division of each centromere into two halves that move to opposite poles ensure that the genetic material is transmitted exactly from parent to daughter cells.

# The Karyotype

## 2.10 General Features of Karyotypes

When a human metaphase cell is photographed and examined, a total of 46 sister chromatid pairs is encountered (Figure 2-14a). Further examination reveals that each of these pairs is of different size from the others and/or differs in the position of its centromere. Some are **metacentric** (M in Figure 2-14a), with the centromere positioned medially so that the four chromatid arms are about equal in length; others are **acrocentric** (A in Figure 2-14a), with two of the arms much longer than the other two; the rest are intermediate between these two extremes.

A complete description of all the chromatid pairs possessed by a given cell type constitutes its **karyotype,** and a karyotype is usually prepared by cutting out individual chromatid pairs from a photograph like Figure 2-14a and arranging them in series according to size. When this is done for humans, it becomes apparent that each type of chromatid pair is

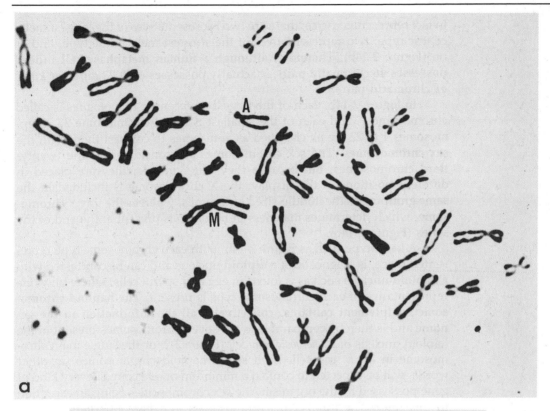

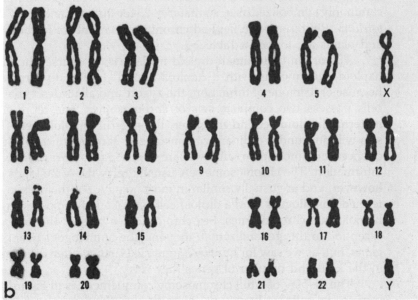

**FIGURE 2-14 Human karyotype.** (a) Mitotic chromosomes of a human male. (b) Same mitotic chromosomes arranged in homologous pairs and numbered. (Courtesy of M. W. Shaw.)

in fact represented twice; there are two representatives of the longest meta-centric type, two representatives of the shortest metacentric type, and so on (Figure 2-14b). Therefore, **although a human metaphase cell indeed possesses 46 chromatid pairs; it actually possesses only 23** *different kinds* **of chromatid pairs.**

In Figure 2-14b, each of the largest metacentric homologues is called chromosome 1, and each of the smallest is called chromosome 22. Chro-mosomes 1 to 22 are all classified as **autosomes** to contrast them with the **sex chromosomes.** The sex chromosomes are considered as the twenty-third chromosomes, but because they differ in size, they are placed in different positions in the display: the **X chromosome** is included in the same grouping as the middle-sized autosomes, whereas the tiny **Y chromo-some,** which determines maleness in humans, is present at the end of the series (Figure 2-14b).

A karyotype such as Figure 2-14b, with each chromosome type repre-sented twice, is diagnostic of a **diploid** nucleus and can be contrasted with a **haploid** nucleus such as is found in egg and sperm cells, where only one representative of each chromosome type is present. **The haploid chromo-some complement contains, collectively, all the information in the ge-nome** and is thereby comparable to the seven chromosomes present in the haploid nucleus of a *Dictyostelium* cell (*Figure 2-9*) or the single main chro-mosome in an *E. coli* cell. Each set of haploid chromosomes, in other words, will be expected to contain a minimum of one copy of every kind of gene possessed by the organism (the sex chromosomes being an exception to this rule). The number of chromosomes in a haploid complement is commonly symbolized as $n$, so that $n = 23$ for humans and $n = 7$ for *Dicty-ostelium*. Representative haploid chromosome numbers for other animals and plants are found in Table 2-2.

The origin of a human diploid nucleus is, of course, the fusion of a haploid egg nucleus with a haploid sperm nucleus to form a **zygote.** Because of its mode of formation, the zygote and all its descendant diploid cells possess two complete sets of chromosomes, one representing the **maternal haploid set** and the other, the **paternal haploid set.** These two sets will not, under ordinary circumstances, carry identical nucleotide se-quences, any more than will two parents ordinarily carry identical genetic information. The chromosome sets usually carry the same kinds of genes, however, and are usually similar in morphology. For this reason, they are termed **homologues,** and a diploid cell is spoken of as possessing pairs of homologous chromosomes. Sex chromosomes present the most common exception to this generalization: they may be homologous over certain re-gions, but, as we saw for humans (Figure 2-14), they often differ in size and in the kind and number of genes they carry.

The display of a full chromosome complement as in Figure 2-14b has come to be called a karyotype, even though the term also has the more general meaning of a description of the complement in any of its guises. Karyotypes are useful in that they permit the rapid recognition of an aber-

**TABLE 2-2  Haploid Chromosome Numbers**

| Common and Scientific Names | Chromosomes | Common and Scientific Names | Chromosomes |
|---|---|---|---|
| Human, *Homo sapiens* | 23 | Frog, *Rana pipiens* | 13 |
| Rhesus monkey, *Macaca mulata* | 21 | Toad, *Bufo americanus* | 11 |
| Horse, *Equus caballus* | 32 | Toad, *Xenopus laevis* | 18 |
| Pig, *Sus scrofa* | 19 | Goldfish, *Carassius auratus* | 50 |
| Cattle, *Bos taurus* | 30 | Housefly, *Musca domestica* | 6 |
| Sheep, *Ovis aries* | 27 | Fruit fly, *Drosophila melanogaster* | 4 |
| Cat, *Felis catus* | 19 | | |
| Dog, *Canis familiaris* | 39 | Garden pea, *Pisum sativum* | 7 |
| Rat, *Rattus norvegicus* | 21 | Bean, *Phaseolus vulgaris* | 11 |
| Mouse, *Mus musculus* | 20 | Tobacco, *Nicotiana tabacum* | 24 |
| Chicken, *Galius domesticus* | ca. 39 | Corn, *Zea mays* | 10 |
| Slime mold, *Dictyostelium discoideum* | 7 | Yeast, *Saccharomyces cerevisiae* | 17 |
| Ciliate, *Paramecium aurelia* | 30–63 | Bread mold, *Neurospora crassa* | 7 |
| Ciliate, *Tetrahymena pyriformis* | 5 | *Chlamydomonas reinhardi* | 16 |

In part from P. L. Altman and D. S. Dittmer, Eds. *Biology Data Book*, Vol. 1, 2nd ed. Bethesda, Md.: Federation of American Societies for Experimental Biology, 1972.

ration in chromosome number or morphology, as we shall see in Chapter 7. They are also of interest in establishing evolutionary relationships between different species. The mouse karyotype, for example, bears little resemblance to the human; it is made up of fewer chromosomes, all of which are small and acrocentric. The various primate karyotypes, in contrast, are all similar to one another (*see Figure 28-3*).

The allotment of the genetic material to more than one major chromosome is a feature that distinguishes eukaryotic organisms from bacteria and viruses. The multichromosome state appears to be a more manageable way to package and distribute the enormous amount of DNA present in the eukaryotic nucleus; this, however, is not its only attribute. As we shall see in Chapter 5, the processes of meiosis and sexual fusion in organisms with multiple chromosomes result in the creation of numerous new combinations of genes. Thus the multichromosomed species possess great potential for diversity and hence for evolutionary flexibility.

## 2.11 Karyotypes Using Banded Chromosomes

It should be apparent with close study of Figure 2-14 that it is virtually impossible to distinguish many of the sister chromatid pairs from one another. For this reason, the construction of karyotypes from such conventionally stained preparations often proves to be both tedious and frustrat-

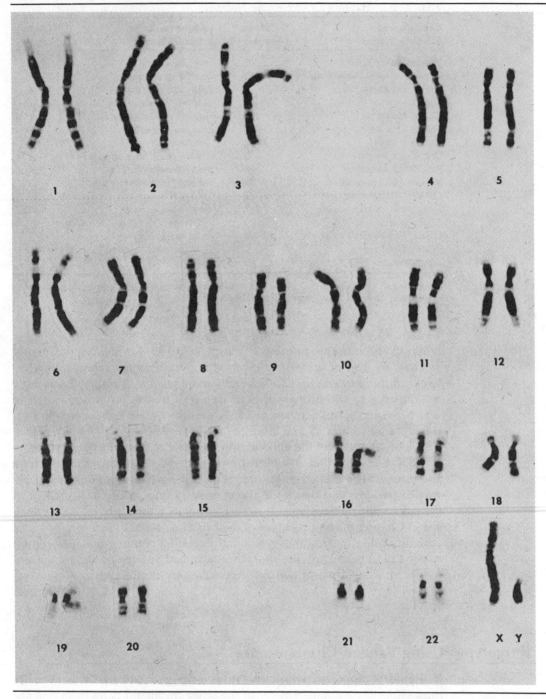

**FIGURE 2-15 Banded human karyotype.** G-banding pattern of metaphase chromosomes from a normal human male. (From J. J. Yunis and M. E. Chandler, *Am. J. Pathol.* **88**:466, 1977.)

ing. Fortunately, in the past ten years staining procedures have been developed that reveal discrete bands along the lengths of sister chromatids. These readily permit the recognition of individual chromatid pairs and the matching of homologues having similar banding patterns. A banded human karyotype is shown in Figure 2-15, and its informational superiority to Figure 2-14 should be obvious. Figure 2-16 shows, for comparison, the shapes and banding patterns in a mouse karyotype.

Of the many staining protocols that have been devised in various laboratories to produce banded chromosomes, the two most generally used are the acid-saline-Giemsa (ASG) technique, which reveals **G bands** (for **Giemsa** stain), and the quinacrine mustard tenchique, which produces fluorescent **Q bands;** technical details for both are described in Box 2.4. G and Q bands have the same locations and are presumed to reveal the same underlying chromosome structures. Although the molecular basis for the staining reactions is not well understood (theories are presented in Section 3.11), the techniques themselves have revolutionized the discipline known as **cytogenetics.**

**BOX 2.4**
Staining and
Banding
Chromosomes

*Feulgen Staining* Cells are subject to a mild hydrolysis in 1 N HCl at 60°C, usually for about ten minutes. This treatment produces a free aldehyde group in deoxyribose molecules. The aldehyde will then react with a chemical known as Schiff's reagent (basic fuchsin bleached with sulfurous acid) to give a deep pink color. The ribose of RNA will not form an aldehyde under these conditions, and the reaction is thus specific for DNA.

*Q Banding* The Q bands (from *quinacrine*) are the fluorescent bands observed after chromosomes are treated with quinacrine mustard, a dye that inserts itself (intercalates) into the DNA duplex, and are then excited with ultraviolet light. The distal ends of each chromatid are not stained by this technique. The Y chromosome becomes brightly fluorescent, both in the interphase nucleus and in metaphase.

*R Banding* The R bands (from *reverse*) are those located in the zones that do not fluoresce with quinacrine mus-

tard, that is, they are between the Q bands. They can be visualized as green, brightly fluorescent bands with acridine orange staining.

*G Banding* The G bands (from *Giemsa*) have the same location as the Q bands and do not require fluorescent microscopy. Many techniques are available, each involving some pretreatment of the chromosomes. In the acid-saline-Giemsa (ASG) technique, for example, cells are incubated in citric acid and NaCl for one hour at 60°C and are then treated with the Giemsa stain, which is a complex mixture of various positively charged dyes that interact with the phosphate groups of DNA. Proteolytic enzyme treatment also reveals these bands without addition of stain.

*C Banding* Chromosomes are exposed first to alkaline and then to acid conditions, a procedure that appears to extract most protein and DNA but leave behind the DNA located in **constitutive heterochromatin** (*see Section 3.12*). This is then visualized with Giemsa stain.

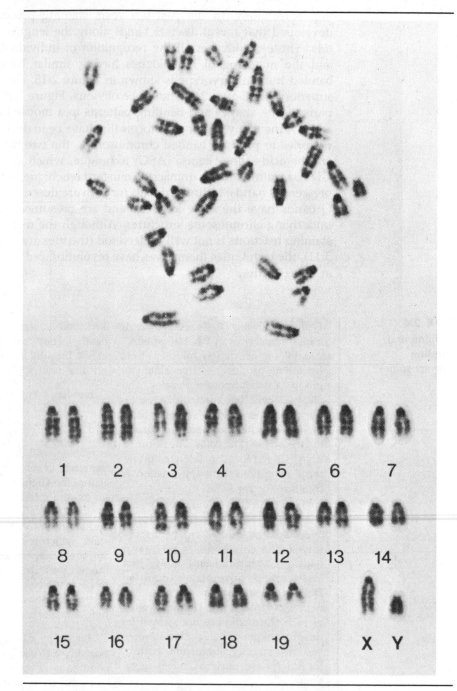

**FIGURE 2-16 Banded mouse karyotype.** G-banding pattern of metaphase chromosomes from the mouse. (From R. A. Buckland et al., *Exptl. Cell Res.* **69:**232, 1971.)

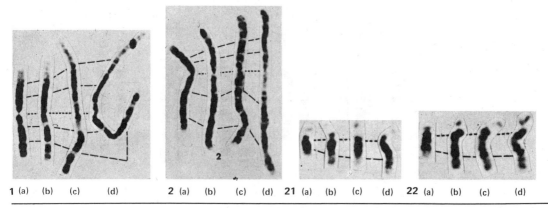

**1** (a)    (b)    (c)        (d)          **2** (a)    (b)      (c)      (d)   **21** (a)    (b)    (c)      (d)   **22** (a)    (b)      (c)          (d)

FIGURE 2-17 **Changes in banding patterns during mitosis.** Human G-banded chromosomes 1, 2, 21, and 22 at mid metaphase (a), early metaphase (b), early prometaphase (c), and late prophase (d), showing the progressive coalescence of the multiple fine bands of late prophase into the thicker and fewer dark bands of metaphase. (From J. J. Yunis, *Science* **191**:1268, 1976.)

At a conference in Paris in 1971, the bands visualized in human mid-metaphase karyotypes by such techniques were assigned a nomenclature in which the letters $p$ and $q$ represent, respectively, the short and long "arms" of a metaphase chromatid; these arms are then subdivided by numbers. A total of 350 bands were defined by the Paris nomenclature. It has recently been found that in late prophase, when mitotic chromosome condensation is less advanced, many more bands can be visualized; these proceed to coalesce with one another to produce the major bands seen in metaphase, as illustrated in Figure 2-17. Thus, in late-prophase karyotypes, it is currently possible to identify at least 2000 bands per human haploid chromosome complement. In Figure 2-18, the Paris Conference numbers are given for the left chromatid of each pair, and the late-prophase numbers are found in the right-hand chromatids. The utility of such band numbers will become evident in Chapters 7 and 15.

## 2.12 Chromosome Purification

An exciting new endeavor in cytogenetics involves the purification of individual chromosome types. In one approach, human chromosomes are liberated from colchicine-blocked cultured cells by mechanical disruption and are stained with the fluorochrome Hoechst 33258. The suspension is then passed through an instrument known as a **fluorescence-activated sorter,** wherein a laser excites the fluorescence of an individual chromosome as it passes through an orifice, and the level of fluorescence emitted determines

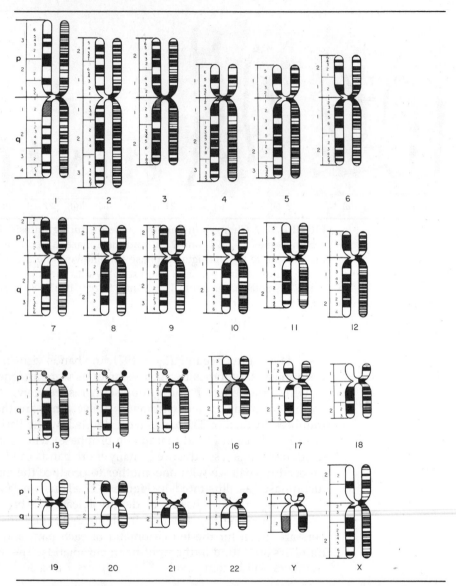

**FIGURE 2-18 Paris Conference nomenclature for human chromosomes.** G-banding patterns at mid metaphase (left chromatid of each chromosome) and at late prophase (right arm). Numbering system for the mid-metaphase bands is that established at the Paris Conference, 1971, with $p$ and $q$ being the short and long arms, respectively. The uppermost band of the first chromosome is designated 1$p$36, the next as 1$p$35, and so on.

into which receptacle that chromosome will be deflected by the sorting component of the instrument. Chromosomes with comparable size and fluorescence (for example, human chromosomes 19, 20, and 21) are all sorted into the same fraction, but 15 relatively pure fractions have been attained, and the methodology is sure to improve. As we will see in Chapter 15, it is currently possible to inject chromosomes from one cell type into another, effecting genetic transformation in much the same way as we saw for *rough* and *smooth* pneumococcal transformation in Chapter 1. The chromosome purification technology should soon allow injections of single chromosome types.

# Atypical Eukaryotic Cell Cycles

## 2.13 Cell Cycle of Ciliated Protozoa

Ciliated protozoa such as *Tetrahymena* and *Paramecium* are atypical in that each cell contains two types of nuclei: a **macronucleus,** which controls the growth and development of the organism, and one or two **micronuclei,** which function only at the time of sexual reproduction. Each micronucleus is diploid, whereas the macronucleus is said to be **hyperpolyploid,** having anywhere from 45 (*Tetrahymena*) to 800 (*Paramecium*) times the amount of DNA in the haploid complement. A macronucleus originates from a micronucleus during the sexual act of conjugation, (*Section 5.14*), proceeds to amplify its chromosome complement, and then transmits this enormous quantity of DNA to daughter macronuclei during asexual reproduction. Since no conventional mitotic configurations are observed during macronuclear division, the division process is said to be **amitotic.** It is not yet established whether a precise allotment of "genome equivalents" is distributed to each daughter during amitosis or whether the process is more random; nor indeed is it clear whether in *Paramecium,* for example, the entire genome is represented 800 times per macronucleus or whether certain chromosomes are represented more abundantly than others.

## 2.14 Cell Cycle of Dipteran Larva Cells

When a eukaryotic microorganism such as yeast is starved of nutrients, it stops dividing, enters a stationary $G_0$ phase, and eventually dies. Several cell types in dipteran larva, the best known being the salivary gland cells of the fruit fly *Drosophila,* also stop mitotic divisions after perhaps 18 hours of larval development, but chromosomal DNA replication and cell growth continue apace. As a result, huge cells are formed, and each comes to contain, by a $2 \longrightarrow 4 \longrightarrow 8 \longrightarrow 16$ doubling series, as much as 1024 times the haploid amount of DNA. The numerous copies of each chromosome type remain associated with one another, moreover, so that each of the eight chromosomes in the nucleus becomes greatly amplified in size. Finally, homologous chromosomes tend to associate with one another, a phenomenon known as **somatic pairing,** so that when the nuclei are broken open and their contents examined with the light microscope, it appears as though four enormous chromosomes are present (Figure 2-19).

The larval cells that carry out this process are said to be **endopolyploid,** and the multistranded chromosomes are termed **polytene chromosomes.** The endopolyploid cells are "terminal" in the sense that they never again divide and are eventually discarded during pupation. They remain, however, very much alive in the sense that they respond to environmental stimuli and produce specific classes of proteins (*Section 9.12*).

For the geneticist, polytene chromosomes are perhaps most fascinating in that they exhibit a series of **bands** of varying widths, each band flanked by lightly staining **interbands,** also of varying widths (Figure 2-19). The bands serve the same "landmark" function as the bands in the ASG-stained chromosomes (*Figure 2-15*), permitting a ready identification of each of the four polytene chromosomes. Even more important, the chro-

**FIGURE 2-19**
**Salivary gland**
**chromosomes** from
*Drosophila*
*melanogaster* as
visualized by
phase-contrast
microscopy. (From
L. Silver and
S. Elgin, *Proc Natl*
*Acad Sci, U.S.*
**73:**423, 1976.)

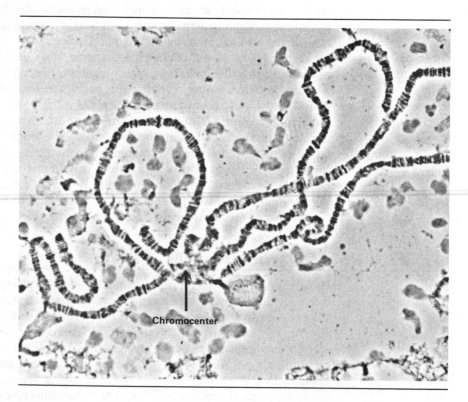

Chromocenter

mosomes are so large and the bands so numerous that it becomes possible to identify very narrow intervals along the length of each chromosome. In 1938, C. Bridges reported that the four polytene chromosomes of this organism, when widely stretched, reveal approximately 5000 to 6000 bands; more recent studies estimate 6900 bands. The question of how these bands relate to genes will be addressed in subsequent chapters of this text.

(a) Matthew Meselson (left) of Harvard University and Frank and Mary Stahl (right and front) of the University of Oregon collaborated in the classical demonstration of semiconservative DNA replication.

(b) Arthur Riggs (City of Hope Medical Center in Duarte, California) and Joel Huberman (Roswell Park Memorial Institute, Buffalo, N.Y.) provided early evidence of discontinuous DNA replication in mammalian cells.

# Questions and Problems

1. Distinguish between the following: kinetochore, centriole, centromere, chromocenter, colchicine, cytokinesis.

2. Cells entering the S phase of their cycle were exposed to $^3$H-thymidine. When these cells reached mitotic metaphase, some were fixed and subjected to autoradiography. Would you expect all the sister chromatids to be labeled, half the sister chromatids to be labeled, or some other labeling pattern? Explain with diagrams such as Figure 2-9b, where the gray DNA strands can be considered $^3$H-labeled and the black strands unlabeled.

3. The unfixed daughter cells resulting from the mitosis described in Problem 2 were allowed to undergo a second cell cycle, this time without any $^3$H label present. When the daughter cells reached mitotic metaphase, some were fixed and subjected to autoradiography. Would you expect all the sister chromatids to be labeled, half the sister chromatids to be labeled, or some other labeling pattern? Explain with diagrams such as Figure 2-9b.

4. When polytene cells are exposed to $^3$H-thymidine pulses during an endoduplication cycle and autoradiographs are made, certain bands are radioactively labeled and others are not. By following the labeling pattern of known marker bands in many different samples, it can be demonstrated that certain bands start replication before others, and that each requires distinctive lengths of time to complete replication. Relate these observations to the autoradiograms shown in Figure 2-8.

5. The DNA in an average polytene chromosome band is about 30 μ in length. In what way is this fact relevant to your answer to Problem 4?

6. If we let X be the minimum amount of genetic material that carries all the information of a species, then in an organism with a diploid chromosome number of 2, how much DNA (X, 2X, 4X, etc.) is found in the following: an egg nucleus; a sister chromatid; a daughter nucleus following mitosis; a homologue following mitosis; a nucleus at the onset of mitotic coiling?

7. Given the haploid chromosome numbers provided in Table 2-2, do these numbers increase with evolutionary advance? Explain.

8. (a) Using the conversion factors provided in Table 2-1 and the estimate of about $1.6 \times 10^{12}$ daltons of DNA per human haploid nucleus, how many genes would be present in the human genome if all its DNA contained genetic information? (b) Various estimates set the maximum number of genes in the human genome at about 50,000. What percent of human DNA would, in such a case, be devoted to carrying genes, and what percent would be involved in other functions? Can you speculate about what some of these other functions might be?

9. Distinguish between two sister chromatids and two homologous chromosomes; distinguish between a diploid and a haploid nucleus.

10. A karyotype of a mitotic nucleus from a female cat shows 76 sister chromatids. What is the diploid and haploid chromosome number of the cat? How many homologous chromosome pairs are present?

11. Related organisms frequently have similar chromosome numbers. Document this statement with several examples from Table 2-2, and note as well some exceptions evident in the Table.

12. Comparing Figures 2-15 and 2-16, do you think it would be easier to construct a mouse or a human karyotype? Explain.

13. In a Meselson-Stahl type of experiment, the following results are observed:

| Generation | Percent of DNA that Is Completely or Partially Hybrid | | |
|---|---|---|---|
| | *Semiconservative* | *Conservative* | *Dispersive* |
| 0 | 0 | | |
| 1 | 100 | | |
| 2 | 50 | | |
| 3 | 25 | | |

Fill in the columns showing predictions for conservative and dispersive replication. Explain your reasoning.

14. A strain of bacteria was cultured in $^{15}N$ ammonium ion for many generations. When the DNA from this organism is sedimented, a single band appears at a density of 1.714. The organism is then removed from its $^{15}N$ source, and replicated in $^{14}N$ for exactly one cell division. Now its DNA bands at two places in equal amounts, 1.709 and 1.705. After a second round of $^{14}N$ replications, a fourth band appears at 1.700. This band becomes more intense after each further cell division in $^{14}N$. Explain these data in terms of the strands' base composition. Hint: is the base composition of each strand of DNA necessarily equal?

15. Diagram how segregation of daughter DNA chromosomes is thought to be accomplished during the *E. coli* cell cycle.

16. Draw a diagram of a replicating *E. coli* chromosome to illustrate Figure 2-5, following the diagrams that illustrate Figure 2-8.

# 3 Molecular Organization of Chromosomes

**A. Introduction**

**B. Protein Structure**
3.1 Amino Acids
3.2 Peptide and Polypeptide Structure
3.3 Polypeptide and Protein Conformations

**C. Chromosomal Proteins**
3.4 Histones
3.5 Nonhistones

**D. The 10 nm Nucleosome Filament of Chromatin**
3.6 Nucleosome Morphology

3.7 Nucleosome Biochemistry Superhelical DNA
3.8 Nucleosome Function

**E. Higher Orders of Chromatin Organization**
3.9 The 25 nm Chromatin Fiber
3.10 Chromatin Loops
3.11 Chromosome Bands Chromomeres
3.12 Mitotic Chromosome Condensation
3.13 Constitutive Heterochromatin

**F. Questions and Problems**

## Introduction

The overall structural organization of chromosomes, discussed in Chapter 2, has a molecular complement: at the level of chromosomal protein and DNA, very regular patterns of organization have been discovered, and these are described in this chapter. After presenting a review of protein structure and describing the **histone** and **nonhistone** proteins found in eukaryotic **chromatin,** we consider how a constant unit of eight histones interacts with chromosomal DNA to create a regular array of "beads" known as **nucleosomes.** We discuss how this topology creates specific **endonuclease cleavage sites,** which allow chromosomal DNA to be clipped into units of defined length, and we describe how nucleosomes serve as "packaging devices" to shorten the overall length of chromosomal DNA in the nucleus.

We then consider how higher orders of chromosome organization might be generated. In particular, we ask how the nucleosome strands interact to form the thicker fibers found in interphase nuclei, and we describe the properties of **constitutive heterochromatin,** which remains condensed throughout interphase. Finally, we investigate how the remaining chromatin fibers coil and condense to produce the banded chromosomes of mitotic nuclei described in Chapter 2.

During the course of this and the next chapter, many biochemical procedures are explained. Since these are utilized in experiments described throughout the text, their mastery at this juncture will greatly clarify the content of later chapters. Readers with a biochemistry background will of course need to give this material less attention than will those without such background.

# Protein Structure

The following sections (Sections 3.1–3.3) give a brief account of the properties of proteins. This information is essential to an understanding of genetics, and is immediately relevant to an understanding of the chromosomal proteins that are described in the subsequent sections of this chapter.

## 3.1 Amino Acids

The basic subunits of proteins are **amino acids.** Figure 3-1 shows the 20 amino acids found in naturally occurring proteins. All of these amino acids except proline have as a common denominator the structure drawn in the shaded region in Figure 3-1. This consists of a carbon atom, known as the **α-carbon,** to which is bonded an amino group (the $\alpha$—**NH$_2$**), a carboxyl group (the $\alpha$—**COOH**), and a proton. At the pH found in most cells, the $\alpha$—NH$_2$ and the $\alpha$—COOH groups of free amino acids exist in solution as —NH$_3^+$ and —COO$^-$. The remaining portion of each amino acid is uniquely distinctive and is referred to as an **R group** (R standing for radical). Thus the R group for alanine is CH$_3$—, that for valine is (CH$_3$)$_2$CH—, and so forth (Figure 3-1). The R groups of amino acids ultimately determine most of the chemical and configurational properties of a protein.

Amino acids are commonly classified according to the chemical properties of their R groups. Amino acids are thus characterized as being either relatively nonpolar (Figure 3-1a) or relatively polar (Figure 3-1b, c, and d). The polar amino acids can be further characterized by the net charge they tend to carry under conditions of physiological pH: most polar amino acids are uncharged (Figure 3-1b), but five are charged (Figure 3-1c and d). Lysine, for example, carries an —NH$_2$ residue in its R group, and this —NH$_2$ group tends to accept protons, becoming —NH$_3^+$. Consequently, lysine is

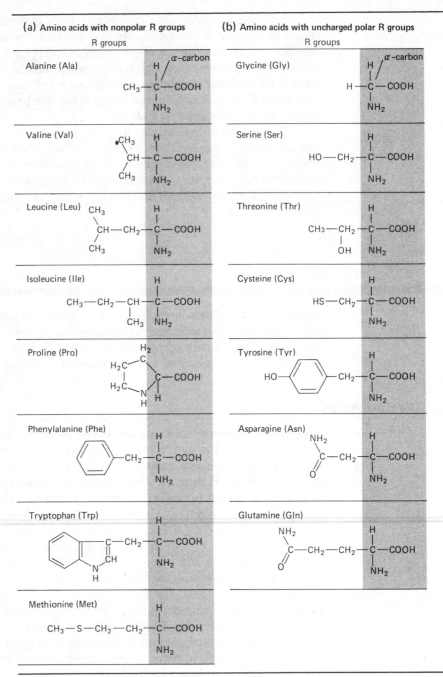

**FIGURE 3-1 Amino acid structure** for the 20 naturally occurring amino acids. The invariant regions are shaded in gray. (From A. L. Lehninger, *Biochemistry*. New York: Worth, 1970.)

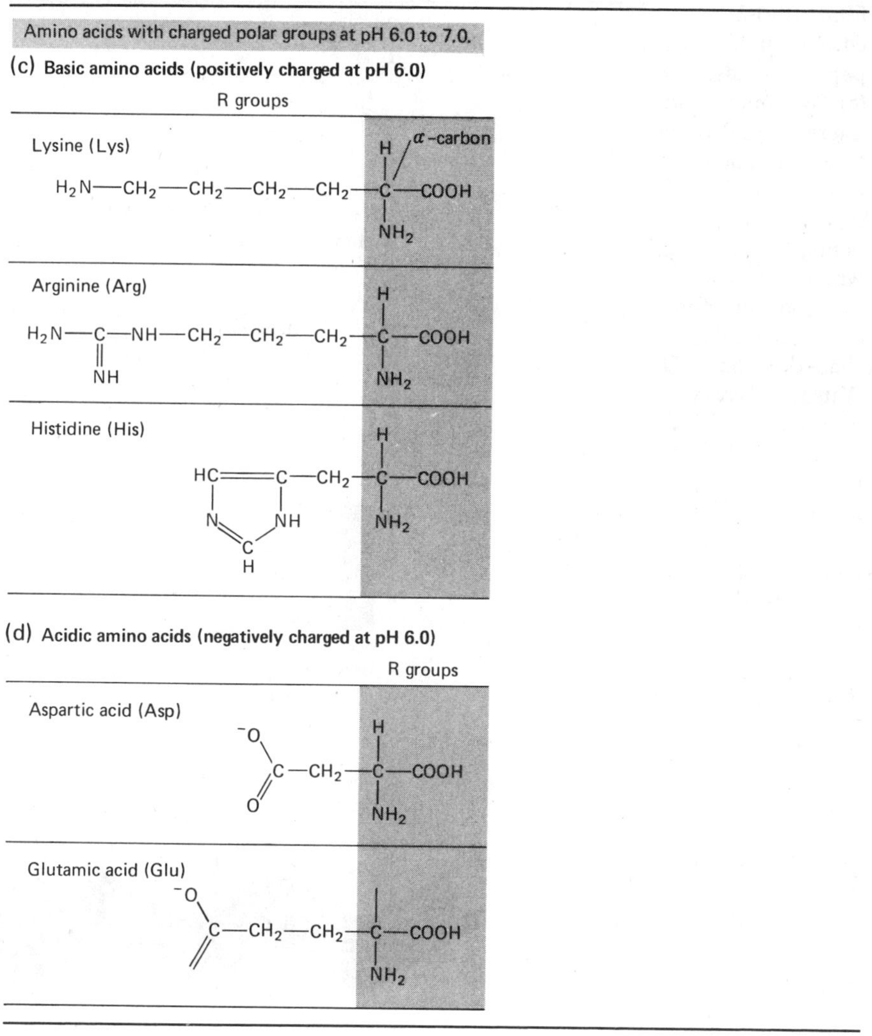

Amino acids with charged polar groups at pH 6.0 to 7.0.

(c) Basic amino acids (positively charged at pH 6.0)

R groups

Lysine (Lys)

$H_2N-CH_2-CH_2-CH_2-CH_2-C-COOH$

H / $\alpha$-carbon

$NH_2$

Arginine (Arg)

$H_2N-C-NH-CH_2-CH_2-CH_2-C-COOH$

$\parallel$
NH

H

$NH_2$

Histidine (His)

$HC=C-CH_2-C-COOH$

N    NH
C
H

H

$NH_2$

(d) Acidic amino acids (negatively charged at pH 6.0)

R groups

Aspartic acid (Asp)

$^-O$
$C-CH_2-C-COOH$
$O$

H

$NH_2$

Glutamic acid (Glu)

$^-O$
$C-CH_2-CH_2-C-COOH$

$NH_2$

called a **basic amino acid;** the two other basic amino acids are arginine and histidine (Figure 3-1c). Similarly, aspartic and glutamic acids tend to lose protons from the —COOH residues located in their R groups and are thus called **acidic amino acids** (Figure 3-1d). Other categories of amino acids, not designated as such in Figure 3-1, include the aromatic amino acids, which possess unsaturated carbon rings in their R groups, and the two sulfur amino acids, which carry a sulfur atom.

## 3.2 Peptide and Polypeptide Structure

Amino acids can become linked in the following way. The $\alpha$—$NH_2$ group of one amino acid interacts with the $\alpha$—COOH of a second, and a total of

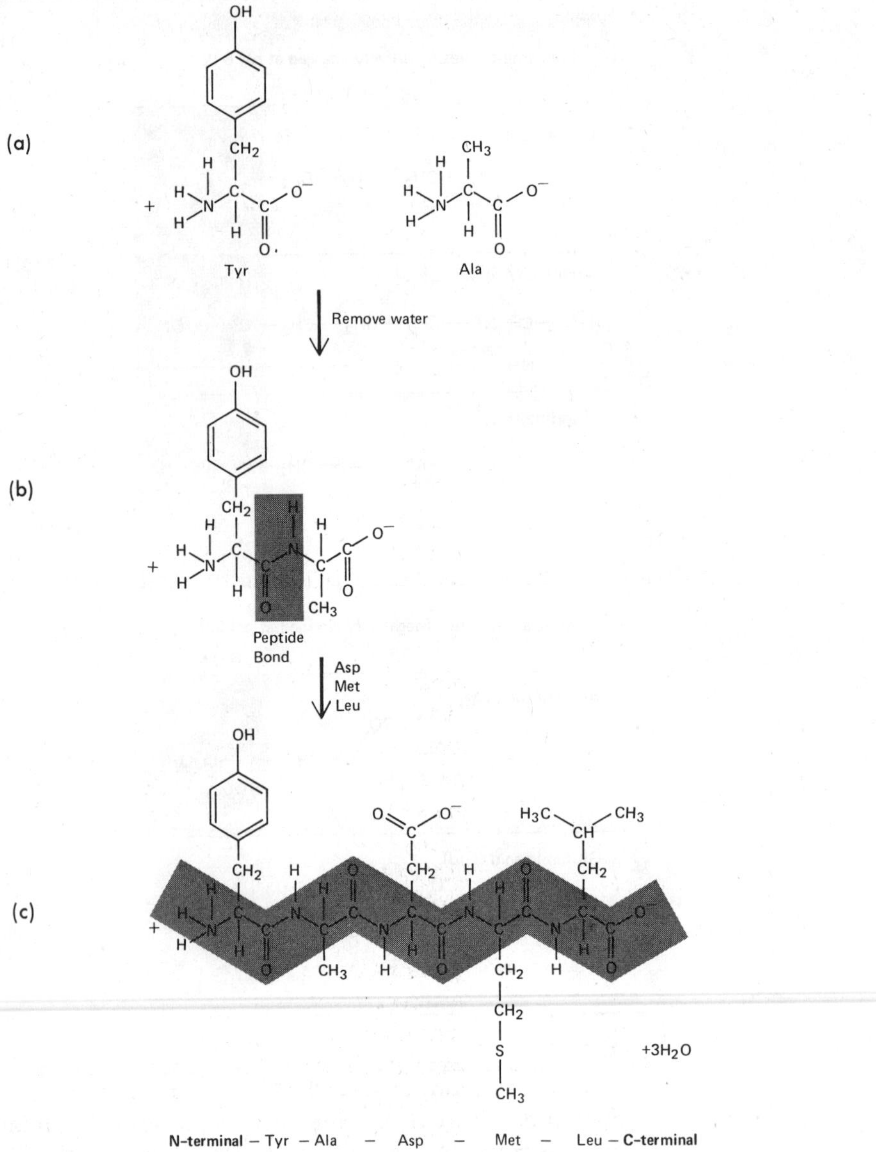

FIGURE 3-2
**Formation of
peptide bonds.**
(a) Two amino acids
interact. (b) Peptide
bond formation
results in a
dipeptide. (c) A
pentapeptide,
where the basic
backbone structure
of the molecule is
shaded. (From J. D.
Watson, *Molecular
Biology of the Gene*,
3rd ed., © 1976 by
J. D. Watson.
Menlo Park, Calif.:
W. A. Benjamin,
Inc.)

one molecule of water is removed, as shown in Figure 3-2a. The resulting
—C—N— linkage is known as a **peptide bond** (Figure 3-2b). When two or
several amino acids become so linked, the resulting polymeric structure is
known as a **peptide** (as, for example, a dipeptide having two amino acids

or a tripeptide having three). A peptide possesses a zigzagging backbone of nitrogen and carbon atoms (Figure 3-2c) that can be analogized with the sugar-phosphate backbone of a polynucleotide, with R groups projecting outward in alternating fashion. A peptide containing many amino acids is called a **polypeptide.**

The amino acid sequence of a purified polypeptide (or peptide) chain can be characterized by several biochemical procedures; here we outline a procedure frequently used in genetic studies. The amino acid at one end of a polypeptide (the **N-terminal amino acid**) possesses a free $\alpha$—$NH_2$ group, whereas the amino acid at the other end possesses a free $\alpha$—COOH group (the **C-terminal amino acid**), as shown in Figure 3-2c. These amino acids can be selectively cleaved from the ends of the chain and can be identified by chromatography, a technique described in Box 3.1. Samples of the polypeptide are then exposed to two different enzymes, **trypsin** and **chymotrypsin.** Each enzyme cleaves peptide bonds only between particular amino acids. Trypsin, for example, will cleave the bond formed between the $\alpha$—COOH of lysine or arginine and the $\alpha$—$NH_2$ of any other amino acid, whereas chymotrypsin generally cleaves the same bond, provided that the $\alpha$—COOH is donated by phenylalanine, tryptophan, or tyrosine rather than by lysine or arginine. Two collections of small peptides result; one set is composed of **tryptic peptides,** the other of **chymotryptic peptides.** The tryptic peptides can then be separated by chromatography (Box 3.1) or by gel electrophoresis (Box 3.2), and each separate peptide class is subjected to an **Edman degradation,** a procedure outlined in Figure 3-3. This yields the complete **amino acid sequence** of the peptide. The process of separation and sequencing is performed similarly with the chymotryptic peptides.

**BOX 3.1**
Chromatography

**Chromatography** is the general term used to describe a variety of techniques that separate molecules according to their size and charge. Molecules are allowed to move through a solid supporting material that is permeated with a solvent; the nature of the material and the solvent can be selected so that separations are effected between molecules having the size range of interest. The term chromatography derives from the fact that in certain early procedures the separated molecules were visualized by appropriate stains. Molecules are now often detected by their ultraviolet absorption or, if radiolabeled, by devices that detect radioisotopes.

Chromatographic procedures that are widely used by molecular geneticists are described in the subsequent paragraphs.

*Paper Chromatography* Amino acids and small peptides are frequently identified by this procedure. A sample is applied to one end of a piece of filter paper, and known amino acids or peptides are applied beside it. The paper is then dipped into a solution containing water and organic solvents (for example, isopropanol or butanol). As the solvent slowly moves up the paper by capillary action, the amino acids (or peptides) in the sample will migrate at characteristic speeds, depending on how strongly they are adsorbed to the water-saturated fibers in the paper relative to their affinity for the more mobile organic solvent. When the solvent front reaches the top of the paper, the paper is dried, sprayed with **ninhydrin** (an amino acid stain), and heated. The amino acids or peptides in the sample appear as purple dots and can be identified by comparison with the positions of the known samples.

*Ion Exchange Chromatography* This technique utilizes columns filled with synthetic resins that carry fixed charge groups. For the analysis of amino acids, the column contains particles with sulfonic acid groups attached, and these acid groups are equilibrated with NaOH so that all are fully charged with $Na^+$ ions. An amino acid mixture is now applied to the top of the column. The amino acids are in solution at pH 3, at which pH they all carry a net positive charge. As they move down the column, particular amino acids have a greater or lesser tendency to displace the $Na^+$ and interact with the sulfonic acid groups. Specifically, the basic amino acids will be bound most tightly (and so move most slowly); the acid amino acids are bound the least tightly (and therefore travel fastest). The column is now washed (eluted) with buffers of increasing pH and $Na^+$ concentration, and because the amino acids move at different rates, they appear in the **effluent** (at the bottom of the column) at different times and can be collected in separate fractions. A device known as an **amino acid analyzer** performs all these operations automatically so that the amino acid composition of a protein can be determined overnight. Ion exchange chromotography can also be used to separate a mixture of polypeptides or proteins that differ in their overall net charge.

*Molecular-Exclusion (-Sieve) Chromatography* This technique is appropriate for separating a mixture of proteins. A column is packed with small particles, one widely used preparation being known as **Sephadex.** Pores exist *between* the particles through which proteins can pass. In addition, pores exist *within* the particles, since each particle consists of a meshwork of long carbohydrate polymers. Therefore, when a protein solution is placed on top of the column and allowed to flow through by gravity, smaller proteins will tend to move into and be retarded by the small intraparticle pores, whereas larger proteins will be able to flow only through the large interparticle pores. As a result, large proteins will appear in the effluent sooner than the small, and proteins can be effectively separated by size.

*Affinity Chromatography* This procedure allows a specific component of interest to be purified from a mixture of components. For example, if one wishes to isolate RNA molecules with long tracts of adenine on one end (poly-A), one first takes a material such as Sephadex, attaches polythymine (poly-T) oligonucleotides to it, and uses it to pack a column. An RNA mixture is then allowed to pass through the column. The poly-A–containing RNA species with an **affinity** for the poly-T will stick to the matrix; the rest will flow through. The retained species is then eluted from the column using a solvent that disrupts the affinity associations. Similar logic applies to an antibody affinity column (*Box 2.3*) wherein a purified antibody, attached to a solid matrix, is used to retain its specific antigen from a mixture of components.

**BOX 3.2**
Gel Electrophoresis

The technique of electrophoresis involves placing molecules in a high-voltage field such that they move toward the positive or the negative pole, depending on their net charge. Movement takes place within a solid substrate permeated with a buffered aqueous solution. The pH of the buffer, the strength of the electric field, the length of time that the field is applied, and the nature of the solid substrate can all be varied, and when the appropriate combination of conditions is chosen, molecules differing only slightly in their net charge can be separated from one another.

**Polyacrylamide gel electrophoresis** is suitable for the separation of peptides and polypeptides, as well as oligonucleotides (short DNA or RNA fragments) and larger polynucleotides. A narrow column of polyacrylamide gel is prepared, and the sample is layered on top. The gel is then placed in an electric field. When low concentrations of polyacrylamide are present in the gel (for example, 5 percent), relatively large polypeptides can move into the gel; at higher concentrations (for example, 12 percent), large molecules may be excluded. Thus the size range of the molecules under examination can be selected. Following electrophoresis, the gel can be stained with a dye such as Coomassie blue so that its component bands are visible, or it can be scanned with ultraviolet light and its absorbance plotted. Molecular weights are obtained by performing standard gel runs with markers of known molecular weight.

Many extensions of this basic technique have been developed. For example, peptides or polypeptides can be labeled with such radioisotopes as $^{125}I$ or $^{14}C$ and subjected to electrophoresis in **slab gels** in which the polyacrylanide gel is molded to be flat and paper-thin. The gel is then dried and sandwiched next to an X ray film for **autoradiography** (*see Box 2.2*). The black lines on the resultant autoradiogram reveal the positions of the labeled macromolecules. A second extension, very valuable for separating a complex array of macromolecules, is called **two-dimensional polyacrylamide electrophoresis:** a radioactively labeled sample is subjected to electrophoresis under one set of conditions (e.g. an 8 percent gel at pH 3.5 in 6 M urea); the gel is then laid on top of a second slab gel (e.g. 16 percent polyacrylamide at neutral pH), and its contents, which were partially separated from one another in the "first dimension," move into the second gel and become even better separated under the second set of conditions.

An important distinction in the protein methodology is whether the gel is run under **nondenaturing** or **denaturing** conditions. In a nondenaturing system, proteins are maintained in their intact, oligomeric configuration, which is necessary when enzyme assays are to be performed directly on the gel. In a denaturing system, proteins are first dissociated into their component polypeptides, usually by exposing them to reagents that disrupt disulfide linkages and to the detergent **SDS** (sodium dodecylsulfate), which disrupts all other noncovalent associations. "SDS gels" are necessary when the molecular weights of individual polypeptide species are to be determined.

For large polynucleotides, gels are commonly composed of agarose rather than polyacrylamide. The position of DNA fragments in the gel is usually determined by staining with **ethidium bromide,** a fluorescent dye that interacts with DNA, and detecting the fluorescence by ultraviolet irradiation.

FIGURE 3-3 The
Edman degradation
to determine the
amino acid
sequence of a
tetrapeptide. (After
A. L. Lehninger,
*Biochemistry*. New
York: Worth, 1970.)

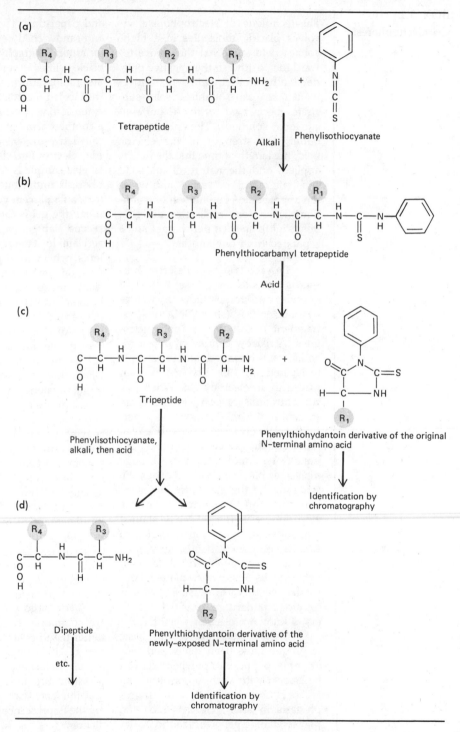

FIGURE 3-3 The Edman degradation to determine the amino acid sequence of a tetrapeptide. (After A. L. Lehninger, *Biochemistry*. New York: Worth, 1970.)

At the end of all of these analyses, the amino acid sequences of numerous tryptic peptides and numerous chymotryptic peptides are known. Because the two enzymes cleave the polypeptide at different positions, many of these amino acid sequences will overlap one another: a tryptic peptide $NH_2$—Ala—Gly—Gly—Lys—COOH may correspond to the chymotryptic fragment $NH_2$—Ala—Gly—Gly—Lys—Ser—Phe—COOH, and so on. By a laborious comparison of these sequences, the overall amino acid sequence for an original polypeptide can eventually be generated, much as will be described for the sequencing of nucleic acids in Chapter 4. An automated apparatus is now available that subjects entire polypeptide chains to a sequential Edman degradation, thereby determining amino acid sequences without enzymatic hydrolysis.

## 3.3 Polypeptide and Protein Conformations

The **primary structure** of a polypeptide chain is simply the sequence of its amino acids. This single chain forms characteristic twists and turns upon

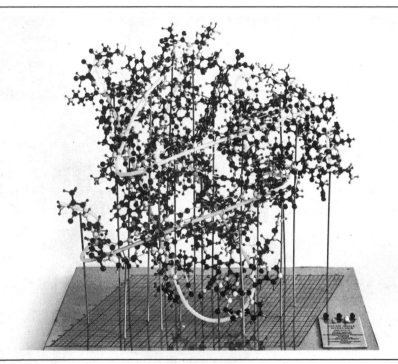

**FIGURE 3-4 Structure of myoglobin** deduced from data derived from high-resolution (2 Å) X ray diffraction. The secondary structure of the polypeptide is formed by the twists and turns taken by the single chain, many of which are α-helical. The tertiary structure of the polypeptide—its overall topology—is indicated by its overall shape in space. (Courtesy of Dr. J. C. Kendrew, Cambridge University.)

itself (the **secondary structure**) and eventaully assumes a characteristic overall shape in space (the **tertiary structure**) (Figure 3-4). Two, three, or many polypeptides may also proceed to associate with one another, the resulting protein being said to possess a **quaternary structure.** Perhaps the best-known example of such an **oligomeric protein** is hemoglobin, which contains four separate polypeptide chains called globins—two α-chains and two β-chains—elaborately fitted together (Figure 3-5).

All of these interactions—secondary, tertiary, and quaternary—are ultimately determined by the primary structure of the polypeptide chain or chains. For example, during the folding process in a single chain, nonpolar R groups of the amino acids tend to orient toward the interior of the poly-peptide chain, whereas polar groups, and particularly charged groups, tend to become localized on the polypeptide's outer surface. Sulfur-con-taining amino acids, when they come into contact, often form **disulfide (S-S) linkages** that tend to stabilize the protein. Since the conformation of a protein ultimately determines its activity, amino acid changes in the pri-mary sequence of a protein may cause it to assume an altered secondary or

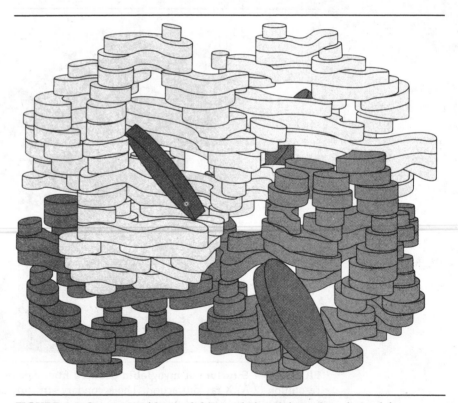

**FIGURE 3-5 Structure of hemoglobin** as deduced from data derived from X ray diffraction. The α-chains are light gray; β-chains are black; heme groups are shown as gray discs. Each heme group binds oxygen, and there are four per molecule. (From "The Hemoglobin Molecule," by M. F. Perutz. Copyright © November 1964 by Scientific American, Inc. All rights reserved.)

tertiary conformation and hence to lose its biological activity. Such amino acid changes, of course, are ultimately elicited by gene mutations; therefore, as we examine "wild" and "mutant" phenotypes throughout this text, we shall in fact usually be referring to the effect exerted by a change in the primary sequence of a polypeptide on the overall topology of a protein.

With this much background in protein structure, we are now in a position to consider the proteins that interact specifically with DNA to form the chromosomes of prokaryotic and eukaryotic cells.

# Chromosomal Proteins

## 3.4 Histones

**Eukaryotic chromatin consists of DNA complexed with histone proteins and a variety of nonhistone proteins.** Histones are very rich (20 to 30 mole-percent) in the basic amino acids lysine and arginine (*Figure 3-1*) and are therefore referred to as **basic proteins.** The lysines and arginines, moreover, tend to cluster toward the N-terminal end so that, at physiological pH, most histones have a positively charged N-terminus.

Five major types of histone are found in most cells, each type differing in its relative content of arginine and lysine (Table 3-1). Within each major type of histone, a number of subtypes are found, all similar to one another. The five types appear to be quite constant from one cell to another and from one tissue to another. In other words, **there is generally no evidence that specific types of histones are found only in specific cell types.** The most prominent exception to this generalization is provided by sperm cells. The DNA of these cells frequently loses most of its histone and becomes associated with other kinds of basic proteins.

**Histones and DNA are present in mammalian chromatin in approximately equal proportions on a weight basis.** This proportion is maintained throughout the mammalian cell cycle (*Section 2.6*) via a strict "coupling" between histone synthesis and DNA replication: histones are synthesized only during the S phase, and an experimentally induced interruption of

**TABLE 3-1  Histones From Calf Thymus**

| Class | Fraction | Lys/Arg Ratio | Total Amino Acids | Molecular Weight |
|---|---|---|---|---|
| Very lysine rich | H1 | 20.0 | 215 | 21,000 |
| Lysine rich | H2A | 1.25 | 129 | 14,500 |
| | H2B | 2.50 | 125 | 13,774 |
| Arginine rich | H3 | 0.72 | 135 | 15,324 |
| | H4 | 0.79 | 102 | 11,282 |

From S. C. R. Elgin and H. Weintraub, *Ann. Rev. Biochem.* **44**:725, 1975.

DNA replication is rapidly followed by a cessation of histone biosynthesis. **Prokaryotes lack the histones found in eukaryotes** and have a relatively low protein/DNA ratio. They do possess basic proteins that associate with their chromosomal DNA, but little is currently known about the roles these proteins might play in the biology of prokaryotic chromosomes.

## 3.5 Nonhistones

The nonhistones are defined as those proteins other than the histones that are isolated together with DNA in purified chromatin. Since these proteins are often more loosely associated with DNA than are the histones, and since most estimates indicate that the nonhistone class includes at least 20 major proteins and hundreds of minor proteins, these proteins are understandably more difficult to purify and study. Interest in these proteins is, nonetheless, extremely high, since a number of reports claim that particular nonhistone proteins either are confined to particular types of cells or else are far more abundant in certain types of tissues than in others. Such tissue specificity suggests, of course, a regulatory role for the nonhistone proteins in determining which sectors of the chromosomes are to express their genetic information, a concept explored more fully in Section 25.18.

A particularly abundant class of eukaryotic nonhistone proteins is the **high-mobility group,** or **HMG,** so called because of their relatively low molecular weight and hence rapid migration through a gel matrix (*Box 3.2*). They resemble histones in their high content of basic amino acids, but they differ in other properties, and their synthesis is not strictly coupled to that of DNA. As detailed in Section 25.18, the HMG proteins are strong candidates as proteins that regulate gene expression.

# The 10 nm Nucleosome Filament of Chromatin

## 3.6 Nucleosome Morphology

**The chromatin of the eukaryotic nucleus is organized into filaments that have a beaded substructure,** each bead known as a **nucleosome.** This organization is best seen if eukaryotic nuclei are gently lysed in a low ionic strength buffer and their contents are spread onto an electron microscope grid. The thinnest chromatin filaments present have a diameter of about 10 nm and, when examined carefully, each proves to consist of closely packed, 10 nm nucleosome units (Figure 3-6). Recalling that a DNA duplex has a diameter of about 20 Å or 2 nm, these beaded 10 nm filaments clearly represent a different form of DNA than the linear polynucleotide chains described in Chapter 1.

X ray diffraction analyses, coupled with the biochemical studies to be outlined in later sections, have generated the molecular model of nucleo-

**FIGURE 3-6
Nucleosomes.**
Chromatin fibers
from a chicken
erythrocyte nucleus
showing
nucleosomes
(arrows) (about
55 × 110 Å) and
connecting strands
(140 Å in length).
The strands become
evident when the
nucleosomes are
stretched during
isolation. (Courtesy
of A. L. Olins.)

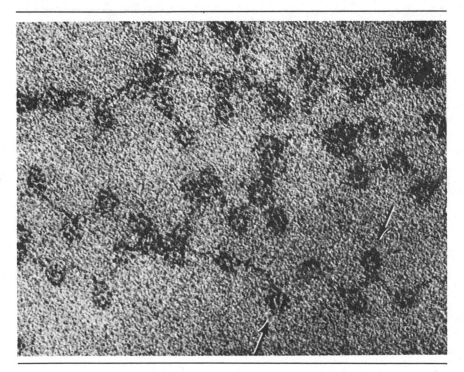

some structure shown in Figure 3-7a. Each nucleosome contains a **core of histone proteins,** around which is wrapped two full turns of DNA duplex. The histones are oriented such that their positively charged N-termini extend outward and interact with the negatively charged phosphate groups of the DNA, and their C-terminal ends interact with one another in the center of the particle.

Prokaryotic DNA is fundamentally different from eukaryotic DNA in that it is not organized into stable nucleosomes. With gentle lysis procedures, *Escherichia coli* chromosomes have been observed to have a beaded substructure, but because of the low protein/DNA ratio in these chromosomes, the beads are not likely to have a morphology homologous to that of nucleosomes. **Therefore, the presence of histones and the histone-mediated packaging of DNA into nucleosomes are unique to the eukaryotic line of evolution.**

## 3.7 Nucleosome Biochemistry

Biochemical studies of nucleosome structure have utilized enzymes known collectively as **deoxyribonucleases,** or **DNAses** (the properties of enzymes are reviewed in Box 3.3). Since the relevant enzymes cut at **internal** sites along a polynucleotide chain, they are known more specifically as *endo*nucleases, the most widely used being **micrococcal endonuclease.** The effect of

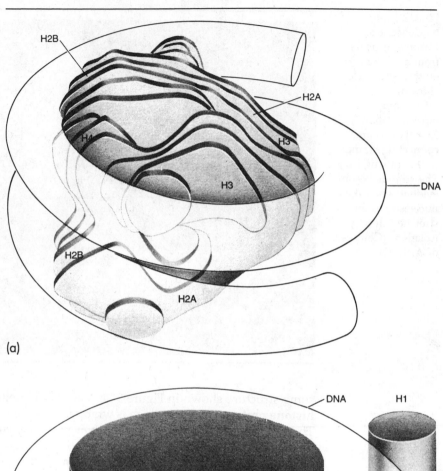

(a)

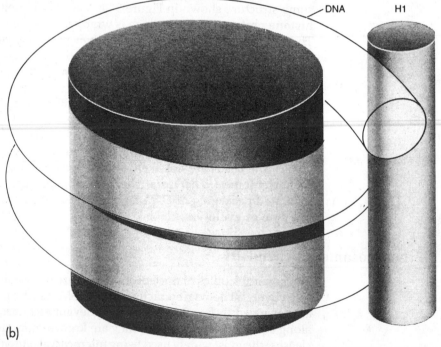

(b)

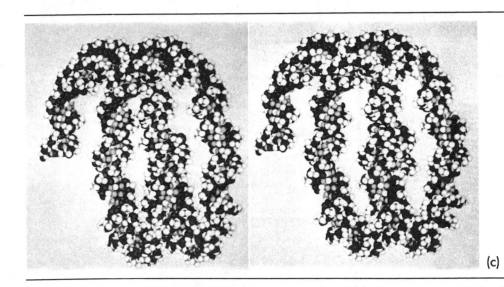

(c)

FIGURE 3-7 Nucleosome models. (a) Nucleosome core particle. One hundred forty-five base pairs of DNA (tube-like structure) wind around a histone core, an octamer of H2A, H2B, H3, and H4. The octamer was modeled from a three-dimensional map based on electron micrographs, and it is seen to form peripheral ridges upon which the DNA is believed to wind. The location of the histones is based on chemical cross-linking data. (b) Chromatosome. One hundred sixty-five base pairs of DNA wrap around the histone core, with the histone H1 located at the sites where the DNA enters and leaves the chromatosome. (Diagrams (a) and (b) from *The Nucleosome* by R. D. Kornberg and A. Klug, copyright © February 1981 by Scientific American, Inc. All rights reserved.) (c) Superhelical DNA. Stereo pair of a computer reconstruction of the superhelical twists adopted by eukaryotic DNA as it wraps around a histone core (facing sideways compared with parts (a) and (b)). Similar twists are adopted by DNA in prokaryotic and phage chromosomes. (See Figure 1-8 for a description of stereo viewing.) (Courtesy of Richard Feldman, NIH.)

enzyme digestion on chromatin structure depends on the length of time the enzyme is allowed to work. If exposure is very brief to allow only the most sensitive sites to be cut, then when the DNA is isolated and subjected to gel electrophoresis (*Box 3.2*), a series of discrete bands is displayed, the smallest being about 200 base pairs in length, the next being 400 base pairs, and so on. In other words, **very nuclease-sensitive phosphodiester bonds occur at intervals of approximately 200 base pairs along the length of chromatin filaments.** When the digested chromatin is examined by electron microscopy, the original 10 nm polynucleosome filaments (Figure 3-8a) are found to have been converted into a collection of mononucleosomes, dinucleosomes, and so on. Thus the endonuclease-sensitive target sites occur *between* nucleosomes, as diagrammed in Figure 3-8b.

**FIGURE 3-8**
**Substructure of the**
**10 nm filament of**
**eukaryotic**
**chromatin** revealed
by endonuclease
digestion.

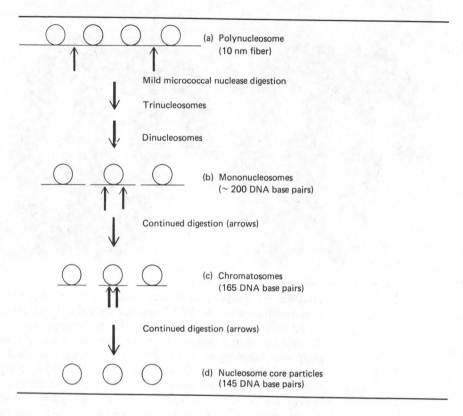

Further digestion with micrococcal endonuclease effects two additional sets of cuts: the mononucleosomes are progressively converted into **chromatosomes** (Figure 3-8c) and then into **nucleosome core particles** (Figure 3-8d), which have no further "dangling" pieces of DNA. Each of these particle types has been subjected to biochemical analysis. The mononucleosomes and the chromatosomes prove to have the same histone composition: each contains one molecule of H1 and a **histone octamer**—two molecules each of histones H2A, H2B, H3, and H4—which together compose the histone core (details of the octamer structure are given in Figure 3-7). The nucleosome core particle, by contrast, contains only the histone octamer: the H1 protein is missing. Therefore, as diagrammed in Figure 3-7b, H1 appears to associate with the base pairs of DNA that enter the chromatosome and the base pairs that leave it, these being the nucleotides digested during the chromatosome ⟶ core particle conversion.

Since chromatosomes contain about 165 base pairs of DNA and mononucleosomes have about 200 base pairs, it follows that there exists additional **linker DNA** between contiguous nucleosomes, as diagrammed in Figures 3-7 and 3-8. The length of linker DNA proves to vary with chromatin source, ranging from 0 to 80 base pairs.

**BOX 3.3**
Properties of
Enzymes Relevant
to Molecular
Genetics

An enzyme is a protein that speeds up the rate of a chemical reaction in a cell. Any biochemical reaction—the synthesis of a large sugar molecule from two smaller molecules, for example—is theoretically capable of occurring at intracellular temperatures and pH in the absence of an enzyme, but it will occur only infrequently. This is because a certain amount of energy, called the **activation energy,** must be supplied to the two molecules so that they can form the proper chemical bonds; the molecules must also be oriented properly with respect to each other so that the bond-forming atoms can react. An enzyme has the effect of allowing molecules to interact more often and more readily so that the activation energy for the reaction is lowered. Enzymes therefore permit biochemical reactions to take place at reasonable rates under intracellular conditions and are said to **catalyze** such reactions. An enzyme is not "used up" as it catalyzes a reaction; instead, it is free to interact with fresh substrates and to catalyze repeated reactions.

Enzymes catalyze biochemical reactions by interacting with the reacting molecules, known as **substrates.** For many enzymes this interaction is highly specific, and a given enzyme will interact only with particular molecules or, on occasion, with slightly modified versions of these molecules. The basis of this specificity is thought to reside in the three-dimensional conformation of the enzyme itself. The polypeptide chain or chains that interact to form an enzyme become oriented to create what is called an **active site.** This site is so constructed that it has a high specific affinity for the substrate molecule or molecules. The diagram below illustrates, quite schematically, the relation between substrates and active sites in the enzyme glucokinase.

Certain enzymes are specific for particular classes of covalent bonds, rather than for particular substrate molecules, and it is these enzymes that are of special interest to the geneticist. Some of them act to catalyze the *breakdown* of a particular bond: the **deoxribonucleases,** for example, catalyze the hydrolysis of the phosphodiester bonds that link nucleotides in a DNA molecule (*Section 1.1*). Other enzymes are commonly involved in the *formation* of particular bonds, an appropriate example being the **DNA polymerases** that promote the formation of DNA polymers. The diagram on the next page illustrates, schematically, the mode of action of a polymerase enzyme. (Diagrams from J. D. Watson, *Molecular Biology of the Gene*, Copyright © 1965 by J. D. Watson. Menlo Park, Calif.: W. A. Benjamin, Inc.)

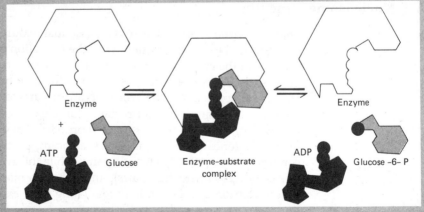

Enzyme    +    ATP    Glucose    ⇌    Enzyme–substrate complex    ⇌    Enzyme    ADP    Glucose –6– P

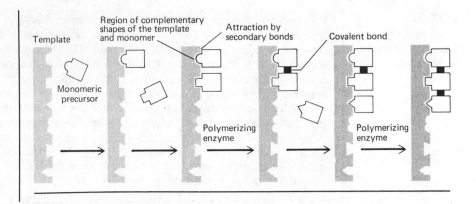

**Superhelical DNA** Before leaving nucleosome structure, it is important to look more closely at the conformation of the DNA itself and introduce the concept of a **superhelical twist** or a **supercoil.** You will recall from Chapter 1 that the two polynucleotide chains of a DNA duplex wind around each other in a right-handed helix, making one complete turn every ten base pairs (*Figure 1-8*). These right-handed twists are, by convention, designated as positive. In addition, **all known natural DNA is superhelical: the axis of the duplex itself follows a helical path in space.** For virtually all DNA, the sense of these superhelical turns is opposite to that of the duplex; therefore, they are called **negative supertwists.** Thus eukaryotic DNA takes negative superhelical twists as it wraps around the "spool" of the histone octet, as illustrated in Figure 3-7c. Available evidence indicates that one or more enzymes participate in creating these twists in the helix and that they are maintained and stabilized by histone interactions.

The superhelical topology of DNA is essential to its replication and, probably, to its recombination as well, a concept elaborated in Sections 8.2 and 17.5.

## 3.8 Nucleosome Function

That nucleosome substructure is an essential feature of eukaryotic chromatin is attested to by its ubiquity: with the exception of the dinoflagellates, it is found in the chromatin of every animal, plant, protist, and fungus examined. A similar conservatism is found in the core histones: the H2A, H2B, H3, and H4 proteins from diverse organisms are very similar in amino acid sequence and can in fact form stable complexes with one another *in vitro.*

What, then, do nucleosomes represent? They are much too small to represent genes (as we shall see, the coding portion of an average gene comprises about 1000 nucleotide pairs, while the average nucleosome comprises only 200 nucleotide pairs), and although many investigators believe that nucleosomes play a role in regulating gene expression (*Section 25.18*), this has yet to be proved. **The most generally accepted "purpose" of nucleosomes, therefore, is to package long, slinky DNA into compact filaments** that are more easily accommodated into the confines of a eukaryotic nucleus. The diameter of a chromatosome "spool" is about 5.5 nm (Figure 3-7b); around this is wound 166 base pairs of DNA, which, when fully extended, would measure about 60 nm. The **packing ratio** effected by nucleosome formation, therefore, is about 10:1. From a genetic perspective, a significant feature of this packing mechanism lies in its topology: **at no point is the DNA buried; instead, it is freely exposed along the entire surface of the "spool,"** available for genetic expression.

**Nucleosomes cannot be static structures.** As we will see in later chapters, the essential processes of DNA replication, transcription, and recombination (*Chapter 1*) all require that the hydrogen-bonded base pairs of a DNA duplex be capable of opening up so that they become accessible to specific enzymes. Therefore, some sort of **nucleosome "breathing"** must occur. In addition, there is evidence for **nucleosome phasing:** base pairs found in one nucleosome at one time may, at a later time, leave that nucleosome and twist around an adjacent nucleosome. Such dynamic events are thought to be stimulated when particular histones become chemically modified or when nonhistone proteins interact with nucleosome components, as will be detailed in the chapters on gene regulation. The important concept to master at this point is that all the "activities" of the eukaryotic genome are dictated by DNA that is organized into nucleosome beads.

# Higher Orders of Chromatin Organization

## 3.9 The 25 nm Chromatin Fiber

The 10 nm nucleosome filaments are seen only if chromatin is exposed to dilute salt solutions. When the ionic strength is increased to physiological levels, each filament self-associates to form a fiber whose diameter is variously estimated at 20 to 30 nm; here we will denote it as a **25 nm fiber.** (Figure 3-9a to c illustrates the relative proportions of DNA, the nucleosome filament, and the chromatin fiber.) The histone H1 apparently mediates this chromatin folding, since 10 nm filaments depleted of H1 fail to associate in an orderly fashion when provided with salt.

Detailed models of the substructure of the 25 nm fiber are shown in Figure 3-10a and b, and the legend to this figure gives additional information on its organization. Formation of the 25 nm fiber brings about a reduction in the length of the DNA by a further factor of 5. Therefore, the overall packing ratio effected by nucleosome-filament and chromatin-fiber formation together is $10 \times 5 = 50:1$. **The 25 nm fiber is thought to represent the fundamental unit of chromatin structure in the interphase nucleus.**

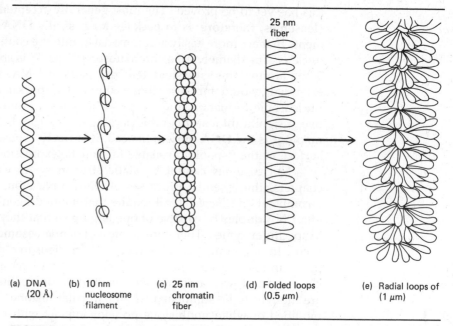

(a) DNA (20 Å)    (b) 10 nm nucleosome filament    (c) 25 nm chromatin fiber    (d) Folded loops (0.5 μm)    (e) Radial loops of (1 μm)

FIGURE 3-9 Levels of chromatin structure. The diameters and configurations of the larger fibers are approximations only. (After M. P. F. Marsden and U. Laemmli, *Cell* 17:849, 1979, and D. E. Comings, *Ann. Rev. Genetics* 12:25, 1978.)

## 3.10 Chromatin Loops

The final level of chromatin organization we will consider is diagrammed in Figure 3-9d and e. Each 25 nm fiber forms a series of loops, each about 0.5 μm in length. The bases of the loops are held close together along a single axis (Figure 3-9d), and the association of many such looped structures creates the **radial-looped fiber** shown in Figure 3-9e.

Chromosome loops can be visualized in preparations such as the one shown in Figure 3-11. The metaphase chromosome has been depleted of all its histones and many nonhistones, allowing its component DNA to spew out. At the top of the figure, each strand is seen to loop and head back toward the central core. The loops depicted here are much longer than the 0.5-μm loops depicted in Figure 3-9, since histone removal has destroyed the primary nucleosome configuration and hence the integrity of the 25 nm fiber. Nevertheless, the micrograph clearly demonstrates a fundamental radial loop organization for the metaphase chromosome, which, many believe, is present as well in the interphase chromosome.

The histone-depleted loops depicted in Figure 3-11 measure some 20 to 30 μm. You will recall (Section 2.7) that the average length of a eukaryotic DNA replicon is also found to be about 30 μm. This correspondence suggests that the higher order loops may have functional significance in DNA replication as well as in chromatin packaging. Specifically, D. S. Coffey and his collaborators have proposed that a complex of replication enzymes is anchored at the base of each loop and that the DNA of a given loop threads its way through the enzyme complex as it is being copied.

## 3.11 Chromosome Bands

Figure 3-9e assumes a uniform density of radial loops along the length of a chromosome, an assumption that may or may not prove to be true for the poorly understood structure of interphase chromatin. Readers of Chapter 2 (*Section 2.11*) will recall, however, that when individual mitotic chromosomes first become visible in early prophase, the distribution of chromatin along their lengths is decidedly nonuniform (*Figure 2-17*): at least 2000 alternating light and dark bands are visualized in the human karyotype by the Giemsa staining procedure, and these appear to coalesce to produce the landmark G bands of metaphase chromosomes. Figure 3-12 diagrams a model of how these mammalian bands might be generated, but the true mechanism is not yet known.

The relationship of genes to the small units that coalesce to form mammalian G bands is at present quite uncertain: some investigators believe these units contain genes, whereas others have proposed that genes instead reside in interband DNA. A similar uncertainty surrounds the relationship between genes and radial loops, and it can safely be said at present that we have little understanding of how genes are organized in the higher order structures of mammalian chromatin.

The bands in dipteran larval chromosomes (*Section 2.4* and *Figure 2-19*) are thought to have an origin similar to the diagrams in Figure 3-12: during successive rounds of endopolyploidization, larval chromosomes apparently retain a partially condensed mitotic configuration and therefore display a nonuniform distribution of chromatin along their lengths. These are then amplified as the chromosomes align side by side, as drawn in Figure 3-13.

There is considerable evidence, detailed in later chapters, that the bands of polytene chromosomes represent units of both DNA replication and transcriptional control. Since an average *Drosophila* band contains about $30 \times 10^3$ base pairs (or 30 **kilobases**) and most eukaryotic genes are found to range in size from 1 to 15 kilobases, each band contains enough DNA for one or several genes. Therefore, the classic adage derived from breeding experiments—"one gene/one band-interband"—will probably give way to a more complex statement: "single dipteran bands may in some cases contain one gene and in other cases more than one gene, plus additional DNA that may serve regulatory or mechanical functions."

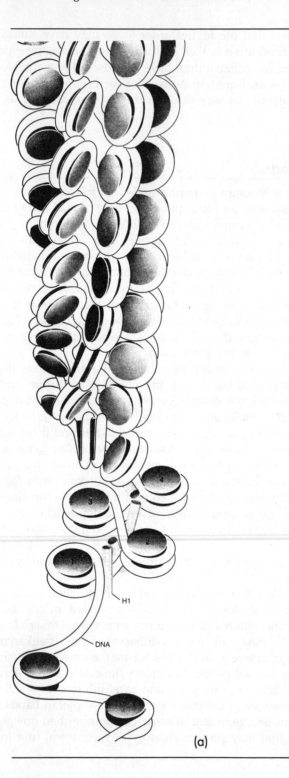

(a)

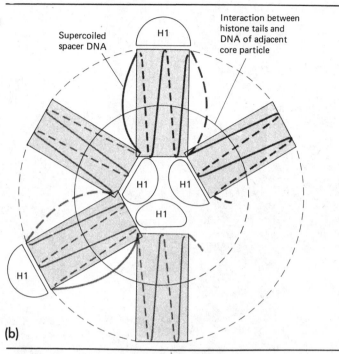

**(b)**

**FIGURE 3-10 Twenty-five nanometer chromatin fiber.** (a) A model showing the formation of the 25 nm fiber in longitudinal section. As salt concentrations are raised to physiological levels (from bottom to top of figure), H1 associates with the linker DNA and the chromatosomes pack together to form the 25 nm fiber which, as drawn here, has a helical pattern, termed a solenoid, with about six chromosomes per turn. (From *The Nucleosome* by R. D. Kornberg and A. Klug, copyright © February 1981 by Scientific American, Inc. All rights reserved.) (b) A model showing the fiber in cross section. Six chromatosomes are arranged radially in the fiber; only five are shown here (gray shading); the sixth would be dropping down to the next level. The DNA is shown associated with the chromatosomes (thin lines) or as linker DNA between chromatosomes (thick lines). The linker DNA loops up between one pair of neighboring chromatosomes (thick solid lines) and loops down between the adjacent pair (thick dashed lines); the chromatosome DNA similarly loops up and down around the chromatosome (thin solid and dashed lines), as detailed in Figure 3-7. Histone H1 is placed at the point of entry and exit of DNA from the chromatosome; because the chromatosomes are packed in a solenoid configuration, then on one chromatosome H1 is in the middle of the fiber, on its neighbor it is on the outside, and on its next nearest neighbor it is back in the center again. (After J. D. McGhee, et al., *Cell* **22**:87, 1980.)

**FIGURE 3-11 Chromosome scaffold.** A metaphase chromosome from the human-derived HeLa cell line. Histones and many nonhistones have been removed by treating with dextran sulfate and heparin, leaving a **scaffold** with the shape of two sister chromatids. A halo of DNA radiates from the scaffold core and forms loops at the periphery (see top of micrograph). Bar = 2 μm. (From Paulson, J. R., and U. K. Laemmli, *Cell* **12**:817–828, 1977.)

**FIGURE 3-12
Model of
chromosome
banding.**

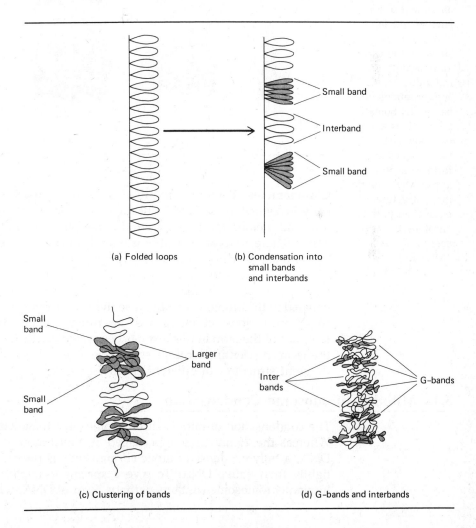

(a) Folded loops

(b) Condensation into
small bands
and interbands

Small band

Interband

Small band

Small band

Small band

Larger band

(c) Clustering of bands

Inter bands

G-bands

(d) G-bands and interbands

**FIGURE 3-13
Model of Polytene
Chromosome
Banding.**
(a) Diagram of an
elemental
chromosome within
a giant polytene
chromosome
showing folded
configurations
alternating with
extended
configurations.
(b) Lateral
alignment of four
such elemental
chromosomes
showing the
amplification of
individual
condensations into
bands. (c) Somatic
pairing between
two polytene
homologues to
form what is
apparently a single
giant structure,
equivalent to the
chromosome arms
seen in *Figure 2-19*
(After E. J. DuPraw
and P. M. M. Rae,
*Nature* **212**:598,
1966.)

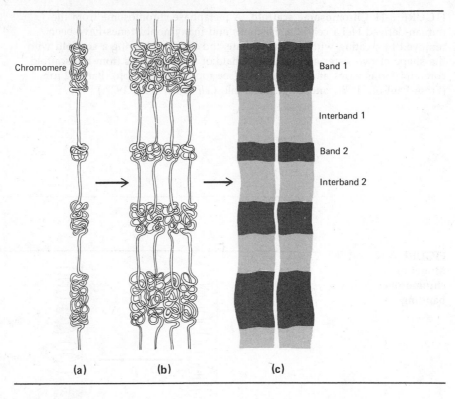

Chromomere

Band 1

Interband 1

Band 2

Interband 2

(a)          (b)          (c)

**Chromomeres**   Before leaving the subject of bands, we should note that
early investigators observed a beaded substructure along condensing mitotic and meiotic chromosomes and named the individual beads **chromomeres.** There exists at present considerable uncertainty about how chromomeres relate to chromosome bands; however, individual dipteran bands
are often claimed to represent laterally amplified chromomeres, and the
small preprophase bands of mitotic chromosomes (*Figure 2-17*) are often
equated with chromomeres. Because there is currently no reason to believe
that these various "chromomeres" are functionally or structurally equivalent, use of the term in this text has been avoided. Readers who encounter
the term in scientific writings should be aware that it may refer to quite
different structures in different chromosome types.

## 3.12  Mitotic Chromosome Condensation

The condensation of mitotic chromosomes is a remarkable phenomenon.
Whereas the 25 nm fiber is packaged 50 times more tightly than native
DNA, a fully condensed mitotic chromosome is packed 5000 times more
tightly than native DNA. To give a specific example, an average-sized
human chromosome contains perhaps 5 cm of DNA, which, as a nucleo-

some filament, is 0.5 cm in length and 10 nm wide. This must condense into a sister chromatid perhaps 1 μm wide and 10 μm long, a feat comparable to packaging a thin strand of material 40 meters long (roughly, half a football field) into a cylinder $1 \times 8$ mm.

Coiling is thought to begin at numerous sites along the length of each 25 nm fiber, causing short segments of the fiber to condense about their own axes. In synchronized cell cycles (*Section 2.6*), wherein most nuclei are induced to enter prophase at about the same time, it is found that histone H1 is specifically phosphorylated at just the time that chromosome condensation is detectable. Since H1 appears to play a key role in chromatin fiber formation (*Figures 3-9b and 3-10*), it is attractive to assign this protein a role in mitotic condensation as well. Changes in nonhistone proteins have also been reported, however, and the condensation process, as yet poorly understood, will likely prove to involve the coordinated activities of a number of nuclear components.

## 3.13  Constitutive Heterochromatin

An important exception to the rule that chromatin condensation occurs only in the mitotic phase of the cell cycle is given by a particular class of chromatin known as **constitutive heterochromatin.** Throughout interphase, constitutive heterochromatin remains condensed, and therefore deeply staining, in contrast to the faintly staining **euchromatin** that fills the rest of the nucleus, and it typically associates closely with the inner membrane of the nuclear envelope (*Figure 2-6*). Blocks of heterochromatin in the interphase nucleus are known as **chromocenters,** and these may fuse to form a single large composite chromocenter (*see, for example, Figure 2-19*). At early prophase, constitutive heterochromatin is considerably more condensed than the remaining chromatin, whereas at late prophase it is often identified as those regions where sister chromatids appear joined together.

In metaphase karyotypes, regions of constitutive heterochromatin are best identified by the **C-banding** technique (*see Box 2.4*), "C" standing for constitutive heterochromatin. These "bands," which are perhaps more properly described as zones, are always localized to particular chromosome sites; in the mouse, for example, C bands are restricted to the centromere regions of every chromosome, whereas in *Drosophila*, the entire Y chromosome, perhaps one third to one half of the X chromosome, and one quarter of each of the large metacentrics (chromosomes 2 and 3) are stained by the C-banding procedure and remain condensed throughout interphase (Figure 3-14). Another favored location for constitutive heterochromatin is at the ends of chromosomes (also known as **telomeres**). Figure 3-15, for example, illustrates the C-banding patterns in two species of rye (*Secale*). Both species display constitutive heterochromatin in their telomeres, but *S. cereale* (cultivated rye) has twice as much as its noncultivated (*S. silvestre*)

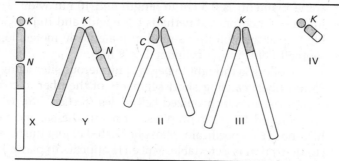

FIGURE 3-14 **Constitutive heterochromatin** in *Drosophila*. Heterochromatin is shaded. K denotes a kinetochore. N and C represent the location of secondary constrictions, constricted sectors within heterochromatin. N has been speculated as the location of nucleolar DNA (Sections 2.5 and 10.7). (From A. J. Hilliker, R. Appels, and A. Schalet, *Cell* **21**:607–619, 1980.)

counterpart, a finding that may give clues about evolutionary relationships between the two.

Constitutive heterochromatin differs from euchromatin not only in its condensation patterns but also in its replication patterns during the cell cycle (*Section 2.8*). In both *Drosophila* and humans, for example, replication

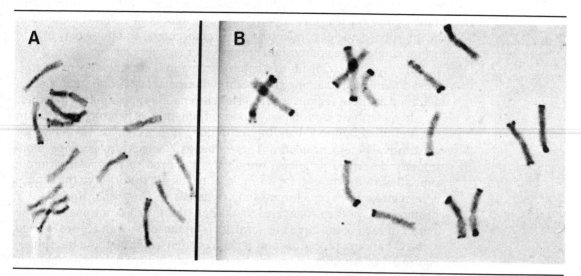

FIGURE 3-15 **Telomeric heterochromatin** in *Secale* species. (a) *S. silvestre* metaphase chromosomes. (b) *S. cereale* metaphase chromosomes. Giemsa C banding was used to visualize the heterochromatin. (From T. Bedbrook et al., *Cell* **19**:545, 1980.)

of constitutive heterochromatin (visualized by autoradiography of ³H-thy-midine–pulsed cells) is restricted to a brief period late in the S phase. Thus, even though it is located in a number of different chromosomes, the whole class of constitutive heterochromatin receives the same kind of differential treatment from the replication enzymes. During the atypical cell cycle that leads to polytene chromosome formation in *Drosophila*, moreover, the **late-replicating** behavior of constitutive heterochromatin gives way to an **under-replicating** behavior: those chromosome segments known to contain constitutive heterochromatin are represented, at most, only a few times in polytene chromosomes, whereas the rest of the genome undergoes as many as ten rounds of replication.

Finally, constitutive heterochromatin differs from euchromatin in its genetic content. Genetic analyses, presented in later chapters, indicate that it contains very **few functional genes.** Moreover, the DNA sequences of constitutive heterochromatin are different from those of euchromatin (*Section 4.10*), and the presence of heterochromatin can affect the expression of genes in contiguous, euchromatic portions of a chromosome (*Section 25.16*). The function of this class of chromatin is as yet a total mystery, but one line of speculation suggests that it may be of great importance to the three-dimensional structure of the interphase nucleus. Thus D. E. Comings has proposed that different chromosomes might be maintained in particular relationships to one another, and to the nuclear envelope, by virtue of associations between their blocks of constitutive heterochromatin, and these relationships might be an important means of controlling gene expression.

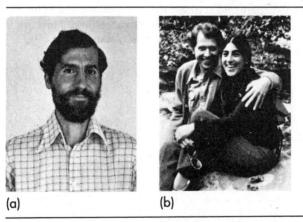

(a)          (b)

(a) Christopher Woodcock (University of Massachusetts, Amherst) and (b) Ada and David Olins (Oak Ridge National Laboratory, Tennessee) were the first to discover the nucleosome substructure of eukaryotic chromatin.

# Questions and Problems

1. Explain why it is more correct to state that the fundamental unit of chromatin is a "string on beads" rather than, as is popularly stated, "beads on a string."

2. Why might histone phosphorylation (the addition of $-PO_4^=$ groups onto histone proteins) alter the structural organization of chromatin?

3. Order the following products of trypsin and chymotrypsin digestion into the sequence of a complete peptide, showing the cleavage sites of each enzyme.

   3 trypsin products: $NH_2$-Phe-Ala-Arg; $NH_2$-Ala-Pro-Gly-Gly;
   $$NH_2\text{-Ala-Gly-Tyr-Lys}$$
   3 chymotrypsin products: $NH_2$-Phe; $NH_2$-Ala-Arg-Ala-Gly-Tyr;
   $$NH_2\text{-Lys-Ala-Pro-Gly-Gly}$$

4. You wish to analyze the proteins myoglobin and hemoglobin and run each on both a nondenaturing and a denaturing (SDS) gel. What kinds of gel patterns would you obtain in each case?

5. Prokaryotic chromosomes have low protein/DNA ratios compared with eukaryotic chromosomes. This fact is used to argue that they lack a nucleosome organization. Explain the reasoning.

6. Design an experiment to learn whether the synthesis of histones and HMG proteins is coupled to the synthesis of DNA in a tissue culture cell line whose cell cycle can be synchronized, meaning that all the cells can be induced to enter $G_1$ at the same time, enter S at the same time, and so on. What result would you expect for each protein type?

7. The eukaryotic genome can be visualized as DNA, as 10-nm filaments, as 25-nm fibers, and as mitotic chromosomes. Describe the structure of each and the conditions and molecules needed to go from one to the next.

8. Diagram the structure, composition, and interconversion of 10-nm filaments, nucleosomes, chromatosomes, and nucleosome core particles.

9. List the eight ways that constitutive heterochromatin differs from euchromatin.

10. Compare and contrast the chromosome bands found in early prophase, metaphase, and larval dipteran chromosomes. How are these thought to relate to genes? To replicons? To chromomeres?

# 4

# Molecular Analysis of Chromosomal DNA and Genetic Engineering

A. **Introduction**

B. **Overall Composition of Genomic DNA**
- 4.1 Amount of DNA (C Value) per Genome
- 4.2 Chemical Composition (GC Content) of Genomic DNA
- 4.3 CsCl Banding Patterns of Genomic DNA: Satellite Bands

C. **Reassociation Kinetics ($C_0t$ Plots) of Genomic DNA**
- 4.4 Dissociation-Reassociation Analysis
- 4.5 Analysis of Reassociation Kinetics

D. **The "Kinetic Classes" of Eukaryotic DNA**
- 4.6 Single-Copy Sequences

- 4.7 Highly Repetitive Sequences
- 4.8 Middle-Repetitive Sequences
- 4.9 Inverted-Repeat Sequences
- 4.10 In Situ Hybridization of DNA Kinetic Classes

E. **Restricting, Sequencing, and Cloning DNA**
- 4.11 Restriction and Modification Enzymes
- 4.12 Sequencing DNA
- 4.13 Cloning Vectors
- 4.14 Preparation of DNA for Cloning
- 4.15 Preparation and Propagation of Recombinant Cloning Vectors
- 4.16 Screening for Recombinant Clones

F. **Questions and Problems**

## Introduction

The previous chapter describes interactions between chromosomal DNA and protein. We now delve one level deeper and look at chromosomal DNA itself, stripped of its protein associations and possessing but a single diagnostic feature: the sequence of nucleotides along its two strands. Within these sequences are, of course, discrete informational units known as genes, and the analysis of genes will occupy most of our attention in this

93

text. In this chapter, however, we will consider important features of chromosomal DNA that can be analyzed independently of gene function. These features not only point up striking differences between prokaryotes and eukaryotes but also have been critical to an understanding of gene function and of molecular evolution.

We begin with entire genomes, asking how much DNA is present per nucleus and the base composition of the DNA. We then describe techniques wherein total genomic DNA is sheared into fragments and the fragments are separated by centrifugation. Such procedures reveal the presence of fragments with base compositions that are markedly different from the genomic average, earning them the name **satellite DNA.** The fragments can also be characterized by a parameter known as **dissociation-reassociation kinetics;** in addition, this approach reveals the existence of discrete classes of DNA, each of which is believed to serve specific genetic functions.

The final sections of the chapter consider available techniques for characterizing specific DNA fragments. We describe the use of **restriction endonucleases** that cut DNA at specific sequences and thereby generate collections of unique fragments that can be separated by gel electrophoresis. We then illustrate how such fragments can be spliced into plasmid or viral chromosomes and cloned in *Escherichia coli* hosts using **recombinant DNA technology,** allowing one to generate large quantities of specific DNA sequences. Finally, we show how the order of nucleotides in these cloned fragments can be determined; this methodology is utilized heavily in the molecular analysis of genes and the DNA sequences that regulate their expression.

# Overall Composition of Genomic DNA

## 4.1 Amount of DNA (C Value) per Genome

The total amount of DNA in a prokaryotic nucleoid or in a eukaryotic haploid nucleus is commonly referred to as its **C value.** The **C value of a eukaryotic nucleus is perhaps 1000 times greater than a bacterium** and 100,000 times greater than a virus; the actual numbers are summarized in Figure 4-1. Figure 4-1 also indicates that eukaryotes vary markedly among themselves with regard to the amount of DNA they possess. Thus a fungus such as yeast contains not much more DNA than *E. coli; Drosophila,* with ~$1.4 \times 10^8$ base pairs per haploid genome, has 70 times the C value of *E. coli;* humans, with $2.87 \times 10^9$ base pairs, have 20 times as much as *Drosophila;* and salamanders have as much as $8 \times 10^{10}$ base pairs per haploid genome. Although it is perhaps not difficult to imagine why humans might possess more DNA than a fungus, it is not at all clear why salamanders should have 30 times more DNA than humans, an enigma we return to in Section 18.6.

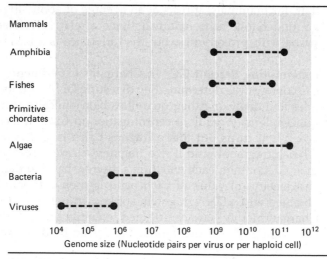

**FIGURE 4-1 C values,** or amount of DNA per cell, in organisms of increasing evolutionary complexity. For each group of organisms, the diagram indicates the maximum and minimum amounts of DNA that have been recorded in various species. For higher forms germ cells are used to calculate the haploid amount. Mammals are all seen to have about the same amount. (Data from R. J. Britten and D. E. Kohne).

Absent from these numbers is a physical sense of their meaning, which is best conveyed by length and weight. An *E. coli* chromosome is about 1 mm long, whereas the DNA in a diploid human nucleus, if strung together in a single strand, would be almost 2 meters long. This leads to the astonishing calculation that all the cells in the human body collectively contain perhaps 25 billion kilometers of DNA (the distance from the earth to the sun is only 58 million kilometers). This much DNA weighs only about 200 grams, or less than half a pound, underscoring how incredibly thin it is.

## 4.2 Chemical Composition (GC Content) of Genomic DNA

Eukaryotic and prokaryotic DNA also differ in their chemical compositions. As we saw in Chapter 1, Chargaff and others have demonstrated that the DNA of a species has a characteristic $G + C/A + T$ ratio. This value can also be expressed as **percent G + C,** and this term is usually shortened to read **percent GC** or the **GC content.** Thus, to say that an organism has a GC content of 30 percent is to say that $G = 15$ percent, $C = 15$ percent, $A = 35$ percent, and $T = 35$ percent of the total DNA. When different species of bacteria are compared, the percent GC varies from as little as 24 to as much as 75 percent (Table 1-3). A similar variability is found in the base composition of various viral DNAs. As one moves up the evolutionary scale, however, the overall GC content becomes less variable so that, for mammals, it rests at an average of about 40 percent GC. It is not yet understood why **mammals have a relatively fixed GC content compared with prokaryotes,** but the difference is a basic one.

**Determining Percent GC**   In Chargaff's experiments (*Table 1-3*), the total GC content was determined by digesting DNA into its component nucleotides and analyzing their quantities biochemically. An alternative method utilizes density gradient centrifugation in CsCl, a technique described in Box 2.1. It turns out that a fragment rich in GC pairs is denser than an AT-rich fragment, and if two fragment classes are sufficiently different in their GC content, each will form a discrete band in the gradient. The buoyant density ($\rho$) value of each band is then compared with the $\rho$ values obtained when DNA fragments of known percent GC (determined by the Chargaff method) are centrifuged to equilibrium; in this way, the percent GC of the DNA in the various bands can be deduced.

## 4.3 CsCl Banding Patterns of Genomic DNA: Satellite Bands

The CsCl banding technique has defined another important parameter of DNA composition, namely, the presence or absence of satellite bands. This is best understood by example.

When a culture of *E. coli* cells is harvested and many millions of *E. coli* chromosomes are isolated by rather gentle procedures, each chromosome breaks at random into fragments averaging about $10^5$ nucleotide pairs in length. These fragments will all move at equilibrium to a single position in a CsCl gradient, forming a band whose $\rho$ value corresponds to 51 percent GC. The formation of a single band in such an experiment means that, overall, the *E. coli* genome has a **uniform base composition:** its random fragmentation does not yield certain chromosome fragments that have significantly different base compositions from other fragments.

When the DNA of most mammalian cells is subjected to the same analysis, on the other hand, a different picture is obtained. The bulk of the DNA fragments equilibrate at a single position in the CsCl gradient, one that corresponds to the percent GC (roughly 40 percent) that is obtained by chemical analysis. In addition to this **main band DNA,** however, **satellite bands** are often (although not always) found, having buoyant densities that are either higher or lower than the main band (Figure 4-2). Some eukaryotes may exhibit only one satellite band, whereas others may exhibit several.

Table 4-1 lists the properties of some nuclear satellite DNAs. It is clear that some are GC-rich and others AT-rich with respect to the main band. Moreover, great variability exists in the percentage of an organism's total DNA that has satellite properties: in the mouse it is perhaps 6 percent of the total, whereas in the crab, up to 30 percent. Variability in the GC content of satellite DNAs appears not to have a significant effect on the total base composition of the mammalian nucleus, since, as we noted earlier, most mammals possess DNA with a total GC content of about 40 percent.

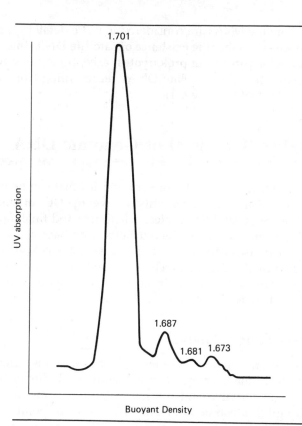

**FIGURE 4-2 Satellite DNA.** Total DNA from *Drosophila melanogaster* yields a main band of DNA with a buoyant density of 1.701 g/cm³ and three lighter satellites. (From S. A. Endow et al., *J. Mol. Biol.* **96:**670, 1975.)

TABLE 4-1   Base Composition of Some Satellite DNAs

| Organism | Main Component (%GC) | Satellite (%GC) | Percent of Total DNA Represented by Satellite |
|---|---|---|---|
| Bacteria | | | |
| *Halobacterium salinarium* | 67 | 58 | 20 |
| *H. cutirubrum* | 68 | 59 | 10 |
| Invertebrates | | | |
| *Cancer antennarius* (crab) | 40 | 3 | 30 |
| *Cancer magister* (crab) | 41 | 3 | 12 |
| *Cancer productus* (crab) | 41 | 3 | 31 |
| *Balanus nubilis* (barnacle) | 47 | 55 | – |
| Vertebrates | | | |
| Mouse | 41 | 34 | 11 |
| Guinea pig | 41 | | |
| Type I | | 39 | 5 |
| Type II | | 45 | 4 |
| Human | 41 | | |
| Type I | | 33 | 0.5 |
| Type II | | 35 | 2 |
| Type III | | 39 | 1.5 |

From B. J. McCarthy, in *Handbook of Cytology*, A. Lima-de-Faria, Ed. Amsterdam: North-Holland, 1969, Table 4. Copyright 1971 by the American Association for the Advancement of Science.

Satellite DNAs are considered in more detail in Sections 4.7 and 4.10. Here we stress that **the existence of satellite DNA does, in general, distinguish eukaryotes from prokaryotes,** exceptions being two species of *Halobacterium* that have satellite DNAs representing from 10 to 20 percent of their total DNA (Table 4-1).

# Reassociation Kinetics (C$_0$t Plots) of Genomic DNA

At this point we can cite three properties that can be used to characterize chromosomal DNA: its **quantity,** its **average GC content,** and the **presence or absence of satellite species.** The fourth and final molecular property of chromosomal DNA considered here is perhaps the most intriguing and most complex. This relates to its **reassociation kinetics.** Again we must describe the biochemical technique used to measure the phenomenon before we can discuss its significance. This technique was first developed and exploited by R. Britten and D. Kohne.

## 4.4 Dissociation-Reassociation Analysis

If a solution containing duplex DNA molecules is either heated or treated with alkali, the hydrogen bonds holding each duplex together become increasingly unstable and the helix starts to come apart (**denaturation** or **melting**), as illustrated in Figure 4-3a. If at this point the solution is cooled or neutralized, the hydrogen bonds will reform and the helix is said to **renature** (Figure 4-3a).

If the solution is instead maintained at high temperature or high alkalinity, the two strands of each helix will separate completely and diffuse away from one another, a phenomenon known as DNA **dissociation** (Figure 4-3b). When a solution of dissociated DNA is incubated under appropriate conditions, the formation of new base pairs eventually causes **reassociation** or **reannealing** of the DNA: pairs of complementary sequences encounter each other in register and form a duplex molecule. In contrast to renaturation, however, the two strands in the new duplexes are very unlikely to have been associated with one another in the original collection of DNA molecules—they almost certainly derive from different duplexes, as illustrated in our example in Figure 4-3b. The term **re**association is therefore somewhat misleading, but it is commonly used nonetheless.

## 4.5 Analysis of Reassociation Kinetics

In a kinetic study of reassociation, one simply dissociates DNA and measures the time course of the reassociation process. The extent of reassociation is estimated by passing samples of the reaction mixture over hydroxylapatite columns: hydroxylapatite crystals, under appropriate ionic

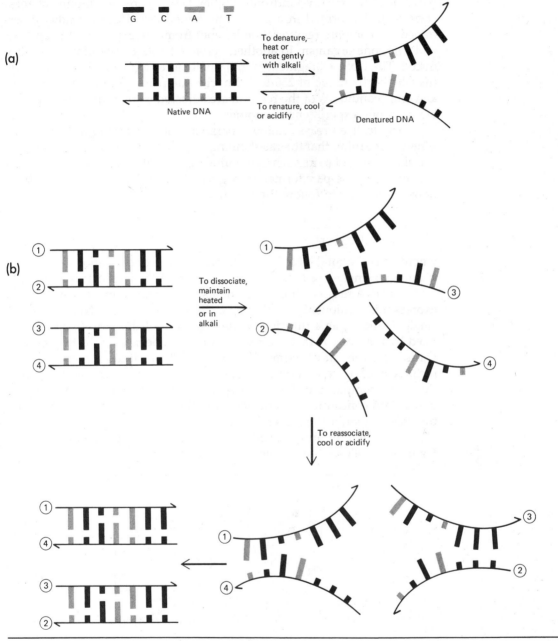

FIGURE 4-3 Denaturation-renaturation (a) and dissociation-reassociation (b) of duplex DNA.

conditions, selectively bind only duplex DNA, and the amount of reassociated DNA formed in a given time period can thus be readily determined. To compare reassociation kinetics from one experiment to another and from one organism to another, certain standard conditions are introduced. The DNA is fragmented into small pieces of relatively uniform size (usually in the range of 250 to 450 nucleotide pairs), and the temperature and ionic conditions of the reaction are adjusted so that they are comparable from one experiment to another.

An idealized reassociation experiment follows **second-order reaction kinetics,** meaning that the rate-limiting step in the reaction is the "correct" collision of strand pairs such that mutual regions of complementarity are in register and base-pair formation begins (*Figure 4-3b*). Second-order reactions are known to follow the equation

$$\frac{C}{C_0} = \frac{1}{1 + kC_0t}$$

where $C_0$ is the total DNA concentration, $C$ the concentration of DNA remaining in single-stranded form at time $t$, and $k$ the reassociation rate constant. In DNA reassociation experiments, it has proved most convenient to express this relationship by plotting $C/C_0$ (the fraction of DNA strands left single stranded) as a function of the logarithm of the term $C_0t$ ($C_0$ multiplied by the time, $t$, that the reassociation has been allowed to proceed). $C_0t$ is a convenient experimental parameter, since it takes account of the reciprocity of concentration and time in second-order reactions. A $C_0t$ **plot,** therefore, compensates for the effect on reaction kinetics of variations in initial DNA concentrations and allows a comparison of reactions that differ by large factors in their overall rates.

A $C_0t$ plot for an idealized DNA reassociation experiment, shown in Figure 4-4, follows the simple sigmoidal curve pattern of a second-order

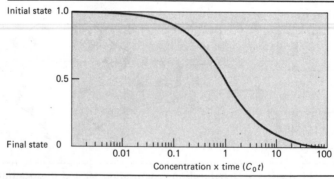

**FIGURE 4-4 Ideal kinetics for DNA reassociation** as seen in a $C_0t$ **plot.** In the initial state all DNA is single stranded; in the final state all DNA has reassociated into duplex form. (After R. J. Britten and D. E. Kohne, *Carnegie Institution of Washington Year Book 66,* Washington D.C., 1967. Courtesy of the Carnegie Institution of Washington.)

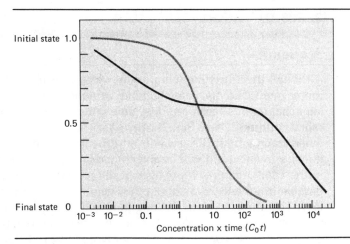

FIGURE 4-5 **Actual kinetics of reassociation** of calf thymus DNA (black curve) and *E. coli* DNA (gray curve), as seen in a $C_0t$ plot in which the DNA is all single stranded in the initial state and double helical in the final state. (After R. J. Britten and D. E. Kohne, *Carnegie Institute of Washington Year Book 66,* Washington D.C., 1967. Courtesy of the Carnegie Institution of Washington.)

reaction. This curve is based on an important assumption, namely, that all the single-stranded DNA fragments in the initial reaction mixture have an equal probability of forming a reassociated duplex. Obviously, if one class of fragments is capable of reassociating much faster than another, then the smooth, S-shaped curve of Figure 4-4 will be replaced by a multiphased curve that reflects the simultaneous occurrence of several different kinds of reactions.

Figure 4-5 shows the actual renaturation kinetics obtained for *E. coli* (gray curve) and calf (black curve) DNA fragments. The *E. coli* $C_0t$ plot has the same symmetrical form as the idealized curve plotted in Figure 4-4. The calf $C_0t$ curve, on the other hand, is clearly skewed, indicating that the population of DNA fragments is heterogeneous with respect to its kinetics of reassociation. In particular, it is evident that some reassociation of calf DNA can occur at very low $C_0t$ values, meaning that a fraction of the calf DNA "finds partners" very rapidly. A second fraction of the calf DNA fragments reassociates with the same kinetics (slope) as the *E. coli* DNA, but does so at much higher $C_0t$ values. Fractions of DNA that reassociate with the same kinetics are said to belong to the same **kinetic class.** Thus the experiment shown in Figure 4-3 indicates that *E. coli* DNA forms a single kinetic class, whereas calf DNA is composed of two kinetic classes.

**The presence of multiple kinetic classes is yet another hallmark of eukaryotic DNA.** In the following sections of this chapter, we consider what is known about the various kinetic classes of eukaryotic DNA that have been analyzed.

# The Kinetic Classes of Eukaryotic DNA

## 4.6 Single-Copy Sequences

Although there are many important exceptions to the rule, most genes are represented but once per haploid genome (a statement documented in later chapters). Therefore, most genes should be found in **single-copy** (also called **unique**) DNA. Single-copy DNA will have the following property: when fragmented, dissociated, and incubated under reassociation conditions, a given strand will be able to reassociate with very few other strands in the DNA mixture. More specifically, only its original partner or a comparable fragment from some other genome will be perfectly complementary. Therefore, a relatively long time will be required for the reassociation reaction to go to completion, since most collisions will produce incorrect pairings. Single-copy DNA, in other words, is predicted to have the kinetic properties of the slowest reassociating kinetic class of DNA in eukaryotic $C_0t$ plots (*Figure 4-5*).

In most eukaryotes, single-copy DNA represents about 70 percent of the chromosomal DNA. Although most genes are believed to reside in this DNA fraction, this does not necessarily mean that 70 percent of the eukaryotic genome contains genes. If all single-copy DNA were to represent genes, then a typical mammalian haploid genome, containing 3 pg of DNA, would contain about half a million genes. Most estimates put the number of genes in a mammalian haploid genome at a maximum of 50,000. Thus the single-copy fraction of eukaryotic DNA appears to contain at least tenfold more DNA than is needed for genetic coding. An analogous excess may be present in the *E. coli* genome as well. Possible functions of this extra single-copy DNA are considered throughout the text.

## 4.7 Highly Repetitive Sequences

At the other extreme from single-copy DNA is the rapidly reassociating DNA that forms duplexes at low $C_0t$ values (*Figure 4-5*). When such rapidly reassociating fragments are isolated and analyzed, they prove to represent the **highly repetitive** kinetic class of eukaryotic DNA.

Highly repetitive DNA typically constitutes from 5 to 10 percent of the eukaryotic genome (but may be absent, as, for example, in the insect *Chironomus*, or more abundant, as in *Drosophila*, where it is 19 percent of the total). In most cases this DNA is composed of a short sequence of bases repeated again and again, with from $10^4$ to $10^7$ repeats per genome. Since this highly reiterated fraction is usually (but not inevitably) somewhat different in its average base composition from the bulk of the DNA in the eukaryotic genome, highly repetitive DNA often bands in a satellite position in a CsCl gradient (*Figure 4-2* and *Table 4-1*). Therefore, **satellite DNA and highly repetitive DNA are often synonymous.**

A number of eukaryotic satellite DNAs have been isolated and sequenced (Table 4-2) using techniques described later in this chapter.

**TABLE 4-2  Sequenced Satellite DNAs**

| Organism | Base Pairs per Repeat | Location | Sequence of One Strand (Satellite Designation) |
|---|---|---|---|
| *Drosophila melanogaster* (fruit fly) | 5 | Arms of Y chromosome and centric heterochromatin of chromosome 2 only; also distal end of 2L | AGAAG (polypurine) ATAAT (1.672) |
| | 7 | | ATATAAT (1.672) |
| | 10 | Centric heterochromatin of all chromosomes and tip of 2L | AATAACATAG (1.686) AGAGAAGAAG (1.705) |
| *Drosophila virilis* | 7 | Centric heterochromatin | ACAAACT (I) ATAAACT (II) ACAAATT (III) |
| *Cancer borealis* (marine crab) | 2 | ? | AT |
| *Pagurus pollicaris* (hermit crab) | 4 | ? | ATCC (I) |
| | 16–43 | ? | (CTG)nCTGCACT (n = 3–12)(II) |
| *Cavia poriella* (guinea pig) | 6 | Centric heterochromatin | CCCTAA (α) |
| *Dipodomys ordii* (kangaroo rat) | 10 | Centric heterochromatin | ACACAGCGGG (HS-β) |

From K. Tartoff, *Ann. Rev. Genet.* **9**:355, 1975.

Guinea pig satellite DNA, for example, has as its basic repeating unit the sequence

```
CCCTAA
| | | | | |
GGGATT
```

which is repeated **in tandem** along long stretches of duplex DNA. Over the course of time, mutations have accumulated within this satellite DNA so that an occasional "module" of the DNA may carry a

```
CGCTAA
| | | | | |
GCGATT
```

sequence, whereas another may contain a

```
CCATAA
| | | | | |
GGTATT
```

sequence. Nonetheless, a basic six-base (hexamer) repeat pattern is conserved. Other satellite DNAs exhibit tetrameric units (as in the hermit crab), heptameric units (in *Drosophila*) and so on (Table 4-2), whereas bovine satellite I DNA appears to repeat only every 1400 bases.

Highly repetitive DNA exhibits rapid reassociation kinetics because the chances that a fruitful collision will occur are obviously very good if from 5 to 10 percent of the DNA in a mixture is of the same sequence.

The chromosomal location of highly repetitive DNA is considered in Section 4.10 of this chapter.

## 4.8 Middle-Repetitive Sequences

**Middle-repetitive DNA** represents perhaps 1 to 30 percent of the eukaryotic genome and contains sequences that are repeated from a few times to hundreds of thousands of times. It is a far more heterogeneous class of DNA than the highly repetitive fraction. When sheared into fragments and

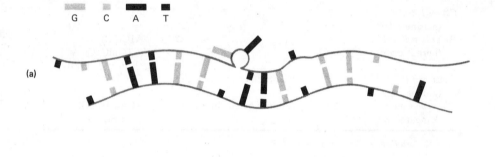

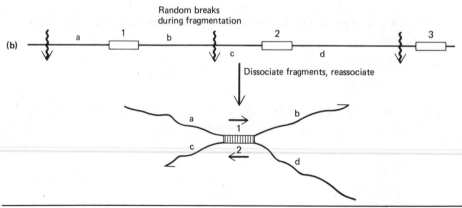

**FIGURE 4-6 Middle-repetitive DNA.** (a) Imperfect DNA duplex formed by middle-repetitive DNA. The two strands are generally complementary, but occasional bases do not match. This reduces the stability of the molecule, as measured by its dissociation temperature. (b) Formation of "double fork" structures in a dissociation-reassociation experiment with middle-repetitive DNA. The three boxes represent middle-repetitive sequences; intervening lines represent unique single-copy sequences. Two single strands resulting from fragmentation and dissociation are shown to have reassociated by their internal, generally homologous middle-repetitive sequences, leaving four dangling strands of flanking single-copy sequences which cannot base-pair with one another.

dissociated, a middle-repetitive strand encounters a **related family** of sequences with which to reassociate. These are not perfectly complementary but are at least sufficiently similar that they can combine to form imperfect duplexes, as illustrated in Figure 4-6a. Such duplexes are more sensitive to heat than are perfectly base-paired duplexes, but they are stable enough to bind to hydroxyapatite and score as reassociated DNA. Therefore, the middle-repetitive class of DNA reassociates with kinetics intermediate between the highly repetitive and the single-copy DNA classes. It is not conspicuous in Figure 4-5, which is why it was not mentioned, but it is in fact detectable in all eukaryotic DNAs when sensitive reassociation experiments are performed.

**Middle-repetitive DNA families are typically dispersed throughout the genome rather than clustered in tandem arrays.** In most eukaryotic genomes, it is organized in the following fashion: a short, middle-repetitive DNA sequence, with an average length of 300 nucleotide pairs, is followed by a single-copy sequence of about 1200 nucleotide pairs, which is followed by another repeat of 300 base pairs, and so forth (Figure 4-6b). This pattern can be detected by several types of experimental protocols, one being to shear DNA into long fragments, dissociate these, allow them to reassociate at $C_0t$ values where reassociation of repetitive DNA is strongly favored, and examine the reassociated DNA by electron microscopy. When this is done, "double fork" structures are found; these consist of a central imperfect duplex region, about 300 nucleotides in length, from which two single strands emerge at either end. The origin of such structures is diagrammed in Figure 4-6b.

Approximately 50 percent of the middle-repetitive DNA of most eukaryotes is interspersed between unique sectors of 1200 base pairs, and R. Britten and E. Davidson have speculated that the middle-repetitive blocks function to control the expression of genes located in the contiguous unique sectors. The remaining middle-repetitive DNA has other patterns of distribution, however, and such species as *Drosophila* display different patterns altogether. Therefore, middle-repetitive DNA is best regarded at present as a heterogeneous class of eukaryotic DNA, some of which may serve regulatory roles. Further information on this class of DNA is given in Section 18.6.

## 4.9 Inverted-Repeat Sequences

The three kinetic classes described thus far include DNA fragments which, when dissociated, reassociate only with another complementary, or partially complementary, fragment. A fourth category of DNA, the **inverted-repeat** class, contains DNA fragments that are **internally complementary,** meaning that they will preferentially reassociate with themselves. Let us examine first a single strand of DNA that is internally complementary (also called **palindromic**):

$$\overbrace{\phantom{AAAAAA}}^{R} \qquad \overbrace{\phantom{AAAAAA}}^{R'}$$

5'. . . . AATGUTCGXXXCGTACATT . . . 3'

(where X is any nucleotide). The regions R and R' are complementary; therefore, when this strand is placed under reassociation conditions, it will rapidly **fold back** or **snap back** upon itself to form a **hairpin**:

5'. . . AATGTAC$G_X$
$\qquad$ | | | | | | | $_X$
3'. . . TTACATGCX$^X$

The same behavior occurs if the DNA is initially double stranded and contains an inverted repeat (also called a region of **twofold** or **dyad symmetry**):

| 5' | R | XXX | R' | 3' |
|---|---|---|---|---|

| 3' | R' | XXX | R | 5' |
|---|---|---|---|---|

In this case, when the strands are dissociated and placed under reassociation conditions, the single strands snap back to form hairpins long before enough time has elapsed for fragment-to-fragment collisions and pairings with complementary strands. Inverted-repeat DNA is therefore isolated in dissociation-reassociation experiments as that fraction of DNA that becomes duplex (that is, binds to hydroxyapatite columns) at zero time (that is, seconds after the beginning of reassociation).

If the sequences denoted XXX are sufficiently long, then reassociation of strands carrying inverted repeats creates a characteristic **stem-and-loop structure** that can be recognized with the electron microscope:

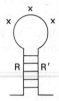

In both the human and *Xenopus* genome, there are an estimated 120,000 such fold-back sequences. They are prominent in the controlling regions of many genes, and at the termini of elements that move from site to site within both prokaryotic and eukaryotic genomes. Detailed examples are given in later chapters of the text.

## 4.10 In Situ Hybridization of DNA Kinetic Classes

Granted the existence of four "kinetic classes" of eukaryotic DNA, one next wishes to learn how these classes are distributed within the chromosomes. An effective technique for addressing this question, called **in situ**

**hybridization,** follows principles developed by S. Spiegelman and his associates for the biochemical technique called **nucleic acid hybridization.**

In a biochemical hybridization experiment, DNA duplexes are dissociated, and the resultant single strands are immoblized on a membrane filter. The filter is then incubated with radioactively labeled, single-stranded DNA or RNA fragments. If the fragments carry base sequences that are homologous to the dissociated DNA, **hybrid** DNA-DNA or DNA-RNA duplexes will form. Where no homology exists, no hybridization will occur. The filter is then washed and exposed to nucleases that destroy single-stranded but not double-stranded nucleic acids. Finally, its level of radioactivity is determined by a device known as a scintillation counter.

*In situ* hybridization follows the same principles, the major modification being that the dissociated DNA is left in place inside the cell nucleus rather than immobilized on a filter. Cells are first preserved by chemical fixatives and mounted on a glass slide suitable for light microscopy; they are treated with ribonuclease to remove RNA and are then exposed to warm alkali (NaOH) to dissociate the DNA. Finally, they are incubated in a solution containing single-stranded, radioactively labeled fragments. These may be DNA fragments collected from a particular band in a CsCl gradient or from a hydroxyapapite column, or they may be highly purified DNA fragments generated by restriction endonucleases (*to be described in Section 4.11*). Alternatively, they can be RNA species or RNA copies (**cRNA**) of particular DNA fragments, the copies being made *in vitro* by using an enzyme known as DNA-dependent RNA polymerase. The temperature and pH of the preparation is adjusted to permit hybridization of the labeled single strands with the dissociated DNA exposed in the cell nuclei. After a suitable period, the slide is washed to remove unhybridized fragments so that only the hybridized, radioactively labeled nucleic acid remains associated with the cellular DNA. When RNA or cRNA is used, the slide is further exposed to ribonucleases, which digest all the RNA molecules that are not part of DNA-RNA complexes. As a last step, the slide is subjected to autoradiography (*Box 2.2*) and stained for light microscopy.

When dividing mouse cells are so exposed to labeled moderately repetitive or single-copy mouse DNA fragments, the radioactivity is distributed uniformly throughout all the chromosomes, indicating that these DNA fractions derive from all parts of the mouse genome. In contrast, when mouse repetitive satellite DNA (or satellite cRNA) is used as the labeled species, the label appears exclusively over each centromere (Figure 4-7). Thus mouse satellite DNA, which constitutes 6 percent of the total DNA of the mouse cell and is composed of repeated tracts of a $\frac{\text{GAAAAATGA}}{\text{CTTTTTACT}}$ module, is neither dispersed throughout the mouse genome nor confined to a single chromosome. Instead, this DNA occupies specific sites in *all* of the mouse chromosomes, with the possible exception of theY chromosome.

*In situ* hybridization of satellite DNAs from a variety of plants and

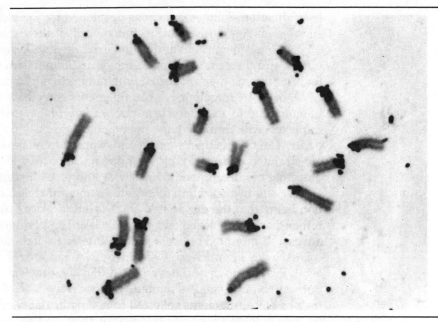

**FIGURE 4-7** *In situ* **hybridization.** Autoradiograph of a mouse metaphase cell hybridized with radioactive cRNA copied *in vitro* from mouse satellite DNA. (From M. L. Pardue and J. G. Gall, *Chromosomes Today* **3:**47, 1971.)

animals has been performed, and the same general pattern seen: hybridization occurs over blocks of DNA in the pericentromeric regions, in portions of the Y chromosome, and often at the telomeres (chromosome ends). These locations should sound familiar to readers of Chapter 3 (*Section 3.13*): all are common locales of **constitutive heterochromatin.** There is no evidence that satellite DNA codes for protein, and most constitutive heterochromatin appears to be genetically inert. Therefore, it is generally assumed that if highly repetitive DNA has any function, it lies in the rather ill-defined category of "chromosome mechanics," possibly mediating chromosome folding at mitosis or the pairing of chromosomes prior to gene recombination. Such speculations are tempered by the fact that chromosomes apparently lacking both constitutive heterochromatin and highly repetitive DNA appear able to fold and pair as well as their satellite-bearing counterparts.

# Restricting, Sequencing, and Cloning DNA

## 4.11 Restriction and Modification Enzymes

Most readers are doubtless aware of the genetic engineering technology that has revolutionized biology during the past ten years, allowing such feats as the insertion of human insulin genes into bacterial genomes. Many of the procedures used in genetic engineering are described in this and subsequent sections. Although it is meaningless to single out any one method as being more important than another, the discovery and exploitation of **restriction endonucleases,** pioneered by W. Arber, D. Nathans, and H. O. Smith, were certainly key to the development of this technology.

We encountered endonucleases in Section 3.7, where we described the sequential degradation of nucleosomes into chromatosomes and core particles. Restriction endonucleases act like micrococcal endonucleases in that they cleave internal phosphodiester bonds in DNA. They differ, however, in that **their sites of attack are restricted to DNA that contains particular nucleotide sequences.**

As an example we can cite the enzyme known as *Eco R*I (restriction enzyme nomenclature begins with an acronym designating the bacterial species from which the enzyme is isolated, in this case, *E. coli*). When *Eco RI* is presented with double-stranded DNA, it recognizes as a substrate only the hexanucleotide sequence

$$5'pXXXpG\downarrow pApApTpTpCpXXX$$
$$XXXpCpTpTpApAp\uparrow GpXXXp5'$$

The lower strand, when read backward, is seen to be the same as the upper strand read forward; thus *Eco R*I is said to recognize a **symmetrical** or **palindromic site,** a symmetry that probably figures in the recognition mechanism. Once this sequence is recognized, the enzyme makes two **staggered nicks** at the sites shown by the arrows. The hydrogen bonds between the four AT nucleotide pairs that lie between these two cuts are not strong enough to hold the two broken chains together. Therefore, the original duplex is cut into two pieces, forming two *Eco R*I **restriction fragments** with complementary, single-stranded "tails":

$$5'pXXXpG \qquad\qquad 5'pApApTpTpCpXXX$$
$$XXXpCpTpTpApAp5' \quad and \qquad GpXXXp5'$$

Table 4-3 presents a catalogue of the most widely used restriction endonucleases, the sequences they recognize, and the nicks they confer (where this is known). Considerable diversity is evident. The most abundant class, like *Eco R*I, recognizes hexanucleotides, but many are available that recognize pentanucleotides or tetranucleotides, and some do not require symmetrical sequences for recognition. The majority of the enzymes make staggered nicks like *Eco R*I, but some (for example, *Hpa* I) make clean breaks that cut the polynucleotide into two **blunt-ended duplexes.**

**The importance of restriction endonucleases lies in their substrate specificity: each enzyme can be used to transform a duplex chromosome into a unique collection of restriction fragments.** As an example we can consider the circular chromosome of the animal tumor virus **SV40.** When exposed to *Eco R*I, a single cut is made in the chromosome (Figure 4-8a), converting it from a circle to a rod. Thus there exists but a single

$$5'GAATTC$$
$$CTTAAG5'$$ sequence in the SV40 chromosome, and this site has to come

**TABLE 4-3   Restriction Endonuclease Sequence Specificities**

| Enzyme | Recognition Sequence |
|---|---|
| **Symmetrical (N = 6)** | |
| *Ava* III | ATGCAT |
| *Bal* I | TGG\|CCA |
| *Bam* HI | G\|GATC̊C |
| *Bcl* I | T\|GATCA |
| *Bgl* II | A\|GATCT |
| *Cla* I | ATCGAT |
| *Eco* RI | G\|AÅTTC |
| *Hind* III | Å\|AGCTT |
| *Hpa* I | GTT\|AAC |
| *Kpn* I | GGTAC\|C |
| *Mst* I | TGCGCA |
| *Pst* I | CTGCA\|G |
| *Pvu* I | CGATCG |
| *Pvu* II | CAG\|CTG |
| *Sma* I | CCC\|GGG |
| *Sac* I | GAGCT\|C |
| *Sac* II | CCGC\|GG |
| *Sal* I | G\|TCGAC |
| *Xba* I | T\|CTAGA |
| *Xho* I | CTC\|GAG |
| **Degenerate symmetrical (N = 6)** | |
| *Acc* I | GT\|(A/C)(G/T)AC |
| *Ava* I | C\|PyCGPuG |
| *Hae* I | (A/T)GG\|CC(T/A) |
| *Hae* II | PuGCGC\|Py |
| *Hgi* AI | G(T/A)GC(T/A)\|C |
| *Hind*II | GTPy\|PuÅC |
| **Symmetrical (N = 5)** | |
| *Asu* I | G\|GNCC |
| *Ava* II | G\|G(A/T)CC |
| *Bbv* I | GC(T/A)GC |
| *Eco* RII | \|CC̊(A/T)GG |
| *Hinf* I | G\|ANTC |
| **Asymmetrical (N = 4, 5)** | |
| *Mnl* I | CCTC cleavage 5 to 10 bases 3′ to site |
| *Hga* I | GACGCNNNNN\| (3′)<br>CTGCTNNNNNNNNNN\| (5′) |
| *Hph* I | GGTGANNNNNNNN (3′)<br>CCACTNNNNNNN\| (5′) |
| *Mbo* II | GAAGANNNNNNNN\| (3′)<br>CTTCTNNNNNNN\| (5′) |
| *Sta* NI | GATGC |
| *Eco* Pl | AGACC (3′) ⎱ cleavage 24 to 26<br>TCTGG (5′) ⎰ bases 3′ to site |

**(continued on next page)**

**(continued)**

Symmetrical ($N = 4$)

| | |
|---|---|
| *Alu* I | AG\|CT |
| *FnuD* II | CG\|CG |
| *Hae* III | GG\|$\overset{*}{C}$C |
| *Hha* I | G$\overset{*}{C}$G\|C |
| *Hpa* II | C\|$\overset{*}{C}$GG |
| *Mbo* I | \|GATC |
| *Taq* I | T\|CGA |

Symmetrical methylated ($N = 4$)

| | |
|---|---|
| *Dpn* I | GmATC |

Symmetrical ($N = 7$)

| | |
|---|---|
| *Eca* I | GGTNACC |

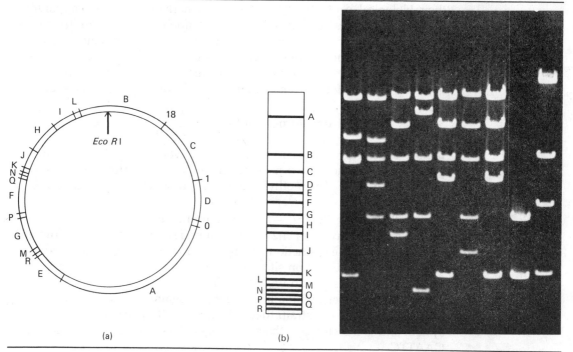

(a)                    (b)

**FIGURE 4-8 Restriction-enzyme analysis of chromosomes.** (a) Circular SV40 chromosome showing the single *Eco* RI cleavage site and the 18 *Hae* III cleavage sites which cut the chromosome into 18 fragments, labeled A–R. (b) Gel electrophoresis of *Hae* III restriction digests of numerous SV40 chromosomes, showing that the 18 fragments migrate according to size to form 18 discrete bands in the gel. (c) Photograph of actual restriction digest gels. The right-hand lane shows four of the fragment classes generated when phage λ is treated with *Hind* III; the remaining lanes show the fragment classes generated when various plasmids of *E. coli* (*Figure 2-2*) are treated with *Bam* HI. Can you find the two that are identical? (Courtesy of Dr. Douglas Berg.)

to serve as an initial landmark in its analysis. A different result is obtained if SV40 chromosomes are exposed to the enzyme *Hae* III: 18 restriction fragments are generated. Since *Hae* III cuts at $\frac{5'GGCC}{CCGG5'}$ sequences, this means that 18 such sequences are present in the SV40 genome. As illustrated in Figure 4-8b, each of the 18 fragments will be of different length and will thus have a distinct molecular weight. Therefore, if an *Hae* III **restriction digest** of SV40 chromosomes is subjected to gel electrophoresis (*Box* 3.4), 18 discrete bands will be detected (Figure 4-8b). Since **every fragment in a particular band has an identical base sequence,** it is possible to cut a given band from the gel, elute its DNA, and subject the DNA to sequence analysis or cloning. These procedures are described in subsequent sections of the chapter.

First, however, we should pause to consider restriction enzymes in their biological context: why are these remarkable enzymes apparently ubiquitous in bacterial cells, and what function do they serve? To answer this question, we must introduce a companion class of enzymes known as **methylating** or **modification enzymes,** which attach methyl groups to the nucleotides found in particular DNA sequences. Thus, for example, a methylating enzyme found in many strains of *E. coli* recognizes the sequence $\frac{5'GAATTC}{CTTAAG5'}$ and transforms it into $\frac{5'\overset{*}{GA}\overset{}{ATTC}}{CTTAAG5'}$, where the asterisks denote methyl groups. The unmethylated version of this sequence is precisely the sequence we just identified as the substrate for the *Eco* RI restriction enzyme. The methylated version, however, is not recognized by *Eco* RI.

Such enzyme "pairs"—a restriction enzyme that cuts a specific sequence and a methylating enzyme that modifies this sequence and thereby "masks" it from the restriction enzyme—are widespread; Table 4-3 shows, with asterisks, methylation sites for other modification enzymes that coexist with known restriction enzymes. These enzyme pairs are believed to provide bacterial cells with a line of defense against foreign DNA. We can first imagine an *E. coli* strain that lacks the *Eco* RI **modification-restriction system** and is infected by a bacteriophage whose chromosome contains a $\frac{GAATTC}{CTTAAG}$ sequence. This phage will encounter no bacterial defenses and will successfully replicate and kill the host. If this same phage instead infects a cell containing the restriction-modification system we are considering, the chromosome will likely be "restricted" the moment it enters the cell, and the severed DNA will be unable to direct a successful infection. Obviously, the infecting chromosome will also be a substrate for the methylating enzymes of the host, but it seems that unmodified phage DNA is usually restricted before it has a chance to be modified. By contrast, host DNA, as it is undergoing replication, is usually modified before it has a chance to be restricted, so that each class of methylating enzyme provides the host with protection from its restriction enzyme counterpart.

## 4.12 Sequencing DNA

The homogeneous collection of restriction fragments that form a single band in a restriction digest gel (*Figure 4-8*) can be eluted out of the gel and subjected to **DNA sequencing.** Two techniques for DNA sequencing are widely used at present, and modifications or new approaches will almost certainly be in use by the time this text appears. We present in Box 4.1 a representative method, one developed by A. Maxam and W. Gilbert. Readers disinterested in the details need only appreciate the fact that it is now possible to determine, accurately and rapidly, the entire sequence of nucleotides along a given DNA strand. When coupled with a technique known as **restriction mapping** (*Section 12.15*), moreover, it becomes possible to determine which sequences lie next to which along the length of a chromosome.

# Cloning DNA

Pure preparations of viral chromosomes, plasmids, and certain kinds of satellite sequences can be isolated by gradient centrifugation in sufficient quantity for restriction and sequencing. Many other kinds of DNA, however, cannot be obtained in pure abundant form without an intermediate step known as **DNA cloning.** The procedure entails the following steps, which are described in detail in subsequent sections.

1. Plasmids or viral chromosomes are constructed in such a way that they will accept the insertion of foreign DNA fragments and retain their ability to replicate. These are known as **cloning vectors.**
2. DNA fragments are prepared for insertion into the vectors.
3. Vectors and fragments are mixed *in vitro,* and the inserted DNA is sealed into the vector to produce **recombinant vectors.** These are introduced into bacterial or eukaryotic hosts, and the host cells are cloned (*Section 2.2*).
4. Individual colonies are screened for the presence of recombinant vectors carrying the DNA fragment of interest. Recombinant vectors are then isolated from the selected host strains, and the fragments of interest, now available in pure abundant form, can be excised from the vector and subjected to biochemical analysis.

## 4.13 Cloning Vectors

Probably the most widely used cloning vector at present is a plasmid known as **pBR322,** whose structure is diagrammed in Figure 4-9. Attentuated forms of various viruses are also very popular. The pBR322 plasmid

BOX 4.1
DNA Sequencing

The DNA sequencing procedure devised by W. Gilbert and A. Maxam is described in this section (see also diagram on facing page). An alternate technique, developed by F. Sanger and his collaborators, is also widely used to determine DNA sequences.

1. A pure preparation of DNA (for example, the DNA eluted from the B band in the *Hae* III restriction-digest gel shown in Figure 4-8b) is labeled enzymatically with $^{32}P$ at the 5' ends of both strands. In the diagram, the label is denoted by asterisks.

2. The duplex B fragments are dissociated (*Section 4.4*) and are separated by electrophoresis into two single-strand fractions known as "heavy" and "light." The basis for this separation lies in the fact that one strand will have more purines than the other and will therefore be heavier than the other. Considered here is the heavy-strand sample; identical procedures can be performed using the light-strand sample to provide a complementary sequence.

3. The heavy-strand sample is divided into four aliquots, one of which will indicate the position of every G in the strand, one of which will indicate every A, one every T, and one every C. These are considered in turn in the next four steps and in the diagram.

   Ⓖ The aliquot is treated with dimethyl sulfate under conditions that result, on the average, in the methylation of one purine per single strand. G is methylated about five times faster than A. Methylation causes the strands to break in the presence of warm alkali, generating $^{32}P$-labeled fragments (fragments 1, 3, and 10 in the diagram). A smaller quantity

of labeled fragments broken at methylated A positions is also generated. Electrophoresis and autoradiography of this sample gives the $G > A$ pattern, with strong bands at guanine breaks and weak bands at adenine breaks.

Ⓐ The purines in this aliquot are methylated as in the previous step, but the sample is then exposed to cold dilute acid, which preferentially causes breaks at A positions. Electrophoresis and autoradiography yields the $A > G$ pattern with dark bands at adenines and light bands at guanines.

Ⓣ This aliquot is treated with hydrazine under conditions that result, on the average, in the hydrazinolysis of one pyrimidine per strand, with T and C equally sensitive. The reacted pyrimidines are rendered susceptible to strand scission by the reagent piperidine. Electrophoresis and autoradiography result in the $C + T$ pattern, with bands of equal intensity corresponding to cleavages at cytosines and thymines.

Ⓒ This aliquot is exposed to hydrazine and piperidine but in the presence of high salt, which preferentially suppresses the hydrazinolysis of thymines. The resultant C pattern displays bands that derive solely from breakages at cytosines.

4. Comparison of the four autoradiographic patterns allows a direct determination of the DNA sequence. In the sample gel in the center of the diagram, for example, the sequence can be directly read as

5' G C G C T C A C T G 3'
  1 2 3 4 5 6 7 8 9 10

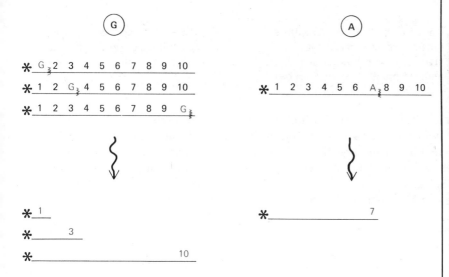

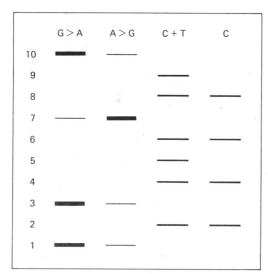

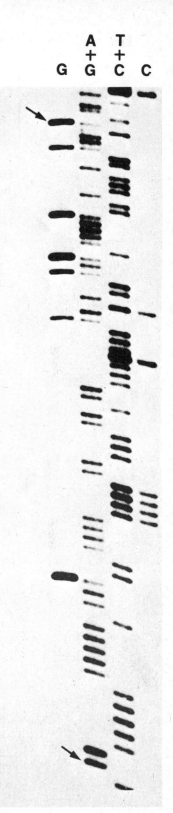

(Courtesy of Dr. Ballas King)

> Knowing this order, adjacent sequences can be pieced together to give long, continuous sequences and, eventually, the order of bases along an entire chromosome (*see Section 12.14*). A photograph of a sequencing gel is shown on page 115.

contains an origin of replication and two relevant genes, one that confers host cells with resistance to the antibiotic ampicillin (Ap^r) and one that confers tetracycline resistance (Tc^r). Within each gene are found a number of restriction enzyme targets (*Section 4.11*); pertinent here is the fact that the Ap^r gene contains a *Pst* I site, whereas Tc^r does not, and that the Tc^r gene contains a *Hin*d III site, whereas Ap^r does not.

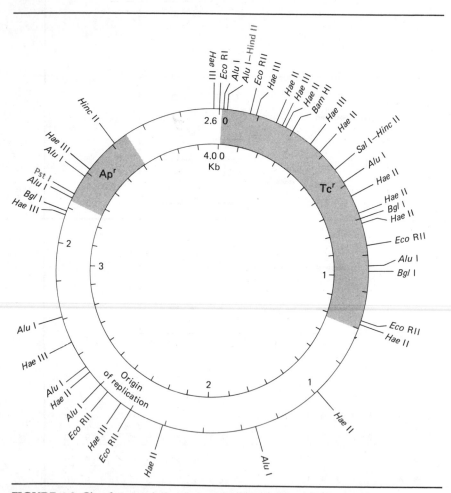

**FIGURE 4-9 Circular restriction map of pBR322.** The relative position of restriction sites are drawn to scale on a circular map divided into units of $1 \times 10^5$ daltons (outer circle) and 0.1 kilobases (inner circle). The molecular weight of the plasmid is $2.6 \times 10^6$. (From F. Bolivar et al., *Gene* **2**:95, 1977.)

Either of these sites can serve as the location for insertion of a foreign DNA fragment; we can consider here the $Tc^r$ locus. Isolated plasmids are treated with *Hin*d III endonucleases, which make staggered nicks at the *Hin*d III site $\frac{5'AAGCTT}{TTCGAA5'}$, splitting each $Tc^r$ gene into two sectors (Figure 4-10a, $Tc^{r'}$ and $Tc^{r''}$) and converting each circular plasmid into a linear chromosome with the two self-complementary ends, as drawn in Figure 4-10a. These **sticky ends** will serve as annealing sites for the foreign DNA, as described next.

## 4.14 Preparation of DNA for Cloning

To prepare DNA fragments suitable for cloning in the restricted pBR plasmid illustrated in Figure 4-10a, a sample of DNA is also treated with the *Hin*d III enzyme. As illustrated in Figure 4-10b, this treatment generates a collection of *Hin*d III fragments, each bearing $\frac{5'pAGCTT \qquad A3'}{A\,3'\ and\ TTCGA'p5'}$ tails, which are capable of annealing with the complementary tails on the restricted pBR vector.

The starting DNA sample is usually in one of two forms. In some cases, the investigator begins with **total genomic DNA** from a particular species—for example, total mouse DNA. Restriction of this sample will yield hundreds of thousands of different *Hin*d III fragments, any one of which would be impossible to resolve as a band in an agarose gel or as a fraction in a gradient. Therefore, the entire fragment collection is mixed with restricted pBR vectors, an experiment known as **shotgun cloning.**

In other cases, the investigator has isolated a very small amount of a purified DNA species and wishes to amplify this sequence for ease in analysis. Suppose, for example, an interesting class of satellite DNA (*Section 4.3*) has been purified by CsCl gradient centrifugation, or a class of restriction fragments has been eluted from a band in an agarose gel (*Figure 4-8*). These fragments may happen to contain *Hin*d III target sites, in which case they can be prepared for cloning by direct exposure to *Hin*d III nuclease. If not, synthetic *Hin*d III sites can be added chemically, and exposure to the nuclease then creates suitable fragments.

## 4.15 Preparation and Propagation of Recombinant Cloning Vectors

To prepare recombinant pBR322 vectors, the *Hin*d III–treated, sticky-ended fragments are simply incubated with the *Hin*d III–treated, sticky-ended plasmids under reassociation conditions, and the two are allowed to hybridize (Figure 4-10c). The four gaps that remain (Figure 4-10c) are then sealed with an enzyme known as **DNA ligase,** whose mechanism of action is diagrammed in Figure 4-11, to create closed circular molecules.

Host *E. coli* cells are next treated with a solution of $CaCl_2$ to facilitate

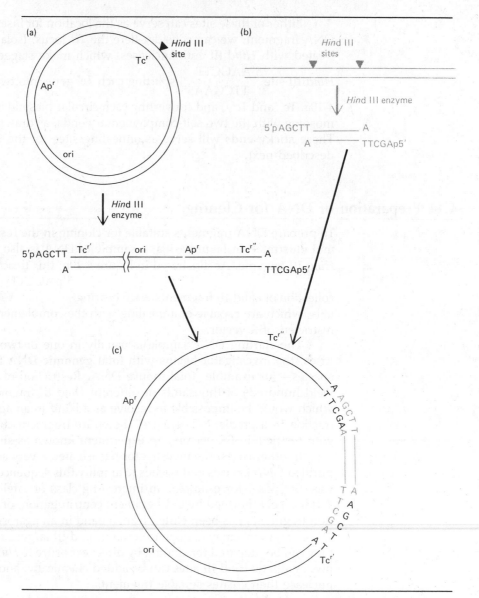

**FIGURE 4-10 Formation of a recombinant plasmid.** (a) The plasmid pBR322 is treated with the *Hind* III restriction enzyme, converting it into a linear molecule with two complementary ends. (b) A class of DNA fragments is created with *Hind* III "sticky ends." (c) The two are incubated together, and the fragments anneal with the vector to form a recombinant plasmid. Fragment-fragment and vector-vector annealing will occur as well, but the resultant molecules will not be selected in the subsequent screen.

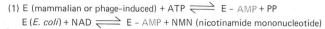

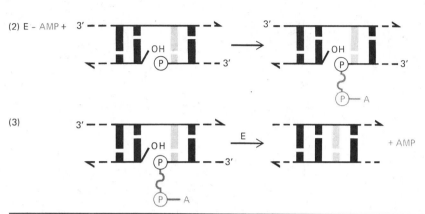

**FIGURE 4-11 DNA ligase** action. (1) The enzyme (E) forms an enzyme-adenylate (E-AMP) complex. (2) The E-AMP reacts with a nicked DNA chain, generating a pyrophosphate bond linking the P terminus of the DNA to the P group of the AMP. (3) The enzyme mediates the formation of a phosphodiester bond with the release of AMP. (Reproduced with permission from "Enzymes in DNA metabolism," *Annual Review of Biochemistry* **38**:795–840. Copyright © 1969 by Annual Reviews, Inc. All rights reserved.)

uptake of DNA. They are then incubated with the recombinant plasmids and plated to a medium containing ampicillin. All *E. coli* cells that have managed to take up a plasmid will carry Ap$^r$ and be resistant to the antibiotic. They will therefore grow to produce colonies, whereas all remaining cells will be killed.

## 4.16 Screening for Recombinant Clones

Of the ampicillin-resistant clones that have taken up plasmids, only some of these can be expected to carry a plasmid in which the foreign DNA fragment is successfully inserted; some plasmids, for example, may simply have circled back on themselves and sealed *via* their own complementary sticky ends. To detect colonies carrying recombinant vectors, advantage is taken of the fact that fragment insertion takes place *within* the Tc$^r$ gene, thereby inactivating it, so that the clones should be ampicillin resistant but tetracycline sensitive. Figure 4-12 illustrates how, by a technique known as **replica plating,** such tetracycline-sensitive colonies are identified. One then analyzes the foreign DNA contained in their recombinant plasmids by isolating the plasmids and treating them with *Hin*d III restriction enzyme: the fragments of interest are released and recovered for analysis.

In the case of the shotgun experiment described earlier, each recombinant clone will carry an unidentified segment of the mouse genome, the

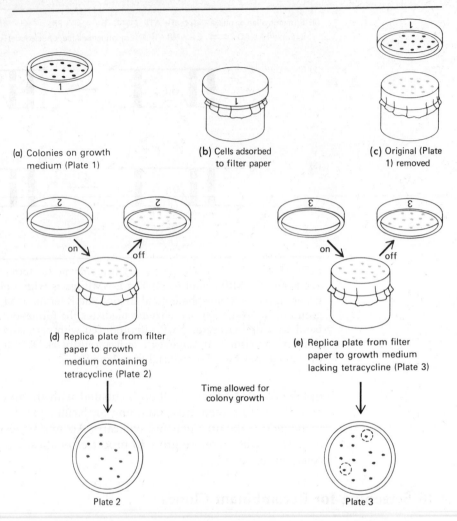

(a) Colonies on growth medium (Plate 1)

(b) Cells adsorbed to filter paper

(c) Original (Plate 1) removed

(d) Replica plate from filter paper to growth medium containing tetracycline (Plate 2)

Time allowed for colony growth

(e) Replica plate from filter paper to growth medium lacking tetracycline (Plate 3)

Plate 2

Plate 3

(f) Comparison of colonies on Plates 2 and 3 reveals those colonies sensitive to tetracycline (circled on Plate 3)

FIGURE 4-12 **Replica-plating** technique, as devised by J. and E. Lederberg, used to screen for tetracycline-sensitive strains of *E. coli*. (a) Cells are plated onto a solidified agar medium containing nutrients, and each forms a colony. (b) An adsorbent material such as filter paper is brought into contact with the agar surface such that some cells adhere to, and are picked up by, the paper (c). The paper is now pressed onto fresh media, and a few cells are transferred (d and e); in the experiment illustrated, plate 2 contains tetracycline, whereas plate 3 lacks the antibiotic. Comparison of these plates reveals cells that are sensitive to tetracycline.

**FIGURE 4-13**
Diagram illustrating
a selection
technique for
identifying an *E.
coli* clone with
DNA homologous
to an RNA probe.
See text for details.

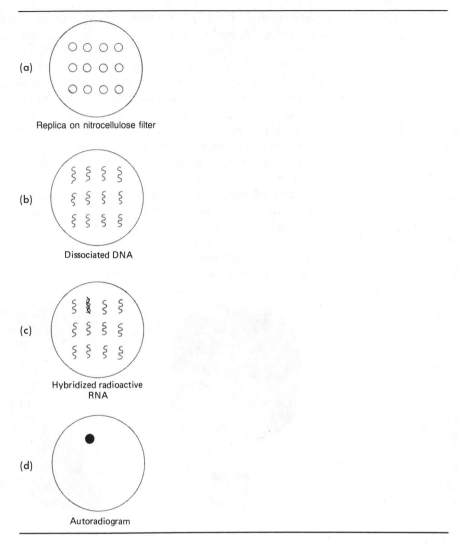

(a)

Replica on nitrocellulose filter

(b)

Dissociated DNA

(c)

Hybridized radioactive
RNA

(d)

Autoradiogram

entire collection of clones being termed a mouse **genomic library.** The final
step is to identify the clone or clones that carry a segment of interest. If, for
example, one wanted to identify the clone carrying all or part of the mouse
insulin gene, how would it be detected among the thousands of recombinants generated in a shotgun cloning experiment? One technique, illustrated in Figure 4-13, is based on the same principle as *in situ* hybridization
(*Section 4.10*). Identified recombinant clones are plated to nutrient agar in a
gridlike fashion and replicated to a nitrocellulose filter (Figure 4-13a). The
picked up cells are lysed *in situ,* and their DNA is dissociated into single
strands (Figure 4-13b). The filter is now placed in contact with a solution
containing radioactively labeled messenger RNA (*Section 1.5*) that codes for

mouse insulin. DNA/RNA hybridization ensues (*Section 4.10*), and because the **RNA probe** has a sequence complementary to the insulin gene sequence, it will hybridize to any insulin gene DNA (Figure 4-13c). The filter is now washed to remove all unreacted RNA and is subjected to autoradiography (Figure 4-13d; *Box 2.2*). Any labeled spot on the filter marks the position of a potential clone carrying the insulin gene. This clone is then recovered by matching the spot to the original grid of *E. coli* colonies.

   With procedures comparable to these, human genomic libraries have been successfully screened to find clones carrying DNA sequences for such important molecules as insulin and the antiviral agent interferon. The DNA has also been carefully scrutinized for clues about how such genes are organized and controlled, as we will detail in later chapters. The recombinant strains can also potentially produce large amounts of such important proteins as human insulin and interferon, a technology that has excited the interest of a number of industrial firms. Human insulin derived from recombinant clones is already available commercially for diabetic patients that cannot use the porcine form of the hormone.

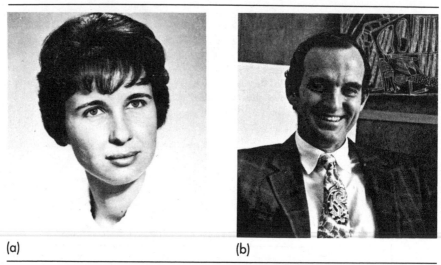

(a)                                                        (b)

(a) Mary Lou Pardue developed, with Professor J. Gall at Yale, the technique of *in situ* hybridization. She is now at the Massachusetts Institute of Technology.
(b) Charles A. Thomas, Jr. utilized reassociation studies to elucidate the structures of several phage chromosomes. He is presently at the Scripps Research Foundation in La Jolla.

# Questions and Problems

1. The DNAs from two different cell types have identical GC content. Does this mean that the two DNAs have the same nucleotide sequence? The same nucleotide composition? The same length? The same buoyant density? The same reassociation kinetics? Explain your reasoning.

2. Does a fast-reassociating class of DNA necessarily have a satellite buoyant density? Explain your answer.

3. If the satellite fraction of guinea pig DNA is sheared into fragments, dissociated, reassociated, and examined with the electron microscope, what would you expect to see? What would you expect of a slow-reassociating fraction in a similar experiment?

4. Outline *in situ* hybridization experiments you would perform to determine whether the Type I human satellite DNA (33 percent GC) is similar in sequence to the repetitive satellite DNA from mouse (34 percent GC).

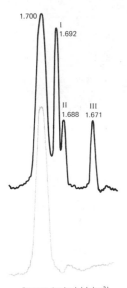

1.700

I
1.692

II    III
1.688   1.671

Buoyant density (ρ) (g/cm³)

5. DNA isolated from *Drosphila virilis* and subjected to CsCl density gradient centrifugation shows the patterns shown on the figure at left, where the black line represents DNA from brain tissue and the gray line represents DNA from salivary glands. (a) How do the satellites from *D. virilis* compare with those of *D. melanogaster* (Figure 4-2)? (b) How would you determine whether any of the *D. virilis* satellites are homologous to the *D. melanogaster* satellites? (c) In what respect would you expect the cell cycles of brain and salivary glands in *D. virilis* to differ? (d) Would you expect *D. virilis* brain cells to have much constitutive heterochromatin? Explain.

6. Why does the single-copy DNA of calf thymus reassociate at a higher $C_0t$ value than the *E. coli* DNA in Figure 4-5? (Hint: recall the relative genome sizes of calf and *E. coli*.)

7. Referring to Figure 3-15, how would you test whether *S. silvestre* and *S. cereale* differ in the sequence of their constitutive heterochromatin DNA as well as in amount?

8. Middle-repetitive human DNA contains foldback DNA sequences that have in common a single *Alu* I restriction site. Describe the experiments needed to document this statement.

9. What is a verbal palindrome? Why is foldback DNA often described as palindromic?

10. Describe how you would clone DNA using the *Pst* I site in the pBR322 plasmid.

11. List four ways that bacteria and mammals differ in their genomic DNA.

12. The sea urchin is known to possess numerous copies of the genes for the five histone proteins. How would you determine whether these genes are dispersed throughout the genome or localized to a single tract of DNA?

13. What is the nucleotide sequence of the DNA shown in Box 4.1, starting with the G marked with an arrow and ending with the A marked with an arrow? Would you expect this DNA to have a satellite buoyant density? Explain.

# 5 The Meiotic Transmission of Chromosomes

**A. Introduction**

**B. Meiosis**

5.1 Key Features of Meiosis

5.2 The Stages of Meiosis

5.3 Leptonema: Lateral Elements Organized

5.4 Zygonema: Synapsis of Homologues

5.5 Pachynema: Crossing Over Occurs

5.6 Diplonema: Desynapsis and Visualization of Chiasmata

5.7 Diakineses: Maximal Chromosome Condensation

5.8 Metaphase I: Independent Alignment of Homologues

5.9 Anaphase I and Telophase I: Separation of Centromeres

5.10 Intrameiotic Interphase: No DNA Replication

5.11 Prophase, Metaphase, Anaphase, and Telophase II: Mitotic-like

**C. Life Cycles of Sexually Reproducing Organisms: Mitosis-Meiosis Alternations**

5.12 Animal Life Cycles

5.13 Higher Plant Life Cycles

5.14 Lower Eukaryote Life Cycles

**D. Meiotic Errors**

5.15 Aneuploidy

5.16 Haploidy and Polyploidy

**E. Questions and Problems**

## Introduction

The transmission of chromosomes from a parent cell to its daughter cells via mitosis (*Chapter 2*) is an asexual process: one parent cell can give rise, through successive mitotic divisions, to a clone of cells that are genetically identical. Such a **vegetative** or **somatic** pattern of cell proliferation is responsible for the growth of multicellular organisms and for the self-propagation of unicellular eukaryotes. Most eukaryotes, however, possess an alternate, **meiotic** mode of chromosome transmission; this mode is inevitably coupled with a **sexual** phase in their life cycle wherein the genes from two different parents come to reside in a single cell. Specifically, meiosis represents the avenue by which haploid **gametes** (ova, sperm, and pollen) arise from diploid cells in the germinal tissues of higher plants and animals. In lower eukaryotes, it is also the avenue by which haploid vegetative cells are generated from diploid zygotes.

This chapter presents a detailed description of meiosis. The life cycles of organisms of genetic interest are then discussed, with emphasis on the alternation of mitotic and meiotic modes of chromosome transmission in each case. Finally, we consider occasional errors that occur or are induced in meiosis, and describe the (often severe) consequences of such errors.

# Meiosis

## 5.1 Key Features of Meiosis

Like mitosis, **meiosis does not involve chromosome replicaton per se:** the meiotic equivalent of the S phase (*Section 2.6*) occurs well before meiosis begins. Recalling that the total amount of DNA in a haploid nucleus is denoted as the C value (*Section 4.1*), therefore, one can say that a diploid cell enters both meiosis and mitosis with a 4C amount of DNA. An important difference between meiosis and mitosis, however, is that meiotic DNA replication is followed by *two* nuclear divisions in succession rather than one. It should be obvious that if a diploid cell (2C) replicates its DNA (4C) and then undergoes two meiotic divisions, the first division will produce two cells with a 2C amount of DNA, whereas the second division will result in four cells that are each haploid, or C, in the amount of DNA they contain. In other words, **the net effect of meiosis is to reduce a cell's chromosome number by half,** usually from an initial 2*n* to a final *n*. This is a key feature of the meiotic process.

As noted in Section 2.10, diploid cells contain two complete sets of chromosomes, one derived originally from a paternal gamete and the other from a maternal gamete. Since meiosis gives rise to such gametes, it follows that the $2n \longrightarrow n$ meiotic reduction must occur in such a way that **each haploid product of meiosis is allotted one complete set of chromosomes** containing all the genetic information pertaining to that species. This complete allotment, then, constitutes the second key feature of the meiotic process.

A third key feature of meiosis is that it is the stage in eukaryotic development in which most **new gene combinations are generated.** These gene combinations come about in two ways. First, the maternally and paternally derived homologous chromosomes that coexist in a 2*n* organism are distributed among the organism's haploid meiotic products in numerous combinations. Thus if we think of a diploid cell as having three sets of homologous chromosomes, 1*M* and 1*P*,, 2*M* and 2*P*, and 3*M* and 3*P*, where *M* stands for maternally derived and *P* stands for paternally derived, the haploid products of the cell will include 1*M*2*M*3*M* cells, 1*M*2*M*3*P* cells,

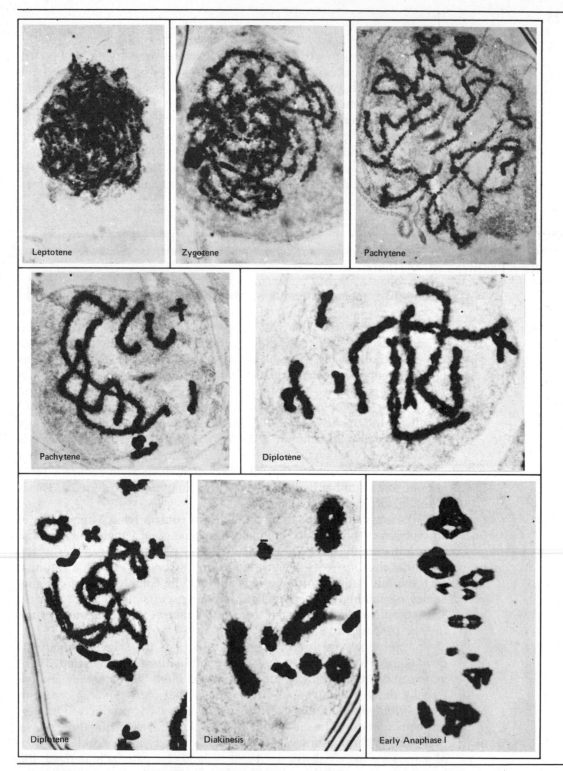

**FIGURE 5-1 Meiosis** in the grasshopper. (Courtesy James L. Walters, University of California, Santa Barbara.)

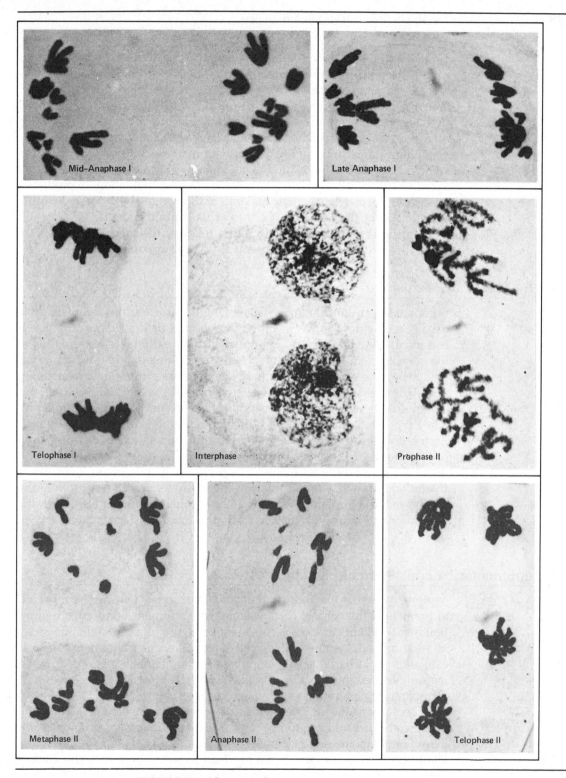

**FIGURE 5-1** (Continued)

1M2P3P cells, and so on (in this case there are eight possible combinations in all). Second, maternally and paternally derived homologous chromosomes frequently take part in genetic exchange during meiotic prophase. When this occurs, a typical haploid product will contain some maternally derived, some paternally derived, and some recombinant chromosomes as members of its complete chromosome set, the recombinant chromosomes containing information derived from both maternal and paternal chromosomes. This cell, with its unique combination of genes, will fuse with a second haploid cell, also carrying a unique gene combination, to produce a 2n diploid zygote. **The genetic makeup of this zygote will clearly be quite different from the makeup of either of its diploid parents.**

In the description of meiosis that follows, the mechanical basis for these three key featues of meiosis will become apparent.

## 5.2 The Stages of Meiosis

Compared with mitosis, meiosis is a lengthy process, the complete cycle usually taking days or weeks rather than hours. Its first stage, prophase I, is particularly complex, and in animals it frequently takes at least four to five days to complete (in contrast to the typical one-half to one hour occupied by mitotic prophase). Prophase I is therefore commonly broken down into five substages: leptonema, zygonema, pachynema, diplonema, and diakinesis (the adjectives corresponding to the first four stages are leptotene, zygotene, pachytene, and diplotene and are often used as nouns). Each of these substages is defined arbitrarily, and all, of course, flow from one to the next. Prophase I is followed by metaphase I, anaphase I, and telophase I, and these are followed by prophase II, metaphase II, anaphase II, and telophase II. A short interphase usually separates the first from the second meiosis in plants. Figure 5-1 shows micrographs of meiosis in the grasshopper. Figure 5-2 diagrams the various meiotic stages that have been defined.

## 5.3 Leptonema: Lateral Elements Organized

Leptonema marks the end of premeiotic interphase. **During the S period of premeiotic interphase, chromosome replication has occurred,** a replication that to all appearances seems identical to the replication that precedes mitosis. Indeed, certain kinds of cells that will ordinarily undergo meiosis can, after an S phase, be so manipulated experimentally that they proceed instead through mitosis. Once a cell enters the leptotene stage, however, it is apparently committed to a meiotic course.

Leptonema means "slender thread," and this stage is heralded by the presence of threadlike chromosomes in their initial phase of meiotic condensation. Each thread in fact represents a pair of sister chromatids that are identical (as in mitosis) and that are held together by a common centromere. The leptotene threads may also exhibit the periodicity that was

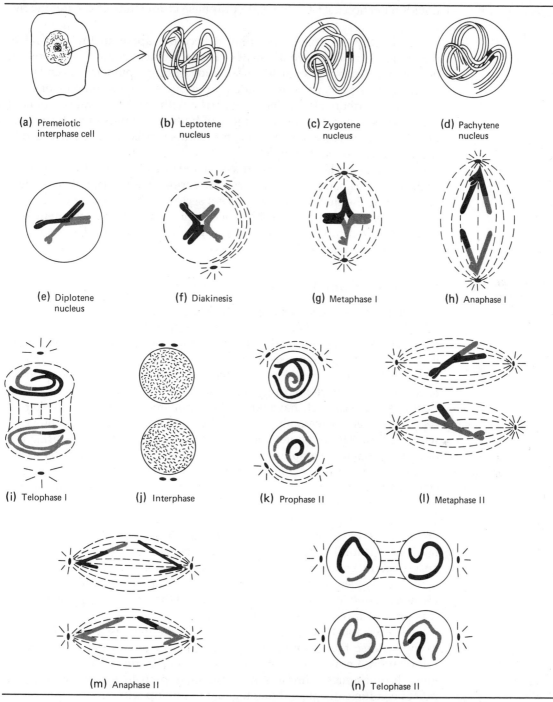

(a) Premeiotic interphase cell

(b) Leptotene nucleus

(c) Zygotene nucleus

(d) Pachytene nucleus

(e) Diplotene nucleus

(f) Diakinesis

(g) Metaphase I

(h) Anaphase I

(i) Telophase I

(j) Interphase

(k) Prophase II

(l) Metaphase II

(m) Anaphase II

(n) Telophase II

**FIGURE 5-2 Diagram of the stages of meiosis** in an animal with one pair of chromosomes ($n = 1$). (After J. McLeish and B. Snoad, *Looking at Chromosomes*. New York: St. Martin's Press, Macmillan & Co., Ltd., 1972.)

noted in Section 2.11 for the early prophase chromatids of the mitotic nucleus.

With the electron microscope, it becomes apparent that a band of material called a **lateral component** or **lateral element** is laid down between or alongside each pair of sister chromatids during leptonema. Each lateral component is perhaps 50 nm wide and appears to be a complex of RNA and protein (ribonucleoprotein) associated with DNA. Lateral component synthesis is apparently a joint activity of both chromatids in each chromatid pair, and DNA from both sister chromatids is believed to be included in the lateral component complex.,

The lateral element material associated with the ends (telomeres) of each chromatid pair becomes associated, during leptonoma, with a particular sector of the nuclear envelope. This sector of the envelope often lacks nuclear pores, may exhibit a thickened region termed an **attachment site,** and is believed to serve as a focal site where homologous chromosomes align with one another.

## 5.4 Zygonema: Synapsis of Homologues

Zygonema (from the Greek *zygon,* a yoke) is defined as the stage in which homologous sets of sister chromatids complete their side-to-side alignment, an association called **synapsis** (from the Greek *syn-,* together, + *apsis,* a joining). Synapsis appears to begin in the telomere-envelope attachment zone described in the preceding section and is widely assumed to initiate with short-lived physical contacts between homologous lateral elements. Such contacts have not yet been documented by electron microscopy, however. Instead, aligned lateral elements are observed to move to within 200 to 300 nm of one another and then participate in the formation of a **synaptonemal complex.** The complex consists of the two lateral elements and a **central region,** approximately 100 nm wide in the mature state, which is bisected by a narrow band, the 20-nm-wide **central component.** A mature synaptonemal complex is illustrated in Figure 5-3. It is not yet clear whether the material of the central region is assembled as a cooperative effort of the synapsing, homologous chromosomes or whether prefabricated central region portions are inserted between two correctly spaced lateral components; both modes of assembly have been reported. It is, however, generally agreed that the central components, like the lateral elements, contain both RNA and protein and that, if DNA is present at all, it is present in very small amounts,.

The establishment of a synaptonemal complex is a crucial genetic event because these complexes mediate the meiotic exchange of genetic information, known as **crossing over** or **recombination,** between homologous chromosomes. Chromosomes that are to participate in recombination must be sufficiently alike (homologous) to line up with one another, gene for gene. The chromosomes then exchange pieces of their DNA in such a way that information is neither gained nor lost from either chromosome.

**FIGURE 5-3**
**Synaptonemal**
**complex** (SC) of the
fungus *Neotiella*
separating two
homologous
chromatid pairs
(Chr 1 and Chr 2).
The lateral elements
(LE) are banded,
which is not the
case for many
organisms. CC,
central component;
CE, central
element. (Courtesy
M. Westergaard
and D.
von Wettstein.)

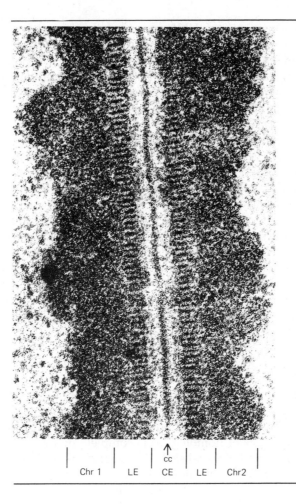

Chr 1 ⎪ LE ⎪ CC CE ⎪ LE ⎪ Chr2

Specifically, if the two chromosomes undergoing pairing are designated A B C D and a b c d, then one recombination event might yield A B c d and a b C D chromosomes, another might yield A b c d and a B C D chromosomes, and so on. In the case of the synaptonemal complex, then, its formation must involve both mutual recognition between sister chromatid pairs and their precise, point-for-point alignment, so that crossing over does not produce chromosomes that transmit too little, or too much, of the requisite genetic information of the species.

As will become clear in later chapters, precise recognition, alignment, and crossing over can occur between chromosomes that are not conjoined by synaptonemal complexes. Why, then, are these complexes ubiquitous in all cells that undergo meiotic synapsis and crossing over, and what role or roles do they perform? One plausible, although not yet proved, explanation is that lateral components serve as scaffolds into which key "recognition sequences" of DNA are carefully threaded during leptonema. At zygonema, such sequences could participate in recognition and alignment, eliminating the staggering problem of how two homologous chromatids could possibly synapse along their entire lengths when each can contain up to half a meter of DNA. Once recognition takes place, central elements would be added between the homologues to stabilize the (often prolonged) synaptic state that develops.

Synaptonemal complexes can be observed only with the electron microscope. With the light microscope, however, a zygotene nucleus can be distinguished in two ways: in many organisms the association of telomeres with the nuclear envelope causes a "bouquet" arrangement of the chromosome strands, and in most organisms, zygotene chromosome strands appear thicker than they do in leptonema. Although this increase in thickness comes partly from continued chromosome condensation, it is primarily the result of sister chromatids at this stage becoming intimately intermeshed with one another so that it becomes impossible to distinguish them as individual entities. For this reason, each synapsed set of homologues, which in fact includes four chromatids, appears to be composed of only two chromosomes (Figure 5-3). Each synapsed homologue set is therefore referred to as a **bivalent**.

## 5.5 Pachynema: Crossing Over Occurs

Pachynema means "thick strand" and denotes the continued shortening and thickening of the bivalents that occur during this stage of meiosis. It is generally agreed that synaptonemal complex formation and the concomitant synapsis of homologues is complete by the onset of pachynema and that the actual physical exchanges that result in chromosomal crossing over occur during the pachytene stage. Detailed discussion of these events at the molecular level will be presented in Chapter 17. The completion of synapsis is accompanied by the dispersal of the "zygotene bouquet" arrangement of chromosome fibers.

## 5.6 Diplonema: Desynapsis and Visualization of Chiasmata

The pachynema-diplonema transition is heralded by **desynapsis:** all 4 chromatids in a bivalent move apart as if repulsed, the result being that the "split" between sister chromatid pairs becomes evident and a separation also appears between the homologues in each set. With the electron microscope, it is seen that most of the synaptonemal complex material is shed during diplonema, and repulsion is presumed to be the consequence of this shedding.

Two kinds of constraints appear to prevent the repulsing chromosomes from separating completely. First, the sister chromatids continue to be held together by centromeres. Second, nonsister chromatids in a bivalent are usually held together at one or several positions along their lengths

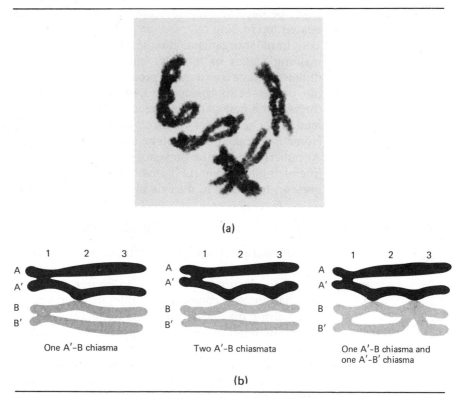

(b)

**FIGURE 5-4** (a) **Chiasmata** visualized during diakinesis in the angiosperm *Trillium erectum.* (Courtesy A. H. Sparrow and R. F. Smith, Brookhaven National Laboratory.) (b) Several chiasma interrelationships among homologous chromatids A-A′ and B-B′.

by regions of apparent contact called **chiasmata** (singular, **chiasma**). Chiasma means a cross, and chromatids connected by a single chiasma typically assume a crosslike configuration when viewed under the light microscope (Figure 5-4a). With the electron microscope, chiasmata are found to include short lengths of synaptonemal complex that have not been shed and continue to hold nonsister chromatids together at comparable positions along their lengths.

Two, three, or all four chromatids in a bivalent may participate in chiasma formation, and a given chromatid may form more than one chiasma. If, for example, we designate the four as A, A′, B, and B′, then chromatid A′ may form a chiasma with chromatid B at one position, while at another position it may exhibit a chiasma with chromatid B or, alternatively, with chromatid B′. These relationships are illustrated in Figure 5-4b.

Chiasmata occur frequently. In a normal meiosis there is at least one chiasma per bivalent, and in a typical human oocyte containing 23 bivalents, perhaps 52 chiasmata are observed. Chiasmata are almost certainly related to crossing over (*see Section 14.3*).

In most organisms the diplotene stage is quickly followed by the remaining stages of meiosis, but in the oocytes of many animals, the diplotene stage is extremely prolonged. The human female fetus, for example, possesses approximately 3,400,000 oocytes in each ovary, and these go through the first stages of meiosis during the fourth to seventh months of fetal life. The oocytes then remain within the ovary in a diplotene stage (sometimes called a **dictyotene** stage), which may last as long as 50 years! At puberty and in the presence of follicle-stimulating hormone (FSH) and luteinizing hormone (LH), one oocyte per menstrual cycle is ovulated and goes on to complete meiosis in the fallopian tubes if fertilization occurs.

## 5.7 Diakinesis: Maximal Chromosome Condensation

Throughout prophase I the chromosomes have continued to shorten, and by diakinesis they appear to be maximally condensed. Chiasmata now appear in electron micrographs to be free of any synaptonemal complex material, instead consisting of uninterrupted spans of chromatin.

During diakinesis chiasmata frequently go through a process known as **terminalization,** in which they appear to move down the chromatids until they reach the ends of the bivalents. The bivalents are thus characteristically interconnected by one or two terminal (or partly terminal) chiasmata as they enter metaphase I. Both the mechanism and the function of terminalization are obscure.

## 5.8 Metaphase I: Independent Alignment of Homologues

Metaphase of the first meiotic division is characterized by spindle formation, as in mitotic metaphase, but the two processes are otherwise different. Each bivalent includes two distinct centromeres, each of which holds two chromatids together. The two centromeres dispose themselves on either side of a plane that is analogous to a metaphase plate, a plane we can imagine bisecting the spindle from left to right (Figure 5-2g).

For meioses involving more than one bivalent (that is, $n > 1$), a key feature of metaphase I is readily visualized by recalling that each bivalent possesses a maternally derived centromere and a paternally derived centromere. For bivalent number 1, the maternal centromere may lie above the metaphase plate and the paternal below. Considering next bivalent number 2, its maternal centromere may also lie above, and its paternal centromere below, the metaphase plate. It is equally probable, however, that the reverse alignment will occur, with the paternal centromere lying above and the maternal centromere lying below the plate. In other words, **the maternal-paternal arrangement of a given set of homologous centromeres with respect to the metaphase plate is *totally independent* of the arrangements assumed by all other sets,** and all possible combinations of arrangements thus occur with equal frequency when large numbers of metaphase I cells are considered. This fact has important genetic consequences, as specified shortly.

# 5.9 Anaphase I and Telophase I: Separation of Centromeres

Bivalents typically continue to be held together by (terminal) chiasmata until anaphase I, at which point these connections are severed and the two centromeres of the bivalent move to opposite poles of the cell. In contrast to anaphase in mitosis, **anaphase I of meiosis involves only a *separation* of independent centromeres. No centromeric divisions occur in anaphase I.**

The nuclear envelope typically disperses during metaphase I and anaphase I. Depending on the organism, the envelope may then reorganize around the two separated sets of homologous chromosomes during telophase I, or the chromosomes may enter directly into the second meiotic division. Similarly, a furrow may or may not divide the cell into two daughter cells.

The germ cells of most female animals take part in **asymmetrical** meiotic divisions, meaning that the metaphase I plate lies near the cell surface rather than in the cell interior. At anaphase I, one set of homologues moves into a tiny bleb of cytoplasm, whereas the other remains behind in the enormous egg cell, known at this stage as an **oocyte.** At telophase I the bleb is separated from the main cell by a furrow, and the resultant tiny cell is called the **first polar body** (Figure 5-5).

# 5.10 Intrameiotic Interphase: No DNA Replication

The interphase that separates the two meiotic divisions may be brief. Its most notable feature is that it includes **no replication of chromosomal DNA,** and thus it is unlike a mitotic interphase or the interphase preceding meiosis I.

# 5.11 Prophase, Metaphase, Anaphase, and Telophase II: Mitotic-like

Cells that bypass a telophase I and enter meiosis II directly do not experience a true prophase II, but otherwise the second meiotic division is morphologically indistinguishable from a mitotic division. **Centromeres** connecting pairs of chromatids move to a metaphase plate, **divide into two halves** (for the first time in the meiotic process), and move to opposite poles at anaphase. At the completion of telophase II, therefore, four haploid cells, or a **tetrad,** have derived from each original diploid cell and each haploid cell now returns to an interphase state.

In animal oocytes, anaphase II and telophase II are again asymmetrical and a **second polar body** is formed (Figure 5-5). The first polar body

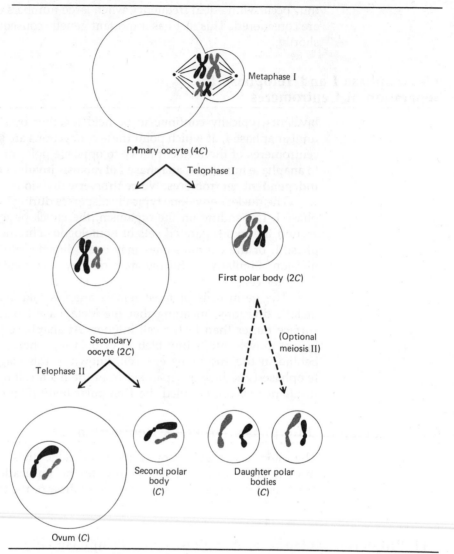

Metaphase I

Primary oocyte (4C)

Telophase I

First polar body (2C)

Secondary
oocyte (2C)

(Optional
meiosis II)

Telophase II

Second polar
body
(C)

Daughter polar
bodies
(C)

Ovum (C)

**FIGURE 5-5 Oogenesis** in an animal showing the formation of polar bodies.

may also undergo a meiosis II so that four haploid products result—three polar bodies and the **ovum;** alternatively, the first polar body may not divide, in which case there will be only three meiotic products, one diploid and two haploid. In either case the haploid ovum is the only viable gamete, in contrast to male animal spermatogenesis in which all four meiotic products are viable.

# Life Cycles of Sexually Reproducing Organisms: Mitosis-Meiosis Alternations

## 5.12 Animal Life Cycles

In animals such as *Drosophila* and humans, the sexual phase of reproduction is experienced exclusively by cells of the **germ line.** These cells become differentiated from the remaining **somatic line** cells early in embryogenesis, and they alone have the potential to undergo meiosis and become gametes. This potential is realized only after the somatic portion of the organism has reached maturity via numerous mitoses. The maturation process may involve several immature stages, as in *Drosophila* (Figure 5-6), or it may occur in the original organism, as in humans.

Gonad differentiation and activity proceeds quite differently for male and female higher animals. In human females, for example, the two original germ line cells of each ovarian primordium undergo approximately 21 mitoses to generate the $3.4 \times 10^6$ ($\approx 2^{21}$) oocytes present in the fetal ovary. As noted earlier, these then enter meiotic prophase, in which they await a one-per-month ovulation during the postpubescent years. In human males

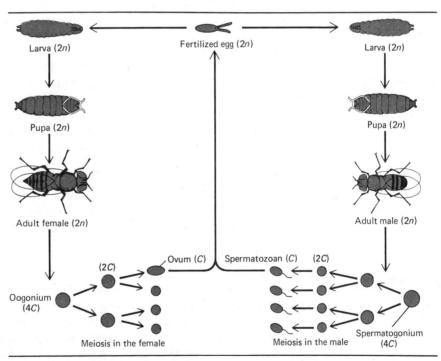

**FIGURE 5-6** Life cycle of *Drosophila*.

the original germ line cells also divide by mitosis to populate the fetal testes. At puberty, however, these give rise to cells called **stem spermatogonia,** which divide by mitosis, on an average of once every 16 days, to produce **primary spermatocytes.** These, in turn, undergo meiosis to generate the millions of spermatozoa produced daily. It can be calculated that 380 rounds of mitosis have occurred in the testis of a 28-year-old man and 540 rounds in a 35-year-old man. There thus exists a very great difference in the number of mitoses undergone by the human male and by the human female germ lines.

## 5.13 Higher Plant Life Cycles

An immediate distinction between an animal and a higher plant is that the latter does not possess a germ line. Instead, among those $2n$ vegetative cells that make up the flowering or **sporophyte** portion of the plant, some are induced to differentiate into **megasporophyte** ($♀$) and **microsporophyte** ($♂$) cells that then undergo meiosis. Meiosis produces haploid ($n$) **spores** (megaspores and microspores) that divide mitotically, sometimes many times, to produce what is called the **gametophyte.** In mosses, the haploid gametophyte is large and leafy, whereas the diploid sporophyte is less conspicuous. In the ferns, the sporophyte is far more prominent, but the gametophyte is still multicellular and green. The gametophyte is most reduced in the angiosperms. As shown in Figure 5-7 for corn, one of the megaspore nuclei undergoes three mitotic divisions to produce an embryo sac containing eight haploid nuclei; one of these then serves as the egg nucleus. The microspore comes to contain three haploid nuclei, two of which serve as sperm nuclei in the pollen grain. At the time of fertilization, one of the sperm nuclei fuses with the egg nucleus to form a $2n$ zygote. The second sperm nucleus fuses with two other nuclei remaining in the embryo sac (the **polar nuclei**) to form the triploid ($3n$) **endosperm,** which helps to nourish the developing embryo. The remaining haploid nuclei play accessory roles in fertilization and seed production.

In comparing higher plants and animals, therefore, it is important to realize that the haploid phase of the organism's life cycle is often far more conspicuous and long-lived in plants than it is in animals. The haploid or gametophyte phase in plants is often a valuable one for genetic studies, as shown in Section 6.8.

## 5.14 Lower Eukaryote Life Cycles

There is no sharp distinction between higher and lower eukaryotes in a taxonomic sense, but geneticists tend to regard the fungi, the algae, and the protozoa as lower versions of eukaryotic cells. The following sections describe the life cycles of four such organisms that have become particularly important in genetic research, namely, *Chlamydomonas, Neurospora, Saccharomyces,* and *Paramecium.*

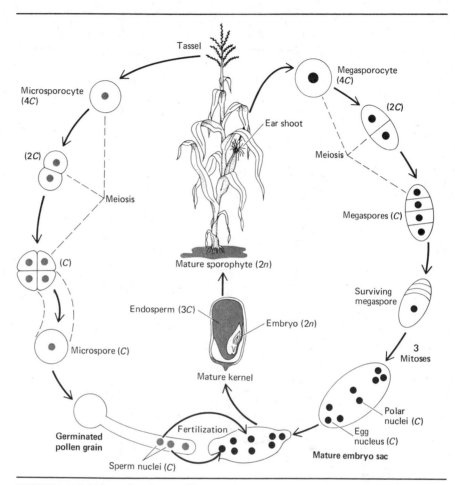

**FIGURE 5-7** Life cycle of corn (*Zea mays*). (From *General Genetics,* Second Edition, by Adrian M. Srb, Ray D. Owen, and Robert S. Edgar. W. H. Freeman and Company. Copyright © 1965.)

*Chlamydomonas* **(Figure 5-8)**   Vegetative cells of *Chlamydomonas reinhardi* are haploid, and each is a free-swimming, flagellated green organism. When the growth medium becomes depleted of nitrogen, these cells undergo certain morphological changes that permit them to mate. No separate germ line is involved in this gametogenic process.

Despite the absence of a separate germ line, there are two sexes of *C. reinhardi:* one called mating type *plus,* the other, *minus.* Gametes of the two sexes appear to be identical under the light microscope, and thus *C. reinhardi* is said to be **isogamous.** Similarly sexed gametes ignore one another, however, whereas *plus* and *minus* gametes undergo rapid pairing and fusion. The two gamete nuclei then fuse, and a diploid zygote results.

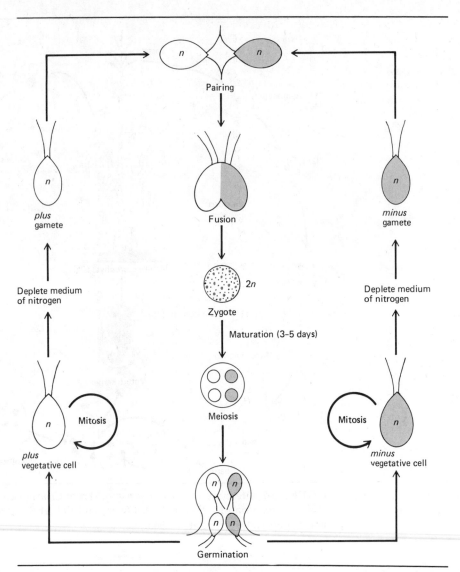

FIGURE 5-8 Life cycle of *Chlamydomonas reinhardi* as it occurs in the laboratory.

After several days of maturation, this zygote undergoes meiosis to yield four haploid products. Each of the meiotic products then divides mitotically to produce four clones, with each clone comprising genotypically identical haploid vegetative cells. Under the appropriate conditions, any one of these haploid vegetative cells can become differentiated into a gamete, and the life cycle of *C. reinhardi* can be repeated.

*Neurospora* **(Figure 5-9)**   Like *C. reinhardi*, the bread mold *Neurospora crassa* also exists in the form of two mating types—in this case, *A* and *a*. Cells of both types can reproduce vegetatively by mitosis, with the daughter cells remaining attached to one another such that long filaments, called **hyphae,** are formed. At the tips of the hyphae, cells called **conidia** will pinch off. Under certain conditions these pinched-off conidia will give rise to new haploid vegetative hyphae. Under other conditions the same conidia may instead take part in sexual reproduction.

During the sexual reproduction process, a conidium from mating type *A* finds its way to a large, specialized hypha, a **protoperithecium,** which develops within a mating-type a organism; conversely, a type *a* co-

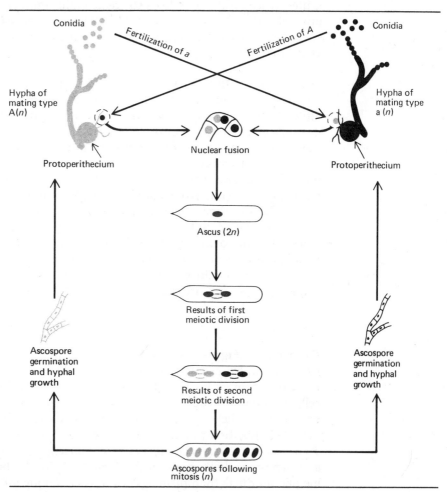

**FIGURE 5-9** Life cycle of *Neurospora*. (From R. Wagner and H. Mitchell, *Genetics and Metabolism*. New York: Wiley, 1955.)

nidium can find its way to an *A* protoperithecium. In either case the two cell types will fuse so that two nuclei will be present in a single cell. Since the protoperithecium donates most of the cytoplasm to this cell, it is considered the female element in *Neurospora* sexuality; the conidium is considered the male element.

The two nuclei do not immediately fuse in the newly formed cell. Instead, each undergoes several mitotic divisions so that a multinucleate cell is formed. Individual nuclei of opposite mating types then fuse to form diploid nuclei, and each of these diploid nuclei becomes walled off in a cell having a thick wall known as an **ascus** (plural, **asci**). Meiosis now occurs to yield four haploid nuclei. These in turn divide mitotically so that each ascus contains eight haploid nuclei, each meiotic product being represented twice. Walls now form to create eight uninucleate cells within the original ascus. Each of these cells is known as an **ascospore.** The ascospores are released when the ascus ruptures, and each can give rise to vegetative hyphae.

**Yeast (Figure 5-10)**   The life cycle of the single-celled fungus *Saccharomyces cerevisiae*, or baker's yeast, is less intricate than that of *Neurospora*, and it differs from both *Neurospora* and *Chlamydomonas* in that individual vegetative cells can be haploid or diploid. Thus vegetative haploid cells of opposite mating type (*a* and *α*) will fuse to form vegetative diploid cells that continue to propagate by mitosis. The cytokinesis that follows a haploid or diploid yeast mitosis is usually unequal; that is, a smaller cell (the "daughter") appears to bud from a larger cell (the "mother"). The two, however, are identical in their nuclear composition.

A vegetative diploid cell of *S. cerevisiae* may also be induced, by nitrogen starvation, to undergo **sporulation,** a complex process that endows the diploid cell with the ability to undergo meiosis. The four resulting meiotic products are contained within a mother cell wall, also called an ascus. Haploid ascospores are then released, each divides to produce a clone of vegetative haploid cells, and the reproductive cycle of the organism can be repeated.

*Paramecium* **(Figures 5-11 and 5-12).**   The life cycles of ciliated protozoa such as *Tetrahymena* and *Paramecium* include some of the most elaborate nuclear behaviors known in eukaryotes. A somatic *Paramecium tetraurelia* cell, as noted in Section 2.13, contains a hyperploid macronucleus and two diploid micronuclei. Sexuality is induced by nutrient starvation and begins with a **conjugation** process between cells of opposite mating type (Figure 5-11a). Conjugal contact triggers macronuclear breakdown (Figure 5-11a) and meiosis in both micronuclei (Figure 5-11b). All but one of the eight meiotic products then disintegrates in each conjugant (Figure 5-11c). The remaining haploid nucleus in each cell undergoes a single mitotic division to give two haploid nuclei (Figure 5-11d). Each of the conjugating cells then

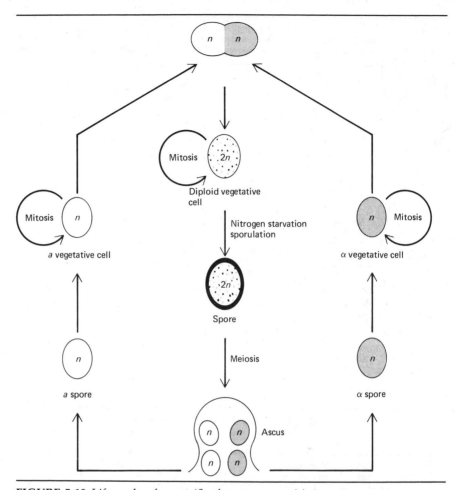

**FIGURE 5-10** Life cycle of yeast (*Saccharomyces cerevisiae*).

donates one of these haploid nuclei to its mate via a bridge that develops to connect them (Figure 5-11e). The two haploid nuclei in each cell fuse to form new diploid micronuclei (Figure 5-11f), and the conjugants separate. After separation, each *Paramecium* experiences two mitotic nuclear divisions (Figure 5-11g) followed by macronuclear development (Figure 5-11h). A final round of mitosis is followed by one cell division (Figure 5-11i).

In addition to the sexual events shown in Figure 5-11, *P. tetraurelia* can also undergo self-fertilization, a process known as **autogamy,** in which meiosis, nuclear breakdown, and a single mitotic division occur as before, but in which the two haploid nuclei in a single cell proceed to fuse together. These events are illustrated in Figure 5-12.

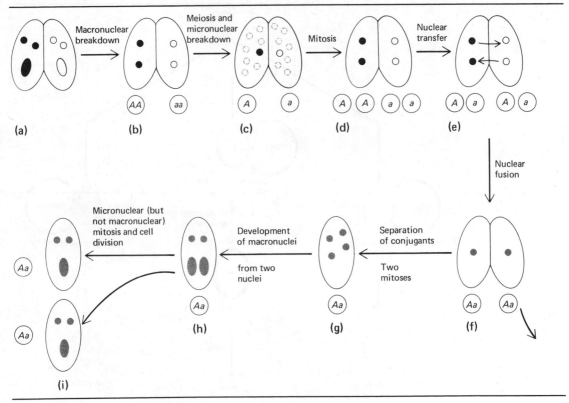

**FIGURE 5-11** Life cycle of *Paramecium aurelia*.

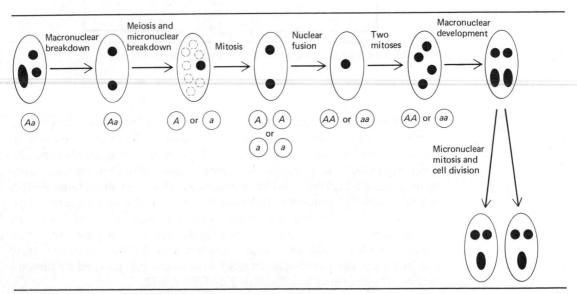

**FIGURE 5-12** Autogamy in *Paramecium aurelia*.

# Meiotic Errors

The importance of meiotic precision is perhaps best appreciated when one considers the consequences of **meiotic errors** that lead to the production of gametes having too many or too few chromosomes. When such gametes undergo fertilization, the resultant zygotes are similarly imbalanced and usually experience severe developmental abnormalities. We first describe the generation of **aneuploid** individuals that have an excess or deficit of a small number of chromosomes, usually only one. We then consider individuals that have lost or gained entire sets of chromosomes.

## 5.15 Aneuploidy

Aneuploids are classified by the number of chromosomes they have gained or lost. Thus if the normal diploid chromosome number of an organism is $2n$, an aneuploid organism might be $2n - 1$ (**monosomic**), $2n + 1$ (**trisomic**), $2n + 2$ (**tetrasomic,** or **double trisomic**), and so on.

Aneuploidy is a common condition. In humans, for example, probably almost a third of all spontaneous abortions involve a fetus with an abnormal number of chromosomes, and among these more than 60 percent are aneuploids, most being either trisomics or monosomics. Some relevant data are summarized in Table 5-1.

Aneuploidy is usually produced by **primary nondisjunction** at meiosis: two homologues fail to disjoin at the first meiotic division, and two of the resulting gametes carry a double dose of the chromosome; the other two gametes lack the chromosome entirely. When these abnormal gametes are fertilized by normal gametes, trisomic and monosomic zygotes result. Figure 5-13 illustrates such events for the homologous X and Y human sex chromosomes, which, although different in morphology (*Figure 2-13*),

**TABLE 5-1  Frequency of Selected Chromosomal Aberrations in Humans**

|  | Frequency of condition in Spontaneous Abortions | Live-Born |
|---|---|---|
| Monosomy X (XO, "Turner's syndrome") | 1/18 | 1/3,000 females |
| Trisomy chromosome 16 | 1/33 | Almost zero |
| Trisomy chromosome 8 | 1/33 | 1/14,500 |
| Trisomy chromosome 21 | 1/40 | 1/600 |
| Trisomy chromosome 18 | 1/200 | 1/4,500 |
| Trisomy, sex chromosomes | | |
| XXY, "Klinefelter syndrome" | Very low | 1/600 males |
| XYY | Very low | 1/1000 males |
| Triploidy | 1/22 | Almost zero |
| Trisomy, X chromosome (XXX) | Very low | 1/1,600 females |

From V. A. McKusick, *Human Genetics.* Englewood Cliffs, N.J.: Prentice-Hall, © 1969, Table 2.3, and P. A. Gerald, *New England J. Med.* **294:**706 (1976).

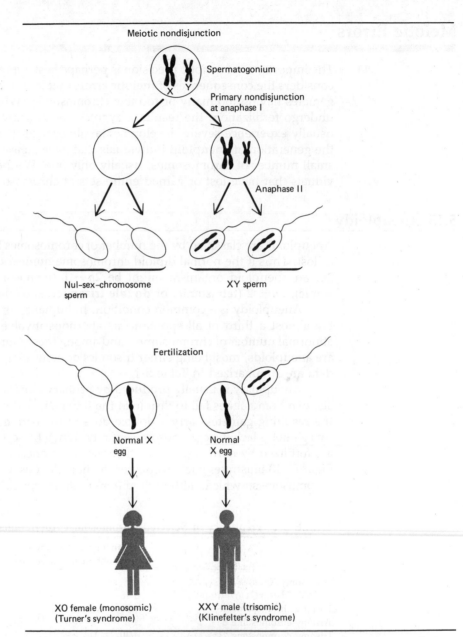

FIGURE 5-13 **Primary nondisjunction** of the X and Y sex chromosomes during human spermatogenesis and the fertilization of the resultant gametes with normal X eggs. Only the sex chromosomes are illustrated; the remaining autosomes carry out meiosis in a normal fashion.

nonetheless ordinarily undergo synapsis at prophase I and separation at anaphase I. When nondisjunction instead occurs, the resultant gametes have either one too many sex chromosomes or no sex chromosomes at all, and humans with disordered sexual development can result (additional information on these human syndromes is presented in Section 6-17 and Table 6-8).

Monosomy is usually lethal: most chromosomes (with the exception of sex chromosomes) must apparently be present in two doses if proper diploid development is to occur. Trisomy, on the other hand, may be less severely deleterious. In the plant *Datura stramonium*, 12 trisomic types have been identified, one for each of the 12 chromosome types, and each plant type has a different morphology (Figure 5-14). In most animals, however, trisomy can be tolerated only for certain chromosomes, the other combina-

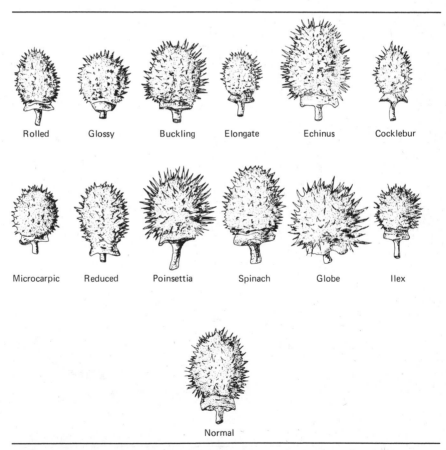

FIGURE 5-14 **Trisomic strains** of the plant *Datura* (Jimson weed). (From E. Sinnott, L. Dunn, and T. Dobzhansky, *Principles of Genetics*. New York: McGraw-Hill, 1958, after A. F. Blakeslee. Used with permission of McGraw-Hill Book Company.)

tions being lethal during the course of embryogenesis. Why this should be so is not known, but the differentiation of a diploid animal zygote is such a delicately balanced process that the presence of an extra chromosome's worth of genetic information might create developmental problems at critical embryonic stages. Supporting this line of reasoning is the observation that live-born human trisomic individuals carry extra copies of chromosomes 8, 13, 18, 21, Y, or X (Table 5-1). The first five of these chromosomes contain large blocks of constitutive heterochromatin, which is presumed to be genetically inert (*Section 3.13*); thus their presence in an extra dose might

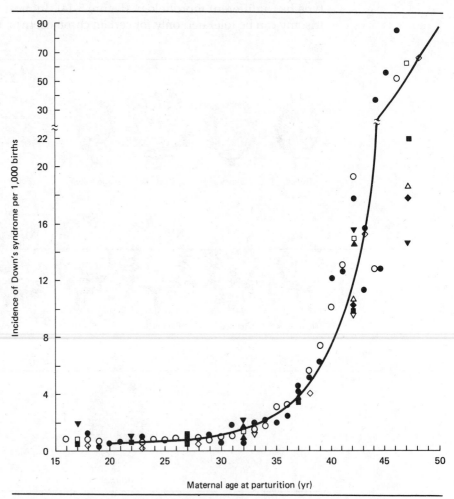

**FIGURE 5-15 Down's syndrome and maternal age.** Incidence of Down's syndrome per 1000 births versus maternal age at parturition. Different studies are represented by different symbols. (From P. H. Crowley, D. K. Gulati, T. L. Hayden, P. Lopez, and R. Dyer, *Nature* **280:**417, 1979.)

do less to upset the postulated balance than an extra dose of fully euchromatic chromosomes. Extra X chromosomes in aneuploid cells also become heterochromatic, as described in Section 25.16.

Trisomy for any of the chromosomes listed in the preceding paragraph except the Y usually produces an abnormal individual, the nature and severity of the syndrome being dependent upon the chromosome involved. The best-known example in humans is trisomy for chromosome 21, which produces **Down's syndrome** (once known by the misnomer Mongolian idiocy). This condition occurs in about 1 in every 600 newborns, and it is clear from Figure 5-15 that its frequency increases significantly when mothers are more than 35 years of age. This suggests that nondisjunction occurs more often in older egg cells, and when it is recalled from Section 5.6 that the human egg cell remains in diplonema from the fetal life of a woman onward, it is perhaps not unexpected that mechanical difficulties become more prevalent by the time the cell is, say, 45 years old. Further information on Down's syndrome is presented in Section 7.9.

Trisomy for the Y chromosome produces XYY males, and these individuals are characteristically taller than average but have normal fertility and produce either X or Y sperm. Initial speculation concerning the existence of an "XYY syndrome" of antisocial tendencies in such men has recently been countered by a large-scale study of XYY Danish men in whom no such syndrome is apparent.

## 5.16 Haploidy and Polyploidy

As stressed earlier in the chapter (*Sections 5.12 to 5.14*), most eukaryotes have either predominantly haploid or predominantly diploid phases in their life cycles. Many haploid organisms go through a brief diploid stage as zygotes, whereas diploid organisms usually produce short-lived haploid gametes. If, however, an egg from a diploid organism is induced to undergo embryogenesis without first being fertilized by a sperm, an aberrant **haploid** individual will result.

Haploid embryos rarely develop normally in animals, just as monosomic organisms rarely survive. In plants, however, haploids frequently reach maturity, although they are usually weak and delicate. They are also usually sterile, a fact that can readily be appreciated in trying to visualize a haploid meiosis in which the pairing of homologues cannot occur and segregation of chromosomes at anaphase I becomes a random process.

**Polyploid** individuals ($2n$ if the original chromosome number was $n$ and $3n$, $4n$, and so on, if the original number was $2n$) are often viable and may be larger than their haploid or diploid counterparts. Polyploidy is common among natural populations of plants, particularly the grasses, and plant breeders frequently create polyploid lines because of their greater vigor and larger flower and fruit size. Even-numbered polyploids, and tetraploid plants in particular, do not usually experience difficulties during meiosis, since each chromosome and its homologue are represented equally. Plants with an uneven chromosome number (triploid, pentaploid, and so on), on the other hand, are usually sterile in that one chromosome set is without homologous partners and (as noted for haploid meioses) aneuploid gametes are produced. Sterility is sometimes desirable for the fruit breeder, since inedible seeds do not form and the plant can still be propagated vegetatively by cuttings. Therefore such commerical plants as the banana and lime are triploids.

Polyploid higher animals are rarely fertile, even when tetraploid, an exception being polyploid silkworms. The difference between plants and animals is believed to rest, at least in part, with the sex chromosomes. Most genera of higher plants do not appear to possess differentiated sex chromosomes and many, although not all, are bisexual (or, in botanical terms, **monoecious**). In higher animals, on the other hand, sex and fertility depend on a critical balance of sex-determining chromosomes, as will be discussed in Section 6.17. Polyploid individuals that have lost this balance can therefore be expected to be sterile.

Although triploids are well known in plants, in *Drosophila*, and even in amphibians, human triploids cannot survive: nearly 1 percent of all human conceptions are triploid, but most die within the first 3 months after fertilization, and the few infants who are born alive die within the neonatal period. As with most aneuploidy, it appears that the very stringent developmental schedule of a human pregnancy is incompatible with such an increase in gene dosage.

Polyploids can arise in a number of ways: a meiotic aberration may result in a failure of reduction division during gametogenesis that affects the entire chromosome complement; an egg may be fertilized by more than one sperm; or chromosomal separation at anaphase may be blocked by **colchicine** or **colcemid,** agents that disrupt microtubules and hence meiotic/mitotic spindle formation. Polyploidy also often accompanies the formation of hybrid plants from two different species, a phenomenon known as **allopolyploidy.** This process will be described in detail in Section 28.2.

(a) Curt Stern was among the first to establish the relationship between chiasmata and genetic exchange. (b) Diter von Wettstein (Carlsberg Laboratory, Copenhagen) is a plant cytogeneticist who, with M. Westergaard, has analyzed the ultrastructure of meiosis.

(a)                                  (b)

# Questions and Problems

1. Examine Figure 5-1 and determine the haploid chromosome number of the grasshopper. Does terminilization of chiasmata occur in this organism?

2. In a diploid organism with a haploid chromosome number of 7, how many sister chromatids are present in its mitotic metaphase nucleus? In its meiotic metaphase I nucleus? In its meiotic metaphase II nucleus?

3. During which stages (or prophase I substages) of meiosis do you expect to find the following (more than one stage may obtain): sister chromatids; synaptonemal complexes; daughter chromosomes; bivalents; chiasmata; spindle microtubules; DNA replication?

4. In a diploid organism with a haploid chromosome number of 3, write out the eight possible combinations of maternal and paternal homologues in its gametes (for example, 1M2M3P). For each case, diagram the alignments of the nonhomologues with respect to the metaphase I plate.

5. For Figure 5-2, imagine that $n = 2$ and that the second bivalent does not undergo crossing over. Show *all possible* metaphase I alignments and telophase II tetrads that could result.

6. Cite the essential differences between mitosis and meiosis.

7. The X and Y sex chromosomes behave like homologues during meiosis. What type(s) of gametes are produced by a male (XY) and a female (XX) with respect to X chromosomes? Does a father transmit copies of his X chromosomes to his sons, daughters, or both? Explain.

8. For each of the three following statements, indicate which (if any) are true for mitosis, for the first meiotic division, and for the second meiotic division. (a) The segregation of centromeres is reductional (that is, centromeres of different parentage always separate). (b) The reciprocal products of a single crossover always segregate (that is, become separated). (c) Chromosomes generally duplicate between this division and the previous one.

9. Explain the following statement: Autogamy in *Paramecium tetraurelia* results in diploid organisms homozygous at all loci.

10. How many bivalents would you expect to find in a human female pachytene nucleus?

11. Life cycles generally represent variations on a central theme, namely, the alternation of meiotic and mitotic divisions between fertilizations (F) or sexual generations. Which of the patterns given below best applies to the following: *Drosphila*, maize, *Chlamydomonas*, *Neurospora*, yeast, and *Paramecium*?

    (a) F $\longrightarrow$ meiosis $\longrightarrow$ mitosis $\longrightarrow$ F
    (b) F $\longrightarrow$ mitosis $\longrightarrow$ meiosis $\longrightarrow$ F
    (c) F $\longrightarrow$ mitosis $\longrightarrow$ meiosis $\longrightarrow$ mitosis $\longrightarrow$ F

12. Nondisjunction can occur during meiotic anaphase I when homologues fail to separate (*Figure 5-13*); it can also occur at meiotic anaphase II when centromeres fail to divide. It is unlikely that both events will occur during the same meiosis. (a) Diagram the consequences of each type of nondisjunction on the sex chro-

mosome consitution of sperm and egg cells formed during spermatogenesis and oogenesis (recall that males are XY and females XX). (b) State whether the following gamete types can be produced by nondisjunction at anaphase I, at anaphase II, or either way: a YY sperm; an XY sperm; an XX egg; a sperm carrying no sex chromosome; an egg carrying no sex chromosome. (c) Describe the kinds of sex chromosomal abnormalities carried by human zygotes formed when the above gametes are fertilized by an X-bearing gamete.

# 6 Mendelian Inheritance of Genes Carried by Autosomes and Sex Chromosomes

A. **Introduction**

B. **Mendel's Experiments**
- 6.1 History of Mendel's Experiments
- 6.2 Mendel's Experiments
- 6.3 Mendel's Interpretation of His Experiments
- 6.4 Genetic Terminology

C. **Segregation of Alleles**
- 6.5 Mendel's First Law and the Segregation of Alleles
- 6.6 Segregation of Alleles in a *Drosophila* Cross and the Test Cross
  The test cross
- 6.7 Segregation in Human Pedigrees
- 6.8 Segregation Analyzed in Pollen
- 6.9 Segregation in Tetrad Analysis
- 6.10 First-Division and Second-Division Segregation
- 6.11 Biased Segregation Ratios
  Zygote lethality
  Meiotic bias

Abnormal gamete development

D. **Independent Assortment**
- 6.12 Mendel's Second Law and the Independent Behavior of Alleles on Nonhomologues
- 6.13 Independent Assortment in a *Drosophila* Cross
- 6.14 The Chi-Square Test
- 6.15 Phenotypic Ratios in Hybrid Crosses
- 6.16 Independent Assortment in Tetrad Analysis

E. **Sex-Linked Inheritance**
- 6.17 The Sex Chromosomes
- 6.18 Patterns of Sex-Linked Gene Transmission
- 6.19 Sex-Linked Inheritance in Birds
- 6.20 Sex-Linked Inheritance in Humans
- 6.21 Sex Linkage and the Chromosome Theory

F. **Questions and Problems**

**153**

# Introduction

This chapter covers the basic principles of **Mendelian genetics,** so-named after Gregor Mendel, a monk who undertook a series of pioneering genetic experiments with the garden pea in the mid–nineteenth century. After presenting Mendel's experiments and defining some essential genetic terminology, we consider the two fundamental laws of Mendelian genetics. We first describe how the behavior of chromosomes at meiosis, behavior unknown to Mendel, in fact underlies the genetic principles Mendel enunciated. We then show how these principles can be deduced from the phenotypes that result from genetic crosses, as Mendel himself deduced them. Finally, we consider the special behavior and expression of the sex chromosomes in Mendelian crosses.

# Mendel's Experiments

## 6.1 History of Mendel's Experiments

Gregor Johann Mendel was born in 1822 in Heinzendorf, which is now in Czechoslovakia. He was of peasant stock and came to study for the priesthood at the Augustinian monastery at Brünn. He failed his first examination for a teaching certificate in natural science, but he then went on to study science at the University of Vienna before returning to Brünn. It was in the garden at the monastery that he grew his peas, and in 1856 he began his genetic experiments. These studies took eight years to perform and resulted in a paper that was published in 1866 in the Proceedings of the Brünn Society for Natural History. The paper was read by few and apparently understood by no one until 1900, long after Mendel's death. Meanwhile, Mendel tried unsuccessfully to repeat his observations with another plant species, *Hieracium*, which, it turns out, forms seeds without a true meiosis. Mendel knew nothing about meiosis, and the failure of these experiments must have represented a great disappointment to him. He also studied several other species of plants, kept bees and mice, and was interested in meteorology. We have, in short, a picture of a man of many and considerable talents whose major work was not appreciated and who undoubtedly came to regard himself as a scientific failure. By 1871, when he assumed administrative duties as abbot of the Brünn monastery, most of his productive scientific work seems to have ceased. He died in 1884.

## 6.2 Mendel's Experiments

Mendel chose to work with the edible pea, *Pisum sativuum*, since seeds for numerous varieties were available from the local seedsman. He first established that **each parental (P) variety bred true,** that is, gave rise to plants like itself when allowed to self-fertilize. He then selected varieties that displayed differences for particular traits. Thus, for example, one variety had a *round* seed shape, whereas another had a *wrinkled* seed shape; one variety had a *yellow* seed color, whereas another had a *green* seed color; and so on. In all he studied seven different characters of the pea, and for each he had two parental varieties.

Mendel next made pairwise crosses between each of his seven sets of parental plants, taking pollen from one and applying it to the stigma of the other. When he examined their offspring, which he called the **first filial (F₁) generation,** he discovered that they were always like *one* of the parents. When peas with *round* seeds were crossed with peas with *wrinkled* seeds, for example, the F₁ offspring always produced *round* seeds. Mendel proposed to designate the traits expressed in the F₁ as **dominant** and the traits not expressed in the F₁ as **recessive**. Thus his full list of traits had the following relationships:

| Dominant | Recessive |
|---|---|
| *round* seeds | *wrinkled* seeds |
| *yellow* seeds | *green* seeds |
| *colored* seed coat | *white* seed coat |
| *inflated* pod | *wrinkled* pod |
| *green* pod | *yellow* pod |
| *axial* flowers | *terminal* flowers |
| *long* stem | *short* stem |

Mendel now allowed each F₁ plant to self-pollinate, meaning that each stigma was fertilized by its own pollen, and the resultant **second filial (F₂) generation** proved to include plants that displayed either the dominant or the recessive trait. The key operation performed by Mendel at this point was to count the number of plants of each type. His results were as follows:

| P | F₁ | F₂ | Ratio |
|---|---|---|---|
| *round* × *wrinkled* | *round* | 5474 *round*<br>1850 *wrinkled* | 2.96∶1 |
| *yellow* × *green* | *yellow* | 6022 *yellow*<br>2001 *green* | 3.01∶1 |
| *colored* × *white* | *colored* | 705 *colored*<br>224 *white* | 3.15∶1 |
| *inflated* × *wrinkled* | *inflated* | 882 *inflated*<br>299 *wrinkled* | 2.95∶1 |
| *green pods* × *yellow pods* | *green* | 428 *green*<br>152 *yellow* | 2.82∶1 |
| *axial* × *terminal* | *axial* | 651 *axial*<br>207 *terminal* | 3.14∶1 |
| *long* × *short* | *long* | 787 *long*<br>277 *short* | 2.84∶1 |

It is clear that in every cross, **a ratio of three dominant to one recessive** was obtained in the $F_2$. Thus the inheritance of each alternate trait obeyed a nonrandom, quantitative rule.

Mendel's next step was to see what happened if crosses were performed between plants that differed by two traits. He chose a variety having the dominant *round* and *yellow* seeds as one parent and a variety with recessive *wrinkled* and *green* seeds as the other parent. All of the $F_1$ plants proved to have *round* and *yellow* seeds, once again demonstrating the phenomenon of dominance. When these were self-pollinated, the $F_2$ plants were found in the following proportions:

| | |
|---|---|
| 315 *round, yellow* | 101 *wrinkled, yellow* |
| 108 *round, green* | 32 *wrinkled, green* |

If you add up these numbers, you will find a 3:1 ratio between each dominant and recessive trait (for example, 315 + 108 are *round*, 101 + 32 are *wrinkled* = 3.18:1), so that each set of traits is behaving exactly the way it did in the previous crosses. Because two sets of traits are present, however, they show up in new combinations, namely, *round* and *green* seeds and *wrinkled* and *yellow* seeds, as though each trait were being inherited independently of the other. Once again, moreover, the numbers indicate that inheritance follows very specific rules: in this case, for every one *wrinkled, green* plant there are 3 (101 ÷ 32) *wrinkled, yellow* plants, 3 (108 ÷ 32) *round, green* plants, and 9 (315 ÷ 32) *round, yellow* plants. Thus, in general, **a ratio of 9:3:3:1** is observed in the $F_2$ when two traits are being followed.

## 6.3 Mendel's Interpretation of His Experiments

Mendel correctly interpreted the significance of the ratios obtained in his crosses by postulating that **each pollen grain and each egg carries only one determinant for each trait** (e.g., a pollen grain will carry either *round* or *wrinkled*, never both). He proposed that the original *round* parental plants were true breeding for *round* because they produced only *one* type of gamete: both their pollen and eggs carried exclusively the *round* (R) determinant. Similarly, he proposed that each parental *wrinkled* plant produced only *wrinkled* (r)- bearing gametes. When crossed together, therefore, their $F_1$ offspring would acquire both the R and the r determinants and could be designated *Rr*. These *Rr* plants, he proposed, would produce *two* types of gametes, in equal numbers: those carrying R alone and those carrying r alone. When allowed to self-fertilize, therefore, the following combinations would be expected to occur with equal probability:

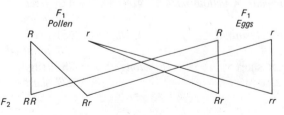

Recalling that *round* (*R*) is dominant to wrinkled (*r*), this gives an $F_2$ ratio of 3 *round* : 1 *wrinkled,* just as Mendel observed experimentally.

These same assumptions, Mendel realized, could account as well for the 9:3:3:1 ratios obtained with two traits. In this case, the parental *round, yellow* plant would again produce only one type of gamete, but this time it would carry both the *round* and the *yellow* determinants and would be designated *RY;* similarly, the parental *wrinkled green* plants would produce only *ry* gametes. When crossed together, therefore, their $F_1$ offspring would be *RrYy.* The $F_1$, he proposed, would again produce equal numbers of gametes carrying *R* alone and *r* alone. In addition, each gamete would have either *Y* alone or *y* alone. Therefore, the $F_1$ would produce four classes of gametes: *RY, Ry, rY,* and *ry.* The results of self-fertilization in this case involve $4 \times 4 = 16$ possible combinations; these are displayed in the following chart, where each box in the grid represents the $F_2$ plant resulting from the union of the egg on the left and the pollen on the top.

**$F_1$ Pollen**

|            | *RY*   | *Ry*   | *rY*   | *ry*   |
|------------|--------|--------|--------|--------|
| *RY*       | *RRYY* | *RRYy* | *RrYY* | *RrYy* |
| *rY*       | *RrYY* | *RrYy* | *rrYY* | *rrYy* |
| *Ry*       | *RRYy* | *RRyy* | *RrYy* | *Rryy* |
| *ry*       | *RrYy* | *Rryy* | *rrYy* | *rryy* |

**$F_1$ Eggs** (labels at left of rows)

As an exercise, write out the seed shape and color of each of these 16 combinations, remembering the effects of dominance, and show how they display a 9:3:3:1 relationship.

## 6.4 Genetic Terminology

With Mendel's experiments in mind, we can define some important genetic terms.

Readers of Chapter 2 should by now feel comfortable with the concept that a diploid cell contains two homologues of each chromosome type possessed by the species (*Figure 2-15*), one of these homologues deriving from one gamete (for example, an egg) and the other homologue deriving from the other gamete (for example, a pollen grain). **Each homologue in a pair contains the same kinds of genes and in the same order,** the physical location of any particular gene in a chromosome being called its **locus.** Thus we can say that in the garden pea, there exists a gene locus that determines seed shape, a locus that determines seed color, and so on. However, if we examine the two homologues in any one nucleus, the genes at a particular locus need not be identical. Specifically, in an $Rr$ pea plant, one seed-shape locus will carry the *round* ($R$) gene, whereas the seed-shape locus on its homologous chromosome will carry the *wrinkled* ($r$). The $R$ and $r$ genes are said to be **alleles** (a term that derives from the Greek *allelon,* of one another), indicating that they are two alternative genes that can occupy the same chromosomal locus.

An organism such as an $Rr$ pea plant, with two alleles at a particular locus, is said to be **heterozygous** at that locus (*hetero* meaning different). By contrast, its true-breeding *round* parent has a copy of the $R$ gene at both of its seed-shape loci; this $RR$ parent is said to be **homozygous** (*homo* meaning the same) at the locus. Similarly, the true-breeding *wrinkled* parental line used in Mendel's crosses is also homozygous, carrying an identical $r$ gene at each seed-shape locus.

We should note here that other systems of genetic notation are also used. In many cases, the allele most often encountered at a locus in natural populations is said to be the **"wild-type"** gene. For example, if we examined peas growing in the wild and determined that most carried the *round* allele at the seed-shape locus, we might designate *round* as wild type and symbolize it as + or as $r^+$. In this system, the less common allele, which is often but not always recessive, might be designated $r$ or $r^-$. Therefore, a homozygote would be designated +/+ or $r/r$, whereas the heterozygote would be +/$r$; alternatively, homozygotes might be $r^+/r^+$ and $r^-/r^-$, with heterozygotes $r^+/r^-$.

All such notation systems summarize the genetic makeup, or **genotype,** of an organism, a term that is traditionally contrasted with phenotype. The **phenotype** derives from the Greek *phain,* to appear, and connotes the particular expression of a gene—wrinkled seed, brown eye, amino acid–requiring—each particular feature being known as a **trait.** Phenotype is also used as a collective term to refer to the sum of the traits that characterize an organism: thus the standard phenotype of humans might be said to include ten fingers, a characteristic growth rate, and several vitamin requirements. The genotype of any individual organism is relatively stable throughout its lifetime, whereas its phenotype will vary, depending on its state of development, environmental influences, and so on. The genotype of a mammal, for example, includes a number of genes that do not find expression in the phenotype until sexual maturity. The phenotype, in short, includes all the products or manifestations of the organism's genes—the amino acid sequence of its proteins, its enzyme activities, its appearance, and even its behavior—whereas the genotype is synonymous with the collection of genes in the organism's chromosome or chromosomes.

In a haploid cell, with only one copy of each gene per nucleus, the presence of a given phenotypic trait can at once be attributed to the presence of a particular gene. Thus if a *C. reinhardi* cell behaves in mating as a *plus* gamete, then it must have the mating-type *plus* ($mt^+$) genotype. Such conclusions cannot immediately be drawn for diploid cells, however, because of the existence of **dominant** and **recessive** genes. A diploid cell possesses two alleles at every chromosomal locus. If one of these alleles turns out to specify a nonfunctional protein, whereas the other specifies a normal protein, then enough normal protein may be produced to endow the cell with a normal phenotype, and the normal allele is said to be dominant to the abnormal allele. Many other kinds of gene-gene interactions can lead to dominant/recessive relationships between alleles, as described in later chapters. For present purposes, **the important feature of a recessive gene is that it will contribute to the phenotype only if it is present in homozygous form.** Thus we saw in Mendel's experiments that both *RR* and *Rr* peas had the *round* seed phenotype and that only *rr* plants could produce *wrinkled* seeds. Similarly, for the recessive *bn* (*brown eyes*) gene and its + allele in *Drosophila*, a fly will develop a *brown-eyed* phenotype only if it has the homozygous (*bn/bn*) genotype; a +/+ homozygote and a +/*bn* heterozygote will both develop wild-type (reddish) eyes, the + allele in the heterozygote dictating the synthesis of sufficient amounts of red pigment that a normal eye color is generated.

With this much background in Mendelian crosses and in genetic terminology, we can relate the Mendelian transmission of genes to the meiotic transmission of chromosomes.

# Segregation of Alleles

## 6.5 Mendel's First Law and the Segregation of Alleles

You will recall from Figure 5-2 that during meiosis the maternal and paternal members of a given bivalent align themselves above and below the metaphase I plate and segregate at anaphase I. If we imagine that we are now watching meiosis in the true-breeding *round* parent of Mendel's crosses, then one of the maternal chromosomes will carry gene *R* and its paternal homologue will also carry gene *R*, and all the products of meiosis will be *R*-bearing gametes, as Mendel deduced. If we instead imagine we are watching meiosis in the heterozygous $F_1$ *round* plants produced from the fertilization of an *R* egg by an *r* sperm, then the maternal chromosome will carry *R* and the paternal homologue will carry its *r* allele, as illustrated in Figure 6-1. When such a cell undergoes meiosis, half the meiotic products will carry *R* and half will carry *r*, as drawn in Figure 6-1. This is, again, the outcome Mendel deduced without knowing the details of meiosis. The

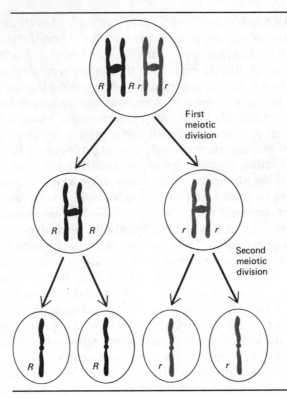

FIGURE 6-1 The **segregation of alleles** $R$ and $r$ during meiosis. Maternal chromosomes are in black and paternal are in gray.

separation of homologues at meiosis, therefore, is responsible for the pattern of inheritance observed when heterozygotes are inbred, namely, that **alleles segregate into separate gametes.** This phenomenon is often called the **Principle of Segregation of Alleles** or **Mendel's First Law.**

In the sections that follow, we will illustrate how the segregation of alleles is demonstrated in a variety of organisms.

## 6.6 Segregation of Alleles in a *Drosophila* Cross and the Test Cross

Genetic analysis of animals is based on the same assumption made by Mendel for the garden pea, namely, that **all the gametes produced by the male and female parents have an equal chance to participate in fertilization.** Thus it is assumed that among the four products of a female meiosis, any one may happen to become the egg cell with equal probability, (the other three becoming polar bodies [*Figure 5-5*]). Similarly, it is assumed that any of the meiotic products of spermatogenesis has an equal probability of fertilizing the ovum in a given cross.

Granted these assumptions (we shall note later the unusual circumstances in which they do not hold), then if we look at a sizable number of zygotes, these progeny should derive from a **random sampling** of all meiotic products. It can thus be expected that the patterns of gene distribution found in these offspring will reflect the patterns of gene transmission followed during meiosis.

To demonstrate the segregation of alleles in an animal such as *Drosophila*, it is first necessary to construct organisms heterozygous for the alleles of interest. Therefore, organisms homozygous for one gene are first mated with organisms homozygous for its allele, exactly as Mendel did with his pea plants. For example, we might begin with a stock of true-breeding wild-type $(+/+)$ *Drosophila* females with normal wings and a stock of true-breeding *vestigial* $(vg/vg)$ males, *vestigial* being a recessive trait wherein the wings are small and rumpled. As diagrammed in Figure 6-2, the females will produce only $+$-bearing eggs, whereas the males will produce only $vg$-bearing sperm. Their $F_1$ progeny will therefore all have a

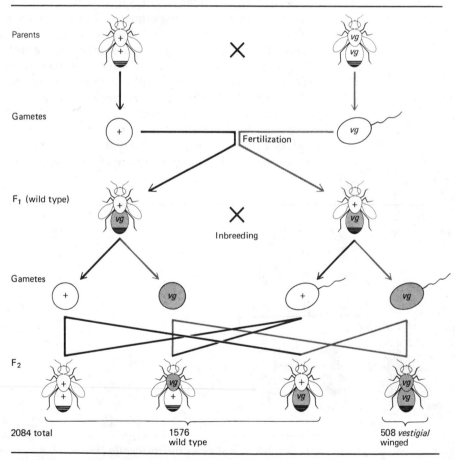

**FIGURE 6-2 Inbreeding $F_1$ heterozygotes.** The $F_2$ shows a 3:1 ratio of phenotypes and a 1:2:1 ratio of genotypes.

heterozygous $+/vg$ genotype and, because of dominance, all will have a wild phenotype (Figure 6-2).

The segregation of alleles during meiosis can now be inferred from the results of inbreeding these $F_1$ flies, just as Mendel self-fertilized his $F_1$ peas. In the present case, each fly, whether male or female, should produce equal numbers of +- and $vg$-carrying gametes. Thus there are four possible unions, as diagrammed in Figure 6-2: + with +, + with $vg$, $vg$ with +, and $vg$ with $vg$, just as with the garden pea self-fertilization. In the $F_2$, therefore, one quarter of the progeny should be $+/+$, one half $+/vg$, and one quarter $vg/vg$, so that the **ratio of genotypes** is 1:2:1 (Figure 6-2). Because of dominance, the **ratio of phenotypes** will be the familiar **3:1 ratio** encountered by Mendel. Figure 6-2 gives actual data that confirm the predicted 3:1 ratio.

**The Test Cross**   A more direct way to follow the transmission of the + and $vg$ alleles in the $F_1$ heterozygotes is to perform a **test cross,** a form of cross not utilized in Mendel's experiments. In a test cross, **organisms are mated with organisms that are homozygous recessive at all loci of interest.** In the present case, with only one locus of interest, the test-cross organisms are *vestigial* ($vg/vg$) flies. Such recessive flies contribute only "neutral" or "silent" $vg$ genes; therefore, **all the genes contributed by the $F_1$ gametes will be expressed in the test-cross progeny.** The results of such a test cross are shown in Figure 6-3. Among 984 progeny, approximately half are phenotypically wild type and half are *vestigial*. Therefore, half the gametes produced by the $F_1$ flies must carry the + gene and half the $vg$ gene. Since we assume that a random sampling of gametes participated in forming the

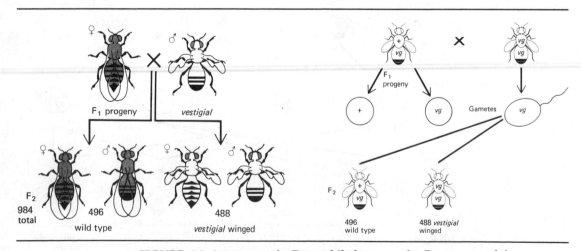

**FIGURE 6-3 A test cross in *Drosophila*** between the $F_1$ progeny of the cross illustrated in Figure 6-2 and *vestigial* flies. The phenotypes are shown on the left and the genotypes on the right.

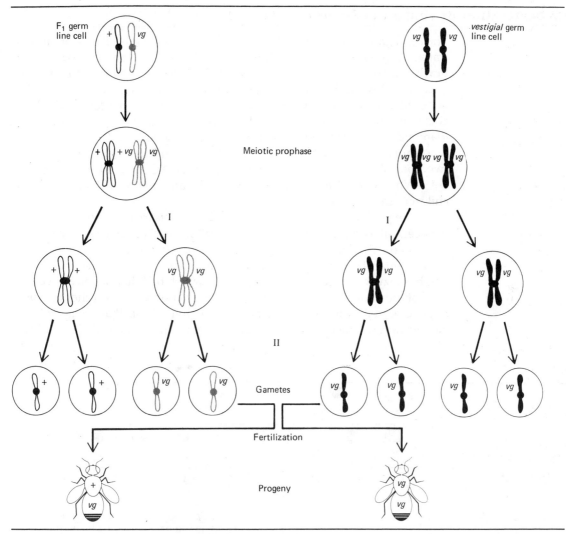

**FIGURE 6-4 Meiotic events in a test cross.** The test cross of Figure 6-3, where I and II indicate the two meiotic divisions.

984 progeny scored, we can conclude that the two alleles segregated in a 1:1 ratio during meiosis. Figure 6-4 diagrams the meiotic events that can be inferred from the test cross we have just analyzed.

In Figure 6-3, the $F_1$ flies are illustrated as females and the test-cross flies as males. In a test cross, moreover, **the females must be collected as virgins,** soon after they emerge from their pupae. If this precaution is not taken and the females are allowed to mate with their $F_1$ brothers, the $F_2$ progeny will display the 3:1 ratio of an $F_1$ inbreeding (*Figure 6-2*) rather than the 1:1 ratio expected of a test cross.

## 6.7 Segregation in Human Pedigrees

The segregation of alleles is also well exhibited in the inheritance of human traits. A classic example is the case of the skeletal abnormality called **brachydactyly,** in which the fingers are very short. A **pedigree** for brachydactyly, shown in Figure 6-5, represents one of the first studies of single-gene inheritance in humans. In a pedigree, females are represented by circles and males by squares; when the sex of an individual is unknown, a diamond-shaped symbol is used. A black symbol represents an individual who expresses the trait. When it is relevant to indicate the phenotypes of two spouses, their marriage is symbolized as ○─□ and their children as ○┬□. When information on a spouse is irrelevant or unavailable, a person's children are indicated as ♀ or ♂.

The pedigree in Figure 6-5 shows that an initial marriage (generation I) between an affected female and a normal male produced eight children (generation II), four of whom were brachydactylous and four of whom were normal. This ratio is reminiscent of a test cross between heterozygous and homozygous recessive individuals, in which half the progeny have the phenotype of one parent and half have the phenotype of the other because of gene segregation. Thus we can adopt the hypothesis that the gene "*B*," conferring the brachydactylous trait, is dominant to its + allele and that

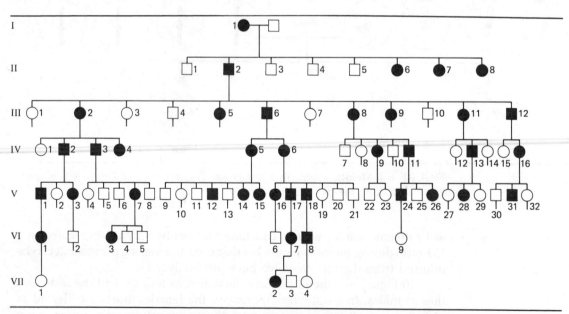

**FIGURE 6-5 Pedigree of human brachydactyly.** (After Farabee, Papers of the Peabody Museum, Harvard University, Vol. 3, 1905, updated by V. A. McKusick, *Human Genetics,* 2nd ed., © 1972. Adapted by permission of Prentice-Hall, Inc., Englewood Cliffs, N.J.)

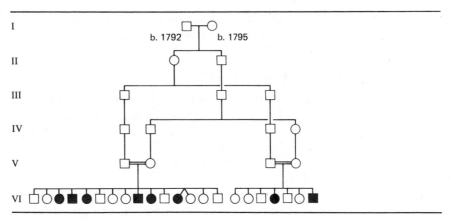

**FIGURE 6-6 Pedigree of microcephaly in humans.** Double marital lines indicate consanguinity. (From V. A. McKusick, *Human Genetics*, 2nd ed., © 1972. Reprinted by permission of Prentice-Hall, Inc., Englewood Cliffs, New Jersey.)

the original mother was a $B/+$ heterozygote. This hypothesis is confirmed by subsequent generations in the pedigree. Whenever brachydactylous individuals have children (we can assume that their spouses are normal, since brachydactyly is a rare affliction), approximately half their offspring are brachydactylous and half are normal, generation after generation.

We can now ask what the outcome would be if two heterozygous brachydactylous individuals from this lineage were to marry and have children. Using the same arguments offered for $+/vg \times +/vg$ crosses in *Drosophila* (*Figure 6-2*), we would predict a 3:1 ratio of affected to nonafflicted. Moreover, we would predict that one of every three brachydactylous individuals would be a $B/B$ homozygote (you should convince yourself of these ratios by making diagrams such as those in Figure 6-2). Such a $B/B$ homozygous individual would, of course, produce only $B$ gametes, and thus all of her or his children would be brachydactylous, regardless of the genotype of the spouse.

A second example of allelic segregation in humans is given by the trait of *microcephaly*, in which the affected individual has a small head and is mentally retarded. The pedigree for this trait (Figure 6-6) differs strikingly from the pedigree for brachydactyly (Figure 6-5): the only microcephalous individuals derive from **consanguineous marriages** (for example, marriages between first cousins), as symbolized by double marital lines. In such cases we can suspect strongly that a rare recessive rather than a dominant gene is responsible for the trait. Homozygotes for this gene should be exceptionally rare and should arise most commonly in consanguineous marriages in which two individuals have a statistically better chance of being heterozygous for the same pair of alleles. It is, in fact, this increased chance of expressing rare recessive (and often deleterious) genes by homozygosity that argues against consanguineous marriages, a concept we consider again in Section 26.7.

## 6.8 Segregation Analyzed in Pollen

Pollen sometimes travels long distances before it makes contact with the pistil of a flower, and consequently it often has a stable and relatively long-lived existence. Since each pollen grain typically contains two identical haploid nuclei (*Figure 5-7*), the phenotype of the grain will reflect its genotype. The segregation of alleles can thus be demonstrated directly in pollen.

We can, for example, consider the genes that determine the carbohydrate composition of maize pollen. The gene *Wx* dictates the synthesis of starch (amylose), and starch-containing pollen grains turn blue when treated with iodine. Pollen carrying the mutant allele *wx*, however, cannot synthesize amylose, and these grains turn a reddish-brown in iodine. In one experiment in which 6919 grains from a heterozygous *Wx/wx* plant were examined, 3437 grains were found to stain blue and 3482 to stain red, giving a 1:1 ratio.

## 6.9 Segregation in Tetrad Analysis

Tetrad analysis represents a powerful and direct way to demonstrate the genetic transmission of alleles. Asci of yeast (*Figure 5-10*) or *Neurospora* (*Figure 5-9*) or the mature zygotes of *Chlamydomonas* (*Figure 5-8*) can be manipulated individually to liberate their four (or eight) haploid products, and these products can be separated. With *Chlamydomonas*, for example, perhaps ten zygotes are aligned to one side in a Petri dish that contains agar and a growth medium, as diagrammed in Figure 6-7a. After meiosis has been completed, each zygote wall bursts open, and the four individual meiotic products (the **tetrad**) can be manipulated into a line, as shown in Figure 6-7b. Each of these haploid cells then reproduces itself mitotically to form a clone of genetically identical haploid cells (Figure 6-7c). If necessary, one or more **replica plates** (*Figure 4-12*) can be made from this original master plate so that the growth properties of the meiotic products can be tested on a number of different media.

Figure 6-7 diagrams a cross between wild-type gametes that do not require arginine and gametes that derive from an arginine-requiring strain. The cross is thus written + X *arg*. The growth medium of the master plate is a minimal medium supplemented with arginine so that all the meiotic products are able to survive (Figure 6-7b and c). After colonies have formed, replicas are made from the master plate to plates containing minimal medium and, as a control, to plates of arginine-supplemented medium. After about 72 hours, colonies are visible on both kinds of medium. On the arginine-supplemented plate, each of the original tetrads is again represented by four colonies. In contrast, on the minimal-medium plates

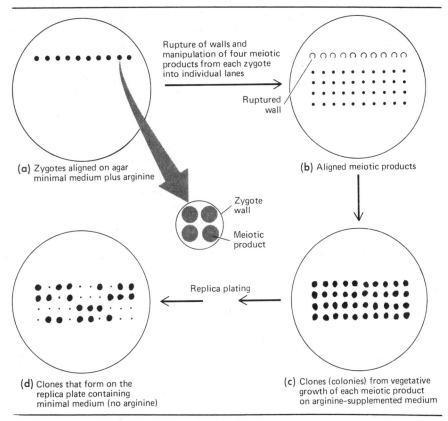

**FIGURE 6-7 Segregation in meiotic tetrads:** tetrad analysis of the cross *arg* × + in *Chlamydomonas*.

colonies develop from only two of the four meiotic products of each tetrad (Figure 6-7d). Microscopic examination of these plates will show that cells from the other two meiotic products were indeed transferred during replica plating but that these cells failed to grow on minimal medium. **Each tetrad, therefore, consists of *equal numbers of two kinds of cells:* half require** arginine for growth and thus carry the *arg* gene, whereas half do not and thus carry its + allele. The outcome predicted by a knowledge of meiosis is therefore confirmed by genetic testing.

## 6.10 First-Division and Second-Division Segregation

In Figure 6-1, the *R* and *r* alleles are drawn segregating at the anaphase of the first meiotic division. **Segregation may in fact be delayed until the second meiotic division if one or the other homologue has participated in a crossover.** This outcome is best illustrated using tetrad analysis of *Neurospora*.

In *Neurospora* (*Figure 5-9*) the ascus is sufficiently narrow that neither nuclei nor ascospores are normally able to move about within it. Consequently, the alignment of the eight ascospores in a mature ascus reflects the positions of the various spindles that mediated the two meiotic and the one (postmeiotic) mitotic division within the ascus. Thus, if we follow the fate of the four chromatids making up any one bivalent in prophase I, as we have done in Figure 6-8, the first pair of ascospores would derive from one chromatid and the second pair from its sister, whereas the third and fourth pairs would derive from the two sister chromatids of the homologous chromosome set. For this reason, *Neurospora* is said to produce **ordered tetrads.**

When a cross is made between wild-type *Neurospora*, which produces pink conidia, and a mutant strain called *albino* (*al*), which has white conidia, tetrad analysis reveals that there is indeed a one-to-one segregation of wild type and *albino* alleles. When, however, the ascospores are removed from the ascus in order, two types of ascus are found. In one type

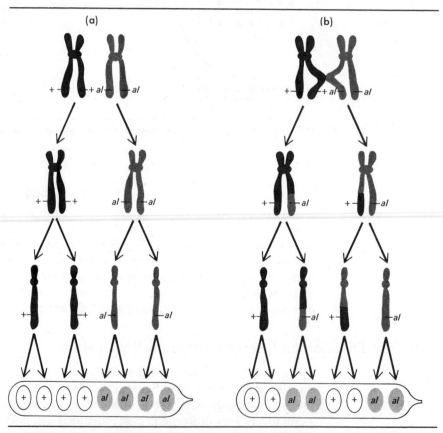

**FIGURE 6-8 Segregation in ordered tetrads of *Neurospora*.** (a) First-division segregation. (b) Second-division segregation.

the order of genotypes (and phenotypes) within the ascus is +, +, *al*, *al*; in the other it is +, *al*, +, *al*. Segregation in both cases is one to one, but the sequence of wild-type and mutant products within the ascus has changed.

These two patterns can be understood by examining Figure 6-8. The sequence +, +, *al*, *al* (Figure 6-8a) results from the segregation of alleles at the first meiotic division or, more precisely, from segregation of the homologous centromeres at anaphase I. The sequence +, *al*, +, *al* results whenever a crossover occurs between two of the homologous chromatids during prophase I, provided that the exchange takes place anywhere between the position of the centromere and the locus of *al* and its + allele. Because of this crossover, the + and *al* alleles are held together by a common centromere and are unable to segregate at anaphase I, as diagrammed in Figure 6-8b. It is only at anaphase II, when the centromeres of sister chromatids divide and the sister chromatids separate, that the two alleles can reside in separate nuclei.

## 6.11 Biased Segregation Ratios

The preceding sections have amply demonstrated that for a diploid $B/b$ heterozygote the $B$ and $b$ alleles should be transmitted meiotically in a 1:1 ratio. Any marked or consistent discrepancy from this ratio—for example, a 2:1 ratio in favor of the $b$ allele—could have one of three general causes.

**Zygote Lethality**  Zygotes carrying the $B$ allele could suffer developmental difficulties and thus be less likely to mature into viable progeny. This is the most frequently encountered factor responsible for distorted segregation ratios and is often referred to as a **marker effect:** the allele selected as a gene "marker" for a particular cross proves to influence adversely the viability of the offspring, and this is reflected in the genetic data.

**Meiotic Bias**  A consistent abnormality in meiosis could preferentially exclude the $B$ allele from the gametes. This situation is particularly applicable to female flower parts and female animals in which only one of the four meiotic products becomes a functional egg. A well-studied case is known in maize. When maize is heterozygous for a heterochromatic "knob" on an abnormal chromosome 10 ($K10/k10$), as many as 70 percent of the functional megaspores are found to carry the knob. It has been proposed that the knobbed region has **neocentromere** activity, meaning that it associates with spindle fibers prematurely and thus experiences a precocious migration to one pole at both meiosis I and II. For this reason, the knobbed chromosome is thought to have a better chance of reaching the basal megaspore, the cell destined to become the functional egg (*Figure 5-7*), than its knobless homologue.

**Abnormal Gamete Development**  The third potential cause of biased segregation ratios is that a $b$-carrying sperm experiences markedly greater success than its $B$ counterpart in reaching an egg and/or fertilizing it. This applies particularly to species in which the genotype of the gamete can affect its success in fertilization. In the case of pollen, for example, the genotype of the gamete is known to be expressed during the complex process of pollen tube formation and nuclear migration. Therefore, pollen grains carrying gene *b* might well direct a faster rate of pollen tube growth than grains of genotype *B,* and an excess of *b*-carrying embryos might thereby appear among the offspring.

In the case of *Drosophila,* however, there is strong evidence that the genotype of the sperm is not expressed at all during transmission and fertilization. Appropriate genetic crosses can be made to yield sperm that carry only the tiny chromosome IV. If these fertilize eggs that carry compensatory extra chromosomes I, II, and III, normal flies will result. It is believed, therefore, that when one type of animal sperm functions better than another type, the difference must be conferred on the sperm during early stages of sperm production, perhaps even when the sperm cell precursor is still heterozygous.

As a specific example of this phenomenon, we can examine the **segregation distorter** locus in *D. melanogaster,* found in or near the constitutive heterochromatin in the right arm of chromosome II (*Figure 3-14*). In males heterozygous for *SD* and its + allele, virtually all of the progeny inherit *SD.* Moreover, if the *SD*-bearing chromosome II carries other genetic markers, these are preferentially transmitted as well. It seems that a gene product is dictated by the *SD* locus during early stages of meiosis, when the spermatocytes are still +/*SD* heterozygotes (*Figure 5-6*), which has an adverse effect on the fertilizing competence of mature gametes carrying the + allele but is not deleterious to gametes carrying *SD.* In other words, the *SD* chromosome has the remarkable property of effectively "killing off" its "rival" wild-type chromosome.

Several laboratories are trying to engineer analogous chromosomes for use in insect control. If, for example, a mosquito X chromosome that had lethal effects on Y chromosomes could be constructed, its introduction into a malaria-carrying mosquito population could result in the elimination of males and, therefore, in sterilization of the population.

# Independent Assortment

## 6.12 Mendel's Second Law and the Independent Behavior of Alleles on Nonhomologues

The Principle of Segregation applies to homologous chromosomes. Consider now the behavior of two sets of nonhomologous chromosomes during meiosis. We noted in Section 5.8 that the different pairs of homologous chromosomes arrange themselves on the metaphase I equator in an independent manner and remain independent throughout meiosis. As a consequence, genes that are located on nonhomologous chromosomes (in other words, genes that are not linked) undergo an **independent assortment** during meiosis.

If we return to Mendel's experiments and imagine we are watching meiosis in the heterozygous *round yellow* plants produced from the fertilization of an *RY* egg by an *ry* sperm, then two maternal chromosomes will carry *R* and *Y*, respectively, and two paternal homologues will carry their *r* and *y* alleles, as illustrated in Figure 6-9. At metaphase I, the two sets of

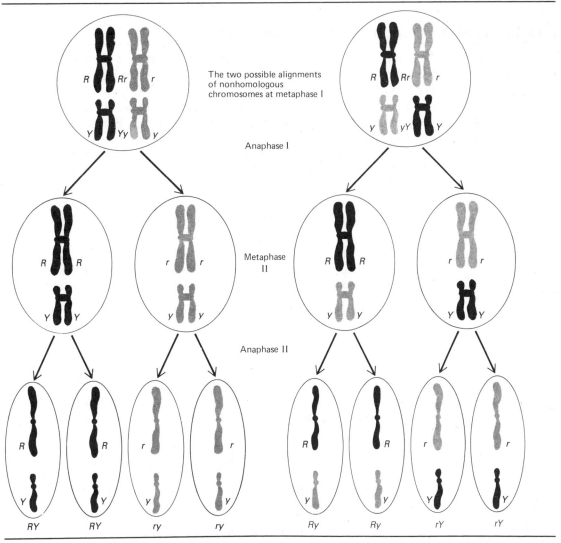

The two possible alignments of nonhomologous chromosomes at metaphase I

Anaphase I

Metaphase II

Anaphase II

**FIGURE 6-9 The independent assortment of nonhomologous chromosomes during meiosis,** with maternal chromosomes in black and paternal in gray.

homologues may happen to line up such that both paternal chromosomes are on one side and both maternal chromosomes are on the other side of the metaphase plate, as drawn in the cell on the left of Figure 6-9. Alternatively, the chromosomes may assume a staggered arrangement, as shown in the cell on the right. As a result, one quarter of the meiotic products from many such meioses will carry $R$ and $Y$, one quarter $r$ and $y$, one quarter $R$ and $y$, and one quarter $r$ and $Y$ (Figure 6-9), just as Mendel deduced without a knowledge of meiosis. This outcome is known as the **Principle of Independent Assortment,** and it represents the **Second Law** to emerge from Mendel's studies.

## 6.13 Independent Assortment in a *Drosophila* Cross

To demonstrate independent assortment in an animal such as *Drosophila*, we follow the same general steps taken to demonstrate allelic segregation. We first construct heterozygous $F_1$ flies, in this case, flies heterozygous for *two* sets of unliked alleles. Then either we can examine allelic transmission patterns directly in a test cross, or we can inbreed the $F_1$ and deduce transmission patterns from the $F_2$ progeny phenotypes.

As an example, we can cross female *Drosophila* homozygous for *vestigial* to males homozygous for the body color mutation *ebony* ($e$). The $F_1$ is wild in phenotype. We have already seen that $vg$ is recessive, and the occurrence of a wild-type $F_1$ in the present cross indicates that $e$ is recessive as well. The $F_1$, therefore, is heterozygous for both $vg$ and $e$ and has the genotype $\dfrac{+\quad +}{vg\quad e}$

If a test cross is to be performed, we follow the rule for the design of a test cross (*Section 6.6*) and choose flies homozygous recessive at the two loci of interest, namely, $\dfrac{vg\quad e}{vg\quad e}$. When these *vestigial ebony* flies are crossed to the $F_1$, we get the results shown in Figure 6-10: four different kinds of progeny are obtained in essentially equal numbers—wild-type, *vestigial*, *ebony*, and *vestigial ebony*. This is indeed the outcome expected if we assume, as in Figure 6-10, independent assortment and the equal production of four types of $F_1$ gametes: $vg$ +, + $e$, + +, and $vg$ $e$. (To prove this to yourself, go through the steps drawn in Figure 6-9, substituting + and $vg$ for the $R/r$ pair and + and $e$ for the $Y/y$ pair.)

If the $F_1$ is instead inbred, four different phenotypes are again found among the progeny flies, but this time they appear in quite different proportions. Thus when the $F_1$ $\dfrac{+\quad +}{vg\quad e}$ are crossed among themselves, the $F_2$ data, shown in Table 6-1, indicate a phenotypic ratio of approximately 9/16 wild type, 3/16 *vestigial*, 3/16 *ebony*, and 1/16 *vestigial ebony*, or, **9:3:3:1**, exactly as in Mendel's crosses.

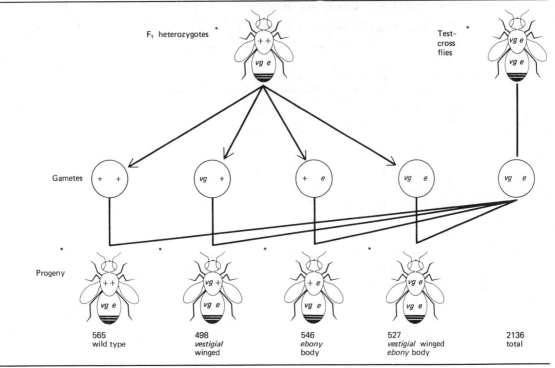

FIGURE 6-10 Test cross demonstrating independent assortment in *Drosophila.*

As we saw when we considered Mendel's data, the most foolproof method for analyzing such a cross is to construct what is known as a **Punnett square.** As illustrated in Table 6-2, all possible egg genotypes are listed on one axis of a square, and all possible sperm (or pollen) genotypes are displayed on a second axis. All possible egg-sperm "fertilizations" are then made, giving a "checkerboard" of expected progeny genotypes. In the present case, allelic segregation and independent assortment predict the

TABLE 6-1 Phenotypes Obtained
by Inbreeding Double

Heterozygotes $\left( \dfrac{+}{vg}\ \dfrac{+}{e} \right)$

|  | Phenotype | Number |
|---|---|---|
|  | *wild type* | 2834 |
|  | *vestigial* | 920 |
|  | *ebony* | 951 |
|  | *vestigial ebony* | 287 |
| Total |  | 4992 |

TABLE 6-2   Genotypes Obtained by Inbreeding Double Heterozygotes $\left(\dfrac{+\ \ +}{vg\ \ e}\right)$

| Eggs | Sperm | | | |
|---|---|---|---|---|
| | $+\ +$ | $+\ e$ | $vg\ +$ | $vg\ e$ |
| $(+\ +)$ | $\dfrac{+\ \ +}{+\ \ +}$ | $\dfrac{+\ \ e}{+\ \ +}$ | $\dfrac{vg\ \ +}{+\ \ +}$ | $\dfrac{vg\ \ e}{+\ \ +}$ |
| $(+\ e)$ | $\dfrac{+\ \ +}{+\ \ e}$ | $\dfrac{+\ \ e}{+\ \ e}$ | $\dfrac{vg\ \ +}{+\ \ e}$ | $\dfrac{vg\ \ e}{+\ \ e}$ |
| $(vg\ +)$ | $\dfrac{+\ \ +}{vg\ \ +}$ | $\dfrac{+\ \ e}{vg\ \ +}$ | $\dfrac{vg\ \ +}{vg\ \ +}$ | $\dfrac{vg\ \ e}{vg\ \ +}$ |
| $(vg\ e)$ | $\dfrac{+\ \ +}{vg\ \ e}$ | $\dfrac{+\ \ e}{vg\ \ e}$ | $\dfrac{vg\ \ +}{vg\ \ e}$ | $\dfrac{vg\ \ e}{vg\ \ e}$ |

presence of four types of eggs and four types of sperm in the $F_1$ gamete pool (if necessary, prove this to yourself with meiosis diagrams); therefore, there are 16 possible genotypic combinations. Closer inspection of Table 6-2 reveals that some of these genotypes are repeated more than once and that nine different kinds of genotypes are present in all. You should find these in the table and should come up with the 9:3:3:1 phenotypic ratio obtained in the actual cross.

To prove that the nine predicted genotypes in fact exist among the $F_2$, one must resort to test crosses between individual $F_2$ flies and $\dfrac{vg\ \ e}{vg\ \ e}$ flies. Table 6-2 predicts that if single $F_2$ flies with a wild phenotype are subjected to such test crosses, 1/9 will prove to be homozygous for both wild-type alleles, 2/9 will be heterozygous for $vg$, 2/9 will be heterozygous for $e$, and 4/9 will be heterozygous for both $vg$ and $e$. Similar predictions can be made regarding the genotypes of the $F_2$ *vestigial* and *ebony* flies; you should work these out yourself.

## 6.14 The Chi-Square Test

The foregoing paragraphs have offered certain predictions that are based on the hypothesis of independent assortment. We now wish to learn whether the data obtained from either test crosses or $F_1$ inbreeding are indeed in agreement with the predictions and thus with the hypothesis. As an example, we can ask whether the $F_2$ data in Table 6-1, obtained by inbreeding $\dfrac{+\ \ +}{vg\ \ e}$ flies, fit a ratio of 9:3:3:1. We can determine the "goodness of fit" of the data to the predicted ratio by a simple statistical test known as the **chi-square ($\chi^2$) test.**

The $\chi^2$ test involves, first, determining a predicted ratio and, second, establishing how closely the observed data fit this ratio. The $\chi^2$ test is made by ascertaining the probability that the deviation of the observed ratio from the predicted ratio is due to chance and not to some other factor such as experimental conditions, biased sampling, or even the wrong hypothesis. The usual statistical procedure establishes an arbitrary criterion for what degree of deviation is considered a significant deviation from that which would be expected from chance alone. If the **probability** ($P$) of obtaining the observed ratio is equal to or less than 5 in 100 ($P \leq 0.05$), the deviation between the expected and the observed ratio is considered significant and not simply attributable to chance. If the probability is 1 in 100 or less ($P \leq 0.01$), the deviation is highly significant, and some nonchance factor is almost certainly operating. When the $P$ value is greater than 0.05, the deviation is not considered statistically significant and can be expected on the basis of chance alone.

A $\chi^2$ test for the data given in Table 6-1 is worked out as follows:

1. *The hypothesis:* That the observed phenotypic ratio is in accord with a predicted phenotypic ratio of 9/16 *wild-type*:3/16 *vestigial*:3/16, *ebony*:1/16 *vestigial ebony.*

2. *Expected or predicted ratio:* By dividing the total number of 4992 by 16 the expected number of *vestigial ebony* flies is determined to be 312. The expected number of wild-type flies is therefore 9 × 312, or 2808, and the expected number of *vestigial* and *ebony* flies is accordingly 3 × 312, or 936, each.

3. *Deviation between observed and expected ratios and the calculation of $\chi^2$.*

| | Wild type | vestigial | ebony | vestigial ebony | Total |
|---|---|---|---|---|---|
| Observed (O) | 2834 | 920 | 951 | 287 | 4992 |
| Expected (E) | 2808 | 936 | 936 | 312 | 4992 |
| Deviation ($D = O - E$) | 26 | 16 | 15 | 25 | |
| $D^2$ | 676 | 256 | 225 | 625 | |
| $D^2/E$ | 0.24 | 0.27 | 0.24 | 2.00 | |
| | $\Sigma(D^2/E) = \chi^2 = 2.75$ | | | | |

4. *Calculating degrees of freedom:* To determine a $P$ value from a $\chi^2$ value, it is essential to establish the number of **degrees of freedom (df).** The degrees of freedom in an analysis involving $n$ classes is usually equal to $n - 1$. What this means is that if a given total number of individuals (4992 in the foregoing example) is divided into $n$ classes (four phenotypic classes in the foregoing example), then once we have calculated the numbers expected in three of these classes, the fourth number is set. It follows that there are only three degrees of freedom in the analysis. Specifically, when we add the expected values of three of the genotypic classes (2808 + 936 + 936) and get 4680, the expected value for the fourth class *must* equal 312 if the total is to be 4992.

**TABLE 6-3   Table of $\chi^2$ (Chi-Square)**

| df | P = 0.99 | 0.98 | 0.95 | 0.90 | 0.80 | 0.70 | 0.5 | 0.30 | 0.20 | 0.10 | 0.05 | 0.02 | 0.01 |
|----|----------|------|------|------|------|------|-----|------|------|------|------|------|------|
| 1  | 0.000157 | 0.00628 | 0.00393 | 0.0158 | 0.0642 | 0.148 | 0.455 | 1.074 | 1.642 | 2.706 | 3.841 | 5.412 | 6.635 |
| 2  | 0.0201 | 0.0404 | 0.103 | 0.211 | 0.446 | 0.713 | 1.386 | 2.408 | 3.219 | 4.605 | 5.991 | 7.824 | 9.210 |
| 3  | 0.115 | 0.185 | 0.352 | 0.584 | 1.005 | 1.424 | 2.366 | 3.665 | 4.642 | 6.251 | 7.816 | 9.837 | 11.345 |
| 4  | 0.297 | 0.429 | 0.711 | 1.064 | 1.649 | 2.195 | 3.357 | 4.878 | 5.989 | 7.779 | 9.488 | 11.668 | 13.277 |
| 5  | 0.554 | 0.752 | 1.145 | 1.610 | 2.343 | 3.000 | 4.351 | 6.064 | 7.289 | 9.236 | 11.070 | 13.388 | 15.086 |
| 6  | 0.872 | 1.134 | 1.635 | 2.204 | 3.070 | 3.828 | 5.348 | 7.321 | 8.558 | 10.645 | 12.592 | 15.033 | 16.812 |
| 7  | 1.239 | 1.564 | 2.167 | 2.833 | 3.822 | 4.671 | 6.346 | 8.383 | 9.803 | 12.017 | 14.067 | 16.622 | 18.475 |
| 8  | 1.646 | 2.032 | 2.733 | 3.490 | 4.594 | 5.527 | 7.344 | 9.524 | 11.030 | 13.362 | 15.507 | 18.168 | 20.090 |
| 9  | 2.088 | 2.532 | 3.325 | 4.168 | 5.380 | 6.393 | 8.343 | 10.656 | 12.242 | 14.684 | 16.919 | 19.679 | 21.666 |
| 10 | 2.558 | 3.059 | 3.940 | 4.865 | 6.179 | 7.267 | 9.342 | 11.781 | 13.442 | 15.987 | 18.307 | 21.161 | 23.209 |

Abridged from Table II of Fisher and Yates: *Statistical Tables for Biological, Agricultural and Medical Research* (1953); published by Longman and by permission of the authors and publishers.

5. *The P value:* Tables of $\chi^2$ values have been prepared from which it is possible to determine the corresponding *P* values, given a particular number of degrees of freedom. A part of such a table is given in Table 6-3, and it is seen that when $\chi^2 = 2.75$ and there are three degrees of freedom, the *P* value is greater than 0.30 and less than 0.50 ($P = 0.50 - 0.30$). Thus from 30 to 50 times out of 100 we could expect chance deviations of the magnitude observed. Since the value of *P* is clearly greater than 0.05, the observed results are in good agreement with those to be expected for the independent assortment of two pairs of alleles.

## 6.15  Phenotypic Ratios in Hybrid Crosses

The Punnett square (*Table 6-2*) is the most accurate way to determine genotypic ratios in a cross, but there is a simpler method for determining phenotypic ratios. When doubly heterozygous individuals such as $\dfrac{+}{vg}\,\dfrac{+}{e}$ are inbred and we follow *vg* and its + allele alone, we expect the genes to segregate in a 1:1 ratio such that the phenotypic ratio will be 3 wild-type: 1 *vestigial*. The same is true for *e* and its + allele. Therefore, assuming that the two gene pairs assort independently and that all genotypes have equal viability, the probability of their appearance will be as follows:

3 of 4 $F_2$ flies should have wild-type wings and
1 of 4 $F_2$ flies should have *vestigial* wings;
3 of 4 $F_2$ flies should have wild-type body color and
1 of 4 $F_2$ flies should have *ebony* body color

A rule established by probability theory states that **the probability that independent events will occur simultaneously is the product of their separate probabilities.** Thus the probability that various combinations will occur is as follows:

wild-type wing and wild-type body color should be $3/4 \times 3/4$
=9/16;

wild-type wing and *ebony* body color should be $3/4 \times 1/4 = 3/16$;
*vestigial* wing and wild-type body color should be $1/4 \times 3/4 = 3/16$;
*vestigial* wing and *ebony* body color should be $1/4 \times 1/4 = 1/16$.

The preceding is often referred to as a **dihybrid cross** and 9:3:3:1 as a
**dihybrid ratio.** We can now use this method to determine expected phenotypic ratios in a **trihybrid cross** wherein inbreeding individuals are heterozygous for **three** unlinked pairs of alleles. For example, true-breeding pea
plants that have *yellow* and *round* seeds and *red* flowers can be crossed with
true-breeding plants that have *green* and *wrinkled* seeds and *white* flowers,
the corresponding traits being allelic. The $F_1$ all have *yellow, round* seeds
and *red* flowers, which indicates that these are the dominant traits. When
these plants are self-fertilized, then, following our previous reasoning,
three of four should have *yellow* seeds while one of four should have green
seeds; three of four should have *round* seeds; and so on. By multiplying
these probabilities, we find that the combinations should occur with the
following frequency:

| | |
|---|---|
| *round-yellow-red:* | $3/4 \times 3/4 \times 3/4 = 27/64$ |
| *round-yellow-white:* | $3/4 \times 3/4 \times 1/4 = 9/64$ |
| *round-green-red:* | $3/4 \times 1/4 \times 3/4 = 9/64$ |
| *wrinkled-yellow-red:* | $1/4 \times 3/4 \times 3/4 = 9/64$ |
| *round-green-white:* | $3/4 \times 1/4 \times 1/4 = 3/64$ |
| *wrinkled-yellow-white:* | $1/4 \times 3/4 \times 1/4 = 3/64$ |
| *wrinkled-green-red:* | $1/4 \times 1/4 \times 3/4 = 3/64$ |
| *wrinkled-green-white:* | $1/4 \times 1/4 \times 1/4 = 1/64$ |

In other words, the predicted **trihybrid ratio** is 27:9:9:9:3:3:3:1.

Summing these classes, we see that for every 64 individuals in the $F_2$,
27 will be expected to exhibit all three dominant traits, 27 will show two
dominant traits, 9 will show one dominant trait, and 1 will exhibit all three
recessive traits. The only individuals in the $F_2$ whose genotype is known by
inspection are the *wrinkled-green-white* plants: these plants must be homozygous recessive, a state usually symbolized by plant geneticists as *rr yy cc*,
where the *c* locus controls pigment formation in the flower. In addition to
the *wrinkled-green-white*, seven other phenotypic classes also emerge, and
26 different genotypes are represented within them. The genotypes can be
derived by a Punnett square analysis, assuming that equal numbers of
*RYC, RYc, RyC, Ryc, rYC, rYc, ryC,* and *ryc* are present in both eggs and
pollen. The resulting genotypes are shown in Table 6-4. Clearly, the genetic
analysis becomes increasingly tedious as more independently assorting alleles are followed in a cross.

Following these lines of reasoning, we can derive a general relationship for the number of gene combinations expected when diploids that are
heterozygous for numerous unlinked genes are crossed. This is shown in

**TABLE 6-4  Phenotypic Ratios in a Trihybrid Cross:** The theoretical number of individuals, with their genotypes and breeding behavior, expected in $F_2$ from a cross of a round, yellow-seeded, red-flowered variety of pea with a wrinkled, green-seeded, white-flowered pea.

| Number of Individuals | Genotype Class | Phenotype Class | Ratio of Phenotypes | Breeding Behavior when Self-Fertilized |
|---|---|---|---|---|
| 1 | RR YY CC | | | Breeds true |
| 2 | Rr YY CC | | | Segregates *round-wrinkled*, 3:1 |
| 2 | RR Yy CC | | | Segregates *yellow-green*, 3:1 |
| 2 | RR YY Cc | Round | | Segregates *red-white*, 3:1 |
| 4 | Rr Yy CC | Yellow<br>Red | 27 | Segregates *round-wrinkled, yellow-green*,<br>9:3:3:1 |
| 4 | Rr YY Cc | | | Segregates *round-wrinkled, red-white*, 9:3:3:1 |
| 4 | RR Yy Cc | | | Segregates *yellow-green, red-white*, 9:3:3:1 |
| 8 | Rr Yy Cc | | | Segregates *round-wrinkled, yellow-green, red-white*,<br>27:9:9:9:3:3:3:1 |
| 1 | RR YY cc | Round | | Breeds true |
| 2 | RR Yy cc | Yellow | 9 | Segregates *yellow-green*, 3:1 |
| 2 | Rr YY cc | White | | Segregates *round-wrinkled*, 3:1 |
| 4 | Rr Yy cc | | | Segregates *round-wrinkled, yellow-green*,<br>9:3:3:1 |
| 1 | RR yy CC | Round | | Breeds true |
| 2 | RR yy Cc | Green | 9 | Segregates *red-white*, 3:1 |
| 2 | Rr yy CC | Red | | Segregates *round-wrinkled*, 3:1 |
| 4 | Rr yy Cc | | | Segregates *round-wrinkled, red-white*, 9:3:3:1 |
| 1 | rr YY CC | Wrinkled | | Breeds true |
| 2 | rr Yy CC | Yellow | 9 | Segregates *yellow-green*, 3:1 |
| 2 | rr YY Cc | Red | | Segregates *red-white*, 3:1 |
| 4 | rr Yy Cc | | | Segregates *yellow-green, red-white*, 9:3:3:1 |
| 1 | rr yy CC | Wrinkled | | Breeds true |
| 2 | rr yy Cc | Green<br>Red | 3 | Segregates *red-white*, 3:1 |
| 1 | rr YY cc | Wrinkled | | Breeds true |
| 2 | rr Yy cc | Yellow<br>White | 3 | Segregates *yellow-green*, 3:1 |
| 1 | RR yy cc | Round | | Breeds true |
| 2 | Rr yy cc | Green<br>White | 3 | Segregates *round-wrinkled*, 3:1 |
| $\frac{1}{64}$ | rr yy cc | Wrinkled<br>Green<br>White | 1 | Breeds true |

From E. W. Sinnott, L. C. Dunn, and T. Dobzhansky, *Principles of Genetics*. New York: McGraw-Hill, 1958, Table 6.1.

**TABLE 6-5   General Hybrid Cross Patterns.** The relation between the number of gene loci involved in a cross and the number of phenotypic and genotypic classes in $F_2$

| Number of Gene Loci Involved in the Cross | Number of Visibly Different $F_2$ Classes of Individuals if Dominance Is Complete | Number of Different Kinds of Gametes Formed by the $F_1$ Hybrid | Number of Genotypically Different Combinations | Number of Possible Combinations of $F_1$ Gametes |
|---|---|---|---|---|
| 1 | 2 | 2 | 3 | 4 |
| 2 | 4 | 4 | 9 | 16 |
| 3 | 8 | 8 | 27 | 64 |
| 4 | 16 | 16 | 81 | 256 |
| $n$ | $2^n$ | $2^n$ | $3^n$ | $4^n$ |

From E. W. Sinnott, L. C. Dunn, and T. Dobzhansky, *Principles of Genetics*. New York: McGraw-Hill, 1958, Table 6.2.

Table 6-5. The major limit placed on such calculations is the number of chromosome pairs ($n$) possessed by a given species, so that $n$ cannot be greater than 4 for *D. melanogaster* nor greater than 7 for the pea. Nonetheless, it is clear that **the independent assortment of genes can yield enormous genetic variability** in sexually reproducing organisms and that this variability is in no way dependent on crossing over between homologous chromosomes. On the assumption that the bringing together of new gene combinations is of selective advantage to a species, the organization of genes into multiple linkage groups and their random assortment during meiosis is indeed beneficial to eukaryotes.

## 6.16 Independent Assortment in Tetrad Analysis

To illustrate independent assortment in tetrad analysis, we can follow a cross between an *arginine*-requiring strain (*arg*) and an *acetate*-requiring strain (*ac*) of *Chlamydomonas*. The meiotic products are separated (*Figure 6-7*) and allowed to form colonies on a medium containing both arginine and acetate. The colonies are then replica plated to four different kinds of media: minimal medium, minimal plus arginine, minimal plus acetate, and minimal plus arginine and acetate.

The cross of the two mutant strains is symbolized as *arg* + × + *ac*, and we assume at the outset that the *arg* and *ac* loci are in different chromosomes so that they undergo independent assortment. Four kinds of meiotic products, and therefore four kinds of haploid vegetative cells, should result from the cross: *arg* +, + *ac*, + +, and *arg ac*. These four can be distinguished from one another by the media on which they will or will not grow (Table 6-6).

Tetrad analysis of the cross gives the results shown in Table 6-7. **Three classes of tetrads are found.** The first consists of colonies derived from two sorts of meiotic products, *arg* + and + *ac*. Since two of the products have the genotype (and phenotype) of one parent and two have the genotype (and phenotype) of the other, this class of tetrad is called a

**TABLE 6-6   Independent Assortment in *Chlamydomonas*.** Types of progeny produced in the cross of *arg +* × *+ ac* and the media on which they will grow

| Type of Progeny | Minimal | Minimal with Arginine | Minimal with Acetate | Minimal with Arginine and Acetate |
|---|---|---|---|---|
| *arg ac* | − | − | − | + |
| *+ ac* | − | − | + | + |
| *arg +* | − | + | − | + |
| *+ +* | + | + | + | + |

Growth is indicated by a plus sign and absence of growth by a minus sign.

**TABLE 6-7   Tetrad Classes** Produced in the Cross *arg +* × *+ ac* in *Chlamydomonas*

| | Tetrad Class | | |
|---|---|---|---|
| | PD | NPD | T |
| | *arg +* | *arg ac* | *arg +* |
| | *arg +* | *arg ac* | *+ +* |
| | *+ ac* | *+ +* | *arg ac* |
| | *+ ac* | *+ +* | *+ ac* |
| Number | 71 | 69 | 95        Total = 235 |

**parental ditype,** or **PD.** Note that each pair of alleles has segregated in a one-to-one fashion.

The second class of tetrad also shows a one-to-one segregation for each pair of alleles, but it consists of two new genotype combinations: two are *arg ac* and two are *+ +*. This tetrad, with two types of product each genotypically different from either of the original parents, is called a **nonparental ditype,** or **NPD.**

Four different genotypes are found in the third class of tetrad. Two are the parental combinations (*arg +* and *+ ac*), and two are the nonparental types (*arg ac* and *+ +*). A tetrad of this sort is called a **tetratype,** or **T.** Again each pair of alleles segregates one to one.

The origin of the PD and NPD tetrads relates directly to the independent alignment of the chromosomes at metaphase I, as illustrated in Figure 6-11a. If the occurrence of the two alternative alignments is equally probable, the ratio of PD/NPD tetrads in any cross involving unlinked genes should be equal to one. The data given in Table 6-7 show that there are 71 PD and 69 NPD tetrads, confirming the prediction. **When a cross yields roughly equal numbers of PD and NPD tetrads, the pairs of alleles being followed can be assumed to assort independently of one another.** In other words, it can be concluded that each set of alleles resides in a different set of homologous chromosomes, the assumption we in fact made about the *arg* and *ac* loci at the outset.

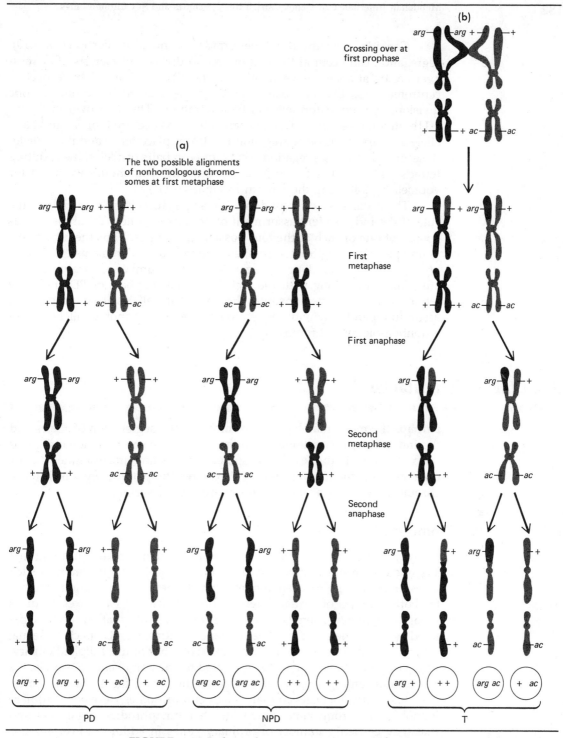

**FIGURE 6-11 Independent assortment by tetrad analysis.** (a) Independent assortment of two pairs of alleles in tetrads of *Chlamydomonas,* yielding PD and NPD tetrads. (b) Independent assortment of two pairs of alleles in tetrads of *Chlamydomonas* where a single exchange has occurred, yielding a T tetrad.

181

Consider next the tetratype tetrads. As noted earlier (*Section 6.10*), segregation can occur at the first or second division of meiosis. If a cross-over occurs at a site between either of the allelic loci and their respective centromeres so that one pair of alleles segregates at the second meiotic division, the tetrad formed will be a tetratype. This is shown in Figure 6-11b, in which second-division segregation has occurred for *arg* and its + allele and first-division segregation has taken place for *ac* and its + allele. (If second-division segregation occurs for both pairs of alleles, the resulting tetrads will be either PD or NPD, an outcome you should work out for yourself by following the pattern in Figure 6-11.)

The occurrence of second-division segregation in no way violates the rule of the independent assortment of nonlinked genes. Tetratype tetrads consist of one of each of the four possible genotypes. Thus the ratio of the genotypes emerging from a cross, as seen in Table 6-7, remains 1:1:1:1 for *arg* +, + *ac*, + +, and *arg ac*, regardless of the number of tetratypes present. This ratio, along with the fact that equal numbers of PD and NPD tetrads are present, demonstrates that each of the alleles of a given pair will assort independently with respect to any member of another allelic pair in a nonhomologous chromosome.

# Sex-Linked Inheritance

Perhaps the most unambiguous examples of the segregation of alleles and the independent assortment of unlinked genes come by following genes that are located in the sex chromosomes. This section first examines the sex chromosomes themselves and then explores the patterns by which these chromosomes are transmitted from parent to offspring.

## 6.17 The Sex Chromosomes

Many higher plants and certain animals are hermaphroditic, meaning that any one organism can give rise to both male and female gametes; in such cases the organism in question must carry the genetic information required for both avenues of sexual differentiation. When the two sexes occur as separate organisms, some of the relevant genetic information can be separated as well. The actual number of genes controlling all aspects of sexual differentiation is unknown for any eukaryote, but in most diploid eukaryotes, key genes have come to be restricted to one pair of chromosomes, the **sex chromosomes,** usually referred to as the **X** and the **Y** chromosomes.

It is generally believed that the sex chromosomes were once ordinary homologous chromosomes occurring in hermaphroditic organisms and that they carried a number of ordinary genes in addition to certain sex-determining genes. Subsequent evolution led to the modification of one homologue. In many cases a physical modification of one of these once homologous chromosomes can be visualized with the light microscope: one of the two sex chromosomes has become much smaller than the other (for example, the human Y chromosome, *Figure 2-14*) and is often highly heterochromatic. In some cases, including that of *D. melanogaster*, the heterochromatic Y chromosome is longer than its X homologue and may have a different shape. For *Drosophila*, the Y chromosome has a hook at one end and is symbolized as ——→. It is important to bear in mind that the postulated transformation affected one homologue only; **modern X chromosomes continue to carry a full complement of ordinary, sex-unrelated genes.**

In *Drosophila*, males and females are most readily told apart by the appearance of their abdominal segments: the female abdomen has a tapered end and a different pattern of stripes than the male abdomen, as seen in Figure 5-6. The sex of a fly is determined, in part, by the number of X chromosomes that individual possesses. Normally, a female fly will have two X chromosomes, whereas a male will have only one X chromosome plus one copy of the heterochromatic Y chromosome. Females are therefore typed as XX and males as XY. An occasional abnormal fertilization will give rise to a fly with one X but no Y chromosome, and such an XO individual, although sterile, is phenotypically a male. Flies with two X chromosomes and a Y chromosome have also been obtained, and they appear as normal females. **The Y chromosome in *Drosophila* therefore plays no apparent role in the determination of sex:** in addition to an apparent contribution made by autosomal genes, femaleness results when two X chromosomes are present and maleness results when one X is present. Indeed, except for its genes that are somehow necessary for producing fertile sperm, the Y chromosome in *Drosophila* appears to carry very few active genes at all.

In mammals normal females are also XX and males XY. In contrast to *Drosophila*, however, **it is the presence of the small, heterochromatic Y chromosome that determines maleness** and its absence that determines femaleness. In humans, mechanical errors in the separation of homologues during meiosis, described in Section 5.15, can produce exceptional XO individuals that have an X but no Y chromosome; these are found to be sterile females exhibiting **Turner Syndrome** (Table 6-8); XXY exceptions, on the other hand, are sterile males exhibiting **Klinefelter Syndrome** (Table 6-8). Similarly, XXX humans are phenotypically female, and XXXY are phenotypically male. It is not yet known how many male-determining genes reside in the Y chromosome, but present evidence suggests that there may be only a few, perhaps distributed along the length of the chromosome in multiple copies. An extra dosage of the Y chromosome, as in XYY persons, produces normally fertile males (Table 6-8).

A third kind of sex chromosome organization is found in birds (and also in butterflies): the male carries the two large chromosomes, called Z, and the female carries one Z and one smaller chromosome called W. Males thus are ZZ and females are ZW.

**TABLE 6-8    Sex Chromosome Abnormalities in Humans**

| Sex Chromosome Constitution | Syndrome | Frequency | Sex Phenotype | Fertility | Characteristic Features (Some Are Not Invariable) |
|---|---|---|---|---|---|
| XO | Turner | One in 3000 "female" births | Female | − | Short-stature, webbed neck, low-set ears, broad chest, wide-spaced nipples and underdeveloped breasts, small uterus, abortive ovary development |
| XXY | Klinefelter | One in 600 "male" births | Male | − | Long legs, small testes, sparse body hair, female-like breast development |
| XYY | | At least one in 2000 male births | Male | + | Unusually tall |
| XXX | | One in 1400 "female" births | Female | Reduced | Mentally retarded |

V. A. McKusick, *Human Genetics*, 2nd ed., © 1972. Adapted by permission of Prentice-Hall, Inc., Englewood Cliffs, N.J.

## 6.18 Patterns of Sex-Linked Gene Transmission

In animals with X and Y chromosomes, the female is termed the **homogametic** and the male the **heterogametic** sex, meaning that as far as sex chromosomes are concerned, females produce only X-carrying eggs, whereas males produce both X- and Y-carrying sperm. Because offspring that are XX are female and those that are XY are male, the inheritance of the X chromosome will follow a specific pattern (Figure 6-12): **a male will transmit his X chromosome to only his female offspring, whereas a female will transmit her X chromosomes to both her male and her female offspring.** It follows that **a male will always inherit his X chromosome from his mother,** since his father must have contributed the Y. Moreover, a male can transmit the information in his X chromosome to his grandchildren only through his daughters.

Any genetic traits that are transmitted by way of this specific pattern are said to be **sex linked.** As an example, we can follow the inheritance of the *white* phenotype in *Drosophila*. Affected flies have no eye pigments and carry a mutation at the *w* locus. When a cross is made between wild-type females and *w* males, all of the $F_1$ flies are wild-type, indicating that *w* is recessive to +. When the $F_1$ flies are inbred, three quarters of the resulting flies possess the wild phenotype and one quarter have the *white* phenotype. This result appears to be a straightforward example of the segregation of a pair of alleles, *w* and +. When, however, the $F_2$ flies are classified for both eye color and sex, it is found that all of the females are wild type, whereas half of the males are wild type and the other half have *white* eyes (Figure 6-13a).

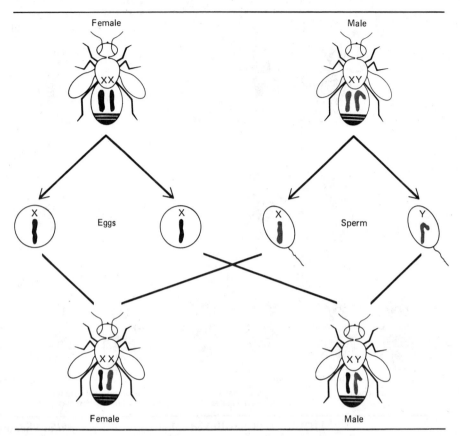

**FIGURE 6-12 Inheritance of X and Y chromosomes in *Drosophila*.**
Chromosomes transmitted by the female are black; those transmitted by the
male are gray. (After E. Altenberg, *Genetics*. New York: Holt, Rinehart and
Winston, 1957.)

We can now examine the **reciprocal cross:** *white* females crossed to
wild-type males (Figure 6-13b). In this case the $F_1$ is no longer all of the
wild type; instead, we find wild-type females and *white* males. Moreover,
when the $F_1$ is inbred, half the progeny are wild in phenotype and half
have *white* eyes. Equal numbers of males and females are represented in
both phenotypic classes (Figure 6-13b).

**Whenever reciprocal crosses give markedly different $F_1$ and $F_2$ phe-
notypic ratios, the trait being studied is likely to be sex linked.** Figures
6-14a and b illustrate the transmission of *w* and its wild-type allele by
assuming that *w* is located in the X chromosome, and in both reciprocal crosses this assumption readily accounts for the observed phenotypic
ratios.

By looking more closely at Figure 6-14, it becomes clear that the presence of the Y chromosome is without effect on the eye color phenotype.

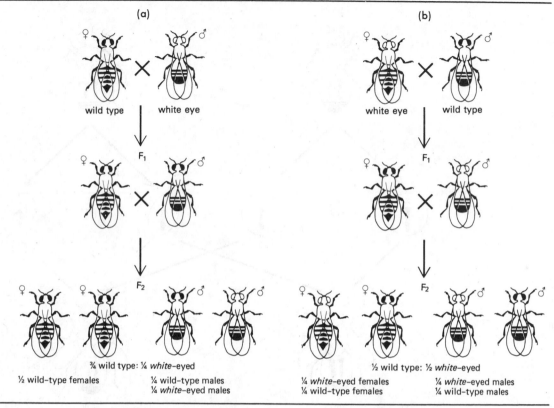

**FIGURE 6-13 Sex-linked inheritance in *Drosophila*: Phenotypes.** Reciprocal crosses are shown in (a) and (b).

Thus in cross *a*, in which all of the original female parents produce +-bearing eggs, all of the $F_1$ males have normal eye color. In cross *b*, in which the female parents produce *w*-bearing gametes, all the $F_1$ males have *white* eye color. The Y does not appear to carry any gene that modifies the expression of either the dominant + or the recessive *w* allele. It therefore behaves as a genetically inert chromosome, which is what we would expect, since, as noted in Section 3.13, constitutive heterochromatin carries very few genes. For this reason, it is inappropriate to regard *Drosophila* males as being either homozygous or heterozygous for most of the genes on their X chromosome. Instead, the male flies are referred to as being **hemizygous,** meaning that they possess only one set of these genes. This set they inherit exclusively from their mothers, and **their Y chromosomes are "silent" as far as the expression of the sex-linked genes is concerned.**

Figure 6-14 a and b shows the predicted genotypes of $F_1$ and $F_2$ flies in the reciprocal crosses we have been considering. We expect, for example,

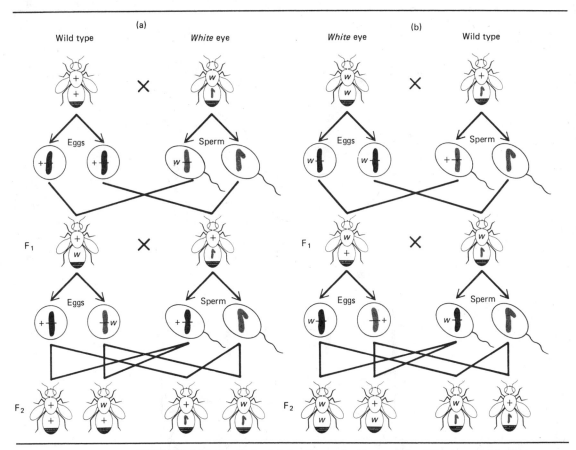

**FIGURE 6-14 Sex-linked inheritance in *Drosophila*: Genotypes.** (a) Crosses correspond to Figure 6-13a. (b) Crosses correspond to Figure 6-13b. Maternally contributed chromosomes (and copies) are black; paternal are gray.

that one half of the F₂ females in both crosses should be heterozygous for *w*. This prediction can be verified by crossing individual F₂ females to *white*-eyed males; you can work this out yourself with diagrams. (Is it in fact necessary to use *white*-eyed males in such test crosses or would males of wild phenotype serve the same purpose?)

    As a second example of sex-linked inheritance in *Drosophila*, we can consider the gene mutation *Bar* (*B*). In females homozygous for *B*, the eyes are highly reduced in size, causing them to look like narrow bars. A *Bar*-eyed male has a similarly reduced eye size. When a *Bar*-eyed female is crossed with a normal male, the sons all have *Bar* eyes, whereas the females have an intermediate phenotype: the eyes are neither normal nor decidedly bar-shaped. As diagrammed in Figure 6-15, this is the expected

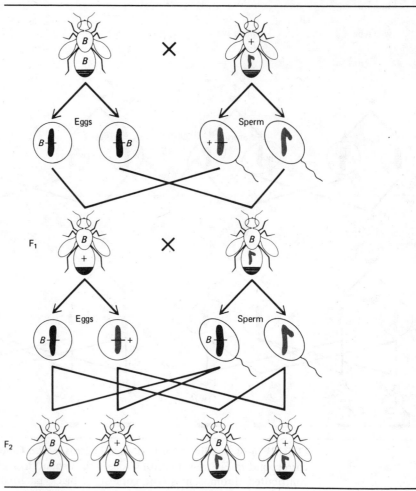

**FIGURE 6-15 The sex-linked inheritance of** *Bar* **eye** in *Drosophila*. Maternal chromosomes are black, paternal are gray.

outcome if the *Bar* locus resides in the X chromosome and if the *Bar* gene is **semi-dominant** to its + allele so that $B/+$ flies have an intermediate size. This deduction is supported by the result of inbreeding the $F_1$: the male progeny of the $B/+$ females are either wild type or *Bar* in equal numbers (Figure 6-15).

The *Bar* mutation is particularly instructive in that it can be identified cytologically. When salivary gland chromosome preparations are made of $B/B$ or $B/Y$ larvae and the X chromosomes are examined, a segment known as 16A, which contains several bands, is found to be present in double dose compared with a normal X chromosome; this is illustrated in Figure 6-16. When salivary gland chromosomes are instead prepared from the "intermediate" $F_1$ females, one X chromosome shows the 16A duplication

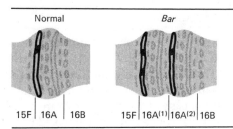

**FIGURE 6-16 Normal** and *Bar* **polytene X chromosomes of** *Drosophila*, showing duplicated 16A segment in the *Bar* chromosome.

15F | 16A | 16B

15F | 16A(1) | 16A(2) | 16B

whereas the other X does not, as would be expected if they were $B/+$. Thus **the inheritance of a particular X chromosome can be correlated with the inheritance of a sex-linked trait,** strongly supporting the thesis that genes on the X chromosome are responsible for generating these traits.

Sex-linked genes not only segregate but also assort independently from genes located on autosomes. A problem that illustrates this fact is included at the end of this chapter.

## 6.19 Sex-Linked Inheritance in Birds

In *Drosophila* and in mammals the heterogametic sex is the male, whereas, as noted earlier, the female is the heterogametic (ZW) sex in birds and butterflies. Sex-linked inheritance in birds follows the pattern to be expected from this reversal of chromosomal determinants, as shown in Figure 6-17 for the dominant gene *B* (*Barred* feathers) in chickens.

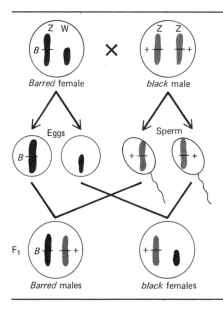

**FIGURE 6-17 Sex-linked inheritance in chickens** (*Barred* vs. *black* feathers). Maternal chromosomes are black; paternal are gray. (After E. Altenberg, *Genetics*. New York: Holt, Rinehart and Winston, 1951.)

## 6.20 Sex-Linked Inheritance in Humans

We saw earlier, in the case of brachydactyly and microcephaly, that human traits are inherited in a Mendelian fashion. The construction of the brachydactyly pedigree (*Figure 6-7*) was straightforward in that brachydactyly is caused by a dominant gene mutation and thus all carriers express the mutant phenotype. In the case of the gene for a rare autosomal recessive trait, on the other hand, its presence is usually masked by a dominant allele, and it is only in the relatively rare cases in which two heterozygous carriers have children (as in consanguineous marriages) that the gene is permitted expression in homozygous individuals (*Figure 6-8*). Moreover, the standard method of ascertaining the genotypes of heterozygotes, the test cross, cannot be feasibly applied to humans. There is thus only limited pedigree data on the inheritance of recessive human autosomal genes.

In contrast, there is a wealth of familial data on the inheritance of recessive sex-linked genes, since **a woman effectively subjects her two X chromosomes to a test cross every time she bears sons.** Her mate contributes only the silent Y chromosome to their male offspring, and thus the

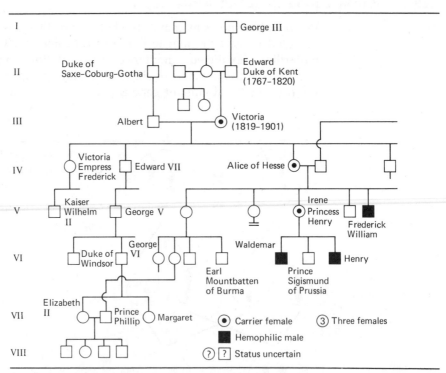

**FIGURE 6-18 Hemophilia pedigree.** The descendants of Queen Victoria showing carriers and afflicted males possessing the X-linked gene conferring the disease hemophilia. (From V. A. McKusick, *Human Genetics*, 2nd ed., © 1972. Adapted by permission of Prentice-Hall, Inc., Englewood Cliffs, N.J.)

phenotypes of her sons will directly reflect the genetic constitution of one or the other of her two X chromosomes. Studies of sex-linked inheritance in humans have amply demonstrated that patterns of gene transmission in humans follow those first recognized in flies.

Certainly the most famous case of human sex-linked inheritance is that of hemophilia A. Individuals suffering from this disorder cannot synthesize a normal blood protein called antihemophilic globulin (AHG), a substance required for the formation of thromboplastin and thus for normal blood clotting. (*Until recently, this condition created severe disability, but excellent transfusion therapy is now available.*) The AHG gene is located in the X chromosome, as demonstrated by the inheritance of a mutant AHG gene in the royal families of Europe during the last century (Figure 6-18). Queen Victoria was probably the original carrier of the mutant gene, since the disease was unknown in preceding generations of highly inbred royalty, and thus the gene probably arose as a new mutation in either the egg or the sperm from which she ultimately developed. One of her nine children, a son, was hemophilic, or a "bleeder," and two daughters proved to be

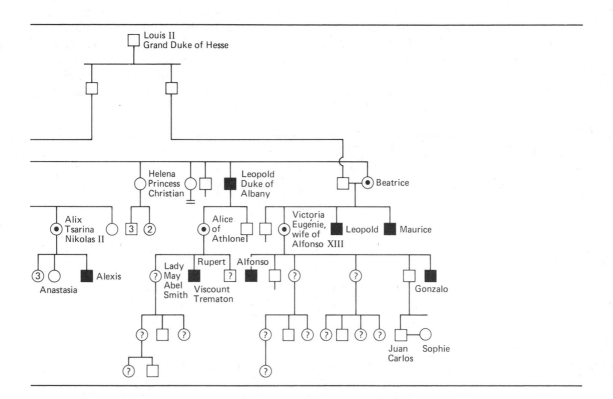

carriers. One daughter, Alice, passed the gene on to her daughter, Alix, who became Empress Alexandra of Russia and bore a hemophilic son, the Tsarevich Alexis. His illness was of considerable importance, for it was a major factor in the ascent of Rasputin and the unpopularity of the royal family at the time of the Russian Revolution.

A frequently occurring form of red-green color blindness is also a sex-linked recessive characteristic in humans. A pedigree is given in Figure 6-19. In the first generation a normal woman who had a color-blind brother married a color-blind man. Of their three children, one male and one female were color blind and one female was normal. The fact that a color-blind daughter was born to this couple at once indicates that the mother must have been a heterozygous carrier, as shown in Figure 6-19. The son then married a woman with normal color vision, an all their children were likewise normal (although the daughters were carriers). The son's color-blind sister (who must be homozygous for the gene) married a man with normal color vision: all of their sons were color blind, whereas their one daughter had normal vision but was a carrier.

Table 6-9 presents a list of the known X-borne mutations in humans. It is of interest in this regard that at least five of the genes that are sex linked in humans are also sex linked in a number of other placental mammals, including the mouse, whose karyotype is strikingly different from that of the human (*Figure 2-16*). This and other lines of evidence suggest that once a certian chromosome in some ancestral mammalian population became entrusted with the all-important role of sex determination, very

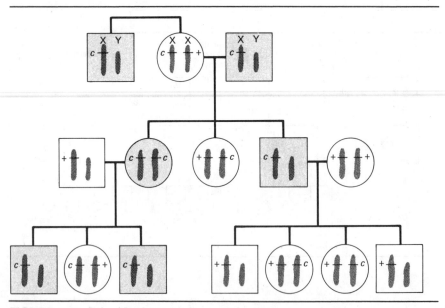

FIGURE 6-19 Color-blindness pedigree in humans.

## TABLE 6-9   Human Sex-Linked Traits

Addison's disease (adrenal insufficiency) (one form)
Agammaglobulinemia (immunoglobulin insufficiency)
Albinism (various forms)
Albright's osteodystrophy
Aldrich syndrome (immunological disorder)
Amelogenesis imperfecta (tooth malstructure)
Anemia, hypochromic
Angiokeratoma (Fabry's disease) (kidney disfunction)
Cerebellar ataxia
Cerebral sclerosis, Scholz type
Charcot-Marie-Tooth peroneal atrophy
Choroideremia (progressive blindness)
Cleft palate
Color blindness (complete and several partial types)
Deafness
Diabetes insipidus (several types)
Dyskeratosis
Ectodermal dysplasia
Ehlers-Danlos syndrome (bruising tendency)
Endocardial fibroelastosis
Faciogenital dysplasia
Fibrin-stabilizing factor deficiency
Focal dermal hypoplasia
Glucose-6-phosphate dehydrogenase deficiency
Glycogen storage disease
Granulomatous disease
Hemophilia A
Hemophilia B (Christmas disease)
Hydrocephalus
Hypomagnesemic tetany
Hypophosphatemia (vitamin D–resistant rickets)
Hypoxanthine guanine phosphoribosyl transferase deficiency
Ichthyosis (epidermal scaling)
Incontinentia pigmentii
Iris, hypoplasis of
Keratosis follicularis spinulosa
Lowe's oculocerebrorenal syndrome
Macular dystrophy
Megalocornea
Menkes syndrome (kinky hair disease)
Mental deficiency
Microphthalmia
Mucopolysaccharidosis II (Hunter syndrome)
Muscular dystrophy, progressive Becker type (onset between age 30 and 40)
Muscular dystrophy, Duchenne type (onset in childhood)
Muscular dystrophy, tardive, Dreifuss type (not lethal)
Night blindness
Norrie's disease (blindness)
Nystagmus (one type)
Ophthalmoplegia (myopia)
Oral-facial-digital syndrome
Parkinsonism (one type)
Pelizaeus-Merzbacher disease (cerebral sclerosis)

Phosphoglycerate kinase deficiency
Pituitary dwarfism
Pseudohermaphroditism, male
Reticuloendotheliosis
Retinitis pigmentosa (blindness)
Retinoschisis
Spastic paraplegia
Spinal and bulbar muscular atrophy
Spinal ataxia
Spondyloepiphyseal dysplasia, late
Testicular feminization syndrome
Thrombocytopenia
Thyroxine-binding globulin reduction
Van den Bosch syndrome (mental deficiency)
XG blood group system
XM system (macroglobulin serum protein)

From V. A. McKusick, *Mendelian Inheritance in Man*, 3rd ed. Baltimore: Johns Hopkins Press, 1971.

little gene rearrangement within this chromosome was tolerated by successive evolutionary lines, even though extensive divergence occurred for most other chromosomes. In other words, **the X chromosome is probably a very ancient chromosome.**

## 6.21 Sex Linkage and the Chromosome Theory

The development in the early part of this century of the **chromosomal theory of inheritance,** which proposes that genes reside in chromosomes, made elegant use of sex linkage. We have already considered the example of the *Bar* mutation in Section 6.19. L. V. Morgan carried this demonstration still further in studies of exceptional strains of *Drosophila* in which the females transmit sex-linked traits to their daughters rather than to their sons. Specifically, when the females are wild type and are crossed to *Bar*-eyed males, the $F_1$ consists solely of wild-type females and *Bar*-eyed males, a result that is the reverse of the normal situation (convince yourself of this with diagrams).

Morgan examined the chromosomes of the exceptional females and found that their two X chromosomes, instead of being separate as they are normally, were permanently joined at or near their centromeres. This attachment, she further observed, prevented the two chromosomes from separating from one another at metaphase. Finally, she noted that females with such **attached X** chromosomes ($\widehat{XX}$) also carried a Y chromosome.

By knowing the karyotype of the exceptional females, we can interpret their unusual pattern of gene transmission. First, it is clear that they should produce two types of egg in equal numbers: $\widehat{XX}$ eggs and Y eggs, as shown in Figure 6-20. When the females are crossed with normal males, therefore, four types of zygote are possible.

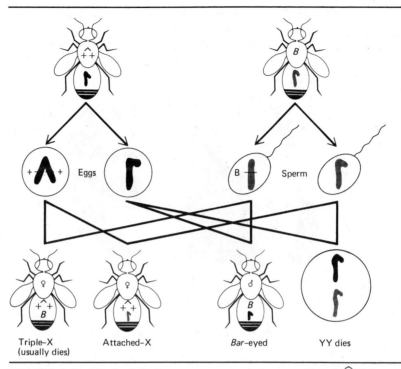

**FIGURE 6-20 Attached X chromosome cross.** An exceptional X͡XY *Drosophila* female (black chromosomes) is crossed to a *Bar*-eyed male (gray chromosomes).

1. Fertilization of an X͡X egg by a Y-bearing sperm will produce an X͡XY zygote, which should develop into a normal female (recall that the Y plays no role in sex determination in flies).
2. Fertilization of an X͡X egg by an X-bearing sperm will produce an X͡XX zygote, which is usually inviable or matures into a peculiar female fly with distinctive characteristics.
3. Fertilization of a Y egg by an X-bearing sperm will produce a normal XY male.
4. Fertilization of a Y egg by a Y-bearing sperm will produce a YY zygote, inviable because the missing X chromosome carries genes that are essential for survival.

These possibilities are summarized in Figure 6-20.

Thus it is clear that among the four possible types of offspring only two, namely X͡XY females and XY males, are viable. In a cross between wild X͡XY females and *Bar*-eyed males, therefore, only wild females and *Bar*-eyed males are expected, and this is what Morgan observed. In other words, these experiments again show a direct relationship between a visible chromosomal abnormality and the transmission of specific hereditary traits, and they therefore offer an unambiguous demonstration that chromosomes are indeed the vehicles of gene transmission: their behavior at meiosis determines the patterns of inheritance.

(a)            (b)

Thomas Hunt Morgan (a) and Lilian Vaughan Morgan (b) were pioneers in establishing the genetics of *Drosophila melangaster.* (Photographs were generously provided by the Morgan family.)

# Questions and Problems

1. (a) A mutant strain of *Chlamydomonas* having paralyzed flagella (*pf*) was crossed to the motile, wild-type strain and 100 tetrads were analyzed. Each tetrad gave a one-to-one ratio of paralysis to motility. Explain this result in terms of the behavior of a pair of homologous chromosomes during meiosis. (b) Instead of analysis of 100 tetrads, as in (a), assume that one meiotic product was recovered at random from each of the 100 tetrads. What would be the ratio of paralyzed to motile?

2. A cross was made between two *Drosophila* that had the wild phenotype. Their progeny were found to have the same phenotype. A sample was taken of 200 of the progeny, each of which was crossed with a fly having a purple eye color. Half of the crosses gave only wild-type flies and the other half gave 50 percent wild-type and 50 percent purple-eyed progeny. What were the genotypes of the original pair of wild-type flies?

3. A cross was made between wild-type (+) *Neurospora* and a mutant strain that cannot synthesize the vitamin thiamin (*thi*). Tetrad analysis showed a one-to-one segregation for +: *thi*. It was also found that 20 percent of the tetrads showed second-division segregation. How were these tetrads recognized, and what are the events in meiosis that lead to second-division segregation?

4. Human albinism is the absence of pigment from the hair, skin, and eyes. Determine from the pedigree given below whether it is a dominant or a recessive gene and whether it is sex linked.

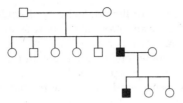

5. A marriage between two brachydactylous heterozygotes, as postulated in the text, has in fact never been reported, and thus it is not known whether *B/B* homozygotes would be brachydactylous or would instead suffer from more severe disabilities. If *B/B* homozygotes are in fact inviable and die at an early embryonic stage, then what proportion of pregnancies would you expect to end in abortion in the proposed mating? Of the couple's viable children, what would be their expected genotypes and the proportions of each?

6. List the nine different genotypes emerging from the cross summarized in Table 6-2.

7. In *Chlamydomonas* a mutant gene *arg* assorts independently of a mutant gene *pf*. What types of tetrads are expected from the cross *arg* + $\times$ + *pf*? How many different genotypes will there be and in what ratio?

8. The tetrads formed in *Neurospora* are ordered as a result of meiosis in the ascus. Assume that one mating type of this organism carries the mutant gene *a* and the wild-type allele of mutant gene *b*, whereas the other mating type carries the wild-type allele of the mutant gene *a* and the mutant gene *b*. A cross, *a* + $\times$ + *b*, gave the following kinds and numbers of ordered tetrads:

| A | B | C | D | E | F |
|---|---|---|---|---|---|
| *a* + | *a b* | *a* + | *a b* | *a* + | + + |
| *a* + | *a b* | *a b* | *a* + | + + | *a* + |
| + *b* | + + | + + | + *b* | *a b* | + *b* |
| + *b* | + + | + *b* | + + | + *b* | *a b* |
| 134 | 132 | 105 | 108 | 11 | 10 |

(a) Show how the data reveal that *a* and *b* are assorting independently.

(b) Classify each tetrad by name and show with diagrams the *simplest* origin for each. Indicate for each tetrad whether segregation occurs at either the first or second division for the *a* locus and *b* locus.

(c) Are all the expected tetrad classes present? If not, which are missing and in what proportions would you expect them to appear?

(d) Instead of analyzing 500 tetrads, assume that one ascospore was recovered at random from each of 1000 asci. How many genotypes would you find and how many of each kind?

9. A cross between two wild-type flies gave progeny all of which were wild type. When they were test-crossed to $\dfrac{vg \quad e}{vg \quad e}$ flies, the following results were obtained:

(a) 1/4 of the test crosses gave wild type, *vestigial ebony, vestigial,* and *ebony* in a 1:1:1:1 ratio.

(b) 1/4 of the test crosses gave all wild type.

(c) 1/4 of the test crosses gave *vestigial* and wild type in a 1:1 ratio.

(d) 1/4 of the test crosses gave *ebony* and wild type in a 1:1 ratio.

What are the genotypes of the original pair of wild-type flies?

10. Assume that the gene $r$ and its + allele show semidominance with respect to flower color such that $+/+$ is red, $+/r$ is pink, and $r/r$ is white and that the gene $s$ and its allele show semidominance with regard to seed color such that $+/+$ has red-black seeds, $+/s$ pink seeds, and $s/s$ white seeds. Give the phenotypes and their ratios expected among the progeny in the following crosses:

(a) $\dfrac{+}{+}\ \dfrac{s}{s} \times \dfrac{r}{r}\ \dfrac{+}{+}$

(b) $\dfrac{+}{r}\ \dfrac{+}{s} \times \dfrac{+}{r}\ \dfrac{+}{s}$

(c) $\dfrac{+}{r}\ \dfrac{+}{s} \times \dfrac{r}{r}\ \dfrac{s}{s}$

(d) $\dfrac{r}{r}\ \dfrac{+}{s} \times \dfrac{+}{r}\ \dfrac{s}{s}$

11. The two homologues drawn below undergo a single crossover between the centromere and the $a$ locus. Explain why you could detect the occurrence of this crossover in *Neurospora* but not in *Chlamydomonas*.

12. Thalassemia is a type of human anemia rather common in Mediterranean populations, but relatively rare in other peoples. The disease occurs in two forms, minor and major; the latter is much more severe. Persons with thalassemia major are homozygous for an aberrant recessive gene; mildly affected persons (with thalassemia minor) are heterozygous; persons normal in this regard are homozygous for the normal allele. The following four questions relate to this situation:

   (a) A man with thalassemia minor marries a normal woman. With respect to thalassemia, what types of children, and in what proportions, may they expect (let $t$ = the allele for thalassemia minor and $T$ = its normal allele)?

   (b) Both father and mother in a particular family have thalassemia minor. What is the chance that their baby will be severely affected? Mildly affected? Normal? Diagram the possible germ-cell unions in this family.

   (c) An infant has thalassemia major. According to the information so far given, what possiblities might you expect to find if you checked the infant's parents for anemia?

   (d) Thalassemia major is almost always fatal in childhood. How does this fact modify your answer to (c)?

13. In humans the most frequent type of albinism (itself quite rare) is inherited as a simple recessive characteristic. Standard symbols are $C$ = normal pigmentation; $c$ = albino. Assume that the genes for thalassemia and albinism assort independently.

(a) A husband and wife, both normally pigmented and neither with severe anemia, have an albino child who dies in infancy of thalassemia major. What are the probable genotypes of the parents?

(b) If these people have another child, what are its chances of being phenotypically normal with respect to pigmentation? Of having entirely normal (that is, nonthalassemic) blood? Of being phenotypically normal in both regards? Of being homozygous for the normal alleles of both genes?

14. The following is a pedigree of a fairly common sex-linked trait.

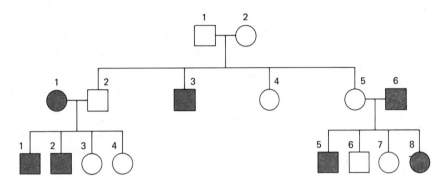

(a) Is the gene that causes the trait dominant or recessive? Explain.

(b) Does this data prove that the gene is sex linked?

(c) Designate the genotype for I 1, 2; II 2, 5, 6; III 1, 8.

15. A cross was made between two wild-type *Drosophila*. Their progeny were 187 *raspberry* eye color males, 194 wild-type males, and 400 wild-type females. Is *ras* a sex-linked gene? Explain. What are the parental genotypes? What are the genotypes of the $F_1$ wild-type females and what is their ratio?

16. A cross was made between a female heterozygous for the recessive genes *ct* (*cut* wings) and *se* (*sepia* eye color) and a *sepia* male. Among their female progeny the phenotypes were 1/2 wild type and 1/2 *sepia*. Among their male progeny the phenotypes were 1/4 wild type, 1/4 *cut*, 1/4 *sepia*, and 1/4 *cut* and *sepia*. Are either of the genes sex linked? What are the genotypes of the parents and their offspring?

17. If a woman having normal color vision has a color-blind father, what is the probability that her sons will be color blind if she marries a man with normal color vision? What genotypes are possible among her male and female offspring? What is the probability of her having a color-blind child if she marries a color-blind man, and is this probability different depending on whether the child is a male or a female?

18. Neither Tsar Nicholas II nor his wife, Empress Alexandra, nor their daughter, Princess Anastasia, had the disease hemophilia. However, their son, the Tsarevich Alexius, did have the disease. Can one automatically assume that Anastasia was a carrier? Why or why not?

19. How could you distinguish from pedigree studies between a sex-linked recessive factor and a dominant autosomal factor expressed only in males?

20. In *Drosophila*, the gene for *red* eye is dominant to its *white* allele and the gene for *long* wing is dominant to its *vestigial* allele. The $+/w$ locus is on the X chromosome; the $+/vg$ locus is not on a sex chromosome. Two *red*-eyed *long*-winged flies bred together produced the following offspring:

Females:   3/4 *red long,* 1/4 *red vestigial*
Males:   3/8 *red long,* 3/8 *white long,* 1/8 *red vestigial,* 1/8 *white vestigial*

What are the genotypes of the parents?

21. Diagram first- and second-division segregation when a maternal chromosome carries genes *F* and *g* and its paternal homologue carries alleles *f* and *G*.

22. For Figure 5-4b, imagine that the paternal and maternal homologues are heterozygous at the loci marked 1, 2, and 3. (a) Assign hypothetical genotypes to the four original sister chromatids at loci 1 to 3. (b) Assuming each chiasma generates recombinant chromatids, diagram the genotypes associated with centromeres A, A', B, and B' at the onset of metaphase I for each of the three bivalents shown in Figure 5-4b.

23. A pair of alleles is always observed to segregate at the first meiotic division. How might this observation be explained?

24. A woman's maternal grandfather is hemophilic, but she is normal, as are her parents and her husband. Would you expect her sons to be hemophilic? If so, in what proportions?

25. The following is a pedigree for a rare sex-linked human trait. Indicate with a dot those females who are carriers and with a P those females who are potential carriers.

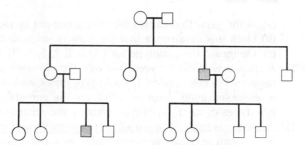

26. A son born to phenotypically normal parents has Klinefelter syndrome (*Section 6.17*) and hemophilia. In which meiotic division of which parent did nondisjunction of the X chromosome (*Section 5.15*) most likely occur? Explain and state your assumptions.

# 7

# Mutation: Induction and Detection of Mutant Organisms and Chromosomes

**A. Introduction**
7.1 Definitions and Terms

**B. Screening Procedures**
7.2 Visible Mutations
7.3 Nutritional Mutations
7.4 Lethal Mutations
7.5 Conditional Mutations
7.6 Mutations to Resistance and Dependency
Fluctuation test

**C. Characterizing Mutant Karyotypes**
7.7 Deficiencies and Duplications
7.8 Inversions
7.9 Translocations and Partial Trisomy

**D. Mutagens, Clastogens, and Carcinogens**
7.10 Generating Mutant Organisms
7.11 Testing Mutagens
Specific locus test
Ames test
Sister chromatid exchange test
Chromosome aberration test
7.12 Carcinogenesis
7.13 Birth Defects

**E. Questions and Problems**

## Introduction

Mutations—heritable changes in the sequence or number of nucleotides in a genome—are "the stuff" of both genetic analysis and evolution. In the genetic crosses followed in the previous chapter, mutant alleles were indispensable: by serving as markers for particular traits, they in fact defined the gene loci of interest. In evolution, mutant genes are thought to provide the raw material upon which natural selection can act to provide fitter organisms for particular environments (*Chapter 27*).

The present chapter considers mutations from a phenomenological point of view. We describe how mutant organisms are **screened** from the bulk of wild-type organisms in a population. We show how mutant cells

201

and organisms are characterized, paying particular attention to analyses of mutant karyotypes. We distinguish **spontaneous** from **induced** mutations and describe how **mutagens** are used to increase the yield of mutant organisms. Finally, we introduce an intensively active field in which **environmental mutagens** are being screened for their potential activity as **carcinogens.** In the following chapter, mutagenic mechanisms are considered at the molecular level.

# 7.1 Definitions and Terms

As summarized in Figure 7-1, mutations are commonly classified under three broad categories.

1. **Point mutations** affect small regions of a chromosome. For example, the **substitution** of one nucleotide for another or the **deletion** or **addition** of one or several bases may bring about point mutations.
2. **Chromosomal mutations** affect larger regions of a chromosome and occur because of some kind of breakage. Examples include **inversions,** wherein a section of a chromosome is inserted backward, and **translocations,** wherein a piece of one chromosome attaches onto another.
3. **Genomic mutations** affect the number of chromosomes present. Included are cases known as aneuploidy, wherein the genome comes to gain or lose a single chromosome, and cases known as polyploidy, wherein the overall chromosome number is doubled or tripled. The generation of $2n + 1$ **trisomy** and $2n - 1$ **monosomy** during errors of meiosis has already been described in Sections 5.15 and 5.16.

In defining a mutant organism, reference is usually made to some "normal" standard with which the mutant organism is compared and contrasted. As noted in Section 6.4, geneticists have adopted the term **wild type** to refer to such a standard. The term wild type originally indicated the sort of organism found in nature—in the wild—as opposed to the type produced in the laboratory or kept under domestication. In fact, however, the term is often applied to the stock maintained in the laboratory from which mutant stocks are derived. With years of laboratory existence the original stock may change; hence its wild-type appellation is often quite arbitrary. Moreover, different laboratories working with the same species may use different wild-type strains. The wild-type term is thus often loosely defined, but it is nonetheless useful.

With wild type as a reference, point mutations are commonly classified as being **forward mutations** if they proceed in the wild-type $\longrightarrow$ mutant direction and as **reverse mutations** or **reversions** if they proceed in the mutant $\longrightarrow$ wild-type direction. Reversion was encountered in Chapter 1 as the process that caused *rough* pneumococcus to resume the capsule characteristics of the *smooth* strain from which it derived. Strictly speaking, a mutant virus or organism is said to have undergone a reversion when a second mutation restores exactly the base sequence of its parental

POINT MUTATIONS AND THEIR CORRECTION

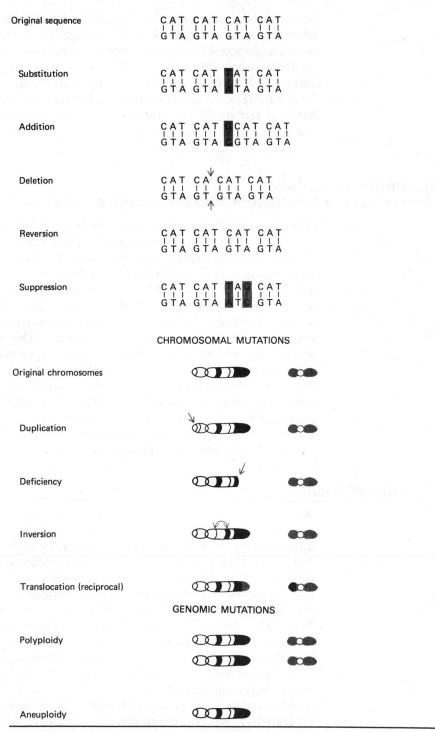

FIGURE 7-1 Point, chromosomal, and genomic mutations.

strain. Thus if the parent is CATCAT and the mutant is CAACAT, the revertant will be CATCAT. Reversion is often, however, mimicked by **suppression.** Suppression results when a second mutation occurs at a site in the chromosome different from the first mutation site and is in some way able to mask, or suppress, the phenotypic expression of the first mutation. Thus suppression will produce an organism that appears to be reverted but is in fact doubly mutant. Reversion and suppression can be distinguished from one another in genetic crosses, as described in later chapters.

# Screening Procedures

Screening procedures are essential for two related types of study. In the first, the geneticist wishes to isolate a mutant strain with a particular phenotype. Since the mutation process is random, the geneticist is in most cases unable to control which gene will become mutant (exceptions to this rule are given in Section 8.7). Techniques must therefore be devised to screen a large number of potentially mutant organisms for the mutant phenotype of interest. In the second, the geneticist wishes to determine whether a given chemical or agent is mutagenic. Organisms of one phenotype are therefore exposed to the putative mutagen, and their descendents are screened for mutants. In this case the important features of the mutant phenotype are that it be readily detectable and the result of a single mutational event, making it possible to calculate how many **mutational hits** are produced by a particular dose of mutagen.

The screening procedures described in the following sections are used for one or both of these purposes.

## 7.2 Visible Mutations

**Visible** mutations affect a morphological trait, and screening is done by inspection. The rare white-eyed fly among a group of red-eyed flies seen under a low-power microscope can be readily recognized and set aside, and thus visibles constitute ideal kinds of mutations for many sorts of genetic studies. In diploid organisms, a visible mutation is often recessive, so that it is expressed only in homozygotes. Some mutations, however, affect the phenotype of heterozygotes, an example being the *B* duplication in *Drosophila*, which leads to the formation of *Bar*-shaped eyes (*Figure 6-16*). Recessive visibles will also, of course, be detected in male diploid organisms if the mutant gene resides in the X chromosome (*Section 6.18*).

Haploid microorganisms can be screened for changes in their colony morphology, and these mutations can also be classified as visibles. Wild-type *Neurospora*, for example, will produce the long, fluffy hyphae depicted in Figure 7-2a, whereas the strain *abn-1* produces minute hyphae under the same growth conditions (Figure 7-2b). Similarly, yeast colonies may be

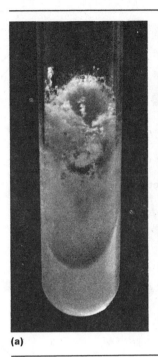 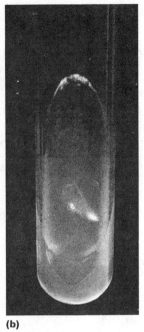

**FIGURE 7-2 Visible mutations in *Neurospora*.** (a) Wild-type hyphae. (b) Small hyphae produced by the mutant strain *abn-1*. (Courtesy of Dr. D. J. L. Luck).

(a)         (b)

classified as normal or as *petite*, and *C. reinhardi* may lose, by mutation, the ability to synthesize chlorophyll in the dark and will form yellow colonies instead of green.

Visible mutations are encountered even in the case of viruses. A Petri plate containing agar and growth medium (**nutrient agar**) can be prepared and inoculated with *E. coli* to provide an even growth of bacteria that will cover the agar surface—a **bacterial "lawn."** If a dilute suspension of wild-type phage T2 (*Section 1.8 and Figure 1-13*) is mixed with the bacteria at the time of inoculation, each virus will encounter, infect, and eventually lyse a single bacterium, releasing about 100 progeny. The progeny can, in turn, infect 100 adjacent bacteria, and so on, so that a clear, bacteria-free **plaque** soon forms within the lawn (Figure 7-3). Certain mutant strains of T2 lyse the bacterial cells so rapidly that during the time a wild-type ($r^+$) phage produces a small plaque with a characteristic fuzzy edge, these rapid-lysis or *r* mutants form a large, clear plaque (Figure 7-3). A single *r*-type phage can therefore be distinguished from many thousands of wild-type phages by the abnormally large plaque it (and its progeny) produces.

## 7.3 Nutritional Mutations

A **nutritional** or **biochemical** mutation affects an organism's ability to produce a molecule (for example, an amino acid) essential for growth. Nutri-

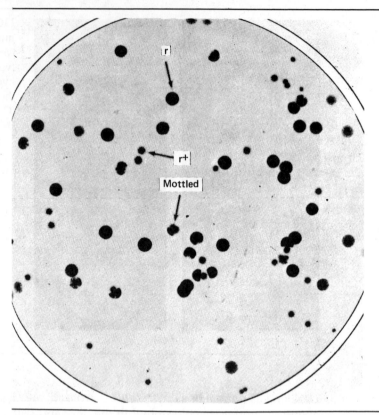

**FIGURE 7-3 Visible mutations of phage.** Plaques produced by wild-type ($r^+$) and mutant ($r$) T2 phages. The mottled plaques are produced when both $r$ and $r^+$ phages grow simultaneously in the same plaque; phages that produce such plaques are described in Section 17.8. (From *Molecular Biology of Bacterial Viruses* by Gunther S. Stent. W. H. Freeman and Company. Copyright © 1963.)

tional mutations are most commonly studied in organisms such as *E. coli*, *Chlamydomonas*, *Neurospora*, and yeast, which can grow in the laboratory on well-defined media. Mutant strains that require a nutritional supplement for growth are termed **auxotrophic**, in contrast to **prototrophic** wild-type or revertant strains. The replica plating technique, described in Figure 4-12, is commonly used to detect auxotrophic microorganisms: a population is first grown on nutrient agar supplemented with the molecule in question; colonies are then replicated to media that lack the supplement, allowing the detection of nongrowing isolates (*see, for example, Figure 6-7*).

Screening for mutant strains of microorganisms with metabolic abnormalities can sometimes be accomplished by the use of special media. Thus *E. coli* cells can be grown on a medium containing the sugar lactose and the reddish-brown indicator dye **2,3,5-triphenyl tetrazolium chloride.** Colonies that are able to ferment lactose ($lac^+$) cause the pH of the medium to drop and thereby bleach the dye so that the colonies appear white, whereas mutant *lac⁻* colonies that cannot ferment lactose from their surroundings appear reddish-brown.

**Enrichment for Auxotrophs** Screening for rare mutant organisms in a population is facilitated considerably when an **enrichment procedure** is devised to increase the proportion of mutant to wild-type organisms present. Two examples can be given. In the first, auxotrophs of *E. coli* or other Gram-negative bacteria can be selectively enriched by placing mutagenized cells in an unsupplemented medium that contains the antibiotic penicillin. Penicillin prevents the formation of cell walls; therefore, wild-type cells that attempt to divide in the presence of penicillin are promptly killed. Auxotrophs, on the other hand, cannot divide in the unsupplemented medium and thus remain alive (although they are slowly starving). The culture is then plated on supplemented, penicillin-free medium, and colonies are allowed to form. These are tested by replica plating (*Figure 4-12*) to determine the specific nutritional requirements they carry.

A second enrichment procedure, utilized both for bacteria and for cultured eukaryotic cells, employs **5-bromodeoxyuracil (5-BUdR)**, an analogue of thymine (as seen in Figure 1-1, thymine is 5-methyluracil) with a bromine atom in place of the methyl group on carbon 5 (Figure 7-4). A mutagenized culture is incubated in minimal medium and presented with 5-BUdR. Wild-type cells incorporate the analogue into their daughter chromosomes, whereas mutant, nongrowing cells do not. The culture is then exposed to near ultraviolet (uv) light (313 nm). 5-BUdR molecules absorb light at this wavelength much more strongly than do the usual bases; therefore, cells that have incorporated 5-BUdR into their chromosomes are much more likely to be photolyzed (literally, "light-broken") by uv exposure than are 5-BUdR–free cells. Thus the mutants survive, whereas the wild-type cells die, the key feature of a successful nutritional enrichment procedure.

## 7.4 Lethal Mutations

A gene that has undergone a **lethal mutation** is typically incapable of producing an active form of an indispensable protein. In haploid organisms this means that the mutant itself or its immediate mitotic progeny will not

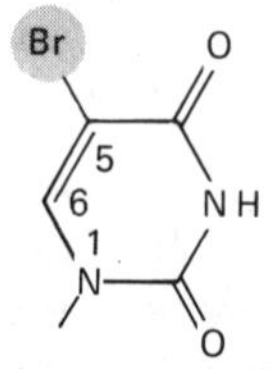

FIGURE 7-4 5-bromodeoxyuracil (5-BUdR).

survive. Mutations in "indispensable" genes must therefore be isolated in haploids as **conditional lethals,** as described in Section 7.5.

Lethal mutations in diploids possess the ability to kill the organism directly or to prevent it from reproducing (**genetic death**); the latter organisms are more often known as **steriles.** A dominant lethal or sterile mutation in a diploid organism cannot be maintained past one generation; however, this is not true of recessive lethals. A new germ-line recessive lethal mutation in a diploid organism will not affect the viability of the original parent, nor will it affect the viability of the next generation, since the gamete carrying the lethal mutation ($l$) is almost certain to fuse with a gamete carrying its dominant allele (+) to produce a viable $+/l$ heterozygote. Indeed, the recessive lethal can be maintained indefinitely in the heterozygous condition by appropriate crosses. It is only when $+/l$ heterozygotes are crossed that the presence of the lethal is detected as an inviability for one quarter of the progeny (those with an $l/l$ genotype). If the lethal gene is carried by the X chromosome, moreover, a female with a $+/l$ genotype will produce only half the expected number of sons. You should work these ratios out yourself using the procedures outlined in Sections 6.6 and 6.18.

**Sex-linked Recessive Lethal (SLRL) Selection**    Screening for sex-linked recessive lethal mutations in *Drosophila* can be facilitated by the use of stocks carrying special chromosomal mutations. One such chromosome is the **Basc X** chromosome of *D. melanogaster,* so named because it carries the dominant *Bar* allele *B,* the recessive apricot eye color allele $w^a$, and an inversion involving the *scute* (*sc*) region of the chromosome. The *scute* inversion effectively prevents meiotic crossing over in the left arm of the X chromosome, for reasons detailed in Section 7.8. Therefore, females heterozygous for a *Basc* X chromosome do not generate eggs carrying recombinant X chromosomes, and any progeny fly with *Bar* eyes can be automatically assumed to carry an intact *Basc* chromosome.

The *Basc* chromosome can be used to screen for recessive lethal mutations in the *D. melanogaster* X chromosome in the fashion diagrammed in Figure 7-5. Wild-type males are mutagenized and crossed with females homozygous for *Basc* (Figure 7-5a). The $F_1$ females are collected as virgins (they are not allowed to mate with their brothers) and are individually crossed with wild-type males (Figure 7-5b). Those females possessing an X chromosome that carries a recessive lethal can be identified by their distinctive male progeny: none of their sons will be wild type; all their sons will be *Bar, apricot* (Figure 7-5c).

## 7.5 Conditional Mutations

A **conditional mutation** affects an organism's phenotype under **restrictive** growth conditions but is without such an effect under **permissive** growth conditions; the wild-type version of this gene, by contrast, is expressed equivalently under both growth conditions. The most widely employed

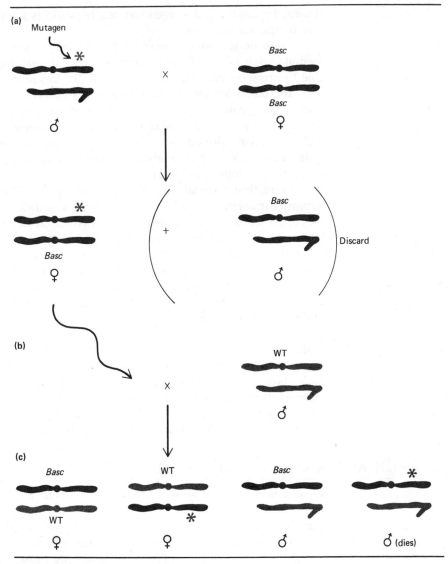

**FIGURE 7-5** *Basc* **chromosome** of *Drosophila* used to screen for a recessive lethal mutation (gray star). Y chromosome is symbolized by ⟶.

environmental variable is temperature. Thus a conditional mutant may exhibit a phenotype at a 37°C permissive temperature but fail to exhibit the phenotype if shifted to a 47°C restrictive temperature, whereas the wild type displays the trait at both temperatures. Strains sensitive to temperature **increase** are termed **temperature sensitive** (*ts*). For **cold-sensitive** strains, by contrast, the restrictive temperature is **lower** than the permissive temperature.

Conditional mutants may be distinguished by other sets of environmental variables as well. For example, a conditional phage mutant may be able to grow on one bacterial strain but fail to grow on another.

To screen for temperature-sensitive strains, a mutagenized population is first grown at permissive temperature. The environment is then warmed to the restrictive temperature, and the population is screened for organisms with altered phenotype. Finally, the selected mutants are retested at permissive temperature to confirm that the phenotypic change is induced by temperature shift.

**Conditional mutations allow the analysis of "essential genes" in haploid organisms** whose mutational defects cannot be corrected by suitable supplements to the growth medium. An organism carrying such a mutation can be maintained and cloned under permissive conditions; then, when the effects of the mutation are to be studied, the strain is shifted to restrictive conditions. Conditional mutants become particularly important for identifying elusive gene products: these products should similarly be functional under permissive conditions and nonfunctional (or nonexistent) under restrictive conditions. If, for example, we wish to identify the DNA polymerase that replicates a bacterial chromosome, we would first select *ts* mutants that cannot synthesize DNA, and therefore cannot grow, at restrictive temperatures. We would then make cell extracts of these mutants under the two growth conditions and look for a strain in which the polymerase fraction is active in the permissive-temperature sample but inactive in the high-temperature sample. The polymerase species that exhibits this behavior could then be purified from the extracts.

## 7.6 Mutations to Resistance and Dependency

Mutations are widely known that confer **resistance** to a drug, to infection by a virus, or to other agents that are normally toxic. Less commonly, mutant organisms will exhibit a **dependency** on such agents for growth. Screening for these mutant strains is a simple matter: a population of cells or organisms is exposed to the agent, and only those that are resistant, or those that are both resistant and dependent, will survive. The dependent strains can subsequently be recognized by their inability to grow in the absence of the agent.

Analysis of many drug-resistant strains has revealed the existence of three general classes: (1) strains that have become impermeable to the drug; (2) strains that have acquired an enzyme to inactivate the drug once it enters the cell; and (3) strains whose intracellular targets have become resistant to the effects of the drug. The third class is of particular interest. Many drugs act by inhibiting essential enzymes such as polymerases or by blocking the function of essential organelles such as ribosomes. Thus a comparison of resistant and sensitive strains often yields important information about how these enzymes or organelles function. For example, the ribosomes from a streptomycin-resistant mutant strain of *E. coli* are found to contain a mutant version of the ribosomal protein S12: its electrophoretic mobility is altered, as is its binding of the antibiotic. Thus S12 is immediately identified as a target for the drug.

**The Fluctuation Test**   When toxic—and potentially mutagenic—agents are used to screen for mutant strains, a fundamental question often arises: did the mutations pre-exist in the population of cells before exposure to the selective agent, or did the agent itself cause the mutations? If, for example, a sample of allegedly streptomycin-sensitive *E. coli* cells is plated on a medium containing streptomycin and some streptomycin-resistant colonies grow up, does this mean that some streptomycin-resistant cells were in fact present in the original sample, or were mutations to resistance induced by the streptomycin in the plating medium?

Such a question is best answered by a **fluctuation test,** first devised by S. Luria and M. Delbrück. In the experiments we describe, Luria and Delbrück were interested in learning whether the presence of the bacteriophage T1 elicited the formation of T1-resistant *E. coli.* They first inoculated a few cells into a single flask containing liquid culture medium, allowed the culture to grow until it contained millions of cells, and then plated ten 0.05 ml samples of the culture on agar media containing T1 phages. As seen in Table 7-1, column a, the number of phage-resistant colonies that arose on each plate was about the same—an average, or **mean,** of about 51 with a variance of 27. (The variance describes the way values are dispersed about a mean.)

Luria and Delbrück next inoculated ten *independent* culture flasks with a few cells, allowed each culture to grow, and plated one 0.05-ml sample from each culture to the T1-containing medium. In this case the results were very different (Table 7-1, column b). The average number of resistant colonies per sample happened to be about the same as in the control, but the variance was now 3498. The individual sample values, in other words, fluctuated greatly.

From such results one can argue as follows. If mutation to resistance is indeed caused by exposure to the phages on the plate, the bacteria in each plated sample should have the same chance of experiencing a mutational event, and the variance in the second experiment (Table 7-1, column b) should be similar to the variance in the control experiment (Table 7-1, column a). Since this is not the observed result, the hypothesis of a virally induced mutation to resistance is not supported. If, on the other hand, mutation to resistance is a random event that can occur at any time during the growth period in liquid culture, the marked fluctuation observed in the second experiment is precisely what one would predict: in one flask a mutation might happen to "hit" a bacterium early in the growth period, so that a large clone of resistant descendents will appear in the final sample, whereas in a second flask a mutation might not happen to hit any bacteria

TABLE 7-1  **Fluctuation Test.** Comparison of the
number of phage-resistant bacteria in different
samples of the same culture and in samples from
independent cultures

| Sample No. | (a) Samples from Same Culture | (b) Samples from Independent Cultures |
|---|---|---|
| | *Resistant Bacteria* | |
| 1 | 46 | 30 |
| 2 | 56 | 10 |
| 3 | 52 | 40 |
| 4 | 48 | 45 |
| 5 | 65 | 183 |
| 6 | 44 | 12 |
| 7 | 49 | 173 |
| 8 | 51 | 23 |
| 9 | 56 | 57 |
| 10 | 47 | 51 |
| Average per sample | 51.4 | 62 |
| Variance* | 27 | 3498 |

*The variance $V = \dfrac{\Sigma D^2}{n-1}$ where $D$ = the amount by which each value
differs from the mean, and $n$ = the number of values tabulated (*Section
21.13*).

After S. E. Luria and M. Delbrück, *Genetics* **28**:491, 1943, and
W. Hayes, *The Genetics of Bacteria and Their Viruses*, 2nd ed. Oxford:
Blackwell, 1968.

until late in the growth period and only a few resistant bacteria will be
present in the final sample.

In summary, then, a fluctuation test can have one of two outcomes. If
exposure to the selective agent is causing mutations, then this mutagenic
activity will dominate the experiment: samples exposed from a single culture will be mutated at the same rate as samples exposed from independent cultures, and the frequency of mutant organisms per culture will be
uniform in both cases. If exposure to the selective agent is not causing
mutations, on the other hand, then the occurrence of spontaneous mutation will dominate the experiment: in this case, the random nature of spontaneous mutations will be reflected in a wide **fluctuation** in the number of
mutant organisms in independent cultures compared with single cultures.

# Characterizing Mutant Karyotypes

In the remaining sections of this chapter, we assume that a screening procedure has been used and that a strain with a heritable mutant phenotype

of interest has been isolated. One would next like to identify the portion of the genome that carries the mutation. For organisms such as *Drosophila,* mice, and humans, it is often possible to learn, by studying a mutant karyotype, whether a piece of a chromosome has been lost or moved to another location, information that becomes enormously useful in understanding the nature and locus of the mutation. Obviously, this approach cannot be used for prokaryotes or for those eukaryotes that have tiny, indistinguishable chromosomes; the nature and loci of their mutations must in all cases be deduced by genetic crosses and molecular analyses.

## 7.7 Deficiencies and Duplications

A **deficiency** describes a chromosomal mutation in which a detectable sector of chromatin is deleted from a chromosome; a **duplication** refers to the addition of contiguous extra copies of a chromosomal sector (*Figure 7-1*). The occurrence of either aberration can sometimes be verified by examining chromosomes directly with the light microscope: the chromosome appears either longer or shorter than normal. Additions and deletions can also be detected in polytene chromosomes of the Diptera (*Section 2.14*), in which one or several bands may be missing or duplicated. Finally, it is possible to detect even very small deficiencies and duplications in G-banded or Q-banded chromosomes (*Section 2.11*). Such analyses have, for example, shown an association between the presence of a deficiency in the short arm of human chromosome 5 and the **cri du chat** syndrome, which includes severe mental retardation and a characteristic catlike cry in very young infants. It has also been demonstrated that Down's syndrome is sometimes produced when individuals are heterozygous for a duplication of band q22 in human chromosome 21. In other words, Down's syndrome can result from 21q22 triplication alone, although most affected individuals are instead trisomic for the entire chromosome 21 (*Section 5.15*).

Malignant cells are sometimes found to possess specific deficiencies or duplications. The white blood cells of patients with chronic myelogenous leukemia frequently exhibit a karyotype in which a portion of the long arm of chromosome 22 is deleted (producing what is known as the **Philadelphia chromosome**) and the missing piece is translocated to the end of the long arm of chromosome 9. Patients with a variety of lymphomas often display a duplication of bands in the long arm of chromosome 14. Whether such changes are a cause or a result of the malignancy is, however, unknown.

The presence of a deficiency or a duplication is particularly obvious in a dipteran organism heterozygous for the aberration. As noted in Section 2.14, somatic pairing occurs between homologous polytene chromosomes in the Diptera. Since pairing cannot take place along a deleted region in the heterozygote, the undeleted homologue simply buckles, as illustrated in Figure 7-6a and diagrammed in Figure 7-7. A dipteran heterozygous for a duplication will also exhibit a buckled region; in this case, however, the

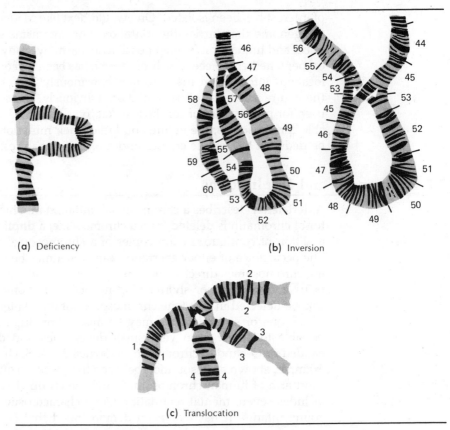

(a) Deficiency

(b) Inversion

(c) Translocation

**FIGURE 7-6 Chromosomal mutation heterozygotes,** showing polytene chromosomes of *Drosophila*. Numbers designate particular bands on particular chromosomes. (After T. S. Painter, *J. Hered.* **25**:464–476, 1934.)

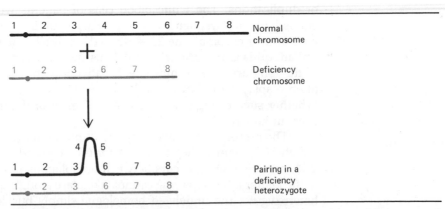

**FIGURE 7-7 Deficiency heterozygote,** showing aberrant pairing with its normal homologue.

buckled region will contain bands that are duplicates of adjacent regions of the chromosome, the adjacent regions being properly paired with the homologue.

**Duplications and deficiencies produce a change in the amount of genetic material present in a chromosome.** Higher organisms homozygous for a deficiency are usually inviable (exceptions include certain strains of corn), whereas organisms heterozygous for a deficiency are frequently viable. These results indicate that a single dose of certain genes may be sufficient for development and self-maintenance, whereas the total loss of those genes cannot be supported.

Small duplications, on the other hand, are not generally as harmful as deficiencies and may, under selective pressures, even become advantageous. A timely example concerns the drug **methotrexate,** widely used in cancer chemotherapy. Methotrexate acts by inhibiting dihydrofolate reductase, an important enzyme in purine biosynthesis and hence essential for DNA replication in the malignant tissue. It has long been known that with continued chemotherapy, higher and higher doses of methotrexate must be used to inhibit tumor growth, and eventually the drug must be discontinued because of its overall toxicity. R. Shimke and his collaborators have discovered that this methotrexate resistance develops by a selective duplication of dihydrofolate reductase genes; as a result, the cell produces more and more enzyme, and more and more methotrexate is required to inhibit activity. When the drug is discontinued and such "methotrexate pressure" is eliminated, the amplified genes very often revert to their original copy number. Similar duplication events are believed to account for many examples of resistance to pesticides and antibiotics.

Duplications are important evolutionary substrates. An organism carrying a duplication has, in effect, an extra portion of genetic material that is not essential to its development and reproduction. Mutations in this extra genetic material are thus much less likely to be deleterious than are mutations in its essential genes. Moreover, the duplicated region can undergo repeated mutation without being subjected to the usual negative selective pressures, and in this sense it enjoys a kind of independent existence within the genome. Should accumulated mutations produce a gene whose protein product is in some way beneficial to an organism, the organism will be provided with a selective advantage, and the new gene—for such it can be called—may become increasingly prevalent in a population. Specific examples of new genes produced by duplications are given in Sections 19.7 and 19.8.

## 7.8 Inversions

**An inversion changes the arrangement of the genetic material in a chromosome rather than the amount.** It is produced when a portion of the chromosome is broken, the broken piece assumes a reversed position, and the break is repaired (*Figure 7-1*). Should the inversion involve an internal segment of a chromosome, as it usually and perhaps always does, two breaks must occur, one at either end of the inverted segment.

An inversion is detected cytologically as a reversal in the band sequence of a polytene chromosome or a banded mammalian chromosome. It also causes a distinctive **looped configuration** in polytene or meiotic cells heterozygous for the inversion: synapsis between the inverted portion of a chromosome and its homologue can occur only if one chromosome loops back to invert itself in the affected region (Figure 7-8). Thus in pachytene preparations of inversion heterozygotes, the length and position of an inversion can frequently be estimated by examining the size of the loops that are present. Similar loops also form during somatic pairing of dipteran polytene chromosomes, in which the inverted region can be identified quite precisely by its banding pattern (Figure 7-6b). It is not uncommon for a second inversion to occur within a single, or simple, inversion, in which case a **secondary loop** will form at synapsis. In some inversion-carrying strains, therefore, extremely complicated synaptic patterns can be found.

An inversion may be **pericentric** (containing a centromere) or **paracentric** (not containing a centromere). If pericentric, it may produce a chromosome quite different from the original in appearance. A metacentric chromosome, for example, can be transformed into an acrocentric chromosome by a pericentric inversion that includes unequal lengths of the chromosome to the right and left of the centromere, as illustrated in Figure 7-9. Much evolution of the karyotype has presumably occurred by such chromosomal alterations.

**Inversions as Crossover Suppressors**    We noted when we considered the *Basc* X chromosome (*Figure 7-5*) that inversions tend to suppress meiotic

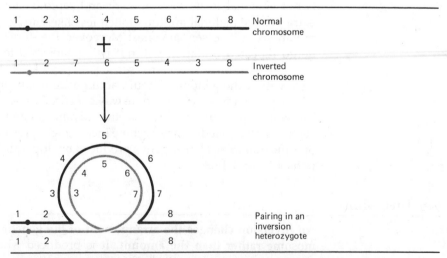

**FIGURE 7-8 Inversion heterozygote,** showing pairing with its normal homologue.

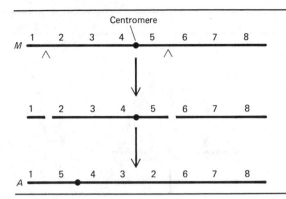

**FIGURE 7-9 Pericentric inversion** transforming a metacentric chromosome (M) into an acrocentric chromosome (A).

recombination. There are two reasons for this suppression. First, the looped synaptic configuration may itself discourage the establishment of chiasmata. Second, when crossovers do occur within inversion synapses, three kinds of products are generated (Figure 7-10): two nonrecombinant chromosomes, one **dicentric** chromosome, and one **acentric fragment.** The dicentric tends to snap into two unequal lengths as it attempts subsequent anaphase separations, and the cells that inherit such broken chromosomes are usually inviable. The acentric fragment cannot associate with spindle microtubules at all and is eventually lost from daughter cells. Therefore, only the nonrecombinant chromosomes are inherited, meaning that **the genes contained within an inversion are transmitted together as a package.** This **crossover suppression** effect of inversions appears to be utilized by *Drosophila*, at least, in adapting to different kinds of environmental situations (*Section 27.4*).

## 7.9 Translocations and Partial Trisomy

A second process that alters the arrangement of genetic material is **translocation:** a portion of one chromosome, or in some cases an entire chromosome, becomes physically associated with (usually) a nonhomologous chromosome (Figure 7-11). Translocations are commonly (but not invariably) **reciprocal,** meaning that each chromosome both donates and receives a piece of chromosomal material (*Figures 7-1 and 7-11*).

As with inversions, translocations can be recognized either as altered banding patterns in mitotic chromosomes or by the unusual synaptic patterns they assume. Thus an organism heterozygous for a translocation will exhibit, in meiotic or polytene cells, a **cross-shaped configuration** at the time of synapsis as the translocated chromosomal regions attempt to pair. Figure 7-6c illustrates the conformation assumed by reciprocal translocation chromosomes in polytene cells, and a detailed view of how such configurations are established is given in Figure 7-11.

Humans occasionally carry translocations; we can consider here two

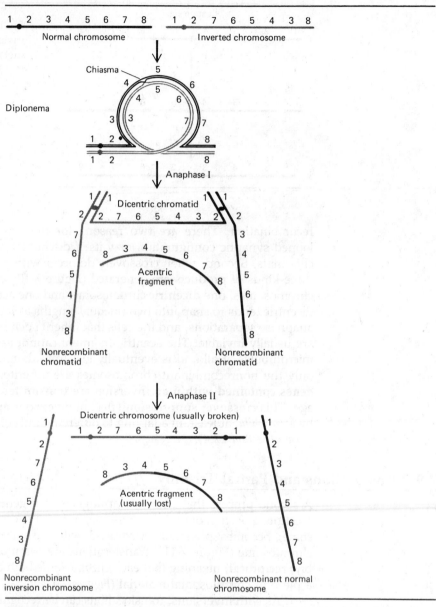

**FIGURE 7-10 Inversion crossover.** A normal and an inverted chromatid cross over in an inverted region that does not contain a centromere (a paracentric inversion).

examples. In the first, a woman carries a normal chromosome 14 and a normal chromosome 21, plus a translocation chromosome, designated t(14;21), that arose as a fusion between 14 and 21. This woman is called a **balanced translocation carrier** because she continues to carry a full diploid

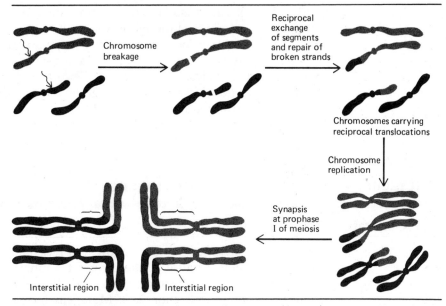

FIGURE 7-11 **Reciprocal translocation** between nonhomologues and its effect on meiotic synapsis.

complement of chromosomes, albeit a rearranged complement, and is phenotypically normal. When meiosis occurs in such carriers, however, errors frequently arise, presumably because the process of synaptic pairing is asymmetric and spindle orientation can become abnormal. (Convince yourself of this with diagrams.) As a result, this woman is likely to produce eggs carrying 14 but not 21, 21 but not 14, 21 + t(14;21) and so on. If one of her 21 + t(14;21) eggs fuses with a normal 14 + 21 sperm, the resultant child will inherit three copies of chromosome 21 (21 + 21 + t(14;21) and will display **translocation Down's syndrome.** An estimated 10 percent of individuals with Down's syndrome carry such translocation chromosomes.

In the second example, a man possesses a normal chromosome 4 and a normal chromosome 9, but a sector of the p arm (*Section 2.12*) of his second chromosome 4 is translocated to his second chromosome 9, meaning that he carries a deficiency chromosome we can designate as $4p^-$ and a translocation chromosome we can designate as t(9;p4). Again, this rearrangement of the genome does not affect his phenotype, and he is again referred to as a balanced translocation carrier. During spermatogenesis, however, roughly half of his sperm will come to carry the $4p^-$ deficiency chromosome rather than its normal homologue, and the resultant $4p^-/4$ offspring will display severe mental retardation and physical handicaps. Moreover, as in our first example, meiotic errors will not infrequently generate 4 + t(9;p4) sperm which, when combined with a normal 4 + 9 egg, will produce a child with an extra segment of chromosome 4. This child is said to be a **partial trisomic** for chromosome 4 and, once again, severe retardation and physical disability will occur.

A balanced translocation carrier is usually unaware of his or her condition until the birth of an aneuploid child or a series of miscarriages occurs and karyotypic analysis is performed. Once a couple learns that one of them carries a translocation, they can go on to produce normal children if they consent to amniocentesis (*Section 19.1*) and to the therapeutic abortion of any aneuploid fetus detected.

# Mutagens, Clastogens, and Carcinogens

## 7.10 Generating Mutant Organisms

**Spontaneous mutations** occur at a low, but measurable, rate in all organisms, presumably because of the error rates inherent in the replication and transmission of a genome; a detailed consideration of spontaneous mutation is found in Section 26.3. **Mutagens** increase mutation rates by stimulating the occurrence of such errors; **clastogens** form a subclass of mutagens that stimulate chromosome breakage. The mode of action of specific mutagens will be detailed in Chapter 8. Here we describe two important considerations in a mutagenesis experiment.

The first is the choice of mutagenic agent. Radiation (X rays, uv, or γ rays) is often effective in inducing chromosomal mutations, as are such chemicals as mitomycin C and nitrogen mustard. A variety of other chemicals (for example, 5-BUdR, aminopurine, nitrous acid, and ethylnitrosourea) are more active in inducing point mutations. It is not possible, however, to predict with certainty the type of mutation elicited by any mutagen: all are highly reactive agents with a variety of effects on a genome. Nor are experiments with one organism necessarily applicable to another: the compound diethylnitrosamine, for example, is a potent mutagen for *Drosophila* but is virtually ineffective in the mouse.

The second is the choice of dose. As mutagens interact with chromosomes, many of the modifications they induce will be **inactivating,** meaning that the altered genome is so crippled that it cannot be replicated and/or transmitted to progeny. If, in a mutagenesis experiment, the dose of mutagen is increased in the hope of obtaining large numbers of mutant organisms, the number of inactivating "hits" will increase as well, to the point at which most of the organisms are killed by the treatment. In practice, therefore, test organisms are first treated with a range of mutagen exposures, and the number of surviving progeny is plotted against dose (an **inactivation curve**). From such a plot is chosen a dose that gives some reasonable number of survivors, and this dose is used to induce mutations.

## 7.11 Testing Mutagens

If one's goal is not to obtain particular kinds of mutants for genetic studies but rather to test the mutagenic/clastogenic potential of a particular compound, then it becomes important to subject the compound to a highly standardized **mutation test.** In such tests, baseline spontaneous mutation rates are first carefully calculated, after which the levels of mutations stimulated by the agent are assessed. We describe here four that are widely used.

**The Mouse Specific Locus Test**   L. B. Russell and W. L. Russell designed this test to analyze **forward mutation rates** (wild type → mutant) in the mouse. They constructed a mouse tester strain homozygous for the seven recessive visible point mutations shown in Figure 7-12. Each mutant allele has a clear-cut effect on the mouse phenotype when homozygous, and none has important effects on viability. To establish spontaneous mutation rates at these loci, males that were homozygous wild type at each of these loci were crossed to females of the tester strain, and a total of 531,500 offspring were examined. Of these, 28 progeny proved to be phenotypically mutant (and therefore homozygous recessive) for at least one trait,

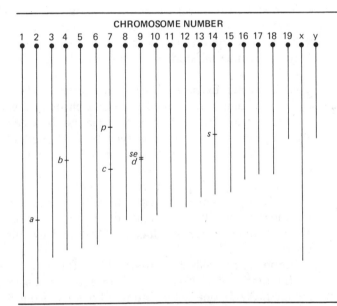

FIGURE 7-12 **Specific locus test.** Mouse karyotype, showing approximate positions of specific loci used in mutagenesis experiments. Abbreviations: *a,* nonagouti; *b,* brown; *c,* chinchilla; *p,* pink-eye dilution; *d,* dilute; *se,* short ear; and *s,* piebald.

giving an average of four spontaneous mutations per locus or about eight spontaneous mutations per locus per million tested male gametes.

This baseline rate can now be compared with rates obtained when the male wild-type mice are first exposed to mutagenic agents. The most potent chemical mutagen tested, ethylnitrosourea, can yield as many as 630 mutations per locus per million tested male gametes, an 80-fold stimulation. The most effective acute dose of X rays (600 R) can stimulate the rate 16-fold, whereas 600 R of chronic γ radiation stimulates the rate about 6-fold.

**The Ames Test**  B. Ames designed this test to screen environmental chemicals for their mutagenic effects. The test utilizes a series of *Salmonella typhimurium* strains designed to be maximally sensitive to the mutagenic action of test compounds. (1) Each strain carries a cell wall mutation that permits most chemicals to enter the cell. (2) They carry a mutation known as *uvrB* that eliminates most of the organism's ability to repair DNA damage once it is formed (repair is described in detail in Chapter 8). (3) They carry the plasmid pKM101, which, for as yet unknown reasons, stimulates the conversion of DNA damage into heritable mutations. (4) Finally, each carries a different *his⁻* mutation so that it is auxotrophic for histidine.

To test the mutagenic potential of a compound, these *his⁻* strains are simply mixed with the agent before plating. An increase in prototrophic colonies then signals a stimulation of the reversion mutation rate and indicates that the compound is acting as a mutagen. Applications of the Ames test to environmental studies are presented in Section 7.12.

**Sister Chromatid Exchange Test**  An alternate way to assess the mutagenicity of a compound, called the **sister chromatid exchange (SCE) test,** involves the cytological detection of exchanges between mitotic sister chromatids (a phenomenon first observed by J. H. Taylor).

In an SCE assay, cultured cells are allowed to undergo two rounds of mitosis in the presence of 5-BUdR (*Figure 7-4*). During the second cell cycle, a potential mutagen is presented to experimental but not to control cells. At the time of the second mitosis, all cells are blocked in metaphase with colchicine (*Section 2.9*). They are then treated with a fluorescent compound or fluorochrome known as **Hoechst 33258** and examined with a fluorescence microscope.

In the control cells it is observed that the two sister chromatids fluoresce differently (Figure 7-13a): as a consequence of semiconservative replication (*Section 2.3*), one sister chromatid possesses one 5-BUdR–containing strand, whereas the other chromatid possesses two such strands (convince yourself of this with diagrams), and the Hoechst reagent proves to fluoresce less strongly when bound to the doubly halogenated chromatid than when bound to the singly halogenated chromatid. In addition, many of the control cells contain sister chromatid pairs that have a checkerboard pattern (Figure 7-13a, arrows), which indicate that exchanges have occurred

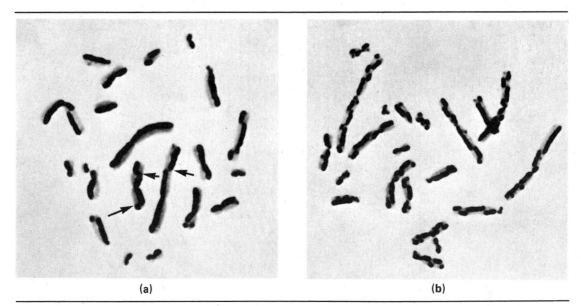

(a)                                              (b)

FIGURE 7-13 Sister Chromatid exchanges. (a) Metaphase cell from the Chinese hamster ovary (CHO) line after two rounds of replication in the presence of 5-BUdR followed by staining with Hoechst 33258 plus Giemsa. Only one chromatid of each sister chromatid pair is stained. Arrows point to several sister chromatid exchanges (SCEs); a total of 12 SCEs are present. (b) As in Figure 7-13a, but exposed to nitrogen mustard ($HN_2$) at $3 \times 10^{-6}$ M for the two cell cycles before sampling. The frequency of SCEs has increased approximately tenfold. (From P. Perry and H. J. Evans, *Nature* **258**:121, 1975.)

between them. Although the molecular basis for such sister chromatid exchanges is poorly understood, they unquestionably require the recombinational exchange of chromosomal material.

When treated cells are now examined, the SCE frequency is frequently enchanced (Figure 7-13b). Figure 7-14 summarizes data on the induction of SCEs by various mutagens. It is evident that mitomycin C and MNNG are particularly effective. The ability to generate SCEs usually correlates well with the overall mutagenicity of a given reagent, whereas, somewhat surprisingly, SCE generation correlates less well with the ability of a reagent to act as a clastogen.

**Chromosome Aberration Tests**  To date, the ability of a reagent to act as a clastogen requires tedious estimates of chromosome breakage frequencies. For humans, peripheral lymphocytes are cultured and blocked in mitotic metaphase with colchicine; hundreds of cells are then scored for the presence of such aberrations as dicentrics, acentric fragments, or chromosomes of abnormal length. Standard control estimates reveal that about 1 cell in

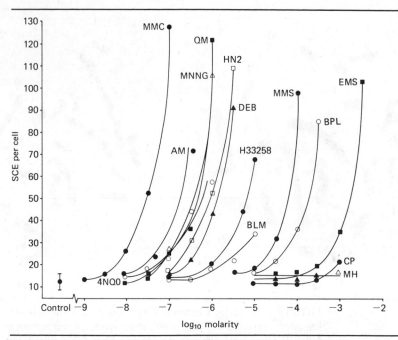

**FIGURE 7-14 Dose-response curves for SCE incidence** in cultured Chinese hamster ovary (CHO) cells against log of initial concentration. MMC, mitomycin C; AM, adriamycin; MNNG, N-methyl-N-nitro-N-nitrosoguanidine; QM, quinacrine mustard; 4NQO, 4-nitroquinoline 1-oxide; HN2, nitrogen mustard; DEB, diepoxy-butane; H33258, Hoechst 33258; BLM, bleomycin; MMS, methylmethane sulfonate; BPL, β-propiolactone; EMS, ethyl methane sulfonate; CP, cyclophosphamide; MH, maleic hydrazide. (From P. Perry and H. J. Evans, *Nature* **258**:121–125, 1975.)

1000 displays at least one such abnormality, a rate that is increased about 20-fold after *in vitro* exposure to 30 R of X rays.

## 7.12 Carcinogenesis

**Carcinogens** are agents that cause normal cells to become malignant such that they divide uncontrollably and possess the ability to spread and establish cell lines in inappropriate parts of the body (**metastasis**). Well-established carcinogens for humans include vinyl chloride and asbestos; some component(s) of cigarette smoke are also clearly carcinogenic. Present estimates hold that 80 to 90 percent of all human cancers are generated by carcinogens present in the human environment, and about one fourth of the current human population is expected to develop cancer at some time in their lives. Therefore, identifying carcinogenic agents and understanding their mode of action are key goals of health-related research.

The relevance of carcinogenesis to genetics lies in the fact that although potent mutagens are not necessarily strong carcinogens (MMS being one example), carcinogens are usually mutagens. This fact allows at least two interpretations, and since "cancer" is actually many hundreds of different types of related diseases, each interpretation may prove to be correct for different malignancies. One theory holds that carcinogens are such highly reactive molecules that even though their cancer-generating activity is unrelated to mutagenesis, mutations are likely to be generated once they enter a cell. The alternate **somatic mutation theory of cancer** holds that mutations are themselves able to precipitate malignancy.

One line of support for the somatic mutation theory of cancer comes from the phenotype of persons suffering from a group of diseases collectively known as **xeroderma pigmentosum (XP)**. As will be detailed in Chapter 8, XP persons are unable to repair ultraviolet-induced damage to their DNA and are presumably therefore highly susceptible to mutation induced by sunlight. It turns out that XP persons are also highly susceptible to numerous forms of skin cancer (squamous cell carcinomas, malignant melanomas, and many others), suggesting that the accumulation of sunlight-induced mutations eventually transforms a skin cell into a malignant cell.

Regardless of theory, the fact that carcinogens are usually mutagens is a strong rationale for screening environmental chemicals for their mutagenicity. Those found to be mutagenic are then given high priority for direct carcinogenicity tests. Since an adequate animal cancer test costs about $550,000 per chemical and takes about three years, an animal screen of the 1000 new chemicals synthesized each year would be prohibitively costly. Although many quick and inexpensive screening tests have been devised, the Ames test (*Section 7.11*) is currently the most widely used, being employed in more than 2000 government, industrial, and university laboratories around the world. An important modification of the Ames test in such assays is that the chemical to be screened is incubated in the presence of a rat liver extract, since many agents prove not to be carcinogenic to mammals until they have been taken up by the body and metabolized by liver enzymes known as **mixed function oxidases.**

The Ames test has been applied to thousands of chemicals. Representative results are shown in Table 7-2. About 95 percent of compounds known to be carcinogenic in animals are found to revert the *Salmonella* mutants, and many compounds shown to be mutagenic have been or are being tested for carcinogenicity. As an example, we can cite an undergraduate laboratory exercise at the University of California at Berkeley in 1975 wherein several hundred commercial products were subjected to the Ames test. It was found that 89% of the major types of hair dye sold in the United States contained mutagenic compounds known as nitrophenylenediamines, an alarming discovery considering that about 25 million Americans dye their hair. A subsequent epidemiological study suggested a considera-

TABLE 7-2   Carcinogens as Mutagens in the Ames Test

| Group of Compounds | Carcinogens Detected as Bacterial Mutagens | Noncarcinogens Not Mutagenic to Bacteria | Compounds of Uncertain Carcinogenicity Detected as Mutagens |
|---|---|---|---|
| A  Aromatic amines, etc. | 23/25 | 10/12 | 5/7 |
| B  Alkyl halides, etc. | 17/20 | 1/3 | 1/1 |
| C  Polycyclic aromatics | 26/27 | 7/9 | 1/1 |
| D  Esters, epoxides, carbamates, etc. | 13/18 | 5/9 | 0/1 |
| E  Nitro aromatics and heterocycles | 28/28 | 1/4 | 0/2 |
| F  Miscellaneous organics | 1/6 | 13/13 | 0/1 |
| G  Nitrosamines | 20/21 | 2/2 | 1/1 |
| H  Fungal toxins and antibiotics | 8/9 | 5/5 | — |
| I  Mixtures (cigarette smoke condensate) | 1/1 | — | — |
| J  Miscellaneous heterocycles | 1/4 | 7/7 | — |
| K  Miscellaneous nitrogen compounds | 7/9 | 2/4 | — |
| L  Azo dyes and diazo compounds | 11/11 | 2/3 | 3/3 |
| M  Common laboratory biochemicals | — | 46/46 | — |
| Total | 157/178 | 101/117 | 11/17 |

From J. McCann, E. Choi, E. Yamasaki, and B. N. Ames, *Proc. Nat. Acad. Sci.* **72:**5135, 1975.

ble excess of breast cancer in postmenopausal women who have dyed their hair for long periods, and the cosmetic industry has responded by modifying the dye composition of most of their products.

Although chemicals produced in industrial laboratories are a major source of environmental mutagens, they are not the only source. A number of products produced by plants and by natural fermentations are also mutagenic and potentially carcinogenic. In many cases these products are **glycosides,** meaning that they carry sugar groups. Such compounds prove not to be mutagenic as long as the sugar groups are present, and the sugar groups survive exposure to the mixed function oxidases of the liver. They are, however, cleaved by the microorganisms present in the intestinal flora, meaning that high levels of sugar-free mutagenic compounds can potentially accumulate in the large intestine. B. Ames and colleagues have recently prepared a cell-free extract of human feces called **fecalase,** which contains many of these **bacterial glycosidase** enzymes, and have asked which natural products become mutagenic in an Ames test after exposure to fecalase. They found that red wine, red grape juice, and tea are so converted, whereas white wine and coffee are not. These findings may well be pertinent to understanding the etiology of colon cancer.

## 7.13  Birth Defects

Cancer poses an unquestioned health hazard, but perhaps even more sobering is the question of how environmental mutagens might affect the world's human gene pool. Some 5 to 10 percent of all children are currently born with genetic defects, and as many as 50 percent of all human conceptions are thought to end in spontaneous abortion. Although meiotic errors (*Sections 5.15 and 5.16*), faulty fertilization, and unsuccessful implantation can account for many of these abortions, the underlying causes of these difficulties may in many cases be genetically determined. The detection and elimination of environmental mutagens thus represents an urgent task for our generation if future generations are to flourish.

(a)                    (b)                    (c)

(a) L. J. Stadler and (b) H. J. Muller independently discovered the mutagenic effects of X rays in plants and animals, respectively. (c) Bruce Ames (University of California, Berkeley) developed the "Ames test" to screen for mutagenic agents.

# Questions and Problems

1. What is the ratio of phenotypes and genotypes expected in the progeny of a cross between two organisms heterozygous for the same lethal gene?
2. Cite six human phenotypic traits that you would classify as wild type (that is, highly invariant) and six traits that commonly appear in a number of variant forms in the human population.
3. Devise an experiment utilizing a colony-forming haploid microorganism such as *C. reinhardi* that would allow you to screen for a mutant strain carrying two mutations, one leading to an acetate (carbohydrate) requirement and the other to a thiamin (vitamin) requirement.
4. Why are the $F_1$ female flies in Figure 7-5 not allowed to inbreed? How would this change the experiment?
5. You wish to isolate a large deficiency in the *a* (*apricot* eye) region of the *D. melanogaster* X chromosome. Show how you would do this utilizing the *Basc* chromosome. Does it matter whether or not the deficiency is lethal?

6. (a) When Chinese hamster ovary cells are cultured in undiluted Burpee's Cola and then examined, the average cell exhibits 20 SCEs. On the basis of this test, would you recommend that the cola be removed from the market (refer to Figure 7-13 in considering this question)? (b) Male strains of mice were fed a steady diet of Burpee's Cola and subjected to the specific locus test. Of 65,548 offspring, 40 showed mutations at one of the 7 loci. What is the mutation rate per locus per $10^6$ gametes? On the basis of this test, would you recommend that the cola be removed from the market?

7. A normal woman and man have a child with severe birth defects and mental retardation. Banded karyotyping reveals that the mother is heterozygous for a reciprocal translocation between portions of chromosomes 8 and 19. The father's karyotype is normal. (a) Draw the possible karyotype(s) of the afflicted child. (b) Why is the woman normal and the child abnormal? (c) What proportion of normal:abnormal children would this couple expect to have? State the assumptions you are making in answering this question.

8. An enrichment procedure known as **tritium suicide** involves exposing mutagenized cells to $^3H$-thymine in a minimal medium, harvesting the cells, and storing them in the cold for weeks to allow the radioactive decay within the tritium-containing cells to "burn them out." What kind or kinds of mutations would such a procedure enrich for?

9. Indicate the proper method or methods to isolate $his^-$, streptomycin-resistant ($str^r$), and streptomycin-independent ($str^i$) strains from a mutagenized culture of *E. coli*. Explain the defects of the improper method or methods.

   1. Grow on minimal medium with penicillin added $\longrightarrow$ plate to minimal medium with histidine and streptomycin added $\longrightarrow$ transfer to his-supplemented, streptomycin-free medium.

   2. Grow on minimal medium + streptomycin $\longrightarrow$ plate to minimal medium $\longrightarrow$ transfer to plate with penicillin and all nutritional substances added, with the exception of histidine.

   3. Grow on minimal medium with $^3H$-thymine added $\longrightarrow$ harvest all the cells and store them in the cold for weeks $\longrightarrow$ plate the cells to minimal medium with histidine added $\longrightarrow$ transfer to his-supplemented penicillin-free medium.

   4. Grow on complete medium lacking histidine but with penicillin added $\longrightarrow$ plate to his-supplemented, streptomycin-added minimal medium $\longrightarrow$ replica plate to minimal medium.

10. In the process of studying a stock of Florida *Drosophila* with a high frequency of sex-linked lethal mutations, Demerec inbred the stock for two generations. As a consequence, a mutability factor (a recessive mutator gene on the second chromosome of *Drosophila* associated with a high frequency of sex-linked recessive lethal mutations) located on the second chromosome became homozygous in some of the $F_2$ flies. Demerec therefore expected a higher frequency of visible mutations (in addition to lethal mutations) in the $F_3$ generation than had been found in the original stock. The visible mutations actually observed in the $F_3$ generation among 15,000 individuals were

| | |
|---|---|
| *yellow*—24 times | *black*—2 times |
| *forked*—3 times | *blistered*—1 time |
| *lozenge*—2 times | *dwarfish*—1 time |
| *vermilion*—2 times | *curled wing*—1 time |

Knowing that these genes are scattered throughout the entire length of the X chromosome, indicate the conclusions that are consistent with the above data and explain your choices.
(a) The mutability factor is affecting the entire X chromosome equally, as the mutations are randomly distributed. (b) The mutability factor is affecting the entire chromosome equally, but some genes contained more sites for lethal mutations than others. (c) The region of the chromosome controlling the *yellow* phenotype is much larger than the other regions scored. (d) The mutability factor cannot be affecting the whole chromosome equally, because among 15,000 progeny, one would certainly have seen more loci represented. (e) List one other conclusion that is consistent with the data presented.

11. A *Drosophila* female is heterozygous for the pericentric inversion shown below. Diagram the four chromosomes generated if a crossover occurs between positions 3 and 4 following meiotic synapsis.

12. A *Drosophila* strain is homozygous for a reciprocal translocation involving noncentromeric portions of chromosomes 2 and 3. (a) Diagram schematically chromosomes 2 and 3 of a normal strain and the translocation strain. Assign arbitrary genes (2A, 2B, etc.) to each chromosome so the effect of the translocation is clear. (b) Diagram the gametes produced by each strain. (c) Diagram synapsis during meiotic prophase in an $F_1$ fly produced by crossing these two strains (*recall Figures 7-10 and 7-11*).

13. Diploid karyotypes are often denoted in a shorthand that describes the total number of autosomes and sex chromosomes present. Thus a normal human female would be 44A, XX and a normal male 44A, XY. (a) How would you denote a person with Turner's syndrome? Klinefelter's syndrome? Down's syndrome? (b) A diploid species with the karyotype male 10A, XY; female 10A, XX, developed a stable tetraploid subpopulation. Give the karyotype of this population. (c) Demonstrate how this population might yield equal numbers of male and female offspring without producing large numbers of inviable zygotes. (d) Would it give fertile offspring in matings with the members of the original population? Why?

14. A woman with a normal karyotype marries a man who is a translocation heterozygote for chromosome 21, the translocation occurring with chromosome 10.
(a) Draw the man's and the woman's karyotype with respect to chromosomes 10, 21, and 10/21.
(b) Draw the synaptic pairing pattern expected of these chromosomes during meiosis in the man.
(c) What are the six possible types of sperm this man can produce with respect to chromosomes 10, 21, and 10/21 (recall the tendency to chromosome imbalance described in Section 7.9)?
(d) Assuming that these sperm fertilize normal eggs, what is the expected fate of each type of zygote?

15. Diagram somatic pairing in the salivary gland cells of a dipteran heterozygous for a duplication.

# 8

# DNA Replication and Repair Mechanisms and Their Contribution to Mutagenesis

A. Introduction

B. DNA Replication Mechanisms
  8.1 Initiation of DNA Replication
  8.2 Chain Elongation and
    Excision-Joining
    DNA polymerase III
    DNA polymerase I
    DNA ligase
    Opening and stabilizing the
     helix

C. Direct Mutagenesis Mechanisms
  8.3 Chemical Alteration of
    Nucleotide Structure
    Nitrous acid
    Hydroxylamine
    Alkylating agents
    Heat
  8.4 Additional Lesions Caused by
    Chemical Mutagens

8.5 Base Analogues
8.6 Irradiation
8.7 Site-Directed Mutagenesis

D. Repair and Misrepair Mechanisms
  8.8 DNA Damage that Requires
    Repair
    Pyrimidime dimers
    Apurinic sites
    Bulky adducts
  8.9 Photoreactivation
  8.10 Excision Repair
  8.11 Base Addition and Deletion
  8.12 SOS Repair
  8.13 Mutator and Antimutator
    Mutations
  8.14 Repair-Deficient Human
    Disorders and Cancer

E. Questions and Problems

## Introduction

Chromosomal DNA is perpetually active in most cells. Some of this activity relates, of course, to the expression of the genes themselves, an activity considered in the next chapter. At any one time, however, most of the genome is not engaged in gene expression, but it continues to participate in three additional activities: **replication, repair,** and **recombination.**

230

Mutations can be thought of as arising in two different ways: some are the consequence of errors made during replication and/or repair, and others arise during the course of recombination. This chapter considers mutagenic mechanisms in the context of replication and repair. Mechanisms that depend on recombination are presented in Chapter 17.

That DNA replication, repair, and mutation are intimately related to one another is apparent when one considers the following.

1. From nucleic acid chemistry, one can calculate how frequently the alternate tautomeric forms of the nucleotides (*Section 1.1*) should arise. One can also calculate the stability of base pairs formed from non-Watson-Crick associations. These calculations reveal that "nonconventional" tautomers and "illegitimate" base pairs should arise sufficiently often that **one base-pair substitution should occur for every $10^4$ to $10^5$ base pairs replicated.**

2. From experimental observations made on *E. coli*, it is estimated that **one base-pair substitution in fact occurs for every $10^9$ to $10^{10}$ base pairs replicated.**

Therefore, during each round of DNA replication, many erroneous bases must be inserted into daughter strands and subsequently removed before they are "immortalized" as base-pair substitutions or mutations. In addition, the onslaught of environmental mutagens (*Section 7.12*), which potentially increases the error/mutation factor still more, must be countered by effective repair strategies to keep "spontaneous" mutation rates as low as they are.

Molecular mechanisms for chromosomal DNA replication have been elegantly worked out for *Escherichia coli* and its phages. Many of these replication enzymes and mechanical processes turn out to operate as well during repair and recombination. We therefore begin with an account of DNA replication that emphasizes information relevant to mutagenesis, focusing largely on *E. coli*. This information is then used in presenting a detailed account of how mutagens damage DNA, how cells try (and usually succeed) to repair the damage, and how mutations result if the damage is either unrepaired or repaired in an erroneous fashion.

# DNA Replication Mechanisms

The broad outlines of DNA replication were presented in Chapter 2. At a molecular level, DNA replication can be thought of in three stages: (1) **initiation** of replication; (2) **elongation** of the daughter strands, which occurs in discontinuous spurts of synthesis; and (3) a complex **excision-joining** reaction to proofread, patch up, and join together the various pieces. These are considered in the following sections.

## 8.1 Initiation of DNA Replication

The bidirectional replication of the *E. coli* chromosome (*Figure 2-5*) always begins at the same unique site, the **replication origin** (*oriC*), which, following the convention used to designate position on the *E. coli* chromosome, is located at 86 minutes (Figure 8-1). The *oriC* region has been cloned and sequenced by recombinant DNA technology (*Chapter 4*). It contains about 300 base pairs with many inverted repeats (*Section 4.9*), and it is thought to assume an elaborate secondary structure with multiple loops that signal the binding of initiation proteins.

The genes encoding essential initiation proteins are identified by conditional *ts* mutations (*Section 7.5*) that confer a specific phenotype; when mutant cells are shifted from permissive to nonpermissive temperature, ongoing DNA replication continues but no new rounds of replication are initiated. The *dnaA, dnaB,* and *dnaC* mutants exhibit this phenotype, and each has been found to be defective in proteins that interact with replication origins.

## 8.2 Chain Elongation and Excision-Joining

The enzymes required for the elongation of daughter strands of DNA may well also function during initiation, but their modes of action are more easily studied during the elongation process. The properties and genetic origin of these enzymes are described in the following paragraphs.

**DNA Polymerase III**   DNA polymerase III holoenzyme is the true **replicase** in *E. coli*, a replicase being defined as a DNA polymerase that acts

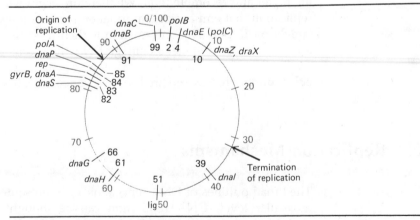

**FIGURE 8-1 Genes involved in DNA replication** in *E. coli.* Positions on the map are given in *minutes,* each minute representing the length of DNA transferred during one minute of sexual conjugation. The total map covers 100 minutes; each minute therefore corresponds to about 130 μm of DNA. Details of how such maps are constructed are given in Chapter 13.

within the replication fork to copy parental DNA strands into daughter strands. The enzyme has six polypeptide subunits, three of which are specified by the *dnaE, dnaX,* and *dnaZ* genes (Figure 8-1). The analogous replicase in eukaryotic cells is known as DNA polymerase α.

The polymerization reaction catalyzed by replicase enzymes is diagrammed in Figure 8-2 and described in the legend to Figure 8-2. The essence of the reaction is that the replicase recognizes a base on the template DNA strand, mediates its association with a complementary nucleoside-5'-triphosphate monomer from the cytoplasm according to Watson-Crick base-pairing rules (*Chapter 1*), and attaches the monomer to the growing daughter strand by a 3', 5' phosphodiester bond, releasing the terminal diphosphate in the process.

Two features of the *E. coli* (and all other) DNA replicase have important consequences for the mode of DNA replication. First, all DNA replicases thus far characterized are **incapable of initiating DNA synthesis:** they can only add nucleotides onto the 3'-OH end of a DNA strand that is hydrogen-bonded to a template strand, as drawn in Figure 8-2. This raises an obvious question: how do the first nucleotides in a daughter strand get laid down? In other words, how does the initiation of DNA polymerization occur? This question, which has plagued biochemists for many years, appears to have been answered by studies of *in vitro* DNA replication conducted by A. Kornberg and associates and by S. Wickner. They find that an enzyme known as **primase,** the product of the *dnaG* gene (Figure 8-1), acts to make a short **primer strand of RNA.** This primer is then elongated by the DNA polymerase III in the fashion depicted in Figure 8-2. Once initiation is accomplished, the primer sequences are enzymatically removed.

The second important feature of all DNA replicases is that they **can add deoxyribonucleotides only to the 3'-OH end** of a hydrogen-bonded primer, never to its 5'-P end. This creates a conceptual dilemma. As stressed in Chapter 1, the two strands of a DNA duplex are oriented in opposite directions (*Figure 1-5*); we now learn that DNA polymerases can elongate chains only from the 3'-OH end and therefore in a 5' ⟶ 3' direction. Thus a single growing point in the *E. coli* chromosome contains two parent strands of opposite polarity being replicated at the same time and in the same direction by an enzyme capable only of unidirectional synthesis. How is this accomplished?

Figure 8-3 diagrams a current view of how elongation occurs. The parent helix locally unwinds and DNA replication proceeds, in the allowed 5' ⟶ 3' direction, along the upper strand (Figure 8-3a) to create a **leading daughter strand.** A primer fragment (shown as a jagged line in Figure 8-3b) is next synthesized opposite the lower strand, and this is elongated, again in the allowed 5' ⟶ 3' direction (Figure 8-3c), to produce a short **lagging daughter strand.** The parental helix, meanwhile, unwinds further so that its upper strand is copied continuously while the lower strand dictates a second round of primer synthesis (Figure 8-3d) and the generation of another short piece of DNA (Figure 8-3e). These pieces of DNA, it should be noted, are known as **Okazaki fragments** after R. Oka-

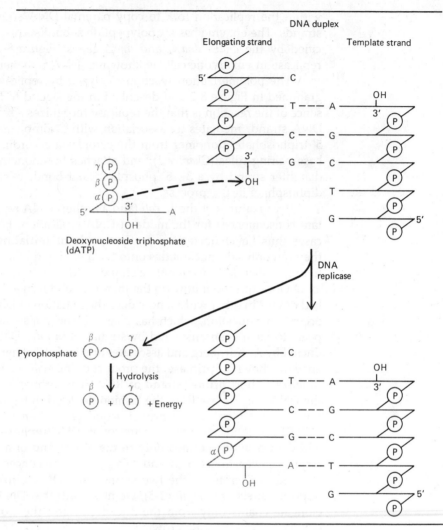

**FIGURE 8-2 DNA polymerization.** A dATP nucleotide is shown being added to an elongating strand opposite a thymine (T) base in the template strand, a Watson-Crick base recognition event catalyzed by the DNA replicase. The α phosphate of the dATP forms a 3′,5′ phosphodiester bond with the free 3′-OH of the elongating strand, and a molecule of **pyrophosphate (P~P)** is simultaneously released. DNA polymerization is potentially reversible, but because one of the reaction products (pyrophosphate) is readily removed by enzyme-catalyzed hydrolysis, the reaction is ordinarily driven toward polymerization. The resultant polymerization will always proceed in a net 5′ ⟶ 3′ direction, meaning that the nucleotide at the 3′ end is always the most recently added to the chain.

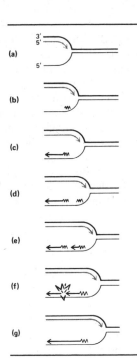

**FIGURE 8-3 The DNA growing point.** The leading strand, depicted in gray, is shown being synthesized continuously; certain experiments suggest that this strand is also synthesized discontinuously, but the matter is not yet settled. The lagging strand is drawn in black; RNA primers are indicated as jagged lines. See text for details. (After D. Dressler, in *Control Processes in Virus Multiplication*, D. C. Burke and W. C. Russell, Eds. Cambridge: Cambridge University Press, 1975.)

zaki, who first reported that discontinuous pieces of DNA, perhaps 1000 nucleotides long, are generated during DNA replication. Finally, primer sequences are degraded (Figure 8-3f), and the discontinuous pieces associated with the lagging strand are joined together enzymatically (Figure 8-3g).

**DNA Polymerase I** DNA polymerase I, specified by the *polA* gene (Figure 8-1), is a single polypeptide but has three enzymatic activities: it polymerizes DNA chains from 3'-OH primers, although at a slower rate than DNA polymerase III; and it has two exonuclease activities, one of which allows it to digest DNA strands in a 5' $\longrightarrow$ 3' direction and the other of which catalyzes a 3' $\longrightarrow$ 5' digestion (see Figure 8-4 and its legend for a description of exonucleases). Although the polymerase activity of this enzyme is undoubtedly utilized by the cell, its two exonuclease activities appear to make the more vital contributions. The analogous eukaryotic enzyme is known as DNA polymerase β.

One function ascribed to DNA polymerase I is the **digestion of the RNA primers** from the ends of Okazaki pieces (Figure 8-3f). In the wake of the advancing replication fork, a DNA polymerase I is thought to recognize a primer–daughter strand juncture, chew up the primer sequences using its 5' $\longrightarrow$ 3' exonuclease activity, and perhaps fill in the gap it creates with deoxyribonucleotides using its polymerase activity.

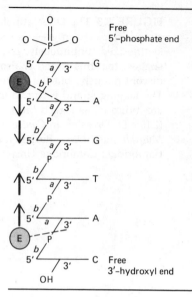

**FIGURE 8-4 Exonuclease action.** An exonuclease must begin its attack from a **free end** of a polynucleotide. Therefore, depending on the specificity of the enzyme, either an exonuclease will begin at a free 3-OH end of a polynucleotide and progressively cleave the bonds on the 3-OH side of the phosphodiester backbone (labeled *a*) or it will begin at a free 5'-P end and cleave the polynucleotide from the 5' side (labeled *b*). A 3'-exonuclease is depicted in light gray and a 5'-exonuclease in dark gray.

DNA polymerase I is also thought to serve as a **proofreader** or **editor** in the wake of the replication fork. Suppose that, during leading strand synthesis (Figure 8-3a), a replicase mistakenly pairs a C opposite a template A. As a result, the 3'-OH end of the leading strand is no longer perfectly hydrogen bonded to the template strand, and the replicase refuses to elongate the strand further. This impasse is apparently resolved by DNA polymerase I, which recognizes the A-C mismatch, clips out the incorrect C with its $3' \longrightarrow 5'$ exonuclease activity, and regenerates a correctly basepaired 3'-OH terminus for the replicase to elongate.

About 400 molecules of polymerase I are present in an *E. coli* cell, in contrast to only about 10 molecules of polymerase III. In part because of this difference in abundance, polymerase I was considered to be the true replicase for many years, and its *in vitro* properties were intensively studied, notably by A. Kornberg and his associates. In 1969, however, P. DeLucia and J. Cairns isolated a mutant strain of *E. coli, polA,* that contained no active form of the "Kornberg enzyme" and appeared capable of normal growth and DNA replication; a search for replicase activity in *polA* strains subsequently revealed the existence of polymerase III. It has since been found that the $5' \longrightarrow 3'$ exonuclease activity of DNA polymerase I is not abolished by the *polA* mutation and that if this exonuclease activity is made temperature sensitive by a second mutation, cells cannot survive at the nonpermissive temperature. Therefore, **the exonuclease activity of DNA polymerase I is critical to the *E. coli* cell.**

**DNA Ligase**   The existence of Okazaki fragments can be demonstrated by exposing cells to [3]H-thymidine for short periods of time, isolating their

DNA, and showing that most of the radioactivity is found in low-molecular-weight DNA strands. If after the short "pulse" of $^3$H-thymidine, the cells are washed, given unlabeled thymidine, and allowed to continue replication for an additional short period (a so-called **pulse-chase experiment**), the label is now found in high-molecular-weight strands of DNA. This joining together of Okazaki pieces to form continuous daughter strands is effected by the enzyme **DNA ligase,** which is specified in *E. coli* by the *lig* gene (Figure 8-1). Mutants defective in ligase activity join Okazaki pieces very slowly. The mechanism of DNA ligase action has already been considered in Section 4.15 (*Figure 4-11*).

**Opening and Stabilizing the Helix** DNA replication clearly requires that a duplex parental molecule be "unwound" so that its internal bases are available to the replication enzymes. Unwinding activity in *E. coli* appears to be mediated by a "helicase" known as the **rep protein,** identified first in the *rep* mutant strain (Figure 8-1), which hydrolyzes ATP while it actively forces the DNA helix apart. Such unwinding of the positive twists in the helix is expected to generate compensatory positive supercoils (*see Section 3.7*) in the parental duplex ahead of the replication fork. These supercoils are removed by the action of **DNA gyrase,** an enzyme jointly specified by the genes *gyrA* and *gyrB* (Figure 8-1). DNA gyrase introduces negative supertwists into DNA, thus compensating for the positive twists introduced by uncoiling, by a concerted breakage and resealing of the DNA helix.

Once opened up, the exposed single strands are stabilized by a third protein known as either the **single-strand binding protein** (SSB) or the **helix-destabilizing protein,** encoded by the *ssb-1* gene (Figure 8-1). The SSB exhibits no interaction with duplex DNA, nor does it actively force open a double helix. Once a helix has opened to reveal its single strands, however, a molecule of SSB protein readily associates with the single-stranded region; additional proteins proceed to bind much more readily than the first (a phenomenon known as cooperativity) so that the single-stranded state is rapidly stabilized by a "coating" of unwinding proteins. The SSB protein is thought to be responsible for holding open a double helix in advance of the replication fork. A single SSB protein is long enough to cover 8 nucleotides of DNA, meaning that the 400 SSB proteins present in each cell can stabilize substantial lengths of both replication forks.

# Direct Mutagenesis Mechanisms

We noted at the start of the chapter that the potential for mutation arises at each Watson-Crick base-pairing event during DNA replication, but that since replicases are so accurate and "proofreading" mechanisms so efficient, the basal error rate is in fact very low. During **direct mutagenesis,** the system is effectively fooled or subverted: nucleotides may be modified, for example, or ambiguous nucleotides may be provided, the result being that mistakes either "slip past" the proofreading enzymes or swamp them out.

## 8.3 Chemical Alteration of Nucleotide Structure

Most direct-acting mutagens transform the chemical structure of a nucleotide, giving it the base-pairing properties of another nucleotide. The resultant **mispairing** will give rise to two sorts of mutation, transitional and transversional. In the case of a **transition,** a purine is replaced by another purine or a pyrimidine is replaced by another pyrimidine (the structures of purines and pyrimidines are found in Figure 1-1). Consider, for example, a particular adenine on one strand of a DNA duplex: a mutagen may transform this into the modified molecule A* which has the base-pairing properties of guanine (Figure 8-5a). At replication, then, this A* will mispair with cytosine (Figure 8-5b), and the complementary daughter strand will carry a cytosine at this position. One round of replication later, the cytosine will pair with a guanine; thus the granddaughter strand, which should be identical to the original, will carry the purine guanine where it should carry the purine adenine, and an A $\longrightarrow$ G transition is said to have occurred (Figure 8-5c). Similar sequences of events produce G $\longrightarrow$ A, T $\longrightarrow$ C, and C $\longrightarrow$ T transitions.

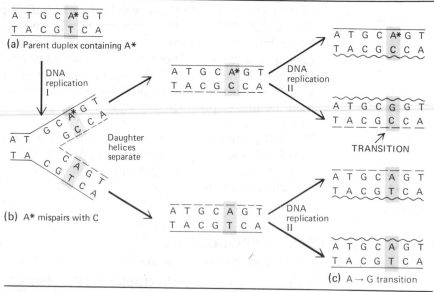

FIGURE 8-5 **Transition induced by chemical modification.** The mispairing of a modified adenine (A*) with cytosine produces an A $\longrightarrow$ G transition.

The second kind of mutation that can result from chemical modification is a **transversion,** in which a purine is replaced by a pyrimidine or vice versa (for example, A $\longrightarrow$ C, G $\longrightarrow$ C, and so on). Of the many transversion-inducing mispairings one might try to visualize, however, most fit very poorly within the confines of a double helix; in the case of purine-purine mispairs, for example, the sugar-phosphate backbone would need to undergo considerable distortion to accommodate both large bases. A few mutagens are, nonetheless, believed to induce transversions directly, although their mode of action remains a matter of speculation.

Permanent alterations in nucleotide structure can be effected by a variety of agents. Of these, the four most widely cited are nitrous acid ($HNO_2$), hydroxylamine ($NH_2OH$), a group of compounds known as the alkylating agents, and heat. Figure 8-6 summarizes what is known of these agents' mutagenic effects on nucleotide structure.

**Nitrous Acid**   Nitrous acid deaminates nucleotides, removing amino (—$NH_2$) groups and substituting instead a keto group (=O). As shown in Figure 8-6a, deamination of cytosine yields uracil and deamination of adenine yields an unusual base, **hypoxanthine (HX).** Uracil is seen in Figure 8-6a to pair with adenine (recall that uracil replaces thymine in RNA), and thus the deamination of cytosine can produce a C $\longrightarrow$ T transition. Hypoxanthine resembles guanine at positions 1 and 6 on its ring and can therefore pair with cytosine, as shown in Figure 8-6a. Again a transition results, this time in the A $\longrightarrow$ G direction.

**Hydroxylamine**   Hydroxylamine reacts only with the pyrimidines of DNA, and probably only its reaction with cytosine is mutagenic. The way in which hydroxylamine is thought to alter cytosine to produce mutations is shown in Figure 8-6b. Note that hydroxylamine attacks the amino group in position 4 of cytosine, converting it to a hydroxylimine (=N—OH); the resulting base ($N^4$-hydroxycytosine) preferentially pairs with adenine to produce a C $\longrightarrow$ T transition.

**Alkylating Agents**   The alkylating agents are a diverse group of highly reactive chemicals that introduce alkyl groups ($CH_3$—, $CH_3CH_2$—, and so on) into nucleotides at numerous positions. Of the various alkylated products, $O^6$-alkylguanine and $O^4$-alkylthymine are the most likely to undergo mispairing. As drawn in Figure 8-6c, $O^6$-alkylguanine can pair with thymine to generate G $\longrightarrow$ A transitions, whereas $O^4$-alkylthymine can pair with guanine to produce T $\longrightarrow$ C transitions.

Alkylating agents are the largest group of mutagens and include mustard gas, epoxides, dimethyl- and diethylsulfonate, methyl- and ethylmethane sulfonate (**MMS** and **EMS**), and N-methyl-N'-nitro-N-nitrosoguanidine (**MNNG**). Most alkylating agents prove to exert their mutagenic effects by triggering misrepair, as described in Sections 8.8 and 8.12.

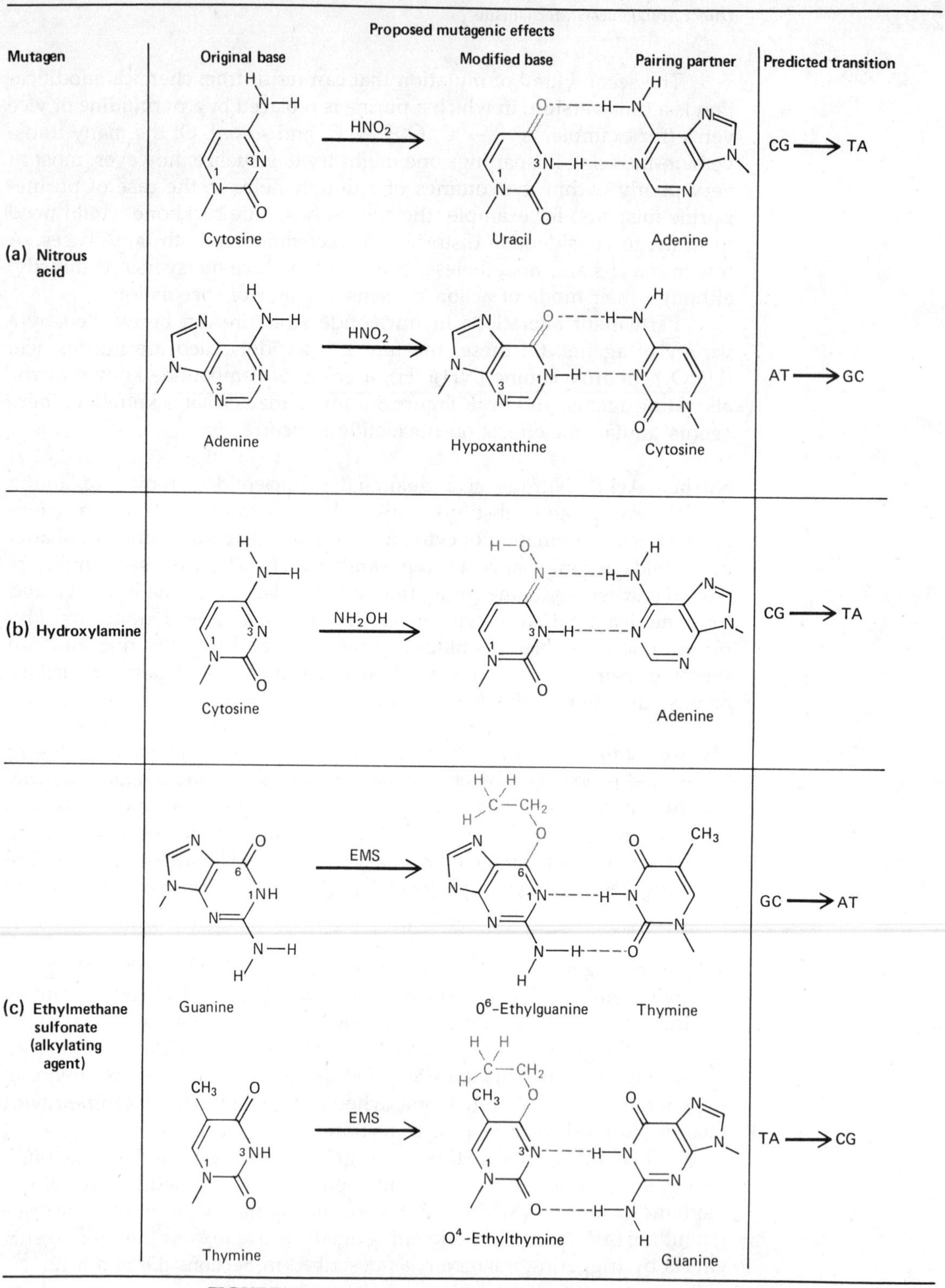

FIGURE 8-6 Mutagens that modify the chemical structure of bases.

**Heat**  Heat has only recently been accorded attention as an important mutagen. It clearly brings about the deamination of cytosine to form uracil, much as does nitrous acid, thereby bringing about C $\longrightarrow$ T transitions; heat also causes G $\longrightarrow$ C transversions by an as yet unknown mechanism. It has been estimated that over 100 heat-induced errors occur in a typical human cell each day, but the vast majority of these are no doubt later repaired.

## 8.4  Additional Lesions Caused by Chemical Mutagens

Chemical mutagens and heat are capable of producing mispairing modifications in specific nucleotides that give rise to mutations directly, as shown in the preceding section. These agents, and radiations as well, also produce a variety of additional alterations in nucleotides and nucleic acids. These can be classified as nonhereditary, inactivating, and premutational alterations.

**Nonhereditary** alterations in nucleotide structure are without effect either on DNA replication or on the transfer of genetic information. Thus the major reaction product of many alkylating agents is 7-alkylguanine, a nucleotide that appears to behave exactly like guanine. This means that measurements of 7-alkylguanine formation are poorly correlated with the mutagenic potential of an alkylating agent.

**Inactivating** alterations, as already noted in Section 7.10, prevent the transmission of the altered genome from parent to offspring. For example, the alkylation of A and C is often inactivating; hydroxylamine decomposes, producing lethal peroxides that damage nucleic acids indiscriminately, ultimately killing the cell directly; nitrous acid and certain alkylating agents may **cross-link** the strands of a DNA duplex so that they cannot separate for replication; and so on. These inactivating events should not be confused with lethal mutations. A lethal mutation is potentially heritable, provided some way is found to keep the mutant daughter organism alive, whereas in an inactivating event, a daughter cell or virus is never even formed.

**Premutational lesions** describe regions of DNA that are sufficiently altered or damaged that under most circumstances they produce inactivation. Such lesions may, however, be repaired by the cell before, during, or after replication. In some cases, the repair is faultless, but in other cases, incorrect nucleotides are inserted into the damaged strand or the daughter strand. This, of course, produces a mutation, and the damage eliciting such **error-prone repair** (**misrepair**) is correctly termed a premutational lesion. Misrepair is discussed in detail in Section 8.12.

## 8.5  Base Analogues

Mutagens known as **base analogues** have chemical structures analogous to naturally occurring bases but carry critical modifications. They are mutagenic only if they are presented to cells at the time of chromosome replication. The two most widely used analogues are 5-bromodeoxyuridine, which we have encountered several times in the text (*Sections 7.3 and 7.11* and *Figure 7-4*), and **2-aminopurine,** an analogue of adenine. 5-BUdR is an analogue of thymine and usually pairs with adenine. The bromine atom in 5-BUdR so alters the charge distribution of the molecule, however, that it tautomerizes to a 5-BUdR* form quite frequently, in which case it possesses the hydrogen-bonding properties of cytosine. Figure 8-7a shows how such shifts can generate G $\longrightarrow$ A transitions; Figure 8-7b shows the induction of an A $\longrightarrow$ G transition.

In addition to stimulating base-pairing ambiguities, base analogues have recently been found to induce mutations by a second mechanism, namely, by causing **imbalances in the deoxynucleoside triphosphate pools of a cell.** Specifically, 5-BUdR inhibits the enzyme that converts cytidine diphosphate (the ribonucleotide) into deoxycytidine diphosphate. As a result, the intracellular supply of dCTP rapidly falls, and at the time of the next DNA replication, template guanines encounter a dearth of cytosines. The void is filled by the available tautomerized forms of 5-BUdR* having the temporary properties of cytosine. The ensuing misincorporations stimulate the frequency of G $\longrightarrow$ A transitions, as in Figure 8-7b.

## 8.6 Irradiation

X rays, gamma ($\gamma$) rays, and ultraviolet (uv) light are widely used mutagens that generate a broad spectrum of lesions in DNA. X rays, for example, induce the intracellular formation of **free radicals** (molecules containing an atom with an unpaired electron), particularly when oxygen is present. These highly reactive compounds can bring about numerous chemical changes in individual bases. They can also wreak havoc with chromosome structure, causing single-strand or double-strand breaks in the DNA. The broken ends may reanneal in a grossly incorrect manner to produce deficiencies or duplications (*Section 7.7*); alternatively, heterologous broken pieces may unite to produce translocations (*Section 7.9*).

The graph in Figure 8-8 demonstrates an important property of X rays: their mutagenic effects are directly proportional to their dosage (measured in **roentgen units, r**), at least in the low-dosage range. X ray dosage is cumulative, meaning that exposure to several low doses over a long period is fully as mutagenic as a single exposure at a moderate dose. It is for this reason that all unnecessary exposure to X rays is to be avoided.

The most important mutagenic effect of both uv and ionizing irradiation is believed to be the stimulation of misrepair, considered in Section 8.12.

## 8.7 Site-Directed Mutagenesis

Recombinant-DNA technology (*Chapter 4*) has opened new doors for the practice of chemical mutagenesis: it is now possible to clone a specific ge-

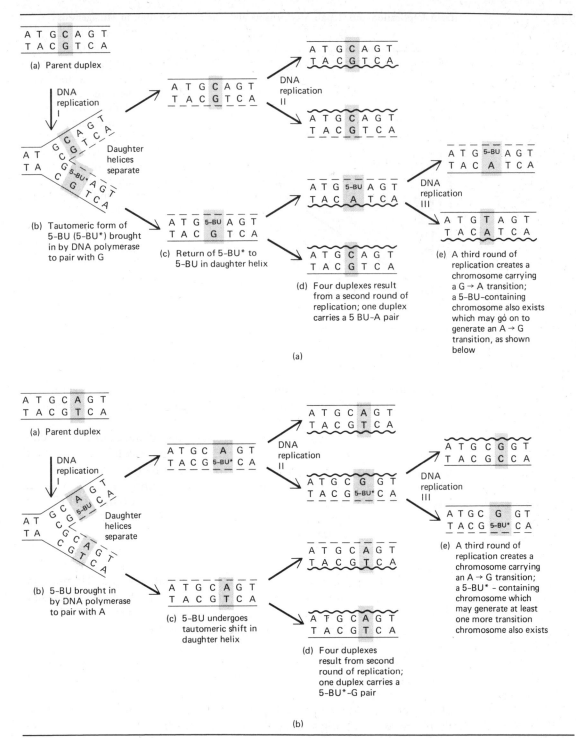

FIGURE 8-7 Transitions induced by base analogues. The analogue 5-BUdR (abbreviated 5-BU) undergoes reversible tautomeric shifts to 5-BU* and creates (a) G ⟶ A transitions and (b) A ⟶ G transitions.

243

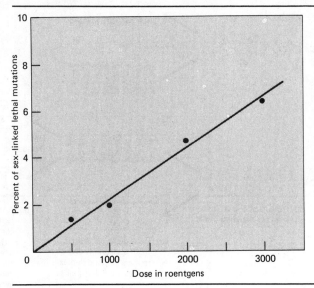

**FIGURE 8-8 X ray mutagenesis.** Increase in sex-linked lethal mutations in *Drosophila* with increasing X ray dosage. Roentgen units are based on the number of ionizations produced in 1 cu cm of air under standard conditions.

netic sequence, introduce a mutagenic lesion directly and specifically into this sequence, and integrate the altered sequence into the genome of a nonmutagenized organism.

As an example, we can cite mutagenesis of a cloned gene specifying the β-chain of rabbit hemoglobin (*Figure 3-5*). The recombinant plasmid was first treated with the restriction enzyme *Eco* RI (*Section 4.11*), which creates a staggered cut in the region encoding amino acids 121 and 122 (Figure 8-9a). The plasmid was next exposed to DNA polymerase I (*Section 8.2*), which, when presented with such a nicked duplex, catalyzes a reaction known as **nick translation:** the exonuclease of the enzyme moves the nick along in a 5′ ⟶ 3′ direction while its polymerase activity copies the exposed template strand (Figure 8-9b). The polymerase was provided, however, with a nucleoside triphosphate pool that lacked dTTP and contained the analogue $N^4$-hydroxy dCTP (C—OH), which the polymerase is "fooled" into using as thymidine. Therefore, a C—OH was inserted opposite each adenine during the copying reaction (Figure 8-9b). The nucleoside triphosphate pool was also lacking in GTP. Therefore, the nick translation reaction stopped when the polymerase came to its first cytosine (Figure 8-9b). The plasmid was now ligated and recloned in *E. coli*. During replication in the host, the C—OH residues were often mistaken for C, and a G was often inserted in the daughter strand (Figure 8-9c). The net effect, therefore, was to bring about AT ⟶ GC transitions in this limited region of the gene. Similar strategies have been used to introduce mutations into selected sectors of cloned viral genes, and the practice holds great promise for future genetic analyses.

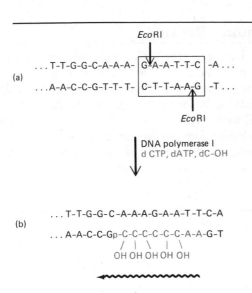

**FIGURE 8-9 Site-directed mutagenesis.** Nick translation of only the lower *Eco*RI site is shown. dC-OH, N⁴-hydroxy dCTP, an analogue used as thymidine; A/G signifies that either an A or a G may occupy the position. See text for details. (After W. Müller, H. Weber, F. Meyer, and C. Weissmann, *J. Mol. Biol.* **124**:343, 1978.)

# Repair and Misrepair Mechanisms

During the course of its lifetime, chromosomal DNA is subjected to a variety of damage: endogenous nucleases may nick it; breaks may occur in it during packaging into a phage head or during mitotic segregation; endogenous cellular chemicals and heat, or uv light and even white light from the sun, may alter it; and noxious environmental chemicals may interact with it. Therefore, even without the added stress of deliberate mutagenesis, the integrity and survival of a genome is critically dependent on the existence of DNA repair mechanisms.

This section begins by describing the three most common kinds of DNA damage that require repair: **pyrimidine dimers, apurinic sites,** and **bulky adducts.** We then describe the systems that have evolved to handle them. Some of these function with remarkable efficiency, earning them their designation as **error-free repair systems.** Other systems are decidedly **error prone** and produce mistakes that are frequently immortalized as mutations.

## 8.8 DNA Damage that Requires Repair

**Pyrimidine Dimers**   Ultraviolet radiation produces several effects on DNA, one being the formation of chemical bonds between two adjacent pyrimidine molecules in a polynucleotide, particularly between adjacent thymine residues, as shown in Figure 8-10a. As the two residues associate, or **dimerize,** their position in the DNA helix becomes so displaced that they can no longer form hydrogen bonds with the opposing purines, and thus the regularity of the helix becomes distorted. Such dimers are inactivating (and lethal) unless repaired.

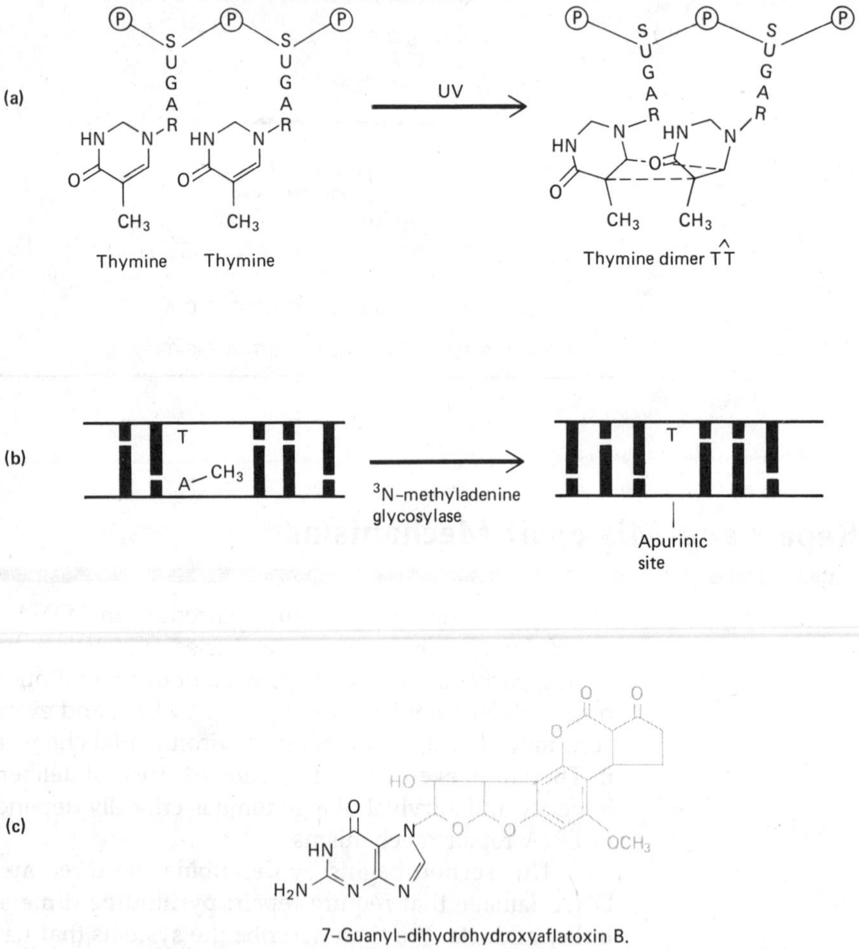

**FIGURE 8-10 DNA damage.** (a) Thymine dimer produced by uv light. (b) $^{3}$N-methyladenine produced by a methylating agent and its removal by a specific glycosylase to form an apurinic site. (c) Bulky adduct formed by the carcinogen aflatoxin B with guanine.

**Apurinic (AP) Sites.** In Section 8.3 we noted that alkylating agents could introduce alkyl groups onto certain bases ($O^6$-alkylguanine and $O^4$-alkylthymine) such that they were likely to undergo mispairing at the time of DNA replication. Alkylating agents also produce modified bases that are not prone to mispairing, the two most common being $N^7$-alkylguanine and $N^3$-alkyladenine. In addition, methylating agents present in the cell under physiological conditions generate significant levels of $N^7$-methylguanine and $N^3$-methyladenine. All of these modified purines are more labile than their unmodified counterparts so that the glycosylic bond that joins them to their sugar (*Section 1.1*) tends to hydrolyze. This hydrolysis may occur spontaneously or may be mediated by specific **glycosylases** present in the cell. The result is that the free base is liberated and an **apurinic (AP) site** is created in the DNA. As illustrated in Figure 8-10b, such AP sites do not disrupt the continuity of the sugar-phosphate backbone. They must, however, be repaired if accurate DNA replication is to occur.

**Bulky Adducts**  Many of the more potent carcinogens, such as aflatoxin B and benzo[a]pyrene, produce what are known as **bulky adducts**—large chemical groups that form covalent associations with DNA. As illustrated in Figure 8-10c, for example, aflatoxin B forms an adduct at the N-7 position of guanine. The presence of this group is thought to destabilize the bond connecting the guanine to its sugar-phosphate backbone. Therefore, the base tends to be released, again leaving a stable apurinic site, which must be repaired if it is not to lead to mutation.

## 8.9 Photoreactivation

A major line of repair for pyrimidine dimers, known as **photoreactivation,** requires the activity of **photoreactivating enzymes,** which have been found in both bacteria and eukaryotes (including humans). These enzymes convert thymine dimers (abbreviated $\widehat{TT}$) into two thymine monomers (TT) and thereby eliminate the lesion from the parental strand. The enzymes are so named because, although they can associate with dimers in the dark, they must absorb a photon of visible light before they can bring about monomerization (Figure 8-11).

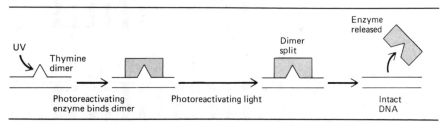

**FIGURE 8-11 Photoreactivation** of a thymine dimer. (From E. M. Witkin, Bact. Rev. **40**:869, 1976.)

## 8.10 Excision Repair

In 1964, R. Setlow and W. Carrier proposed that a damaged DNA duplex could be repaired by four enzymatic steps, collectively termed **excision repair**, which are drawn in Figure 8–12 b to d. (1) An **endonuclease** activity (*Section 4.11*) first makes a single-stranded nick in strand A. (2) An **exonuclease** activity (*Figure 8-4*) enters the nick and digests a portion of strand A. (3) A **DNA polymerase** activity (*Figure 8-2*) fills the resultant gap with a sequence complementary to strand B. (4) A **ligase** activity (*Figure 4-11*) seals the newly copied segment to strand A. All excision repair systems are believed to utilize Steps 2 to 4 in an equivalent fashion. Each system, however, is unique in the first step, being specific for the kind of damage recognized and the kind of endonuclease used to make the initial single-stranded incision.

The best characterized systems utilize two enzyme activities to bring about the first step. The damage is initially recognized by a battery of

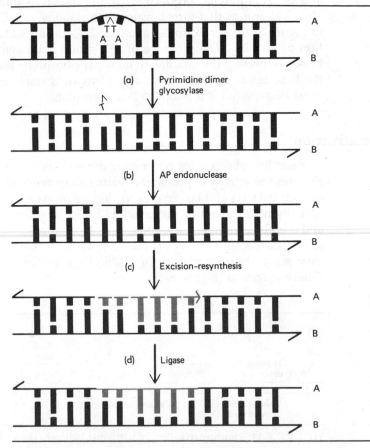

FIGURE 8-12 **Excision repair** of a thymine dimer.

**glycosylases** that clip damaged bases out of DNA duplexes. We have already described (*Section 8.8*) the glycosylase that removes $N^3$-methyl-adenine from DNA. In addition, there exists a **pyrimidine dimer glycosylase,** which cuts the glycosylic bond between the 5'-thymine of the dimer and its sugar (Figure 8-12a); a **uracil glycosylase,** which removes misincorporated uracils (these should be found only in RNA); and a variety of glycosylases that recognize other forms of alkylated bases. Operating in conjunction with these glycosylases are **AP endonucleases.** These apparently recognize the "holes" left in the DNA duplex when purines or pyrimidines are removed, and they respond by making a cut across the phosphodiester linkage (Figure 8-12b).

Excision repair can be monitored directly as the removal of damaged segments of DNA *in vitro*. In eukaryotes, its occurrence is most often detected as **unscheduled DNA synthesis:** cells in the $G_0$, $G_1$, or $G_2$ stage of the cell cycle (*Section 2.6*), which are expected to be inactive in chromosomal replication, are placed in the dark, exposed to a mutagen, and then pulsed with $^3$H-thymidine and subjected to autoradiography (*Box 2.2*); isotope incorporation signals excision repair. As detailed in Section 8.14, certain human diseases are believed to be caused by defective excision repair systems.

## 8.11 Base Addition and Deletion

Excision repair is remarkably efficient and conceptually error free: thousands of dimers have been shown to be excised from an *E. coli* chromosome without error. The risk of introducing mistakes during such complex repair pathways must always be present, however, and may contribute to an important class of mutations consisting of **deletions** (ABDEFGH) and **additions** (ABXCDEFGH).

Deletions and additions are known to occur spontaneously; they are also elicited by various mutagens. One model to account for the "spontaneous" generation of these lesions is shown and explained in Figure 8-13. Essentially, the model proposes that following strand digestion during a routine repair process, either the digested strand or the intact strand may "slip" or "buckle" and form base pairs with inappropriate sectors of its complement. The remaining repair steps then create strands that are either too long (Figure 8-13a) or too short (Figure 8-13b).

Additions and deletions are also efficiently induced by a class of mutagens known as **acridines,** the most widely used of which is **proflavin** (Figure 8-14). Acridines are flat, aromatic molecules that interact with DNA in such a way that they become wedged, or **intercalated,** between the stacked bases of a double helix. Streisinger has proposed that acridines may stabilize the postulated buckled-out regions of DNA that lead to the formation of additions and deletions (Figure 8-13), but their precise mutagenic mode is unknown.

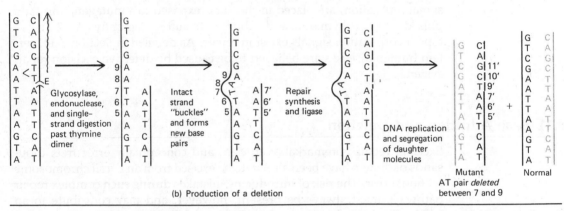

**(a) Production of an addition**

**(b) Production of a deletion**

**FIGURE 8-13 Addition** (a) and **deletion** (b) produced during excision repair. Thymine dimers are shown as T̂T̂. (After G. Streisinger et al., *Cold Spring Harbor Symposia on Quantitative Biology* **31**:77, 1966.)

Additions and deletions often produce an effect on the "reading" of the genetic message known as a **frameshift.** Frameshift mutations are described in detail in Section 11.7, where we consider experiments by Crick and his associates that provided important insights into the nature of the genetic code.

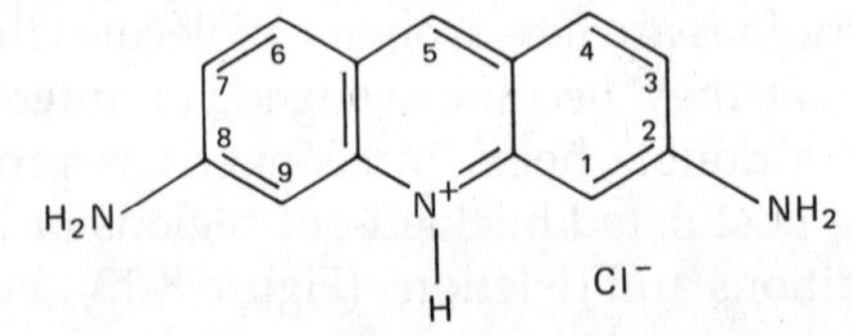

**FIGURE 8-14 Proflavin,** an acridine mutagen.

Proflavin (chloride): (2, 8–diamino acridine hydrochloride)

# 8.12 SOS Repair

Many of the genes that participate in *E. coli* DNA repair, including most of those shown in Figure 8-15, are said to be **constitutive:** their protein products are present at comparable levels at all times. Other proteins important for repair, however, are **inducible:** effective levels of these proteins are synthesized by the cell only when a particular mode of repair is indicated. We describe here the properties of the inducible **SOS repair system,** first studied by E. Witkin, which is elicited in *E. coli* when the cell is unable to replicate its DNA and survival is in jeopardy. It should be noted that this novel repair pathway is but one of several features of a pancellular **SOS response** that includes many other adaptations designed to cope with the crisis at hand.

The SOS response can potentially be elicited by any of the forms of DNA damage described in Section 8.8. Let us assume that chromosome replication has stalled, and the SOS response is induced, because one of the parental template strands carries a thymine dimer that has escaped correction by constitutive repair systems (Figure 8-16a). The replicase will stall at the dimer in the replication fork, and because the parental helix has already unwound, classic excision-resynthesis repair pathways are useless: the lone damaged strand lacks a complementary partner for a repair polymerase to copy (Figure 8-16b). The SOS repair solution is to abandon the usually stringent base-pairing requirements and somehow polymerize the daughter strand past the dimer, adding nucleotides that are not dictated by the parental strand (Figure 8-16c). This "solution" leaves the parental lesion unexcised and is very likely to generate a mutant base sequence in the daughter strand, but it is apparently more adaptive than certain death.

Two gene products figure prominently in the SOS response. The first is a protein called **LexA,** the product of the *lexA* gene (Figure 8-15). LexA is responsible for maintaining the SOS system in a repressed state until DNA damage occurs. Damage somehow overcomes this repression and permits expression of several genes, notably *recA*. The **RecA** protein is a protease,

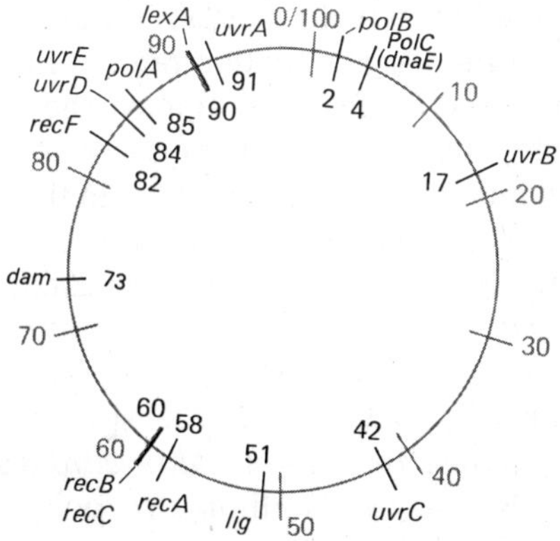

FIGURE 8-15 Genes involved in repair in *E. coli.*

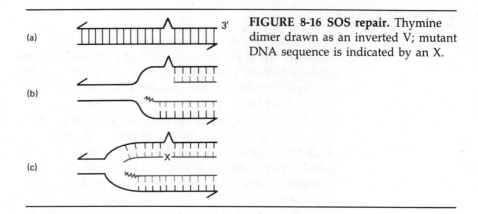

**FIGURE 8-16 SOS repair.** Thymine dimer drawn as an inverted V; mutant DNA sequence is indicated by an X.

and one of its targets is LexA, meaning that once RecA synthesis begins, LexA levels drop rapidly and the SOS-specific genes are permitted free expression. These genes dictate a battery of enzymes that bring about the untemplated, error-prone DNA repair characteristic of the SOS response. Within four hours after the "damage signal" is removed, RecA levels drop, LexA levels rise, and the SOS functions return to their repressed state.

Error-prone repair pathways that replicate past premutational lesions (*Section 8.4*) are found in yeast as well as *E. coli* and are likely to exist in higher eukaryotes as well. Many may prove to employ bypass mechanisms that are quite different from those utilized in SOS repair. For the geneticist, the important point is that these last-ditch efforts to save the genome are likely to be major sources of base substitutions (both transitions and transversions), deletions, and additions.

## 8.13 Mutator and Antimutator Mutations

Two interesting classes of mutations have the effect of altering the spontaneous mutation frequency typical for a particular species. **Mutator mutations** enhance this basal rate. In phage T4, for example, the phage-specified replicase is encoded by gene 43. Certain mutations in gene 43 have the effect of increasing spontaneous mutation rates at many T4 loci, producing deletions, additions, transitions, and transversions. When the mutant replicases are isolated, their ratio of exonuclease to polymerase activity is typically found to be much lower than the ratio in the wild-type enzyme. In other words, the ability of the mutant enzymes to excise base mispairings in the T4 replication fork is defective. Similar mutator mutations occur in the *dnaE* gene of *E. coli,* and numerous mutator genes have been described in yeast and in other eukaryotes.

**Antimutator mutations** that decrease spontaneous mutation rates have also been reported in gene 43 of T4. *In vitro* testing of the resulting enzymes often reveals that the "antimutator polymerases" have a high exonuclease/polymerase ratio compared with their wild-type counterparts, meaning that *in vivo*, these enzymes are unusually efficient in mispair recognition and excision.

The preceding cases should not be taken to mean that the exonuclease activities of DNA polymerases are the sole guardians of accurate DNA replication. The *thy⁻* mutation in Chinese hamster ovary cells, for example, results in greatly depleted dTTP pools and elevated dCTP pools. This imbalance in deoxynucleoside triphosphate pools exerts mutator activity, possibly by promoting mispairing (recall the 5-BUdR effect described in Section 8.4) or perhaps by stimulating an SOS response.

## 8.14 Repair-Deficient Human Disorders and Cancer

We conclude this chapter with a brief description of four autosomal recessive human disorders that have two features in common: afflicted persons are prone to developing cancer, and they exhibit defects in their ability to repair damaged DNA. The clinical manifestations of each syndrome (Table 8-1) are complex, and many are not easily related to repair deficiencies, but the relationships may become more apparent with future study.

Three of the disorders—**ataxia-telangiectasia (AT)**, the **Bloom syndrome (BS)**, and **Fanconi's anemia (FA)**—can be classed together as **chromosome breakage syndromes,** although the specific types of chromo-

**TABLE 8-1   Repair-Deficient Human Disorders**

| Disorder | Syndrome |
| --- | --- |
| Ataxia-telangiectasia (Louis-Bar syndrome) | Cerebellar ataxia; telangiectasia; neurological deterioration; often immune deficiency; hypersensitivity to ionizing radiation (may be blocked at early stage of excision of radioproducts); high frequency of spontaneous chromosome aberrations; homozygotes and heterozygotes cancer-prone, especially lymphoreticular tumors |
| Bloom's syndrome | Growth retardation; telangiectatic erythema (exposed areas); hypersensitivity to sun and to UV; high frequency of spontaneous chromosome aberrations, especially quadriradials, and sister chromatid exchanges |
| Fanconi's anemia or pancytopenia | Bone marrow deficiency; anatomical defects; growth retardation; high frequency of spontaneous chromosome aberrations; cultured cells show hypersensitivity to DNA cross-linking agents; homozygotes and heterozygotes cancer-prone |
| Xeroderma pigmentosum (classic XP) | Hypersensitivity to sun and to uv; sun-induced keratoses, skin carcinomas and melanomas; occasionally neurological abnormalities; probably defective in an early stage of excision repair; at least seven types |
| Xeroderma pigmentosum (XP variant) | Similar to XP, but not defective in excision; may be defective in SOS-type repair |

From P. Hanawalt, et al., *Ann. Rev. Biochem.* 48:783, 1979.

some aberrations encountered are different in each case. FA cells are unique in their hypersensitivity to cross-linking agents such as mitomycin C; AT cells are uniquely hypersensitive to ionizing radiation; and BS cells show a tenfold greater incidence of sister chromatid exchanges (*Section 7.11*) than normal, AT, or FA cells. Thus one can conclude that human cells possess a variety of repair pathways capable of being subverted by mutation and can speculate that cancer susceptibility is enhanced if repair is defective.

The fourth disorder, **xeroderma pigmentosum (XP),** is in fact at least eight different diseases. Seven of these (XP-A through XP-G), referred to as **classic XP,** all exhibit defective excision repair of uv-induced thymine dimers. As a result, the cells are very sensitive to uv irradiation and show enhanced rates of uv-induced mutations but can repair other kinds of DNA damage quite normally. The obvious interpretation of these observations is that classic XP cells are defective in human enzymes analogous to pyrimidine dimer glycosylases or AP endonucleases (*Section 8.10* and *Figure 8-12*), a hypothesis that is under intensive investigation.

The eighth type of XP, known as **XP variant,** produces similar kinds of skin lesions, but the cells excise uv-induced dimers in an apparently normal fashion. The prevalent hypothesis is that XP variant cells are defective in some form of repair that acts in the replication fork, perhaps analogous to SOS repair (*Section 8.12*).

The coupling of repair defects with malignant diseases brings us once again to the correlation between DNA damage and carcinogenesis (*Sections 7.11 and 8.12*). It is probably fair to state that at the present time, it is in this area of human health-related research that molecular genetics is making its most far-reaching contributions.

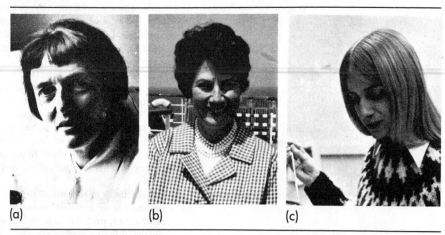

(a)               (b)             (c)

(a) Charlotte Auerbach performed early studies on the mutagenic effects of certain chemicals. She is now at the University of Edinburgh. (b) Evelyn Witkin, of Rutgers University, discovered misrepair mutagenesis in *E. coli.* (c) Sue Wickner (National Institutes of Health) uses mutant strains of *E. coli* to dissect DNA replication *in vitro.*

# Questions and Problems

1. After treatment with uv, *C. reinhardi* must be left in the dark for at least 12 hours in order to recover a significant number of mutant organisms. Can you estimate the length of the *C. reinhardi* cell cycle from this information? Explain your reasoning.

2. When *C. reinhardi* cells are exposed to a chemical mutagen at a particular concentration, 70 percent of the cells cannot form colonies. Describe the kind or kinds of effects these cells have suffered.

3. What molecular similarities exist between adenine and cytosine that render them susceptible to the same mutagens? What is similar about thymine and guanine in the same regard?

4. Explain why mutagenic transitions are easier to explain at a molecular level than are transversions.

5. Diagram a sequence of molecular events leading to the reversion of the addition shown in Figure 8-13a and a reversion of the deletion shown in Figure 8-13b.

6. A portion of a gene contains the sequence $\frac{\text{CATTG}}{\text{GTAAC}}$. For each of the mutagens listed below, show a probable series of events by which this sequence would become mutated after one or two rounds of replication: (a) nitrous acid; (b) hydroxylamine; (c) EMS; (d) uv irradiation.

7. Describe how you would set up a screening procedure for temperature-sensitive *dna* strains of *E. coli* using the 5-BUdR enrichment procedure (*Section 7.3*).

8. It is reported that the alkylating agent MNNG selectively mutates replicating regions of bacterial chromosomes, producing double or multiple mutations in close proximity to one another. MNNG is highly reactive in alkylating thiol (sulfur-containing) groups in proteins. Devise a hypothesis on the mechanism of MNNG mutagenesis based on these observations.

9. The purines and pyrimidines of DNA have the stable chemical structures shown in Figure 1-1, and the accuracy of Watson-Crick base-pair formation depends on the stability of these configurations. Watson and Crick proposed a theory of mutagenesis wherein, on rare and short-lived occasions, certain hydrogen atoms in a purine or pyrimidine ring would migrate to new unstable positions (a **tautomeric shift**). The resultant tautomers are drawn below.

Rare enol form of thymine (T*)     Rare imino form of cytosine (C*)     Rare imino form of adenine (A*)     Rare enol form of guanine (G*)

(a) Show with diagrams how T* forms base-pairs with guanine, C* with adenine, A* with cytosine, and G* with thymine (use Figure 1-6 as a model). (b) Show with diagrams similar to Figure 8-5 the mutagenic effect if a cytosine undergoes a tautomeric shift during DNA replication I but resumes its stable configuration at DNA replication II. (c) Would you expect tautomeric shifts to produce transitions? Transversions? Deletions/additions? Explain.

10. Because base analogues can induce both G $\longrightarrow$ A and A $\longrightarrow$ G transitions, they are able to cause reversions of the mutations they induce. (a) Would you expect hydroxylamine to be able to cause reversions in hydroxylamine-induced mutations? Explain. (b) Would you expect 5BUdR to be able to cause reversions in hydroxylamine-induced mutations? Explain.

11. The drug adenine arabinoside is able to combat such herpes virus infections as chicken pox, shingles, and herpes encephalitis without killing infected humans. How do you think the drug might act? How could you test this prediction with adenine-arabinoside–resistant herpes mutants?

12. Phage lysates of T2 and T4 were uv irradiated separately, and the surviving fraction was determined at 10-second intervals, on assay plates incubated either in the dark or in the light (permitting photoreactivation [PhR].) The results are given below.

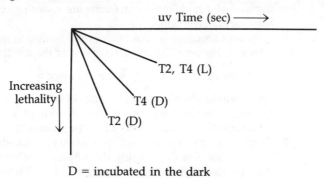

D = incubated in the dark
L = incubated in the light

Which is the best explanation? Explain your choice.

A. PhR is less efficient in T2 than in T4.

B. T4 has a repair mechanism that eliminates damage not suseceptible to PhR. T2 lacks this mechanism.

C. T4 has a repair mechanism that eliminates the same class of damage as PhR. T2 lacks this mechanism.

D. The T2 genome is larger than the T4 genome; hence it is more susceptible to uv.

13. Additions and deletions are frequently found to occur in regions of the chromosome that contain tandemly repeated DNA sequences. Using diagrams similar to Figure 8-13, explain why such sequences would favor this mode of mutagenesis.

14. You have two strains of bacteria, one carrying a deletion of the *recA* gene, the other carrying a mutation that makes the Lex A protein resistant to protease. Would you expect these strains to be more or less prone to uv-induced killing than wild-type strains? Explain.

15. Describe what you would expect to happen to DNA replication when an *E. coli* strain carrying a temperature-senstive *dnaG* mutation is shifted to nonpermissive temperature.

16. Would you expect Okazaki fragments to continue being synthesized for some minutes after a shift to nonpermissive temperatures in rapidly growing cultures of *dnaA* strains? *dna C-D* strains? *dnaG* strains? *dnaE* strains? a strain carrying a temperature-sensitive *lig* mutation? Explain for each case.

17. Distinguish between the activities of the rep protein, DNA gyrase, and the SSB protein in opening up a parental duplex for replication.

18. The DNA polymerase II enzyme of *E. coli*, encoded by the *polB* gene, has both exonuclease and polymerization activities. How might you use mutant strains to learn what role this enzyme plays in the cell?

# 9

# Genes and Gene Transcripts: General Features

A. Introduction

B. General Features of Transcription
9.1 Overview of Transcription
9.2 Defining Genes and Gene Transcripts
9.3 RNA Polymerases
9.4 Transcription of the Sense Strand of a Gene
9.5 General Structure of a Sense Strand and its Transcript

C. Anatomy of Structural Genes
9.6 Leader Sequences
9.7 Promoter Sequences
9.8 Terminator Sequences and Poly-A Tails

D. "Split Genes" in Eukaryotes

9.9 Intervening Sequences (Introns) and hnRNA
9.10 Splicing Out Intervening Sequences
Small nuclear RNA
9.11 "Purpose" and Evolution of Intervening Sequences

E. Visualizing Structural Gene Transcription
9.12 Lampbrush Chromosomes and Polytene Puffs
9.13 EM Images: Multiple Initiations and RNP Particles

F. Questions and Problems

## Introduction

This chapter and the chapters that follow are broadly concerned with a key activity of the genetic material: its ability to direct protein synthesis. This process is conveniently thought of as occurring in two major stages, **transcription** and **translation,** which are summarized diagrammatically in Figure 9-1. Many of the details of this figure will become meaningful as the chapters progress.

The present chapter describes the overall features of transcription and of genes that code for the amino acid sequence of proteins in prokaryotes and eukaryotes. The following chapter presents studies of specific genes and provides a more detailed picture of the principles set forth here. The genes contained in the DNA of mitochondria and chloroplasts, which have unique properties, are considered in Chapter 16.

# General Features of Transcription

## 9.1 Overview of Transcription

The transcription of genes is superficially similar to DNA replication (*Figure 8-2*): a polymerase enzyme matches complementary nucleotides along a DNA template according to the Watson-Crick base-pairing rules, and the bases are then polymerized to form a polynucleotide copy of the original strand. The polynucleotide transcripts are usually single-stranded RNA molecules; the enzyme that forms these copies is thus called **DNA-dependent RNA polymerase** (abbreviated as **RNA polymerase** in this text), and ribonucleotides, rather than deoxyribonucleotides, are the substrates for the polymerization reaction. The base uracil replaces thymine in RNA, as noted in Chapter 1, but otherwise the DNA strand is faithfully copied by the polymerase enzyme. As with DNA replication, polymerization proceeds in a 5' $\longrightarrow$ 3' direction and the transcript has opposite polarity to the template, so that the sequence 3' TACAAC 5' in DNA is transcribed as 5' AUGUUG 3' in RNA. Chain elongation proceeds at about 40 to 50 nucleotides per second (in *Escherichia coli* at 37°C), presumably in concert with a "melting" of the helix in front of the polymerase and a closing of the helix in its wake.

The "purpose" of transcription is to allow gene expression. Genes must remain in chromosomes, where their replication, repair, and transmission are assured, yet they must also be able to direct the activities of the cell, most notably protein synthesis. They therefore produce transcripts that diffuse away from the chromosomes and participate in protein synthesis. The transcripts do not, therefore, remain hydrogen bonded to the template the way daughter chromosomes do; instead, they "peel off" the chromosome, as illustrated in Figure 9-1. They are not, moreover, endowed with the same kind of permanence as the original genetic material; instead, they are typically degraded by cellular ribonucleases once their functional usefulness has been spent.

## 9.2 Defining Genes and Gene Transcripts

A **structural gene** is defined as a sequence of nucleotides that specifies the amino acid sequence (that is, the structure) of a polypeptide. Associated with the beginning and end of each structural gene are nucleotide sequences known as **controlling elements**, which are involved in the regulation of transcription. The interactions between controlling elements, RNA polymerase enzymes, and other regulatory proteins are considered in Chapters 22 through 25. Here it is sufficient to note the existence of two ubiquitous types of controlling elements: a **promoter** sequence, to which the RNA polymerase must bind if it is to transcribe the gene into mRNA, and a **terminator** sequence, which signals that the polymerase should dissociate from the template.

We must note at this point a special arrangement of structural genes known as an **operon.** In viruses and bacteria, groups of two or more related structural genes often share a single promoter and a single terminator and can thus be denoted as promoter . . . gene 1 . . . gene 2 . . . gene *n* . . . terminator. Additional controlling elements, such as **operators,** are usually associated with operons; these are described in detail in Chapter 22. In the case of operons, then, several genes will be copied into a single long RNA transcript.

The initial transcript of most (and perhaps all) structural genes and operons must be modified before it acquires full biological activity. Such **post-transcriptional processing** may involve adding nucleotide sequences onto the original transcript, cleaving sequences from the original transcript, or modifying existing bases by, for example, methylation. The final **processed** transcript of a structural gene is known as **messenger RNA (mRNA);** for operons, the transcript is sometimes specified as a multigenic mRNA. The mRNA molecules are then translated into polypeptide chains under the aegis of the **translation apparatus** (*Figure 9-1* and *Chapter 11*) of the cell.

Structural genes represent most of the genes in a chromosome, and when "genes" are considered in this and subsequent chapters, the term will usually refer to structural genes. Bacteria and eukaryotes possess, in addition, several kinds of genes whose RNA transcripts serve not as messenger intermediaries but as the final products of gene expression. Included here are the genes that specify **ribosomal RNA (rRNA)** and **transfer RNA (tRNA)** molecules, both of which perform critical functions in the translation of mRNA (Figure 9-1). The rRNA and tRNA genes are associated with promoter and terminator controlling elements, and their initial RNA transcripts undergo an elaborate processing before they assume their biologically active form. The tRNA and rRNA transcripts are not, however, translated into polypeptide chains. The tRNA and rRNA genes are mentioned here because, as is evident in the next chapter, their analysis has figured prominently in our present understanding of gene structure.

## 9.3 RNA Polymerases

An RNA polymerase enzyme must have the following properties. (1) It must be able to recognize promoter controlling elements in the double-stranded state of DNA. (2) It must be able to "burrow into" the DNA duplex at the proper promoter region and unwind the initial sequence of

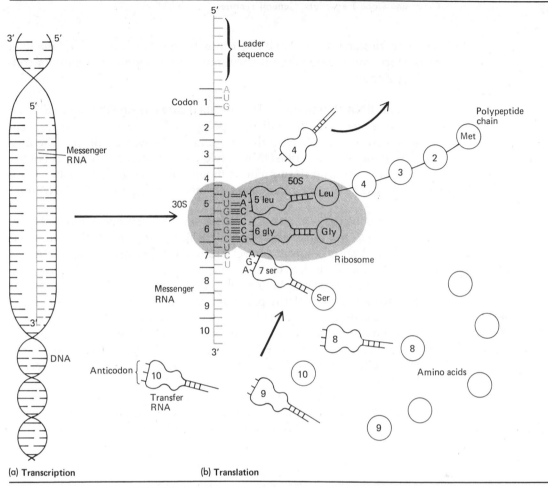

(a) Transcription  (b) Translation

**FIGURE 9-1 Transcription and translation.** (a) During transcription a strand of the DNA helix serves as a template for the synthesis of a complementary RNA copy or transcript (in this case, messenger RNA). This then "peels off" the DNA template. (b) During translation, three classes of RNA interact with a variety of enzymes and proteins to generate the formation of a new polypeptide chain. Ribosomal RNA is a component of the ribosomes that serve as a kind of scaffolding for the process of polypeptide synthesis. The ribosomes contain a large (50S) and a small (30S) subunit. Transfer RNA interacts with amino acids and mediates their correct insertion into the growing polypeptide chain. Messenger RNA carries the information contained in a gene to the ribosome. The information is encoded as groups of three nucleotides, with each specifying a particular amino acid. Each codon is recognized by a complementary anticodon on a transfer RNA molecule that has previously associated with that particular amino acid. In the figure most of the amino acids are represented by numbered circles; the amino acid glycine has just been bound to its site on the ribosome by the corresponding transfer RNA. It will form a peptide bond with leucine, thereby extending the growing polypeptide chain. The ribosome then moves the length of the codon along the messenger RNA and so comes in position to bind the transfer RNA carrying serine. (From *Ribosomes* by M. Nomura, copyright © October 1969 by Scientific American, Inc. All rights reserved.)

the gene for transciption. (3) It must copy the gene accurately. (4) Finally, it must stop transcribing when it encounters and recognizes terminator controlling elements.

**The *E. coli* RNA Polymerase** The so-called **core enzyme** of the *E. coli* RNA polymerase possesses three distinct polypeptides—$\beta$, $\beta'$, and $\alpha$; each enzyme has two copies of $\alpha$, so that the unit can be written **$\beta\beta'\alpha_2$**. The core enzyme alone can transcribe DNA into RNA. Its $\beta$ subunit, the target of the antitranscription drug **rifampicin,** is thought to catalyze RNA polymerization, whereas its $\beta'$ subunit appears essential for binding to the DNA template.

*In vivo,* the core enzyme normally associates with one of two polypeptides known as **$\sigma$ (sigma)** and **nusA**. The $\beta\beta'\alpha_2$ - $\sigma$ form of RNA polymerase initiates transcription at correct promoter sites along *E. coli* DNA and copies only one strand, whereas the core enzyme lacking $\sigma$ begins transcription at random along the template and copies both strands. Thus **the sigma factor is essential for promoter recognition. The nusA protein is similarly essential for correct termination:** $\beta\beta'\alpha_2$-nusA complexes recognize terminator sequences. There thus appears to be a cycle during transcription: as drawn in Figure 9-2, $\sigma$ associates with the core enzyme prior to initiation, after which $\sigma$ dissociates and nusA associates to allow correct termination.

The genes for the $\beta$ and $\beta'$ subunits of the *E. coli* enzyme, denoted *rpoB* and *rpoC* (for *RNA polymerase*), are located at 88.5 minutes on the genetic map (Figure 9-3) and are transcribed together as part of a larger

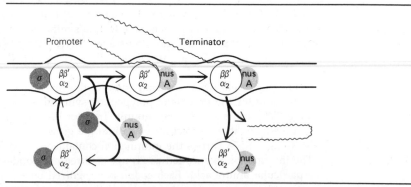

**FIGURE 9-2 Initiation-termination cycle** of transcription. The core enzyme $\beta\beta'\alpha_2$ (M = 400,000) associates with a $\sigma$ polypeptide (M = 95,000) prior to initiation of transcription at the promoter. Sigma then dissociates, and the nusA polypeptide (M = 69,000) adds on and travels with the polymerase to the correct terminator sequence. Sigma competes effectively with nusA for polymerase binding once the enzyme is released from the DNA. (From J. Greenblatt and J. Li, Cell **24**:421–428, 1981.)

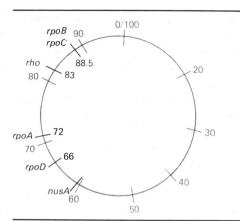

**FIGURE 9-3 Genes involved in DNA transcription** in *E. coli.*

operon (*Section 10.2*). The gene for α (*rpoA*) is not, however, found in this grouping but at 72 minutes (Figure 9-3), whereas the *rpoD* gene for σ maps at 66 minutes and the *nusA* gene maps at 61 minutes. Therefore, this multi-subunit enzyme is specified by genes that are generally well separated from one another in the *E. coli* genome.

**Eukaryotic RNA Polymerases**   Three RNA polymerases, **I, II,** and **III,** are present in eukaryotic nuclei. As summarized in Table 9-1, these enzymes transcribe different classes of genes: polymerase I, for example, resides in the nucleolus, is specialized for the synthesis of the large rRNA species known as 18S and 28S (described in Chapter 10), and is resistant to the inhibitor **α-amanitin** (derived from the mushroom *Amanita phalloides*), whereas polymerase II is highly sensitive to α-amanitin. All three types of enzyme have multiple polypeptide subunits and all appear to require factors analogous to sigma and nusA protein for accurate transcription.

## 9.4 Transcription of the Sense Strand of a Gene

The statement that the sequence of nucleotides in a gene encodes the amino acid sequence of a polypeptide refers to only one strand, the **sense strand,** of a duplex gene. Thus if the DNA base sequence 3' . . . . .

**TABLE 9-1   RNA Polymerases from Eukaryotes**

| Enzyme Class | Subnuclear Localization | Cellular Gene Transcripts | Viral Gene Transcripts |
|---|---|---|---|
| I | nucleolus | 18S, 28S rRNAs | None identified |
| II | nucleoplasm | hnRNA, mRNA | mRNA precursors |
| III | nucleoplasm | tRNAs, 5SRNA | Low-molecular-weight RNAs |

ATGTTTCAGACA . . . . 5′ codes for a sequence of several amino acids in a protein, then the base sequence of the complementary DNA strand, 5′ . . . TACAAAGTCTGT . . . 3′, is dictated by the Watson-Crick base-pairing rules followed by DNA polymerase enzymes and not by any dictum that it, too, should contain a "meaningful" sequence. In fact, the sole function of this **antisense strand** is to generate a complementary sense strand for use by the next generation of a virus or cell. (The sense strand of a gene will, for its part, give rise to an antisense strand during DNA replication.) By contrast, **promoters and terminators function as double-stranded structures.** Therefore, a gene can be depicted as

gene

---

promoter                                    terminator

---

antigene

with only the sense strand of the gene being transcribed.

In the case of many small viruses, all of the genes are located on one strand of the chromosome, and this strand alone is transcribed. In other cases, an example being the virus T4, **one strand of the duplex is the sense strand for certain genes, whereas the other strand bears the sense sequences for other genes** (experiments detailing this conclusion are described in Chapter 23). This means that in visualizing the chromosome of T4 we can think of a given single strand of DNA as having long sequences of bases containing meaningful information, followed by long sequences of bases that do not code directly for protein and presumably code for nothing at all.

## 9.5 General Structure of a Sense Strand and Its Transcript

We can now focus more closely on a sector of a chromosome that carries a structural gene. Since the chromosome is transcribed in a 5′ ⟶ 3′ direction and since the resultant mRNA is translated in a 5′ ⟶ 3′ direction (*Figure 9-1*), it follows that the beginning of the gene will lie to the 3′ end of the sense strand and the conclusion of the gene will lie at the 5′ end:

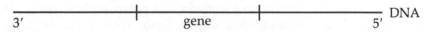

As indicated in Figure 9-1 and detailed fully in Chapter 11, the first amino acid in a polypeptide chain is usually methionine, which is specified in mRNA by the sequence 5′-AUG-3′ (the **initiating codon for translation**). We can therefore be more specific:

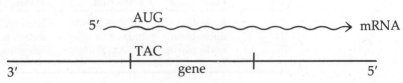

You will note that the AUG is not drawn at the 5′ end of the mRNA. The nucleotides that lie between the 5′ end of the mRNA and the AUG of the gene constitute what is best called a **5′ noncoding sequence** but what is more commonly called a **leader sequence.** The initial nucleotide at the 5′-end of an RNA transcript is usually found to be a purine (5′-pppA or 5′-pppG, generally denoted 5′pppPu), meaning that the DNA will usually carry an **initiating pyrimidine (Py) for transcription** (C or T). Incorporating this information, we have

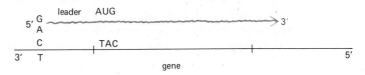

Finally, we must add the **promoter,** defined as the site at which an RNA polymerase attaches to DNA and initiates transcription, and the **terminator,** where transcription is concluded. A **3′ noncoding sequence** or a **trailer** sequence may be transcribed between the conclusion of the gene and the terminator sequence.

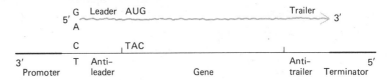

The above diagram of a DNA strand and its mRNA transcript allows us to imagine an RNA polymerase enzyme binding to the promoter at the left and "flowing" into the gene. Such a picture of transcription has led to the useful terms **upstream** and **downstream:** the promoter is said to be upstream from the gene; the antitrailer sequence is said to be downstream from the gene; and so on.

We must at this point deal with two matters of notation. In the preceding paragraphs, the mRNA was drawn with the 5′ end on the left and the 3′ end on the right. This convention is very logical, for the mRNA is translated in a 5′ ⟶ 3′ direction and the amino acid sequence of its protein product can therefore be read in the usual left-to-right fashion. The difficulty comes with choosing a convention to display the sequence of a gene. The sense strand sequence of a gene can be read from left to right only if the 3′ end of the strand is placed on the left, as we have done in the preceding paragraphs. Most biologists, however, prefer to retain the 5′ ⟶ 3′ orientation of the message even when they are contemplating the structure of a gene. Therefore, we will usually adopt the standard convention of depicting the sequence of a gene as the sequence of its **nontranscribed (antisense)** strand, a sequence identical to that of its mRNA product except, of course, that thymines replace uracils.

The second matter of notation concerns the numbering of nucleotides in a sequence. In many presentations of the structure of a gene, the initiating 5' purine on the antisense strand is denoted as Number 1; nucleotides upstream are given negative numbers (−1, −2, and so forth), and downstream nucleotides are given positive numbers. This convention persists for most prokaryotic genes, and it is the one we will generally follow in this text. We should note, however, that the A of the ATG initiating codon for translation is sometimes denoted Number 1.

In the sections that follow, we examine structural genes in more detail, using information derived principally from genes that have been cloned and sequenced as described in Chapter 4. Table 9-2 summarizes sequences that have been identified as critical for gene expression; these will be explained during the course of the chapter.

**TABLE 9-2   DNA Sequences Important for Gene Expression†**

| Name | Consensus Sequence (5' ⟶ 3') | Location and Function |
|---|---|---|
| A. Ribosome-binding sequence (Shine-Dalgarno sequence) | AGGA | Prokaryote leader close to AUG start; binds to 3'AUUCCUCCA . . . 5' of the 16S rRNA |
| B. Cap sequence | PyPyA*TTCPu | Eukaryote leader terminus; signals addition of m$^7$G cap to the A (asterisk) in mRNA |
| C. −35 region | TTGACA | E. coli promoter; upstream binding site for RNA polymerase |
| D. Pribnow box (−10 region) | TATAAT | E. coli promoter; midstream binding site for RNA polymerase |
| E. Goldberg-Hogness box (TATA box) (−30 region) | TATAAAAA | Possible eukaryotic promoter sequence |
| F. Terminator | Inverted repeat followed by T cluster | Prokaryotic terminator; signal for polymerase and transcript to dissociate |
| G. Intron-exon boundaries | Exon I-AG\|GTAAGTA-(intron)-PyTTTTTTTCTTNCAG\|G-Exon II | Eukaryotic structural gene; introns perhaps excised at sites denoted by vertical lines |

†Presented are "**consensus sequences**," each representing the most prevalent version of a particular element (often determined by computer analysis of many sequences). The sequence found in a particular gene may well be slightly different from the consensus sequence.

# Anatomy of Structural Genes

## 9.6 Leader Sequences

The length of mRNA leaders varies considerably, but all appear to have the same major function: **leader sequences establish appropriate associations with a ribosome so that translation will initiate efficiently at the AUG "start" codon.** This is accomplished differently for prokaryotes and eukaryotes.

For prokaryotes and their phages, efficient translation is found to depend on the presence, in the leader region, of a so-called **ribosome binding sequence** (Table 9-2A), also known as **Shine-Dalgarno** sequence after the two investigators who first noticed it. This is found about 5 to 10 base pairs upstream from the AUG codon, and it proves to be homologous to the 3'-OH terminus of 16S rRNA, the major RNA component of the small ribosomal subunit (*Figure 9-1*). It has therefore been proposed that as this portion of the leader peels off the DNA template, it forms base pairs with the complementary rRNA sequence in the ribosome and thereby assumes the correct mRNA-ribosome orientation.

For eukaryotes and their viruses, the original initiating pyrimidine in the leader is usually a 5'-pppA. This acquires a critical modification shortly after it is transcribed: a 7-methyl guanosine is added enzymatically to form an **$m^7G(5')pppA$** terminus known as a **cap.** The cap is recognized in the cytoplasm by a **cap-binding protein,** which, in turn, mediates the correct mRNA-ribosome associations. When the nucleotide sequence of cloned eukaryotic genes is examined, a **cap sequence** (Table 9-2B) can be identified which specifies that this modification should occur.

## 9.7 Promoter Sequences

**Prokaryotic Promoters**  A promoter sequence is classically positioned upstream from the initiating purine, the idea being that the RNA polymerase must bind to and open up the DNA ahead of the site at which the initial nucleotide will be copied. In fact, most (although not all) RNA polymerases are large, multisubunit proteins; therefore, even as one end of the enzyme is binding to an upstream recognition signal, the other end may be making contact with downstream transcribed regions.

These steric concepts apply in considering the approaches taken to identify promoter sequences. The technique most widely used, known as **footprinting,** involves isolating chromosome fragments derived from the 5' region of a gene and incubating these with purified RNA polymerases under *in vitro* conditions in which transcription cannot proceed. The mixture is then exposed to pancreatic DNase, which digests all the DNA except that which is protected by its association with RNA polymerase (see

Figure 9-4). These "polymerase-protected sites," or "footprints," are then isolated and subjected to DNA sequencing (*Box 4.1*). More detailed information can then be obtained by asking which bases in these DNA sequences are protected from chemical modification (e.g., methylation) by RNA polymerase contacts.

For *E. coli* RNA polymerase, such approaches have yielded the results summarized in Figure 9-5. The contact sites are seen to cluster in three regions: the **−35 region;** the **Pribnow box** (again, named after the scientist who first noticed it) centering at −10; and the region surrounding the initiating purine at +1. Space-filling models reveal that upstream from the Pribnow box, **RNA polymerase contacts are made only on one side of the DNA helix.** Thus if we refer to Figure 1-8 and call the visible parts of the DNA helix the front side and the parts that go behind the paper the back side, then RNA polymerase can be visualized as reaching in from the front

**FIGURE 9-4 RNA polymerases binding to promoters.** Depicted is the complete chromosome of the *E. coli* phage T7; three *E. coli* RNA polymerase molecules (arrows) have bound to the three early T7 promoters (A1, A2, and A3). (Courtesy of Dr. Th. Koller.)

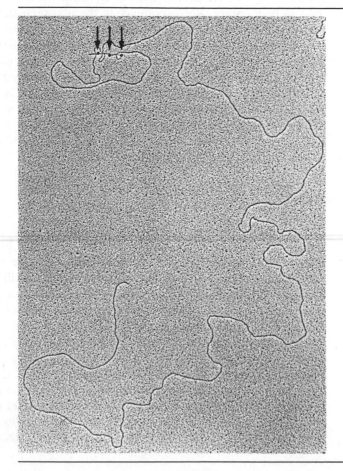

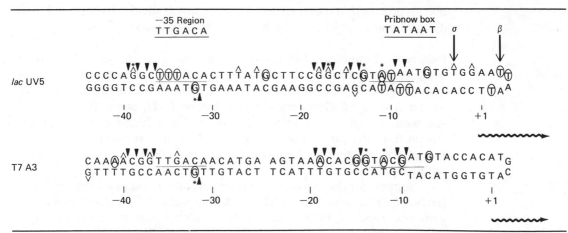

**FIGURE 9-5 RNA polymerase contacts** with the *lac* UV5 promoters of *E. coli* and the A3 promoter of T7 (*Figure 9-4*). The figure shows RNA polymerase contacts with phosphates and purines in the *lac* UV5 and T7 A3 promoters, plus thymine contacts in the *lac* promoter. The unwound regions in both, represented by a separation of the strands, are also shown. Base pair positions are numbered relative to the start of transcription at +1, and the two sequences are aligned with respect to their Pribnow boxes and −35 regions (underlined) with the consensus sequences indicated at the top. The location of the sigma and beta subunits of RNA polymerase, identified by photochemical cross-linking to bromouracil-substituted *lac* UV5 promoter DNA, are indicated. (▼) Ethylated phosphates that interfere with binding of the polymerase; (○,∧) purines (G,A) that the polymerase protects from or enhances to dimethyl sufate attack, respectively; (○,∧) thymines (substituted with bromouracil) that the polymerase protects from or opens to photochemical cleavage, respectively; (*) methylated purines that interfere with binding of the polymerase. (From U. Siebenlist, R.B. Simpson, and W. Gilbert, *Cell* **20:** 269, 1980.)

side to make contacts with those bases that are accessible from a frontal position.

The importance of the −35 and Pribnow box regions for *E. coli* promoter function is supported by two other kinds of observations. First, at least 50 prokaryotic promoters have, to date, been sequenced, and all contain similar sequences in these two locations (Table 9-2C and D). Second, a number of **promoter mutations** have been identified that either block or enhance the transcription of particular prokaryotic genes; sequence analysis of these mutant promoters usually reveals base substitutions or deletions in one of these two regions.

While essential recognition and binding events clearly occur at −35 and at −10 in *E. coli*, the unwinding of the DNA strands and the selection of the sense strand for initiation of transcription begins in the middle of the Pribnow box and extends just past the +1 site (Figure 9-5). Again thinking in three dimensions, we can think of the *E. coli* RNA polymerase as an oblong protein that, having bound to recognition sites by its back end, uses its front end to pry open the helix and initiate transcription.

**Eukaryotic Promoters**   Much recent work has involved attempts to identify eukaryotic promoters. Comparisons of DNA sequences upstream from various sequenced eukaryotic genes have consistently uncovered the presence of a so-called **Goldberg-Hogness box** or **TATA box** with the consensus sequence shown in Table 9-2E. Since this is found about 30 bp upstream from the cap sequence and is AT rich, it may serve a polymerase recognition function analogous to the AT-rich −35 or −10 sequence of *E. coli* promoters.

Support for this concept comes from *in vitro* mutagenesis experiments (*Section 8.7*) in which the TATAAAAA sequence of a cloned eukaryotic gene was replaced by a TAGAAAAA sequence. Specific transcription of this mutant DNA in an *in vitro* system was drastically descreased compared with the normal gene, in much the same way that one sees for mutations in the *E. coli* promoter sequences. Additional *in vitro* mutagenesis studies, however, have shown that the deletion of sequences upstream from the Goldberg-Hogness box can also disrupt transcription, and certain animal virus genes can be transcribed efficiently in eukaryotic cells even though they lack a Goldberg-Hogness sequence. It appears, therefore, that **eukaryotic genes utilize a variety of promoters,** a variability that may be important in determining which gene is expressed in which cell. This concept is explored more fully in Chapters 24 and 25.

Since the DNA associated with nucleosomes is twisted into a constrained, superhelical configuration, the DNA of eukaryotic genes somehow "relaxes" so as to become more accessible to RNA polymerases. Evidence has been obtained for changes in the conformation of transcribed DNA (*Section 25.18*), but it is not yet possible to describe these changes in molecular terms.

## 9.8 Terminator Sequences and Poly-A Tails

Prokaryotic terminator sequences share two characteristics (Table 9-2F): (1) a string of Ts in an AT-rich region directly precedes the site at which transcription is terminated; and (2) an inverted-repeat sequence (*Section 4.9*) is found upstream, lying within a somewhat GC-rich region. A current model for how these sequences signal termination is detailed in Figure 9-6. Briefly, the model proposes that once RNA polymerase has transcribed the inverted-repeat sequence and the series of Ts (Figure 9-6a), the RNA transcript forms a stem-and-loop structure (*Section 4.9*) *via* its self-complementary inverted-repeat sector (Figure 9-6b) and dissociates from the template (Figure 9-6c). This event, it is proposed, somehow causes the polymerase to undergo a change in configuration such that it can no longer elongate the 3' terminus, and the enzyme and transcript both dissociate from the template.

As noted in Section 9.3, termination in *E. coli* is effected by the $\beta\beta'\alpha_2$-

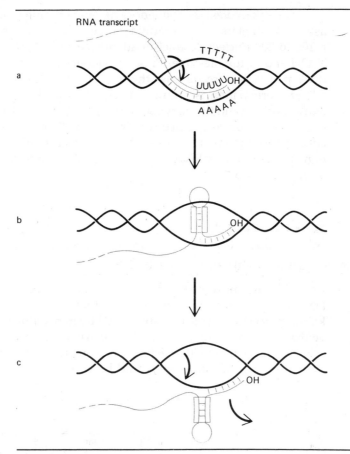

RNA transcript

a

b

c

**FIGURE 9-6 Termination of transcription by RNA polymerase.** The polymerase molecule has been omitted for clarity; the enzyme would be expected to span at least 10 to 20 bp on either side of the termination point. The inverted-repeat sectors are depicted as two blocks on the RNA strand. (a) RNA polymerase has reached the position at which it will terminate. The transcript is still hydrogen bonded to the template DNA strand through the distal half of the potential hairpin. Formation of the hairpin (an intramolecular event in the RNA) should displace the transcript from the template. (b) Hairpin formation in the transcript restores the potential of the DNA helix to reanneal. At this point, polymerase shifts rapidly into a configuration in which it is unable to elongate the transcript further. (c) Transcript displaced, possibly by competition from the nontemplate strand of DNA, followed by mRNA release. (From P. J. Farnham, and T. Platt, *Cell* **20**:739, 1980.)

nusA form of the RNA polymerase. Release of the transcript from the template also appears to require the protein **rho,** specified by the *rho* gene (*Figure 9-3*). Rho protein has ATPase activity and binds to both DNA and RNA, but its ability to facilitate termination is not yet well understood at a molecular level.

For eukaryotes, the sequences signaling termination are not yet established. A hallmark of most termination events, however, is the addition of a 100 to 200 nucleotide tract of adenosine residues (a **poly-A tail**) to the 3'-OH end of the message. Present evidence suggests that the eukaryotic RNA polymerase reads past the site at which the poly-A tract will be added. Termination is followed by an endonucleolytic cut at the polyadenylation site and the enzymatic addition of the tail. The "purpose" of the tails is as yet unclear, since certain eukaryotic mRNAs are readily translated without them. Their existence, in any case, permits the ready isolation of eukaryotic mRNA by affinity chromatography using poly-T columns (*see Box 3.1*).

# "Split Genes" in Eukaryotes

## 9.9 Intervening Sequences (Introns) and hnRNA

Certainly the most unexpected discovery to emerge from recombinant DNA and sequencing technology was that **the coding region of most eukaryotic structural genes is interrupted by noncoding sectors,** called **intervening sequences,** or **introns.** Thus if we consider a gene whose amino acid coding sequences (the **e**xpressed sequences, or **exons**) are interrupted by two introns, its structure would be written as

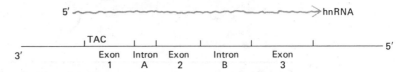

Intervening sequences are extremely diverse. Some eukaryotic genes lack them altogether; at the other extreme, the gene for α-collagen contains 52. They also range in length from 10 to 10,000 base pairs. Exons are similarly diverse: the gene encoding the chicken protein ovomucoid, for example, is split into 14 exons; the shortest of these is only 20 nucleotides, whereas the longest is 181 nucleotides. Specific examples of introns and exons are given in Chapter 10; here we focus on the more general question of how they are transcribed and processed.

The primary transcript of the "split gene" drawn above is perhaps most appropriately called **precursor mRNA (pre-mRNA)** because, before it can be translated by ribosomes, the introns must be removed by steps that we will consider in Section 9.10. The name most commonly given to this transcript, however, is **heterogeneous nuclear RNA,** or **hnRNA,** a term that has an interesting history.

Some years prior to the discovery of intervening sequences, it had been established that the eukaryotic nucleus contains enormous quantitites of RNA molecules that share several properties expected of mRNA: they are heterogeneous in size; they hybridize with a considerable fraction of the genome in DNA-RNA hybridization experiments (*Section 4.10*); and they usually carry poly-A tails on their 3′ ends. Unlike mRNA, however, they appear never to leave the nucleus, and their half-life (30 minutes or less) is much shorter than most mRNA (on the order of 7 to 24 hours). Moreover, when analyzed by **sucrose gradient centrifugation** (Box 9.1), a technique that estimates the size of macromolecules in terms of their sedimentation coefficient (expressed in **Svedberg units** or **S values**), this RNA is found to vary in size from 10S to 200S, whereas most mRNA ranges between 6S and 30S. In short, the puzzling conclusion was reached that the eukaryotic nucleus synthesizes numerous, enormous, and heterogeneous RNA molecules that appear to be degraded shortly after they are produced and do not engage in protein synthesis.

**BOX 9.1**
Sucrose Gradient
Centrifugation

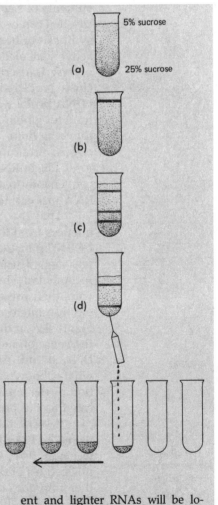

This technique is described as a method for estimating the size of RNA molecules. It can also be used for other species such as proteins, ribosomal subunits, or ribosomes.

a. A centrifuge tube containing a continuous gradient of sucrose concentration ranging from 5 percent at the top to 25 percent at the bottom (other ranges may also be chosen) is prepared.

b. A small sample of RNA is carefully layered on top of the gradient, and the tube is centrifuged at high speeds for several hours. Like-sized RNA molecules tend to move with similar velocities.

c. Centrifugation is stopped. RNAs of similar molecular weights are located in discrete regions of the gradient.

d. The bottom of the tube is punctured with a hypodermic needle, and successive samples are collected with a fraction collector. The absorbance of each sample is then measured at 260 nm, an ultraviolet absorption maximum for nucleic acids. Samples with little RNA will show little absorbance; those with more RNA will show more absorbance. A plot of these absorbances generates a profile of RNA concentrations within the gradient. In such a profile, heavier RNAs will be found near the bottom of the gradient and lighter RNAs will be located near the top. This size information is expressed by calculating a sedimentation coefficient for each species in terms of Svedberg (S) units, heavier species having larger S values than light species.

This technique differs from the CsCl density gradient procedure we have described for DNA (*Box 2.1*) in two respects. (1) The sucrose gradient is *preformed*, whereas the CsCl gradient is produced only after many hours of centrifugation; therefore, the sucrose gradient approach is faster. (2) The CsCl gradient contains an *equilibrium* distribution of macromolecules, whereas the sucrose gradient contains molecules that are in the process of moving to the bottom of the tube; the centrifugation is stopped while they are still in transit.

The concept that hnRNA might in fact be pre-mRNA was given strong support by studies of mouse β-globin, one of the polypeptide constituents of hemoglobin (*Figure 3-5*). The mRNA for β-globin is a 10S species that can be readily purified from erythroid cells engaged in the massive synthesis of hemoglobin. This mRNA was then copied *in vitro* into a radioactively labeled complementary DNA (cDNA) species using the enzyme reverse transcriptase (which makes single-stranded DNA copies of RNA chains) and radiolabeled deoxyribonucleotide triphosphates. Finally, the cDNA probe was used in DNA-RNA hybridization assays (*Section 4-10*) to hunt for globin-like sequences in the hnRNA of erythroid cell nuclei. When this was done, a 15S hnRNA species was found to hybridize with the cDNA, indicating the following progression: globin genomic DNA $\longrightarrow$ 15S hnRNA $\longrightarrow$ 10SmRNA. It thus appeared that a 15S $\longrightarrow$ 10S conversion occurred, and it was proposed that this was accomplished by RNA processing enzymes.

That the 15S $\longrightarrow$ 10S reduction is accomplished by removing sequences from the *interior* of the transcript was convincingly demonstrated by S. Tighlman, P. Leder, and their colleagues. The mouse β-globin gene was cloned using a λ phage chromosome as a vector (*Section 4.13*). The recombinant chromosomes were first partially denatured (*Figure 4-3*); they were then presented with the 15S globin hnRNA and examined with the electron microscope. The resultant images resemble those diagrammed in Figure 9-7: in the globin gene region, the 15S hnRNA has hybridized with the sense strand of the gene, preventing the reannealing of the antisense DNA strand. As a result, the antisense DNA strand bulges out, forming what is known as an R-loop because it is caused by RNA-DNA hybridization of the complementary strand. Thus the 15S hnRNA is clearly implicated as the primary transcript of the β-globin gene.

A very different result came when this experiment was repeated using the 10S mRNA species rather than the 15S hnRNA species. As drawn in Figure 9-7b, two R-loops are present, between which is a sector of 645 nucleotides of duplex DNA that is not homologous to the 10S species and therefore does not participate in R-loop formation. In other words, during the 15S $\longrightarrow$ 10S processing, sequences in the middle of the 15S transcript are excised, and the broken pieces are resealed to form the 10S species.

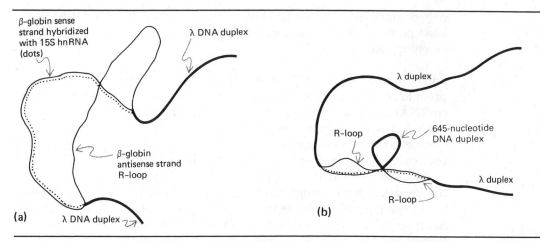

FIGURE 9-7 **R-loop mapping of mouse β-globin DNA** (see text for details). A smaller, 116-bp intron is present in this gene as well as the 645-bp intron shown. (From S. Tilghman, P. Curtis, D. Tiemeier, P. Leder, and C. Weissman, *Proceedings of the National Academy of Science* U.S. **75**:1309, 1978.)

## 9.10 Splicing Out Intervening Sequences

The cell must use great precision in excising introns from hnRNA and splicing the cut ends together so that a correct mRNA molecule is generated for translation. DNA sequence analysis has been performed on the introns present in a large number of cloned structural genes, and "consensus sequences" at the intron-exon margins have been identified (Table 9-2G) that may serve as recognition signals for excision-splicing enzymes. The most invariant feature of these sequences has come to be known as the **GT-AG Rule:** as depicted in boldface in the consensus sequence of Table 9-2G, the location of the 5′ cut carries a GT (or a GU in the RNA transcript), and the location of the 3′ cut carries an AG.

Support for the importance of this consensus sequence comes from studies of a mutation in the human gene for the β-chain of hemoglobin. DNA was obtained from a human homozygous for this gene who suffers from a disease known as $\beta^+$-thalassemia (defective production of β-globin), and sequence analysis revealed that the mutant β-globin DNA differed from the wild type at only one position: a G $\longrightarrow$ A transition had occurred within the small intron of the gene. The new A created an **AG** couple, and this was surrounded by the same consensus nucleotides that are located at the natural 3′ splice junction of this intron. In other words, **the mutation has created a new 3′ splice junction in the middle of the small intron.** It is proposed, therefore, that many of the transcripts from this gene may be spliced at the new, incorrect junction rather than at the correct junction and that the abnormally spliced mRNA cannot be translated properly. As a result, the total amount of β-globin synthesis is severely reduced.

**Small Nuclear RNA (snRNA): A Possible Splicing Mediator**  Splicing of pre-mRNA may be mediated by a class of stable, small nuclear RNAs (**snRNA**) found in all eukaryotes. The most abundant of these, called **U-1 snRNA**, has a 5'-terminal sequence complementary to the consensus sequence at the two putative splice junctions, as shown in Figure 9-8. It has been proposed, therefore, that the snRNA holds the two ends of the intron in the double-stranded configuration drawn in Figure 9-8 and that it is this configuration that is recognized by a specific ribonuclease-splicing activity.

Experimental evidence for the participation of U-1 in gene splicing has recently been obtained by V. W. Yang and associates. They took advantage of the fact that the infection cycle of adenovirus, which occurs in human nuclei, depends on the correct splicing of viral RNA transcripts. They isolated nuclei of virally infected cells and incubated them with antibodies (*Box 2.3*) directed against U-1. The antibodies had no effect on such parameters as overall transcription or polyadenylation of the adenoviral RNA molecules, but they were extremely effective at blocking the splicing reactions.

## 9.11 "Purpose" and Evolution of Intervening Sequences

There is as yet no definitive answer to the following obvious question: why do eukaryotes "bother" to have introns that they first transcribe and then excise? Several observations are compatible with the theory that introns serve no purpose.

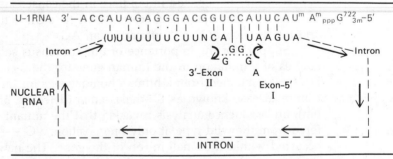

**FIGURE 9-8 Splicing out an intervening sequence.** Model showing how base pairing of U-1 snRNA to the consensus sequence at an exon I-intron-exon II region of hnRNA could form an **RNA splicing complex,** with dots indicating the predicted points of cutting and splicing. (From J. Rogers, and R. Wall, *Proc. Nat. Acad. Sci.* **77:**1877, 1980.)

1. Eukaryotic genes lacking introns (e.g., the genes for histones and actin) are expressed as well as intron-containing genes.
2. Mutations induced within most introns have no obvious effect on the phenotype.
3. Certain rat strains have two nonallelic genes for the hormone insulin, one containing a single intron and the other containing two, and both are expressed equivalently.

On the other hand, there are indications that at least some introns have functional roles.

1. *In vitro* deletion of an intron within a structural gene of the animal virus SV40 blocks the expression of the gene.
2. An intron in a yeast mitochondrial gene has the puzzling property of encoding an RNA splicing enzyme necessary for removing its own sequences from the gene transcript (Section 16.6 describes this intron in more detail). Although this property in itself is hardly a satsifactory "reason" to have introns, the splicing enzyme might play additional roles in mitochondrial gene expression.

Perhaps the most interesting theory is W. Gilbert's proposal that introns might have been important in eukaryotic evolution. He suggests that two (or more) functional genes, originally separated in the genome, might be brought together to create a new gene interrupted by one (or more) introns. Each exon in the new gene would encode a **domain** in the polypeptide product whose function corresponds to the function of the original gene, and the organism would be able to express this "patchwork gene" because it is able to splice out introns. Supporting this notion is the fact that each of the three exons in the globin gene appears to play a specific functional role in hemoglobin physiology: the central exon, for example, specifies the heme-binding sector of the polypeptide, whereas the right-hand exon specifies residues critical for intersubunit bonding. In short, the Gilbert proposal suggests that most present-day introns may be **evolutionary relics,** once important for creating new eukaryotic genes and persistent simply because cells have no way to get rid of them.

# Visualizing Structural Gene Transcription

## 9.12 Lampbrush Chromosomes and Polytene Puffs

Structural gene transcription is ordinarily impossible to visualize with the light microscope, but early cytogeneticists discovered two eukaryotic cell types in which the process could be detected. The first are vertebrate and invertebrate oocytes in the diplotene stage of meiotic prophase (*Section 5.6*). The chromosomes of such cells, you will recall, are partially condensed, but to accomplish transcription they send out long loops of

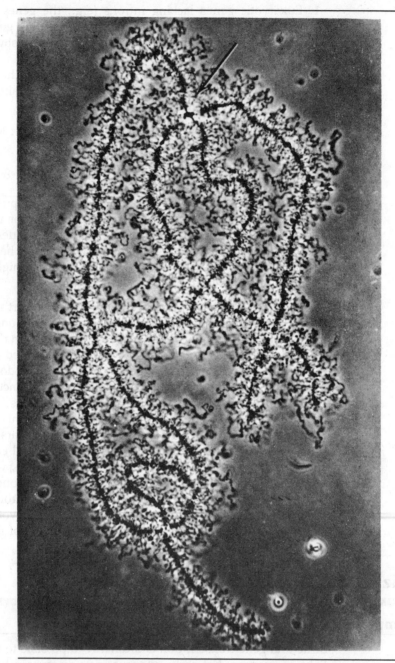

**FIGURE 9-9 Lampbrush chromosomes** from a *Triturus* oocyte. Two homologues are present, each being made up of two chromatids that cannot be individually resolved here. Extending from each chromatid pair is an array of lateral loops that reminded early cytologists of a lampbrush. Each loop is believed to be a single strand of nucleosomes surrounded by newly transcribed RNA. (From J. Gall.)

nucleosome-packaged DNA (*Section 3.6*) from their central axes, earning them the designation **lampbrush chromosomes** (Figure 9-9). Closely associated with each loop are large hnRNA primary transcripts, which are presumed to function, in some unknown way, during egg maturation and/or embryonic development.

The second cell types wherein transcription is manifest are the larval cells of many dipterans, where polytene chromosomes reside (*Section 2.14*). During certain periods of larval life, or in response to such stimuli as hormones, specific polytene chromosome bands lose their compact form and appear to separate into their component strands, forming what is aptly termed a **puff** (Figure 9-10). Associated with the puffs are large hnRNA transcripts, which, in several cases, have been shown to be processed into smaller mRNA species that dictate the synthesis of protein products known to be elicited by the particular stimulus applied. Details of the puffing response are presented in Sections 24.4 and 24.6.

## 9.13 EM Images: Multiple Initiations and RNP Particles

The electron microscope (EM) has afforded the most detailed images of structural gene transcription. Figure 9-11 shows the process in *E. coli*. Ribosomes are seen to associate with each mRNA even as it peels off the tem-

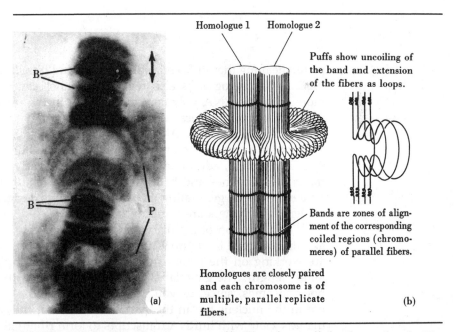

FIGURE 9-10 (a) **Two chromosome puffs** (P) and unpuffed polytene bands (B) from the salivary gland of *Chironomus*. (From U. Clever.) (b) Schematic representation of chromosomal puffing. (From A. B. Novikoff and E. Holtzman, *Cells and Organelles.* New York: Holt, Rinehart and Winston, 1970, Figure IV-25.)

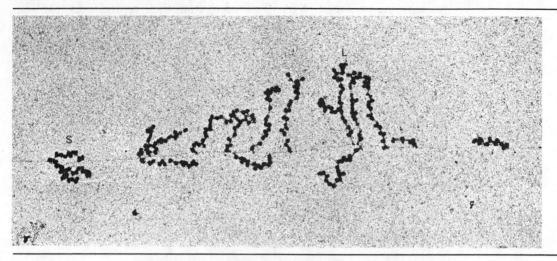

**FIGURE 9-11 Structural gene transcription in *E. coli*.** The faint left-to-right strand is chromosomal DNA. Ribosomes (globular structures) associate with the mRNA transcripts that peel off the DNA. The promoter for this gene is apparently near the far left. (From O. L. Miller, B. A. Hamkalo, and C. A. Thomas, Jr., *Science* **169**:392, 1970.)

plate, meaning that in *E. coli,* **transcription and translation are tightly coupled.** Such micrographs also reveal that genes can be "worked on" by more than one RNA polymerase at a time. In the present case, the promoter is apparently located to the far left, and at the time the cell was broken open, some RNA polymerase had traversed almost the entire length of the gene, generating the long transcripts designated *L* in Figure 9-11. Meanwhile, additional polymerases had clearly attached to the promoter at later times and had traversed shorter lengths of the gene at the time of cell lysis, generating the short transcripts designated *S*. Intermediate-length polymers are found in between.

Transcription of a eukaryotic structural gene is seen in Figure 9-12a and diagrammed in Figure 9-12b. In this case, at least 12 RNA polymerases are working on the gene to generate the 12 transcripts of progressively longer length. The globular structures associating with the transcripts are not ribosomes, since, as you will recall, translation in eukaryotes occurs not in the nucleus but in the cytoplasm. Instead, they represent proteins that associate with hnRNA transcripts to form **ribonucleoprotein particles (RNPs).**

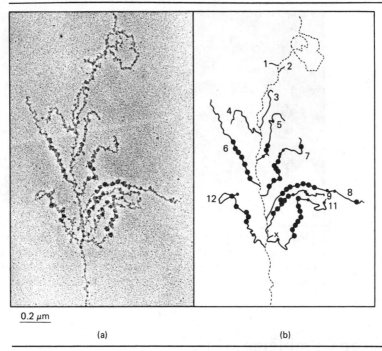

0.2 μm

(a)                                      (b)

**FIGURE 9-12** (a) **Structural gene transcription in *D. melanogaster* embryos.** (b) Interpretive tracing of micrograph shown in (a). Dotted line represents chromatin strand; solid lines represent transcripts; and solid circles represent globular RNP structures. Fibril No. 10 (designated x) has apparently been broken. (From A. L. Beyer, O. L. Miller, Jr., and S. L. McKnight, *Cell* **20:**75, 1980.)

RNPs are also found in the cytoplasm, where they contain mRNA rather than hnRNA. Comparison of the two RNP types shows that the proteins associated with hnRNA in the nucleus are quite different from those associated with mRNA in the cytoplasm. Therefore, some of the nuclear RNP proteins may play specific roles in hnRNA $\longrightarrow$ mRNA processing and then dissociate from their products prior to cytoplasmic export. It has also been reported that the cytoplasmic RNP proteins associated with messages that are translated on rough ER (*Section 2.5*) are different from those that are translated on free ribosomes. Therefore, some of the cytoplasmic RNP proteins may specify the "cytoplasmic address" to which their associated mRNAs are to be delivered.

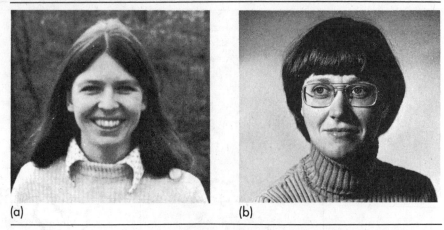

(a) Joan Steitz (Yale University) studies the role of small nuclear RNA in the processing of mRNA. (b) Shirley Tilghman analyzed intron splicing in collaboration with Dr. Philip Leder at the NIH, and presently studies eukaryotic gene structure at the Institute for Cancer Research, Philadelphia.

# Questions and Problems

1. Compare Figures 8-1 and 9-3. How would you characterize the relative complexity of transcription and replication based on a comparison of the two maps? Speculate why transcription lacks an elaborate error-correcting mechanism (Hint: compare the genetic consequences of a single transcription error *versus* a single replication error).

2. The promoter for the wild-type *lac* operon of *E. coli* is termed a low-level promoter in that RNA polymerases bind to it only very infrequently unless a regulatory protein known as CAP is present. Several mutations have been isolated that convert this into a high-level, CAP-independent promoter, one being the *uv5* promoter. The Pribnow box sequence of the wild-type *lac* promoter region reads 3'ATACAAC5'; the *uv5* promoter reads 3'ATATTAC5'. (a) What mutation or mutations have occurred to produce the *uv5* promoter? (b) Write out the duplex sequences of the two promoters. Which would be more readily "melted" (denatured) by a polymerase? Explain. How might this difference explain the different properties of the two promoters?

3. Is a cDNA copy of an mRNA molecule equivalent to the transcribed strand of a gene? If not, how would you distinguish them?

4. What might you expect to find if you compared the restriction fragments generated from β-globin gene DNA, from 15S cDNA, and from 10S cDNA by a given restriction enzyme (assume the cDNA to be duplex)?

5. Phage SP8 has a duplex DNA chromosome and infects *B. subtilis*. Native and dissociated (*Section 4.4*) SP8 DNA was run on a CsCl gradient, giving the profile shown in (a) below. The heavy (*H*) and light (*L*) strands (*Box 4.1*) were then separated from one another and purified into two fractions, (b) and (c). De-

scribe how you would use DNA/RNA hybridization (*Section 4.10*) to determine whether the *H* or *L* strands are the sense strands of SP8. What outcome would you expect if the *H* strand is sense for some genes and the *L* strand is sense for other genes?

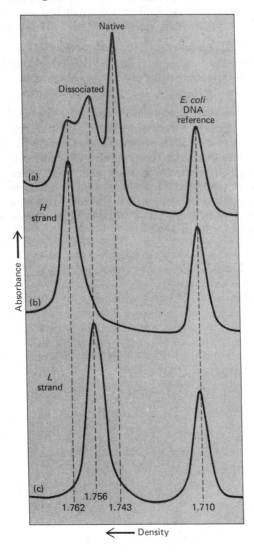

6. Bacteria lack the enzymes for excising and splicing introns; they also fail to recognize eukaryotic promoters or to translate most mRNAs that lack Shine-Dalgarno sequences. If you wanted to construct a clone of *E. coli* capable of synthesizing the β-hemoglobin protein, how would you solve these problems using recombinant DNA technology?

7. You have isolated some rifampicin-resistant mutants of *E. coli*. Cite three possible lesions they might carry (*Section 7.6*). How would you distinguish between these possibilities?

8. Figures 9-11 and 9-12a are superficially similar in that each shows a group of wavy strands studded with particles. In what ways are the two images in fact similar, and in what ways do they differ?

9. For each of the terms listed below pertaining to prokaryotes, an analogous term applies to eukaryotes. Find the four "matched sets" and then describe, for each set, how the terms are similar and how they are different.

1. Poly (dT) track
2. Rifampicin
3. Shine-Dalgarno sequence
4. Pribnow box

5. α-amanitin
6. Goldberg-Hogness box
7. Cap sequence
8. Terminator sequence

10. The *E. coli* RNA polymerase binds to promoters A1, A2, and A3 of the T7 genome (Figure 9-4); transcription of a gene associated with one of these promoters yields a T7 RNA polymerase that is active in the remaining phases of the phage infection cycle. Footprinting and sequencing experiments suggest that the phage enzyme is much smaller than the host enzyme and that it recognizes different promoters. What experimental results would lead to this conclusion?

11. Which of the following could represent the sense strand of a gene for the following mRNA: 5'AUAGGCGAU3'? (More than one may apply.) What is wrong with the others?

1. 3'UAUCCGCUA5'
2. 5'ATCGCCTAT3'
3. 5'TATCCGCTA3'
4. 3'TATCCGCTA5'

# 10

# Genes and Gene Transcripts: Specific Genes

A. **Introduction**

B. **Transfer RNA Genes**
    10.1  General Properties of
          Transfer RNA
    10.2  Genes for tRNA in
          *Escherichia coli*
    10.3  Genes for tRNA in
          Eukaryotes
    10.4  tRNA Processing

C. **Ribosomal RNA Genes**
    10.5  General Properties of
          Ribosomal RNA

10.6  Genes for rRNA in *E. coli*
10.7  Genes for 18S and 28S rRNA
       in Eukaryotes
10.8  Genes for 5S rRNA in
       Eukaryotes

D. **Structural Eukaryotic Genes**
    10.9  Insulin Genes
    10.10 Conalbumin Genes
    10.11 Histone Genes

E. **Questions and Problems**

## Introduction

The preceding chapter describes structural genes in general terms so that their important features can be emphasized. We now examine five specific examples so that the diversity of mechanisms used to transcribe and process individual genes can be appreciated. Two of our examples, the genes for tRNA and for rRNA, are not translated into protein at all, but their transcripts play key biological roles, and they have been very actively investigated. The three eukaryotic structural gene examples were chosen to illustrate three degrees of intron involvement with eukaryotic gene structure: the insulin gene with two introns, the conalbumin gene with 17, and the histone genes with none. Additional eukaryotic and prokaryotic structural genes are described in subsequent chapters; our intent here is to give a sense of what genes are really like.

285

# Transfer RNA Genes

## 10.1 General Properties of Transfer RNA

Transfer RNA molecules play a key role in protein synthesis, bringing amino acids to the ribosome and aligning them in the order dictated by the sequence of codons in an mRNA molecule. The overall reaction is drawn in Figure 9-1, which should be reviewed, and details will be presented in Chapter 11.

The structure of mature tRNA molecules is drawn in Figure 10-1. Each tRNA folds into a characteristic L-shaped molecule (Figure 10-1a) that can be represented in two dimensions as a **cloverleaf**-shaped structure (Figure 10-1b). In order to consider the transcription and maturation of these molecules, two properties of tRNA must be appreciated.

1. Each tRNA species associates with a specific type of amino acid at its 3' **acceptor end** (Figure 10-1). Since there are 20 amino acids used in protein synthesis, this means a minimum of 20 tRNAs. It turns out, however, that most amino acids are recognized by several different tRNA types called **isoacceptor tRNAs;** thus, for example, an organism may possess two species of tyrosine tRNA, $tRNA_1^{Tyr}$ and $tRNA_2^{Tyr}$, which have different sequences but which both bind tyrosines at their acceptor ends.

2. Once charged with its appropriate amino acid, each tRNA molecule is in a configuration to recognize a specific **codon** in mRNA by its prominent **anticodon** (Figures 9-1 and 10-1). Thus once a particular $tRNA^{Tyr}$ has bound a tyrosine, it seeks out in the message a 3-nucleotide codon specifying tyrosine (for example, 5'-UAC-3'), using its 3'-AUG-5' anticodon as a probe and forming a $\begin{smallmatrix} 5'\text{-UAC-}3' \\ | \; | \; | \\ 3'\text{-AUG-}5' \end{smallmatrix}$ set of base pairs.

The primary structure of a tRNA transcript, and the mature structure of the tRNA molecule, must therefore include features for both specific amino-acid recognition and specific codon-anticodon interactions.

## 10.2 Genes for tRNA in *Escherichia coli*

At least 40 different kinds of tRNA are present in an *E. coli* cell; these are specified by a diverse array of genes. We consider this array in some detail

**FIGURE 10-1 tRNA Structure.** (a) Three-dimensional configuration of yeast phenylalanine tRNA deduced by X-ray crystallography. (From S. H. Kim, et al., *Science*, **179**, 285–288, 1973, copyright the American Association for the Advancement of Science.) (b) Two-dimensional cloverleaf configuration of a charged yeast alanyl-tRNA (From J. D. Watson, *Molecular Biology of the Gene*, 3rd ed., copyright © 1976 by J. D. Watson; W. A. Benjamin, Inc., Menlo Park, California.) Details on the maturation of these molecules are presented in Section 10.4.

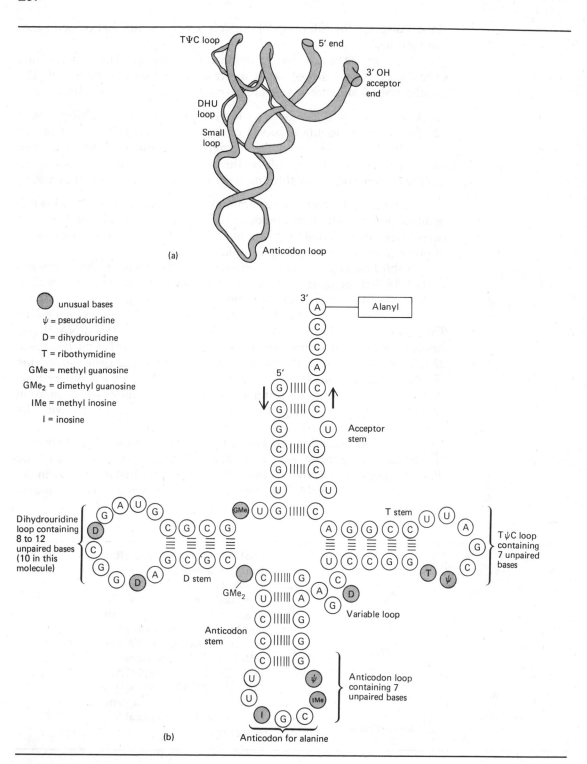

(a)

TΨC loop

5' end

3' OH acceptor end

DHU loop

Small loop

Anticodon loop

unusual bases

ψ = pseudouridine

D = dihydrouridine

T = ribothymidine

GMe = methyl guanosine

GMe₂ = dimethyl guanosine

IMe = methyl inosine

I = inosine

Alanyl

Acceptor stem

Dihydrouridine loop containing 8 to 12 unpaired bases (10 in this molecule)

D stem

GMe₂

T stem

TΨC loop containing 7 unpaired bases

Variable loop

Anticodon stem

Anticodon loop containing 7 unpaired bases

Anticodon for alanine

(b)

since it displays the full range of possible arrangements that related genes can assume.

The simplest case is that an amino acid is recognized by only one kind of tRNA (a **"solo" tRNA**), as is the case for tryptophan in *E. coli*. The genetic specification of a solo tRNA could occur in one of three ways.

1. A **single copy** of the gene could be present per *E. coli* chromosome.
2. Two or more identical copies of the gene could lie adjacent to one another in the chromosome, forming a **redundant tandem cluster**.
3. Two or more identical copies of the gene could lie in different sectors of the chromosome, constituting a set of **unlinked redundant genes.**

The alternative, and more common, case is that an amino acid is recognized by (at least) two nonidentical but isoacceptor tRNAs. In some cases, these are encoded by genes that lie next to one another in a cluster; in other cases, the genes map to unlinked loci.

A third alternative is also common in *E. coli*, namely, that the genes for tRNAs that recognize very different amino acids are nonetheless adjacent to one another (a **nonredundant tandem cluster**).

We can give two examples to illustrate the actual intricacy that exists. The first concerns the two isoacceptor tRNA$^{Tyr}$ species. The tRNA$_1^{Tyr}$ species is encoded by a redundant tandem cluster, called *tyrT*, which maps at 27 min (Figure 10-2). The cluster contains two identical genes specifying tRNA$_1^{Tyr}$, located side by side. The tRNA$_2^{Tyr}$ species, which differs from tRNA$_1^{Tyr}$ at two nucleotide positions, is specified by *tyrU*, which maps at 88 min. The *tyrU* gene is part of a nonredundant gene cluster with the order *thrU-tyrU-glyT-thrT*.

The second example concerns the three isoacceptor tRNA$^{Gly}$ species. The location of gene(s) encoding tRNA$_1^{Gly}$ is not yet established. The tRNA$_2^{Gly}$ species is encoded by the *glyT* gene that, we just noted, lies in the 88 min cluster. The tRNA$_3^{Gly}$ species is specified by genes in two locations.

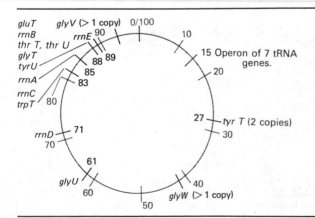

**FIGURE 10-2 tRNA and rRNA (*rrn*) genes** in the genetic map of *E. coli*. Position 15 has been identified genetically as the locus of suppressor (*SupB,E*) mutations and by cloning the locus of an operon carrying genes for the following tRNA molecules: tRNA$_M^{Met}$, tRNA$^{Leu}$, tRNA$_1^{Gln}$, tRNA$_1^{Gln}$, tRNA$_M^{Met}$, tRNA$_2^{Gln}$, tRNA$_1^{Gln}$. The duplicated genes in this operon have identical sequences.

At 94 min there exists a redundant tandem cluster of two or three identical copies of a tRNA$_3^{Gly}$ sequence called *glyV*, and at 42 min there exists a single copy of a gene called *glyW*, which also specifies tRNA$_3^{Gly}$ (Figure 10-2). The glyV and glyW loci, therefore, constitute a set of unlinked redundant genes.

In all cases where tRNA genes have been cloned and sequenced, the genetic unit is preceded upstream by a canonical "Pribnow box" (*Section 9.7* and *Table 9-2*), the recognition site for RNA polymerase. Each primary transcript is an oversized precursor molecule, which, like pre-mRNA molecules, must undergo processing. In the case of solo tRNA genes, the primary transcripts have the structure 5'-leader-tRNA-trailer-3'. In the more common case of tandem clusters, the entire unit is preceded by a single promoter and copied into a single long transcript, meaning that these genetic units can be formally classified as **tRNA operons** (*Section 9.2*). A stretch of nucleotides called **spacer DNA** occupies the interval in the operon between one tRNA sequence and the next; since this spacer is included in the transcript, it is often denoted as **transcribed spacer.** The pre-tRNA transcripts from tRNA operons, therefore, have the general structure 5'-leader-(tRNA-spacer)$_n$-tRNA-trailer-3', where the transcribed spacer is also fated for processing.

**Mapping tRNA Genes**  Map positions for tRNA genes (Figure 10-2) have been determined in three ways. In some cases a tRNA gene will undergo mutation and give rise to **suppressor tRNAs** that have anomalous properties in protein synthesis, as described in Section 11.16. The map position of the gene giving rise to this suppressor phenotype can then be determined by genetic crosses (*Chapter 13*).

The chromosomal location of other tRNA genes has been deduced from their presence in **specialized transducing phages.** The construction of these phages is detailed in Section 13.13. Here it is sufficient to know that when bacteriophages such as lambda (λ) infect *E. coli*, their chromosomes are spliced into the host chromosome at one stage of the infection cycle and, at a later stage, are spliced out again. The excision process is occasionally faulty such that contiguous *E. coli* genes are erroneously "picked up" along with the phage chromosome. The resultant specialized transducing chromosomes resemble recombinant λ vectors (*Figure 9-7*) in that they contain both λ genes and heterologous DNA that can be transduced (carried over) to subsequent cells by infection. As an example we can consider the transducing phage λ*rif*$^d$*18*. The phage was selected because it had picked up the *rif* gene (now called *rpoB*) of *E. coli*, which codes for the β subunit of RNA polymerase (*Section 9.3*) and maps at 88 min (*Figure 9-3*). When λ*rif*$^d$*18* chromosomes were cleaved with restriction endonucleases (*Section 4.11*), one of the resultant fragment classes proved to contain a totally unexpected sequence for tRNA$_2^{Glu}$, while another contained the *tyrU* gene noted previously. These tRNA genes, therefore, are at once localized close to *rpoB* on the genetic map.

The third method for locating tRNA gene positions is by recombinant DNA technology. In this case, as with specialized transducing phages, the presence of a tRNA is often discovered next to a cloned gene whose map position is already known from genetic crosses.

## 10.3  Genes for tRNA in Eukaryotes

In eukaryotes, estimates of tRNA gene redundancy have been made by hybridizing (*Section 4.10*) eukaryotic DNA and a total-tRNA preparation from the same species. It is found, for example, that about 0.08 percent of the yeast genome will hybridize with tRNA. Knowing the yeast genome size, this percentage extrapolates to about 400 tRNA genes in the yeast genome, or an average of from 5 to 7 sequences for each of the 60 tRNA sequences present in the cell. By similar calculations, *Drosophila* is believed to carry an average of 10 copies of each of about 60 tRNA sequences, and the toad *Xenopus laevis* is believed to carry 200 copies of each gene per haploid genome.

The tRNA genes of eukaryotes can be mapped by *in situ* hybridization (*Section 4.10*) using radiolabeled tRNAs as probes. Figure 10-3 illustrates

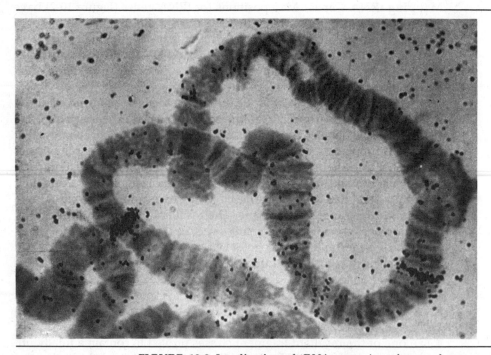

**FIGURE 10-3 Localization of tRNA genes** in polytene chromosomes of *D. melanogaster*. *In situ* hybridization of radiolabeled $tRNA_{3b}^{val}$. (From T. A. Grigliatti).

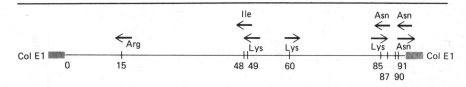

**FIGURE 10-4 Eukaryotic tRNA gene cluster.** A 9.34 kb fragment, derived from region 42A on chromosome 2R of *D. melanogaster*, was cloned in plasmid Col El. Arrows indicate transcription direction of the eight tRNA genes; genes with opposite directions of transcription are encoded on opposite strands. (From Hovemann, B., S. Sharp, H. Yamada, and D. Söll, *Cell* **19**:889, 1980.)

results for the $tRNA_{3b}^{Val}$ of *Drosophila*: at least two nonclustered redundant genes clearly exist for this species. Similar studies indicate that most tRNA genes occur in clusters and that some clusters contain more than one species of tRNA gene (i.e., nonredundant tandem clusters). Thus the overall picture seen for *E. coli* is repeated in eukaryotic genomes.

The detailed organization of eukaryotic tRNA genes has been most effectively analyzed by recombinant DNA technology. For example, a hybrid plasmid carrying a 9.34 kb fragment of *Drosophila* DNA has been found to carry a cluster of eight tRNA genes, whose organization is diagrammed in Figure 10-4. Particularly interesting is the fact that five of these genes are encoded by one strand of the DNA and three by the other strand, much as we saw for the chromosome of T4 (*Section 9.4*). Moreover, the strand location and, therefore, the direction of transcription is different for two redundant genes encoding the same species of isoacceptor $tRNA_2^{Lys}$. The evolution and the "purpose" of these arrangements is at present most unclear.

Perhaps the most interesting feature of eukaryotic tRNA genes is that **their promoters lie within the genes themselves.** Figure 10-5 illustrates the DNA sequence of a $tRNA_1^{Met}$ gene cloned from a *Xenopus laevis* tRNA gene cluster. Transcription is initiated several bases upstream from the structural gene and terminates at a poly-T cluster downstream from the 3' end. In this respect, the unit resembles an "orthodox" eukaryotic structural gene. When the sequences upstream from the 5' leader of many such tRNA genes are analyzed, however, no "consensus sequence" such as the Goldberg-Hogness box (*Table 9-2*) can be found. Moreover, if these upstream sequences are deleted by *in vitro* mutagenesis (*Section 8.7*), the resultant truncated gene continues to be transcribed. M. Birnsteil and his colleagues therefore examined the effect of creating deletions and point mutations in the gene itself. They discovered that two separate sequences within the gene are essential for polymerase recognition; as indicated in Figure 10-5, one of these occurs at positions 8 to 13, and the second occurs at positions 51 to 72. It therefore appears that RNA polymerase III, the enzyme that transcribes tRNA genes (*Table 9-1*), detects two distinct sequences within

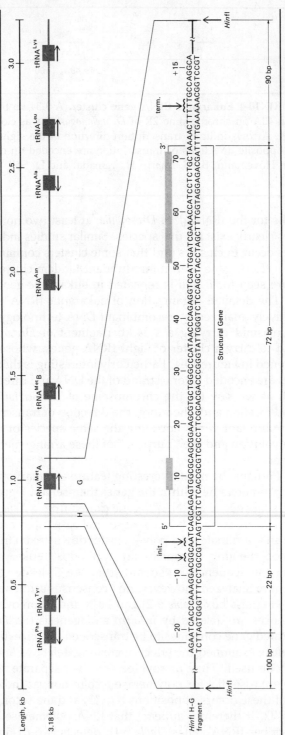

**FIGURE 10-5 The tRNA$_i^{Met}$ Gene of *X. laevis*,** with the location of its **split promoter** indicated as gray bars. The gene is contained within a 3.18 kb DNA segment that carries a total of eight tRNA genes. Note that, as with the *Drosophila* gene cluster (*Figure 10-4*), transcription is in opposite directions for different genes. (From H. Hofstetter, A. Kressmann, and M. L. Birnstiel, *Cell* **24**:573–585, 1981.)

the genomic DNA (which may well lie near one another as a consequence of secondary DNA folding); the polymerase then "reaches back" and begins copying the gene from the 5' end.

# 10.4 tRNA Processing

The **pre-tRNA** transcripts of both single genes and operons undergo an elaborate series of processing steps before acquiring their final, mature tRNA configurations. In *E. coli*, processing begins with endonucleolytic cuts that eliminate leader, spacer, and trailer sequences. Temperature-sensitive mutants that produce defective **RNase P** enzymes are unable to process normally, and large pre-tRNAs therefore accumulate in the cell. From analysis of the processing of such molecules *in vitro* it appears that the RNase P enzyme recognizes, and cuts, regions of leader and spacer sequences that are internally complementary and therefore loop back on themselves to form double-stranded stems (*recall Figure 9-8*). A second enzyme, **RNase D,** is involved in removing trailer sequences by exonuclease digestion from the 3' ends.

The next stage in tRNA processing, which occurs on either the precursor molecules or the cleaved products, is termed **nucleoside modification:** certain uridines, for example, are reduced to dihydrouridine or rearranged into a form known as pseudouridine; certain adenosines are deaminated to yield inosine; and methyl groups are added at certain positions to yield 5-methyl-cytosine or 1-methyl-guanosine. The structures of some of these **minor bases** are given in Figure 10-6; at least 50 have been described to date. The importance of nucleoside modification becomes apparent in the phenotype of the *hisT* mutant of *Salmonella*. The *hisT* mutation affects the production of a pseudouridylation enzyme; the consequent absence of pseudouridine in key positions on many tRNA molecules results in the defective biosynthesis of many proteins in *hisT* strains.

In the final tRNA processing step, a specific **nucleotidyl transferase** enzyme adds the sequence 5'CCA-OH3' onto the 3'-terminus generated by RNase D cleavage. This step is not required in cases in which the CCA sequence is an internal part of the transcript and becomes exposed after RNase D action. The final CCA-terminus (*Figure 10-1*) is essential for the amino acid acceptor function of all tRNA molecules.

Eukaryotic tRNA molecules are processed by enzymes analogous to those in *E. coli:* nucleases trim the leader and trailer; bases are modified; and CCA termini are added, all within the nucleus. In addition, some (but not all) eukaryotic tRNAs have short (14–62 bp) intervening sequences (*Section 9.9*) on the 3' side of the anticodon that must be excised before the molecules are active. Two mutant strains of yeast, *rna*1 and 1*os*1, cannot process out these introns and accumulate oversized tRNA precursors in the nucleus. When the pre-tRNAs are presented with wild-type yeast extracts, normal processing takes place, allowing biochemical analysis of the reaction. The nucleotide sequences that are cut do not conform to the

**FIGURE 10-6**
**Minor bases** found in tRNA.

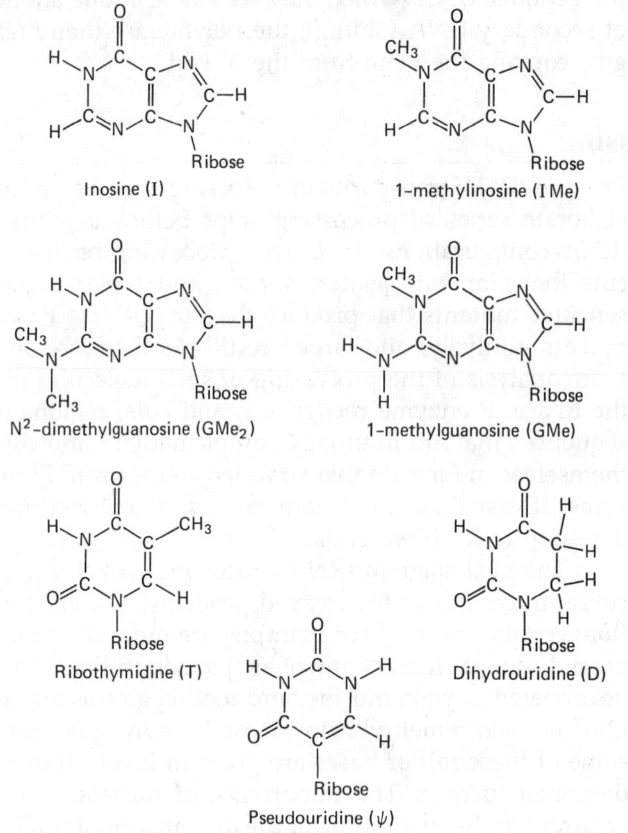

Inosine (I)

1-methylinosine (I Me)

$N^2$-dimethylguanosine (GMe$_2$)

1-methylguanosine (GMe)

Ribothymidine (T)

Dihydrouridine (D)

Pseudouridine ($\psi$)

GT-AG rule established for hnRNA introns (*Section 9.10*); it is thus probable that unique tRNA splicing enzymes are involved in this processing step.

# Ribosomal RNA Genes

## 10.5 General Properties of Ribosomal RNA

Ribosomes contain three major classes of ribosomal RNA (rRNA) plus at least 50 proteins and serve as essential scaffolds for the translation of mRNA into protein (*Figure 9-1* and *Chapter 11*). In prokaryotes, the **23S** and **5S rRNAs** (*see Box 9.1*) reside in the large ribosomal subunit (which sediments at 50S), while the **16S rRNA** resides in the small 30S ribosomal subunit. In eukaryotes, the large ribosomal subunit contains a **28S** and a **5S** species and the small subunit contains an **18S** rRNA. Partly because the eukaryotic rRNAs are larger, their intact ribosomes are also larger (**80S**) than prokaryotic ribosomes (**70S**).

## 10.6 Genes for rRNA in *E. coli*

Genes for rRNA (*rrn*) in *E. coli* are present in seven positions in the *E. coli* genome (*Figure 10-2*), and the entire DNA sequence of one such locus has been determined. Each *rrn* locus specifies a 30S **pre-rRNA** transcript containing 16S, (tRNA), 23S, and 5S sequences (a tRNA sequence usually—and perhaps always—lies between 16S and 23S); therefore, each *rrn* locus is properly considered an operon (*Section 9.2*). Processing of the primary transcript involves initial cuts in the transcribed spacer, followed by secondary trimming of the leader and trailer sequences, much as we saw for pre-tRNA. Details are given in Figure 10-7.

About 0.4 percent of an *E. coli* chromosome is devoted to carrying rRNA sequences, whereas rRNA accounts for 40 to 50 percent of the RNA synthesized in rapidly growing cells. A visual explanation for this discrepancy in percentages is seen in Figure 10-8: the rRNA genes are in a state of perpetual transcription, with polymerases packed along the DNA and strands of rRNA of progressively longer lengths coming off the template (compare with the more sparsely transcribed structural gene, Figure 9-11). Such dense packing is apparently accomplished, at least in part, by the presence of two promoters at the start of each *rrn* operon.

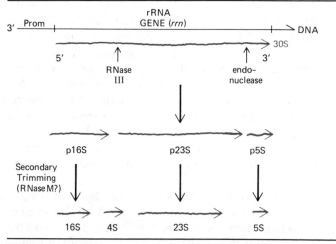

**FIGURE 10-7 rRNA processing in E. coli.** A cleavage site for RNase III exists in the transcribed spacer between the 16S and 28S sequences. Normally this cut is made during the course of transcription so that a full-length 30S transcript is rarely formed; the full-length transcript is, however, recovered in an RNase III–deficient mutant strain (*rnaC⁻*). Secondary trimming may involve a second enzyme, RNase M or maturase, and generates a 4S tRNA from either the p16S or the p23S precursor. Nucleoside modifications, notably methylations, occur at an early stage of processing for the 23S molecule and later for the 16S species.

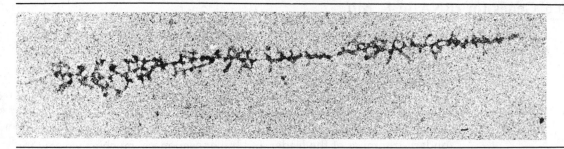

**FIGURE 10-8 rRNA transcription in E. coli.** Sixty molecules of 30S pre-rRNA can be generated per minute from each gene. (From O. L. Miller, *Int. Rev. Cytol.* **33**:1, 1972.)

## 10.7 Genes for 18S and 28S rRNA in Eukaryotes

Each eukaryotic nucleus possesses 1 or more nucleoli (*Figure 2-6*); these represent sites where newly synthesized rRNA accumulates and becomes associated with the ribosomal proteins that are synthesized in the cytoplasm and then migrate back into the nucleolus for assembly. Experiments by D. Brown and J. Gurdon were among the first to establish that **the nucleolus is not only a storage site for rRNA but also the site of rRNA synthesis.**

The Brown and Gurdon experiments focused on the mutation *0-nu* (for *zero-nucleolus*) in *Xenopus laevis*. Normal *Xenopus* cells (+/+) contain 2 nucleoli whereas heterozygous toads (+/0-nu) exhibit only one nucleolus in each of their nuclei. Such heterozygotes are viable, but when they are crossed with one another about one quarter of their progeny die at the swimming larva stage. You will recall from Chapter 6 that whenever two heterozygotes are crossed one quarter of their progeny are expected to be homozygous for a mutant allele. Therefore we can predict that the inviable progeny in the *X. laevis* crosses represent *0-nu/0-nu* homozygotes. This expectation is confirmed when cytological specimens are prepared from embryo tail tips: about one quarter of the embryos exhibit no nucleoli in their nuclei.

The swimming larva stage of *X. laevis* embryogenesis is preceded by a period when the embryos have used up the store of ribosomes inherited from their mothers and are called on to synthesize their own rRNA. Brown and Gurdon therefore compared the ability of homozygous and heterozygous larvae to synthesize rRNA during this period. They exposed embryos to $^{14}CO_2$ and examined the amount of radioactivity incorporated into rRNA. The results are shown in Figure 10-9, in which the absorbance at 260 nm is shown in gray and the radioactivity is shown in black. It is clear that the control larvae have much more 28S and 18S RNA than the anucleolate mutants, as measured by absorbance. Even more striking is the fact that the anucleolate embryos incorporate virtually no radioactivity into ei-

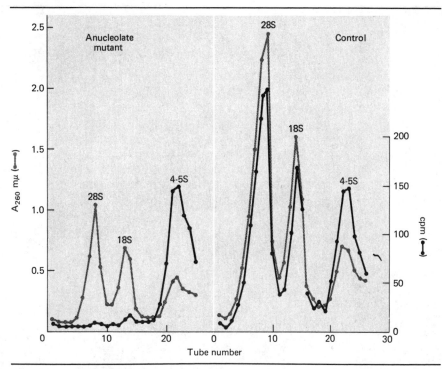

**FIGURE 10-9 Brown and Gurdon experiment.** Sucrose gradient centrifugation of total RNA isolated from anucleolate (*0-nu/0-nu*) embryos and control (*+/+* and *+/0-nu*) embryos. The bulk RNA is given by the absorbance at 260 nm (gray tracing). The RNA synthesized between the neurula and muscular response stage is represented by the radioactivity measurements (black tracing). (From D. Brown and J. Gurdon, *Proc. Natl. Acad. Sci. U.S.* **51**:139, 1964.)

ther species of rRNA, indicating that the 28S and 18S rRNA that they do possess is of maternal origin.

The mutation carried by *0-nu* toads proves to be a deletion of the **nucleolar organizer (NO),** a sector of chromosome 12 in *Xenopus laevis*. This sector can be shown by *in situ* hybridization studies with 18S and 28S rRNAs to be the exclusive site of 18S and 28S gene sequences. RNA-DNA hybridization studies reveal that 500 copies of the 28S and 18S sequences are present in each *X. laevis* nucleolar organizer. A similar level of NO **gene redundancy** is found for all other eukaryotes examined, but there is some variation with regard to the distribution of these genes: *Drosophila* resembles *Xenopus* in having but one NO per haploid genome (in the X and Y chromosomes); humans have 5 NOs (in the short arms of the 5 acrocentric D and G chromosomes); whereas the field vole (a rodent) has genes for 18S and 25S rRNA in almost all of its chromosomes.

Transcription of nucleolar organizer DNA is visualized in Figure 10-10. Segments of the NO DNA are seen to be in an active state of transcrip-

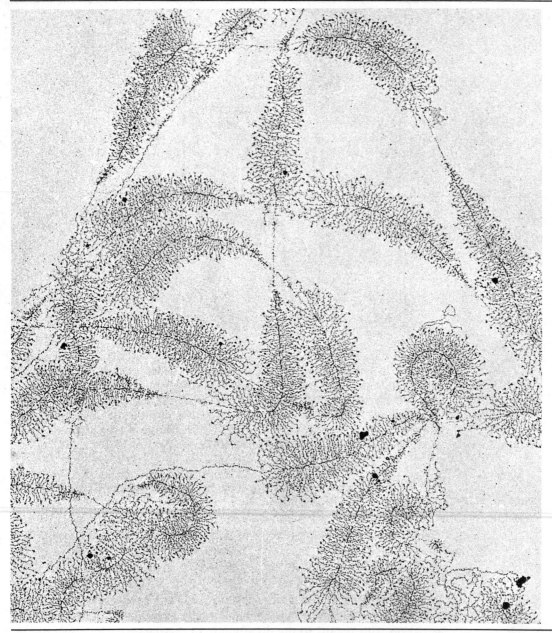

**FIGURE 10-10 rRNA transcription in the nucleolus** of *Triturus*. DNA is represented by long axial strands, RNA by the "feathers." (From O. L. Miller and B. R. Beatty, *Science* **164**:955, 1969.)

tion in which multiple initiation sites produce multiple rRNA strands of progressively longer lengths (*compare with Figure 10-8*). Each such "feather region" is flanked on either side by DNA that is clearly not being transcribed and is appropriately termed **nontranscribed spacer.**

NO DNA from *Xenopus* can be isolated as a satellite band in CsCl gradients (*Section 4.3*) and has been subjected to extensive molecular analysis, notably in the laboratory of D. Brown. Figure 10-11 diagrams the structure of this DNA. The 18S and 28S sequences are copied as a single large 40S pre-rRNA transcript; processing of this transcript is detailed in Figure 10-11 and its legend. All 500 copies of the rRNA gene in the DNA appear to be virtually identical to one another. The intervening nontranscribed spacers differ little in sequence but vary in lengths, even along the same piece of NO DNA (Figure 10-11); this is because they are virtually repetitive, with the modular units varying in their repetition frequency from one spacer to the next. Thus the nontranscribed spacer in NO DNA bears a strong structural resemblance to the nontranscribed satellite DNAs found in constitutive heterochromatin (*Section 4.7*). Since the function of both classes of DNA is unknown, it cannot be said whether or not this resemblance is fortuitous.

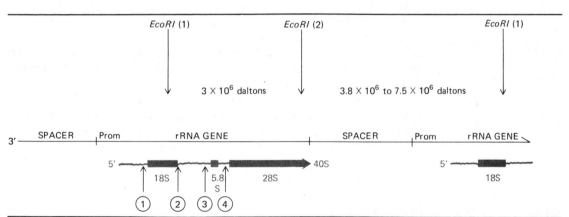

**FIGURE 10-11 rRNA gene organization and processing** in *Xenopus*. Each gene is separated from the next by a nontranscribed spacer of variable length (as determined by the size of the indicated *Eco*RI restriction fragments). The 40S pre-rRNA transcript (45S for mammals) contains a 5′ leader sequence that is cleaved from the molecule at step ①. Cuts ② and ③ in the transcribed spacer occur next, followed by cut ④ adjacent to the 28S block. Secondary trimming produces the final 18S, 28S, and 5.8S rRNAs; all three then associate with ribosomal proteins. Some eukaryotes (including *Drosophila* and *Tetrahymena*) carry an intervening sequence in the 28S rRNA coding sequence that must also be spliced out. (After P. Wellaner, *et al.*, *J. Mol. Biol.* **105**:461, 1976, and R. Perry, *Ann. Rev. Biochem.* **45**:605, 1976.)

## 10.8 Genes for 5S rRNA in Eukaryotes

You will recall that in *E. coli*, 5S rRNA derives from the cleavage of the large *rrn* gene transcript (*Figure 10-7*). In the simple eukaryotes such as yeast, 5S genes map close to the larger rRNA genes, but they are transcribed by a different enzyme (RNA polymerase III rather than polymerase I, as summarized in Table 9-1) and are thus independent transcripts. In the complex eukaryotes, 5S RNA genes not only are transcribed independently but also map far from the other RNA genes (the 5.8S RNA shown in Figure 10-11 is a distinct species).

The independent location of the 5S genes is suggested in Figure 10-9, where the *0-nu* deletion in *X. laevis* has abolished 18S and 28S transcription but does not affect "4S RNA" synthesis (a combination of both tRNA and 5S rRNA). *In situ* hybridization studies (*Section 4.10*) make the point more clearly: when labeled 5S rRNA is applied to *X. laevis* chromosomes, the NO regions are unreactive, and binding instead occurs over the telomeres of most of the 18 chromosomes. Similarly, 5S genes are found in many of the human chromosomes that do not carry nucleolar organizers.

The 5S genes in *X. laevis* are very similar to the NO genes in their organization: highly conserved transcribed sequences, 120 nucleotides long, alternate with nontranscribed spacers, about 600 nucleotides long, that contain repeating modules of a simple sequence. An estimated 20,000 copies of 5S genes are present in the *X. laevis* genome; 2000 copies are believed to be present in the human haploid complement.

Knowing that 5S genes are transcribed by RNA polymerase III, the same enzyme that copies tRNA genes, it is perhaps not surprising that D. Brown and colleagues have found that **the 5S promoter is "internal."** Unlike the tRNA promotor, it is apparently not split into two parts; instead, it lies right in the middle of the gene. This was demonstrated by showing that 5S DNA clones carrying residues 50 to 83 are actively transcribed even if sectors of the flanking sequences (1–49 and 84–terminus) have been deleted; by contrast, clones that carry deletions extending into the 50 to 83 residue region are transcriptionally inactive.

# Structural Eukaryotic Genes

## 10.9 Insulin Genes

The hormone insulin is composed of two polypeptide chains (A and B) and a C peptide, joined by disulfide bridges (*Section 3.3*). It plays a key role in regulating sugar, lipid, and amino acid metabolism. Its deficiency or malfunction can generate the disease **diabetes mellitus,** which may affect as much as 5 percent of the human population and which is presently the third leading cause of death in the United States.

Most vertebrates have only one insulin gene per haploid genome, but a few species, including the laboratory rat, have two nonallelic genes, called I and II, each specifying a distinct protein. The primary transcript of each insulin gene encodes a 110 amino acid polypeptide, called **pre-proinsulin,** with the structure diagrammed in Figure 10-12a. The pre-region consists of 24 amino acids that compose what is called the **signal sequence;** this specifies that the protein is to be sequestered within membranes for secretion, and it is cleaved off the polypeptide soon after translation begins. The remaining **proinsulin** molecule folds up to allow disulfide bridge formation, and proteolytic enzymes then cleave the internal basic amino acids (Figure 10-12a) to form the two chains of mature insulin.

Figure 10-12b diagrams the structure of the rat pre-proinsulin gene II, which is seen to be interrupted by two introns (small capital letters in Figure 10-12b). The gene I version lacks the second intron completely and contains seven amino acid substitutions in its coding sequence; these are indicated below the line in Figure 10-12b. The human insulin gene, which has also been sequenced, contains two introns at almost identical positions to the rate gene II introns, but they differ significantly in length. Thus the three insulin sequences, schematized in Figure 10-13, illustrate two important properties of introns: their number can vary, and their length can vary.

## 10.10 Conalbumin Genes

The gene for chicken conalbumin, a major egg white protein, exemplifies a prominent group of eukaryotic genes, those split by multiple introns. Figure 10-14 illustrates the structure of this gene as visualized by R-loop mapping (*Section 9.9*). The cloned genomic DNA, about 10,600 bp in length, encodes an mRNA of about 2500 bp. The gene is split into no less than 17 exons, labeled 1 to 17, by 16 introns, labeled A to P. It is certainly difficult to imagine why so many introns are present, particularly when related chicken egg white proteins, such as ovomucoid and ovalbumin, are encoded, respectively, by genes split by only 6 and 8 introns.

We should note here another type of variable, namely, the length of introns. Although the introns of the various egg white protein genes vary only between 50 and 200 bp, introns can become enormous. The present record is held by the mouse gene for dihydrofolate reductase, which is 42 kb in length, specifies a 1.6 kb mRNA, and contains five introns. Thus each intron must average 8000 nucleotides in length, which is at least ten times longer than a "typical" intron.

## 10.11 Histone Genes

As noted in Section 3.8, the five histones (*Table 3-1*) that associate with eukaryotic DNA are among the most highly conserved proteins known. The primary amino acid sequence of a given type of histone is virtually

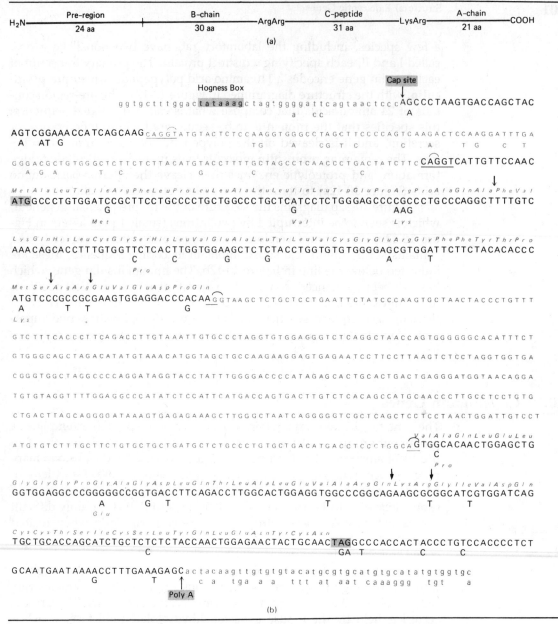

FIGURE 10-12 Pre-proinsulin gene in the rat. (a) Structure of pre-proinsulin polypeptide showing positions of mature B, C, and A chains. (b) Structure of antisense strand of pre-proinsulin II gene; nucleotide substitutions in the gene I sequence are indicated in the second line. Large caps: mature mRNA sequences. Small caps: two introns. Lower case: gene-flanking sequences. Nucleotide redundancies at intron-exon margins are underlined; GT and AG dinucleotides at margins are overlined. Initiation and termination codons and important regulatory regions are boxed. Arrows indicate, in order, the boundaries of the pre-region and the peptides B, C, and A. (From Lomedico, P., N. Rosenthal, A. Efstratiadis, W. Gilbert, R. Kolodner, and R. Tizard. *Cell* **18**:545–558, 1979.)

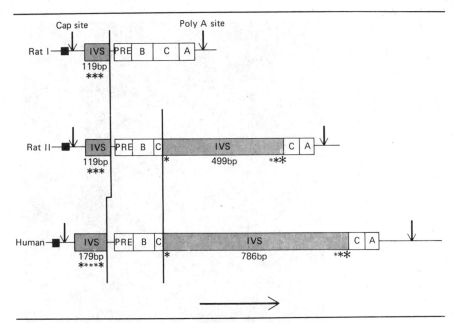

**FIGURE 10-13 Human and rat insulin genes.** The topology of the two rat insulin genes (I and II) and the single human insulin gene is displayed. The coding sequences for the peptide chains (pre-, B, C, and A) of pre-proinsulin are represented by the clear boxes. Intervening sequences (IVS) are distinguished by the shaded areas, with the length of each intervening sequence indicated below. The extent and position of nucleotide homology between intervening sequences of the two species is represented by the size of the asterisks—the larger the asterisk the greater the homology. Vertical lines describe the positions at which intervening sequences occur. The internal intervening sequence in the rat II and the human gene occurs in exactly the same position (valine 39 of the C-peptide region) whereas the position of the intervening sequence located in the 5'-untranslated region is displaced by three nucleotides in the human gene as compared with both rat genes. Also indicated are the sites for polyadenylation and capping (shown by arrows) as well as the Hogness box (the small black box). (From Bell, G. J., R. L. Pictet, W. J. Rutler, B. Cordell, E. Tischer, and H. M. Goodman, *Nature* **284**:26, 1980.)

identical throughout the eukaryotic kingdom, presumably because very precise interactions are necessary for the formation of nucleosomes (*Section 3.7*).

In all eukaryotes that have been examined to date, the five histone proteins are encoded by five structural genes that are transcribed separately but are clustered together. Figure 10-15 illustrates this arrangement for the histone genes of the newt. Figure 10-15 also shows that histone H2B is encoded on the opposite strand, and transcribed in the opposite direction, from the other four histones. Thus the unit is similar in several respects to a eukaryotic tRNA gene cluster (*Figure 10-4*).

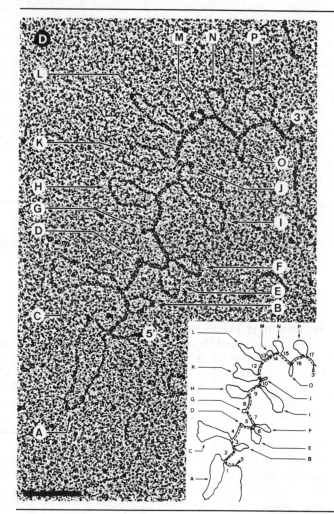

**FIGURE 10-14 Conalbumin mRNA hybridized to cloned conalbumin gene** under conditions that favor RNA-DNA hybrids. In the line drawing, the dashed line represents the RNA and the solid line the DNA. Capital letters refer to intronic sequences, and numbers to the exonic RNA-DNA hybrid sequences (not all of the exons are numbered). The 5' and 3' ends of the con-mRNA molecules are indicated with arrows. (From M. Cochet, F. Gannon, R. Hen, L. Maroteaux, F. Perrin, and P. Chambon, *Nature* **282**:567–574, 1979.)

Histone gene clusters have been analyzed from a variety of species, and considerable variability has been found. Thus in *Drosophila*, the gene order is H1-H3-H4-H2A-H2B, and genes H1, H3, and H2A are transcribed right to left (*compare with the newt, Figure 10-15*). The length of the nontranscribed spacer DNA between each gene also varies considerably from one

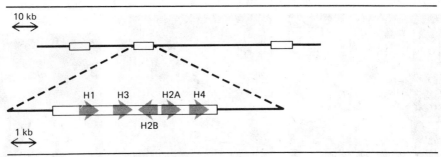

**FIGURE 10-15 Histone gene cluster** from the newt. The upper part of the figure shows histone gene clusters (boxes) interspersed with spacer sequences (single line). The lower part of the figure shows an enlargement of one histone gene cluster, with arrows indicating the direction of transcription. (From E. C. Stephenson, H. P. Erba, and J. G. Gall, *Cell* **24:**639–647, 1981.)

species to the next. Finally, there are striking differences in the number of such clusters per genome. In most organisms, histone clusters are tandemly repeated as illustrated at the top of Figure 10-15, with tracts of spacer DNA in between each cluster, much as we saw for the eukaryotic rRNA genes (*Figure 10-11*). The number of repeats varies from ten in the chicken up to several hundred in the sea urchin; a "rationale" for such repetitiveness can be found in the unusually high demand for histone biosynthesis during early sea urchin embryogenesis. The clusters have been mapped by *in situ* hybridization to unique chromosomal locations (for example, region 39D-E of *D. melanogaster* chromosome 2 and band q34 of human chromosome 7); therefore, there is no evidence for a dispersal of clusters in the fashion of the tRNA and rRNA genes.

The histone genes present the geneticist with several intriguing riddles. Granted that all H3 structural genes, for example, are highly conserved in their nucleotide sequence, how has this conservation been maintained while the gene has apparently been quite free, during the course of evolution, to change the DNA strand on which it is coded and to change its relative order within a cluster? How, moreover, is sequence homogeneity maintained within a genome? If 299 "good" copies of H3 exist in a sea urchin nucleus and a mutation occurs in the three hundredth gene, is this mutation lethal to the organism or is it tolerated? Is it immediately corrected in somatic cells, or is it corrected in the germ line so that only "pure" histone sequences are transmitted from parent to offspring? These questions will be evaluated in Section 19.7.

(a)　　　　　　　　　　　(b)　　　　　　　　　　　(c)

(a) John Gurdon (Medical Research Council, Cambridge, England) and
(b) Donald Brown (Carnegie Institute, Baltimore) study gene structure and
regulation in *Xenopus* and collaborated on studies of ribosomal RNA genes.
(c) Barbara Hamkalo and Oscar Miller, Jr. collaborated on studies of eukaryotic
gene transcription. Dr. Hamkalo is presently at the University of California,
Irvine, and Dr. Miller at the University of Virginia.

# Questions and Problems

1. In what respect do the histone genes resemble the mammalian X chromosome
   (*Section 6.17*)? How might this similarity be explained?
2. A series of strains of *D. melanogaster* carry recessive mutations known as *bobbed*,
   which, in homozygous form, cause slow development, reduced growth and
   fertility, and poor viability. Some *bobbed* mutations have a stronger effect than
   others on development and viability. Describe DNA/RNA hybridization experiments that would ascertain whether the various *bobbed* mutations represent
   more or less extensive NO deletions.
3. It has been proposed that the DNA carrying rRNA sequences in *E. coli* is in a
   "melted" state at all times. What considerations have led to this suggestion?
4. The cellular RNAse activity responsible for the short lifetime of bacterial mRNA
   acts against free, single-stranded RNA. Why are bacterial rRNA and tRNA
   relatively resistant to this nuclease?
5. The X and Y chromosomes of *D. melanogaster* each carry an NO. Mutant X
   chromosomes are known that carry either an NO duplication or an NO deficiency. (a) Diagram the karyotypes of flies you would construct with one, two,
   three, and four "doses" of the NO. What is the sex in each case (*Section 6.17*)?
   (b) Would you expect a difference in the ability of DNA extracted from these 4
   strains to hybridize with 18S rRNA? 28S rRNA? 5S rRNA? If so, what would the
   differences be?

6. A common property of tandem-repeat genes is that they undergo meiotic synapsis "out of register" and then undergo crossing over to produce a deficiency and a duplication chromosome. Diagram how such an event could generate the X chromosomes described in Problem 5.

7. The following arrangements (a to e) for tRNA genes are found on three bacterial chromosomes. For each, describe the probable structure of the primary transcript and whether the chromosome as a whole displays redundant or nonredundant tandem clusters, unlinked redundant genes, single-copy genes, or some other arrangement.

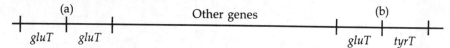

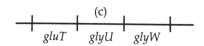

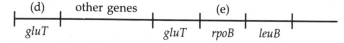

8. (a) Describe how transcript (a) in Problem 7 would be processed to yield mature tRNA$_2^{Glu}$ molecules. Include in your description the necessary enzymes. (b) What would be different if the transcript derived from a eukaryotic chromosome?

9. Two types of spacer DNA are described in this chapter. What are the distinctive properties of each?

10. What do eukaryotic tRNA and 5S RNA genes have in common? How are they different?

11. You are told that certain RNA samples have the following properties: (a) a larger precursor size than final size; (b) an additional sequence added onto the original transcript; (c) a homogenous molecular weight; (d) a short lifetime. In each case state whether the RNA could reasonably be tRNA, mRNA, rRNA, and/or hnRNA (one or more than one may apply), and give your reasons.

# 11

# Structural Gene Expression and the Genetic Code

**A. Introduction**

**B. Protein Synthesis**
   11.1  An Overview of Protein Synthesis
   11.2  Charging tRNA Molecules
   11.3  Initiating Protein Synthesis
   11.4  Elongation of the Polypeptide Chain
   11.5  Termination of Polypeptide Synthesis
   11.6  Genetic Specification of the Translation Apparatus

**C. "Cracking" the Genetic Code**
   11.7  The Triplet Nature of the Code
   11.8  The Degeneracy of the Code
   11.9  Deciphering the Code: *In Vitro* Codon Assignments
   11.10 Patterns to the Genetic Code and the Wobble Hypothesis

   11.11 *In Vivo* Codon Assignments and Codon Preference
   11.12 Initiator Codons
   11.13 Universality of the Code and the Mitochondrial Exceptions

**D. Nonsense Mutations and Chain Termination**
   11.14 Properties of Nonsense Mutations
   11.15 Establishing the Chain-Termination Codons

**E. Suppressor Mutations**
   11.16 General Nature of Suppressors
   11.17 Nonsense Suppressor Genes in *E. coli*
   11.18 Nonsense Suppressor Genes in Yeast

**F. Questions and Problems**

## Introduction

This chapter begins with a general account of the mechanics of protein synthesis, with emphasis on features important to an understanding of genetic phenomena. The rest of the chapter considers the genetic code used in translating mRNA into protein. We describe the key experiments that led to the "cracking" of the code, and we consider important genetic signals such as the codons that specify the "start" and "stop" of translation. Included in this presentation is a consideration of translation-defective mutants, particularly **suppressor mutations** that change the codon recognition properties of particular tRNA molecules.

# Protein Synthesis

A group of ribosomes translating eukaryotic mRNA is illustrated in Figure 11-1. We have already presented certain features of this translation process in Chapter 9, features that are summarized in Figure 9-1b. In the sections that follow, we first review the overall process of synthesis and then give a detailed analysis of individual events occurring during synthesis. In the course of these descriptions, for the sake of completeness, certain statements are made about the genetic code, evidence for which appears later in the chapter.

## 11.1 An Overview of Protein Synthesis

A polypeptide is synthesized in a stepwise fashion, starting with the N-terminal amino acid and ending with the C-terminal amino acid. In a like manner the mRNA dictating the sequence of the polypeptide is read by the protein-synthesizing apparatus in a stepwise fashion, one codon at a time, starting at or near the 5′ end and working toward the 3′ end. Each codon in an mRNA molecule consists of three nucleotides (a **triplet**). A given codon

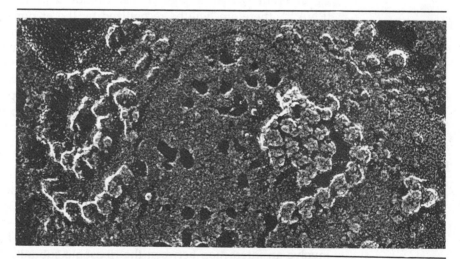

**FIGURE 11-1 Polyribosomes.** Two spirals of ribosomes sitting on the surface of endoplasmic reticulum (the membrane has been perforated with holes during the preparative procedure). (Courtesy of Dr. John E. Heuser).

is complementary, in the Watson-Crick sense, to an anticodon carried by a tRNA molecule. Each tRNA is also charged with its specific amino acid to form an aminoacyl-tRNA. As succeeding codons of the mRNA are exposed to a specific ribosomal site, successive aminoacyl-tRNA molecules participate in an anticodon-codon recognition event that is presumed to involve the familiar hydrogen bond pairing of adenine with uracil and guanine with cytosine. Thus a codon UUG (written in a 5′ ⟶ 3′ direction) on mRNA will be recognized by the AAC (written in a 3′ ⟶ 5′ direction) anticodon of a tRNA molecule, as illustrated in Figure 9-1. The UUG codon specifies the amino acid leucine, and the tRNA bearing the AAC anticodon is charged with a leucine molecule. The next codon in the mRNA illustrated in Figure 9-1 is a 5′—GGC—3′ codon specifying glycine. A glycyl-tRNA will pair with this codon via its 3′—CCG—5′ anticodon, and so forth. As successive amino acids are thus brought into position on the ribosomes, peptide bonds form between them and a polypeptide is synthesized (*Figure 9-1*).

## 11.2 Charging tRNA Molecules

The processing of primary tRNA transcripts, detailed in Section 10.4, produces mature tRNA molecules that fold into the characteristic L-shaped structures depicted in Figure 10-1. The nucleotide modifications made during tRNA processing create **loops** of non–base-paired residues; the remaining nucleotides form stable base-paired **stems.**

Although the overall shape of all mature tRNA molecules is remarkably similar, each is different in nucleotide sequence and thus assumes subtly different spatial configurations. Unique aspects of the structure of tRNA$^{Ala}$, therefore, will be recognized by the enzyme alanyl-tRNA synthetase, which proceeds to "charge" the tRNA with an alanine residue in the fashion diagrammed in Figure 11-2. A tRNA$^{Tyr}$, meanwhile, is recognized and "charged" with tyrosine by a tyrosyl-tRNA synthetase enzyme, and so on. The resultant **aminoacyl tRNAs** have two features critical to the subsequent steps of protein synthesis: at one end is exposed the anticodon required to recognize the appropriate codon in mRNA; at the acceptor end is the amino acid specified by that anticodon-codon couple.

## 11.3 Initiating Protein Synthesis

The first event in polypeptide synthesis is that the mRNA leader (*Section 9.6*) binds to the small ribosomal subunit (Figure 11-3a). In prokaryotes, binding is evidently mediated by the Shine-Dalgarno sequence (*Table 9-2*) encoded within the leader sequence. In eukaryotes, binding appears to involve an interaction between the ribosome subunit, the m$^7$G(5′) cap at the start of the message (*Section 9.6*), and a factor known as cap-binding protein. Binding is followed by a "scan" along the leader sequence until the initiator codon is reached.

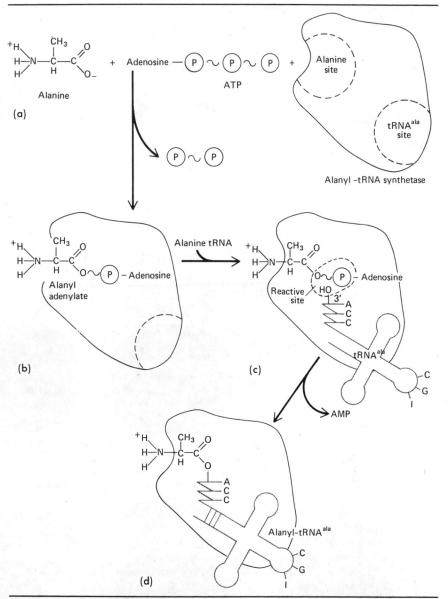

**FIGURE 11-2 Aminoacyl-tRNA synthetase activity.** (a) Alanine interacts with the alanine site of an alanyl-tRNA synthetase enzyme in the presence of ATP to form (b) an adenylated residue. This associates (c) with the 3'—OH of the AAC terminus of an alanine tRNA, sequestered in the tRNA$^{Ala}$ site of the synthetase. The AMP is then hydrolyzed (d) to form an ester bond between the tRNA$^{Ala}$ and the alanine molecule. The resultant alanyl-tRNA is released from the enzyme to participate in protein synthesis.

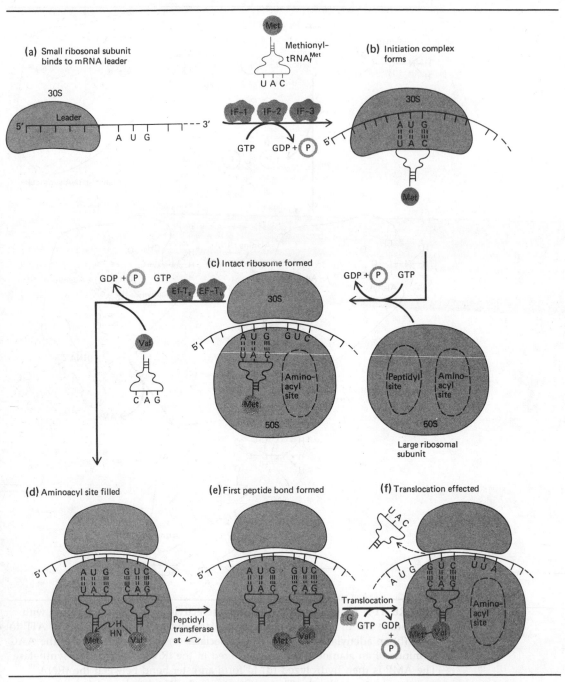

**FIGURE 11-3 Initiation and elongation of polypeptide synthesis,** with molecular interactions often indicated in highly schematic form. (See text for details.)

The **initiator codon,** which is usually AUG but is sometimes GUG, is recognized by the **initiator tRNA** known as **tRNA$_f^{Met}$.** This tRNA is charged with methionine which, in prokaryotes but not eukaryotes, subsequently becomes formylated, explaining the tRNA$_f^{Met}$ terminology. Therefore, the first amino acid to be incorporated into a polypeptide is either methionine or N-formyl-methionine; this may or may not be cleaved off the N-terminus during post-translational processing of the polypeptide product.

Pairing of the (N-formyl) methionyl-tRNA$_f^{Met}$ anticodon with the initiator codon and the association with the small ribosomal subunit requires at least three protein **initiation factors** (IF-1, IF-2, and IF-3) and the hydrolysis of a molecule of GTP. The resultant **initiation complex** (Figure 11-3b) is primed to combine with the large ribosomal subunit to form an intact 70S (or 80S in eukaryotes) ribosome.

The large ribosomal subunit possesses two sites, a **peptidyl site** and an **aminoacyl site,** whose actual spatial configurations are unknown. As the two ribosomal subunits associate (Figure 11-3c), the methionyl-tRNA$_f^{Met}$ becomes bound to the peptidyl site in a reaction that also requires GTP; at the same time, the mRNA becomes sequestered in a groove formed by the apposition of the two ribosomal subunits.

## 11.4 Elongation of the Polypeptide Chain

The alignment of the initiator codon of mRNA with the anticodon of the methionyl-tRNA$_f^{Met}$ in the peptidyl site of the ribosome fixes the alignment of the next codon of the mRNA, and consequently, the alignment of the aminoacyl site. In Figure 11-3d, the next codon specifies valine, and a valyl-tRNA$^{Val}$ is seen to occupy the aminoacyl site, paired with the mRNA codon. This association again requires GTP and is mediated in *Escherichia coli* by a complex of two **elongation factors** (EF-T$_s$ and EF-T$_u$).

At this point the peptidyl site still contains a methionyl group covalently linked to its tRNA$_f^{Met}$ by the α—COOH group of methionine. A displacement now occurs so that the α—COOH group of methionine leaves the tRNA linkage and forms a peptide bond with the α—NH$_2$ of the valyl-tRNA sitting in the adjacent aminoacyl site (Figure 11-3e). This exchange of one bond for another, called the **peptidyl transferase reaction,** is mediated by peptidyl transferase, an enzyme bound to the larger ribosomal subunit. At the completion of this step the tRNA$^{Val}$ carries a dipeptide residue (NH$_2$-met-val-), while the tRNA$_f^{Met}$ has given up its amino acid (Figure 11-3e).

The next step in polypeptide synthesis, called **translocation,** involves two simultaneous events that are mediated by GTP hydrolysis, by the elongation factor EF-G, and in all likelihood by a pronounced conformational change in the structure of the ribosome. In the first event the dipeptidyl-tRNA (in Figure 11-3e, the dipeptidyl-tRNA$^{Val}$) moves from the aminoacyl to the peptidyl site on the ribosome, dislodging the empty tRNA$_f^{Met}$ in the process (Figure 11-3f). The second and simultaneous event is that the whole ribosome moves along the mRNA a distance equivalent to three nucleotides (one codon). As a result, the valine codon remains associated with the anticodon of the dipeptidyl-tRNA$^{Val}$, and a new codon moves into register with the aminoacyl site (Figure 11-3f). In Figure 11-3f the new codon specifies leucine, and the cycle described by steps (c), (d), and (e) can now repeat itself: leucyl-tRNA$^{Leu}$ will move into and bind to the aminoacyl site; a peptidyl transferase reaction will occur; a translocation of the tri-peptidyl-tRNA$^{Leu}$ to the peptidyl site will take place; and a fourth codon will be exposed to the aminoacyl site. As the ribosome moves in a 5' $\longrightarrow$ 3' direction along the mRNA strand, 8 to 15 amino acids will be incorporated into the growing polypeptide every second. Since the correct aminoacyl-tRNAs apparently find the correct codons by a random trial-and-error diffusion process, this rate of chain elongation seems truly remarkable.

After a ribosome has translated perhaps 25 codons of an mRNA, the 5' end of the mRNA becomes free to form a second initiation complex and a second ribosome begins moving along the mRNA, mediating the synthesis of a second polypeptide chain. A third ribosome follows, and so on. The resulting structure, called a **polyribosome** or **polysome,** consists of an mRNA molecule that is being simultaneously translated by several ribosomes into several polypeptide chains. Polysomes from *E. coli* are visualized in Figure 9-11.

In eukaryotes, polyribosomes that are synthesizing proteins destined for secretion become associated with the membrane systems of the outer nuclear envelope and the endoplasmic reticulum, often in beautiful whorls (*Figure 11-1*). As noted in Section 10.9, these proteins carry an N-terminal **signal sequence** of hydrophobic amino acids that is thought to mediate the membrane association.

## 11.5 Termination of Polypeptide Synthesis

The synthesis of a protein concludes when a ribosome encounters one or more **terminator codons** in the mRNA. A terminator codon is not recognized by the anticodons of any of the normally occurring aminoacyl-tRNAs, and its presence in the aminoacyl site of the ribosome therefore precludes the addition of any further amino acids to the C-terminus of the chain.

In *E. coli* and its phages, and probably in all eukaryotic cytoplasms as well, the RNA triplets UAA, UGA, and UAG all function as terminator codons, a fact we explore fully in the final sections of this chapter. When a terminator codon moves into an aminoacyl site (Figure 11-4a), the following events are thought to occur. The terminator codon first interacts with one of two **release factors:** RF-1, which appears to recognize UAA or UAG, or RF-2, which recognizes UAA or UGA. The resultant RF · terminator codon · ribosome complex effectively blocks further chain elongation (Figure 11-4b). With the aminoacyl site so clogged, the completed polypeptide

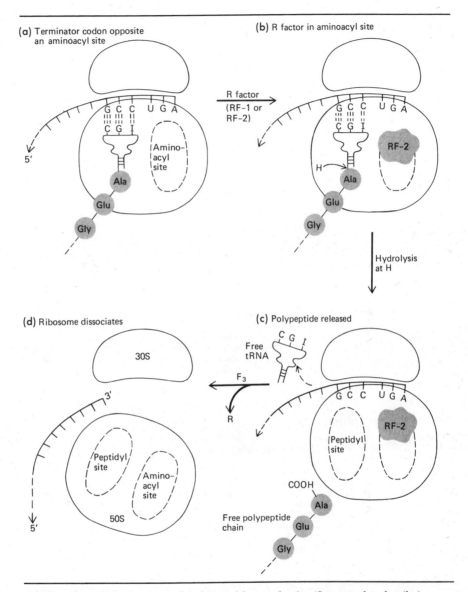

**FIGURE 11-4 Termination of polypeptide synthesis.** (See text for details.)

remains esterified to the final tRNA occupying the peptidyl site. This linkage is then broken by hydrolysis in a reaction mediated by still another protein factor, and both a free tRNA molecule and a complete polypeptide are released from the ribosome (Figure 11-4c). The ribosome then dissociates into its large and small subunits (Figure 11-4d), which are now free to form new initiation complexes and participate in another round of polypeptide synthesis.

## 11.6 Genetic Specification of the Translation Apparatus

The translation apparatus in *E. coli* consists of more than 140 distinct macromolecular species, including at least 60 types of tRNA, 20 types of aminoacyl-tRNA synthetase, 30 proteins in the large ribosomal subunit (proteins L1–L30), 19 proteins in the small ribosomal subunit (S1–S19), 3 rRNAs (23S, 16S, and 5S), and 9 factors for initiation (IF), elongation (EF), and release (RF). These multiple components must fit together and interact in a highly specific manner because mutations affecting their structure often prove to be highly deleterious or lethal: cells are unable to translate accurately their mRNA into protein.

Figure 11-5 indicates some of the mapped translation apparatus genes in *E. coli*. Although a number of genes are dispersed, two conspicuous clusters at 72 and 88.5 minutes carry a large number of the relevant genes.

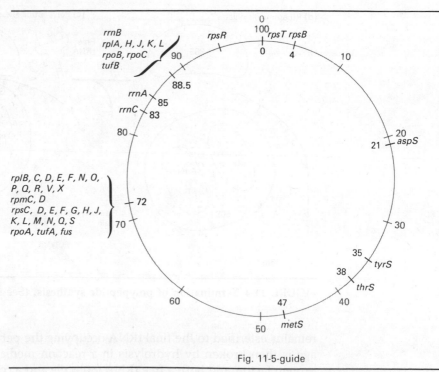

Fig. 11-5-guide

**FIGURE 11-5 Genes specifying components of the translational apparatus** in *E. coli* (*see also Figures 9-3 and 10-2*). Abbreviations: *rrn*, ribosomal RNA gene cluster; *rpl*, ribosomal protein of the large subunit; *rpm*, the *rpl* genes continued past *rplZ*; *rps*, ribosomal protein of the small subunit; *rpo*, RNA polymerase; *tuf*, EF-T$_u$ gene; *fus*, EF-G gene; *aaS*, aminoacyl-tRNA synthetase genes.

# "Cracking" the Genetic Code

By the late 1950s it was clear to a number of investigators that the ultimate products of genes were polypeptides, that the synthesis of a polypeptide involved the ordering of amino acids into a defined sequence, and that this information might be encoded by the sequence of nucleotides in a gene. In this and the following sections we present a largely historical account of how the genetic code was deciphered in the early 1960s. The experimental approaches taken involved a brilliant mix of genetics and biochemistry, and the results unquestionably revolutionized the biological sciences.

## 11.7 The Triplet Nature of the Code

Those trying to deduce the nature of the genetic code quickly realized that if a single nucleotide in an mRNA specified a single amino acid in a protein, proteins could contain only 4 amino acids, whereas in fact they contain 20. Similarly, a doublet code made up of all possible pairs of the 4 nucleotides could generate only 16 combinations, still not enough to specify the 20 amino acids. Therefore, the simplest code that could be envisioned as biologically useful was a triplet code. When all triple combinations of the four nucleotides are made, $4^3$, or 64, different sequences or codons are possible. Since, in theory, only 20 are necessary the possibility was entertained that the code is **degenerate,** meaning that an amino acid can be specified by more than one triplet. Of course, another theoretical possibility is that as many as 44 of the putative triplets are never found in natural mRNAs, or that they occur but do not specify amino acids. In the experiments described below, F. Crick, L. Barnett, S. Brenner, and R. Watts-Tobin were able to establish both the triplet nature and the degeneracy of the code. Since these experiments illustrate several important genetic principles, they are described in detail.

In their experiments, Crick and coworkers made use of $r$-mutant strains of the T4 phage that exhibit two distinct phenotypes: they cannot grow on strain $K(\lambda)$ of $E. coli$, and they bring about a rapid atypical lysis of $E. coli$ strain B so that their plaque morphology is distinctive (*Figure 7-3*). The wild-type allele of the $r$ gene dictates the synthesis of a T4 protein that inserts into the $E. coli$ cell membrane during the course of infection, and the insertion of this protein is apparently necessary for a normal lytic response.

It is possible to classify $r$ strains of T4 as being **leaky** or **nonleaky.** A leaky strain is capable of partial wild-type function; in this case it is capable of producing plaques on $E. coli$ strain B that somewhat resemble wild-type plaques. Mutagenesis of T4 with base analogs such as 5-BUdR (*Figure 7-4*), which characteristically generate GC $\rightleftharpoons$ AT transitions (*Section 8.5*), typically produces leaky $r$ strains. In contrast, a nonleaky $r$ strain shows no growth on $E. coli$ strain $K(\lambda)$ and produces unmistakably large, clear $r$ plaques on strain B. The $r$ strains induced by acridine dyes (*Figure 8-14*), which generate deletions or insertions of one or a few nucleotides in the T4 chromosome (*Section 8.11*), are characteristically nonleaky.

Crick and coworkers pursued these observations with a series of carefully designed experiments that established molecular distinctions between acridine-induced and base analog–induced mutations. They began with a nonleaky acridine-induced *r* strain that they designated *FCO*. During successive cycles of infection of *E. coli* B with this T4 strain they occasionally came upon a plaque that was indistinguishable, or nearly so, from a wild-type plaque. Moreover, the phages isolated from these plaques could successfully infect strain *K*(λ). At first sight, therefore, these phages seemed to represent strains in which the *FCO* mutation had reverted to wild type, but Crick and his associates were able to demonstrate that each new strain was in fact a **pseudo-wild** strain carrying *two* mutations, the original *FCO* mutation plus a second mutation. They found that if they infected *E. coli* cells simultaneously with both a new phage strain and a wild-type strain, as drawn in Figure 11-6, they could isolate rare recombinant strains carrying either *FCO* alone or a new mutation alone (details on how such phage recombination occurs are given in the next chapter). This outcome would not have occurred had *FCO* undergone reversion. Therefore each new strain was in fact doubly mutant, and each could generate a pseudo-wild lytic response only because it carried, in addition to its original *FCO* mutation, a second mutation that acted to **suppress** the effect of the original.

When Crick and his coworkers studied the properties of the recombinant phages that carried only the new suppressor mutations (Figure 11-6), they found in almost all cases that each produced a nonleaky *r* phenotype. In other words, each suppressor mutation, acting alone, generated a phage with the phenotype of an acridine-induced *r* strain.

These observations led Crick and his coworkers to make the following proposals. They first postulated that acridines produce *either* deletions *or* insertions in DNA. Specifically, they suggested that the original *FCO* strain carried an insertion, designated as +, and that each of the *FCO* suppressor mutations consisted of a spontaneous deletion in the *r* gene, designated as −. They next argued that if the reading of a genetic message begins from a fixed point and proceeds sequentially from that point, one codon at a time, a + mutation would have the effect of shifting the reading frame of a transcript (a **frameshift mutation**) in a forward direction; for example, if the correct reading frame is GAG · GAG · GAG · GAG ·. . . , the insertion of a C would produce the reading GAG · GAC · GGA · GGA ·. . . .

Similarly, a − mutation involving the deletion of an A would shift the frame in a reverse direction and produce a GAG · G↓GG · AGG · AGG ·. . . message. In either case it is clear that a single deletion or insertion, particularly if it occurs near the beginning of a gene, would lead to a disruption in the reading of most of the genetic message and a nonfunctional (nonleaky) protein would almost certainly be synthesized. In con-

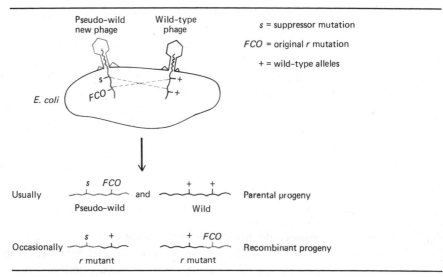

FIGURE 11-6 Suppressor mutations in pseudo-wild strains of T4. The new strain is allowed to infect *E. coli* along with wild-type phages (a mixed infection). Occasional recombination events, indicated by dashed lines, generate recombinant progeny carrying either *FCO* alone or the new suppressor mutation(s) alone.

trast, they argued, a base analog-induced transitional mutation in a gene would disrupt only a single base in an mRNA, resulting, at most, in the insertion of a single incorrect amino acid into a polypeptide. This mutant polypeptide might well still be capable of partial functioning; that is, it might well be able to sustain a leaky phenotype.

Finally, Crick and his coworkers proposed that the − mutations were able to exert their suppressor effects because they were of opposite sign from the *FCO* mutation. They argued that when both + and − mutations are included within nearby portions of the same gene they have the effect of canceling each other out: the reading frame of the transcript from such a doubly mutant gene would first be shifted out of and then back into the correct phase. Looking at our GAG polynucleotide, the insertion of a C ( ↑ ) followed by the deletion of an A ( ↓ ) would have this effect:

$$\text{GAG} \cdot \text{GAC} \cdot \text{GGA} \cdot \text{GGA} \cdot \text{GG} \downarrow \text{G} \cdot \text{GAG} \cdot \text{GAG} \dots$$
$$\uparrow$$

In this sequence, only four triplets are wrong instead of the entire message, and a pseudo-wild phenotype might well be achieved by the resultant protein.

Based on these arguments, Crick and his coworkers designed and performed a series of experiments that established the reading frame used in translating messenger RNA. They first infected *E. coli* B cells with the − suppressor strains of T4 that they had collected. These strains were, as we have said, of the *r* phenotype, but pseudo-wild plaques were occasionally encountered, much as in the experiments with *FC0*. These plaques again proved to be formed by suppressed strains carrying double mutations, but in this case the new second mutations were suppressors of − mutations and were thus of the + class. In this way Crick and his coworkers eventually accumulated a store of about 80 different *r* strains, all designated (+) or (−).

Crosses were then made between the various + and − strains. Most combinations of + and − produced recombinants with a pseudo-wild phenotype, just as the original combination of *FC0* and its − suppressor gave pseudo-wild progeny (Figure 11-6). In other words, most − mutations were found to be suppressors of + mutations and vice versa. In contrast, phages carrying combinations of any two + mutations or any two − mutations still had the nonleaky *r* phenotype. Thus, two mutations of the same sign could not suppress each other.

The final step was to create strains that carried **three** mutations of the same sign, + + + or − − −. All exhibited a pseudo-wild phenotype. Thus three mutations of the same sign were found to suppress one another where two could not. The simplest interpretation of this result is to propose that the **coding ratio** is three: if each codon consists of a triplet of nucleotides, three insertions or three deletions would have the effect of restoring the correct reading frame. Again, by example with poly · GAG, three + + + insertions of C would produce

GAG · GAG · GAC· GGA · GCG · ACG · GAG · GAG.
                  ↑               ↑      ↑

Four incorrect codons are included in this message, but the bulk is back in phase.

## 11.8 The Degeneracy of the Code

Besides indicating that a genetic message is read in successive frames of three nucleotides, the Crick experiments suggested that the code is degenerate. You will recall that many of the + − combinations in T4 were found to yield a pseudo-wild phenotype. This indicates that a reasonably functional protein was often synthesized by such doubly mutant strains and suggests that the wrong triplets found between a + and a − mutation (or between + + + or − − − mutations) *do* code for amino acids, even though they are probably in most cases the wrong amino acids. Since the wrong triplets are created by random mutational events, they presumably include most of the 64 possible triplet combinations of the 4 nucleotides, suggesting that many of the 20 amino acids are specified by more than 1 triplet.

A more direct demonstration that the code is degenerate was made shortly after the 1961 publication of the Crick paper, when several laboratories succeeded in deciphering the genetic code by biochemical means.

## 11.9 Deciphering the Code: *In Vitro* Codon Assignments

Three approaches were taken to establish, *in vitro*, which nucleotide triplets specify which amino acids. This work was done most notably in the laboratories of M. Nirenberg and G. Khorana. In the first approach synthetic polyribonucleotides that contained one, two, or three different types of base were constructed. They were then presented to a complete **cell-free protein-synthesis system** containing ribosomes, the full complement of aminoacyl-tRNAs, and the protein factors necessary for polypeptide synthesis. The composition of the polypeptides synthesized by the *in vitro* system was then analyzed. In cases in which the artificial messenger contained only one repeating nucleotide, the results were straightforward. Poly U, for example, was found to direct the synthesis of polyphenylalanine. Assuming a triplet code, this experiment indicated that UUU is a codon for phenylalanine. Similarly, poly C directs the synthesis of polyproline and poly A, that of lysine, indicating that CCC codes for proline and AAA for lysine.

When mixed copolymers are made, one can calculate all the possible triplets they contain assuming that the bases are incorporated randomly into the molecule; for example, when poly AC is synthesized from a mixture containing equal proportions of A and C, eight triplets should occur: CAC, AAC, AAA, ACA, ACC, CCC, CCA, and CAA. Using poly AC it was found that six amino acids were incorporated into polypeptides: asparagine, glutamine, histidine, lysine, threonine, and proline. The next step was to vary the ratio of A and C, and it was found that when the polymer contained more A than C the ratio of asparagine to histidine in the polypeptide increased. By such experiments the *composition* of the bases in a certain triplet, although not their absolute order, could often be deduced.

The second approach taken was similar to the first except that methods were devised to synthesize polyribonucleotides of a known sequence; for example, a regular copolymer of C and U (CUCUCU . . .) was constructed and was found to direct the synthesis of regularly alternating co-polypeptides of leucine and serine. Polymers of repeating tri- and tetranucleotides were also elaborated and the polypeptides they specified were analyzed. In this way amino acid assignments could readily be made for a number of triplets.

We should note at this point that magnesium concentrations must be kept artificially high in these *in vitro* experiments if the translation of artificial messengers is to occur. High magnesium concentrations allow initiation without an initiation codon, and an AUG or GUG codon will, of course, be absent from most of the artificial polyribonucleotides that might be synthesized *in vitro*. The artificial initiation process occurs at random along a synthetic messenger, reading in all possible frames: thus a regularly repeating poly AAG stimulates the synthesis of three kinds of homo-polypeptide: polyarginine (AGA), polylysine (AAG), and polyglutamic acid (GAA).

A third and quite different approach to solving the genetic code was made by synthesizing trinucleotides of known base sequences, associating them with ribosomes, and testing the ability of the resulting complexes to stimulate the binding of specific aminoacyl tRNAs. For example, the trinucleotide GCC was found to be active in promoting the binding of alanyl-tRNA$^{Ala}$ but *not* of any other aminoacyl-tRNA, and it was therefore concluded that GCC is a codon for alanine.

None of these *in vitro* methods is in itself without ambiguity, but taken together they established the codon assignments for the amino acids with considerable certainty. The full genetic code so derived is summarized in Tables 11-1 and 11-2. Each codon is written as it would appear in mRNA, reading in a 5' $\longrightarrow$ 3' direction; the corresponding codons in DNA will, of course, be both complementary to these mRNA codons and written in the reverse order on a 5' $\longrightarrow$ 3' strand (*Section 9.5*). Similarly, the bases in the corresponding tRNA anticodons will be both complementary and antipolar to the mRNA codons.

An examination of Tables 11-1 and 11-2 reveals several features of the genetic code. First, the code is indeed seen to be degenerate, there being, for example, six codons that specify serine. Only two amino acids are represented by a single codon: tryptophan and methionine. Methionine, as we have learned, is signaled by the special codon AUG, which also specifies the initiation of translation. The three codons UAA, UAG, and UGA are often called **nonsense codons** because they fail to stimulate aminoacyl-tRNA binding. They serve, as already noted, a punctuation function as

**TABLE 11-1   The Genetic Code**

| First Position (5' End of mRNA) (Read Down) | Second Position (Read Across) | | | | Third Position (3' End) (Read Down) |
|---|---|---|---|---|---|
| | U | C | A | G | |
| U | Phe | Ser | Tyr | Cys | U |
| | Phe | Ser | Tyr | Cys | C |
| | Leu | Ser | Stop | Stop | A |
| | Leu | Ser | Stop | Trp | G |
| C | Leu | Pro | His | Arg | U |
| | Leu | Pro | His | Arg | C |
| | Leu | Pro | Gln | Arg | A |
| | Leu | Pro | Gln | Arg | G |
| A | Ile | Thr | Asn | Ser | U |
| | Ile | Thr | Asn | Ser | C |
| | Ile | Thr | Lys | Arg | A |
| | Met (start) | Thr | Lys | Arg | G |
| G | Val | Ala | Asp | Gly | U |
| | Val | Ala | Asp | Gly | C |
| | Val | Ala | Glu | Gly | A |
| | Val (start) | Ala | Glu | Gly | G |

**TABLE 11-2** **The Genetic Code,** showing the RNA codons specifying each amino acid.

| GCA GCC GCG GCU | AGA AGG CGA CGC CGG CGU | GAC GAU | AAC AAU | UGC UGU | GAA GAG | CAA CAG | GGA GGC GGG GGU | CAC CAU | AUA AUC AUU | UUA UUG CUA CUC CUG CUU | AAA AAG |
|---|---|---|---|---|---|---|---|---|---|---|---|
| Ala | Arg | Asp | Asn | Cys | Glu | Gln | Gly | His | Ile | Leu | Lys |

| AUG | UUC UUU | CCA CCC CCG CCU | AGC AGU UCA UCC UCG UCU | ACA AC ACG ACU | UGG | UAC UAU | GUA GUC GUG GUU | UAA UAG UGA |
|---|---|---|---|---|---|---|---|---|
| Met | Phe | Pro | Ser | Thr | Trp | Tyr | Val | Stop |

chain-terminator signals, and are considered more extensively in Section 11.14.

## 11.10 Patterns to the Genetic Code and the Wobble Hypothesis

Two notable patterns emerge when the genetic code is tabulated as in Table 11-1. First, amino acids with similar structural properties tend to have related codons. Thus the aspartic acid codons (GAU and GAC) are similar to the glutamic acid codons (GAA and GAG); similarly, the codons for the aromatic amino acids phenylalanine (UUU, UUC), tyrosine (UAU, UAC), and tryptophan (UGG) all begin with uracil. This feature of the code is thought to have evolved to minimize the consequences of mistakes made during translation or of mutagenic base substitutions. Thus, if an amino acid in a protein is erroneously replaced by one with similar properties, the protein may still be functional.

The second pattern to the genetic code is that for many of the synonym codons specifying the same amino acid the first two bases of the triplet are constant, whereas the third can vary; for example, all codons starting with CC specify proline (CCU, CCC, CCA, and CCG) and all codons starting with AC specify threonine. This flexibility in the third nucleotide of a codon may well help to minimize the consequences of errors, and F. Crick has offered a molecular explanation for its occurrence. He suggests that as an aminoacyl-tRNA molecule is lining up to form base pairs with an mRNA codon on a ribosome, the *initial* codon-anticodon interaction is between the nucleotide at the 5' end of the codon and the nucleotide at the 3' end of the anticodon, while the second interaction takes place between the two middle nucleotides. He proposes that once correct base pairs have formed at these two positions some **wobble** will be permissible at the third

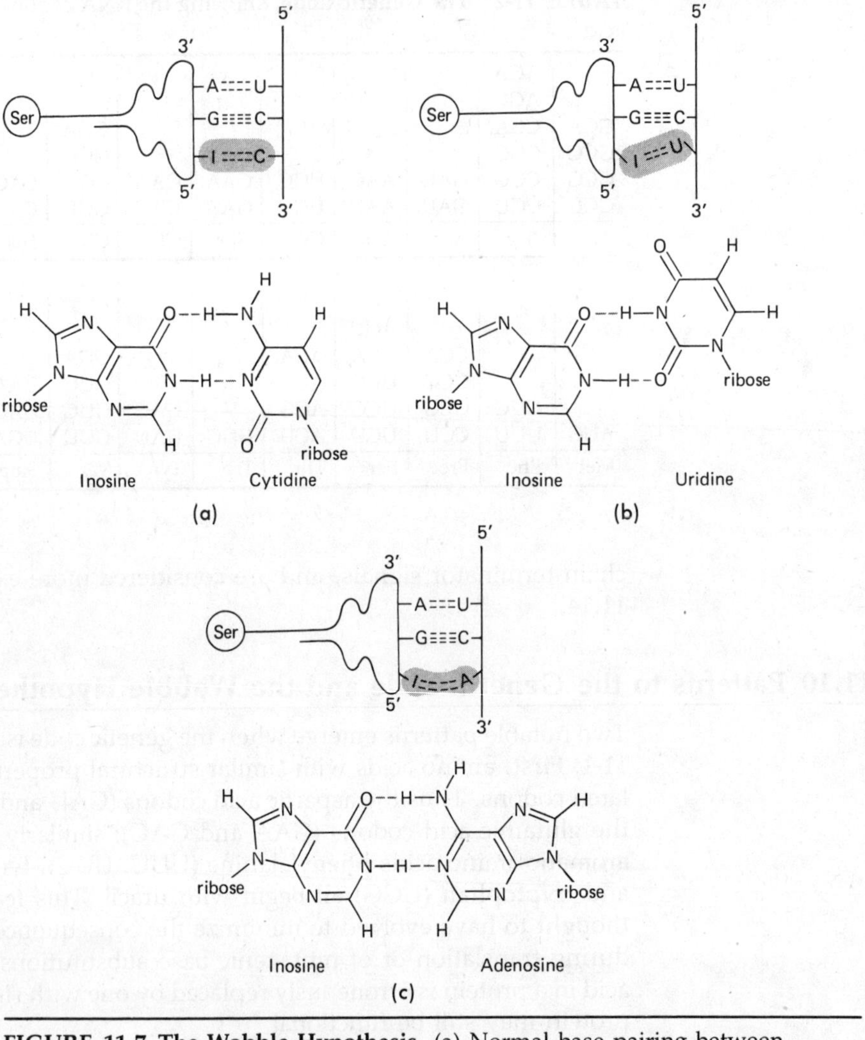

FIGURE 11-7 The Wobble Hypothesis. (a) Normal base pairing between inosine (I) and cytidine in the third position of anticodon-codon recognition (shaded). (b) Wobble base pairing between I and uridine. (c) Wobble base pairing between I and adenosine.

position. For example, the purine inosine (I) is similar to guanine and will normally pair with cytosine (C), as shown in Figure 11-7a; when it is in the third position of an anticodon, however, it may be free to shift its position so that it can form base pairs with a U (Figure 11-7b) or an A (Figure 11-7c) in the third position of the codon, such pairs being called **wobble base pairs.** Therefore, as illustrated in Figure 11-7, a seryl-tRNA$^{Ser}$ molecule with the anticodon 3'—AGI—5' will be able to interact with the serine codons UCC, UCU, and UCA. Similarly, a U at the wobble position will be able to pair with an A or a G.

Because of the proposed wobble base-pairing, one tRNA species is thought to be able to recognize more than one codon for the same amino acid. As noted in Section 10.1, moreover, several distinct **isoacceptor tRNA** species often exist for a given amino acid. In some cases these distinct species respond to the same codon or codon family, but in other cases they will respond only to different codons. In short, there is no simple one-to-one relationship between the number of amino acid codons that exist in a cell and the number of tRNA species possessed by the cell.

## 11.11 *In Vivo* Codon Assignments and Codon Preference

The code summarized in Table 11-1 was established with synthetic molecules and cell-free protein-synthesizing systems. Experiments designed to determine whether this same code is also used *in vivo* soon followed. One approach compared the amino acid sequences of proteins from wild-type and mutant strains of various organisms.

Early experiments by C. Yanofsky and coworkers, for example, focused on the enzyme tryptophan synthetase of *E. coli*. In one chymotryptic peptide (*Section 3.2*) derived from the wild-type enzyme, position 8 was occupied by glycine. Several base analog–induced mutant strains had substitutions for glycine at position 8, and the new amino acids were always either arginine or glutamate, as shown in Figure 11-8a. When the arginine-containing strains were again treated with base analogs, revertants were recovered that had glycine position 8. New mutant strains were also found, and these carried either serine, isoleucine, or threonine at position 8 (Figure 11-8a). In contrast, when the glutamate-containing

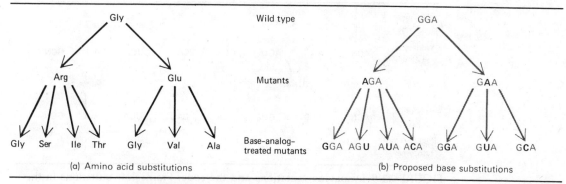

**FIGURE 11-8 Tryptophan synthetase mutations and the code.** (a) Amino acid substitutions observed at position 8 of a chymotryptic peptide from tryptophan synthetase A of *E. coli*, as described in the text. (b) Proposed base substitutions (boldface) leading to the amino acid substitutions seen in (a).

strains were again mutagenized with base analogs, new mutant strains carrying either valine or alanine at position 8 were obtained in addition to the glycine revertants (Figure 11-8a). By assuming that the original glycine at position 8 was coded for by GGA and that each mutational event involves a single base change, it was found that the observed amino acid substitutions could be explained by a unique set of codons (Figure 11-8b), all selected from the *in vitro* codon assignments. This outcome strongly suggests (but does not prove) that the *in vitro* code is followed *in vivo*.

Similar experiments, performed by E. Terzaghi, G. Streisinger, A. Tsugita, and their associates, concerned the enzyme lysozyme, which is specified by T4 and digests bacterial cell walls. In this case the complete amino acid sequence of the enzyme was known, and comparisons were thus possible between the wild-type sequence and the sequence specified by pseudo-wild strains that carried two acridine-induced, + − frameshift mutations. One pseudo-wild enzyme was found to carry a cluster of five amino acids that differed from those in the wild-type enzyme, as shown in Figure 11-9, thus confirming the prediction of Crick and his colleagues that pseudo-wild strains should dictate the synthesis of proteins that carry several wrong amino acids in tandem. When *in vitro* codon assignments were written out for the five wild-type amino acids, it was found that if a single nucleotide were first deleted and then inserted, the *in vitro* codon assignments for the five mutant amino acids could be generated, as illustrated in Figure 11-9. The amino acid sequences of several other pseudo-wild strains can be similarly made to fit the *in vitro* code by assuming simple frameshift events.

The most unambiguous validation of the *in vitro* code has been made possible by comparing the amino acid sequences of various proteins with the nucleotide sequences of the genes that specify them (*Box 4.1*). Starting

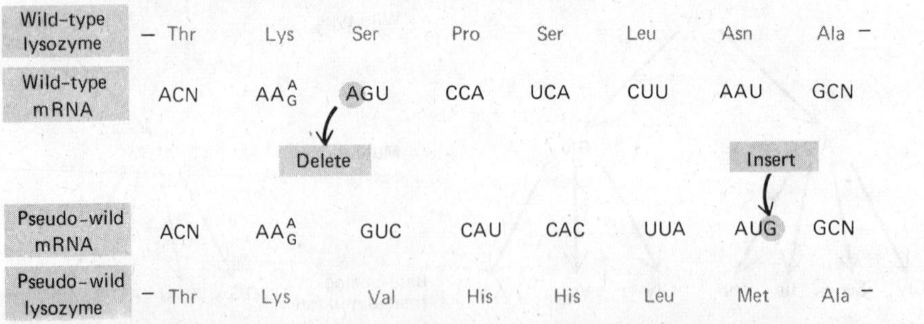

| Wild-type lysozyme | — Thr | Lys | Ser | Pro | Ser | Leu | Asn | Ala — |
|---|---|---|---|---|---|---|---|---|
| Wild-type mRNA | ACN | AA$_G^A$ | AGU | CCA | UCA | CUU | AAU | GCN |
| | | | Delete | | | | Insert | |
| Pseudo-wild mRNA | ACN | AA$_G^A$ | GUC | CAU | CAC | UUA | AUG | GCN |
| Pseudo-wild lysozyme | — Thr | Lys | Val | His | His | Leu | Met | Ala — |

**FIGURE 11-9 Lysozyme mutations and the code.** Frameshift mutations in the lysozyme gene of T4 phage and their effect on the primary structure of the enzyme. The lysine codon is shown as AA$_G^A$ to indicate the possible variability at the third position. N = any nucleotide. (After C. Yanofsky, *Ann. Rev. Gen.* **1:**117, 1967.)

from initiator codons, one can mark off successive triplets and determine which amino acids they should specify according to the *in vitro* code; this theoretical sequence is then compared with the known sequence of amino acids in the protein. A number of "proofs" of the *in vitro* code have been made in this fashion (see, for example, the insulin gene, Figure 10-10), and these data provide clear evidence for the existence of **colinearity** between the sequence of nucleotides in a gene and the sequence of amino acids in a polypeptide.

Sequence analyses of prokaryotic and eukaryotic genes have disclosed the phenomenon of **codon preference**. Given the choice of six leucine codons (*Table 11-2*), for example, the *E. coli* phage φX174 uses the CUU codon 12 times as often as it uses the CUA codon, with the other four codons given intermediate preference. For the animal virus SV40, by contrast, the most popular leucine codons are UUG and UUA and the least popular is CUC. Codons of the type NCG (N for any nucleotide) are heavily used by φX174 but are little used by vertebrates in general. Presumably these differences reflect various factors, including a cell's tRNA composition and perhaps structural features of the mRNAs themselves.

## 11.12 Initiator Codons

Sequencing studies have revealed that GUG as well as AUG can serve as an initiator codon *in vivo*. The *in vitro* code (*Table 11-1*) states that GUG should code for valine, yet the N-terminal amino acid inserted in response to an initiator GUG is (N-formyl)-methionine. This aberrant finding is explained by the observation that methionyl-tRNA$_f^{Met}$ appears to obey unique codon-anticodon recognition rules. You will recall that most aminoacyl-tRNAs are thought to make initial contact with the first nucleotide in the triplet codon, the third nucleotide being the "wobble base" (*Figure 11-7*); methionyl-tRNA$_f^{Met}$, on the other hand, makes initial contact with the third nucleotide (G) of the codon, the first nucleotide becoming the "wobble base." Therefore, methionine is inserted in response to AUG and GUG codons (and perhaps UUG and CUG codons as well).

Why doesn't N-formyl methionyl-tRNA$_f^{Met}$ recognize internal AUG and GUG codons and insert N-formyl methionine in these positions? It appears that the elongation-factor complex EF-T$_s$ · EF-T$_u$ (*Figure 11-3*) is unable to interact with N-formyl methionyl-tRNA$_f^{Met}$ and therefore never brings this species into the aminoacyl site of the bacterial ribosome. Hence, internal AUGs and GUGs are accessible only to methionyl-tRNA$^{Met}$ and valyl-tRNA$^{Val}$ species. As is evident in Figure 11-10, tRNA$_f^{Met}$ and tRNA$^{Met}$ have the same 5'-CAU-3' anticodon but are otherwise very different in their base sequence; presumably, therefore, they assume quite different tertiary structures and are recognized by quite different sets of proteins in the translation apparatus.

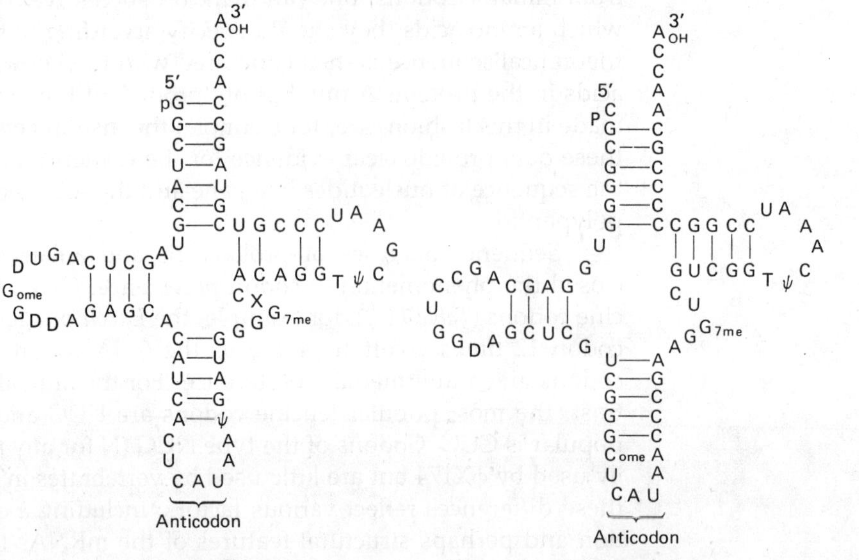

FIGURE 11-10 (a) tRNA$_f^{Met}$ and (b) tRNA$^{Met}$. Abbreviations are the same as in Figure 10-1b plus G$_{7me}$, 7-methylguanylic acid; G$_{ome}$, 2'-0-methylguanylic acid; C$_{ome}$, 2'-0-methylcytidylic acid. X represents an unidentified nucleotide.

## 11.13 Universality of the Code and the Mitochondrial Exceptions

That the genetic code is universal has been demonstrated in a number of ways. In one approach, synthetic trinucleotides of known sequence are complexed with *E. coli* ribosomes and their binding affinity for *E. coli* aminoacyl-tRNAs is compared with their affinity for aminoacyl-tRNAs from other organisms. In general, such studies demonstrate that different types of organisms are similar, if not identical, in the codons that their various aminoacyl-tRNAs recognize, although some may have little or no aminoacyl-tRNA that corresponds to certain codons. One can also present an *E. coli in vitro* protein-synthesizing system with, for example, purified mRNA from poliovirus (which is normally translated by human cells) and recover normal polio viral coat protein.

It is speculated, therefore, that the code became fixed in its present form when the first kinds of cellular life evolved, from one to three billion years ago. Whether the code originated essentially all at once or whether it began with only a few meaningful codons and gradually expanded to include all 64 codons is not, of course, known, but apparently once it was established, it specified such essential biological information that any further variations did not survive evolutionary pressures.

The known exceptions to the rule of a universal code are given by mitochondria. As we noted in Chapter 2 and will describe at length in

Chapter 16, mitochondria possess their own chromosomes, transcribe their own tRNA and rRNA species, and use these to translate a small number of mRNAs into mitochondria-specific proteins. One of these proteins, subunit II of mammalian mitochondrial cytochrome c oxidase, was sequenced and found to carry tryptophans at positions 65, 106, and 163. The mitochondrial gene for subunit II was subsequently cloned and sequenced, and all three tryptophan positions were found to be occupied by a 5′TGA3′ triplet in the DNA. This triplet is transcribed as a 5′UGA3′ codon in mRNA that, as we noted in Section 11.5, should signal the termination of polypeptide synthesis rather than the insertion of a tryptophan. The molecular basis for the switch was soon established: mitochondria were found to contain a tRNA species that is charged with tryptophan but that carries a 5′UCA3′ anticodon rather than the 5′CCA3′ anticodon present on the tRNA$^{Trp}$ species of wild-type prokaryotes and eukaryotes. As diagrammed in Figure 11-11, the mitochondrial anticodon is able to recognize 5′UGA3′ whereas the "usual" anticodon cannot.

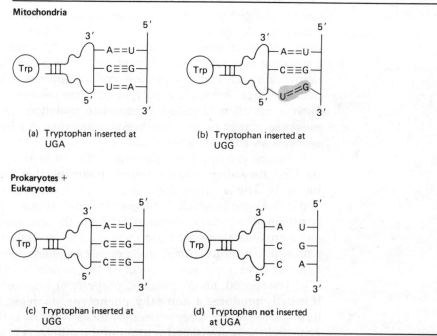

(a) Tryptophan inserted at UGA

(b) Tryptophan inserted at UGG

(c) Tryptophan inserted at UGG

(d) Tryptophan not inserted at UGA

**FIGURE 11-11 Mitochondrial exception to the universal code.** The UCA anticodon of the mitochondrial tRNA$^{Trp}$ forms normal base pairs with the UGA codon (a); its U at the wobble position also allows the formation of a wobble base pair with the UGG codon (b). The tRNA$^{Trp}_{CCA}$ found in all wild-type prokaryotes and eukaryotes can recognize the UGG codon as well (c), but the C at the wobble position CCA disallows base-pair formation with the UGA triplet (d).

Although yeast and human mitochondria both read the UGA codon as tryptophan rather than "stop," a "universal mitochondrial code" clearly does not exist. Yeast mitochondria, for example, are unique in reading the CUX family of codons as threonine rather than leucine (*Table 11-1*), whereas human mitochondria are unique in reading AUA as methionine rather than isoleucine. Moreover, corn mitochondria reportedly read UGA as "stop" rather than as tryptophan. In general, therefore, it can be said that mitochondrial genetic systems have followed quite independent evolutionary pathways, a concept expanded in Chapter 16.

# Nonsense Mutations and Chain Termination

## 11.14 Properties of Nonsense Mutations

As stated earlier in the chapter, three RNA codons are not recognized by any of the aminoacyl-tRNAs normally present in cells. These are UAG (also called *amber* because an investigator who studies the properties of this codon belonged to the Bernstein family, and Bernstein means "amber" in German); UAA (also called *ochre*), and UGA (*opal*). One or more of these three codons is used profitably by the cell to signal the natural end of translation of a particular polypeptide (*Section 11.5*). If, however, a mutation occurs such that one of these codons appears in the interior of a genetic message, incomplete polypeptides are released from the ribosome. Such a mutation is called a **nonsense** mutation, as contrasted to the **missense** mutations we have been considering in which a mutant codon specifies an alternate amino acid.

Strains carrying nonsense mutations can be recognized by four criteria. First, **they often produce peptide fragments** that can be detected experimentally. The length of the fragment will depend, of course, on the position in the gene in which the nonsense codon appears. A given codon may be **proximal** to (near) or **distal** to (far from) the promoter region of its gene, and a proximal nonsense codon will exert its effect near the N-terminus of a protein to produce a small polypeptide fragment, whereas a distal mutation will have a less severe effect on the protein length.

The second, and correlative, property of a nonsense mutation is that **it usually produces a nonleaky phenotype,** as might be expected for an incomplete protein. Only when the mutation occurs in the distal part of the gene and the untranslated portion of the mRNA is not essential does the polypeptide product show any activity at all. Frequently, so little of the protein is synthesized that extracts of the mutant strains will not even cross react in the precipitin test with antibody made against the wild-type protein (the precipitin test is described in Box 2.3.)

The third characteristic of nonsense mutations, called **polarity,** is observed when a nonsense mutation lies in a gene proximal to the promoter of an operon (*Section 9.2*). In such a case, the nonsense mutation prevents expression not only of the promoter-proximal gene but also of all genes in the operon that are distal to the mutation. Such polar effects have been found to be "relieved" when *E. coli* cells also carry a mutation in the *rho* gene, which specifies the rho protein involved in the termination of transcription (*Section 9.8*). It appears, therefore, that when an *E. coli* ribosome falls off a message at a nonsense codon, rho blocks further transcription of that message and no genes distal to the nonsense codon are copied. Mutant versions of rho perform this activity less efficiently, and polarity is thereby less severe in those *rho⁻* strains that carry nonsense mutations in an operon.

The fourth feature of a nonsense mutation is that **its phenotypic expression is usually conditional.** Certain genes in bacteria and other cells suppress the expression of nonsense codons in a manner quite distinct from the suppression of polarity. The molecular basis of nonsense suppression is discussed in Section 11.16; meanwhile its basis may occur to you.

## 11.15 Establishing the Chain-Termination Codons

The behavior of UAG, UAA, and UGA triplets *in vitro* made them likely candidates as the sources of nonsense mutations. It was observed, for example, that neither UAG, UAA, nor UGA was active as a trinucleotide in promoting the binding of aminoacyl-tRNAs to ribosomes. Moreover, when synthetic polynucleotides were synthesized such that any one of these triplet combinations appeared in the message, these polynucleotides were characteristically translated into short peptide fragments rather than long polypeptides. Such observations indicated—but did not prove—that these three codons were involved both in nonsense chain termination and in natural chain termination.

The approaches taken to analyze nonsense mutations *in vivo* resemble the experiments described in Section 11.11 for the tryptophan synthetase gene. Thus A. Garen and colleagues mutagenized *E. coli* with base analogs, and strains were isolated that exhibited no alkaline phosphatase activity whatsoever, these being likely to be carrying nonsense mutations in the alkaline phosphatase gene. When these strains were again mutagenized with base analogs, "cured" strains having at least partial alkaline phosphatase enzyme activities were recovered. The amino acid sequences of the wild-type and the cured alkaline phosphatases were then compared. It was found that the wild-type protein had a tryptophan (coded for by UGG) in a certain position and that the cured proteins from several strains of *E. coli* carried a spectrum of seven amino acid substitutions in its place, including the original tryptophan. By writing out the codons for these seven amino acids it could be deduced that the original nonsense mutation must have converted the UGG to a UAG, for the substituted codons could all be related to UAG by single nucleotide changes. This result is diagrammed in Figure 11-12. The study of other cured strains led, by similar reasoning, to

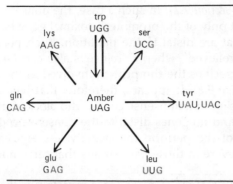

FIGURE 11-12 Nonsense mutations and their intragenic suppression. The relationship between seven codons that specify amino acids and the UAG *amber* nonsense codon. Each differs by a single base, indicated in boldface. (After A. Garen et al., *Science* **160**:152, 12 April, 1968. Copyright 1968 by the American Association for the Advancement of Science.)

the implication of UAA and UGA as nonsense codons. The reader is reminded here of the one exception to this generalization. As detailed in Section 11.13 and Figure 11-11, the UGA codon specifies tryptophan insertion in mitochondria, meaning that only UAG and UAA function as mitochondrial termination codons.

# Suppressor Mutations

## 11.16 General Nature of Suppressors

Suppressor mutations were defined in Section 7.1 as second mutations that restore the original (wild) phenotype of a mutant strain. Such a second mutation may occur within the gene carrying the first mutation (**intragenic**) or may lie in a different gene (**intergenic**). We have already encountered several examples of intragenic suppression in the course of this chapter: base substitutions in *E. coli* have been shown to convert missense or nonsense codons into other missense codons that restore protein activity (*Figures 11-9 and 11-12*), while a suppressor frameshift mutation occurring close to the position of a first mutation can produce a functional lysozyme or *r* gene product in T4 (*Figures 11-6 and 11-10*).

Intergenic suppression can originate in two different ways; one indirect and one direct. During **indirect intergenic suppression** a second gene mutation has the effect of circumventing the expression of the original mutant phenotype without in any way altering the original mutant gene product. For example, certain mutant strains of *Neurospora* are known to have an abnormally high sensitivity to intracellular arginine. When these strains also carry the *s* suppressor gene, they synthesize very little arginine and thus the arginine sensitivity, although still present, is not exhibited. Additional examples of genes that affect the expression of other genes are given in later chapters. Of more immediate interest here is the phenomenon of direct intergenic suppression.

**Direct intergenic suppression** is most easily defined after an example is given to illustrate it. We first imagine an *E. coli* cell carrying an *amber* mutation in its tryptophan synthetase gene such that the mRNA for tryptophan synthetase carries a UAG codon near its proximal end. Such a cell, as we have learned, would synthesize no tryptophan synthetase and would normally be inviable. Now let us imagine that a second mutation occurs, this time in a gene that specifies tRNA$^{Leu}$. This mutation has no effect on the affinity of the tRNA$^{Leu}$ molecule for leucine, and thus leucyl-tRNA$^{Leu}$ complexes form normally. The mutation does, however, change the anticodon of the tRNA$^{Leu}$ so that instead of being 3'—AAC—5' and pairing with a 5'—UUG—3' leucine codon, it is 3'—AUC—5'. This mutant tRNA has the capacity to recognize a 5'—UAG—3' nonsense codon and insert leucine in response to it. Therefore, when the mutation in the tRNA gene, called an *amber* **suppressor mutation,** is included in the cell along with the *amber* mutation in the tryptophan synthetase gene, a tryptophan synthetase of normal size will be synthesized. This will contain one wrong amino acid (leucine), but it will usually have at least some activity and the cell can survive.

In general, then, direct intergenic suppressor mutations can be said to affect genes whose products function in the translation of mRNA into protein. The mutant suppressor molecules permit a mutant codon to be translated incorrectly but beneficially. Most often the suppressor molecules are tRNA molecules, as in the foregoing example. In *E. coli* nine different tRNA genes (*Figure 10-2*) have been observed to mutate so that their tRNA transcripts perform a mutant activity: each inserts its normal amino acid in response to a nonsense triplet. As in the preceding example, the amino acid inserted is usually wrong and the protein synthesized in the presence of the suppressor is therefore usually still mutant; however, it is almost always able to function more effectively than the original nonsense fragment.

## 11.17 Nonsense Suppressor Genes in *E. coli*

One of the first isolated and most intensively studied **suppressor strains** of *E. coli* carries the mutant gene known as *su3*. A cell carrying *su3* plus an *amber* mutation in its alkaline phosphatase gene will demonstrate partial or pseudo-wild enzymatic activity. Similarly, if this cell is infected with a T4 strain carrying an amber mutation in a head protein gene, some infective progeny phages will be produced.

H. Goodman and colleagues analyzed the tRNA present in wild-type and *su3* cells. Both proved to have identical amounts of tRNA$_2^{Tyr}$. The wild-type cells also possessed the isoacceptor tRNA$_1^{Tyr}$, whereas *su3* cells produced a reduced amount of tRNA$_1^{Tyr}$ and, in addition, a new tyrosine-inserting tRNA species not present in the wild type cells. When the sequence of this new tRNA was determined, it was found to be identical to tRNA$_1^{Tyr}$ except at the anticodon: thus, whereas the wild-type anticodon reads 5'-G*UA-3' (where G* is a modified guanine), the mutant anticodon reads 5'-CUA-3'. The mutant tRNA is thus able to read 5'-UAG-3' codons, insert tyrosine, and eliminate the chain-terminating effects of nonsense triplets, exactly as in our earlier hypothetical example.

The *su3* mutation maps to 27 min on the *E. coli* chromosome (*Figure 10-2*), thus localizing the gene for tRNA$_1^{Tyr}$. Further analysis has revealed that two closely linked copies of this gene (*tyrT*) occupy the locus; therefore, *su3 E. coli* cells carry one *tyrT* and one *tyrTsu3* gene, explaining how they are able to•synthesize both normal and suppressor versions of tRNA$_1^{Tyr}$.

Other *amber* suppressors in *E. coli* insert glutamine, leucine, lysine, or serine in response to UAG. The codons for these amino acids all differ from UAG by a single base (*Table 11-1*), and in a number of these strains a suppressor tRNA with a mutant anticodon has been identified. Ochre and opal suppressors are also known in *E. coli*, as are suppressor tRNAs with base changes in nonanticodon regions of the molecule. For example, one suppressor tRNA carries a mutation in the dihydrouridine loop (*Figure 10-1*), which changes its translation properties.

Strains of *E. coli* carrying efficient *amber*-suppressor genes are capable of the same rates of growth as wild-type cells, a puzzling and unexplained fact since one would expect that in such strains many natural chain-terminator codons would be suppressed and overly long proteins would be synthesized by **read-through** into ordinarily nontranslated sectors of mRNA molecules such as trailer or spacer sequences.

**Temperature-Sensitive Amber Suppressor**   Since a nonsense mutation typically abolishes all protein activity, *amber* mutations cannot ordinarily be isolated in essential genes: if a nonsuppressor strain is used to select for such a mutation then the mutant cell will die, whereas if a suppressor strain is used the cell will not express a mutant phenotype during the screening procedure. This dilemma is circumvented by the *SupD$^{ts}$* mutant of *E. coli*, which produces an efficient serine-inserting amber suppressor at 14°C but has virtually no suppressor activity at 43°C.

## 11.18 Nonsense Suppressor Genes in Yeast

Of the eukaryotes, yeast is presently the best characterized with respect to intergenic nonsense suppressors. Suppressor genes can be detected by their ability to suppress nutritional nonsense mutants and by their ability to restore near-normal cytochrome *c* levels to strains carrying proximal *amber* or *ochre* mutations in the *CYC1* structural gene. Using the cytochrome *c* assay, F. Sherman and colleagues were able to identify eight loci in the yeast genome that, when mutated, exert suppressor activity. Alleles at a given locus, moreover, specify either *amber* or *ochre* suppressors: thus the *SUP2-a* allele of the *SUP-2* locus encodes an *amber* suppressor, while the *SUP2-o* allele encodes an *ochre* suppressor. Molecular studies have shown that the mechanism of suppression is analogous to that in bacteria so that, for example, the *SUP5-a* gene specifies a tRNA$^{Tyr}$ species with a mutant anticodon (5'-CTA-3') that inserts tyrosine in response to UAG.

**Yeast Suppression of a Human Nonsense Mutation**   J. C. Chang and Y. W. Kan have recently demonstrated that humans also undergo nonsense mutations and that these can be suppressed *in vitro* by yeast suppressor tRNAs. They analyzed the cDNA (*Section 4.14*) from a patient with a β-globin defect known as β°-thalassemia; this disease differs from β⁺-thalassemia (*Section 9.10*) in that β-globin synthesis is nil in the homozygous state. The mutant gene proved to have undergone an A ⟶ T transversion such that the codon **A**AG, specifying lysine at amino acid position 17 of the β-globin chain, was converted to an *amber* T**AG** (UAG). Chang and Kan then isolated mRNA from this patient and presented it to a cell-free protein synthesis system (*Section 11.9*) suitable for eukaryotic messages. Although the patient's mRNA could dictate α-globin synthesis in a normal fashion, no β-globin synthesis was observed unless they added yeast suppressor tRNAs that were known to insert serine in response to *amber* codons. In the presence of the yeast tRNA, a β-globin species was synthesized in which serine now occupied amino acid position 17.

(a)                                      (b)                                      (c)

(a) Masayasu Nomura (University of Wisconsin) studies the genetic specification and assembly of ribosomes. (b) Charles Yanofsky (Stanford University) has demonstrated relationships between amino acid substitutions and the genetic code in *E. coli*. (c) Luigi Gorini contributed much to our understanding of the molecular nature of suppressor mutations.

# Questions and Problems

1. In what ways is a polyribosome similar to an *rrn* gene (*Section 10.6*) in the process of being transcribed?

2. Crick et al. found that they could mixedly infect *E. coli* cells with a + and a − *r* strain and could recover + − recombinants. In most cases these had a pseudo-wild phenotype, but several continued to exhibit an *r* phenotype, particularly in cases where both the + and − mutations occurred near the beginning of the *r* gene. Give an explanation for this result, using diagrams.

3. The following repeating polyribonucleotides were used in a cell-free system to direct amino acid incorporation into polypeptides: (a) poly AC; (b) poly UAC; (c) poly AAG; (d) poly GAU; (e) poly UAUC. Which amino acids would you expect to be incorporated in each case?

4. When U occupies the third (5′) position of an anticodon, wobble base pairing can occur with either A or G in the codon, whereas a C in this position can pair only with a G. Predict whether the following mutant anticodons can suppress *amber* (5′-UAG-3′) or both *amber* and *ochre* (5′-UAA-3′) codons: (a) 5′-UUA-3′; (b) 5′-CUA-3′.

5. The RNA chromosome of the R17 bacteriophage, which contains three genes, also serves as phage messenger RNA. The R17 coat protein, specified by the middle gene in the chromosome, has the C-terminal amino acid sequence Asn-Ser-Gly-Ile-Tyr. The R17 replicase, specified by the gene at the 3′ end of the chromosome, has the N-terminal sequence Met-Ser-Lys-Thr-Thr-Lys.

   The following R17 RNA sequence is found to be protected by ribosomes from ribonuclease digestion:

   | 1 | 2 | 3 | 4 | 5 | 6 | 7 | 8 | 9 | 10 | 11 | 12 | 13 | 14 | |
   |---|---|---|---|---|---|---|---|---|----|----|----|----|----|--|
   | 5′ . . . AAC | UCC | GGU | AUC | UAC | UAA | UAG | AUG | CCG | GCC | AUU | CAA | ACA | UGA | ~ |

   | 15 | 16 | 17 | 18 | 19 | 20 |
   |----|----|----|----|----|----|
   | GGA | UUA | CCC | AUG | UCG | AAG . . . 3′ |

   (a) Which codon specifies the C-terminus of the coat protein? (b) What is significant about the two codons that follow the codon identified in (a)? Would you expect an *ochre* suppressor mutation to confuse the termination of coat protein synthesis? Explain. (c) Explain how the AUG number 18 and not the AUG number 8 is identified as the initiator codon for replicase synthesis. (d) If AUG number 8 were to initiate protein synthesis, write out the sequence of amino acids that would be polymerized.

6. A bacterial operon has the gene sequence promoter-*a b c d*-terminator, where genes *a-d* specify polypeptides A-D, respectively. (a) Write out a general structure for the mRNA transcribed from such an operon, assuming that it has not undergone processing. Indicate the polarity of the molecule and include the following features: leader, trailer, and transcribed spacer sequences; initiator and terminator codons; initial purine. (b) The C-terminus of polypeptide C is found to be . . . Thr-Arg-Arg; the N-terminus of polypeptide D is found to be Met-Leu-Ser. The sequence of a ribosome-protected fragment of the operon mRNA is found to be . . . ACUCGCCGCUGAUGCUAUCA . . . . Define and

comment on the reading frames used to translate the *c* and *d* sequences in this mRNA. How long is the transcribed spacer?

7. When "multigenic" mRNAs transcribed from bacterial operons are translated, equimolar quantities of each polypeptide chain encoded by the message are synthesized in the case of the *trp* (tryptophan-synthesis) and *gal* (galactose-utilization) operon mRNAs of *E. coli*. When *lac* operon mRNA is translated, however, the enzyme β-galactosidase (specified by the first coding sequence at the 5′ end of the message) is synthesized three to five times more frequently than the transacetylase enzyme (specified by the third and final coding sequence at the 3′ end of the message). In what region(s) of the *lac* operon mRNA would you look for sequences that might explain this discrepancy?

8. A mutation called *glyUsuA36* in *E. coli* converts a tRNA$^{Gly}$, which recognizes the GGA glycine codon, into a tRNA$^{Gly}$, which recognizes the AGA arginine codon. (a) Write out the probable anticodons for these two tRNAs and specify the nature of the mutational event. (b) The missense mutation *trpA36* of *E. coli*, which alters amino acid number 211 in the active site of the tryptophan synthetase enzyme, is suppressed in *glyUsuA36* strains. What amino acid would you expect to find at position number 211 of tryptophan synthetase in wild-type cells? In *trpA36* cells? In cells carrying both *trpA36* and *glyUsuA36*?

9. Wild-type yeast are sensitive to the inhibitor canavanine (*can$^s$*). The *ochre* mutation *canl-100* causes yeast to be canavanine-resistant (*can$^r$*). When *canl-100* cells are mutagenized, two types of *can$^s$* strains can be recovered which we can call A strains and B strains. The mutations carried by the A strains map to the *canl-100* locus; the mutations carried by the B strains are not linked to *canl-100*. (a) Describe the probable nature of the type A and the type B mutations. (b) Describe how you would use the new *can$^s$* strains to isolate temperature-sensitive *ochre* suppressors.

10. Describe how you would screen for an *amber* mutation in an essential gene of *E. coli* using the *SupD$^{ts}$* strain. Assume you have an antiserum available that is directed against the protein product of this gene.

11. A suppressor tRNA is discovered that will suppress certain frameshift mutations if they are of the +1 addition variety but not if they are +2 or −1 or −2. Describe a likely conformation assumed by its anticodon loop.

12. A mutation changes a 3′TCT5′ DNA sequence to an ACT, which creates a mutant phenotype in strain Y of *E. coli*. Strain Y is now mutagenized, and three strains (A, B, and C) are recovered that have an apparently wild phenotype. Sequence analysis of this position of the gene reveals that strain A carries a GCT sequence, strain B a TCT sequence, and strain C a TTT sequence. Explain why each of these strains has an apparently wild phenotype.

13. A nonsense suppressor mutation is found to map to the *rpsR* gene of *E. coli* (*Figure 11-5*). Speculate on how the product of this gene might function in translation.

14. A temperature-sensitive suppressor mutation is found to relieve UAA and UAG nonsense mutations but is without effect on UGA mutants. tRNA extracts from the suppressor strain demonstrate no suppressor activity in an *in vitro* translation assay, nor do washed ribosomes, but the soluble protein fraction of the cells is highly active. (a) Design an *in vitro* translation assay that could be used for the assays described. (b) What component(s) of the translation system are likely to be affected by this mutation? Explain your reasoning.

15. Is iso-1-cytochrome *c* dispensable for growth in yeast? Explain.

16. Transversions can arise by four general mutational pathways (AT ⟶ CG, AT ⟶ TA, GC ⟶ CG, GC ⟶ TA) and transitions can arise by two mutational pathways (GC ⟶ AT, AT ⟶ GC) (*Section 8.3*). You wish to learn which pathway(s) are followed by a new mutagen. Outline how you would accomplish this using the following tester strains: a serine-missense T4 mutant; a tyrosine-missense T4 mutant; a leucine-missense T4 mutant; and *amber*, *ochre*, and *opal* suppressor strains of *E. coli*. The diagram below will assist your answer.

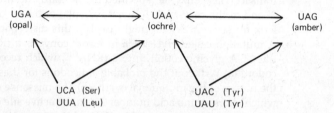

17. It is observed that heat induces reversion of *amber* but not *ochre* mutations. Explain how this observation can be used to argue that heat causes guanine transversions.

18. Sequencing of the ϕX174 chromosome has allowed a determination of the frequency with which the 64 possible codons are actually used in the phage genome. The results are tabulated below.

| Phe | TTT | 39 | Ser | TCT | 35 | Tyr | TAT | 36 | Cys | TGT | 12 |
|-----|-----|----|-----|-----|----|-----|-----|----|-----|-----|----|
|     | TTC | 26 |     | TCC | 9  |     | TAC | 15 |     | TGC | 10 |
| Leu | TTA | 19 |     | TCA | 16 | Ter | TAA | 3  | Ter | TGA | 5  |
|     | TTG | 26 |     | TCG | 14 |     | TAG | 0  | Trp | TGG | 16 |
| Leu | CTT | 36 | Pro | CCT | 34 | His | CAT | 16 | Arg | CGT | 40 |
|     | CTC | 15 |     | CCC | 6  |     | CAC | 7  |     | CGC | 29 |
|     | CTA | 3  |     | CCA | 6  | Gln | CAA | 27 |     | CGA | 4  |
|     | CTG | 24 |     | CCG | 21 |     | CAG | 34 |     | CGG | 8  |
| Ile | ATT | 45 | Thr | ACT | 40 | Asn | AAT | 37 | Ser | AGT | 9  |
|     | ATC | 12 |     | ACC | 18 |     | AAC | 25 |     | AGC | 5  |
|     | ATA | 2  |     | ACA | 13 | Lys | AAA | 47 | Arg | AGA | 6  |
| Met | ATG | 42 |     | ACG | 19 |     | AAG | 31 |     | AGG | 1  |
| Val | GTT | 53 | Ala | GCT | 64 | Asp | GAT | 44 | Gly | GGT | 38 |
|     | GTC | 14 |     | GCC | 17 |     | GAC | 35 |     | GGC | 28 |
|     | GTA | 10 |     | GCA | 12 | Glu | GAA | 27 |     | GGA | 13 |
|     | GTG | 11 |     | GCG | 12 |     | GAG | 34 |     | GGG | 3  |

(a) Are the codons used randomly or nonrandomly? Explain with examples. (b) The "most popular" codon for each amino acid has a distinctive feature. What is it? How might this relate to the wobble hypothesis? (c) Referring to Table 1-3, would you expect that a preference for particular codons might prove to be a distinctive feature of a species? How might an analysis of the tRNA species present in a cell help to document your answer?

19. A hemoglobin variant called **Constant Spring** is known in humans. Whereas the normal length of the hemoglobin α-polypeptide (*Figure 3-5*) is 141 amino acids, the length of the Constant Spring α-chain is 172 amino acids. (a) Describe how this phenotype could arise either via a point or a frameshift mutation. (b)

Amino acid 142 in Hb-Constant Spring is found to be glutamine. Three other types of 172-residue mutant α-hemoglobins have also been sequenced, and they are found to carry Glu, Ser, or Lys at position 142. Show how this pattern indicates that the natural chain-terminator codon of the α-hemoglobin gene is UAA. (c) The amino acid sequence of residues 143 to 172 in all four hemoglobins is found to be the same. Does this argue for the occurrence of frameshift or point mutations? Explain.

20. You wish to devise an *in vitro* system wherein a cloned eukaryotic gene is first transcribed and then translated. What enzymes, factors, organelles, and so on would you include in the incubation medium, and what role would each play?

21. Explain why a frameshift mutation does not lead to the incorrect expression of all the genes on a chromosome.

22. You wish to learn which mRNA sequences are "shielded" from nuclease digestion when mRNA first associates with ribosomes. Outline how you would perform this experiment. Which nucleotides would you predict would be shielded in prokaryotes? Eukaryotes?

23. In the iso-l-cytochrome *c* of yeast, alanine usually occupies amino acid position 12. A mutant strain is found to carry proline, and further mutagenesis produces either alanine revertants or serine or threonine mutants. What base changes are occurring in the DNA?

24. The hemoglobin variants listed below have undergone the indicated amino acid substitutions. For each, show one point mutation that could have generated the substitution, and indicate whether a transition or a transversion is required.

| Hemoglobin Variant | Affected Chain | Position | Amino Acid Substitution |
|---|---|---|---|
| a) Hikari | β | 61 | Lys ⟶ Asn |
| b) I | α | 16 | Lys ⟶ Glu |
| c) D Ibadan | β | 87 | Thr ⟶ Lys |
| d) G Philadelphia | α | 68 | Asn ⟶ Lys |
| e) O Indonesia | α | 116 | Glu ⟶ Lys |
| f) G Chinese | α | 30 | Glu ⟶ Gln |
| g) San José | β | 7 | Glu ⟶ Gly |
| h) G Galveston | β | 43 | Glu ⟶ Ala |
| i) S | β | 6 | Glu ⟶ Val |
| j) Mexico | α | 54 | Gln ⟶ Glu |
| k) Shimonoseki | α | 54 | Gln ⟶ Arg |
| l) K Ibadan | β | 46 | Gly ⟶ Glu |
| m) Norfolk | α | 57 | Gly ⟶ Asp |
| n) Seattle | β | 76 | Ala ⟶ Glu |
| o) M Milwaukee | β | 67 | Val ⟶ Glu |
| p) L Ferrara | α | 47 | Asp ⟶ Gly |
| q) G Accra | β | 79 | Asp ⟶ Asn |
| r) M Boston | α | 58 | His ⟶ Tyr |
| s) Kenwood | β | 95 | Lys ⟶ Glu |
| t) Zurich | β | 63 | His ⟶ Arg |
| u) Horse hemoglobin | α | 24 | Phe ⟶ Tyr |

# 12 Mapping Viral Chromosomes

**A. Introduction**

**B. Viral Infection Cycles**
12.1 Lytic Infection Cycles
12.2 Lysogenic Infection Cycles
12.3 DNA Tumor Virus Cycles

**C. Complementation Analysis**
12.4 The Complementation Test
The cistron
12.5 Intragenic Complementation

**D. Recombination-Frequency Mapping**
12.6 Defining Gene, Marker, and Locus
12.7 The General Case for Mapping by Recombination Analysis
12.8 One-Factor Cross
12.9 Two-Factor Cross
12.10 Three-Factor Cross
Using double recombinants to determine marker order
Deriving map distances from a three-factor cross
Circular genomes in three-factor crosses: The T4 map

**E. Deletion Mapping**
12.11 General Properties of Deletions
12.12 Deletions that Foreshorten Map Distances

12.13 Deletion Mapping of the *rII* Locus of T4
12.14 Deletion Mapping by Specialized Transduction

**F. Mapping Without Recombination**
12.15 Physical Maps
Denaturation mapping
Transcription R-loop mapping
Heteroduplex mapping
12.16 Restriction Mapping and Sequencing DNA Chromosomes: The SV40 Genome
12.17 Sequencing RNA Chromosomes: The MS2 Genome
12.18 Overlapping Genes

**G. Approaches to Solving Mapping and Complementation Problems**
12.19 Two-Factor Phage Cross Problem
12.20 Three-Factor Phage Cross Problem
12.21 Phage Complementation Problem
12.22 Deletion Mapping Problem

**H. Questions and Problems**

# Introduction

**A genetic map provides a summary of a species' genetic information.** It indicates the order and position of genes along a chromosome, with neighboring genes **closely linked** and widely separated genes **loosely linked.** It reveals whether the genes are ordered in a circular array, as for many viruses and most bacteria, or whether two ends of the chromosome exist. It displays whether functionally related genes are grouped together in operon-like clusters or whether they are dispersed. And finally, it can often indicate the kind of mutation a chromosome has undergone, with a large deletion or rearrangement giving very different results in a mapping experiment from a point mutation.

In this chapter we describe strategies used to map viral chromosomes. We first describe the two major kinds of bacteriophage infection cycles, known as **lytic** and **lysogenic.** We then go through the steps taken to analyze a collection of phage mutant strains. They are first subjected to **complementation analysis** to determine whether their mutations reside in the same or different genes. They are then subjected to **mixed infection** in order to analyze their patterns of recombination with one another, leading to the construction of **recombination-frequency maps.** Any deletion mutant strains are instead used for the construction of **deletion maps.** Finally, we consider mapping strategies that depend not on recombination analysis but rather on the properties of the chromosomes themselves. These include **physical mapping** techniques and the powerful new approaches of **restriction mapping** and **nucleotide sequencing.**

# Viral Infection Cycles

It is meaningless to speak of a virus as a prokaryote or a eukaryote since a virus has no cellular structure; instead, viruses are identified according to the type of cell they infect, with bacteriophages infecting bacteria and animal and plant viruses infecting eukaryotic animals and plants. Representative species are shown in Figure 12-1. Each virus contains a single type of nucleic acid—either DNA or RNA—and a protein coat that serves in the infection process and in protecting the virion from the external environment. The nucleic acid contains genetic information necessary for the construction of new viruses but little or none of the apparatus necessary for such construction (for example, ribosomes or enzymes). Viruses must therefore infect living cells and divert these cells' biosynthetic apparatus into the business of synthesizing virus progeny.

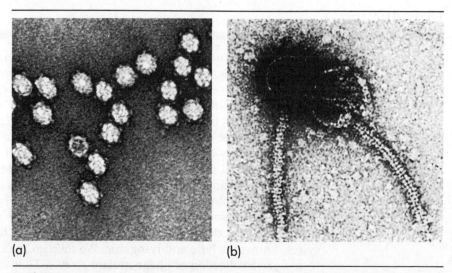

(a)                                              (b)

**FIGURE 12-1** (a) ΦX174 bacteriophages. (Courtesy of J. Finch.) (b) Two λ
bacteriophages. (Courtesy of M. Schnös.) (c) T2 bacteriophage showing its
duplex DNA chromosome (released by osmotic shock) with two free ends.
(From A. K. Kleinschmidt et al., *Biochim. Biophys. Acta* **61**:857, 1962.)

## 12.1 Lytic Infection Cycles

A **lytic infection cycle** causes the death of the host cell. Many **virulent**
bacteriophages can only conduct lytic cycles; these include the close relatives **T2** and **T4** and the close relatives **T3** and **T7**. Other important phages,
known as **temperate,** can either conduct a lytic cycle or else a **lysogenic
cycle,** which does not lead to the immediate death of the host. Lysogeny is
described in the next section.

Lytic cycles follow the same general course; a T2 infection sequence is
illustrated in Figure 12-2, and the T7 cycle is diagrammed in Figure 12-3.
The phage first adsorbs to the wall of its bacterial host (Figure 12-2, *arrow*)
and injects its chromosome into the host cytoplasm (recall the diagrams of
a T2 phage in Figure 1-13 and our discussion of the Hershey-Chase experiments in Section 1.8). Three sets of genes are then sequentially expressed:
the **early gene** products inhibit *Escherichia coli* RNA synthesis; the **DNA
metabolism gene** products include enzymes to replicate the phage chromosome and nucleases to digest the *E. coli* chromosomes so that free nucleotides are available for phage DNA synthesis; and the **late gene** products
include phage coat, tail, and assembly proteins to package the newly replicated chromosomes into mature viruses. A final late gene product is the
enzyme **lysozyme,** which digests the bacterial cell walls. Perhaps 250 new
phages are released in what is called the **burst,** and these are then free to
find new bacterial hosts.

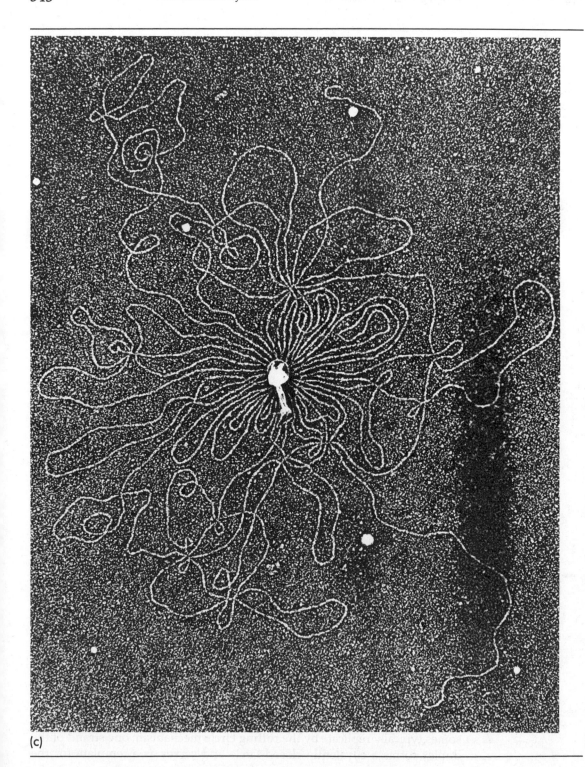

(c)

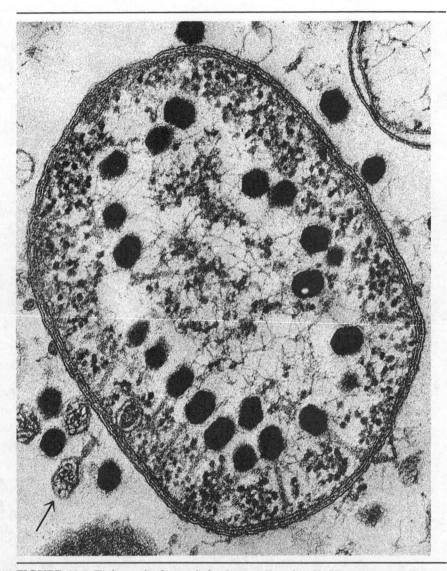

**FIGURE 12-2 T2 bacteriophages infecting an *E. coli* cell.** Hexagons are phage heads; small black particles are ribosomes; thin fibrous material is DNA. Arrow points to phage adsorbed to outer cell wall. (From Simon, L. D., *Virology* **38**:287, 1969.)

The chromosomes of most virulent and temperate phages are DNA duplexes that range in length from 12 μm (T7) to 56 μm (T2). (Phages with single-stranded genomes are considered in Section 12.18.) In late stages of infection, when rapid production of phages is in progress, these chromosomes typically replicate by the **rolling circle** mechanism: one strand of the duplex forms a covalently closed circle and serves as a template for the

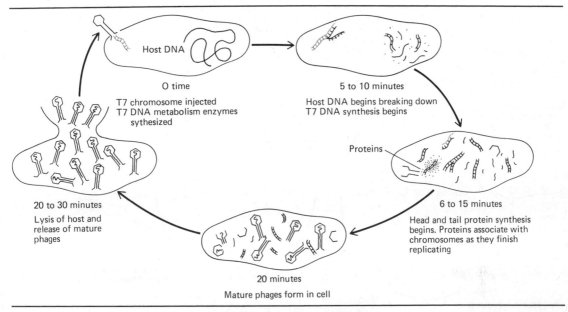

**FIGURE 12-3 The T7 life cycle** at 30°C.

generation of enormously long DNA molecules, called **concatemers,** which contain many genomes' worth of phage sequences. The concatemers are then clipped into chromosome-length rods by the so-called **headful mechanism:** DNA is pulled into a forming phage head until there is no more room, at which point it is cleaved.

## 12.2 Lysogenic Infection Cycles

Temperate phages such as λ and φ80 can lyse their host via the same general route as phage T7; details are given in Figure 12-4a. Under certain conditions of infection, however, the injected chromosome does not express its virulent genes, nor is it replicated as an independent entity. Instead, it dictates the synthesis of a **repressor** molecule that inhibits the expression of virulent genes, and it becomes physically inserted into a special region of the host chromosome, in which state it is replicated along with the host chromosome for many generations and is called a **prophage** (Figure 12-4b). The bacterium carrying such a prophage is called a **lysogen,** and the act of repressor synthesis and prophage integration is called **lysogeny.** (Details of this process are considered in Section 18.1.)

An *E. coli* carrying a λ prophage is said to be lysogenic for λ (abbreviated *E. coli* **(λ)**) and usually grows at the same rate as a normal bacterium. Very infrequently (perhaps once in 10,000 divisions), levels of repressor drop below threshold levels, whereupon the prophage is released from the host chromosome, directs a lytic cycle, and lyses its host (Figure 12-4b).

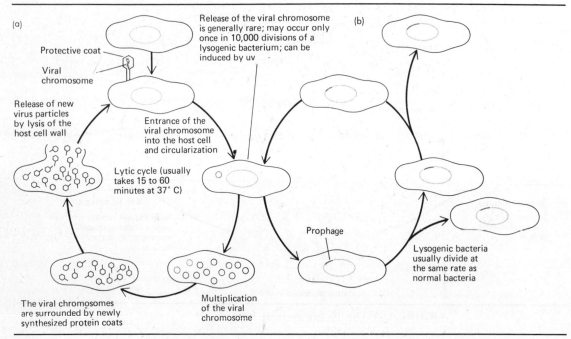

(a)

Release of the viral chromosome is generally rare; may occur only once in 10,000 divisions of a lysogenic bacterium; can be induced by uv

(b)

Protective coat

Viral chromosome

Release of new virus particles by lysis of the host cell wall

Entrance of the viral chromosome into the host cell and circularization

Lytic cycle (usually takes 15 to 60 minutes at 37° C)

Prophage

Lysogenic bacteria usually divide at the same rate as normal bacteria

The viral chromosomes are surrounded by newly synthesized protein coats

Multiplication of the viral chromosome

**FIGURE 12-4 The λ life cycle,** showing its lytic (left) and lysogenic (right) cycles. Phage chromosomes are depicted in dark gray and black and host chromosome in light gray. (From J. D. Watson, *Molecular Biology of the Gene*, 3rd ed. Copyright © 1976 by J. D. Watson; Menlo Park, Calif.: W. A. Benjamin, Inc.)

Repressor levels can be induced to drop by irradiating lysogenic bacteria with ultraviolet light or by exposing them to certain chemicals, such manipulations being known as **induction.** There are thus two occasions when a temperate phage may initiate a lytic cycle, one immediately after it infects its host and the other after it leaves the lysogenic state.

The strategic advantage to lysogeny should be evident: once λ DNA sequences are introduced into the *E. coli* chromosome, they are transmitted to subsequent generations of *E. coli* with each cell division, so that billions of cells harboring λ prophages are generated in a short period of time from a single lysogenic cell. The transmission of phage information in space and time is thus not solely dependent on the infectious meeting of virus with host. As the host invades new habitats, the phage genome effectively goes along for the ride.

## 12.3 DNA Tumor Virus Cycles

The final variety of infectious cycle considered here is exhibited by the tumor virus **SV40** and its close relative **polyoma,** both of which infect animal cells and can cause tumors. Like λ, SV40 can conduct two types of infections, one leading to host death and the other to a physical insertion of viral DNA into the host genome. Thus if SV40 virions are incubated with cultured monkey cells, they will carry out a **productive infection** that is formally analogous to a lytic cycle in bacteriophages. Several hours after infection, SV40-specific molecules are detected, of which the **T-antigen** and **t-antigen** are most significant: the former is a protein kinase required for viral DNA replication and for the resultant transcription of early viral genes. The latter is believed to be important for transformation. Some 12 to 20 hours after infection, SV40 chromosome replication commences; three viral structural proteins then begin to appear in the nucleus of the host cell, where they assemble to form mature progeny viruses. By two days after infection approximately 10,000 new virus particles have been released. The host cell is killed in the process, although a lytic "burst" does not necessarily occur. Details on the regulation of this cycle are found in Section 23.10.

If rather than being presented to monkey cells, SV40 is presented instead to cultured mouse cells, an alternate mode of infection is possible. Following infection, T-antigen appears in the host cell nucleus but no viral DNA or capsid protein synthesis occurs. After 18 to 24 hours, the host cell itself divides a few times and acquires novel properties: it can grow without being attached to a solid substratum (it acquires **anchorage independence**); it will proliferate in culture media depleted of its usually high levels of calf serum (it acquires a so-called **serum independence**); and it can be caused to clump to other cells when presented with carbohydrate-binding lectin proteins such as concanavalin A (it acquires **lectin agglutinability**). Most mouse cells so altered will return to normal after a few days and are said to have been abortively transformed, but some remain anchorage-independent, serum-independent, and lectin-agglutinable and are said to be **stably transformed.** If such transformed cells are then injected into the peritoneal cavity of a young mouse, they will divide uncontrollably and eventually kill the animal. In other words, stably transformed cells have the potential to cause tumors.

The SV40 chromosome in a stable tranformant remains in the nucleus of its host cell and appears to become covalently linked to the host's chromosomal DNA, as described in detail in Section 18.3. It continues to dictate the synthesis of T-antigen, but late gene expression almost never occurs, apparently because mouse ribosomes cannot be diverted into synthesizing late proteins. In certain respects, therefore, the transformed state is reminiscent of the lysogenic stage in phage λ. If a transformed cell is now fused with a "permissive" monkey cell so that their two cytoplasms blend together (cell-fusion techniques are discussed in Chapter 15), then mature infectious virions can be produced, showing that a complete copy of the virus chromosome must have been present in the mouse transformant.

# Complementation Analysis

## 12.4 The Complementation Test

Any genetic analysis invariably begins with a collection of mutant strains, generated as described in Chapter 7, which carry mutations that mark particular genes. Before the positions of the mutations are mapped within the genome, it is first important to ascertain **how many mutant genes are represented in the collection.** To use a phage example, suppose we have selected for *rapid-lysis* (*r*) mutant phages of T4, which, as described in Sections 7.2 and 11.7, cannot conduct a lytic cycle in *E. coli* strain $K(\lambda)$ and produce abnormal plaques (*Figure 7-3*) when they lyse *E. coli* strain *B* cells. Our ultimate goal is to locate their mutations in the T4 chromosome, but we first need to know how many genes are marked by the mutations at hand. It is possible that only one gene exists in T4 that, when mutated, causes the *r* phenotype, but it is also possible that two or more such genes exist, in which case some of the mutants in the collection may carry defects in one gene and other mutants may carry defects in a second or third gene. To distinguish among these possibilities, a series of **complementation tests** is performed.

The principle behind a complementation test is as follows. If two homologous chromosomes carry mutations within different genes—let us say one chromosome is mutant in gene A and the other chromosome in gene B—and these two chromosomes come to reside in the same cell, one chromosome will direct the synthesis of a mutant A gene product but a normal B product and the other will direct the synthesis of a normal B product but a mutant A product. **The important point is that the cell will come to contain both normal A and normal B products** and thus normal function can occur (Figure 12-5a). In such a case the two mutant chromosomes are said to **complement** each other, and it is inferred that the two mutations lie within the boundaries of two different genes. If, on the other hand, the two chromosomes carry mutations in the same gene (for example, both carry mutations in gene B, although not necessarily at identical positions with B), then neither chromosome will be able to direct the synthesis of a normal B product, and even when both chromosomes come to reside in the same cell, normal B function will not be observed. The two mutant chromosomes in this case do not complement each other, and it can be inferred that the mutations are allelic, that is, that they lie within the boundaries of the same gene.

The actual design of a complementation test depends on the organism and on the kinds of mutations being studied. For the *r* mutant strains of T4, a simple test is to grow a culture of *E. coli* K (λ) cells and add equal numbers of two of the *r* mutant strains in the collection (a **mixed infection**). After allowing sufficient time for a single lytic cycle to occur, the lysate is plated onto a lawn of *E. coli* B bacteria, and the number of *r* plaques is scored. It is found that if the two strains can complement each other, 100 to

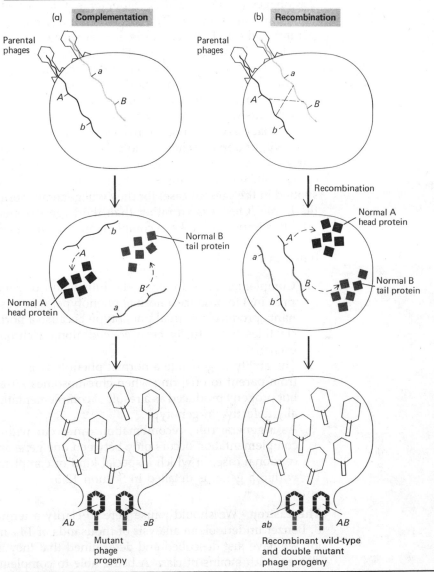

**FIGURE 12-5 Salient differences between** (a) **complementation and**
(b) **recombination** as illustrated for two mutant phage strains. One strain, *A b*,
carries a normal gene for the A head protein but a defective tail protein gene *b*.
The other strain, *a B*, carries a normal B tail protein gene but an abnormal head
protein gene *a*. During complementation, normal A and B proteins are
synthesized under the direction of the complementing chromosomes, and
normally constructed phages emerge that are mutant (*A b* and *a B*) in genotype.
During recombination, normal proteins are also synthesized and normally
constructed phages emerge, this time carrying normal (*A B*) genotypes. It
should be obvious that the phenomenon of complementation will operate in a
cell in which recombinant chromosomes have also arisen.

250 phages (a normal burst) will emerge per infected cell, whereas if the two strains are noncomplementary then fewer than 0.1 phage will emerge per infected cell. In other words, for every cell that releases a normal burst of at least 100 phage progeny, 999 cells produce no phage progeny at all.

The few successful bursts that do occur in a mixed infection between noncomplementing strains could, of course, take place because one of the two mutations in question has undergone reversion. They could also take place because recombination has occurred: two noncomplementing (allelic) strains that carry mutations at two close, but distinct, sites within the same gene will undergo recombination with a very low frequency to produce phages with a normal nucleotide sequence that can direct a normal lytic cycle. Both the revertants and the recombinants, however, can be easily spotted in the present case, for they will generate normal-sized plaques on the *E. coli* B tester lawn rather than the large *r* plaques.

It is important to keep complementation and recombination distinct. They are two quite different processes, as summarized below and illustrated in Figure 12-5.

1. Complementation involves the interaction of gene **products** (usually proteins) to produce a normal phenotype, whereas wild-type recombinants produce a normal phenotype because a normal sequence of nucleotides has actually been created along a chromosome via genetic exchange.
2. The ability to generate a normal phenotype is genetically transmitted from parent to offspring when chromosomes arise by recombination, but progeny produced as a result of complementation remain individually defective in genotype.
3. As a general rule, recombination can occur within a gene, whereas complementation occurs only between one gene and another. The exceptional case, in which a special kind of complementation can occur within a gene, is detailed in Section 12.5.

**The Cistron**   We should pause here to clarify a term. In the late 1950s, S. Benzer undertook an analysis of thousands of T4 *r* mutants in the fashion we have just described and determined that they indeed fell into two classes, with strains of class A being able to complement the function(s) missing in strains of class B, and vice versa. Instead of concluding that the *r* phenotype is controlled by the genes A and B, however, Benzer denoted the functional units as *cistron A* and *cistron B,* and the term "cistron" has since become widely used as a synonym for a gene.

A cistron derives its name from the **cis-trans test** for complementarity. If two mutations are located in the same chromosome, this corresponds to a *cis* configuration. A wild-type chromosome should be able to complement the missing functions of such a doubly mutant chromosome, regardless of whether the two mutations lie in the same or different genes; thus, complementation should be observed in mixed infections involving *cis* chromosomes and wild-type chromosomes. When the two mutations are instead in the *trans* configuration, each is located on a separate chromosome. Complementation in mixed infections involving two such *trans* chromosomes will occur *only* if the two mutations are located in different genes, or, if one prefers, in different cistrons. In practice, complementation studies usually do not include the *cis* part of the test, since it requires the construction of doubly mutant strains. The *trans* part of the test is identical to the complementation test we have just described.

## 12.5 Intragenic Complementation

There is one major exception to the rule that **complementation occurs only between one gene and another.** The exception, known as **intragenic complementation,** is illustrated schematically in Figure 12-6. In the example two identical polypeptides, both called α and both specified by the same gene, assume opposite orientations with respect to each other and so form

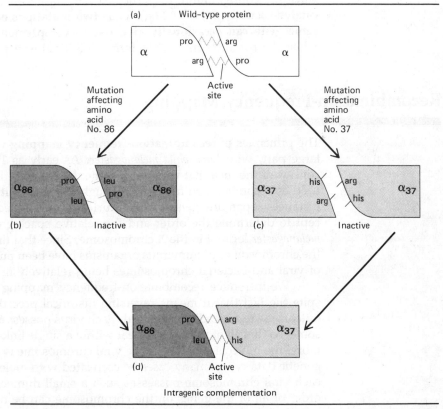

**FIGURE 12-6 Intragenic complementation** between two different mutant subunits of a dimeric protein. (See text for details.)

an enzyme αα (Figure 12-6a). In wild-type cells the region of the enzyme important to its catalytic activity (the active site) contains an arginine and a proline contributed by one α subunit and an arginine and a proline contributed by the other α subunit, the interaction of all four amino acids being essential to catalytic activity. The arginine, we can imagine, resides at position No. 86 on the α polypeptide and the proline at position No. 37. We now suppose that a GCG ⟶ GAG mutation occurs and that leucine rather than arginine appears at position No. 86 in one mutant strain. This mutant polypeptide, which we can call $\alpha_{86}$, will not form an active enzyme when it interacts with a second $\alpha_{86}$ polypeptide, perhaps because the four amino acids cannot interact in such a way as to form an active catalytic site (Figure 12-6b). We can similarly suppose that a second strain has undergone a GGG ⟶ GTG mutation and produces an $\alpha_{37}$ polypeptide in which histidine instead of proline appears at position No. 37. Again, $\alpha_{37}\alpha_{37}$ dimers are found to be inactive (Figure 12-6c). When, however, the two mutant polypeptides are allowed to interact in a single cell during a complementation test, then $\alpha_{86}\alpha_{37}$ **heterodimers** can form, and the presence of the proline, leucine, arginine, and histidine residues in the active site results in a configuration, or perhaps a net charge, that is compatible with catalytic activity (Figure 12-6d). Thus **two mutations occurring within the same gene can occasionally give rise to complementary gene products when the products interact to form a multisubunit protein.**

# Recombination-Frequency Mapping

The principles of **recombination-frequency mapping** were established, in large part, with *Drosophila melanogaster*. As early as 1911, T. H. Morgan conceived the idea that variations in the degree of gene linkage, as measured by recombination frequencies, might correlate with differences in the distances separating genes. A. H. Sturtevant proceeded to utilize this concept to determine the order and the relative spacing of five genes of *D. melanogaster* located in the X chromosome. Since that time genetic maps of the chromosomes of numerous organisms have been published, with maps of viral and bacterial chromosomes being relatively recent.

We introduce recombination-frequency mapping with viruses—despite the fact that it means reversing historical precedents—because the outcomes are so readily visualized. Each virus possesses only one chromosome, so that all of its genes appear within a single linkage group. Furthermore, the molecular nature of the viral chromosome is well known so that genetic data can in many cases be correlated with molecular data. Finally, each viral chromosome possesses such a small number of genes that the order of all of the genes on the chromosome can be determined.

## 12.6 Defining Gene, Marker, and Locus

Genetic mapping studies frequently make use of the terms **gene, marker, and locus;** these should be clearly distinguished before we proceed.

**Gene** We have been defining a gene in molecular terms throughout the text ( *for example, Section 9.2*) and have presented in Section 12.4 an operational definition of a gene as a unit of complementation. By now you should also feel comfortable with the concept of a gene as existing in a number of different forms, one usually denoted wild type and the others mutant. A gene cannot be mapped, and, indeed, its existence cannot ordinarily be known, until a strain carrying a mutation in that gene is somehow procured. Thus we might rightly infer that the chromosome of a bacteriophage must contain genes specifying bacteriophage head protein, but until it is possible to devise a means of obtaining mutant phages that are unable to direct the synthesis of this head protein, it will be difficult to learn much about the chromosomal location or the properties of the normal gene.

**Marker** A genetic marker is a mutation that marks the existence of a given gene. The marker is not, of course, synonymous with the gene itself, any more than a substituted base could be considered synonymous with a gene. One reason that the two are often confused is that the *name* given to a marker often becomes loosely used as the *name* of the gene; for example, when a mutant strain of phage that is defective in head protein synthesis is isolated, the mutation it carries might arbitrarily be denoted *hp* for *head protein*. In this case, then, *hp* would serve as a genetic marker. Subsequent genetic analysis might localize the marker at a certain position on the phage's map, leading to the conclusion that a gene specifying the structure of head protein resides at that position. Unfortunately this gene is often called *hp* when it should strictly be called "the gene marked by the *hp* mutation." The latter terminology is obviously more cumbersome and the former is therefore often used, even though it can result in misunderstanding.

**Locus** "Locus" is a convenient term that denotes a physical position within a chromosome. Thus, if two markers are said to "map to the same locus," this indicates that they occupy similar positions. The term can be used quite generally to denote a certain chromosomal region or it can be used more specifically to refer to small sectors of a chromosome. The term **gene locus** is also used to denote that portion of a chromosome's nucleotide sequence that constitutes a certain gene. Two **allelic genes** (*defined in Section 6.1*) will, of course, occupy the same chromosomal (and gene) locus.

## 12.7 The General Case for Mapping by Recombination Analysis

For phage recombination-frequency mapping, a culture of bacteria is subjected to **mixed infection** by two or more phage strains carrying different genetic markers—for example, by one strain with the chromosome constitution $a+$ and another with $+b$. During the ensuing lytic cycle exchanges will occasionally occur so that recombinant $+ +$ and $ab$ progeny phages appear in the burst that follows lysis of the host cell (Figure 12-5b). In an analogous process known as **transfection**, specially treated bacterial cells can be caused to take up "naked" copies of two types of phage chromosomes, thus bypassing the need for the phage injection apparatus; such chromosomes also proceed to replicate and recombine within the host.

The principle underlying genetic mapping in such situations is very simple. Granted the assumption that genetic exchanges will occur at all points along a chromosome's length, then **the farther apart two genes are on a chromosome, the more likely it is that they will be separated by a genetic exchange.** To give an extreme example, we can assume that the $a$ and $b$ loci lie at opposite ends of a chromosome, as shown in Figure 12-7a. In this case almost any exchange occurring in an internal position on the chromosome will have the effect of bringing the markers into new combinations. The observed **recombination frequency** between the markers will

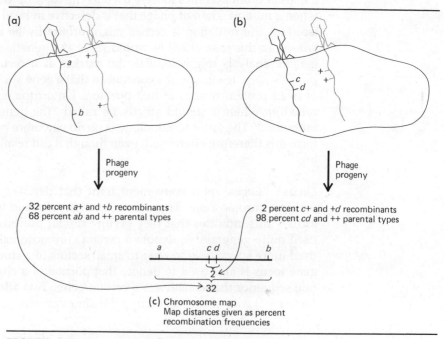

(a)

(b)

Phage progeny

Phage progeny

32 percent $a+$ and $+b$ recombinants
68 percent $ab$ and $++$ parental types

2 percent $c+$ and $+d$ recombinants
98 percent $cd$ and $++$ parental types

(c) Chromosome map
Map distances given as percent
recombination frequencies

FIGURE 12-7 **Recombination-frequency mapping** in phage crosses. Note that the placement of $c$ and $d$ between $a$ and $b$ cannot be assumed on the basis of the two crosses shown alone; other crosses must also be performed.

therefore be large. In contrast, two other loci, *c* and *d*, might lie right next to each other in the chromosome, in which case there will be a small probability that any particular exchange will happen to occur in the interval separating the two markers. The observed recombination frequency for these two markers will, therefore, be small (Figure 12-7b). By comparing the recombination frequencies obtained with a number of different marker combinations (for example, *ab* × + + crosses and *cd* × + + crosses), the four markers can be arranged in a linear order (Figure 12-9c) such that their relative distances from one another correspond to their recombinational distances.

This basic principle forms the basis of mapping by recombination frequency in all organisms. The one aspect of the cross that is varied is the number of markers manipulated at any one time. In the next three sections we consider cases in which one, two, or three loci are considered, these being respectively known as **one-factor**, **two-factor**, and **three-factor** crosses.

## 12.8 One-Factor Cross

**In a one-factor cross, a mutant strain is crossed to a wild-type strain from which it differs by a single phenotypic trait.** As an example we can cross a newly isolated mutant λ strain that creates *minute* plaques with a wild-type λ strain. A dense suspension of *E. coli* is incubated with an even denser suspension of the two types of phage, so that on the average, about 10 phages (five of each type) infect each cell at the same time. Stated more formally, the **multiplicity of infection** (m.o.i.) is adjusted to have a value of approximately 10; $10^9$ phages and $10^8$ bacteria are present. Phage multiplication is then allowed to proceed until lysis occurs; the progeny phages in the lysate are diluted to appropriate concentrations for plating onto bacterial lawns; and the morphologies of the plaques that form on the lawns are individually scored.

**The major purpose of a one-factor cross is to determine whether the mutant phenotype is controlled by a single genetic marker that is suitable for mapping studies.** Thus, in an actual cross of this kind, 4390 plaques were examined; 2050 proved to have the *minute* phenotype while 2340 had the wild phenotype. No additional kinds of plaque morphology emerged even though the phages had ample opportunity, given the high multiplicity of infection, to undergo genetic exchange. Therefore, we can conclude that the *minute* trait is determined by a single mutant locus to which we can assign the marker notation *mi*. We can further conclude that the *mi* strain is suitable for mapping: the equivalent numbers of *minute* and wild-type plaques among the progeny indicate that the *mi* mutation confers no selective lethality and that the mutant and wild plaque types can be distinguished from one another without ambiguity.

Let us explore what the result would be for such a cross if the *minute* phenotype is actually the composite effect of *two* gene mutations, *a* and *b*, affecting plaque formation. When such a *minute* strain is crossed to wild type at high multiplicity, we would predict that genetic exchange would occasionally separate *a* and *b* so that they would come to reside in different progeny chromosomes. In other words, the original *minute* strain might be the double mutant *a b*, in which case recombination during a cross with + + chromosomes would be expected to yield *a*+ and +*b* progeny. It is usually the case for viruses and haploid organisms that recombinant progeny carrying only one mutation will have phenotypes different from either their doubly mutant parent or their wild-type parent. Thus we would expect, in addition to wild and *minute* plaques, plaques with novel properties; intermediate-sized plaques might turn up, for example, or plaques with a distinctively new morphology. When such additional plaque types appear, it is clear that the two strains differ by mutations at more than one locus, and that each mutation and its unique phenotypic effect must be monitored during subsequent crosses.

## 12.9 Two-Factor Cross

The hypothetical cross just outlined was, in fact, a **two-factor cross**, in which a doubly mutant parent, *a b*, was crossed with a wild-type parent, + +. When, as in this case, both mutant genes are carried by one parent and their wild-type alleles are carried by the other parent, the cross is said to be **in coupling**—the two markers are coupled in the same chromosome. In genetic shorthand the cross is written + + × *a b*, and recombination between the *a* and *b* loci will produce singly mutant progeny designated +*b* and *a*+. The alternative in a two-factor cross is that the markers be **in repulsion,** in which case each parent carries only one of the two markers being followed in the cross (+*b* × *a*+).

**The major purpose of a two-factor cross is to establish the recombination frequency between two genetic markers.** We can return to our actual example from phage λ and outline a cross between the original *minute* strain and a second strain that creates *clear* plaques, a strain demonstrated in one-factor crosses to carry a single plaque morphology marker designated *c*. The *minute* parent is in this case given the genetic notaton *mi* + to indicate that its chromosome is wild type (at least with respect to plaque morphology) at all but the *mi* locus, while the *clear* parent is designated + *c*. The cross is therefore written as

$$mi + \times + c.$$

Any recombination between the two markers will yield wild-type (+ +) or doubly mutant (*mi c*) progeny. The results of this cross reveal the presence of four plaque phenotypes in the following proportions:

| Plaque Morphology | Number Scored |
| --- | --- |
| *minute* | 1205 |
| *clear* | 1213 |
| wild-type | 84 |
| *minute* and *clear* | 75 |

To determine the **percent recombination** $R$ between $mi$ and $c$, we apply the following expression:

$$R = \frac{\textbf{sum of two recombinant types}}{\textbf{sum of all types}} \times \textbf{100}$$

Therefore:

$$R_{mi,c} = \frac{84 + 75}{1205 + 1213 + 84 + 75} \times 100$$

$$= \frac{159}{2577} \times 100$$

$$= 0.0617 \times 100$$

$$= 6.2$$

In other words, the $mi$ and $c$ loci are separated from one another by recombinaton during a phage cross about 6 percent of the time.

Table 12-1 summarizes a number of two-factor λ crosses involving the $mi$ and $c$ strains plus two additional plaque morphology markers called *cocarde* (*co*), and small (*s*), where all four markers have been established by complementation tests to define distinct genes. In the first section of the

**TABLE 12-1   Two-Factor Crosses Involving Phage λ**

Numbers of progeny plaques, classified according to genotype, are given opposite each cross. Percent recombination is

$$\frac{\text{sum of two recombinants}}{\text{sum of all types}} \times 100$$

In the upper section several examples of the same cross are given to show the amount of variation between experiments. In the lower sections, only total numbers, obtained by adding together the results of individual experiments, are given.

| Parents | Number of Experiments | Progeny | | | | Recombination, % |
|---|---|---|---|---|---|---|
| co + × + mi | 1 | 5162 co + | 6510 + mi | 311 + + | 341 co mi | 5.3 |
| | 1 | 459 | 398 | 17 | 25 | 4.7 |
| | 1 | 720 | 672 | 44 | 46 | 6.1 |
| co mi × + + | 1 | 36 | 30 | 795 | 620 | 4.4 |
| | 1 | 74 | 56 | 1005 | 956 | 6.2 |
| s + × + co | 2 | 7101 s + | 5851 + co | 145 + + | 169 s co | 2.4 |
| s co × + + | 2 | 46 | 53 | 1615 | 1774 | 2.8 |
| s + × + mi | 1 | 647 s + | 502 + mi | 65 + + | 56 s mi | 9.5 |
| s mi × + + | 3 | 1024 | 1155 | 13083 | 13253 | 7.6 |
| s + × + c | 1 | 808 s + | 566 + c | 19 + + | 20 s c | 2.8 |
| c + × + mi | 1 | 1213 c + | 1205 + mi | 84 + + | 75 c mi | 6.2 |
| c + × + co | 3 | 6000 c + | 6000 + co | 14 + + | − c co* | 0.1 |

*Type *c co* not distinguishable from *c +*.

From A. D. Kaiser, *Virology* 1:424, 1955.

table, the *co* and *mi* markers are seen to recombine about 5 percent of the time, whether they are initially in coupling or in repulsion. The second section shows that *co* and *s* are closer together, recombining only 2.4 to 2.8 percent of the time; *s* and *mi*, by contrast, are very far apart (third section), recombining an average of 8.5 percent of the time.

The final step is to construct a genetic map from such data. We begin by averaging the *R* values obtained in the various crosses in Table 12-1. By letting each average *R* value represent the map distance between two markers, we obtain the following:

| Markers | Map Distance |
|---------|--------------|
| *co–mi* | 5.3 |
| *co–s*  | 2.6 |
| *s–mi*  | 8.5 |
| *s–c*   | 2.8 |
| *c–mi*  | 6.2 |
| *c–co*  | 0.1 |

We next identify the two most distant markers, which are clearly *s* and *mi*, and place these at either end of the map, arbitrarily placing *s* at the left and *mi* at the right:

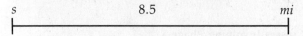

We proceed to identify the next-to-longest interval, which is defined by *c* and *mi*, and locate the *c* marker 6.2 units away from *mi*:

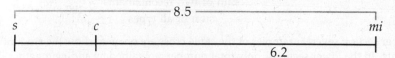

If distances were perfectly additive, we would expect the *s* to *c* interval to be 8.5 − 6.2 = 2.3 units. Looking at the averaged data, we find the experimental values to be a close 2.8 units. We can therefore write our skeletal map with some confidence as

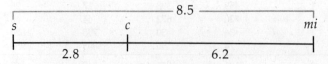

The final task is to position *co* with respect to these three markers. Again we choose the longest interval, which is that interval separating *co* and *mi*. By placing the *co* locus 5.3 units away from *mi*, we expect 6.2 − 5.3 = 0.9 map units to separate *co* from *c*. The data call for a 0.1 interval, which is again in the right range. Similarly, we would expect 8.5 − 5.3 = 3.2 units to separate *co* from *s*, which is similar to the 2.6 value found.

We can therefore write our final map as

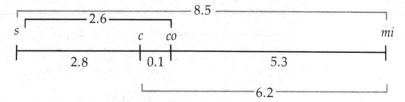

With such a map, we can now ask whether *s* and *mi* in fact define the two ends of the λ chromosome or whether the four strains we have selected define only a partial sector of the chromosome. This question can only be answered by isolating and crossing additional mutant strains, an activity that has occupied numerous geneticists since A. D. Kaiser's pioneering studies (*Table 12-1*) in 1955. The modern λ map is shown in Figure 12-8a; some of the data leading to the generation of this map is presented in later sections of the chapter and in the problem sets. It is clear from Figure 12-8a that our four markers are all located in the right half of the chromosome and that other mutations were essential to define the entire genome.

## 12.10 Three-Factor Cross

**A three-factor cross simultaneously provides the map order and the recombination frequencies between three markers.** It is therefore a more efficient way to obtain mapping data than is a multiple series of two-factor crosses as described previously. A three-factor cross also indicates the frequency of double exchanges, in which two chromosomes participate in not one but two recombinational events.

In a three-factor cross, parental chromosomes are marked at three positions. Using our phage λ markers, we might perform the cross

*s co mi* × + + +

Mixed infections at high multiplicities are again performed, and analysis of progeny phenotypes reveals the presence of eight plaque types in the proportions given in Table 12-2. The first two phenotypes are clearly parental;

**TABLE 12-2   Three-Factor Cross Involving Phage λ.** Parents are *s, co, mi,* and + + +

| Plaque Morphology | Number Scored | Class | Class Total |
|---|---|---|---|
| *small cocarde minute* | 975 ⎫ | | |
| wild type | 924 ⎭ | Parental | 1899 |
| *small* | 30 ⎫ | | |
| *cocarde minute* | 32 ⎭ | Single I | 62 |
| *small cocarde* | 61 ⎫ | | |
| *minute* | 51 ⎭ | Single II | 112 |
| *small minute* | 5 ⎫ | | |
| *cocarde* | 13 ⎭ | Double | 18 |
| Total progeny | 2091 | | |

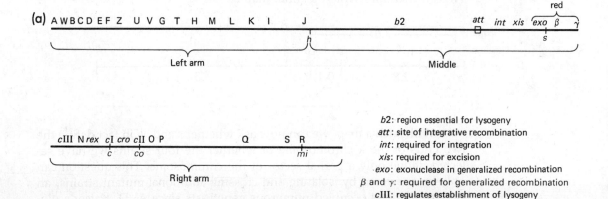

**(a)** A W B C D E F Z  U V G T  H  M  L  K  I      J                    b2              att   int xis  exo β  γ

                                                                                                            red

Left arm                                    Middle

cIII N rex  cI cro cII O P          Q      S R
         c     co              mi

Right arm

b2: region essential for lysogeny
att: site of integrative recombination
int: required for integration
xis: required for excision
exo: exonuclease in generalized recombination
β and γ: required for generalized recombination
cIII: regulates establishment of lysogeny
N: regulates gene transcription
rex: blocks growth of T4rII mutants on λ lysogens
cI: produces repressor essential for lysogeny
cro: regulates gene transcription
cII: regulates establishment of lysogeny
O: phage DNA replication
P: phage DNA replication
Q: regulates gene transcription
S: required for cell lysis
R: produces λ lysozyme

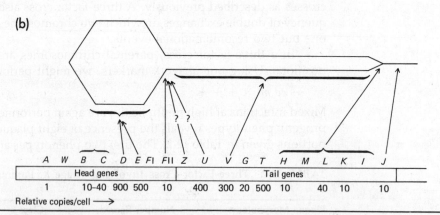

**(b)**

| A | W | B | C | D | E | Fl | Fll | Z | U | V | G | T | H | M | L | K | I | J |
|---|---|---|---|---|---|----|-----|---|---|---|---|---|---|---|---|---|---|---|
| Head genes | | | | | | | Tail genes | | | | | | | | | | | |
| 1 | | 10–40 | 900 | 500 | | 10 | | 400 | 300 | 20 | 500 | 10 | | 40 | 10 | | 10 | |

Relative copies/cell ⟶

FIGURE 12-8 λ Genetic map. (a) Entire λ chromosome, with the markers used in the original Kaiser crosses shown below the line. The chromosome measures about 15 map units and contains 48,100 base pairs of DNA. (b) Left arm of the λ chromosome, showing schematically the order of the morphogenetic genes and the likely points of action of those gene products, where known. The approximate relative number of copies of the various gene products made during a productive infection are shown below the map. (With permission from S. Casjens and R. Hendrix, *J. Mol. Biol.* **90**:20, 1974. Copyright by Academic Press Inc. [London] Ltd.)

the remaining six are recombinant. Since in this instance we already know, from our two-factor crosses, that the map order of these markers is *s co mi* as written, we can proceed to analyze the origin of the six recombinant types.

The *small* and the *cocarde minute* phages arise every time a single crossover occurs between the *s* and the *co* loci, which we can designate as Interval I:

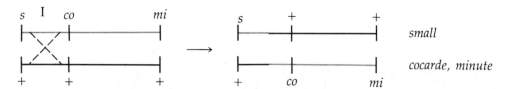

These two **reciprocal chromosome types** should be present in the progeny with about equal frequency, and this is borne out by the Table 12-2 data: 30 *small* and 32 *cocarde minute* plaques were scored. Thus these can be grouped together in the **Single I** progeny class, which comprises 62 of the 2091 plaques scored.

The *small cocarde* and the *minute* plaques arise every time a single crossover occurs between the *co* and the *mi* loci, which we can designate as Interval II:

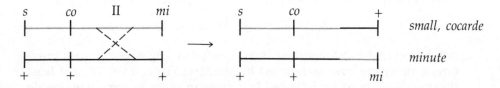

These two reciprocal chromosome types are also seen in Table 12-2 to be present in about equal numbers; together they form the **Single II** progeny class, representing 112 of the total.

The two final plaque types in Table 12-2, *small minute* and *cocarde*, arise as a consequence of double crossovers, one in Interval I and the other in Interval II:

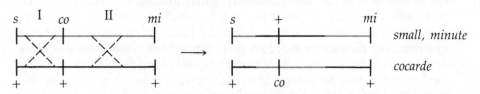

Again, these are present in roughly equivalent numbers (Table 12-2) and represent the **Double** class of progeny, 18 of the total.

Double-crossover chromosomes are unquestionably present in the progeny of two-factor crosses as well as of three-factor crosses. They can-

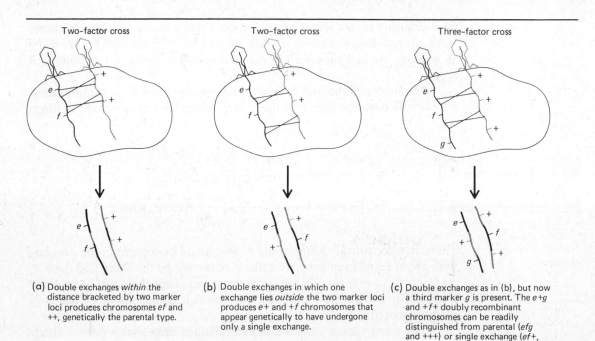

Two-factor cross

Two-factor cross

Three-factor cross

(a) Double exchanges *within* the distance bracketed by two marker loci produces chromosomes *ef* and ++, genetically the parental type.

(b) Double exchanges in which one exchange lies *outside* the two marker loci produces *e*+ and +*f* chromosomes that appear genetically to have undergone only a single exchange.

(c) Double exchanges as in (b), but now a third marker *g* is present. The *e*+*g* and +*f*+ doubly recombinant chromosomes can be readily distinguished from parental (*efg* and +++) or single exchange (*ef*+, ++*g*, +*fg*, and *e*++) chromosomes.

**FIGURE 12-9 Double exchanges** in two-factor crosses (a) and (b) and in a three-factor cross (c) where one exchange occurs between each marker pair.

not, however, be distinquished from parental or single-recombinant progeny, a statement best understood by studying Figure 12-9. Thus **at least three markers must be followed in a cross in order to recognize double recombinants.**

**Using Double Recombinants to Determine Marker Order**   We can now state an important generalization about three-factor crosses: the **double-exchange class of progeny will always be the rarest class.** This rule derives from a basic tenet of probability theory: **the possibility that two independent events will occur simultaneously is the product of their individual probabilities.** A familiar example of this principle occurs with coin-flipping: given that the probability of flipping heads or tails is 50-50 (0.5) for one coin, the probability that two coins flipped simultaneously will come up heads is $(0.5) \times (0.5) = 0.25$, or one quarter of the double flips. Thus, if the recombination frequency between markers $a$ and $b$ is 0.05 (so that $R = 0.05 \times 100 = 5$) and the recombination frequency between $b$ and $c$ is 0.03 (so that $R = 3$), the expected probability that *two* exchanges will occur between $a$ and $c$ in a given cross is the product of the two individual frequencies; that is, $(0.05)(0.03) = 0.0015$ (so that $R = 0.15$). Clearly, this product will always be a much smaller number than either individual frequency alone, an outcome that makes intuitive sense: it seems logical that any one chromosome would be far more likely to participate in one exchange than in two; more likely to take part in two than in three; and so forth, just as it would be a better bet to wager that a given flip will be heads than that when two coins are flipped both will be heads.

The infrequent nature of double exchanges can be profitably applied to three-factor crosses when gene order is being investigated, for of the six possible recombinant classes among the progeny **the two rarest classes can be taken as the doubly recombinant classes.** Once these two classes are recognized, the map order of the three markers in the cross can be deduced directly. Suppose, for example, a three-factor cross is performed between a *cocarde minute* strain and a *clear* strain (where the marker order is not known in advance), and the two rarest progeny types prove to be *cocarde clear* and *minute*. The most foolproof way to proceed is to write out all the possibilities and then select the correct one, in the following manner:

Possible Marker Order    Expected Double Recombinants

| | | | Possible Marker Order | | | | Expected Double Recombinants |
|---|---|---|---|---|---|---|---|
| (1) | *co* | *mi* | + | | *co* | + | + |
| | + | + | *c* | | + | *mi* | *c* |
| (2) | *co* | + | *mi* | | *co* | *c* | *mi* |
| | + | *c* | + | | + | + | + |
| (3) | + | *co* | *mi* | | + | + | *mi* |
| | *c* | + | + | | *c* | *co* | + |

Of the three possible marker orders, only (3) yields expected double recombinants whose genotypes match those of the two rarest types in the actual cross; therefore, a *c co mi* marker order can be determined. Try this one yourself: from a cross *c co$_2$ + $\times$ + + mi*, the rarest progeny are *c mi +* and *co$_2$ + +*. What is the relative order of *c*, *co$_2$*, and *mi*?

**Deriving Map Distances from a Three-Factor Cross**    Data from three-factor crosses can be used not only to order genetic markers but also to calculate map distances between markers. This is done as follows:

1. Classify the progeny of the cross as Parental, Single I, Single II, and Double, as done in Table 12-2.
2. Count up the number of progeny in each class and the grand total scored, again as done in Table 12-2.
3. Calculate the **frequency** of each class of recombinant progeny as follows:

| Recombinant Class | Frequency | Percent of Total Progeny |
|---|---|---|
| Single I | 62/2091 = 0.029 | 2.9 |
| Single II | 112/2091 = 0.053 | 5.3 |
| Double | 18/2091 = 0.0086 | 0.86 |

4. For simplicity, designate the three percentages so obtained as $\alpha$, $\beta$, and $\gamma$. Thus

$\alpha$ = percent of total progeny of the Single I class = 2.9
$\beta$ = percent of total progeny of the Single II class = 5.3
$\gamma$ = percent of total progeny of the Double class = 0.86

5. The recombination frequency values can now be calculated directly; for generality, we will let $a$, $b$, and $c$ represent $s$, $co$, and $mi$. The markers $a$ and $b$ are clearly separated in the Single I progeny ($a + +$ and $+ b c$). They are also separated in the Double progeny ($+ b +$ and $a + c$). Therefore,

$$R_{ab} = \alpha + \gamma$$

Similarly,

$$R_{bc} = \beta + \gamma$$

The outside markers, $a$ and $c$, are not separated from one another in the Double class. Therefore,

$$R_{ac} = \alpha + \beta$$

Work out for yourself how these relationships yield the map

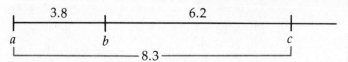

6. Examination of this map reveals an apparent anomaly: the distance between the two outside markers (8.3) is less than the sum of the $a$-$b$ distance plus the $b$-$c$ distance (3.8 + 6.2 = 10). That the two numbers are not the same is due to the occurrence of double crossovers. If two crossovers occur in the $a$-$c$ interval then, of course, the one cancels the other out and no recombination appears to have occurred when the $R_{ac}$ frequency is being calculated; each of these crossovers is, however, likely to be scored when $R_{ab}$ and $R_{bc}$ frequencies are being determined. **In general, therefore, as genes lie increasingly farther apart on a chromosome, their apparent distance in two-factor map intervals becomes increasingly foreshortened by the occurrence of double crossovers.**

**Circular Genomes in Three-Factor Crosses: The T4 Map** Distinctive linkage patterns are obtained when three-factor crosses are performed with phages T4 (and T2). Early T4 mapping studies by G. Streisinger established the linear map of markers shown in Figure 12-10a. This map showed no signs of being unusual as long as the markers followed in the various crosses were fairly closely linked. Aberrant results were obtained, however, when three-factor crosses involved markers placed at the two "ends" of this linear map; for example, when the cross $r67$ $h42 + \times + + ac41$ was performed, the rarest recombinant classes were found to be $r67 + +$ and

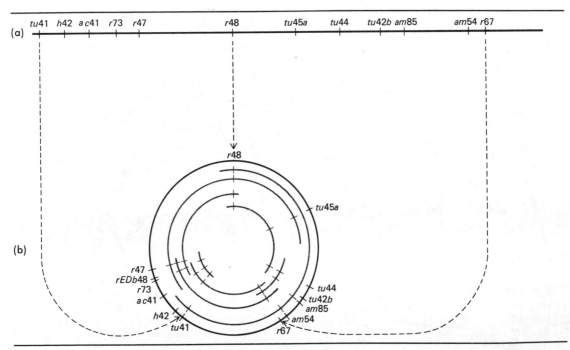

FIGURE 12-10 Mapping T4. (a) A linear sequence of genetic markers in T4 established by recombinational analysis. (b) A circular genetic map of T4. The markers used for each three- or four-factor cross are connected by an arc. (After G. Streisinger, R. Edgar, and G. Denhardt, *Proc. Natl. Acad. Sci. U.S.* **51**:775, 1964.)

+ *h*42 *ac*41, indicating a map order of *r*67-*h*42-*ac*41, whereas the map shown in Figure 12-10a predicted that the *r* marker would be more closely linked to *ac* than to *h*. To resolve this dilemma Streisinger and coworkers proposed that the ends of the linear array in fact closed on themselves to form a circle. Additional crosses, shown by the various arcs in Figure 12-10b confirmed this explanation: all the markers could be assigned a unique order on a circular map. A more recent map of the T4 chromosome is found in Figure 12-11.

The T2 and T4 chromosomes are, in fact, physically rod-shaped in the virion. The circular map results from the fact that the sequence of nucleotides in all T4 chromosomes is **terminally repetitious,** meaning that the genetic "text" begins over again at the end of the chromosome. Thus a T4 chromosome from one phage particle might read ABCDEFGAB, with the terminal repetition representing 2000 to 6000 nucleotide pairs. The DNA from a second T4 particle may be terminally repetitious for different parts of the chromosome; for example, CDEFGABCD. A third chromosome might read FGABCDEFG, and so on. All of these combinations occur with equal frequency in a T4 phage population. Stated more formally, each

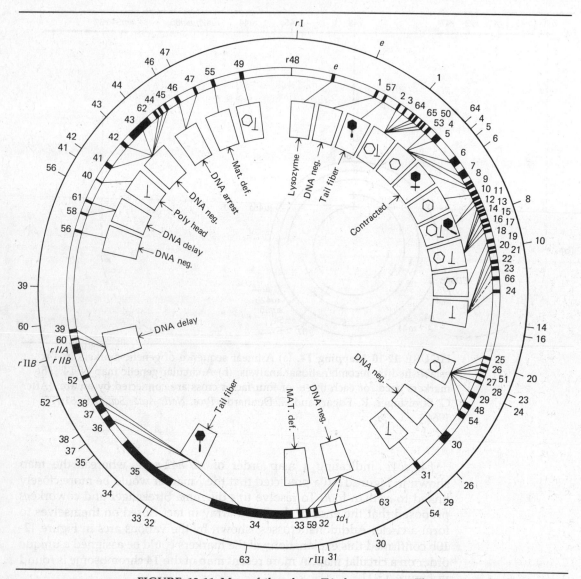

**FIGURE 12-11 Map of the phage T4 chromosome.** The inner circle shows the function of genes and their recombination distances. The outer circle represents proposed physical distances between markers, determined by isolating fragments of T4 DNA that have been prematurely packaged into phage particles and analyzing the markers that they contain. (From G. Mosig, *Adv. Genetics* 15:1, 1970.)

phage chromosome is a different **circular permutation** of a common nucleotide sequence, ABCDEFG, and each is terminally repetitious for the particular region of the sequence in which the chromosome happens to begin. For this reason, when a large population of T4 is examined, as in a mapping experiment, it appears as if any one gene in the sequence is adjacent to the next gene in the sequence, and no ends are detected.

The topography of the T4 chromosome is the consequence of the mode of T4 maturation that, as described in Section 12.1, utilizes the **concatemer/headful** mechanism. You will recall that long reiterated concatemers of T4 DNA spin out into the infected cell. It turns out that the lengths of DNA that are cut off from the concatemers and packaged into T4 headfuls are somewhat longer than a genome's worth of T4 DNA. As a result, each chromosome carries a terminal repetition and a circularly permuted set is generated.

# Deletion Mapping

## 12.11 General Properties of Deletions

As detailed in Section 8.10, a deletion chromosome lacks at least one nucleotide, and more commonly a stretch of nucleotides. By definition, the sector of a chromosome lost by a deletion has specific properties in genetic crosses:

1. The deletion, if extensive, will shorten map distances between outside markers.
2. The deleted sector is not available to participate in recombination, since the requisite nucleotides for pairing and crossing over are simply not available.
3. The deleted sector cannot ordinarily be reverted.
4. The deleted sector cannot complement defects in other genes.
5. The deletion cannot ordinarily be corrected by a suppressor mutation.

## 12.12 Deletions That Foreshorten Map Distances

As a first example of a deletion chromosome we can consider the mutant strain $\lambda b2$, so named because of the $b2$ defect in the central portion of the phage $\lambda$ genome. Phenotypically, $\lambda b2$ is able to direct a lytic cycle but can establish only **abortive lysogeny,** meaning that the prophage does not establish a stable state of integration in the host chromosome and thus does not become replicated and passed on to future bacterial generations. Three-factor crosses involving $\lambda b2$ indicate that $b2$ lies between $h$ and $c$ on the $\lambda$ map (*Figure 12-8a*). The $\lambda b2$ chromosome has also been shown by CsCl density gradient centrifugation (*Box 2.1*) to contain 17 percent less DNA than the wild-type chromosome, which suggests that the $b2$ marker represents a large deletion lying between $h$ and $c$.

If $b2$ is in fact a large deletion, then its presence should physically shorten the distance between $h$ and $c$. Therefore, recombination frequencies between these two markers in a $\lambda b2$ strain should be *lower* than those in a strain that does not carry $b2$. More specifically, since the interval between *h* and *c* represents roughly 34 percent of the λ chromosome, the removal of 17 percent of the chromosome in this region should shorten the interval by half, and recombination between *h* and *c* should occur only half as often in λ*b2* as in wild-type phage. This is exactly what is observed. In the cross *h b2* + × + *b2 c* the percent of recombination between *h* and *c* is 2.2, whereas it is 5 in the cross *h* + + × + + *c*.

## 12.13 Deletion Mapping of the *rII* Locus of T4

We can next consider how deletion chromosomes are used as tools in mapping studies, using as an example the analysis of the *rII* locus of phage T4. The *rII* locus is defined by *r* mutations that, as detailed in Section 12.4 of this chapter, fall into two complementation groups or cistrons, A and B, and therefore define two genes *rIIA* and *rIIB*. Both genes were found, by mixed-infection crosses with other T4 mutant strains, to lie close to one another in the same sector of the genetic map. In the late 1950s, therefore, S. Benzer and his collaborators set out to order a collection of thousands of *rII* mutations, the suspicion being that many of these would serve to define the order of mutations **within** the A and B genes.

Benzer had in his collection several *rII* strains that were observed never to revert to wild type, even after mutagenesis. These were therefore excellent candidates for deletion mutants. Benzer proceeded to cross various deletion strains to strains carrying revertible point mutations in either the A or B genes. As diagrammed in Figure 12-12, he was able to identify deletions that covered a large portion of gene A by their inability to form wild-type recombinants when mixedly infected with *rIIA* point mutants. Other deletion strains could not form wild-type recombinants when crossed with *rIIB* point mutants, while still others behaved as if they carried deletions that covered both genes.

Benzer reasoned that **if two strains carrying deletions that overlapped one another were crossed, then wild-type recombinants would never be found among the progeny, whereas if the two crossed strains carried deletions that did not overlap, then wild-type recombinants would occasionally form.** Figure 12-13 illustrates this principle of **mapping by overlapping deletions** and Figure 12-14 shows the deletions Benzer actually studied, each deletion being identified by a particular number (for example, 1272 and 1241). The ends of the various deletions in the collection delimit 47 different segments of the A and B genes, shown at the bottom of Figure 12-14 as *A1a*, *A1b1*, and so on. The ''big seven''deletions shown at the top of Figure 12-14 cover particularly large regions of the *rII* locus.

With such a collection of deletion-mutant strains it becomes a relatively simple matter to map the *rII* genes. A strain carrying a point mutation at an unknown site is crossed first with the seven strains carrying the ''big seven'' deletions. If, for example, the point mutation lies in segment *A3e*, it will give no wild-type recombinants in crosses with strains *r1272*, *r1241*, or *J3*, but it will give wild-type recombinants with the remaining

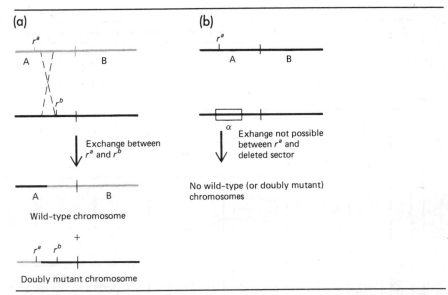

FIGURE 12-12 **Behavior of deletions in genetic crosses.** A phage carrying a point mutation ($r^a$) in the $A$ gene of the $rII$ region is crossed (a) with a chromosome carrying a second point mutation ($r^b$) in the $A$ gene or (b) with a phage carrying a deletion ($\alpha$) in the $A$ gene. Wild-type recombinants can emerge in cross (a) but, because no nucleotides are available in the $\alpha$ chromosome for the $r^a$ region to pair and exchange with, a normal wild-type sequence cannot be restored by recombination in cross (b).

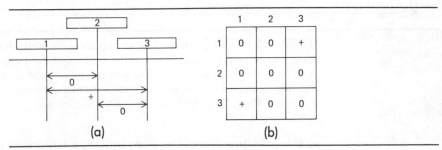

FIGURE 12-13 **Mapping by overlapping deletions.** (a) Three deletion mutations are shown, each affecting a different segment of a gene. Mutant strains carrying deletions number 1 and number 3 can recombine to produce wild-type recombinants (a result symbolized by the + below the arrow connecting number 1 and number 3), whereas neither strain can produce wild-type recombinants when crossed with strains carrying deletion number 2 (a result symbolized by the arrows labeled 0). (b) The same data as in (a) displayed in matrix form. The parental strains (numbers 1, 2, and 3) are drawn on the horizontal and vertical axes, and the result of a cross between any pair (for example, number 1 × number 3) is found in their intersecting box (in this case, a +). (From S. Benzer, in *The Chemical Basis of Heredity*, W. D. McElroy and B. Glass, Eds. Baltimore: Johns Hopkins Univ. Press, 1957.)

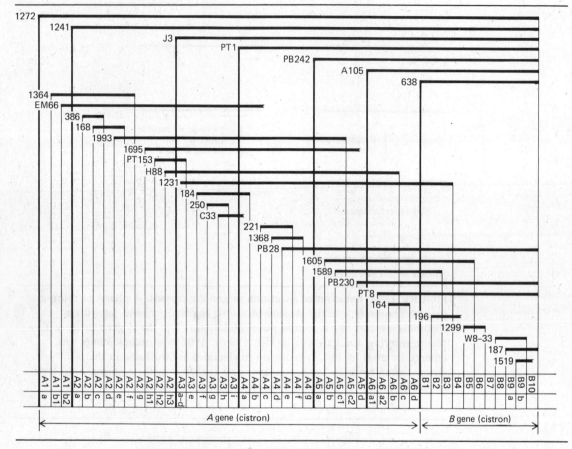

**FIGURE 12-14 *rII* Deletions.** The *rII* region of T4 is divided up into 47 small segments (shown as small boxes at the bottom of the figure). Some deletion ends have not been used to define a segment and are drawn fluted. The *A* and *B* genes, defined by independent complementation tests, are indicated. (From S. Benzer, *Proc. Natl. Acad. Sci. U.S.* **47:**410, 1961.)

strains. The position of the point mutation within the region bracketed by the *J3* and *PT1* deletions can then be determined by crosses with the appropriate small deletion strains, in this case *r1231*, *r184*, *r250*, and *C33* (Figure 12-14). The *A3e* mutant strain should not yield wild-type recombinants with *r1231* but should with the others.

Having localized the mutation to segment *A3e* by means of only 11 crosses, Benzer could now subject the strain to two-factor crosses with other strains that also carried point mutations within the *A3e* segment, looking for rare recombinants of the sort illustrated in Figure 12-12a. In this way, he was able to construct the first **gene fine-structure map,** shown in Figure 12-15.

In addition to generating this map, Benzer went on to note with what

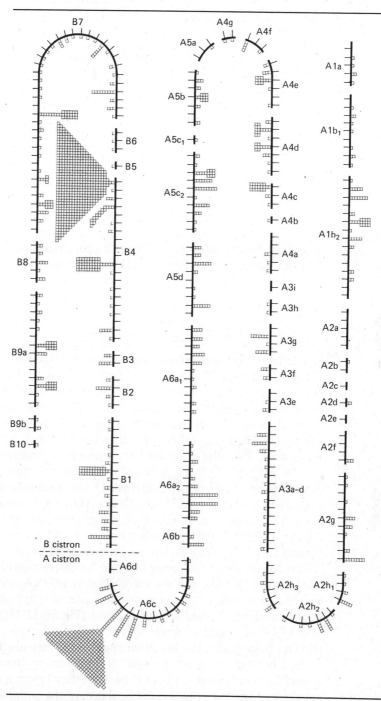

**FIGURE 12-15 The *rIIA* and *rIIB* genes of T4.** Each square identifies the location in the chromosome of an independently occurring mutation. Where more than one mutation occurs, the number of squares indicate the number of mutations. (After S. Benzer, *Proc. Natl. Acad. Sci. U.S.* **47**:410, 1961.)

frequency the various sites underwent mutation, scoring both $r^+ \longrightarrow r$ forward mutations and $r \longrightarrow r^+$ reversions. These results are included in Figure 12-15, in which each square represents one spontaneous mutational occurrence at each indicated site. It is obvious that the genes exhibit a number of **hot spots** in which spontaneous mutations occur preferentially. More than 500 mutational events are scored at the most prominent hot spot, whereas only one mutation is noted at many other sites.

Benzer collected similar data on the topography of mutations elicited by various mutagens, and the profiles proved to be different both from one another and from those generated for spontaneous mutations. Such evidence gave early support to the notion that different mutagens preferentially alter different spectra of nucleotides. Why there should be particularly "hot" mutational sites within a gene has not yet been explained, although one popular theory holds that the reactive nucleotides are those thrust into prominence by the particular tertiary structure assumed by a chromosome.

## 12.14 Deletion Mapping by Specialized Transduction

**Specialized transduction** is a mode of gene transfer carried out by temperate phages such as $\lambda$. You will recall from Section 12.2 and Figure 12-4 that during lysogeny a $\lambda$ chromosome integrates into the *E. coli* chromosome, using mechanisms that are considered in detail in Section 18.1. Integration normally occurs between a sector of the $\lambda$ chromosome known as *attP* and a fixed invariant site, called *attB*, that lies at 17 min in the *E. coli* chromosome map. Thus the integrated $\lambda$ prophage is usually flanked on the one side by the *galK*, *T*, and *E* genes of the *E. coli* galactose operon and on the other side by a block of genes involved with the synthesis of biotin, *bioA*, *B*, *F*, *C*, and *D*, as drawn in Figure 12-16a. When $\lambda$ prophages are induced to leave this chromosomal site, they usually interact at the two *att* regions (Figure 12-16b); they then splice out an intact $\lambda$ chromosome and leave behind an intact *E. coli* chromosome, as diagrammed in Figure 12-16c. Occasionally, however, an error occurs: the interaction does not involve *att* sites (Figure 12-16d), and the piece of circular DNA that emerges from the bacterial chromosome is a genetic and a physical hybrid, carrying bacterial genes covalently linked to phage genes (Figure 12-16e). The released hybrid DNA proceeds to direct a lytic cycle and copies become packaged within phage coats. The resultant **specialized transducing particles** are so named because they are capable of transferring (**transducing**) their acquired bacterial genes to the next bacteria they infect, a transduction that is **specialized** in that only the genes flanking the prophage attachment site are transferred.

The total length of the DNA that is packaged into a specialized transducing particle is roughly the same as that of a normal phage chromosome. A specialized transducing particle carrying a sizable piece of the bacterial genome will therefore usually (although not always) lack a corresponding

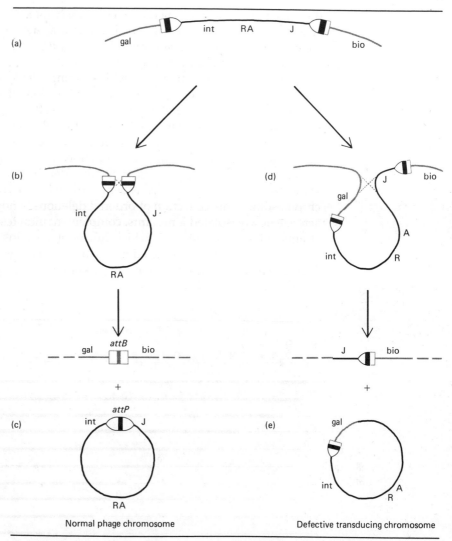

**FIGURE 12-16 Specialized transduction.** An integrated λ prophage is diagrammed in (a), with bacterial DNA in gray, phage DNA in black, and the spindle structures the *att* regions. Normal excision is drawn in (b) and (c); an abnormal excision, producing a λd*gal* transducing chromosome, is drawn in (d) and (e). Such abnormal events occur approximately once for $10^6$ normal excisions.

length of phage genome. The missing phage genes are taken away from the end of the prophage chromosome opposite the one to which the bacterial genes are added. Therefore, as illustrated in Figure 12-16e, a λ particle transducing a certain number of *gal* genes will typically lack a corresponding block of genes in the J region of its chromosome, whereas a *bio*-transducing λ particle will often lack genes in the N region. For this reason such specialized transducing particles are denoted as **defective** in phage genes, so that, for example, λ *dgal* is defective and *gal*-transducing whereas λ*dbio* is defective and *bio*-transducing.

Specialized transduction provides an important method for mapping the bacterial chromosome, as we detail in Section 13.14. Defective transducing chromosomes are of added value in that they represent **a collection of specific deletions that can be used to deletion map the temperate phages themselves.** Crosses are performed, for example, between λ*dgal* phages and strains carrying point mutations at known positions on the λ chromosome. Such analyses allow one to order the deletions in the fashion shown in Figure 12-17, exactly as Benzer did with the *rII* deletions of the T4 chromosome. This collection of ordered deletions is now available to characterize newly isolated λ mutants: complementation tests with the deletion strains will quickly establish which function(s) are lost by the new mutation, and crosses with the deletion strains will readily determine in which segment of the λ chromosome the unknown mutation resides. Indeed,

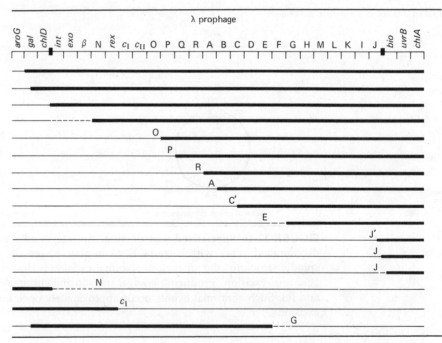

**FIGURE 12-17 Terminal deletion mutations of prophage λ.** The dark solid line represents the deleted segment, and the dashed lines indicate regions whose presence or absence cannot be determined from the genetic analyses that have been done. C′ and J′ indicate that the end points of the deletions lie between known *sus* markers of the C and J genes. (From S. Adhya, P. Cleary, and A. Campbell, *Proc. Natl. Acad. Sci. U.S.* **61**:956, 1968.)

most of the left arm of the λ chromosome (*Figure 12-8b*) was characterized by deletion mapping.

# Mapping Without Recombination

Recombination-frequency and deletion mapping techniques utilize data based on recombination (or on the prevention of recombination by deletion). The remaining sections of the chapter describe ways to generate maps that do not require mixed infection to determine marker order or position.

## 12.15 Physical Maps

A number of mapping procedures are based on the appearance of chromosomes in the electron microscope after various manipulations. We describe here three representative techniques.

**Denaturation Mapping**  In this procedure, a chromosome or DNA fragment is heated just enough to induce AT pairs to come apart or denature (*Section 4.4*) but not enough to denature the more strongly hydrogen-bonded GC pairs. Regions of the DNA containing long AT-rich stretches can thereby be recognized by electron microscopy as "bubbles," and these serve as physical landmarks along the genome (Figure 12-18). The distances separating the bubbles can first be established in preparations of wild-type chromosomes, after which one can analyze the chromosomes of deletion mutants and make accurate estimates of the location and length of various deletions.

**Transcription R-Loop Mapping**  The R-loop procedure was described in Section 9.9 in the context of identifying intervening sequences in eukaryotic genes, but it is also useful to locate transcriptionally active regions of a chromosome. Figure 12-19 illustrates a nearly intact λ chromosome that was incubated with ribonucleotide triphosphates and *E. coli* RNA polymerase *in vitro* for eight minutes and then subjected to destabilizing conditions that favor the formation of RNA-DNA hybrids. The newly synthesized RNA molecules hybridize back to their template DNA and create the R-loop "bubbles" indicated by the arrows (Figure 12-19), thus pinpointing with precision the location of the three major promoters (*Section 9.7*) of the λ genome.

**Heteroduplex Mapping**  This procedure serves to identify regions of non-homology in two otherwise homologous chromosomes or DNA fragments. As an example we can consider two λ chromosomes. One carries the long *b2* deletion described in Section 12.12 plus a distinctive sequence called *b5*.

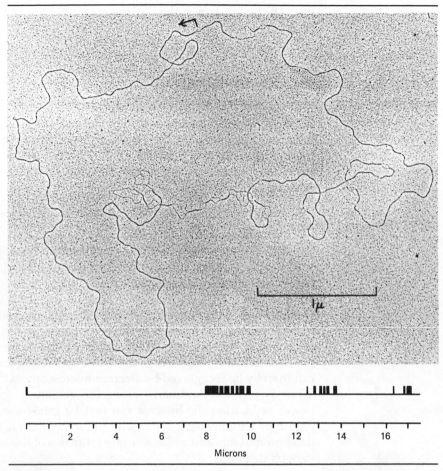

**FIGURE 12-18 Denaturation mapping.** Circular chromosome of phage λ, partially denatured to reveal regions rich in AT base pairs. Below is a denaturation map of this chromosome, with the position of denatured sites (vertical bars) positioned with respect to the arrow in the micrograph. (From R. B. Inman, *J. Mol. Biol.* **51**:61, 1970.)

The other lacks the *b2* deletion (and is thus *b2*⁺) and carries the sequence *i*^λ, which is different from *b5*. To construct a denaturation map, the two chromosome types are first dissociated into their separate polynucleotide strands and are then mixed together under renaturation conditions (*Section 4.4*). Whenever two complementary strands from the two phage types form a hybrid duplex, the duplex exhibits the two regions of discontinuity shown in Figure 12-20: the nonhomologous *b5* and *i*^λ sectors fail to anneal and form a symmetrical bubble, while the *b2*⁺ sector of the nondeleted strand loops out in the region of the deletion to form an asymmetrical bubble. You may recall analogous bulges when deficiency-carrying chro-

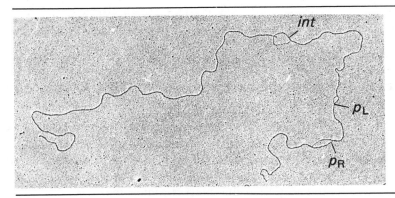

**FIGURE 12-19 Transcription R-loop map** of λ DNA. λgt *WES · λC* DNA was transcribed with *E. coli* RNA polymerase at a polymerase to DNA weight ratio of 2:1 for eight minutes. The $P_R$, $P_L$, and *int* promoters are identified as loops (*arrows*). (From Brack, C., *Proc. Natl. Acad. Sci.* **76**:3164, 1979.)

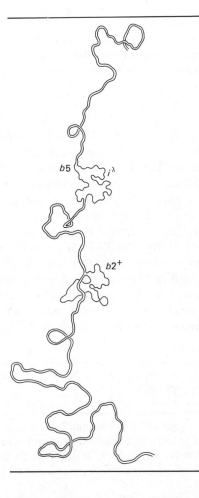

**FIGURE 12-20 Heteroduplex mapping** of λ chromosomes. The first discontinuity in the molecule is formed between two nonhomologous regions called $i^\lambda$ and *b5*. The second discontinuity is the consequence of the *b2* deletion. (From B. Westmoreland, W. Szybalski, and H. Ris, *Science* **153**:1345, March 21, 1969. Copyright 1969 by the American Association for the Advancement of Science.)

mosomes pair with their homologues (*Figures 7-7 and 7-8*). Such maps allow precise measurements of the lengths of deletions and regions of non-homology.

## 12.16 Restriction Mapping and Sequencing DNA Chromosomes: The SV40 Genome

The properties of site-specific restriction endonucleases, described in Section 4.11, render them eminently suited to mapping studies: the selective substrate specificity of each enzyme can be used to transform a duplex chromosome into a collection of homogeneous fragments that are far easier to study than the intact chromosome.

As an example we can consider the mapping of SV40. In Section 4.11, we noted that this animal virus has a circular duplex genome of 5243 base pairs. We also stated that the chromosome has a single *Eco* R1 site and 18 *Hae* III sites and that *Hae* III would cut the chromosome into 18 *Hae* III restriction fragments that could be separated from one another by gel electrophoresis (*Figure 4-8*).

K. J. Danna, G. H. Sack, and D. Nathans were among the first to use restriction enzymes to produce a **restriction map** of SV40. They began with the enzyme *Hind* II from *Haemophilus influenzae* and found that it converted the SV40 chromosome into the 11 fragments shown in Figure 12-21, with Hin-A the largest (slowest migrating) and Hin-K the smallest (fastest migrating) fragments in the gel. A number of approaches were then used to reconstruct the original order of the 11 *Hind* II fragments within the SV40 chromosome. One strategy was to use two restriction endonucleases simultaneously or in succession. For example, if an SV40 *Hind* II digest was exposed to *Eco* RI and then subjected to electrophoresis, fragment *F* was no longer found in its characteristic position on the gel. Instead, two new smaller fragments were present, the summed sizes of which correspond to the size of fragment *F*. Thus fragment *F* was identified as carrying the *Eco* RI cleavage site. Alternatively, the SV40 chromosomes were first digested with the enzyme *Hpa* I, which generates three large fragments: Hpa-A, Hpa-B, and Hpa-C. Each of these was then eluted separately from a gel, exposed to *Hind* II, and rerun by electrophoresis. When, for example, the Hpa-C fragments were exposed to *Hind* II, bands identical in mobility to the Hin-B and Hin-I were found on the resultant gels, indicating that the B

**FIGURE 12-21** *Hind* II digest of SV40 chromosome. Digest of $^{32}$P-labeled chromosomes subjected to slab gel electrophoresis and autoradiography. (With permission from K. J. Danna, G. H. Sack, Jr., and D. Nathans, *J. Mol. Biol.* **78**:363, 1973. Copyright by Academic Press Inc., [London] Ltd.)

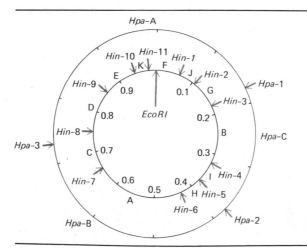

**FIGURE 12-22 Early restriction map of the SV40 genome.** Map units are given as

$$\frac{\text{distance from } EcoRI \text{ site}}{\text{length of SV40 DNA}}$$

in the direction F-J-G-B . . . Arrows illustrate the sites of restriction endonuclease cuts (*Hin*-1, *Hin*-2, *Hpa*-1, . . .) and letters indicate the fragments generated, with the inner circle showing the position of the eleven *Hind* II fragments (A–K) and the outer circle the 3 *Hpa* fragments (A–C). (With permission from K. J. Danna, G. H. Sack, Jr., and D. Nathans, *J. Mol. Biol.* **78:**363, 1973. Copyright by Academic Press Inc., [London] Ltd.)

and I fragments lie next to one another in the chromosome. By such approaches, Nathans and colleagues were able to produce the restriction map of SV40 shown in Figure 12-22, where the single *Eco* RI cleavage site is arbitrarily assigned a "0" position and other sites map between 0 and 1 on the circle.

More detailed physical maps of SV40 have since been generated (Figure 12-23). These have utilized restriction enzymes that recognize short nucleotide sequences (for example, tetranucleotides). Since it is more probable that a particular tetranucleotide will be present in a chromosome than, say, a hexanucleotide, such restriction enzymes cleave the chromosome into more numerous small fragments. The small fragments are again arranged into a restriction map and are then subjected to DNA sequencing (*Box 4.1*). Once the nucleotide sequences of particular fragments are known, **overlapping sequences** can be sought: for example, if one strand of the fragment X generated by *Hae* III has the sequence 5′CCGACTA<u>AGCTTG</u> . . . 3′ and fragment Y generated by *Hind* III has the sequence 5′<u>AGCTTG</u>CTATA . . . 3′, then the underlined region of overlap indicates that fragments X and Y derive from a sector of the chromosome with the sequence 5′CCGACTA<u>AGCTTG</u>CTATA3′, and a map order 5′ X – Y 3′ is deduced.

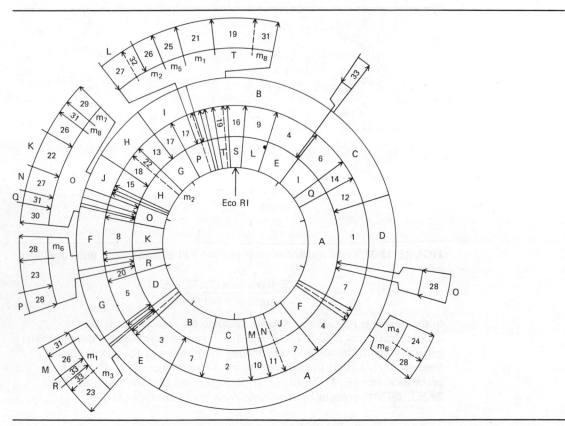

**FIGURE 12-23 Detailed restriction map of SV40.** Inner circle: the *Eco* RI site is shown. Inner ring: the location of the 32 *Alu* (*Arthrobacter luteus*) sites (and fragments) are indicated. *Alu* sites that are also recognized by *Hind* III are shown by dashed arrows. Middle ring: the 50 fragments produced by double digestion of SV40 DNA with *Alu* and *Hae* III (*Haemophilus aegyptius*) enzyme are indicated. These *Alu-Hae* fragments are identified by an arabic numeral; fragments moving with the same mobility have the same number. Outer ring: the 18 *Hae* sites (and fragments) are shown. For clarity, several regions are shown on an expanded scale on the outside. Outward arrows are *Alu* sites (dashed arrows are *Alu* or *Hind* III sites), and inward arrows *Hae* III sites. *Alu* fragments are indicated on the inside and *Hae* fragments on the outside (From R. Yang, A. van de Voorde, and W. Fiers. *Eur. J. Biochem.* **61**:119–138, 1976.)

It should be obvious that a careful comparison of numerous overlapping and sequenced restriction fragments will eventually yield the nucelotide sequence of the entire SV40 chromosome. This has in fact been accomplished in the laboratories of W. Fiers and S. M. Weissman.

How is a restriction map converted into a functional map, with the boundaries of genes located within the sequence? The most widely used approach, known as **Southern Blot Hybridization,** is described in Box 12.1;

**BOX 12.1**
Southern Blot
Hybridization.

This is a widely used technique, developed by E. N. Southern for detecting specific sequences among DNA fragments separated by gel electrophoresis. An agarose gel containing DNA fragments (e.g., a restriction digest, a) is laid over a strip of cellulose nitrate such that the DNA elutes out of the gel and becomes trapped in the cellulose nitrate (b). A $^{32}$P-RNA probe is now used for DNA hybridization, and complementary bands are detected in the resultant autoradiogram (c).

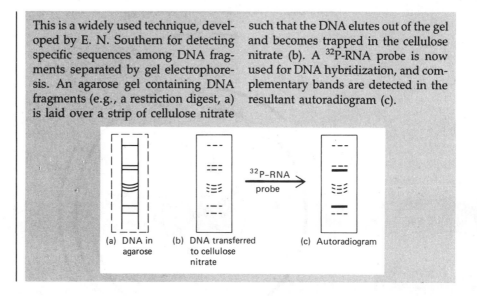

(a) DNA in agarose    (b) DNA transferred to cellulose nitrate    $^{32}$P-RNA probe    (c) Autoradiogram

this allows one to determine which restriction fragments are homologous to particular mRNA species that specify particular gene products. A functional map of the SV40 genome is presented in Figure 12-24, where the chromosome is seen to specify five gene products, namely, the T, t, VP1, VP2, and VP3 proteins. Further information on the coding of these proteins is found in Section 12.18.

## 12.17 Sequencing RNA Chromosomes: The MS2 Genome

The phage **MS2** and its close relatives **R17, Qβ,** and **f2** have small circular genomes made of single-stranded RNA. The sequence of the MS2 chromosome has been determined in W. Fiers' laboratory, using RNA sequencing methods that need not concern us here. When the primary sequence is examined, certain segments are found to exhibit **internal complementarity;** that is, one would expect such sequences to fold back on themselves like a hairpin and form interstrand base pairs, much as we saw for inverted repeats of DNA (*Section 4.9*) and for tRNA (*Figure 10-1*). Each gene is therefore thought to loop and bend like a many-petaled flower, the entire chromosome resembling a bouquet. Such a "flower model" for the secondary structure of the coat protein gene is shown in Figure 12-25; the proposed secondary structures for the larger A-protein (required for phage maturation) and the replicase genes entail even more elaborate configurations.

The folded secondary structure of the MS2 chromosome has several consequences. First, the "packaging problem" of fitting a 1-μm length of RNA into a 25-nm capsid is simplified if the chromosome exists naturally as a compact molecule. Second, the sensitivity of the chromosome to ribo-

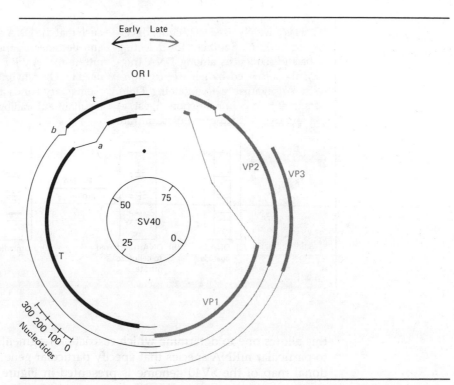

**FIGURE 12-24 SV40 mRNAs,** as transcribed from the circular chromosome of SV40. "Early" (black) and "late" (gray) mRNAs are transcribed in opposite directions from opposite strands, as indicated at the top of the map, from promoters located near the origin of replication (ORI); details on the regulation of this transcription are given in Section 23.10. The two early gene products, T- and t-antigens, are specified by mRNAs created by different splicings of the same RNA transcript, as are the three late gene products, the capsid proteins VP1, VP2, and VP3; details are found in Section 12.18. Narrow lines indicate nontranslated portions of a transcript; V-shaped lines indicate introns that are spliced out of a transcript. The inner circle shows the standard mapping coordinates of the SV40 chromosome, where 0 represents the single *Eco* RI site (*Figure 12-23*). (After Ziff, E. B., *Nature* **287**:491–499, 1980.)

nucleases present in an *E. coli* host cell is greatly reduced if much of the RNA is "tucked inside" the structure. Finally, some of the nucleotide sequences prominently displayed on the exterior of the "bouquet" may play important roles in terms of associating with ribosomes, initiating transcription, seeding capsid proteins, and so on. In other words, RNA phage chromosomes may conduct their infectious cycles not only by virtue of the genes encoded in their chromosomes but also by virtue of the physical topology of the chromosomes themselves.

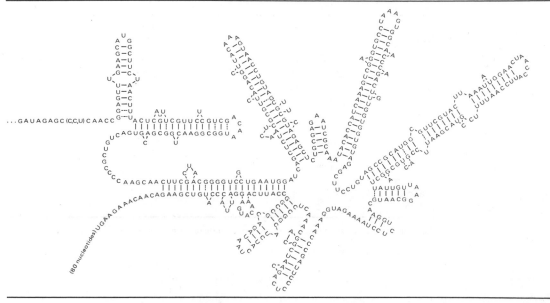

FIGURE 12-25 **The coat protein gene of MS2.** (From W. Fiers *et al.*, *Nature* **237**:83, 1972.)

## 12.18 Overlapping Genes

We conclude this chapter by describing one of the most unexpected results to emerge from sequence analysis of viral chromosomes, namely, the discovery that certain sequences encode more than one gene product.

The first example of such "overlapping genes" was discovered by F. Sanger and colleagues for the phage **φX174,** which has a small circular genome of single-stranded DNA. In this case, certain transcripts were found to encode one gene product when translated in one triplet reading frame and a second gene product when read in a second triplet frame. Specifically, as shown in Table 12-3 and Figure 12-26a, they found that the B gene of φX174 is located within the A gene sequence and the E gene is located within the D gene sequence. In the D-E case, the relevant mRNA sequence is shown in Figure 12-26b, where the E reading frame is shifted from the D frame by one nucleotide (convince yourself of this by counting off triplet units). Similar constructions have since been found in other phages: as diagrammed in Figure 12-26c, two of the three possible reading frames of the MS2 chromosome prove to encode proteins required by the phage.

The animal virus SV40 employs a somewhat different strategy. Shortly after infection it synthesizes a single species of "early mRNA" (Figure 12-24) that can be spliced in two different ways. If the intron labeled *a*

**TABLE 12-3  Summary of the φX174 Genome**

| Gene | Function | Protein Molecular wt | Number of Nucleotides* |
|---|---|---|---|
| A | Replication | 56,000 | 1536 |
| B | Replication or packaging | 13,800 | (360) |
| C | Replication or packaging | 7000 | |
| D | Replication or packaging | 16,800 | 456 |
| E | Lysis | 9900 | (273) |
| J | Capsid | 5000 | 114 |
| F | Capsid | 46,000 | 1275 |
| G | Capsid | 19,000 | 525 |
| H | Capsid | 36,000 | 984 |
| Noncoding and C | | | 485 |
| Total | | | 5375 |

*Values in parentheses are overlapping sequences and therefore not included in the addition to obtain the total length of DNA.

From F. Sanger et al., *Nature* **265**:687, 1977.

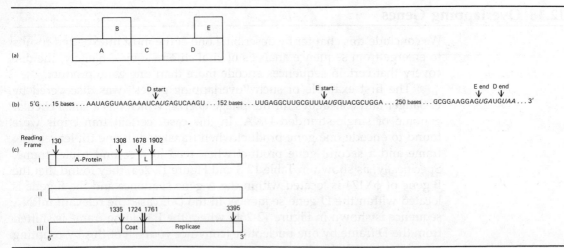

**FIGURE 12-26 Overlapping genes.** (a) Extent of overlap in the φX174 genome. (b) φX174 sequence showing reading-frame shift between genes D and E. (c) Three alternate readings of the MS2 chromosome. Reading-frame I, defined by nucleotide #1 at the 5′ end, allows transcription of the A-protein and L-protein (for host-cell lysis) genes; Frame II, shifted by one nucleotide, encodes no known proteins; Frame III, shifted by two nucleotides, encodes the coat and replicase proteins. (From Sanger, F., et al., *Nature* **265**:687, 1977, and Beremaud, M. N., and T. Blumenthal, *Cell* **18**:257, 1979.)

in Figure 12-24 is spliced out, this removes a "stop" codon, and the resultant mRNA can be translated all the way to its terminus, generating a large
(96 kdalton) polypeptide, the T antigen. If instead the smaller intron labeled *b* in Figure 12-24 is spliced out, then the "stop" codon remains in the
resultant mRNA and translation arrests after only a portion of the molecule
has been read, generating the small (15 kdalton) polypeptide, the t antigen.
Thus both messages have identical 5' and 3' ends, but distinct internal
coding sequences.

The "late mRNA" transcript from the SV40 genome can also be
spliced in two different ways, one generating a 16S mRNA which encodes
the capsid protein VP1, and the other a 19S mRNA, which encodes VP2
and VP3 (Figure 12-24). The 19S transcript contains two AUG "start" codons, one of which initiates VP2 translation and the other VP3 translation.
The 16S transcript, after splicing, continues to carry nucleotide sequences
that overlap sequences in the 19S transcript. However, the message is
translated in a different reading frame, beginning with a unique AUG
codon, so that none of the VPI amino-acid sequence is the same as either
the VP2 or the VP3 amino-acid sequences.

Overlapping genes are thought to have evolved in viruses because
they are under selective pressure to conserve a small genome size.
Whether such parsimony will be detected in other types of chromosomes is
not yet established.

# Approaches to Solving Mapping and Complementation Problems

Here and at the conclusion of the next few chapters we consider approaches to solving genetics problems. It is **not** recommended that these
sample problems be memorized or applied step-by-step to other problems.
The only reliable approach is to understand thoroughly the principles involved and then consider each question as a new puzzle to solve. The
sample problems are instead presented as "warm ups," illustrating the
ways that data are presented and the ways the data can be used.

## 12.19 Two-Factor Phage Cross Problem

**Problem**   When five mutant strains of a bacteriophage are crossed, they
yield the following recombination frequencies (numbers indicate percent
recombination):

| Strain | A | B | C | D | E | Crossed with strain |
|--------|---|---|---|---|---|---------------------|
|        | 0 | 12 | 3 | 8 | 2 | A |
|        |   | 0 | 9 | 4 | 6 | B |
|        |   |   | 0 | 5 | 0 | C |
|        |   |   |   | 0 | 2 | D |
|        |   |   |   |   | 0 | E |

Draw a map showing the location of each mutation. What is unusual about results with strain $E$, and how could they be explained?

**Solution**

(a) Write out all recombination frequencies from the data:

$$R_{AB} = 12 \quad R_{BE} = 6 \quad R_{CE} = 0$$
$$R_{AC} = 3 \quad R_{BD} = 4 \quad R_{DE} = 2$$
$$R_{AD} = 8 \quad R_{BC} = 9$$
$$R_{AE} = 2 \quad R_{CD} = 5$$

(b) Identify A and B as lying farthest apart and put them at either ends of the map:

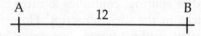

(c) Fill in the remaining distances, working always from large to small recombination frequencies, and omitting any anomalous data:

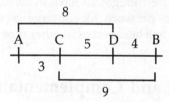

(d) Note that E is anomalous. If $R_{BE}$ is 6, then $R_{AE}$ should also be about 6; instead it is 2. C does not recombine with E at all. Conclusion: E is most likely a deletion extending to either side of C.

## 12.20 Three-Factor Phage Cross Problem

**Problem**　Order the mutations $a$, $b$, and $c$ (where $a\,b\,c$ is not necessarily the marker order) on the basis of the results of the following three phage crosses and give the distances between the markers.

| Cross | Percent Wild Type Among Progeny from the Cross |
|---|---|
| (1) $a\,b + \times + + c$ | 8.75 |
| (2) $a + c \times + b +$ | 3.75 |
| (3) $+ b\,c \times a + +$ | 1.25 |

**Solution**

(a) Note that the data are presented in an unusual fashion: only "percent wild type" is given. The reciprocal "triple-mutant" classes will, of course, figure equally among the progeny and must be included in the calculation of recombination frequencies. Therefore, the values given should first be multiplied by 2.

(b) Cross #3 produces the smallest numbers of + + + phages; this is therefore identified as the double-crossover class. Writing the marker order . as $+ a + \times b + c$ allows the generation of + + + (and $b\ a\ c$) progeny by double crossovers. The marker order is therefore assigned as $b\ a\ c$, and the frequency of doubles, $R_D = 2(1.25) = 2.5\%$ so that $\gamma = 2.5$.

(c) The Single I ($\alpha$) and Single II ($\beta$) classes can now be identified as crosses #2 and #1 respectively. Therefore:

$$\alpha = 2(3.75) = 7.5$$
$$\beta = 2(8.75) = 17.5$$

(d) Recombination frequencies are therefore:

$$R_{ba} = \alpha + \gamma = \phantom{0}7.5 + 2.5 = 10.0$$
$$R_{ac} = \beta + \gamma = 17.5 + 2.5 = 20.0$$
$$R_{bc} = \alpha + \beta = 17.5 + 7.5 = 25.0$$

(e) Draw map as:

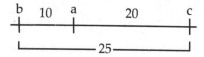

## 12.21 Phage Complementation Problem

**Problem**   Seven mutant strains of T4 can grow individually only in suppressor strains of *E. coli*. All possible pairwise mixed infections of nonsuppressor *E. coli* cells are performed with the seven strains to test for the ability to carry out a complete infection. In the results given, 0 = no phage growth; + = phage growth and lysis.

| 1 | 0 | | | | | | |
|---|---|---|---|---|---|---|---|
| 2 | + | 0 | | | | | |
| 3 | 0 | + | 0 | | | | |
| 4 | 0 | 0 | 0 | 0 | | | |
| 5 | + | + | + | 0 | 0 | | |
| 6 | 0 | + | 0 | 0 | + | 0 | |
| 7 | + | 0 | + | 0 | + | + | 0 |
| | 1 | 2 | 3 | 4 | 5 | 6 | 7 |

What is the minimum number of genes that are represented in this collection of seven mutant strains?

**Solution**

(a) Examine the data for obvious anomalies. Strain 4 clearly cannot complement any other; it must therefore represent a long deletion or, less likely, a multipoint mutation or a dominant mutation inhibiting growth under all conditions of mixed infection.

(b) Proceed to assess the remaining strains. Strain 1 (first row of matrix) appears to be mutant in the same gene (no complementation) as strains 3 and 6. Call this gene A and write

```
     1   3   6
  +--+---+---+--+
           A
```

(the relative order is, of course, arbitrary here). Strains 2, 5, and 7 are not mutant in gene A since they complement A-deficient strains. These are therefore assessed next.

(c) Strain 2 (second row of matrix) cannot complement strain 7. Call them B-deficient mutants:

```
   2   7
 +--+--+--+
      B
```

(d) Strain 5 complements all strains (except, of course, itself and the deletion strain). It is therefore mutant in a third gene C:

```
  5
 +-+--+
    C
```

(e) Three genes, A though C, are represented in the collection.

## 12.22 Deletion Mapping Problem

**Problem** Five point mutants (*a* through *e*) were tested for wild recombinants with the seven deletion mutants drawn below. What is the order of point mutants on this deletion map (+ = recombination, 0 = no recombination)?

deletion map

```
    1      2
   ___    ___
  6      3
 ___    ___
 5      4
___    ___
        7
       ___
```

data — point mutants

| deletions | 1 | 2 | 3 | 4 | 5 | 6 | 7 |
|---|---|---|---|---|---|---|---|
| a | 0 | + | 0 | 0 | + | + | 0 |
| b | + | + | + | 0 | + | 0 | 0 |
| c | + | + | + | + | 0 | 0 | 0 |
| d | 0 | + | + | 0 | + | 0 | 0 |
| e | + | 0 | 0 | 0 | + | + | 0 |

**Solution**

(a) Recognize anomalies. In this case #7 is probably an extensive deletion covering a whole region and is not useful.

(b) Start with small deletion #5 and note that only *c* is unable to recombine with it: *c* . . . . . .

(c) Similarly with #2, only *e* is unable to recombine with it: . . . . . *e*.

(d) Deletion #6 covers *b* and *d* as well as *c*, whereas deletion #1 covers *d*. The map order in this region is therefore *c b d* . . . . .

(e) Deletion #3 covers both *a* and *e*, while deletion #1 covers *a* and *d*. The map order here is therefore . . . *d a e*

(f) The complete order is *c b d a e.*

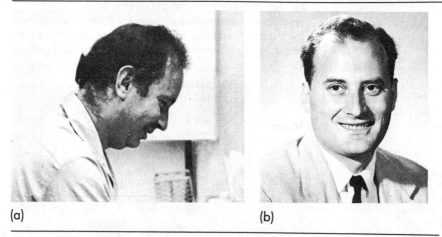

(a)                                                    (b)

(a) George Streisinger (University of Oregon) has made a number of contributions to our present understanding of phage T4 genetics. (b) Seymour Benzer (California Institute of Technology) produced the first fine-structure maps of a phage gene.

# Questions and Problems

1. Given the information in the problem presented in Section 12.20, what percent wild type would you expect from the cross *a* + × + *c*?

2. A three-factor cross between 2 strains of a virus (*abc* and + + +) was performed. The results are given as follows:

| Genotype of Progeny | Number of Plaques |
|---|---|
| + + + | 1200 |
| a b c | 1100 |
| a + + | 280 |
| + b c | 300 |
| a b + | 180 |
| + + c | 160 |
| a + c | 85 |
| + b + | 75 |

Determine the linkage order of these three genes and linkage distance between *a-b*, *b-c*, and *a-c*. Comment on your results.

3. Wild-type phage λ is able to multiply both in strain *A* and strain *B* of *E. coli*; *sus* mutant strains are able to multiply only in *A*, not in *B*. Multiplication in *B* does result after mixed infection of strain *B* by certain pairs of mutants. All 40 possible pairwise mixtures of eight different *sus* strains are tested for the ability to multiply cooperatively in strain *B* with the following results:

| | 1 | 2 | 3 | 4 | 5 | 6 | 7 | 8 |
|---|---|---|---|---|---|---|---|---|
| **1** | 0 | | | | | | | |
| **2** | + | 0 | | | | | | |
| **3** | 0 | + | 0 | | | | | |
| **4** | + | 0 | + | 0 | | | | |
| **5** | 0 | + | 0 | + | 0 | | | |
| **6** | + | 0 | + | 0 | + | 0 | | |
| **7** | + | + | + | + | + | + | 0 | |
| **8** | 0 | 0 | 0 | 0 | 0 | 0 | 0 | 0 |

**Mutant strain** (rows) / **Mutant strain** (columns)

(a) On the basis of these results, classify mutations 1 through 7 into genes.
(b) Give two possible explanations for the behavior of strain 8. How could you distinguish between these alternatives experimentally?

4. Traits *A, B, C, D, E, F, G, H* in T4 were mapped in a series of two-factor crosses. The values turned out to work very consistently, with one exception.

| | A | B | C | D | E | F | G | H |
|---|---|---|---|---|---|---|---|---|
| A | 0 | 10 | 17 | 1.5 | 3 | 6.5 | 10 | 8 |
| B | — | 0 | 0 | 6 | 9 | 0 | 2 | 4 |
| C | — | — | 0 | 11 | 14 | 4 | 7 | 9 |
| D | — | — | — | 0 | 0 | 0.5 | 4 | 2 |
| E | — | — | — | — | 0 | 3.5 | 7 | 5 |
| F | — | — | — | — | — | 0 | 0 | 0 |
| G | — | — | — | — | — | — | 0 | 2 |
| H | — | — | — | — | — | — | — | 0 |

(a) Draw a map (need not be on scale) of this part of the T4 genome
(b) Which recombination value is different from the one predicted from the map?

5. A geneticist performed the following experiments with two suppressor sensitive mutations $p$ and $q$. Plaques caused by $+ +$, $p+$ or $+q$ and mixed yields could be distinguished.

   1. Simultaneously infect nonpermissive host and plate on nonpermissive host.
   2. Simultaneously infect nonpermissive host and plate on permissive host.
   3. Simultaneously infect permissive host and plate on nonpermissive host.
   4. Simultaneously infect permissive host and plate on permissive host.

   Predict the results of the experiments if
   (a) $p$ and $q$ fail to complement and are eight units apart.
   (b) $p$ and $q$ complement and are ten units apart.
   NOTE: Recombination rarely occurs before phage growth is underway.

6. One mutant of T4 is unable to make a head protein ($H^-$), whereas another mutant is unable to make tail fibers ($T^-$), where $H$ and $T$ are known to be closely linked on the phage chromosome. Both of these strains can infect *E. coli* strain $K$, but neither are able to produce progeny phage. However, if strain $K$ is mixedly infected with both ($H^-$) and ($T^-$) strains of T4, phage progeny are produced in approximately the same amounts as cells infected with wild-type phage. What would the genotypes of the progeny phage be? Explain your choice.
   (a) Approximately equal amounts of wild type ($++$) and the double mutant ($H^-T^-$).
   (b) Mostly ($H^-$) and ($T^-$) in approximately equal amounts with a small proportion of wild type ($++$) and double mutant ($H^-T^-$).
   (c) Equal amounts of ($H^-$), ($T^-$), wild type ($++$), and the double mutant ($H^-T^-$).
   (d) Approximately equal amounts of wild type ($++$) and the double mutant ($H^-T^-$), with small amounts of ($H^-$) and ($T^-$).

7. Suppose the genome of T4 phage is represented as *ABCDEF*, and because of the circularly permuted DNA of T4, a typical sample from several particles yields sequences such as *ABCDEFAB*, *CDEFABCD*, *EFABCDEF*, and so on.

   Now suppose deletion mutants are formed by deleting region *CD*, that these mutants are allowed to infect *E. coli* cells, and that the DNA of the progeny phage is then examined. Which of the sequences below ought to be found if the "headful" hypothesis is correct? Explain your choice.
   (a) ABEFAB, EFABEF, FABEFA
   (b) ABEFABEF, EFABEFAB, FABEFABE
   (c) ABEFABEFA, EFABEFABE, FABEFABEF

8. Two-factor crosses involving strains $a$, $b$, $c$, $d$, and $e$ of a bacteriophage give the following results:

| Cross | Recombination Frequency (%) |
|---|---|
| $a \times b$ | $<10^{-3}$ |
| $a \times c$ | 2.0 |
| $a \times d$ | 3.0 |
| $a \times e$ | 1.0 |
| $b \times c$ | $<10^{-3}$ |
| $b \times d$ | 1.0 |
| $b \times e$ | 0.8 |
| $c \times d$ | 1.1 |
| $c \times e$ | 2.8 |
| $d \times e$ | 3.8 |

Order the markers along an unbranched genetic map.

9. The following recombination frequencies are measured for two-factor bacteriophage crosses:

$$a\,b^+ \times a^+\,b \qquad 3.0\%$$
$$a\,c^+ \times a^+\,c \qquad 2.0\%$$
$$b\,c^+ \times b^+\,c \qquad 1.5\%$$

(a) What map order is suggested for the mutations $a$, $b$, and $c$? Why are the distances not additive?

(b) Consider the three-factor cross $a\,b^+\,c \times a^+b\,c^+$. Which two recombinant types do you expect to be rarest?

(c) Compute the recombination frequencies expected for each of the three reciprocal pairs of recombinant types emerging from the three-factor cross in (b).

10. Four mutant T4 strains are tested for complementation in *E. coli*. We get these results (+ = complementation, 0 = no complementation):

|   | 1 | 2 | 3 | 4 |
|---|---|---|---|---|
| 1 | 0 | 0 | + | + |
| 2 | 0 | 0 | 0 | + |
| 3 | + | 0 | 0 | + |
| 4 | + | + | + | 0 |

(a) Explain the above results in terms of the functional relationships between mutations 1 to 4.

(b) Draw a map showing the relative locations of these mutations. What ambiguity remains?

11. A cross between two λ strains gives the following results:

| | | | |
|---|---|---|---|
| + | + | + | 38 |
| $a$ | $b$ | $c$ | 23 |
| $a$ | + | + | 273 |
| + | $b$ | $c$ | 318    Total: 12,324 |
| $a$ | $b$ | + | 112 |
| + | + | $c$ | 121 |
| $a$ | + | $c$ | 6,389 |
| + | $b$ | + | 5,050 |

(a) What are the genotypes of the parental phages?

(b) What is the gene order?

(c) What are the map distances between the genes?

12. A three-factor cross yields the following kinds and numbers of progeny:

| | | | | | | | | |
|---|---|---|---|---|---|---|---|---|
| + | + | + | 235 | | $p$ | $q$ | $r$ | 270 |
| $p$ | $q$ | + | 62 | | $p$ | + | + | 7 |
| + | $q$ | + | 40 | | $p$ | + | $r$ | 48 |
| + | $q$ | $r$ | 4 | | + | + | $r$ | 60 |

Total: 726

(a) What are the genotypes of the parental phages in this cross?

(b) What is the gene order?

(c) What are the map distances between the genes?

13. Four T4 phage deletion strains are tested for recombination by pairwise crossing in *E. coli*. We get these results (+ = recombination, 0 = no recombination):

|   | 1 | 2 | 3 | 4 |
|---|---|---|---|---|
| 1 | 0 | + | + | + |
| 2 | + | 0 | 0 | + |
| 3 | + | 0 | 0 | 0 |
| 4 | + | + | 0 | 0 |

(a) Explain the above results in terms of a genetic map.

(b) A mutant phage, *m*, known to contain *no* deletions, recombines with mutants 1 to 4 as follows:

$$\begin{array}{cccc} 1 & 2 & 3 & 4 \\ m \quad 0 & + & 0 & + \end{array}$$

Explain the nature of the mutation in phage *m* and show its position on the map.

14. The ends of 6 deletion strains (*C, D, E, F, G,* and *H*) define 11 separate regions of the *rII* locus of phage T4, as shown below.

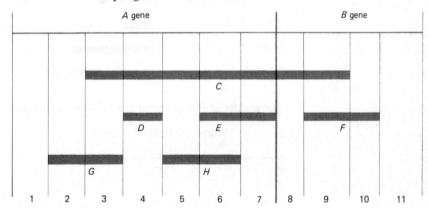

When *E. coli* cells are mixedly infected with pairwise combinations of these phages plus four strains carrying mutations *K, L, M,* and *N*, the following results are obtained:

0 = no lysis
C = complementation
R = recombination

|   | C | D | E | F | G | H | K |
|---|---|---|---|---|---|---|---|
| K | 0 | R | R | C | R | 0 | 0 |
| L | 0 | C | C | R | C | C | C |
| M | 0 | 0 | 0 | C | R | 0 | 0 |
| N | 0 | R | 0 | R | R | 0 | 0 |

(a) In which regions of the map can you place mutations *K, L, M,* and *N*?

(b) What is the nature of mutations *M* and *N*? Give your reasoning.

**15.** The $b_2$ sector of the $\lambda$ chromosome is not marked by *sus* mutations. Why?

**16.** The $\lambda$ chromosome exhibits a clustering of functionally related genes (*Figure 12-8b*), with the left arm *morphological*, the right arm largely *regulative*, and the middle region *integrative* and *recombinational*. Is the T4 chromosome (Figure 12-10) similarly "logical"?

**17.** You isolate a mutant strain of $\lambda$ that can be induced to leave the lysogenic state and direct a lytic cycle by shifting the temperature to 42°C. How might you explain this phenotype?

**18.** A culture of *E. coli B* was mixedly infected with two *rIIB* mutants. An aliquot of the *B* lysate was diluted by $10^2$ and a 0.1-ml sample was plated on *E. coli* strain *K* lysogenic for $\lambda$. A second aliquot of the *B* lysate was diluted by $10^5$, and a 0.1 ml sample was plated on *E. coli* strain *B* bacteria. A total of 20 plaques were counted on the $K(\lambda)$ plates, and 20 plaques were also counted on the *B* plates. What is the map distance between these two mutations?

**19.** Benzer isolated a mutant phage strain Z with a single point mutation somewhere in the *rIIA* locus. He crossed mutant Z with three other deletion mutant strains *A, B,* and *C* which were deleted for the various regions of the *rIIA* gene indicated below by solid bars.

Normal *rIIA* locus

Strain *A*   $a \quad b \quad\quad c \quad d \quad\quad e$

Strain *B*   $b \quad\quad c \quad d \quad\quad e$

Strain *C*   $c \quad d \quad\quad e$

$d \quad\quad e$

The results of this series of crosses were as follows:

| | Wild-type Recombinants/ 10,000 Progeny |
|---|---|
| Z × A | 140 |
| Z × B | 35 |
| Z × C | 0 |

Benzer then crossed mutant strain Z with three strains I, II, and III possessing *point* mutations at sites I, II, and III, respectively, with the following results:

| | Wild-type Recombinants/ 10,000 Progeny |
|---|---|
| Z × I | 2 |
| Z × II | 8 |
| Z × III | 4 |

From this information, in what region ($a-b$, $b-c$, $c-d$, or $d-e$) of the *rIIA* locus is the mutation in strain Z located? Which of the three strains carrying point mutations have their mutant sites closest to that in strain Z, and what is the map distance separating them from Z?

**20.** The deletion mutants *C, D, E, F,* and *G* map in the following regions of the T4 *rII* genes.

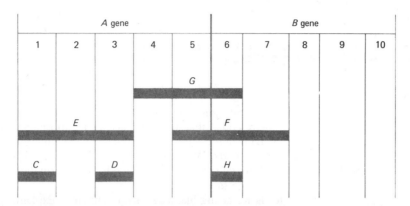

The following data for the unmapped mutants $R$, $S$, $T$, and $U$ were obtained. (0 = no lysis; R = recombination; C = complementation.)

|   | C | D | E | F | G | H |
|---|---|---|---|---|---|---|
| R | R | R | 0 | R | R | C |
| S | R | R | R | R | 0 | C |
| T | C | C | C | 0 | R | R |
| U | C | C | C | R | R | R |

Assign map positions to the four mutations.

21. An enrichment procedure (*Section 7.3*) designed to isolate nonreplicating phage mutants is known as $^{32}$P **suicide:** phages are grown in the presence of $^{32}$P, and the lysates are stored for several weeks, during which the decay of β-particles produces lethal breaks in the phosphodiester backbone of the DNA. The efficiency of killing per radioactive disintegration can be expressed by a constant $k$. For the T series phages, k = 0.1, whereas for φX174, k = 1.0. Why are the two values different?

22. (a) When a collection of rod-shaped T4 chromosomes is subjected to a dissociation-reassociation experiment (*Section 4.4*) and the reassociated molecules are examined with the electron microscope, most are found to be in the form of circles. Explain this result, using diagrams similar to those in Figure 4-3.

(b) This experiment was repeated using the rod-shaped chromosomes of T7, and almost all the resultant duplexes were rod shaped. What does this experiment reveal about the T7 chromosome?

23. For the φX174 genome, answer the following:

(a) What is the expected C-terminal amino acid of the D protein? The E protein?

(b) Would you expect an *amber* mutation in gene E to affect the translation of gene D? Why or why not?

(c) Speculate why this arrangement of genes might have evolved.

24. The λ chromosome is replicated as a circle. During packaging, the following sequence in the replication product, known as **cos,** is recognized by the enzyme **ter** (for **ter**minus-generating activity), which makes the two single-strand cuts shown by the arrows.

**(b)**

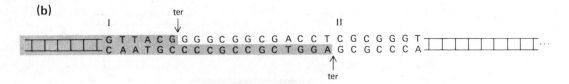

These cuts generate rod-shaped λ chromosomes with two single-stranded "sticky ends" that are then packaged into phage heads.

(a) Draw a rod-shaped λ chromosome showing the nucleotide sequences of its two ends. Why are these ends called "sticky"?

(b) Illustrate how these ends mediate the formation of a circular λ chromosome at the time of replication.

(c) Is ter acting like a restriction enzyme? Explain.

(d) A mutation in λ creates a cos sequence at a second site in the genome. What would be the probable effect of such a mutation? Could this mutant be temperature sensitive? Explain?

25. The $sus_{20}$ and $sus_9$ strains of λ are defective in gene C and gene G, respectively. When nonsuppressor strains of *E. coli,* lysogenic for either mutant phage, are induced with uv, each phage will dictate an abortive lytic cycle wherein all components are synthesized normally except the C protein or the G protein. When such induced cells are lysed artificially and the lysate is plated onto a permissive bacterial lawn, no plaques will form because no infective phage particles have been produced. If, however, the $sus_{20}$ lysate is first incubated for several hours with the $sus_9$ lysate and the mixture is then plated, numerous plaques form on the permissive lawn.

(a) Explain why this phenomenon is called *in vitro* complementation.

Pairwise mixtures of various *sus* lysates were made, and the following result was obtained (0 = no complementation, + = complementation, letters = genes carrying a *sus* mutation).

|   | A | B | C | D | E | G | H | I | J | K | L | M |
|---|---|---|---|---|---|---|---|---|---|---|---|---|
| A | 0 | 0 | 0 | 0 | 0 | + | + | + | + | + | + | + |
| B | — | 0 | 0 | 0 | 0 | + | + | + | + | + | + | + |
| C | — | — | 0 | 0 | 0 | + | + | + | + | + | + | + |
| D | — | — | — | 0 | 0 | + | + | + | + | + | + | + |
| E | — | — | — | — | 0 | + | + | + | + | + | + | + |
| G | — | — | — | — | — | 0 | 0 | 0 | 0 | 0 | 0 | 0 |
| H | — | — | — | — | — | — | 0 | 0 | 0 | 0 | 0 | 0 |
| I | — | — | — | — | — | — | — | 0 | 0 | 0 | 0 | 0 |
| J | — | — | — | — | — | — | — | — | 0 | 0 | 0 | 0 |
| K | — | — | — | — | — | — | — | — | — | 0 | 0 | 0 |
| L | — | — | — | — | — | — | — | — | — | — | 0 | 0 |
| M | — | — | — | — | — | — | — | — | — | — | — | 0 |

(b) Which genes can and cannot complement one another *in vitro*?

(c) Study Figure 12-8b and offer an explanation for these results.

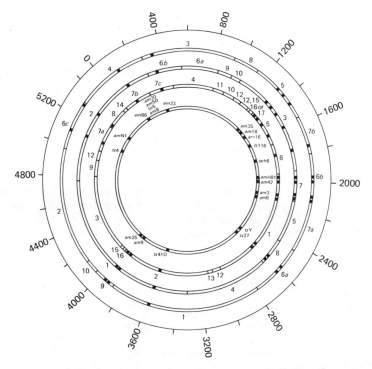

26. The map of φX174 above shows five concentric circles: the outermost circle shows map distance in nucleotides; next come restriction maps of *Hind* II, *Hae* III, and *Alu* I; and the inner circle shows the locations of various nonsense and *ts* mutants mapped by two-factor crosses. The locations of the mutations on the restriction fragments were ascertained by the **marker rescue** technique: individual fragment classes from a wild-type restriction digest are cut out of a gel; each fragment class is incubated with *am*- or *ts*-chromosomes under reassociation conditions (*Section 4.3*) to form partially hybrid chromosomes; and *E. coli* cells are transfected with the hybrids. High titers of wild-type phages arise only when the restriction-fragment class carries the normal allele of the mutant gene.

    (a) Diagram the marker rescue technique using the five largest fragments of a *Hind* II wild-type digest and *am33* mutant chromosomes. Which fragment class would you expect to produce rescue?

    (b) Which fragments would you expect to rescue *am33* in the *Hae* III and *Alu* I digests?

    (c) Locate a fragment class in each digest that would fail to rescue any identified marker.

    (d) Fragment 2 of the *Alu* I digest fails to rescue *am9*, whereas fragment 1 of the *Hae* III digest can do so. How do you interpret this result?

27. A mutant phage contains a small inversion. In a heteroduplex map (Figure 12-20) with the wild-type phage, would you expect to see a discontinuity? If so, would you expect it to resemble the *b5/iλ* region or the *b2* region in Figure 12-20? Explain.

28. Four *sus* mutant strains of phage λ give the following results when crossed with the λ deletion strains (Figure 12-17). (0 = no recombination; + = recombination).

**Deletion Strain**

|  |  | $c_1$ | G | O | P | R | A | C' | E |
|---|---|---|---|---|---|---|---|---|---|
| *sus* | *sus₁* | + | 0 | + | + | + | + | + | + |
| strain | *sus₂* | + | 0 | 0 | 0 | 0 | 0 | 0 | + |
|  | *sus₃* | + | + | 0 | 0 | 0 | 0 | 0 | 0 |
|  | *sus₄* | + | 0 | 0 | 0 | 0 | + | + | + |

(a) In which gene or genes (give all possibilities) does each *sus* mutation lie?
(b) How would you go about giving specific gene and map assignments to each *sus* mutant?

29. A circular duplex DNA chromosome from an animal virus has a molecular weight of $6.5 \times 10^6$. Treatment of this DNA with the restriction enzymes *Hpa*, *Hin*, and *Eco* yields the following gels, where numbers refer to molecular weight $\times 10^6$.

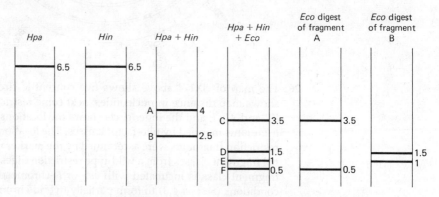

(a) How many *Hpa* sites are present in the chromosome? How many *Hin* sites? How many *Eco* sites?
(b) Draw a restriction map of the chromosome, showing the positions of sectors A to F.
(c) If you omitted *Hpa* and restricted the chromosome with just *Hin* + *Eco*, would you expect to generate fragments C to F? Why or why not? If not, what would you expect to generate?

# 13

# Mapping Bacterial Chromosomes and Plasmids

**A. Introduction**

**B. Molecular Overview of Bacterial Conjugation**
13.1 $F^+ \longrightarrow F^-$ Transfer
13.2 The Formation of *Hfr* Cells and *Hfr* $\longrightarrow F^-$ Transfer

**C. Mapping by Bacterial Conjugation**
13.3 Early Conjugation Studies
13.4 Interrupted Mating
13.5 Differences Between *Hfr* Strains
13.6 Rapid Mapping by Conjugation
13.7 The *E. Coli* Chromosome Map

**D. Bacterial Transformation**
13.8 Molecular Overview of Bacterial Transformation
13.9 Establishing Gene Linkage by Transformation
13.10 Transformation Mapping Using *Bacillus Subtilis*

**E. Generalized Transduction**
13.11 Molecular Overview of Generalized Transduction
13.12 Mapping by Generalized Transduction

**F. Specialized Transduction**
13.13 Transduction of Markers to Host Bacteria
13.14 Mapping Bacteria by Specialized Transduction
13.15 F-Mediated Sexduction

**G. Plasmids**
13.16 General Properties of Plasmids
13.17 Detecting the Presence of a Plasmid
13.18 F Element (F Sex Factor)
13.19 R Plasmids
    Conjugal transfer genes
    Antibiotic resistance genes
    Transposons
    Transposon-plasmid interactions
    R plasmids and public health
13.20 Col Plasmids

**H. Approaches to Solving Mapping Problems**
13.21 Transduction Problem

**I. Questions and Problems**

# Introduction

As we move from the bacteriophage (*Chapter 12*) to the bacterium, we make a major leap in the complexity of both genetic organization and gene transmission. A bacterium possesses perhaps a hundred-fold more DNA than most of its phages (*Figure 2-2*). The DNA of a bacterium, moreover, is found in both the single main chromosome and the relatively small plasmids (*Figure 2-2*), so that frequently the transmission of more than one genetic element from parent to offspring must be followed. Finally, a number of different avenues have evolved whereby the DNA from one bacterial cell can undergo genetic exchange with the DNA from another bacterial cell. Thus the DNA may be transmitted through a specialized sex pilus (**conjugation**); it may enter the cell as a duplex fragment (**transformation**); or it may be carried into and out of the cell by a bacteriophage (**transduction**). This chapter describes each of these modes of bacterial gene transmission and demonstrates how each is used to construct bacterial chromosome maps.

# Molecular Overview of Bacterial Conjugation

## 13.1 $F^+ \longrightarrow F^-$ Transfer

A bacterial cell may contain one or more plasmids, small DNA molecules that usually maintain a distinct existence from the main chromosome and replicate independently of it. Plasmids are considered in some detail in later sections of this chapter and in Chapter 18, but one kind of plasmid, the **sex element** known as **F**, plays a key role in the sexuality of bacteria and is thus appropriately considered here as well.

*Escherichia coli* cells that carry an F element (the term **F factor** and **F agent** are also used) are known as $F^+$, and such cells are found, albeit quite rarely, among natural populations of the bacterium. The F element is made up of at least 15 genes, 9 of which are known to control the elaboration of **F pili**, long appendages that extend from the surface of $F^+$ cells (Figure 13-1). Cells that possess F pili are sensitive to infection by single-stranded RNA phages (*Section 12.17*); $F^-$ cells, which lack an F element and are devoid of pili, are insensitive to such phage infection.

An $F^+$ cell will usually ignore another $F^+$ cell, whereas it will readily establish contact, probably via an F pilus, with an $F^-$ cell. Once contact is made, the pilus is believed to become modified and to serve as a protoplasmic channel between the two cells; in this context it is often referred to as a **conjugation tube.** Under ordinary circumstances the only genetic element transferred through the tube is the F element itself. In a mixture of $F^+$ and $F^-$ cells, each $F^+$ donor will pass a copy of the F element on to an $F^-$

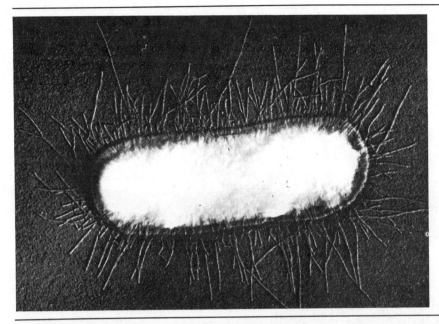

FIGURE 13-1 A bacterium with pili (also called fimbriae) extending from its surface. Each pilus is a hollow cylinder with a 20-Å central hole. (From J. P. Duguid et al., *J. Path. Bact.* **29**:197, 1966.)

recipient, while retaining at least one copy for itself; eventually, then, virtually every cell in such a mixed population becomes an $F^+$ cell.

The transfer of F is thought to proceed by the rolling circle mechanism, which we encountered earlier in considering phage replication (*Section 12.1*). As drawn in Figure 13-2, the 5' end of one strand of the F element is thought to be drawn into the recipient cell, where it is copied in the $5' \longrightarrow 3'$ direction. Meanwhile, the second strand remains in the donor and serves as a rolling template for its own replication.

The infectious transmission of F elements from cell to cell is of interest in itself, but it yields no information about the genes in the main bacterial chromosome. Fortunately for bacterial genetics, the F element infrequently (about once in every 10,000 $F^+$ cells) becomes associated with the main bacterial chromosome in such a way that a copy of the main chromosome, instead of simply the F element, is transferred through the conjugation tube from donor to recipient.

## 13.2 The Formation of *Hfr* Cells and *Hfr* $\longrightarrow$ $F^-$ Transfer

The events that allow transfer of the main bacterial chromosome to an $F^-$ cell are diagrammed in Figure 13-3. In this process, an F element first inserts itself into the bacterial chromosome. The insertion process, which is

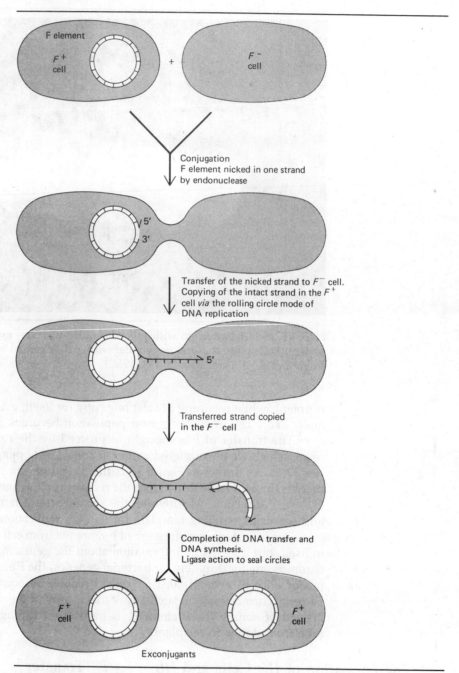

**FIGURE 13-2 Transfer of the F element** from an $F^+$ to an $F^-$ cell during conjugation in *E. coli*.

in some ways analogous to λ chromosome insertion (*Sections 12.2 and 12.14*), causes the circular F element to break at a particular point and to become a linear segment of the bacterial chromosome, as diagrammed in Figure 13-3a. An $F^+$ cell that carries such an integrated F element is known as an *Hfr* cell, for reasons that will become apparent shortly.

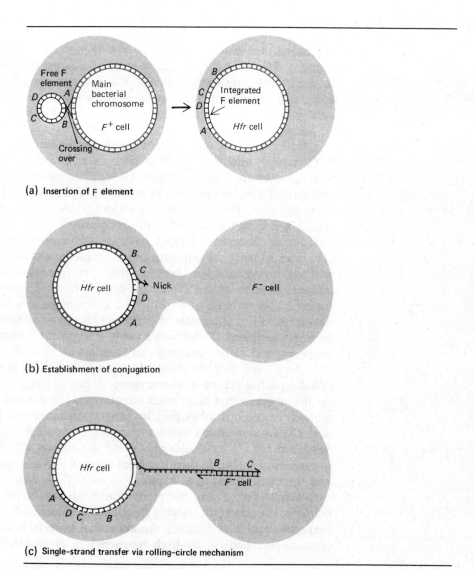

(a) **Insertion of F element**

(b) **Establishment of conjugation**

(c) **Single-strand transfer via rolling-circle mechanism**

**FIGURE 13-3 Integration of the F element** to form an *Hfr* cell and transfer of *Hfr* chromosome to an $F^-$ cell during conjugation in *E. coli*.

The integrated F element is ordinarily replicated passively along with the bacterial chromosome, very much like a λ prophage, and in this way is transmitted from one cell generation to the next. In other words, the integrated F element ordinarily shuts off its ability to replicate independently and behaves as an ordinary segment of the bacterial chromosome. When conjugation is initiated, however, the replication apparatus of the F element is somehow activated. The F element's DNA is nicked in one strand (Figure 13-3b), and the 5' end of the nicked strand is drawn into the conjugation tube, apparently by the impetus of a rolling circle type of DNA replication (Figure 13-3c). This mode of transfer is similar to the round of replication that effects the transfer of an F element in an $F^+ \longrightarrow F^-$ conjugation (*Figure 13-2*). The difference, of course, is that the transferred strand is now covalently linked to the enormous main bacterial chromosome; one strand of this chromosome is therefore drawn into the $F^-$ cell as well (Figure 13-3c) and is copied in the 5' $\longrightarrow$ 3' direction as it enters the $F^-$ recipient.

In Figure 13-3a, the F element carries some arbitrary markers called $A$, $B$, $C$, and $D$, and the nick that initiates rolling circle replication occurs between markers $C$ and $D$ (Figure 13-3b). This means that the F element is split during transfer and only some of its genes enter the $F^-$ cell at the start of conjugation. The remaining F element genes (the $D–A$ segment in Figure 13-3) will be transferred to the recipient cell only after approximately 1300 μm of chromosomal DNA have already passed through the conjugation tube. Usually the entering chromosome breaks at some intermediate position during transfer. Therefore the $F^-$ recipient cell usually inherits an incomplete copy of the F element during $Hfr \times F^-$ conjugation and remains $F^-$ (in contrast to an $F^+ \times F^-$ interaction, in which the $F^-$ cell in converted to $F^+$). If transfer of the entire chromosome is in fact completed, then the recipient $F^-$ cell will inherit a complete F element and the $Hfr$ property, and its descendants can then act as $Hfr$ cells.

An $F^-$ cell that has received only a part of the donor chromosome is called a **partial zygote** or **merozygote**. A partial zygote is initially diploid for those genes that have been transferred, but the cell does not remain diploid. Instead, it takes part in genetic exchanges so that some of the donor DNA is included in the recipient chromosome (such exchanges are described from a molecular viewpoint in Chapter 17). All nonintegrated fragments of DNA are then lost from the cell line in subsequent divisions, and the clone emerges as haploid.

When donor and recipient chromosomes carry different genetic markers, the emergent cells are frequently recombinant. It is for this reason that cells capable of donating chromosomal material to recipient cells have come to be called $Hfr$ for **high frequency of recombination.**

# Mapping by Bacterial Conjugation

## 13.3 Early Conjugation Studies

Many features of bacterial sexuality were worked out in the 1950s by W. Hayes, F. Jacob, J. and E. Lederberg, and E. Wollman well before the molecular details of the process were known. Their studies revealed that recombination would occur when *Hfr* cells of one genotype were crossed with $F^-$ cells of another genotype. We can, in particular, follow a series of crosses performed by Jacob and Wollman.

The *Hfr* strain used in the Jacob-Wollman cross had the following phenotype and genotype: it was capable of synthesizing the amino acids threonine (*thr$^+$*) and leucine (*leu$^+$*); it was sensitive to the metabolic inhibitor sodium azide (*azi-s*), and it was sensitive to the phage $T_1$ ($T_1$-*s*). It could ferment lactose (*lac$^+$*) and galactose (*gal$^+$*) as well as glucose, and it was sensitive to the antibiotic streptomycin (*str-s*). The $F^-$ strain used by Jacob and Wollman had the complementary genotype, that is, *thr$^-$ leu$^-$ azi-r $T_1$-r lac$^-$ gal$^-$ str-r* ($F^-$), where *resistance* is denoted by *-r*. The two cell types were mixed and allowed to interact for 60 minutes. The cells were then plated to various media and allowed to grow.

Jacob and Wollman began by plating the cells in the mating mixture to a minimal medium containing streptomycin. On such a medium all unmated *Hfr* cells, being streptomycin-sensitive, were killed, and any cells that required threonine or leucine, including unmated $F^-$ cells, were unable to grow. This step therefore allowed only $F^-$ cells with the recombinant genotype *thr$^+$ leu$^+$ str-r* to form colonies. Such recombinants emerged with very high frequency—some 10 percent of the total cells plated—meaning that genetic exchange had occurred frequently between the *thr$^+$ leu$^+$* markers contributed by the *Hfr* donor and the *str-r* marker contributed by the $F^-$ recipient. This result, in other words, indicated that the *thr* and *leu* loci are quite loosely linked to the *str* locus.

The next step involved determining the rest of the genotype of the *thr$^+$ leu$^+$ str-r* recombinants. Jacob and Wollman replica-plated (*Figure 4-12*) the recombinant colonies to a minimal medium that contained either sodium azide or bacteriophage $T_1$ and determined the distribution of sensitivity and resistance to these agents among the colonies. They also replica-plated the *thr$^+$ leu$^+$ str-r* recombinants to special indicator media (such as that described in Section 7.2) to determine whether the various strains were able to ferment lactose or galactose. Typically, they found that among the *thr$^+$ leu$^+$ str-r* recombinants, 90 percent carried the *azi-s* marker of the *Hfr* donor, 80 percent carried the $T_1$-*s* marker, 40 percent carried the *lac$^+$* marker, and 25 percent carried the *gal$^+$* marker.

## 13.4 Interrupted Mating

At the time the Jacob-Wollman crosses were performed, nothing was known about what was involved in bacterial conjugation, and thus the results were interpreted in many different ways. That conjugation in fact involves the transfer of a linear *Hfr* chromosome to the *F⁻* cell was established genetically by Jacob and Wollman in their subsequent **interrupted mating** experiments.

In these experiments *Hfr* and *F⁻* cells, having the same genotypes as before, were allowed to conjugate. At specific time intervals samples were taken and agitated in a Waring Blendor to separate mating cells. Samples that had been allowed to mate for 5 minutes, 10 minutes, and so on, were thus obtained. Each sample was then plated to a streptomycin-containing minimal medium to select, as before, recombinant *F⁻* cells that were *thr⁺ leu⁺ str-r*. The complete genotypes of these recombinant cells were then determined by appropriate platings.

The results are shown in Figure 13-4. Focus attention first on the order of appearance of the unselected markers in the recombinant cells. It is seen that *azi-s* is not transferred at all during the first nine minutes of mating, after which it is transferred readily. The $T_1$-*s* marker comes next,

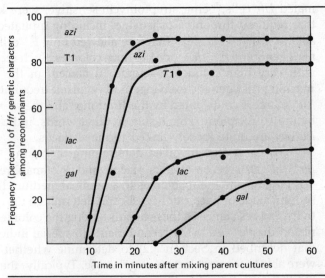

**FIGURE 13-4 Interrupted Mating.** Kinetics of transfer of various unselected donor (*Hfr*) markers (*azi*, $T_1$, *lac*, and *gal*) among *F⁻* recombinants selected for *thr⁺ leu⁺ str'* markers following an interrupted mating experiment. For each marker, a plateau is observed after the frequency of transfer reaches a certain level. This corresponds to the rate of spontaneous cessation observed for that marker in noninterrupted mating experiments. (After F. Jacob and E. L. Wollman, *Sexuality and the Genetics of Bacteria.* New York: Academic Press, 1961.)

first appearing in recombinants at ten minutes. The $lac^+$ marker does not appear until about 18 minutes, whereas the *gal* locus makes its first appearance only after 25 minutes. The most direct way to interpret these results is to draw the donor chromosome as $\overrightarrow{(thr^+, leu^+)}$, with the arrowhead indicating the origin of entry, and to assume that this chromosome moves into the recipient in a linear, orderly fashion. Since *azi-s* begins to enter the recipient cells shortly after the selected $thr^+$ $leu^+$ markers, whereas $gal^+$ does not enter for an additional 16 minutes, it can at once be deduced that *azi-s* must lie much closer to the $thr^+$ $leu^+$ loci than does $gal^+$. Applying this logic to the remaining unselected markers, the order of genes in the donor chromosome becomes $\overrightarrow{gal^+\ lac^+\ T_1\text{-}s\ azi\text{-}s\ (thr^+, leu^+)}$.

You will notice in Figure 13-4 that the percent recombination for each marker reaches a "plateau" value. These values prove to be the same as those obtained in the noninterrupted matings described in Section 13.3, with 90 percent for *azi*, 80 percent for $T_1$, and so on. Work out for yourself how these numbers arise, recalling that "natural" breaks in the donor chromosomes are increasingly likely to occur with increasing conjugation time and assuming that, once inside a recipient, all donated DNA is equally likely to participate in recombination with the $F^-$ chromosome.

## 13.5 Differences Between *Hfr* Strains

The *Hfr* strain used in the foregoing experiments was called *Hfr-H*. As different *Hfr* lines were isolated from $F^+$ populations in various laboratories, it became clear that each had distinctive properties. Thus a second strain, *Hfr-AB311*, when subjected to interrupted mating, transferred its markers to $F^-$ cells in the order $\overrightarrow{thr\ leu\ azi\ T_1\ lac\ gal}$; in this case the order of the markers is the reverse of the *Hfr-H* order. In a third strain, *Hfr-C*, the transmission order proved to $\overrightarrow{gal\ thr\ leu\ azi\ T_1\ lac}$. Here, not only is the orientation the reverse of *Hfr-H* but also the linkages appear to differ from both *Hfr-H* and *Hfr-AB311*: *gal* and *lac* now appear to be at opposite ends of the chromosome, whereas earlier they had appeared next to each other. (Such a situation should sound familiar: we noted in Section 12.10 that linkage relationships established in one phage T4 cross did not always hold for another.)

Once marker transfer data were collected from a number of different *Hfr* strains, it was possible to draw several conclusions about *Hfr* integration and the main bacterial chromosome. First, it became clear that the main chromosome can pass from *Hfr* to $F^-$ cells in either one direction (let us say $A \longrightarrow Z$) or the opposite direction ($Z \longrightarrow A$). This outcome was interpreted, correctly, to mean that $F$ itself can assume one or the other of opposite orientations within the main chromosome at the time it integrates, thus establishing the polarity of transfer.

The genetic data also reveal that the *E. coli* chromosome carries numerous potential sites for F integration, many with opposite orientations, as drawn in Figure 13-5. When a given *F*[+] cell converts to an *Hfr*, integration may occur at any one of these sites. Once integrated, however, the trait is stable for that particular strain: the daughters of an *Hfr* cell inherit an F element integrated at the same site as in the parental chromosome.

Finally, the interrupted mating experiments demonstrate that the *E. coli* chromosome, like the T4 chromosome, is genetically circular. Thus when various *Hfr* maps are compared with one another, they prove to be circular permutations of a common marker order, reading in the same or opposite direction. This result is, of course, one we have come to expect, since we have been assuming a circular *E. coli* chromosome throughout the text.

## 13.6 Rapid Mapping by Conjugation

B. Low has devised a particularly efficient method for conjugation mapping that depends on many of the principles we have described. This method utilizes 15 different *Hfr* strains of *E. coli*; each strain carries a different F integration site, and therefore initiates the transfer of a different seg-

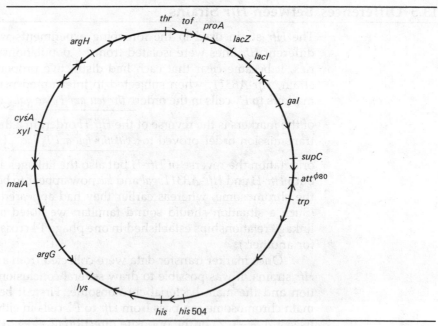

**FIGURE 13-5 Sites of F-element integration** in the chromosome of *E. coli*. Each different point of insertion is symbolized by an arrowhead pointing in the direction of chromosome transfer, so that the gene behind the arrowhead is the first one transferred. (From A. Campbell, *Episomes*. New York: Harper and Row, 1969.)

ment of the main chromosome, and each is sensitive to streptomycin. The 15 different *Hfr* colonies are replica-plated onto a layer of $F^-$ streptomycin-resistant cells that carry an unmapped mutation. The plate is also **recombinant-selective,** meaning, for example, that if the unmapped mutant strain is an acetate auxotroph, the medium is made acetate-free so that only recombinants that receive an $ac^+$ locus will be able to grow.

Cells from each *Hfr* colony proceed to transfer their unique sectors of the main chromosome into the $F^-$ recipients on the plate. Since the probability of recombination decreases linearly with physical distance from the origin of transfer, and since chromosome transfer may cease spontaneously, cells underlying a given *Hfr* strain are most likely to acquire genes transferred early in conjugation and are relatively unlikely to receive genes that are transferred late. After 30 minutes, the plate is sprayed with streptomycin to kill all *Hfr* cells and is then incubated for one to two days. Heavy patches of growth are observed only in those areas where at least one *Hfr* cell has transmitted a wild-type gene that replaces its mutant allele in a recipient mutant cell. Thus, simple inspection of the plate allows one to determine, for example, that *Hfr-AB311* cells are uniquely capable of converting the $F^-$ acetate mutants into prototrophs with high efficiency, and this localizes the acetate-utilization locus close to the F element integration site of the *Hfr-AB311* strain.

## 13.7 The *E. Coli* Chromosome Map

Partial maps of the *E. coli* chromosome have been presented throughout the text; the complete genetic map of *E. coli*, as it is now known, is given in Figure 13-6. The map is divided into minutes, with 100 min representing the time taken for a complete chromosome to pass from *Hfr* donor to $F^-$ recipient during conjugation. This detailed map was not constructed with conjugation data alone, since transduction (*Sections 13.12 and 13.14*) must be used to establish linkage relationships between markers that are close to one another. In practice, therefore, a newly isolated mutant strain of *E. coli* carrying a mutation at some unknown position on the chromosome is first subjected to mapping by interrupted mating to determine the general location of the mutant gene; its precise location is then determined by transduction.

A striking feature of the *E. coli* chromosome is its nonrandom distribution of mapped genes: the 72 min and 83 min regions, for example, are densely populated with genes, whereas the 33 min region is a veritable genetic wasteland. One possibility is that the "empty" areas in fact contain genes that have not yet been marked by mutations, perhaps because we do not yet know how to select for such mutations. The alternative possibility is that the DNA in these regions does not contain genes and instead performs other functions for the cell. One intriguing suggestion is based on the fact that the 33-min "wasteland region" lies almost exactly 180° from the 86 min site where bidirectional chromosome replication in *E. coli* originates (*Figure 2-5*). It has been proposed, therefore, that the 33 min DNA might attach to the membrane at the time of replication and that the late replication of this DNA somehow provides the signal for mesosome in-growth and cell division (*Section 2.2*).

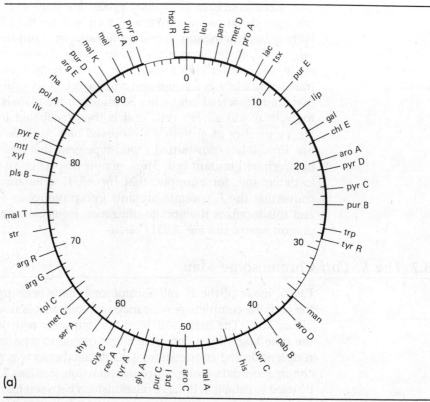

(a)

**FIGURE 13-6 E. coli map.** (a) Circular reference map of *E. coli* K-12. The large numbers refer to map position in minutes, relative to the *thr* locus. The 52 loci were chosen on the basis of greatest accuracy of map location, utility in further mapping studies, and/or familiarity as long-standing landmarks of the *E. coli* K-12 genetic map. The two thin portions of the circle represent the only two map intervals that are not spanned by a continuous series of P1 cotransduction linkages. (b) Linear scale drawings (next pages) representing the circular linkage map of *E. coli* K-12. The time scale of 100 min, beginning arbitrarily with zero at the *thr* locus, is based on the results of interrupted conjugation experiments. Parentheses around a gene symbol indicate that the location of that marker is not well known, sometimes having been determined only within very wide limits. An asterisk indicates that a marker has been mapped more precisely but that its position with respect to adjacent markers is not known. Arrows above genes and operons indicate the direction of messenger RNA transcription of these loci. Genetic symbols are defined in B. J. Bachmann, and K. B. Low, *Microbiol. Rev.* **44:**1, 1980.

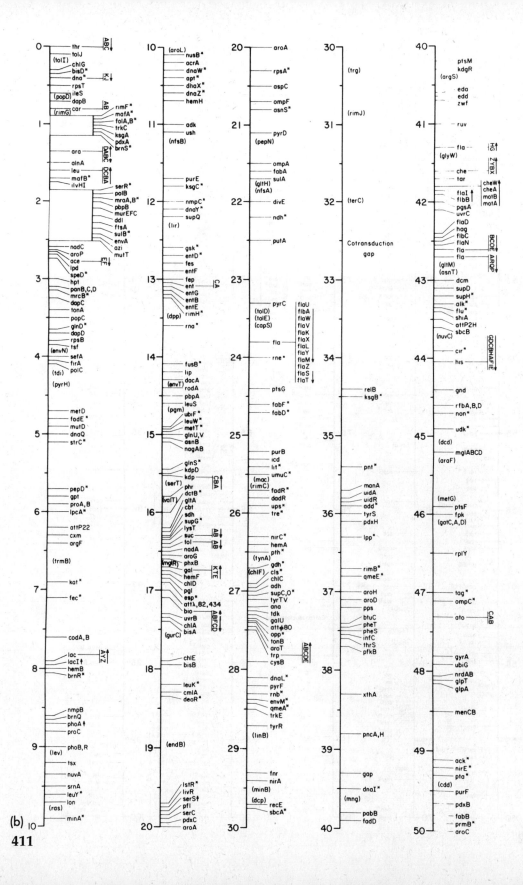

(b)

411

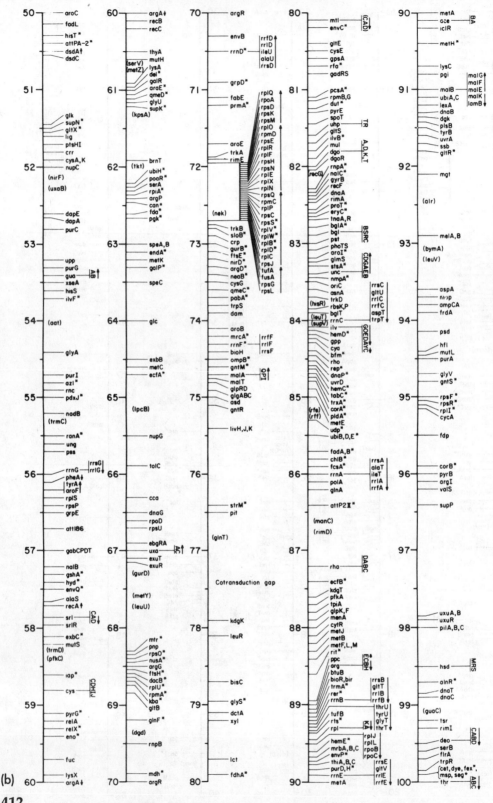

# Bacterial Transformation

## 13.8 Molecular Overview of Bacterial Transformation

Transformation was described in Chapter 1 in the context of classical experiments with *Diplococcus pneumoniae* (pneumococcus), where DNA isolated from *smooth* (virulent) cells was taken up by *rough* (nonvirulent) cells and incorporated in such a way that information for *smooth* capsule formation was transmitted to progeny bacteria (*Figure 1-12*). Transformation is formally analogous to conjugation: in both situations, **donor** DNA is transferred to a **recipient** or **host** cell, where it undergoes genetic exchange with the host chromosome to produce a recombinant bacterium. The processes are distinct, however, in that transforming DNA passes from donor to recipient in a free exposed state rather than through the shield of a conjugation tube. Moreover, in transformation, the recipient cell ordinarily receives random fragments of donor chromosomes, in contrast to the orderly transmission of a fixed gene sequence during conjugation.

A number of bacteria, including *Streptococcus pneumoniae, Hemophilus influenzae,* and *Bacillus subtilis,* are readily induced to undergo transformation in the laboratory, and at least low levels of transformation probably take place in their natural habitats. A bacterium such as *E. coli,* on the other hand, can undergo transformation only under special laboratory conditions. The cells must carry mutations abolishing exonuclease I and V activity; the cells must be treated with highly concentrated $CaCl_2$ solutions to render their membranes permeable to DNA; and they must be exposed to very high concentrations of donor DNA. In the paragraphs that follow, transformation is described as it is believed to occur in the more "natural" situations; the *in vitro* transformation of *E. coli* has, however, important applications for genetic research.

In any large population of bacteria such as *D. pneumoniae,* a few cells are probably always present that are capable of being transformed. A variety of laboratory procedures have, however, been developed to maximize the presence of transformable or **competent** cells so that transformation proceeds with high efficiency. These procedures include growing cells under particular culture conditions, harvesting them at a particular phase in their growth cycle, and so forth. Competent cells appear to elaborate a surface protein called **competence factor,** which is involved in an energy-dependent binding of donor DNA fragments to the cell surface.

DNA fragments from the donor cell must also be in a particular state if high-efficiency transformation is to occur: they must be large, relatively intact, and double-stranded, having a molecular weight in the 0.3 to $8 \times 10^6$ dalton range. The requirement for duplex DNA presumably relates to the way in which donor DNA enters the cell: as the end of a DNA fragment crosses the membrane, an intracellular DNase hydrolyzes one of its strands. The energy for pulling in the intact strand subsequently derives from the hydrolysis of the degraded strand. Mutant strains of *D. pneu-*

*moniae* that lack this DNase activity (also called **DNA translocase**) cannot be transformed.

Once single-stranded DNA fragments have successfully been pulled into a cell, they are capable of inserting into homologous regions of the host chromosome. Obviously, any one piece of DNA will be homologous only to a certain segment of the recipient cell genome, and most of the pulled-in pieces will not be genetically marked. If, however, donor DNA derives from an erythromycin-resistant $ery^r$ strain and the host strain is erythromycin-sensitive ($ery^s$), then an occasional host cell will pull in a donor DNA fragment containing a copy of the $ery^r$ gene. If this donor DNA succeeds in inserting into the $ery^s$ locus of the recipient, then a transformed, erythromycin-resistant cell will emerge. Models for how such insertions might take place are presented in Section 17.2; here we are concerned with how transformation is used to establish genetic linkage.

## 13.9 Establishing Gene Linkage by Transformation

To establish gene linkage by transformation the following considerations are made. If two genes, $E$ and $F$, are so distantly linked that the probability is remote that both will ever be included within the same DNA fragment, then double transformants—cells transformed for both $E$ and $F$—can arise only when a cell happens to receive and integrate two separate pieces of DNA, one carrying gene $E$ and the other gene $F$. On the other hand, if $E$ and $F$ are closely linked, then double transformants can arise either because, as before, a cell receives two separate DNA fragments, one with $E$ and the other with $F$, or else because a cell receives a single fragment carrying both $E$ and $F$.

An experiment is therefore constructed in which a culture of competent cells carrying markers $E$ and $F$ is divided among several flasks and each subculture is presented with progressively more dilute concentrations of DNA fragments derived from donors carrying $E$ and $F$. The recipient cells from each flask are then plated and the numbers of single ($E$ or $F$) and double ($EF$) transformants are scored. Looking first at the single transformants, one finds, as expected, that as the concentration of DNA is decreased, fewer and fewer single transformants appear among the progeny, with the probability of transformation being about the same for either marker. This result is shown in Figure 13-7, Curve A. Looking next at double transformants, one predicts that if the genes $E$ and $F$ are linked, the concentration dependence for producing double transformants (Figure 13-7, Curve B) should have approximately the same slope as the concentration dependence for single transformants (Curve A). If, on the other hand, $E$ and $F$ are unlinked, then, as DNA concentrations decrease, the occurrence of double transformations will decrease with a much greater slope (Figure 13-7, Curve C), since the occurrence of such double transformations will be totally dependent on the rapidly decreasing probability that

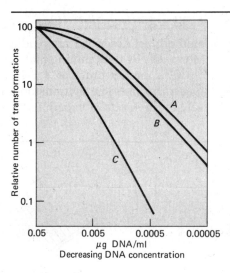

**FIGURE 13-7 Transformation.** Idealized curves illustrating the effect of decreasing concentration of transforming DNA on the relative number of single and double transformants for markers $E$ and $F$. Curve A: Number of single transformants of either the $E$ or $F$ type. Curve B: Number of double transformants of type $EF$ where $E$ and $F$ are closely linked. Curve C: Number of double transformants of type $EF$ when $E$ and $F$ are unlinked and therefore carried by different DNA fragments. (From W. Hayes, *The Genetics of Bacteria and Their Viruses*, 2nd ed. New York: John Wiley, 1968.)

two appropriate fragments will enter the same cell. You will recall from the probability concepts we considered in Section 12.10 that if two events occur independently of one another, the frequency at which they can be expected to occur simultaneously will be the product of their separate frequencies. This same consideration underlies the **principle of establishing linkage by transformation.** The frequency of transformation for gene $E$ alone and for gene $F$ alone is first established. These frequencies are then multiplied together to determine the theoretical frequency at which double $EF$ transformants should arise **if** $E$ transformations and $F$ transformations are independent events. One then examines the frequency with which $EF$ transformants in fact arise. If they arise with the same low frequency as calculated for the theoretical situation, then the genes are considered unlinked. If the number of $EF$ transformants instead greatly exceeds such expectations and in fact approaches the number of individual transformants, then the hypothesis that $E$ and $F$ transformations are independent events is clearly incorrect. Instead, the two genes must often participate in the same transformation event, meaning that they must be closely linked.

Obviously this approach depends on the assumption that the DNA molecules in the transformation mixture have a known, fairly uniform size. When, for example, the mean molecular weight of transforming DNA is $5 \times 10^6$, we may find that genes $E$ and $F$ are linked, whereas gene $G$ is not linked to either $E$ or $F$. If the DNA is then isolated more gently and larger fragments are acquired, linkage between $G$ and either $E$ or $F$ may be observed. A comparison of such overlapping data often allows relative intergenic distances to be deduced.

## 13.10 Transformation Mapping Using *Bacillus Subtilis*

The transformation method just described can readily define whether two markers are linked. Moreover, by determining that marker *A* is linked to *B* and that *B* is linked to *C*, a marker order of *A-B-C* can be deduced. The ordering of genes along a chromosome by transformation proceeds much more rapidly, however, in the kinds of experiments we now describe for *Bacillus subtilis*.

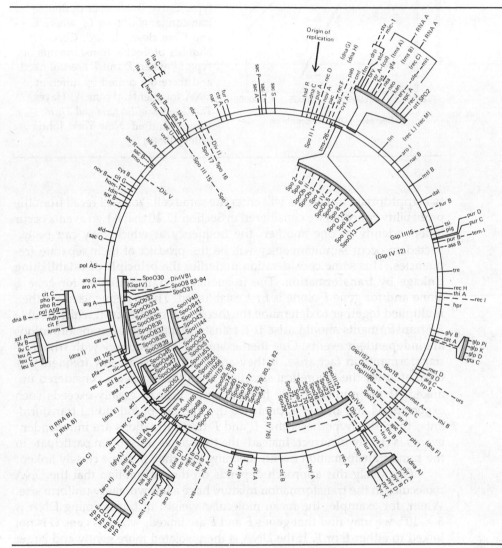

**FIGURE 13-8 B. subtilis map.** Distances are based on transduction frequencies with phage PBS1. (From F. E. Young and G. A. Wilson, in *Spores*, VI, P. Gerhardt, R. N. Costilov, and H. L. Sadoff, Eds. Washington, D.C.: American Society of Microbiologists, 1975.)

The initiation of the chromosome replication cycle in *B. subtilis* can be **synchronized**—meaning that every chromosome in a culture of cells begins being replicated at the same time—by procedures developed by N. Sueoka, H. Yoshikawa, and colleagues. Cells can, for example, be grown in a deuterated ($D_2O$) medium and then transferred to an aqueous ($H_2O$) medium; the transfer itself stimulates a synchronous onset of DNA replication. Such methods of initiating synchronous chromosome replication have been combined with transformation in the following way: the heavy (deuterated) DNA is induced to replicate in light ($H_2O$) medium so that the resulting daughter DNA duplexes are hybrid in density (*Section 2.3*). At intervals during the synchronous replication cycle, cells are removed from the culture and their DNA is extracted, fragmented, and subjected to CsCl density centrifugation (*Box 2.1*) to separate the hybrid DNA fragments from the heavy, unreplicated material. The fragments containing newly replicated DNA can then be used to transform competent recipient cells. By choosing recipient cells that carry numerous auxotrophic markers and donor cells carrying their prototrophic alleles, the genes that are included in the newly replicated DNA at each sampling time can be identified.

The resulting data indicate that synchronous chromosome replication in *B. subtilis* begins at a unique position, such that the gene *purA*, involved with adenine synthesis, appears in early samples of hybrid DNA. At the other extreme the *metB* gene, involved with methionine synthesis, does not appear among the hybrid transforming fragments until the replication cycle is almost complete, thus localizing the *met* gene near the terminus of the replicating chromosome. The time of appearance of intermediate genes can also be scored and a map constructed in which the map units relate marker positions relative to the *met* terminal marker. Thus it is possible not only to establish by transformation that certain markers must be closely linked to one another, but also to map an entire chromosome.

Figure 13-8 shows a recent map of the *B. subtilis* chromosome; the map is seen to be in the form of a single circular linkage group. The relative marker order was established using transformation, but many of the details of the map were provided by transduction analysis, the topic of the next sections.

# Generalized Transduction

## 13.11 Molecular Overview of Generalized Transduction

Generalized transduction is one of three types of transduction known to occur in bacteria; the other two (specialized and F-mediated) are described in subsequent sections. During generalized transduction, a piece of a donor bacterial chromosome becomes incorporated into a phage head and is transduced (carried over) to a recipient bacterium, where it may insert into the recipient genome, much as a DNA fragment becomes incorporated during transformation.

Generalized transduction is performed by phages such as P1 of *E. coli*, P22 of *Salmonella*, and SP01 and PBS1 of *B. subtilis*. Toward the ends of their lytic cycles, while phage chromosomes are being packaged into protein coats, an occasional piece of host DNA, usually of the same length as a phage chromosome, becomes mistakenly packaged in a phage coat and is liberated along with the other phage progeny in the burst. The bacterial DNA in this **transducing particle** can be efficiently injected into a second bacterium, and, since it shares a homologous sequence with a portion of the host chromosome, it occasionally inserts into the host chromosome. When the inserted DNA carries donor genes that differ from recipient markers, recombinants will result.

We might pause to ask why it is that *all* bacteriophages do not mediate generalized transduction; that is, why bacterial rather than phage DNA is not mistakenly packaged within, for example, the coats of wild-type T viruses. The answer appears to be a physiological one. Some generalized transducing phages may be less selective about the DNA incorporated into protein coats. Others may cause the DNA of the host to break into fragments of the appropriate size at just the right time in the transducing phage's lytic cycle. In the case of P22, this is apparently not a matter of chance: a phage-specified protein, coded by gene 3 and most likely the endonuclease responsible for cutting P22 concatamers into "headful" lengths during virion maturation (*Section 12.1*), apparently recognizes particular signals (base sequences?) in the *Salmonella* chromosome as well and cuts these into "headful" lengths. As a result, P22 will transduce certain *Salmonella* markers far more frequently than others, these transduced markers presumably lying between preferred cutting sites. **High-transducing (HT)** mutants of P22 also package phage-length pieces of *Salmonella* DNA into transducing particles, but these now contain a random collection of markers: the marker-specificity of wild type is almost completely abolished. The mutations carried by HT strains, therefore, apparently alter the cutting specificity of the phage endonuclease.

## 13.12 Mapping by Generalized Transduction

Generalized transducing particles are formed quite infrequently. In the analysis of transducing data, therefore, it is not necessary to be concerned (as one is with transformation) about the possibility that an apparent linkage of two markers has arisen because a cell has become doubly recombinant for two separate pieces of DNA. Even if a given bacterium is infected with a multiplicity of infection as high as ten, the probability that one of the infecting particles will prove to be a transducing particle, and that this particle will generate a singly recombinant cell, is very small. Therefore, the probability that two of the ten infecting particles are effective transducing particles becomes as low as 1 in $10^8$, which is of the same order as the probability that one of the genetic markers being followed will undergo a reversion.

It follows that direct inferences can be drawn from generalized trans-dution data: **if two markers are cotransduced with high frequency, they are considered to be closely linked, whereas if two markers are never cotransduced, they are probably separated by a length of DNA that is at least as long as one phage chromosome.** This length is equivalent to perhaps $6 \times 10^7$ daltons of DNA for phage P1 and as much as $1.7 \times 10^8$ daltons for phage PBS1; in contrast, the average DNA fragment in a transformation preparation is in the $0.3 \times 10^6$ to $8 \times 10^6$ dalton range. Thus generalized transduction proves to be not only a more direct approach to establishing genetic linkage than transformation but also a means of examining linkage between both closely-linked and more widely separated markers.

When two markers are followed during generalized transduction as described above, a **two-factor transduction** is said to take place. Such crosses, as they are called, permit the establishment of **relative cotransduction frequencies,** and if marker $A$ is cotransduced with $B$, $B$ is cotransduced with $C$, and $A$ is not cotransduced with $C$, it is concluded that the markers assume an $A$-$B$-$C$ order along the chromosome. A far more informative cross is provided by a **three-factor transduction** that, like the three-factor crosses described in Section 12.10 for bacteriophages, allows both marker order and map distance to be established directly.

As an example of a three-factor transduction cross we can cite a typical experiment in which the donor *E. coli* cells have the genotype $a^+b^+c^+$ and the recipients have the genotype $a^-b^-c^-$. The donor is infected with P1, the P1 progeny are used to infect recipient cells, and the recipient cells are then plated and subjected to selection for the presence of *one* of the donor markers (for example, the presence of $a^+$). Such recombinant recipient cells are said to have been transduced for at least one marker, and they are now tested by replica plating to see whether they are transduced (recombinant) for one or both of the other markers as well.

It is here that we apply reasoning similar to that for three-factor crosses of bacteriophages: **the rarest class of transductants should represent the most unlikely transduction event,** and the most unlikely event is for a recipient marker to be flanked by two donor markers. This requires the occurrence of four crossovers, as shown in Figure 13-9. Therefore, if *abc* is, in fact, the correct gene order, then $a^+b^-c^+$ cells should represent the rarest class of transductants; if *bac* is the correct gene order, then $b^+a^-c^+$ cells should be the rarest; and so on.

**Three-factor transduction data can supply cotransduction frequencies as well as gene order; to derive these frequencies, the total number of times the selected marker appears with an unselected marker in the recipient cells is calculated.** To give some actual data, an experiment was performed by E. Signer, J. Beckwith, and S. Brenner with *trpA$^+$ supC$^+$ pyrF$^+$* donor cells and *trpA$^-$ supC$^-$ pyrF$^-$* recipient cells of *E. coli*, where *trpA* is a

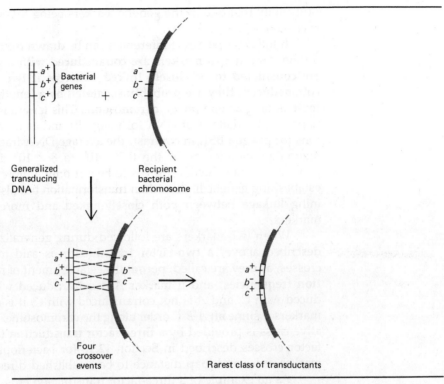

**FIGURE 13-9 Formation of a rare transduced bacterium by four crossovers** between the DNA of a transducing particle and a bacterial chromosome.

gene involved in tryptophan biosynthesis, *supC* is an *ochre*-suppressor gene (*Section 11.17*), and *pyrF* is a gene involved in pyrimidine biosynthesis. P1-mediated transductants for *supC⁺* were initially selected. These transductants could be classed as follows:

|  | Number |
|---|---|
| 1. *supC⁺ trpA⁺ pyrF⁺* | 36 |
| 2. *supC⁺ trpA⁺ pyrF⁻* | 114 |
| 3. *supC⁺ trpA⁻ pyrF⁺* | 0 |
| 4. *supC⁺ trpA⁻ pyrF⁻* | 453 |
|  | 603 total |

The marker order is at once recognized as being *supC trpA pyrF*, based on the genotype of the rarest class of transductants (class 3). The *supC⁺* and *trpA⁺* markers are both transduced in classes (1) and (2) but not in (3) or (4); therefore, the *supC-trpA* cotransduction frequency is calculated as 36 + 114/603 = 150/603 = 0.25. Similarly, the cotransduction frequency for *supC⁺* and *pyrF⁺* is seen to be equivalent to the class (1) frequency, namely, 36/603 = 0.06.

Cotransduction frequency is inversely related to the distance between genes. Thus, if two markers are close together, they will almost always be transduced together and their cotransduction frequency will approach one. If, on the other hand, two markers are never or almost never included in the same piece of transducing DNA, their cotransduction frequency will approach or be zero. A mathematical expression of this relationship is given as

$$d = L(1 - \sqrt[3]{x})$$

where $d$ = the physical distance between two markers on a chromosome,
$L$ = the average length of transducing DNA, and
$x$ = the cotransduction frequency of two markers.

In other words, when the average length of the transducing particle's DNA is known, the molecular distances separating nearby genes can be estimated. When a value for $L$ is not known, it can simply be considered as some constant.

As stated earlier, the kind of resolution that can be achieved with generalized transduction in *E. coli* is difficult to achieve with interrupted mating experiments; on the other hand, a generalized transduction experiment is most easily designed when one has some idea of the general location of a given mutation. In practice, therefore, when a new mutant strain is isolated in *E. coli*, its mutant gene is introduced into *Hfr* strains and its general chromosomal position is determined by conjugation. Its position is then pinpointed by generalized transduction using appropriately marked donor and recipient strains.

# Specialized Transduction

## 13.13 Transduction of Markers to Host Bacteria

The formation of specialized transducing particles was described in detail in Section 12.14, where we considered how λdgal phage chromosomes could be used in deletion-mapping experiments to order λ genes. We now re-direct our attention to the extra **bacterial** genes carried by specialized transducing phages and consider how these are used to analyze **bacterial** chromosomes. If a λ phage lysate containing λdgal$^+$ phages is used to infect gal$^-$ bacteria, there arise transductants that are able to ferment galactose. These transductants are almost invariably lysogenic for λ, and, when they are examined closely, it appears that they are formed in a unique fashion. The entering λdgal$^+$ DNA first circularizes (Figure 13-10a) as in a normal phage λ infection (*Figure 12-4*). Its gal$^+$ region then pairs with the homologous gal$^-$ region in the recipient chromosome (Figure 13-10b), but the donor genes do not displace the recipient genes (in contrast to the situation in transformation or in generalized transduction). Instead, recombination

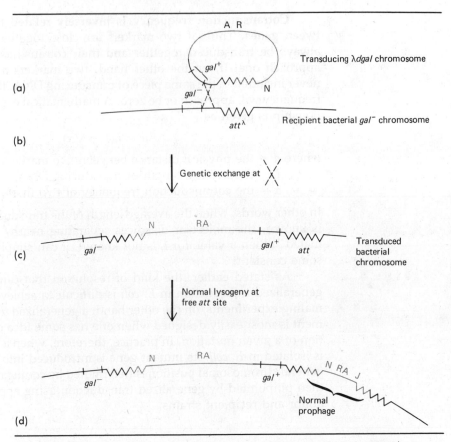

**FIGURE 13-10 Specialized transduction** of a *gal⁻* bacterium by a λ*dgal⁺* transducing particle. Bacterial DNA is represented in black and phage DNA in gray. Zigzag lines represent the *att* DNA normally involved in phage integration and excision (*Sections 12.14 and 18.1*).

occurs such that the λ*dgal⁺* DNA is inserted into the site where pairing took place (Figure 13-10c). Meanwhile, the usual λ attachment site (*att*λ) in the *E. coli* chromosome remains unoccupied, and it ordinarily becomes filled by a second, normal λ phage that infects the cell at the same time as the λ*dgal⁺* particle (Figure 13-10d). Thus the resulting transduced bacterium carries two copies of most (but not all) λ genes and two copies of one or several bacterial genes (Figure 13-10d).

With respect to the bacterial genes, the cell is a **partial diploid**, *gal⁺/ gal⁻*. It and its descendants are able to ferment galactose because the *gal⁺* gene is dominant to the *gal⁻* gene, meaning, in molecular terms, that the normal gene product carries out the activity that comes to define the phenotype, in this case the ability to ferment galactose.

When partial diploids of the sort illustrated in Figure 13-10d are treated with ultraviolet light to induce λ excision, the resulting lysate will contain approximately half normal phages and half λ*dgal* transducing phages. Such a lysate, known as a **high-frequency transduction (HFT)** lysate, will clearly be greatly enriched for bacterial genes involved in galactose utilization.

## 13.14 Mapping Bacteria by Specialized Transduction

Mapping bacteria by specialized transduction is conceptually the simplest of all approaches: **bacterial genes transduced by a specialized transducing phage are identified as closely linked to the bacterial *att* site for that phage.** Thus the *att*$^\lambda$ site at 17 min is identified as closely linked to the *gal* and *bio* genes; the *att*$^{\phi 80}$ site at 27 min is flanked by a cluster of *trp* genes on one side and two suppressor genes, *supC* and *supF*, on the other; and so on.

Selection procedures have also been devised to **place genes of interest next to known *att* sites** so that they become included in specialized transducing particles. Thus the genes of the *E. coli lac* operon, which are not normally linked to any identified *att* site, have been manipulated so that they lie next to the *att*$^{\phi 80}$ site (details are presented in Figure 13-11) or next to the *att*$^\lambda$ site. The resultant *lac*-transducing phages, known as **φ80plac** and **λplac** because they are not defective in phage genes and are therefore **plaque-forming, have played essential roles in the isolation and molecular analysis of the *lac* genes.

## 13.15 F-Mediated Sexduction

The F element of *E. coli* (*Section 13.1*) integrates into a chromosomal site specific for a given *Hfr* strain (*Figure 13-5*). Within any *Hfr* population certain cells are found to be $F^+$, meaning that they transfer F elements (and not main chromosomes) to $F^-$ cells. Such $F^+$ cells are created when integrated F elements undergo excision and resume their status as autonomous plasmids. On rare occasions an error occurs during excision and genes from the main bacterial chromosome become included in the F element, which is now called an **F′ element.** The F′ element is unlike a λ*d* particle in that rarely, if ever, is any of its own F genetic information deleted (perhaps because such deletions would invariably disallow sexual transfer). Instead, the F element is simply expanded in length by the number of main genes added to it. Since the F′ element does not become encased in coat protein, this extra length does not present a packaging problem, and the element continues its autonomous existence within the bacterial cell.

The original *Hfr* cell giving rise to an F′ element remains haploid and is usually viable; its genes are distributed somewhat differently, but all are still present in the cell in a single copy. When, however, its F′ element is infectiously transmitted to an $F^-$ cell of differing genetic constitution, the recipient cell and its descendants become **partial diploids** for the bacterial

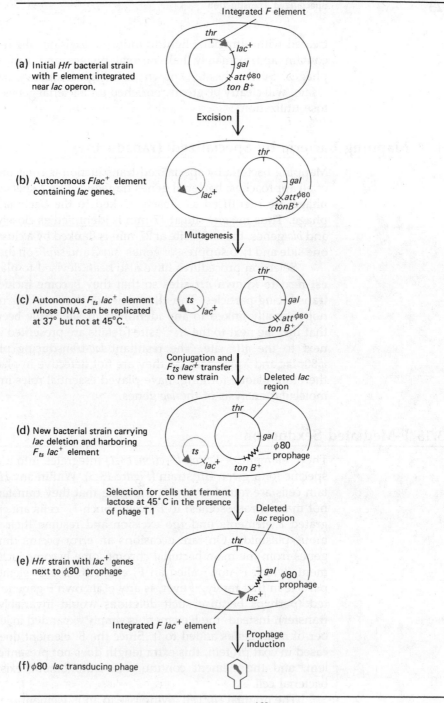

(a) Initial *Hfr* bacterial strain with F element integrated near *lac* operon.

Integrated *F* element

*thr*
*lac*⁺
*gal*
*att*φ80
*ton B*⁺

Excision

(b) Autonomous *Flac*⁺ element containing *lac* genes.

*thr*
*lac*⁺
*gal*
*att*φ80
*tonB*⁺

Mutagenesis

(c) Autonomous *Fₜₛ lac*⁺ element whose DNA can be replicated at 37° but not at 45°C.

*thr*
*ts*
*lac*⁺
*gal*
*att*φ80
*ton B*⁺

Conjugation and *Fₜₛ lac*⁺ transfer to new strain

Deleted *lac* region

(d) New bacterial strain carrying *lac* deletion and harboring *Fₜₛ lac*⁺ element

*thr*
*gal*
φ80 prophage
*ts*
*lac*⁺
*ton B*⁺

Selection for cells that ferment lactose at 45°C in the presence of phage T1

Deleted *lac* region

(e) *Hfr* strain with *lac*⁺ genes next to φ80 prophage

*thr*
*gal*
φ80 prophage
*lac*⁺

Integrated *F lac*⁺ element

Prophage induction

(f) φ80 *lac* transducing phage

**FIGURE 13-11 Placing *lac* genes next to att^φ80.** The two critical features of the selection step (d) are that (1) the temperature sensitivity of the *F lac*⁺ element's replication is lost once the plasmid is integrated into the host chromosome and (2) integration into the gene *tonB*⁺, a gene that confers sensitivity to phage T1, produces a bacterium resistant to phage T1.

genes introduced by the F' element. Continued cycles of replication and infection by F' elements can eventually convert an entire population into F' partial diploids. In addition, the bacterial genes in the F' element will not infrequently take part in exchanges with the recipient's chromosomal genes to produce true recombinants, a phenomenon commonly referred to as **sexduction** or **F-duction.**

Genetic studies of *E. coli* make frequent use of F' elements. Since different *Hfr* strains have different integration/excision sites for F elements, each will give rise to F' elements that carry different arrays of bacterial genes. For a given F' type, moreover, one can determine the frequency with which neighboring genes are "picked up." These frequencies are comparable to cotransduction frequencies and are used in constructing genetic maps in much the same way that cotransduction frequencies are used.

F' elements also allow the study of gene interactions in bacteria. One can ask whether a mutant gene is dominant or recessive to its wild-type allele by bringing them together in partial diploids. One can also use partial diploids to perform a bacterial complementation test, asking whether a mutant bacterial gene carried on the F' element can complement a mutation carried on the bacterial chromosome. Finally, one can ask whether a group of genes must be located in the same physical piece of DNA in order to be transcribed in a controlled manner by placing some genes in an F' element and others in the main chromosome. Such experiments with F' elements are described in Chapter 22.

# Plasmids

## 13.16 General Properties of Plasmids

Plasmids have already been considered in Section 4.13 as vectors utilized in recombinant DNA experiments. Here we consider the genetic properties of naturally occurring plasmids.

Plasmids are autonomous genetic elements that can be found in almost every known type of bacterial cell. Plasmids can have an independent, self-replicating existence quite separate from the main chromosome of the cell. Additionally, most plasmids can integrate into or out of the main chromosome, much like the temperate phage λ (*Section 12.2*); plasmids and temperate phages with such integration properties have come to be called **episomes.** All known bacterial plasmids exist within the cell as covalently closed duplex DNA rings (*Figure 2-2*).

Bacterial plasmids can be classified in several ways. One classification is based on the kind of specialized genetic information a plasmid carries. By this criterion there exist plasmids that carry **antibiotic resistance** genes and are known as **R plasmids,** plasmids that specify proteins known as

**TABLE 13-1   Some Properties Coded by Naturally Occurring Plasmids**

| Property | Exemplified By |
|---|---|
| Fertility—ability to transfer genetic material by conjugation | F, R1, Col1 |
| Production of bacteriocins | CloDF13 (*Enterobacterium cloacae*) ColE1 |
| Antibiotic production | SCP1 plasmid of *Streptomyces coelicolor* |
| Antibiotic resistance | R222 |
| Heavy metal resistance ($Cd^{2+}$, $Hg^{2+}$) | p1258 (*S. aureus*), R6 |
| Ultraviolet resistance | Col1b, R46 |
| Enterotoxin | Ent |
| Virulence factors, hemolysin K88 antigen | ColV, Hly |
| Metabolism of camphor, octane, and so on | Cam, Oct (*Pseudomonas*) |
| Tumorigenicity in plants | T1-plasmid of *Agrobacterium tumefaciens* |
| Restriction/modification | Production of *Eco* RI endonuclease and methylase by plasmid of RY13 |

Plasmids listed are indigenous to *E. coli* unless otherwise indicated. From S. Cohen, *Nature* 263:731, 1976.

**colicins** (to be described shortly) and are called **Col** plasmids, and so forth. Table 13-1 lists some of the properties encoded by naturally occurring plasmids.

A second way to think of plasmids is based on the fact that some are **conjugal** (also called **transmissible**) and others are **nonconjugal (nontransmissible).** The best known example of a conjugal plasmid is the F element, already described in Section 13.1, the genes of which dictate the formation of surface pili and the conjugal transfer of F-element DNA from $F^+$ to $F^-$ cells. It turns out that similar conjugal genes are present in certain R and Col plasmids as well, whereas they are absent in nonconjugal plasmids. Nonconjugal plasmids are therefore ordinarily restricted to an "asexual" parent $\longrightarrow$ daughter mode of transmission. In some cases, however, they may be **cotransferred** or **mobilized** along with particular conjugal plasmids when both of these elements are present in the same cell.

## 13.17 Detecting the Presence of a Plasmid

How can one determine genetically that a bacterial population carries plasmids? When the plasmid has infectious properties, its presence is readily detected by the rapid transmission of a certain trait or traits from cell to cell. These traits, moreover, will exhibit no linkage (in transduction experiments, for example) to any markers located in the main chromosome. Finally, in a clone of cells carrying plasmids, segregants that have lost their plasmids will often arise.

The loss of a plasmid can be spontaneous or induced. In some cases a copy of the plasmid may fail to be transmitted to a daughter cell. Alternatively, exposure to acridines, starvation for thymidine, or X-irradiation selectively abolish the replication of certain plasmids without affecting the replication of the main chromosome. Plasmid loss creates "segregant" clones that are viable but lack those traits determined by plasmid genes. That they have not arisen by mutation is clear from the fact that they arise with high frequency and lose many traits in a single event.

In the following sections we consider F elements, R plasmids, and Col plasmids, focusing first on the traits these plasmids confer on *E. coli* and then on the genetic specification of these traits.

## 13.18  F Element (F Sex Factor)

*E. coli* cells harboring an F or F' element are endowed with a number of phenotypic traits (*Section 13.1*): (1) They possess pili and can transfer their plasmid DNA to recipient ($F^-$) cells. (2) They are sensitive to infection by single-stranded RNA phages and certain single-stranded DNA phages (the so-called **male-specific phages,** since $F^+$ cells are sometimes called "male" *E. coli* ). (3) They resist the growth of such **female-specific phages** as T3 and T7. (4) They exclude the acquisition of additional F elements, first by a **surface-exclusion** (*Sfx*) of F-element uptake and second by an **immunity** mechanism that prevents the retention of any additional F elements that manage to enter. (5) They can be converted into *Hfr* cells when the F element inserts into the main bacterial chromosome.

This rich array of F-controlled traits has permitted the isolation of mutant F and F' elements; some, for example, carry mutations conferring male phage resistance whereas others carry *sfx*$^-$ mutations. Complementation analysis between various point mutations has been performed by forcing the formation of mutant$_1$ F/mutant$_2$ F diploids and looking for a restoration of conjugal function before immunity eliminates one of the F elements. Deletion mapping has been performed using F'*lac* deletion plasmids formed as a consequence of faulty excision (*Section 13.15*). Such analyses, initiated by M. Achtman, N. Willetts, and A. J. Clark and pursued by others, have led to the F element map shown in Figure 13-12.

The upper left quadrant of the F element map contains the *tra* (DNA *tra*nsfer) genes, which cluster together in several operons (Figure 13-12). Some of the *tra* genes specify pilus formation; others are more specifically involved with DNA transfer and surface exclusion. Additional genes on the plasmid include those concerned with incompatibility (*inc*) and replication (*rep, oriV*), and long expanses have yet to be mapped.

Within the unmapped expanses of the F element lie three loci labeled *IS2, IS3,* and γδ (Figure 13-12). These represent entities known as **insertion sequences (ISs),** and we will be considering their properties in detail in Section 18.4. Their relevance here is that **insertion sequences are thought to mediate the episome phenotype of the F element.** Thus the main bacterial chromosome contains integrated IS elements that are homologous to those present in F, and recombination between these sequences permits F integration. Each resultant *Hfr* cell proceeds to generate a clone carrying an F element at a particular chromosomal locus. Since different bacterial strains carry ISs at different locations and in various orientations, F ele-

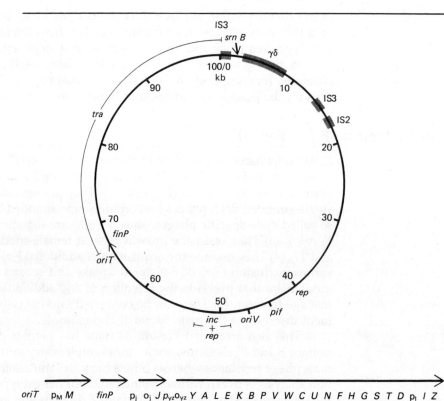

FIGURE 13-12. F element map of E. coli. Distances are given in kilobases (kb). Insertion sequences are shown as heavy lines. The *transfer* (*tra*) genes (below) are expressed as three independent units, one for *traM*, one for *traJ*, and one for the large operon *traY* ⟶ *traZ*. (From N. Willetts and R. Skurray, *Ann. Rev. Genet.* **14**:41, 1980.)

ments will exhibit a strain-dependent insertion behavior, as we have already concluded from genetic data (*Figure 13-5*).

## 13.19 R Plasmids

Conjugal R plasmids can be thought of as possessing two distinct types of information: genes for conjugal transfer and genes for antibiotic resistance. We consider these two groups separately and then consider how they interact with one another.

**Conjugal Transfer Genes**  Conjugal R plasmids contain genes that dictate both the synthesis of pili and the transfer of DNA. Therefore, for the purpose of discussion, we can say that a conjugal R (and Col) plasmids possess *tra* operons, although their component genes appear to differ somewhat from the *tra* genes of the F element. A major distinction between the F and

R plasmids is that the *tra* genes in most conjugal R (and Col) plasmids are highly repressed, so that naturally occurring $R^+$ strains transfer their plasmids at only very low frequency.

**Antibiotic-Resistance Genes**   The drug-resistance genes in R plasmids are symbolized differently than the comparable genes in the *E. coli* chromosome. Thus the chromosomal streptomycin-resistance gene is denoted *str-r* while the plasmid-borne gene is written Sm; other common plasmid genes include the gene for resistance to ampicillin (Ap), chloramphenicol (Cm), kanamycin (Km), sulfonamides (Su), and tetracycline (Tc). The chromosomal and plasmid resistance genes generally express their effects in different ways; for chromosomal genes, resistance is typically the result of an alteration in a ribosomal protein (*Section 7.6*), whereas the plasmid genes typically dictate the synthesis of enzymes that inactivate the antibiotics as they enter the cell. For instance, the Cm gene directs the synthesis of chloramphenicol acetyl transferase, an enzyme that adds an inactivating acetyl group to the chloramphenicol molecule.

Some R plasmids carry only one resistance gene whereas others carry two or more. These plasmids can be mapped by transduction with phage P22 or by restriction/sequencing, so that, for example, the R222 plasmid has been found to have the gene order Su-Sm-Cm.

**Transposons**   The antibiotic-resistance genes of R plasmids are usually contained within larger units known as **transposons (Tn).** Transposons are considered in detail in Section 18.5. For present purposes it is sufficient to know that transposons are short segments of DNA that lack the ability to self-replicate and persist because they insert themselves into a pre-existing chromosome or plasmid and are replicated along with the host DNA. As their name implies, transposons are also able to move (transpose) from one site of insertion to another, so that they can move from one plasmid to another, for example, or from one site to another within the same plasmid. A typical transposon carries one or more genes for antibiotic resistance and two inverted repeats (*Section 4.9*) at its termini; these appear to be essential for inserting the transposon into the recipient replicon, as we detail in Section 18.5.

**Transposon-Plasmid Interactions**   A critical property of R plasmids can now be appreciated, namely, that **they can pick up additional transposons.** Thus if a bacterium harbors an R plasmid carrying the transposon Tn10, which encodes tetracycline-resistance, and if it is coinfected by a plasmid carrying Tn3, which encodes ampicillin resistance, the two plasmids can interact such that **a copy of the Ap transposon is inserted into the Tc-carrying plasmid.** The bacterium will thus carry, and can transmit to its progeny, a Tc-Ap plasmid conferring simultaneous resistance to *two* antibiotics. Such events can of course be repeated so that multiresistance R plasmids are generated, an example of which is diagrammed in Figure 13-13.

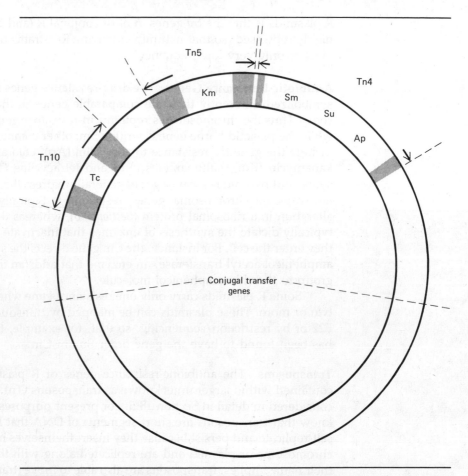

**FIGURE 13-13 Multiresistance R plasmid.** Three transposons are schematically represented within the plasmid: Tn10 carries a gene for tetracycline resistance (Tc), Tn5 a gene for kanamycin resistance (Km), and Tn4 genes for streptomycin (Sm), sulfonilamide (Su), and ampicillin (Ap) resistance. Each Tn is flanked by an inverted repeat, designated by the shaded regions and the arrows that point outward.

**R Plasmids and Public Health**  The mobility of transposons and the ability of a given R plasmid to "pile on" large numbers of resistance genes has major implications for pubic health, since R plasmids can be transmitted not only from cell to cell within a species but also across species lines. Thus, for example, *E. coli, Proteus,* and other nonpathogenic members of the intestinal flora often harbor R plasmids that have permitted the bacteria to survive high doses of antibiotics ingested by their hosts. These R plasmids can subsequently be transferred to infecting pathogenic bacteria such as *Salmonella, Shigella* (which causes dysentery), or *Haemophilus influenzae* type b (which can cause meningitis), so that these organisms also become resistant to the same spectrum of antibiotics.

Even more disturbing is the finding that such transfers take place freely in sewers and in polluted rivers. In Japan, where resistance statistics have been gathered, drug-resistant *Shigella* rose from about 0.2 percent in 1953 to 58 percent in 1965, with most apparently harboring R plasmids. In the same survey, 84 percent of *E. coli* and 90 percent of *Proteus* collected from hospital patients were similarly resistant to antibiotics. There is thus little question that the large-scale and indiscriminate application of antibiotics in agriculture and in medicine has created a pool of drug-resistant enterobacteria that can transmit their resistance genes to infecting pathogens and to other bacteria in polluted environments. Obviously, medical efforts to arrest infection are greatly impeded in such cases, and drug-resistant pathogens may create serious epidemics in such vulnerable locales as hospitals or crowded living areas. The situation would be improved if drug therapy were limited in the future to persons with acute cases of infectious disease and if the use of antibiotics in agriculture were modulated; in many countries of the world, however, such limits are unfortunately not in force at present.

## 13.20 Col Plasmids

Bacterial cells that harbor **Col plasmids** synthesize proteins known as **colicins** that act specifically to kill *E. coli* cells. Such bacteria also synthesize **immunity proteins** that render the Col-harboring cells insensitive to the bactericidal effects of the colicins they produce; this explains how *E. coli* cells are able to carry Col plasmids in the first place. Plasmids specifying other bacteriocins are also known; for example, a **cloacin** encoded by the CloDF3 plasmid is toxic to *Enterobacterium cloacae*; **vibriocins** are directed against *Vibrio cholerae*, and so forth. Within the Col family, moreover, a large spectrum of colicin species is known—colicin lb, colicin E1, and so on—each specified by a distinct Col plasmid.

Of the colicins, E2 and E3 have been most extensively studied. Both are physically similar proteins and compete for the same receptor on their target bacteria; however, their mode of killing is quite different. E2 proves to be an endonuclease that degrades the cell's DNA when it enters a nonimmune target cell. E3, on the other hand, brings about the cleavage of a fragment from the 16S rRNA of *E. coli* ribosomes, thereby destroying the ability of the target cell to synthesize protein.

The sector of a conjugal Col plasmid coding for the colicin and immunity proteins is separable from its conjugal transfer component in much the same way as we saw for R plasmids. This sector can, moreover, pick up antibiotic-resistant genes contained within transposons, and thus confer both resistance and colicinicity to host cells.

# Approaches to Solving Mapping Problems

## 13.21 Transduction Problem

**Problem**   In a P1 transduction experiment, the donor bacteria were $synP^+$ $supM^+$ $trpZ^+$; the recipients were $synP^-$ $supM^-$ $trpZ^-$. Initial selection was for $supM^+$ transductants.

$$48 \text{ of these were } M^+ P^+ Z^+ \ (1)$$
$$120 \text{ of these were } M^+ P^+ Z^- \ (2)$$
$$500 \text{ of these were } M^+ P^- Z^- \ (3)$$
$$0 \text{ of these were } M^+ P^- Z^+ \ (4)$$

a. What is the marker order?
b. What is the cotransduction frequency between $M$ and $P$? $M$ and $Z$?

### Solution

a. Recognize the rarest class, (#4). The $P^-$ recipient marker is in the middle in the order written; the marker order is therefore $M \ P \ Z$.
b. Grand total = 48 + 120 + 500 = 668 transductants.

c. Cotransduction of $M$ and $P = \dfrac{48 + 120}{668} = 0.25$

d. Cotransduction of $M$ and $Z = \dfrac{48}{668} = 0.072$

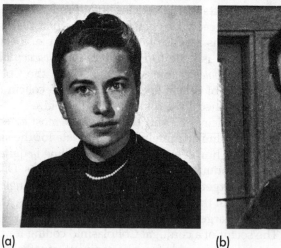

(a)                                                   (b)

(a) Harriet Ephrussi-Taylor performed some of the early bacterial transformation experiments indicating linkage between genetic markers.
(b) Jonathan Beckwith (Harvard Medical School) has utilized transduction and sexduction to elucidate many features of *E. coli* genetics.

# Questions and Problems

1. Given the map of the six deletion mutants shown below, predict the results of crosses involving the five point mutants (a through e) with the six deletions (1 through 6). (Indicate + for recombination, 0, no recombination, in the table.)

(deletions)

|  | 1 | 2 | 3 | 4 | 5 | 6 |
|---|---|---|---|---|---|---|
| a |  |  |  |  |  |  |
| b |  |  |  |  |  |  |
| c |  |  |  |  |  |  |
| d |  |  |  |  |  |  |
| e |  |  |  |  |  |  |

(point)

2. A **complementation map** displays complementation relationships in the same fashion as a deletion map. Construct a complementation map from the following data derived from pairwise mixed infections of phage strains 1 to 5. Some or all of the strains may carry deletions. + for complementation, 0 for no complementation.

|  | 1 | 2 | 3 | 4 | 5 |
|---|---|---|---|---|---|
| 1 | 0 | 0 | 0 | + | + |
| 2 |  | 0 | 0 | 0 | 0 |
| 3 |  |  | 0 | + | 0 |
| 4 |  |  |  | 0 | 0 |
| 5 |  |  |  |  | 0 |

3. The following groups of genes can be cotransduced by phage P1.
   1. A, H, and L
   2. H and S
   3. L and O
   4. M and O
   What is the order of the markers? Explain your logic.

4. Two labs calculated cotransduction frequencies for the genes $A$ and $B$. Lab A found 0.63, the lab B found 0.47. Which lab reported the genes to be closer together?

5. In a generalized transduction experiment, donor *E. coli* cells have the genotype $trpC^+$ $pyrF^-$ $trpA^-$ and recipient cells have the genotype $trpC^-$ $pyrF^+$ $trpA^+$. P1-mediated transductants for $trpC^+$ were selected, and their total genotypes were determined with the following results:

| Genotype | | | Number of Progeny |
|---|---|---|---|
| $trpC^+$ | $pyrF^-$ | $trpA^-$ | 274 |
| $trpC^+$ | $pyrF^+$ | $trpA^-$ | 279 |
| $trpC^+$ | $pyrF^-$ | $trpA^+$ | 2 |
| $trpC^+$ | $pyrF^+$ | $trpA^+$ | 46 |

(a) Determine the order of the three markers. (b) Determine the cotransduction frequencies for $trpC$ and $pyrF$ and for $trpC$ and $trpA$. (c) Calculate physical map distances between these markers assuming the P1 chromosome to be $10\mu$ in length. (d) Calculate similarly the map distances between the $supC$, $trpA$, and $pyrF$ markers from the data in the text and construct a map for the region. (For [c] and [d], you will need to know the following: $\sqrt[3]{0.92} = 0.973$; $\sqrt[3]{0.46} = 0.772$; $\sqrt[3]{0.06} = 0.391$; $\sqrt[3]{0.25} = 0.63$.)

6. Phages $\phi80$ and $\lambda$ are very closely related, such that in dissociation-reassociation experiments, the H strands of $\phi80$ (*Box 4.1*) will form heteroduplex hybrids with the L strands of $\lambda$ and vice versa. A $\lambda lac$ transducing phage was constructed using techniques similar to those diagrammed in Figure 13-11 for making $\phi80lac$ transducing phage. Each phage was shown to carry the contiguous $lacI$, $lacZ$, and $lacY$ genes of the $lac$ operon, but the sense strand (*Section 9.4*) for those genes was found to be the H strand for $\phi80lac$ and the L strand for $\lambda lac$.
   (a) Give a plausible explanation why the genes are encoded on different strands in the two transducing phages (Hint: review Figure 13-5). Do you expect the two sequences to be in the same or opposite orientation? Explain.
   (b) Draw the structure of the H strands of the two phages, showing the $lac$ genes (or their antisense sequences) in the middle of the strands and the terminal phage genes $A$ and $B$ (*Figure 12-8a*) at the 5' and 3' ends respectively.
   (c) Draw the structure of the hybrid that forms when these two H strands are allowed to anneal. Show how single-stranded nuclease treatment of this hybrid would allow you to isolate "pure" $lac$ DNA. (This experiment was in fact performed by J. Beckwith, J. Shapiro, and their colleagues in 1969, the first "isolation of a gene".)

7. A transducing phage carrying the closely linked genes $A^-B^+C^+$ was allowed to infect a recipient *E. coli* with the genotype $A^+B^-C^-$, and wild-type *E. coli* recombinants ($A^+B^+C^+$) were selected. The reciprocal cross ($A^+B^-C^-$ phages and $A^-B^+C^+$ recipients) was also performed, and $A^+B^+C^+$ wild-type recombinants were also selected. The number of wild-type recombinants proved to be equivalent in both crosses. What does this tell you about the possible order of the three genes?

8. Two bacterial strains were obtained with the following genotypes: Hfr, $arg^-$, $leu^+$, $azi$-s, $str$-r and $F^-$ $arg^+$ $leu^-$ $azi$-r $str$-s. You want to mate the two strains and after conjugation, detect and enrich for the $F^-$ recombinant genotype $arg^+$ $leu^+$ $azi$-r. Which of the following media will accomplish this selection? Explain why the remaining media will not work.
   (a) Minimal media with streptomycin.
   (b) Minimal media with leucine and sodium azide.
   (c) Minimal media with sodium azide.
   (d) Enriched media without arginine or leucine, with streptomycin.
   (e) Minimal media with streptomycin and sodium azide.

9. Suppose a strain of E. coli is found which only grows in minimal media when supplemented with methionine. When sexduced with an F' element which is known to carry $ara^+$ $leu^+$ $azi$-$r$ markers, the original strain can grow on unsupplemented media.
   Explain these results. Outline how you would test whether this explanation is correct.

10. Construct a complementation map from the following data derived from pairwise mixed infections of phage strains 1 to 5. Some or all of the strains may carry deletions. + for complementation, − for no complementation.

|   | 1 | 2 | 3 | 4 | 5 |
|---|---|---|---|---|---|
| 1 | 0 | + | + | 0 | 0 |
| 2 |   | 0 | 0 | 0 | + |
| 3 |   |   | 0 | + | 0 |
| 4 |   |   |   | 0 | + |
| 5 |   |   |   |   | 0 |

11. How is the transfer of F similar to phage infection? How is it different?

12. A new mutation is introduced into an Hfr strain so that its general position can be determined by interrupted mating. Outline how you would carry out these steps.

13. The order of markers transferred by four Hfr strains to $F^-$ cells is as follows:

| Hfr Strain | Order of Markers Donated |
|------------|--------------------------|
| 1 | Q  W  D  M  T  → |
| 2 | A  X  P  T  M  → |
| 3 | B  N  C  A  X  → |
| 4 | B  Q  W  D  M  → |

   Draw a map as in Figure 13-5 showing the sites and orientation of insertion of the F element in these four strains.

14. An Hfr strain of E. coli (A) transfers the gal operon early and with high efficiency and the lac operon late and with low efficiency to a $gal^-$ $lac^-$ $F^-$ strain (B) in an $A \times B$ cross. $Gal^+$ recombinants of strain B remain $F^-$. It is possible to isolate from strain A a variant called C that transfers the lac operon but not the gal operon early and with high frequency to strain B. In a $C \times B$ cross, $lac^+$ cells of strain B are generally $F^+$. What is the nature of strain C? How would you design experiments to isolate such a strain?

15. Compare the partial zygotes that result from $Hfr \longrightarrow F^-$ conjugation with the partial diploids that result from $F' \longrightarrow F^-$ conjugation.

16. An F' element including the maltose (mal) gene is discovered in E. coli. The episome marked with $mal^+$ is introduced into a $mal^-$ $F^-$ strain. Most of the resulting cells transfer the original episome to other $F^-$ cells. However, occasional cells appear which transfer the entire E. coli chromosome, beginning with the maltose gene. These cells fall into two types: (a) those that transfer

$mal^+$ very early and $mal^-$ very late and (b) those that transfer $mal^-$ very early and $mal^+$ very late.

Draw a diagram of the interaction between the episome and the chromosome that shows how cells of type (a) are most likely to have arisen. Do the same for the case of cells of type (b).

17. Genes *purB* and *purC* of *B. subtilis* are shown to be linked in transformation experiments. Outline three experimental approaches that could be used to demonstrate whether or not these genes are linked to the marker *lin* (*lincomycin-resistant*, lincomycin being an antibiotic).

18. Phage 105 is a lysogenic phage of *Bacillus subtilis*. It has recently been discovered that the gene order of the prophage map is identical to the gene order of the linear DNA in a phage 105 particle.

    (a) What does this finding imply about the location of the termini of the mature phage DNA and the prophage attachment site?

    (b) Why has it not been possible to isolate specialized transducing phage of phage 105?

# 14

# Mapping Eukaryotic Chromosomes in Sexual Crosses

**A. Introduction**

**B. Classical Studies on Linkage and Recombination**
14.1 Partial Linkage
14.2 The Morgan Crosses and the Sturtevant Map
14.3 Correlating Chiasmata with Gene Recombination

**C. Mapping *Drosophila* in Sexual Crosses**
14.4 Two-Factor Crosses and the Test Cross
14.5 Three-Factor Crosses
14.6 Double Crossovers and Interference

**D. Cytological Mapping**
14.7 Deficiency Mapping
14.8 Cytogenetic Analysis of the *Zeste-White* Region of the *Drosophila* X Chromosome
14.9 Correlating Genetic and Cytological Maps

**E. Linkage Groups and Chromosomes**
14.10 Detecting Linkage Groups
14.11 Assigning Linkage Groups to Chromosomes

**F. Mapping by Tetrad Analysis**
14.12 Mapping Two Linked Genes by Tetrad Analysis
14.13 Mapping Three Linked Genes by Tetrad Analysis
14.14 Gene-Centromere Linkages
14.15 Mapping the *CYC1* Gene of Yeast

**G. Approaches to Solving Mapping Problems in Eukaryotes**
14.16 Multiple Markers in One Cross
14.17 Determining Phenotypic Classes

**H. Questions and Problems**

# Introduction

This chapter parallels the two preceding chapters on mapping viruses and bacteria, but we now focus on nucleate organisms whose chromosomes undergo meiotic segregation and independent assortment (*Chapter 6*) as well as crossing over. We first consider "classical" early experiments demonstrating linkage in eukaryotes and correlating the occurrence of genetic recombination with the occurrence of physical exchanges between chromosomes. We next examine how genetic maps are generated in organisms with a sexual cycle, focusing attention on *Drosophila* and *Chlamydomonas*. In Chapter 15 we go on to describe mapping procedures utilizing "parasexual" organisms and cells and conclude with a description of how human chromosomes are mapped *in vitro*.

# Classical Studies on Linkage and Recombination

## 14.1 Partial Linkage

The studies of Mendel (*Sections 6.1 to 6.3*) on the garden pea, published in 1866, had little impact on the scientific world until 1900 when C. Correns and H. de Vries independently rediscovered his paper in the course of their own investigations on quantitative aspects of gene transmission. At about the same time investigators such as T. Boveri, C. Correns, and particularly W. Sutton were investigating the behavior of chromosomes at meiosis. It was Sutton in 1903 who clearly postulated a relationship between chromosomal behavior at meiosis and the segregation and independent assortment of genes, a relationship that forms the basis of the **chromosome theory of inheritance.**

Sutton and others realized that one gene could not possibly correspond to a whole chromosome: organisms clearly must possess more genes than chromosomes. Thus it was proposed that each chromosome carries many linked genes. Linkage was soon described experimentally, but it was found to be "partial" or "incomplete" linkage. For example, in experiments published in 1905 by W. Bateson, E. Saunders, and R. Punnett (and later expanded by Punnett in 1917) sweet peas with *purple flowers* and *long* pollen grains were crossed to *red*-flowered plants with *round* pollen grains. Nothing was unusual about the $F_1$ progeny: all were *purple* and *long*, showing these to be the dominant traits. When the $F_1$ was inbred and each pair of alleles was examined separately, moreover, each proved to behave as expected of segregating Mendelian genes: *purple* and *red* flowers were present in a 3:1 ratio, as were the *long* and *round* traits.

It was when the traits were considered together that **partial linkage** was seen. Among 6952 $F_2$ plants, 4831 were *purple, long*; 390 were *purple, round*; 393 were *red, long*; and 1338 were *red, round*. In other words, the parental types (*red, round* and *purple, long*) were present in excess of what would be predicted of independently assorting genes. Had the traits been located on independent chromosomes, perhaps 3915 *purple, long;* 1305 *purple, round;* 1305 *red, long;* and 435 *red, round* plants would have been expected in the $F_2$ in accordance with a 9:3:3:1 ratio (*Section 6.12*). The two traits did not, on the other hand, show complete linkage; had they done so, a 3:1 ratio, or 5214 *purple, long* and 1738 *red, round* progeny would have been expected (*Section 6.6*).

A number of theories were offered to explain partial linkage, but the correct interpretation was made in 1911 by T. H. Morgan based on analyses of crosses with *Drosophila melanogaster*.

## 14.2 The Morgan Crosses and the Sturtevant Map

Morgan began with the isolation of two spontaneously mutant male flies, one having white eyes and the other miniature wings. When these flies were crossed with normal females and the $F_1$ progeny were inbred, the mutant traits exhibited the sex-linked pattern described in Section 6.18. Morgan therefore postulated that the recessive genes *w* (*white*) and *m* (*miniature*) were both located in the X chromosome. By appropriate crosses he constructed a stock of female flies homozygous for both *w* and *m* and then set about to demonstrate that exchange could occur between these linked genes.

The results of his experiments are shown in Figure 14-1. The homozygous *white miniature* females were first crossed to wild-type males. In the $F_1$ the males were all *white miniature* and the females were wild in phenotype (and heterozygous in genotype), as is expected for sex-linked traits. When the $F_1$ were inbred, the original parental combinations (*white miniature* and wild type) were the most prominent phenotypes in the $F_2$. As we saw in the sweet pea experiments, **an excess of parental phenotypes in the $F_2$ strongly suggests the presence of linked genes.** The linkage was a relatively loose one in this case: 900/2441, or 36.9 percent, of the progeny were recombinant (either *white* with wild-type wing or *miniature* with wild-type eye), as shown in Figure 14-1.

Morgan extended his studies to a third sex-linked recessive marker, *y*, which confers a *yellow* body instead of the normal gray. In crosses (similar to those of Figure 14-1) involving *yellow*-bodied and *white*-eyed flies, 1.3 percent of the progeny were recombinant.

Morgan interpreted these experiments in the context of a study of salamander meiosis made by F. Janssens in 1909. Janssens had described chiasmata (*Figure 5-4*) and had suggested these might represent sites of physical exchange between maternal and paternal homologues. Morgan therefore proposed that partial linkage was observed whenever two markers in the same chromosome were physically separated from one another by a chiasma (he coined the term **crossing over** for this process). To explain why 36.9 percent of the progeny were recombinant for *w* and *m* and only

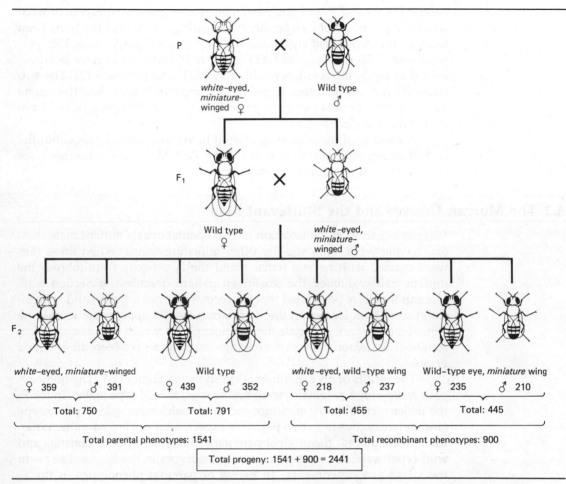

**FIGURE 14-1 Linkage and recombination.** Morgan's crosses involving the markers *white eye* and *miniature* wing in the X chromosome of *D. melanogaster*.

1.3 percent were recombinant for *w* and *y*, Morgan proposed that *w* and *m* were relatively far apart on the X chromosome and therefore had a relatively good chance of being separated by crossing over, whereas *w* and *y* were close together and recombined quite infrequently.

It remained for A. Sturtevant, in 1911, to realize that the recombination frequencies from such two-factor crosses could be used to generate a chromosome map. Two more sex-linked markers, *v* (*vermillion eyes*) and *r* (*rudimentary wings*), were by this time known, and analysis of additional crosses had yielded slightly different recombination frequencies from those reported by Morgan. The five markers were therefore ordered and spaced as shown below, with *y* assigned an arbitrary value of 0 and the other map units expressing distances from *y*.

| 0 | 1.0 | | 30.7 | 33.7 | | 57.6 |
|---|-----|---|------|------|---|------|
| $y$ | $w$ | | $v$ | $m$ | | $r$ |

Such a use of recombination frequencies is, of course, identical to their more recent use in mapping viral and bacterial chromosomes, as described in detail in Chapters 12 and 13. Thus an idea conceived for *D. melanogaster* has since been found to be applicable to all chromosomes that undergo recombination.

## 14.3 Correlating Chiasmata with Gene Recombination

Janssen's hypothesis that chiasmata lead to physical exchange and Morgan's hypothesis that physical exchange leads to gene recombination are basic tenets of classical genetics. Experimental support for these hypotheses rests on two observations: (1) the occurrence of chiasmata can be correlated with the physical exchange of chromosomal material; and (2) the physical exchange of chromosomal material can be independently correlated with the ocurrence of gene recombination. It is concluded, therefore, that chiasmata must be correlated with gene recombination.

**Correlation Between Physical Exchange and Chiasmata**   The correlation between physical exchange of chromosomal material and the occurrence of chiasmata has been demonstrated in the plant *Disporum sessile*. The normal karyotype of *Disporum* is shown in Figure 14-2a, with homologues arranged above and below one another. Focusing on the $A_1$—$A_2$ pair of homologues, we see that $A_1$ carries an additional "tail" of chromatin, called a satellite, so that $A_1$ can be distinguished from $A_2$. The two are therefore called **heteromorphic homologues.** The homologues labeled $b$ are, in this normal strain, identical in morphology. Figure 14-2b shows the chromosome complement of a mutant *Disporum* strain that has undergone a reciprocal translocation (*Section 7.9*) between $A_2$ and one of the $b$ chromosomes: a small piece of material has broken off the $b$ and is attached to the $A_2$, whereas a large piece has broken from $A_2$ and is attached to the $b$. The two translocation chromosomes so formed are called, respectively, $a$ and $B_2$. Thus the karyotype of the mutant *Disporum* is particularly useful for study in that it carries chromosomes $A_1$, $a$, $B_2$, and $b$, all of which can be distinguished from one another.

Synapsis involving these four chromosomes produces the expected cross-shaped configuration (*Figure 7-11*) shown in Figure 14-2c. If no physical exchanges occur between the chromatids, then, when the homologous centromeres separate at anaphase I, each centromere will be expected to carry a symmetrical pair of chromatids, as shown in Figure 14-2d. If, on the other hand, a physical exchange should occur within the lengths between the centromeres and the vertical axis of the cross (the so-called **interstitial regions**), then at anaphase I centromeres will carry chromatids of unequal length, one of the four arms being conspicuously longer than its counterpart. This is shown in Figure 14-2e.

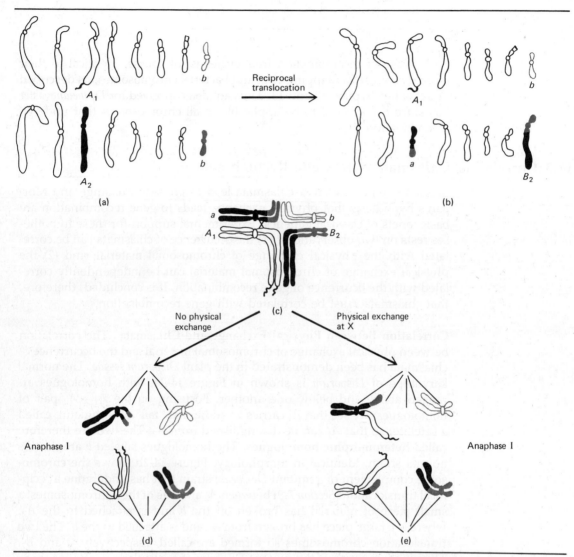

**FIGURE 14-2 Correlating physical exchange with chiasmata** in a translocation strain of *Disporum sessile*. See text for details. (After H. Kayano, *Cytologia* **25**:461–467, 1960).

The experiment therefore involves two sets of observations. Mutant *Disporum* cells are first examined in meiotic prophase I, and the number of chiasmata within one or both of the interstitial regions is counted. The mutant cells in meiotic anaphase I (or II) are then examined, and the number of structures whose chromatid arms are of unequal length is counted. For *Disporum*, both numbers prove to be approximately 30 percent; that is, 30 percent of the synapsing chromosomes have at least one interstitial chiasma and 30 percent of the separating translocation chromosomes have unequal arms. Similarly close values have been obtained in several other cases. Such results do not, of course, *prove* that chiasmata lead to physical exchanges, but they offer good support for the hypothesis.

**Correlation Between Physical Exchange and Recombination**  Experiments that relate the occurrence of physical exchange between chromosomes to the occurrence of genetic recombination are well exemplified by C. Stern's experiments with *Drosophila*, published in 1931 and diagrammed in Figure 14-3. Stern first interbred various strains of flies that carried chromosomal abnormalities and mutant genes. He eventually obtained a stock of female flies that carried heteromorphic X chromosomes. One of these X's carried a translocated segment of a Y chromosome; the other X was broken into two

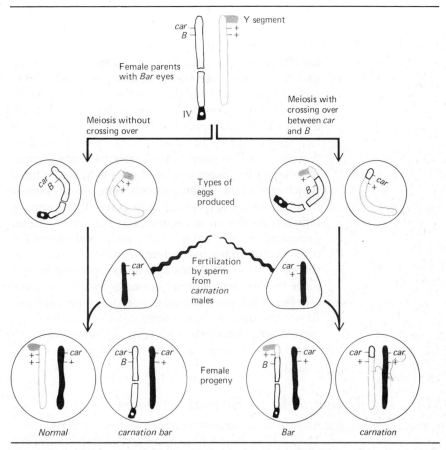

**FIGURE 14-3 Correlation between physical exchange and recombinant progeny** in *Drosophila*, as demonstrated in Stern's experiment with heteromorphic homologues.

segments, one segment remaining attached to its proper X centromere and the other being translocated to the tiny chromosome IV. The homologues, moreover, differed genetically at two different loci. The broken X carried mutations in two genes, called *car* and *B*, which affect the color (*carnation*) and shape (*Bar*) of the eye. The X with the attached Y segment carried normal alleles of each of these genes (both designated as + in Figure 14-3). The *B* gene is dominant to its + allele whereas *car* is recessive.

Such females were crossed with *carnation* males whose X chromosome (recall that males have only one X, whereas females have two) carried *car* and the + allele of *B* (Figure 14-3). Female offspring of the cross, who would have acquired one X from their mother and one (always + *car*) from their father, were then examined.

Four different types of female fly were found among the offspring (Figure 14-3): those with normal eyes, those with both *Bar*-shaped and *carnation*-colored eyes, those with *Bar*-shaped eyes, and those with *carnation*-colored eyes. The genotypes of these females can at once be recognized as $\dfrac{+\ +}{car\ +}$, $\dfrac{car\ B}{car\ +}$, $\dfrac{+\ B}{car\ +}$, and $\dfrac{car\ +}{car\ +}$, as drawn in Figure 14-3. When the chromosome morphologies of these flies were examined, it was found that the normal flies had inherited an intact X chromosome with an attached Y segment from their mothers and that the doubly mutant flies had inherited the broken X. These two classes of progeny can therefore be said to have inherited unaltered versions of the maternal X complement. The two singly mutant flies, in contrast, each carried an X chromosome that was an altered version of the maternal X complement. Thus the flies with *Bar*-shaped but normal-colored eyes exhibited a broken X, one segment of which carried the Y translocation, whereas the *carnation*-eyed flies carried a normal-looking X (plus, in both cases, the normal X inherited from their fathers). As drawn in Figure 14-3, both new chromosome shapes can be generated by postulating that a reciprocal physical exchange of segments occurs between the two heteromorphic X's during meiosis. In other words, the altered chromosomes appear to represent recombinant chromosomes. Since the flies with normal eye color and shape and the flies with *carnation* and *Bar* eyes have parental genotypes, whereas those exhibiting either the *Bar* or the *carnation* trait alone have recombinant genotypes, this experiment clearly demonstrates a correlation between the inheritance of recombinant chromosomes and the inheritance of recombinant sets of genes.

# Mapping *Drosophila* in Sexual Crosses

The sections that follow describe the steps taken in constructing and analyzing genetic crosses in *Drosophila*. Exactly the same principles hold for mapping all other sexually reproducing diploid eukaryotes; the problem section at the end of this chapter includes examples from other organisms.

## 14.4 Two-Factor Crosses and the Test Cross

You will recall from Section 12.9 that the basic design of a two-factor cross in phages is to mixedly infect a bacterium with doubly marked phages ($+ +$ and $a\ b$, for example) and to look for recombinant phages ($+\ b$ and $a\ +$) among the progeny. Exactly the same strategy holds in a eukaryotic two-factor cross, but the existence of diploidy and dominance in eukaryotes necessitate a few additional considerations.

The first step in a two-factor cross in *Drosophila* is to create flies with the general genotype $\dfrac{+\ \ +}{a\ \ b}$, heterozygous at the two loci of interest, where the continuous line connecting the markers denotes their linkage. This is usually accomplished by taking true-breeding $\dfrac{a\ b}{a\ b}$ flies and crossing them to true-breeding $\dfrac{+\ +}{+\ +}$ flies, thereby creating $\dfrac{+\ \ +}{a\ \ b}$ heterozygotes in the $F_1$ (or in the $F_1$ females in the case of sex-linked genes).

The next step is to allow these double heterozygotes to undergo gametogenic meiosis. Gametes produced without recombination will be either $+ +$ or $a\ b$, whereas gametes that are generated by crossing over will be either $+\ b$ or $a\ +$.

The final step in the experiment is to devise some way of telling these gametes apart, so that recombination frequencies can be deduced. This is usually accomplished by a test cross (*Section 6.6*): the organisms under test are mated with organisms that are homozygous recessive at all loci of interest. In this case, one would mate the double heterozygotes to flies that are homozygous recessive $\left(\dfrac{a\ b}{a\ b}\right)$ or hemizygous recessive $\left(\dfrac{a\ b}{\longrightarrow}\right)$, where the $\longrightarrow$ symbol denotes the Y chromosome. Such flies produce gametes that carry either $a\ b$ or "silent" $\longrightarrow$ chromosomes, neither of which will obscure, via dominance, the genetic makeup of the chromosomes contributed by the double heterozygote.

The test cross can be performed in one of two ways. In the 1911 crosses of Morgan described previously, $\dfrac{w\ m}{w\ m}$ flies were crossed to $\dfrac{+\ +}{\longrightarrow}$ males to generate $\dfrac{w\ m}{+\ +}$ female heterozygotes in the $F_1$. In this case, the $F_1$ males were all $\dfrac{w\ m}{\longrightarrow}$ hemizygotes and therefore suitable test-cross organisms, so that the $F_1$ was simply inbred and recombination frequencies were directly deduced from the $F_2$ phenotypes (*Figure 14-1*).

In other cases, the $F_1$ females must be taken from their vials while they are still virgin, to prevent their mating with their brothers, and allowed to mate with test-cross males. This can be illustrated by a cross designed to determine the recombination frequency between two autosomal markers, $g$ (*glass eye*) and $e$ (*ebony* body color). A *glass ebony* × *wild*

**TABLE 14-1 Linkage and Recombination Between *gl* and *e* in *Drosophila***

$$\frac{gl\ e}{+\ +} \times \frac{gl\ e}{gl\ e}$$

| Progeny | Number | Percent |
|---|---|---|
| $\dfrac{gl\ e}{gl\ e}$ | 1031 | 43.4 |
| $\dfrac{+\ +}{gl\ e}$ | 1159 | 48.6<br>Total: 92.0 |
| $\dfrac{gl\ +}{gl\ e}$ | 92 | 3.8 |
| $\dfrac{+\ e}{gl\ e}$ | 99 | 4.2<br>Total: 8.0 |

type cross yields $\dfrac{+\ +}{gl\ e}$ $F_1$ heterozygotes in both females and males. Since the $F_1$ males have the potential to produce $+\ +$ sperm, they are unsuitable for test crossing. The $F_1$ female virgins must therefore be collected and crossed to $\dfrac{gl\ e}{gl\ e}$ males, which yields the results shown in Table 14-1.

The crosses we have outlined utilized markers in coupling $\left(\dfrac{+\ +}{a\ b}\right)$, but the same results should occur if the markers are introduced in repulsion $\left(\dfrac{+\ b}{a\ +}\right)$. Thus the cross *glass* × *ebony* $\left(\dfrac{gl\ +}{gl\ +} \times \dfrac{+\ e}{+\ e}\right)$ will yield wild-type $F_1$ heterozygotes with a $\dfrac{gl\ +}{+\ e}$ genotype, and test-crosses of such females to $\dfrac{gl\ e}{gl\ e}$ males should again reveal that *gl* and *e* remain linked approximately 92 percent of the time and are recombined approximately 8 percent of the time.

It should be noted that female heterozygotes are used in all these analyses because, at least under laboratory conditions, **crossing over does not occur in male *Drosophila*.**

## 14.5 Three-Factor Crosses

As in bacteriophages, two-factor crosses, if performed in enough combinations, can generate both the order and the spacing of markers on a chromosome. Three-factor crosses are preferable, however, since they can reveal marker orders directly by permitting the recognition of those rare recombi-

**TABLE 14-2** Linkage and Recombination between *y, ec,* and *ct* in *Drosophila*

$$\frac{y\ ec\ ct}{+\ +\ +} \times \underrightarrow{y\ ec\ ct}$$

| Progeny Phenotypes | Number | Class | Class Total | Class Frequency |
|---|---|---|---|---|
| *yellow, echinus, cut*<br>wild type | 1071<br>1080 | Parental | 2151 | |
| *yellow*<br>*echinus cut* | 78<br>66 | Single I | 144 | $\alpha = 144/2880 = 0.050$ |
| *yellow echinus*<br>*cut* | 282<br>293 | Single II | 575 | $\beta = 575/2880 = 0.199$ |
| *yellow cut*<br>*echinus* | 4<br>6 | Double | 10 | $\gamma = 10/2880 = 0.0034$ |
| | | GRAND TOTAL | 2880 | |

nant classes that arise as a result of double crossovers (see Section 12.10 to review the principles underlying three-factor crosses).

A typical cross for *D. melanogaster* that involves three sex-linked markers is outlined here. Females homozygous for *yellow* body (*y*), a rough eye known as *echinus* (*ec*), and *cut* wings (*ct*) are crossed, as usual, with wildtype males. The $F_1$ females will thus be phenotypically wild type and heterozygous for all three loci, $\frac{y\ ec\ ct}{+\ +\ +}$. These females are then inbred with their $\underrightarrow{y\ ec\ ct}$ brothers (as noted earlier, these in effect are test crosses). We can predict that the females will produce eight different types of eggs: *y ec ct, + + +, y + +, + ec ct, y ec +, + + ct, y + ct,* and *+ ec +*. We cannot, however, predict the number of each, since the gene loci are linked and therefore do not assort independently of one another.

The results of the cross are shown in Table 14-2. The eight progeny phenotypes fall into four classes, just as in the three-factor phage crosses outlined in Section 12.10. Using the same terminology introduced in Section 12.10, the parental class consists of *yellow echinus cut* and wild-type flies; the single-recombinant-I class consists of flies with the complementary phenotypes *yellow* and *echinus cut;* the single-recombinant-II class consists of the complementary *yellow echinus* and *cut* flies; and the double-recombinant class consists of *yellow cut* and *echinus* flies. The parental class is, as expected, the largest; the double-recombinant class is the smallest. Assuming, as we did for prokaryotic genetics, that the rarest class arises as the result of double crossovers, the marker order can be assigned as *y ec ct.*

To determine map distances between the loci, the recombination frequencies for each class are calculated, as shown in Table 14-2, to yield values for $\alpha$, $\beta$, and $\gamma$. The recombination frequencies between each pair of loci are then determined by familiar relationships:

$$R_{y,ec} = \alpha + \gamma = 5.3 \text{ percent}$$
$$R_{ec,ct} = \beta + \gamma = 20.2 \text{ percent}$$
$$R_{y,ct} = \alpha + \beta = 24.9 \text{ percent}$$

Converting recombination frequencies to map distances by multiplying by 100, we obtain the following map:

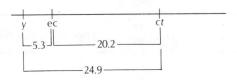

## 14.6 Double Crossovers and Interference

The map drawn above reveals that the $y - ct$ interval of 24.9 map units is smaller than the $y - ec$ plus $ec\text{-}ct$ interval ($5.4 + 20.3 = 25.7$ map units). As with phages (*Section 12.10*), this "foreshortening" of the outside markers is caused by the occurrence of double crossovers. Double crossovers in eukaryotes were considered in Section 5.6. As shown in Figure 5-4, they may occur between the same two homologous chromatids in a bivalent (**two-strand doubles**) or they may involve three or four chromatids. In a **three-strand double,** a chromatid engaged in crossing over with one homologous chromatid at one locus will establish a second chiasma with its second homologous chromatid at a different locus. In a **four-strand double,** two homologous chromatids cross over at one level and the other two at a second level. Each type of double crossover yields a distinct ratio of parental : recombinant chromatids, as illustrated in Figure 14-4.

If we assume that the various double crossover configurations will occur with equal probability, then we would expect the ratio of two- : three- : four-strand double exchanges to be 1:2:1 (the three-stranded events involve two possible acts of three chromatids and are thus expected to occur twice as often). Three-strand events are in fact found to be deficient or two-strand events to be excessive, or both, but the observed ratio does not vary greatly from the ratio predicted. In genetic terminology it is therefore said that there is little if any **chromatid interference:** the fact that a chromatid has formed a chiasma with one homologous chromatid does not interfere with its ability to form a chiasma with its second homologous chromatid.

In contrast, **there is often a marked interference exerted by one chiasma on the probability that a second chiasma will occur between the same two chromatids.** This can be seen by reexamining the data in Table 14-2. The double crossover class $\gamma$ constitutes 0.34 percent of the progeny. As noted in Section 12.10, the probability that two independent events will occur at the same time is given by the product of their individual probabilities. Therefore, the frequency of double recombinants in the $y - ct$ interval

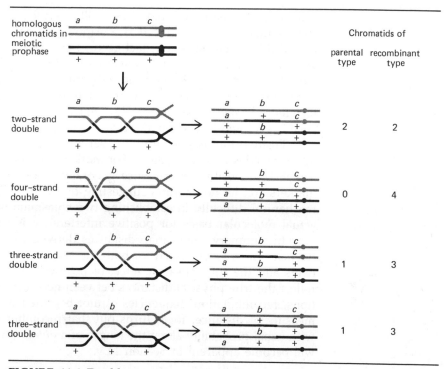

**FIGURE 14-4 Double crossover configurations and the types of recombinant chromosomes they produce.** (After A. M. Srb, R. D. Owen, and R. S. Edgar, *General Genetics*, 2nd ed. San Francisco: W. H. Freeman and Company, 1965.)

would be expected to be the frequency of recombination between *y* and *ec* multiplied by the frequency of recombination between *ec* and *ct*. Thus one would expect (0.053) (0.202) = 0.011 or 1.1 percent of the progeny to be doubly recombinant. The discrepancy between the predicted (1.1 percent) and the observed (0.34 percent) figures indicates that the two crossover events are not independent of each other; in genetic terminology it is said that there is **interference** exerted by one chiasma on the occurrence of a second chiasma.

The nature and extent of interference can be conveyed by an expression known as the **coefficient of coincidence,** usually symbolized as **s,** where

$$ s = \frac{\text{observed doubles}}{\text{expected doubles}} = \frac{\gamma}{(\alpha + \gamma)\,(\beta + \gamma)} $$

If the predicted frequency of double crossovers in an interval is the same as the observed frequency ($\gamma$), then $s = 1$; in this case, the occurrence of one exchange in the interval has no influence on the occurrence of a second, and it is said that there is no interference. If the number of double exchanges is less than predicted, then $s$ will be some fractional number; in the $y$-$ct$ example, $s = 0.34/1.1 = 0.31$, meaning that only 31 percent of the expected double crossovers in fact occur. Since in this case, one crossover appears to inhibit the second, the cross is said to exhibit **positive interference.** (When markers are very close together, particularly within the same gene, double exchanges occur more often than predicted and **negative interference** is said to take place. The molecular basis for negative interference is explored in Section 17.11).

Positive interference is not observed when markers are located on opposite sides of a centromere. For markers located in the same chromosomal arm, positive interference generally increases the closer the two markers are, an observation suggesting that chromatids experience some mechanical difficulty in establishing two chiasmata near each other. The actual molecular basis for positive interference is, however, unknown.

In summary, then, map distances derived from recombination frequencies must be corrected for both the shrinking effect of double crossovers and the compensatory effect of positive interference if they are to reflect the true physical distances between genes. Even with such corrections, recombination frequencies cannot be relied on to give an accurate measure of distance, for it turns out that recombination does not occur with equal probability along the length of a eukaryotic chromosome. This will become apparent in Section 14.9.

# Cytological Mapping

An important and independent view of an organism's genome is gained from physical maps of chromosomes. Dipteran polytene chromosomes have proved to be particularly useful for this, since chromosomal aberrations will frequently alter both their banding patterns and their somatic pairing configurations (*Figure 7-6*), thus allowing the location and the extent of aberrations to be defined with some precision. When such physical disorders can be correlated with particular mutant phenotypes, it becomes possible to construct **cytological maps.**

## 14.7 Deficiency Mapping

**Deficiency mapping** is formally equivalent to the deletion mapping techniques described for viruses (*Section 12.13*), the major difference being that larger segments of chromosomal DNA are typically missing in a deficiency than in a phage deletion (*Section 7.7*). In *Drosophila*, most deficiencies are lethal when carried in the homozygous or hemizygous condition; they must therefore be maintained in heterozygous flies. Where a fly is heterozygous for a deficiency in one chromosome and for a recessive, nonlethal mutation at a comparable position on the homologous chromosome, the recessive mutation will characteristically be expressed in the phenotype, a phenomenon aptly termed **pseudodominance.** It thus becomes possible to correlate the cytological location of a deficiency with the map locations of known genetic markers.

An example of deficiency mapping involves the analysis of the X chromosome of *D. melanogaster.* Figure 14-5a shows the noncentromere end of this chromosome in its polytene state. You will recall that such a chromosome represents an amplified version of two homologous chromosomes that have come together by somatic pairing (*Figure 3-12*). Therefore, if a fly is made heterozygous for an X chromosome carrying a terminal deficiency, its polytene X chromosomes will resemble those shown in Figure 14-5b and 14-5c: that portion of the polytene structure deriving from the normal X will carry a full set of bands, whereas the portion deriving from the deficiency homologue will be foreshortened.

M. Demerec utilized the chromosomes shown in Figure 14-5 to carry out cytological mapping. The deficiency chromosome in Figure 14-5b is seen to lack 11 bands present in the normal chromosome; in Figure 14-5a these bands are represented by the segment marked 260-1. Females carrying the recessive markers *y* (*yellow body*), *ac* (*achaete,* affecting the number and size of body bristles), and *sc* (*scute,* affecting the formation of scutellar bristles) were made heterozygous for this deficiency, and all three recessive genes were expressed, thereby localizing these markers to the X terminus. When such triple mutants were instead made heterozygous for the deficiency shown in Figure 14-5c, a deficiency involving the loss of only the eight bands in the 260-2 region of Figure 14-5a, then only *y* and *ac* were expressed. This outcome localizes the *sc* locus to the few bands present in the second deficient chromosome but absent in the first.

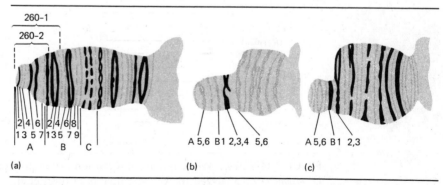

**FIGURE 14-5 Deficiency mapping.** (a) Normal polytene X chromosome of *D. melanogaster.* (b) and (c) Polytene X chromosomes from deficiency heterozygotes. (From A. M. Srb, R. D. Owen, and R. S. Edgar. *General Genetics,* 2nd ed. W. H. Freeman and Company. Copyright © 1965, after Demerec.)

## 14.8 Cytogenetic Analysis of the *Zeste-White* Region of the *Drosophila* X Chromosome

B. Judd and collaborators have performed a particularly detailed genetic and cytogenetic analysis of the region of the X chromosome shown in Figure 14-6, extending between band 3A3 (the locus of *z*, the *zeste* eye marker) to band 3C2 (the locus of *w*, the *white* eye marker). The first goal of the Judd experiments was to isolate *Drosophila* strains carrying mutations in every one of the genes in the *zeste-white* region. To screen for such mutations the Judd group employed the procedure outlined in Figure 14-7. Normal males were treated with mutagens and then crossed with females homozygous for the *Basc* chromosome (*described in Section 7.4*). The $F_1$ female progeny from this cross therefore inherited *Basc* chromosomes from their mothers and X chromosomes from their fathers.

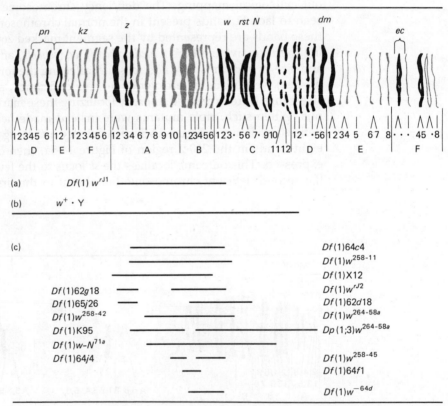

FIGURE 14-6 *Zeste-White* region, the portion of the polytene X chromosome of *D. melanogaster* analyzed in the experiments of Judd et al. The lines below designate (a) the extent of the deficiency in the *Df*(l) $w^{rJ1}$ chromosome used in screening; (b) the extent of the extra genes found in the $w^+Y$ chromosome used in screening; (c) the extents of various deficiencies used in complementation studies and in cytogenic analysis. (From B. Judd, et al., *Genetics* **71**:139, 1972.)

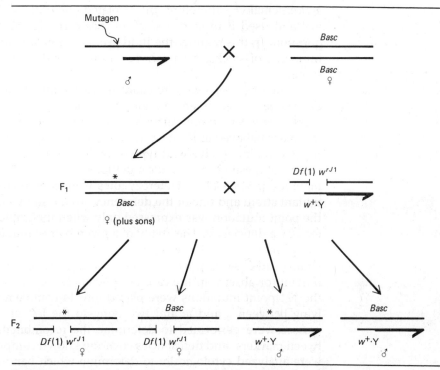

**FIGURE 14-7 Screening for *zeste-white* mutants.** The X chromosome is represented by a thin line, the Y by a thicker line, and a mutation in the region of interest by ∗. (From B. Judd, et al., *Genetics* **71**:139, 1972.)

Because the fathers were mutagenized, certain of the $F_1$ females would be expected to inherit a paternally derived X chromosome carrying a mutation in the *zeste-white* region (X∗). Judd identified these carriers by crossing individual $F_1$ females to males carrying a deficiency in the 3A2-3C2 region of the X chromosome, a deficiency designated $Df(1)w^{rJ1}$ (Figure 14-6a). Such a deficiency is normally lethal in males, but the male strain was constructed to possess a Y chromosome carrying a translocated segment of X chromosomal material that includes all the genes absent in the deficient X chromosome. This chromosome is designated $w^+ \cdot Y$ (Figure 14-6b).

The possible $F_2$ progeny from such a cross are shown in Figure 14-7. Among the $F_2$ females half should carry a *Basc* X chromosome (and can be identified by their *Bar, apricot* eyes) and half should carry the original paternally derived X chromsome. If this chromosome carries a mutation (X∗) in the *zeste-white* region and the mutation does not cause lethality, then the mutation will be expressed in the $F_2$ daughters since it will lie opposite the $Df(1)w^{rJ1}$ deficiency. Should the mutation be a recessive lethal, as proves most often to be the case, then the only viable $F_2$ females will be of the *Bar, apricot* variety. Whenever Judd and his colleagues encountered an all *Bar-*

*apricot*-female $F_2$, therefore, they isolated one of the $X^*/w^+\cdot Y$ males in the vial and used it to start a stock of flies carrying the $X^*$ chromosome in question. In this manner the Judd group was able to amass a collection of hundreds of stocks, each carrying an independently derived $X^*$ chromosome.

The next steps were to characterize the mutations genetically and to determine, by complementation, how many genes they marked. Initial crosses between strains carrying $X^*$ chromosomes revealed that 121 chromosomes behaved as if they carried point mutations (they could, for example, recombine to give wild-type recombinants). The approximate map locations of these 121 *zeste-white* mutations were first estimated by deficiency mapping (*Section 14.7*). Heterozygotes were constructed between a given mutant strain and one of the deficiency strains shown in Figure 14-6c, and the point mutation was expressed only when it occupied a position bracketed by a deficiency. The ability of a given pair of mutant chromosomes to complement each other was then assessed by crossing appropriate mutant strains and observing whether the resulting heterozygotes were normal or mutant (or absent, in the case of noncomplementing lethals). In this way the 121 point mutations were placed into 14 complementation groups, 12 lying between *z* and *w* and two lying to the left of *z*. Finally, pairwise crosses were performed to determine the recombinational distances between markers, and deficiencies covering known complementation groups were analyzed cytologically to determine which bands were missing.

The results of the Judd studies are summarized in Figure 14-8. The 13 *zeste-white* (*zw*) loci between the *z* and *w* loci cover a region of the polytene chromosome estimated to contain 13 bands. The initial Judd results, in other words, suggested a correspondence between the presence of a polytene chromosome band and the presence of a gene. More recent studies from Judd's lab, however, have demonstrated the complexity of the situation. Alternate screening procedures detected the presence of additional nonessential genes in the region, so that nine complementation groups are now estimated to lie in the six-band 3B region (Figure 14-8). On the other hand, there appear to be more bands than genes in the 3C region. Therefore, as noted in Section 3.11, it appears that some genes may not have a banded organization, whereas other genes associate with several bands' worth of DNA (some of which may well have regulatory functions).

## 14.9 Correlating Genetic and Cytological Maps

Over short distances, as in the Judd experiments, there appears to be reasonable correspondence between physical distances in a polytene chromosome and distances on a genetic map (*Figure 14-8*). Over longer distances, however, the correspondence varies considerably, depending on the region of the chromosome being considered. It is seen in Figure 14-9, for example, that the X chromosome of *D. melanogaster* undergoes recombination more often in the center than at either end; markers at the ends there-

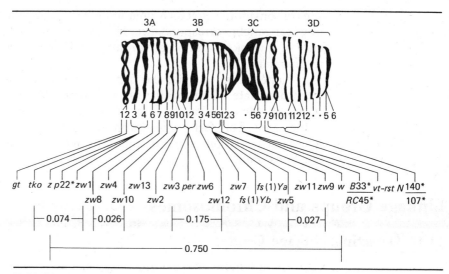

**FIGURE 14-8 Genetic map of the *zeste-white* region.** The 13 *zw* complementation groups are shown along with other markers in the region. Asterisks denote rearrangement break points that do not interrupt known gene functions. Representative map-unit distances are given below the markers. (From R. P. Wagner, B. K. Judd, B. G. Sanders, and R. H. Richardson, *Introduction to Modern Genetics.* New York: John Wiley & Sons, 1980.)

fore appear much closer together on the genetic map than they actually are on the physical chromosome. More specifically, there is as much recombination in the *w-fa* interval, which constitutes perhaps five polytene bands, as there is in the *y-w* interval, which has about 75 bands (Figure 14-9). Recombination is also rare in the immediate vicinity of centromeres, the

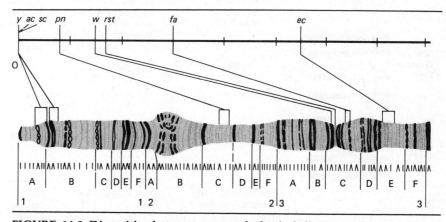

**FIGURE 14-9 Disparities between map and physical distances,** demonstrated by a portion of the left arm of the X chromosome of *D. melanogaster* and the corresponding genetic map. (After C. B. Bridges.)

sites of constitutive heterochromatin (*Section 3.13*), which again distorts the physical-genetic correspondence. Finally, as noted in Section 7.8, the presence of inversions and other chromosomal rearrangements will suppress recombination.

Nonetheless, despite the fact that **distances between markers are often disparate when cytological and genetic maps are compared,** it is invariably the case that the *order* of markers deduced by both approaches is the same. This fact argues impressively for the linear arrangement of genes in chromosomes first proposed from the Morgan/Sturtevant recombinational data.

# Linkage Groups and Chromosomes

## 14.10 Detecting Linkage Groups

As more and more markers become available for study in a given eukaryotic species, a limited number of groups of linked genes, called **linkage groups,** begin to appear. As a general rule, **markers that assort independently are placed in different linkage groups and those that exhibit linkage are placed in the same group.**

The major qualification to this rule becomes apparent when the physical process of meiosis is recalled. As noted in Section 5.6, chiasmata form quite frequently during meiosis, typically at least once per bivalent. Taking as an example the *Drosophila* homologues marked by *gl* and *e* and by their + + alleles (Figure 14-10), if we were to examine 100 such bivalents we might expect to find that 84 have no chiasmata in the *gl-e* interval, while 16

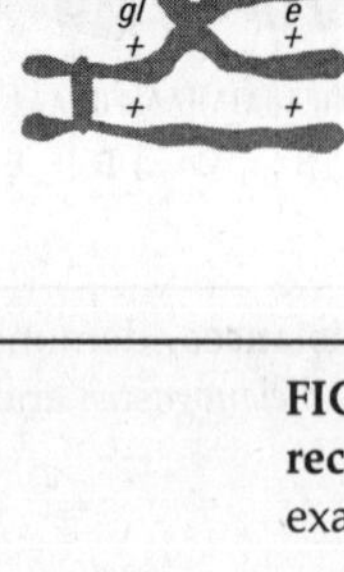

|  | Number of bivalents of each class | Number of chromatids in each class (bivalents × 4) | Number of recombinant chromatids |
|---|---|---|---|
| 1) Non-Crossover Class | 84 | 336 | 0 |
| 2) Crossover Class | 16 | 64 | 32 |
| | Total bivalents = 100 Crossover frequency = 16/100 = 0.16 | Total chromatids = 400 | Recombination frequency = 32/400 = 0.08 |

FIGURE 14-10 The relationship between crossover frequency and recombination frequency, using the *glass ebony* data from Table 14-1 as an example.

have single chiasmata in this interval. It can thus be said that the overall **crossover frequency** in the *gl-e* interval is $16/100 = 0.16$, or 16 percent. By definition, however, only two of the four chromatids in a bivalent will participate in any single exchange event. Therefore, if we consider the $16 \times 4 = 64$ chromatids in the chiasma-bearing bivalents, only $1/2 \times 64 = 32$ chromatids will be $gl +$ or $+ e$ recombinants (Figure 14-10), and of the 400 chromatids present in the 100 bivalents, only $32/400 = 0.08$, or 8 percent, will be recombinant. Thus **the recombination frequency deduced from genetic analysis of random meiotic products will be half the crossover frequency.**

It follows that when two markers on the same chromosome are so far from one another that they happen to be separated by a chiasma in virtually every meiotic bivalent, their crossover frequency will approach 100 percent and their recombination frequency 50 percent. Markers that appear to be linked in 50 percent of the offspring and unlinked in the other 50 percent are indistinguishable from markers that assort independently. In other words, **if two markers recombine about half the time, it is impossible to decide whether they lie in the same or different linkage groups.**

To resolve this question the markers in question must be tested for their linkage to other markers in the relevant linkage group. Thus, for example, if the recombination frequency between *a* and *b* is about 50 percent, between *a* and *c* 20 percent, and between *b* and *c* 30 percent, then *a* and *b* are probably linked; if, however, *b* and *c* also recombine 50 percent of the time, then *a* and *b* are probably not linked.

The number of linkage groups possessed by a diploid organism should correspond to its number of chromosome pairs and thus to its haploid chromosome number. The number of linkage groups in *D. melanogaster* is four, and there are four pairs of chromosomes, the XX or XY pair, the large autosomes II and III, and the tiny autosome IV. Another *Drosophila* species, *D. pseudoobscura*, has five pairs of chromosomes and five linkage groups, corn has 10 linkage groups and a haploid chromosome number of 10, and so on. This correspondence is, of course, further evidence in support of the chromosome theory of inheritance.

## 14.11 Assigning Linkage Groups to Chromosomes

In order to assign a particular linkage group to a particular chromosome, some sort of cytogenetic correlation must be made between the presence of a genetic marker carried by a linkage group and the presence of a distinctive chromosome. This is classically accomplished by chromosomal mutations: the presence of an identifiable deficiency in a chromosome (*Section 14.7*), for example, may allow the expression of a recessive trait in a known linkage group, thereby assigning that linkage group to that chromosome.

The assignment of the four linkage groups in *D. melanogaster* (Figure 14-11) to its four chromosomes was a straighforward affair: the chromosomes are easily distinguished from one another and have the landmark banding patterns in their polytenic state. Assigning the ten linkage groups

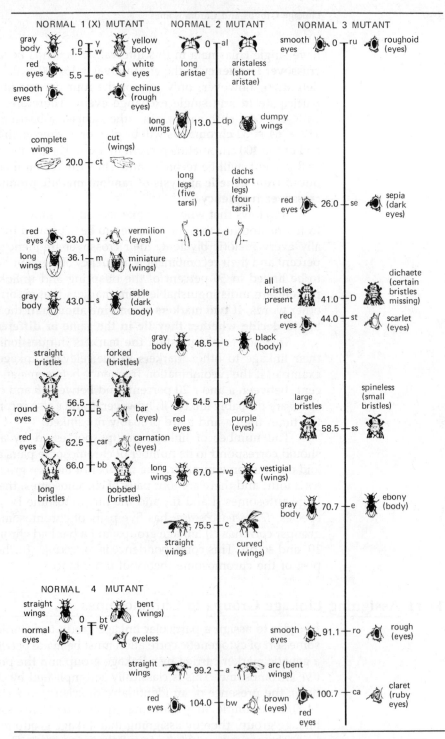

**FIGURE 14-11 Genetic map of *Drosophila melanogaster,*** showing some of the more widely used visible markers. A complete map of this organism presently contains some 1000 gene loci.

of maize (Figure 14-12) to its ten haploid chromosomes was also greatly facilitated by "natural" landmarks: most of the chromosomes of maize bear characteristic arrays of chromomeres and "knobs," and the deletion or translocation of these features can be correlated with the acquisition of particular mutant phenotypes.

In the case of the mouse, 20 linkage groups were defined genetically (Figure 14-13) and 20 chromosomes were known to be present in the haploid karyotype, but it proved almost impossible to tell these small acrocentric chromosomes apart until "banding technology" was introduced (*Section 2.11* and *Figure 2-16*). O. J. Miller, D. A. Miller, and their coworkers proceeded to assign linkage groups to these chromosomes by studying **overlapping translocations.** For example, mouse stock T138Ca was known to carry a translocation that caused markers in linkage group (LG) II to be linked to markers in LG IX, whereas stock RB163H carried a translocation that caused LG II markers to associate with LG XII markers. The karyotype of stock T138Ca revealed a translocation chromosome with the banding patterns of chromosomes 9 and 17; the translocation chromosome of stock RB163H proved to carry bands from both chromosomes 9 and 19. Thus it was deduced that LG II must be in chromosome 9, that LG IX corresponds to chromosome 17, and that LG XII is in chromosome 19.

# Mapping by Tetrad Analysis

As outlined in Section 6.9, many "lower" haploid eukaryotes can be subjected to tetrad analysis, wherein all four products of individual meioses are analyzed separately. Tetrad analysis also permits linkage and mapping studies.

## 14.12 Mapping Two Linked Genes by Tetrad Analysis

When two markers assort independently (that is, when they are not linked), then one expects to find in tetrad analysis a near one-to-one ratio of PD to NPD tetrads (*Section 6.16*). **Linkage between two markers is signaled by a PD:NPD ratio that is considerably greater than unity.** For example, in an *arg* + × + *pab* (para-amino benzoic acid requirement) cross of *Chlamydomonas reinhardi,* the tetrads obtained are as follows:

| PD | NPD | T |
|---|---|---|
| *arg* + | *arg pab* | *arg* + |
| *arg* + | *arg pab* | *arg pab* |
| + *pab* | + + | + *pab* |
| + *pab* | + + | + + |
| 119 | 1 | 71 |

Total Tetrads: 191

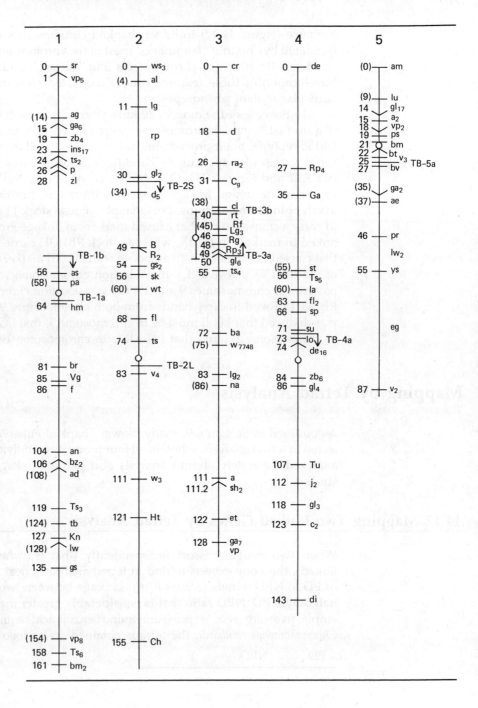

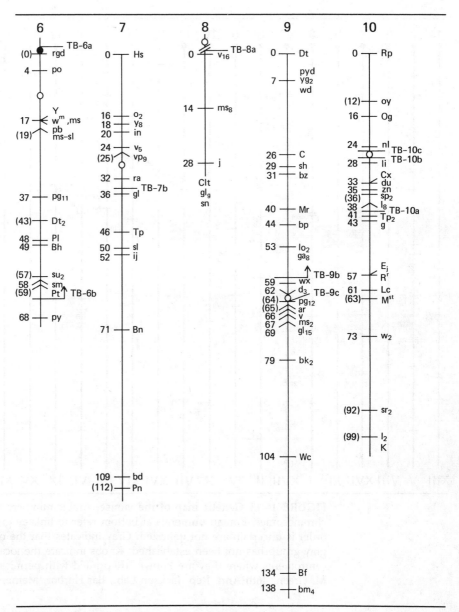

**FIGURE 14-12 Genetic map of corn (*Zea mays*).** Parentheses indicate probable position based on insufficient data, ○ indicates centromere position, and ● indicates organizer. Positions designated TB identify the genetic location of breakpoints of A-B translocations, which generate terminal deficiencies; ——↘ indicates that the TB breakpoint is in that position or is some distance in the direction indicated. (From M. G. Neuffer and E. H. Coe, Jr., in *Handbook of Genetics*, R. C. King, Ed. New York: Plenum Press, 1974.)

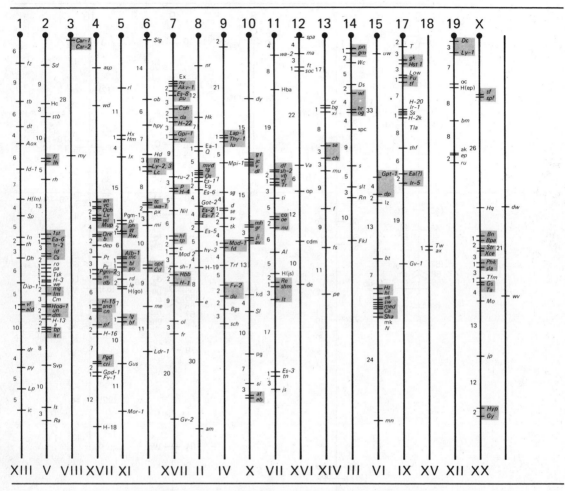

**FIGURE 14-13 Genetic map of the mouse.** Arabic numbers at top refer to chromosomes; Roman numerals at bottom refer to linkage groups. Loci whose order is uncertain are not italicized. Gray indicates that the order within the gray group has not been established. Knobs indicate the locations of the centromeres where they are known. (Reprinted with permission from M. Green, 46th Ann. Rep. Jackson Lab., Bar Harbor, Maine, 1975.)

The unequal ratio of PD to NPD is a major departure from independent assortment and indicates that the *arg* and *pab* loci tend to be inherited in their parental configurations rather than in any nonparental or recombinant configurations.

Assuming that *pab* and *arg* are linked, the origin of the three tetrad classes is as diagrammed in Figure 14-14. The PD tetrads, of course, arise when the parental chromatids are transmitted intact to the meiotic products (Figure 14-14a). A PD tetrad might also arise after a two-strand double

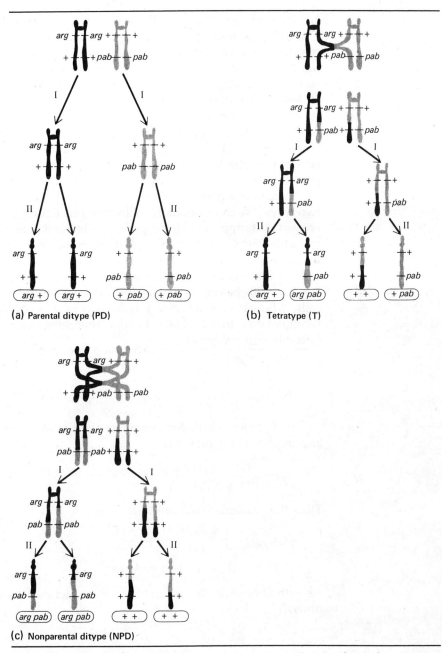

(a) Parental ditype (PD)

(b) Tetratype (T)

(c) Nonparental ditype (NPD)

**FIGURE 14-14 Tetrad analysis of two linked genes.** Origin of tetrads in a cross of *arg* + X + *pab* in *Chlamydomonas*. I and II represent the two meiotic divisions.

crossover (*Figure 14-4*), but these would be expected to be relatively rare. The T tetrads form whenever a single crossover occurs between the *arg* and *pab* loci in two of the four chromatids in a bivalent (Figure 14-14b). Finally, the NPD tetrads, with four recombinant products, must result from the simultaneous occurrence of two crossovers in the *arg pab* interval involving all four chromatids; in other words, the NPD tetrads must result from a four-strand double crossover (Figure 14-14c). Once again the assumption is made that such double crossovers will be rare compared with single crossovers, and this assumption is borne out by the fact that only 0.5 percent of the tetrads in the present cross are of the NPD class, whereas 37.2 percent are of the T class and 62.3 percent are of the PD class.

Recombination frequencies can be calculated from such data in one of two ways. We can essentially follow the procedures set forth for *Drosophila* crosses by supposing that the progeny obtained in the *C. reinhardi* cross are in fact collected as random individuals rather than in their tetrad groupings. Since there are 191 tetrads, this means that there are 4 × 191 or 764 cells in the collection, of which 4 × 1 = 4 belong to the NPD class and 4 × 71 = 284 belong to T. All 4 NPD cells are recombinant *arg pab* or + + cells, whereas only half the T cells (that is, 142 cells) are *arg pab* or + + recombinants (*recall Figure 14-10*). Therefore, using the formula $R = \dfrac{\text{recombinant progeny}}{\text{total progeny}}$, the expression becomes

$$R_{arg,pab} = \frac{4 + 142}{764} = \frac{146}{764} = 0.191 \text{ or } 19.1 \text{ percent.}$$

A simpler calculation considers the tetrads in their class groupings and utilizes the expression

$$R = \frac{NPD + 1/2\,T}{PD + NPD + T}$$

Thus the particular data are treated as

$$R_{arg,pab} = \frac{1 + 1/2\,(71)}{191} = \frac{36.5}{191}$$

$$= 0.191 \text{ or } 19.1 \text{ percent}$$

As with *Drosophila*, this recombination frequency is translated as 19.1 map units.

## 14.13 Mapping Three Linked Genes by Tetrad Analysis

The use of a three-factor cross to determine both marker order and intra-marker map distances for linked genes can be illustrated for *C. reinhardi* by a cross between an *arg + +* strain and a *+ pab thi* strain, where *thi* indicates a requirement for thiamin. The kinds of tetrads obtained are shown in

**TABLE 14-3 Tetrad Analysis of Three Linked Genes.** Data obtained from the cross *arg + +* × *+ pab thi* in *Chlamydomonas*.

| (1) | (2) | (3) | (4) |
|---|---|---|---|
| *arg + +* | *arg + +* | *arg + +* | *arg + +* |
| *arg + +* | *arg pab thi* | *arg + thi* | *arg pab +* |
| *+ pab thi* | *+ + +* | *+ pab +* | *+ + thi* |
| *+ pab thi* | *+ pab thi* | *+ pab thi* | *+ pab thi* |
| 14 | 54 | 103 | 2 |

| (5) | (6) | (7) | (8) |
|---|---|---|---|
| *arg + +* | *arg + thi* | *arg + thi* | *arg + thi* |
| *arg pab thi* | *arg pab +* | *arg pab thi* | *arg + thi* |
| *+ + thi* | *+ + +* | *+ pab +* | *+ pab +* |
| *+ pab +* | *+ pab thi* | *+ + +* | *+ pab +* |
| 6 | 3 | 6 | 2 |

| (9) |
|---|
| *arg pab +* |
| *arg pab thi* |
| *+ + +* |
| *+ + thi* |
| 1 |

Table 14-3. The standard procedure with data of this kind is to classify each tetrad class as a PD, NPD, or T for each of the three pairwise marker combinations possible, namely, *arg pab, arg thi,* and *thi pab.* For example, tetrad (3) in Table 14-3 is a PD with respect to *arg* and *pab* but a T with respect to both *arg thi* and *thi pab.* Tetrad (7) is a T with respect to *arg* and *pab,* NPD with respect to *arg* and *thi,* and T with respect to *thi* and *pab.* The total number of PD, NPD, and T tetrads is then determined for each marker combination, as shown in Table 14-4. The recombination frequencies calculated from these numbers are therefore

$$R_{arg,pab} \frac{1 + 1/2(71)}{191}(100) = 19.1 \text{ percent}$$

$$R_{arg,thi} \frac{8 + 1/2(167)}{191}(100) = 48 \text{ percent}$$

$$R_{thi,pab} \frac{2 + 1/2(121)}{191}(100) = 32.8 \text{ percent}$$

The marker order of *arg pab thi* can be deduced directly from these values: there is more recombination between *arg* and *thi* than between the other two sets of markers, and, by our usual line of reasoning, *arg* and *thi* must therefore lie farthest apart. The marker order can also be deduced from the fact that the rarest tetrad classes (tetrads 4 through 9 in Table 14-3) can all be accounted for by assuming the occurrence of two-, three-, or

**TABLE 14-4**   Analysis of data given in Table 14-3

|                | arg-pab | arg-thi | pab-thi |
|----------------|---------|---------|---------|
| PD             | 119     | 16      | 68      |
| NPD            | 1       | 8       | 2       |
| T              | 71      | 167     | 121     |
| Total          | 191     | 191     | 191     |
| Map distance   | 19.1    | 48.0    | 32.8    |

four-strand double crossovers between chromatids bearing an *arg pab thi* sequence. If, for example, the marker order were instead *arg thi pab*, then tetrads of class 4 would arise by a single crossover and should be relatively frequent.

Extensive studies of *C. reinhardi* by tetrad analysis have disclosed the existence of 16 linkage groups (Figure 14-15). In *Saccharomyces cerevisiae* (yeast), 18 linkage groups have been recognized (Figure 14-16), whereas *Neurospora crassa* has only 7 (Figure 14-17). Most of the *N. crassa* linkage groups have been assigned to their respective chromosomes by correlating the inheritance of genetic translocations with the inheritance of chromosomes having altered morphologies.

## 14.14 Gene-Centromere Linkages

The ordered tetrads of *N. crassa* (*Section 6.10*) are particularly useful for determining the map distance separating a gene and its centromere. Except in certain abnormal cases, the centromeres of homologous chromosomes will always separate at the first meiotic division, meaning that they will always undergo first-division segregation. If a gene *b* is located very near a centromere, then *b* will also segregate from its + allele at the first division, and + + *b b* tetrads will emerge from all + × *b* crosses. If gene *b* instead is located some distance from the centromere so that crossing over takes place within the centromere-*b* interval, then every time an odd number of crossovers occurs, a second-division segregation tetrad will result (+ *b* + *b*). Thus the recombination frequency, and the map distance, between the gene and its centromere is simply half the frequency of second-division segregation tetrads emerging from a cross. The one-half factor again derives from the fact that only two of the four chromatids in the relevant bivalent have participated in any one exchange.

## 14.15 Mapping the *CYC1* Gene of Yeast

Working over a period of 15 years, F. Sherman and his colleagues succeeded in generating a fine-structure map of the *CYC1* gene of yeast. Located in linkage group X (*Figure 14-16*), the *CYC1* gene codes for the small heme protein, iso-1-cytochrome c, which localizes in the mitochondrion and is

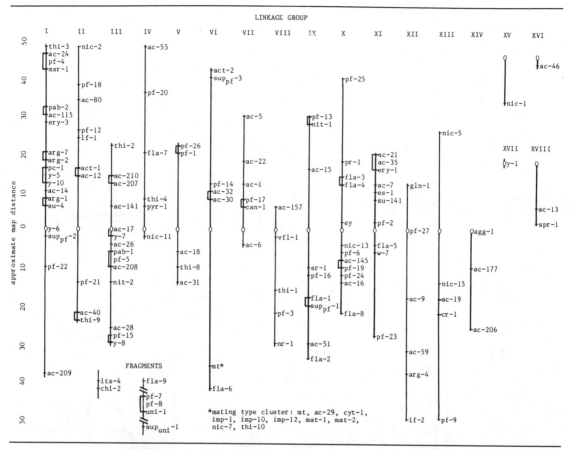

**FIGURE 14-15 Genetic map of *Chlamydomonas reinhardi*.** Abbreviations: *ac*, acetate-requiring; *act*, actidione (cycloheximide)-resistant; *agg*, phototactic aggregation; *arg*, arginine-requiring; *can*, canavanine-resistant; *chl*, chlorophyll-deficient; *cr*, chloroplast ribosome-deficient; *cyt*, abnormal cell division; *ery*, erythromycin-resistant; *es*, suppresses *ery*; *ey*, lacks eyespot; *fla*, ts flagellar defect; *gln*, glutamine; *imp*, mating-defective; *lf*, long flagella; *lts*, light-sensitive; *mat*, maternal inheritance altered; *msr*, methionine sulfoximine resistant; *mt*, mating type; *nic*, nicotinamide-requiring; *nit*, nitrate reductase-deficient; *nr*, neamine-resistant; *pab*, p-aminobenzoic acid-requiring; *pc*, protochlorophyllide; *pf*, paralyzed flagella; *pr*, paromomycin-resistant; *pyr*, pyrithiamine-resistant; *spr*, spectinomycin-resistant; *su*, *sup*, suppressor; *thi*, thiamine-requiring; *uni*, single flagellum; *vfl*, variable number of flagella; *w*, white; *y*, yellow. (From E. Harris, *Chlamydomonas* Genetics Center, Duke University)

essential for respiration. The Sherman experiments can be said to be the historical equivalent of Benzer's experiments with the *rII* gene of T4 (*Section 12.13*) in that they provided the first detailed map of a eukaryotic gene.

Sherman and colleagues screened for *cyc1* mutations by first isolating yeast strains that could not respire, could not use lactate as a carbon

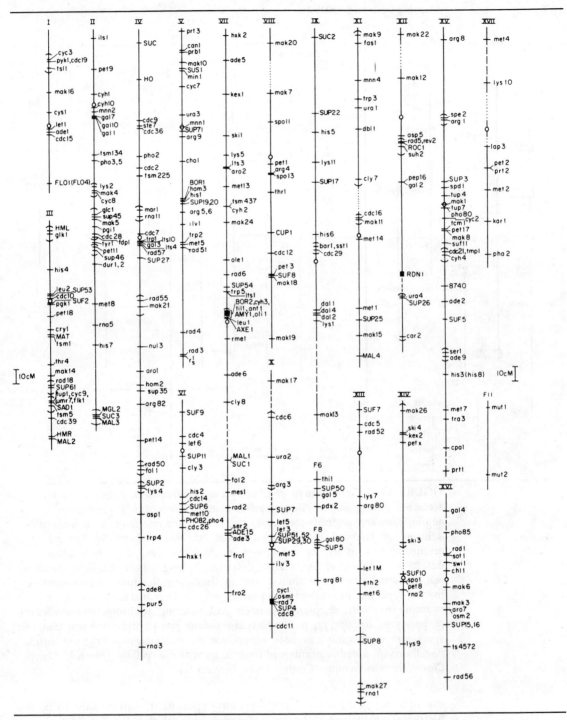

**FIGURE 14-16 Genetic map of *Saccharomyces*.** Solid lines represent linkage distances established by tetrad analysis; dashed and dotted lines (not drawn to scale) show linkage established by mitotic and aneuploid analysis, respectively. A key to the gene symbols can be found in R. K. Mortimer and D. Schild, *Microbiological Reviews* 44:519–571, 1980.

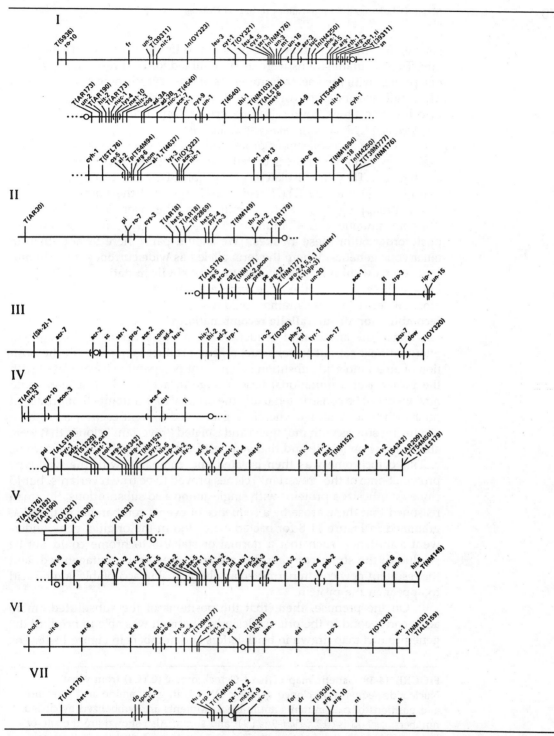

**FIGURE 14-17 Genetic map of *Neurospora crassa*.** Order of markers in parentheses is uncertain. Linkage group I is estimated to be at least 200 map units long, and the total for all seven groups probably exceeds 1000. A key to symbols and a list of unmapped genes is found in Neurospora Newsletter **29** (1982). Courtesy of D. D. Perkins and M. Björkman.

469

source, and that were therefore resistant to the toxic effects of chlorolactate. These mutant strains were then examined for the presence or absence of a protein with the spectral properties of iso-1-cytochrome c and a total of 210 single-site mutants was isolated; these mutants either lacked iso-1-cytochrome c altogether or else produced a nonfunctional cytochrome (*recall Section 11.18 on yeast nonsense mutations*).

Sherman and colleagues proceeded to cross the various *cyc1* mutants against one another and calculated the frequency of *CYC1* wild-type recombinants. They also isolated a collection of strains carrying deletions that extended into the *CYC1* region and subjected the point mutants to deletion mapping (*Section 12.13*). In this way they were able to assign the 210 point mutations to 47 mutational sites in the gene and to unambiguously order 60 of these 47 sites. [We should pause here to note than in eukaryotic genetics, where the term allele has wide currency, the 210 mutant strains would usually be said to carry **allelic mutations, multiple alleles,** or **heteroalleles** (the three terms are synonymous). Similarly, recombination between these strains would usually be spoken of as **heteroallelic recombination** or **interallelic recombination.**]

Since the amino acid sequence of iso-1-cytochrome c was known, Sherman next set out to construct a **gene-product map** wherein the position of an amino acid substition in a mutant polypeptide is correlated with the position of a mutational base change in a gene. Such a map could obviously not be constructed using the original *cyc1* mutants that produced no cytochrome. Sherman therefore proceeded to treat each of these non-producing strains with mutagens and isolated "revertant" clones that were able to grow on lactate and that synthesized a functional iso-1-cytochrome c. This cytochrome was then isolated and compared with the wild-type protein. Some of the "revertant" clones proved to be true revertants, but 15 clones synthesized proteins with single amino acid substitutions. Sherman reasoned that these arose by a sequence of events comparable to that diagrammed in Figure 11-8 for bacteria: a codon in the original gene underwent a mutation such that a normal or stable cytochrome could not be produced; the strain carrying this mutation was again mutagenized, and the resultant new codon permitted an alternate, but acceptable, amino acid to appear in the protein.

On the premise, then, that the positions of the substituted amino acids correspond to the original lesions, Sherman was able to produce the gene-product map drawn in black letters and numbers in Figure 14-18. The

**FIGURE 14-18 Genetic map of Iso-1-Cytochrome c (CYC1) from yeast.** Nucleotide sequence of cloned genomic DNA is in gray; amino acid sequence and positions of *cyc1* mutants are in black. Mutants are symbolized by circled numbers: *cyc1-13* is ⑬ , *cyc1-9* is ⑨; and so on. Also shown are the site of heme attachment, the ε-N trimethyllysine at position 77, and the N-terminal methionine that is excised from the normal protein. (From F. Sherman, et al., *Genetics* **81**:51, 1975, and M. Smith, et al., *Cell* **16**:753, 1979.)

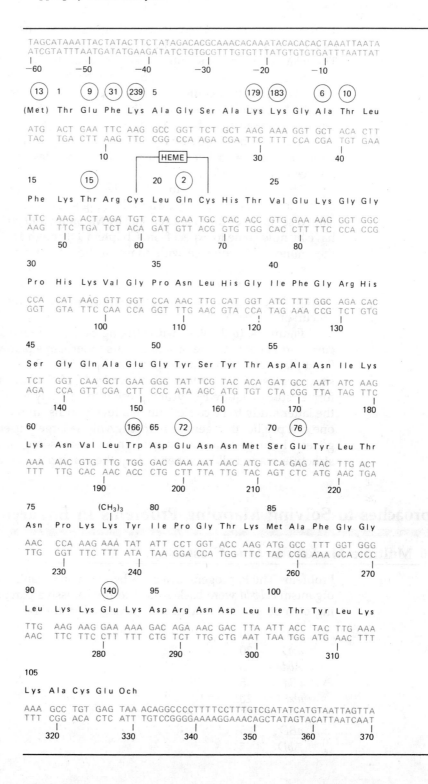

15 mutations are depicted as circled numbers so that, for example, the *cyc1*-13 mutation that affects the initial methionine residue is denoted ⑬. The relative order of the substituted amino acids proved to be identical to the relative order of the 15 mutations established by deletion mapping. In other words, **a strict colinearity was established between the genetic map and the biochemical positions of the amino acids.** Similar colinearity demonstrations were made by C. Yanofsky for the tryptophan synthetase genes of *E. coli,* and these provided strong evidence indeed for the concept of a structural gene as a linear sequence of nucleotides encoding a linear sequence of amino acids.

The gray letters in Figure 14-18 give the complete nucleotide sequence of the iso-1-cytochrome c gene and flanking segments, as determined for a cloned segment of the yeast genome by M. Smith and collaborators. As we have by now come to expect from Chapters 9 through 11, the codons match the amino acid sequence with complete fidelity and confirm all of Sherman's predictions about missense substitutions in the various mutant strains. It is noteworthy that no intervening sequences (*Section 9.9*) are present in this gene, nor is the protein initially synthesized as a longer precursor (*Section 10.9*).

Figure 14-18 displays with striking clarity the revolution that has occurred in genetics in the past five years. Sherman's pioneering studies defined plausible codons at 15 amino acid postions; Smith's sequence analysis provides *all* the codons plus hundreds of flanking nucleotides. The former study required 15 years; present technology would probably allow the latter study to be carried out in a few months. In the future, therefore, one can predict that genetic mapping will be largely used to establish the general location of a mutant gene in a chromosome, and that gene fine-structure maps will be generated by the cloning/sequencing technology.

# Approaches to Solving Mapping Problems in Eukaryotes

## 14.16 Multiple Markers in One Cross

**Problem**  The $F_1$ progeny of a cross between an organism *AABBDD* and an organism *aabbdd* were backcrossed to the recessive parent and the following phenotypic frequencies were observed in the $F_2$:

| | |
|---|---|
| *ABD* | 20 |
| *abD* | 20 |
| *Abd* | 5 |
| *aBd* | 5 |
| *abd* | 20 |
| *ABd* | 20 |
| *aBD* | 5 |
| *AbD* | 5 |

What is the relationship of the three markers $ABD$ to each other (assuming that they are on the same chromosome)?

**Solution**   Problems of this sort are best solved by considering pairs of markers separately. Thus for the $AB$ pair, we can classify the test-cross progeny (ignoring the recessive chromosome) as

$$
\begin{array}{lll}
AB & 20 + 20 = 40 \\
ab & 20 + 20 = 40
\end{array} \Big\} \; 80
$$

$$
\begin{array}{lll}
Ab & 5 + 5 = 10 \\
aB & 5 + 5 = 10
\end{array} \Big\} \; 20
$$

And can draw their linkage as $\underline{A \qquad 20 \qquad B}$
For the $BD$ pair we have

$$
\begin{array}{lll}
BD & 20 + 5 = 25 \\
bd & 5 + 20 = 25
\end{array} \Big\} \; 50
$$
$$
\begin{array}{lll}
Bd & 5 + 20 = 25 \\
bD & 20 + 5 = 25
\end{array} \Big\} \; 50
$$

The genes $B$ and $D$ are not linked genetically, even though they reside in the same chromosome

For the $AD$ pair we have

$$
\begin{array}{lll}
AD & 20 + 5 = 25 \\
ad & 5 + 20 = 25 \\
Ad & 5 + 20 = 25 \\
aD & 20 + 5 = 25
\end{array}
$$

$A$ and $D$ are similarly not linked.

# 14.17 Determining Phenotypic Classes

**Problem**   In snails, pink shell color may be due to the homozygous condition of either or both of the recessive genes $a$ and $b$. Brown shell color is dependent upon the presence of both dominant $A$ and $B$ genes. Pink male snails of genotype $AAbb$ were crossed with pink females of genotype $aaBB$. The $F_1$ brown females were crossed with pink males of genotype $aabb$. There were 710 progeny of which 125 were brown and 585 were pink.

a. What ratio is expected for unlinked genes?
b. Based on the current data, what is the percent recombinants?
c. What is the map distance?
d. In what percent of the tetrads did a crossover between the two genes occur?

**Solution**

1. Write out all available information in genetic shorthand. Begin by assuming partial $AB$ linkage:

Genotype of $F_1$ brown females: $\dfrac{Ab}{aB}$

Genotype of testcross males: $\dfrac{ab}{ab}$

Possible genotypes of testcross progeny:

Parental: $\dfrac{Ab}{ab}$ and $\dfrac{aB}{ab}$

Recombinant: $\dfrac{AB}{ab}$ and $\dfrac{ab}{ab}$

2. Recognize that the 125 brown progeny must correspond to the $\dfrac{AB}{ab}$ recombinant types since these are the only class with both $A$ and $B$ genes.

3. Assume reciprocal recombination, so that there should also be 125 $\dfrac{ab}{ab}$ pink snails among the progeny, giving a total of 250 recombinants.

4. The remaining $585 - 125 = 460$ pink testcross progeny should have the reciprocal parental genotypes, that is,

$$\left.\begin{array}{l} \dfrac{Ab}{ab} = 230 \\[2em] \dfrac{aB}{ab} = 230 \end{array}\right\} \text{460 parental}$$

5. Note that 460 parental + 250 recombinant yields the given total of 710.
6. Now proceed to answer the questions.
   a. For unlinked genes, one would expect equal proportions of $AB$, $ab$, $Ab$, and $aB$ progeny and a ratio of 75 percent pink to 25 percent brown. The excess of pinks (82 percent) signals linkage.

   b. Percent recombination $= \dfrac{250}{710} = 35$ percent.

   c. Map distance = 35 map units.
   d. A crossover occurred in 70 percent of the tetrads.

(a) Burke Judd (University of Texas) studies gene-chromomere relationships in the *zeste-white* region of *Drosophila*. (b) Fred Sherman (University of Rochester) has analyzed the fine structure of the cytochrome-*c* gene in yeast.

# Questions and Problems

1. A *Neurospora* zygote was constructed with the genotype *AaBb*. From ordered tetrad data, how would you tell
   (a) if *A* and *B* are linked.
   (b) the genotype of the parental strains.

2. For the cross

   list all the different kinds of *Neurospora* tetrads that can be obtained from a *single* crossover event. Recall that crossovers can occur between *a* and the centromere, between *a* and *b*, and below *b*. They might also involve different pairs of strands.

3. The following genes are known to be on the same autosome: *d, e, i, k, l,* and *n*. On the basis of the following recombination frequencies, what is the gene order? (If you are right, you will know it).

   | | | | |
   |---|---|---|---|
   | *l-d* | 25% | *n-e* | 6% |
   | *l-i* | 10% | *i-n* | 5% |
   | *d-e* | 4% | *k-l* | 18% |
   | *d-k* | 7% | *l-n* | 15% |

4. (a) In a cross between an $a+c/+b+$ male and an $abc/abc$ female, list the classes of offspring that you would expect to find in the $F_1$ in the order of decreasing frequency, assuming the order of the genes is as written. The $a - b$ distance is 16, and the $b - c$ distance is 6. (b) If you knew that the classes listed in the order of decreasing frequency were

**1.** $a+c/abc$            **3.** $abc/abc$
   $+b+/abc$               $+++/abc$
**2.** $+bc/abc$            **4.** $ab+/abc$
   $a++/abc$               $++c/abc$

what would be the order of the genes?

5. In *Drosophila* three linked genes, $a$, $b$, and $c$, gave the following results when $ac$ females were crossed to $b$ males and the $F_1$ inbred:
$F_1$: the phenotypes were $+$ female and $ac$ males
$F_2$:

| Phenotypes | Number of Females | Number of Males |
|:---:|:---:|:---:|
| $a$ | 49 | 2 |
| $b$ | 0 | 428 |
| $c$ | 49 | 48 |
| $+$ | 451 | 23 |
| $ab$ | 0 | 47 |
| $ac$ | 451 | 428 |
| $bc$ | 0 | 1 |
| $abc$ | 0 | 23 |
| | 1000 | 1000 |

(a) Are the three genes sex linked or autosomal? (b) What are the parental and $F_1$ genotypes? (c) What is the percent recombination between the three genes? (d) What is the sequence of the three genes in their chromosome? (e) What is the coefficient of coincidence and is there any interference?
Hint: there are "hidden" recombinants whose phenotypes are different from their genotypes. Their frequency must be inferred from the data.

6. Two male plants are scored for the phenotypes of the pollen (*review Sections 5.13 and 6.8*) they produce. The possibilities are: *brown* (*br*), *crinkly* (*cr*), *glossy* (*gl*), *green* (*gr*), and *small* (*s*), and the results are:

| Plant Number 1 | | Plant Number 2 | |
|:---|:---:|:---|:---:|
| *Pollen Type* | *Number* | *Pollen Type* | *Number* |
| *br cr gl* | 7 | *gl s +* | 129 |
| *+ + cr* | 330 | *s + +* | 4 |
| *+ gl cr* | 123 | *gr + +* | 138 |
| *+ cr br* | 670 | *+ gl +* | 873 |
| *+ gl br* | 366 | *+. + +* | 337 |
| *br + +* | 99 | *s + gr* | 846 |
| *+ + +* | 4 | *gl + gr* | 2 |
| *+ + gl* | 721 | *gr gl s* | 361 |
| Total | 2320 | Total | 2690 |

You learn from the published literature that

$$R_{gr,br} \cong 0.04$$

(a) Give the genotypes of plant number 1 and plant number 2. (b) Calculate recombination values. (c) Map the genes. (d) What aspect of the data (or the results derived from them) lacks internal consistency?

7. In the cross

$$\begin{array}{c} \text{♀} \qquad \text{♂} \\ Q \parallel q \quad q \parallel \\ \qquad \times \\ k \parallel + \quad k \parallel \end{array}$$

where
$Q$ = dominant mutation, quiet voice
$q$ = normal allele, loud voice
$k$ = recessive mutation, krinkly whiskers
$+$ = normal allele, straight whiskers

The $F_1$ progeny types are

1. Quiet, krinkly
2. Loud, straight
3. Loud, krinkly
4. Quiet, straight

(a) What is the significance of these progeny types resulting from this cross? (b) Design a cross to demonstrate that these genes are sex linked. (c) How could the cross $\dfrac{Qk}{q+} \times \dfrac{Q+}{\underline{\qquad}}$ be used to obtain valid data for mapping studies?

8. Given that four chromatids can cross over in two-, three-, and four-strand double crossovers and your general knowledge of genetics, which of the following statements are true? Explain your answers. (a) The more strands involved in crossing over, the fewer parental-type chromatids and the more recombinant-type chromatids. (b) The same two chromatids will very likely exchange genetic material via crossovers several times. (c) If two chromosomes (four chromatids) work it right, when they get finished exchanging information no resultant chromatid will carry the same information as either parent. Diagram.

9. Suppose mapping studies showed that the gene for sickle cell anemia was near the end of a linkage group, whereas *in situ* hybridization of labeled hemoglobin mRNA from reticulocytes showed a major band of grains halfway between the telomere and the centromere of a single chromosome type. Are these data compatible? Why or why not?

10. Diagram a series of not more than five crosses that will produce homozygous *white miniature* females starting with Morgan's two mutant male *Drosophila*, one *white* and one *miniature*, and a stock of wild-type females (recall that *w* and *m* are sex-linked genes).

11. Diagram the meiotic events occurring in the cross shown in Figure 14-1.

12. If a genetic marker is present in the X and Y chromosomes in the region of their homology, would it behave in crosses as a sex-linked trait? If not, how would its presence be detected in the X and Y chromosomes (assume that crossing over occurs equally in both sexes).

13. Duchenne-type muscular dystrophy is an inherited human disease, and one type of color blindness is sex linked. Two healthy parents with normal color vision have fully normal daughters, half their sons are healthy but color blind, and the other half has normal color vision but Duchenne's disease. (a) Is

Duchenne's disease an autosomal or a sex-linked trait? (b) Diagram the probable genotypes of the parents. (c) Could such parents ever have a normal son? If so, under what circumstances?

14. In *Drosophila*, the markers *a*, *b*, and *c* are crossed in all three possible pairwise combinations to give the following recombination frequencies:

$$R_{a,b} = 0.06$$
$$R_{a,c} = 0.10$$
$$R_{b,c} = 0.06$$

What is the expected frequency of *a* + *c* eggs produced by a female with an $\dfrac{a \quad b \quad c}{+ \ + \ +}$ genotype?

15. The recessive genes *an* (*anther* ear), *br* (*brachytic*), and *f* (*fine* stripe) all lie in chromosome 1 of maize. When a plant heterozygous for these markers is test-crossed with a homozygous recessive plant, the following results are obtained (data from R. A. Emerson):

| Test-Cross Progeny | Numbers |
| --- | --- |
| wild type | 88 |
| fine | 21 |
| brachytic | 2 |
| brachytic fine | 339 |
| anther | 355 |
| anther fine | 2 |
| anther brachytic | 17 |
| anther brachytic fine | 55 |
| | 879 |

Determine the sequence of the three genes and their map distances.

16. Calculate the coefficient of coincidence for the phage cross given in Section 12.20.

17. In *Drosophila* the sex-linked genes *cut* (*ct*), *lozenge eyes* (*lz*), and *forked* bristles (*f*) are the following map distances apart: *cut* to *lozenge*, 7.7 units, and *lozenge* to *forked*, 29.0 units. Assuming that there is no interference, what are the expected numbers of genotypes out of 1000 flies recovered from the cross $\dfrac{ct\ lz\ f}{+++} \times \dfrac{ct\ lz\ f}{\longrightarrow}$? What effect will a coefficient of coincidence of 0.5 have on your answer?

18. (a) In the absence of interference, what is the theoretical maximum recombination frequency (in percent) between two markers? (b) Can the total genetic map length of a chromosome ever exceed this value? Explain. (c) If a certain pair of homologues exhibits an average of 2.5 chiasmata per meiosis, what is the expected genetic map length of the entire chromosome?

19. Female corn flowers were made heterozygous for markers *c* (*colorless* aleurone), *wx* (*waxy* endosperm) and *sh* (*shrunken* endosperm) and were pollinated by male flowers homozygous for these markers. Derive a linkage map for these three loci from the following data, where traits are wild type unless denoted as mutant.

| Progeny Phenotype | Number |
|---|---|
| wild type | 17,959 |
| *colorless* | 524 |
| *waxy* | 4455 |
| *shrunken* | 20 |
| *colorless, waxy* | 12 |
| *colorless, shrunken* | 4654 |
| *waxy, shrunken* | 509 |
| *colorless, waxy, shrunken* | 17,699 |

**20.** An analysis of tetrads in a cross of *Chlamydomonas* involving the loci *pf* (*paralyzed flagella*) and *nic* (requirement for the vitamin nicotinamide) gave the following results: PD = 70, NPD = 2, and T = 28. Calculate the linkage relationship between *pf* and *nic*.

**21.** In *Chlamydomonas* the genes designated here as *a, b,* and *c* are linked. The following are the tetrads recovered from a cross involving these three loci.

| (1) | (2) | (3) | (4) | (5) |
|---|---|---|---|---|
| *a b c* | *a b c* | *a b c* | *a b c* | *a + +* |
| *a + +* | *a + c* | *a b c* | *a + +* | *a b +* |
| *+ b +* | *+ b +* | *+ + +* | *+ b c* | *+ + c* |
| *+ + c* | *+ + +* | *+ + +* | *+ + +* | *+ b c* |

| Number of tetrads | | | | |
|---|---|---|---|---|
| 10 | 10 | 440 | 160 | 10 |

| (6) | (7) |
|---|---|
| *a b +* | *a b c* |
| *a + c* | *a b +* |
| *+ b c* | *+ + c* |
| *+ + +* | *+ + +* |

|  |  |
|---|---|
| 10 | 360 |

(a) What are the genotypes of the parents? (b) Determine the map distances between *a, b,* and *c*. (c) Determine the simplest origin of each of the seven classes of tetrads. (d) What is the coefficient of coincidence?

**22.** Analysis of ordered tetrads in *Neurospora* from a cross *a b c* × + + + gave the following tetrad types:

| (1) | (2) | (3) | (4) | (5) | (6) |
|---|---|---|---|---|---|
| *a b c* | *a b +* | *a b c* | *a b +* | *a b c* | *a b +* |
| *a b c* | *a b +* | *+ b c* | *+ b +* | *a + c* | *a + +* |
| *+ + +* | *+ + c* | *a + +* | *a + c* | *+ b +* | *+ b c* |
| *+ + +* | *+ + c* | *+ + +* | *+ + c* | *+ + +* | *+ + c* |

| Number of tetrads | | | | | |
|---|---|---|---|---|---|
| 300 | 300 | 100 | 100 | 100 | 100 |

(a) Are any of the genes linked? If so, which ones? (b) If any of the genes are linked, how far apart are they? (c) Calculate the distance between each gene and its centromere. (d) Construct a genetic map from the analysis of the tetrad data.

23. In *Neurospora* the second-division segregation frequencies of *ser*, *thr*, and *arg* are found from the study of ordered tetrads to be 10 percent, 10 percent, and 16 percent, respectively.

From the cross $a + + + + \times A$ *ser thr arg ad* the following recombination frequencies are found:

$R_{ad,ser} = R_{ad,arg} = R_{ad,thr} = 50$ percent
$R_{ser,thr} = 9.8$ percent
$R_{thr,arg} = 12.6$ percent
$R_{a,ser} = R_{a,thr} = R_{a,arg} = R_{a,ad} = 50$ percent

Draw the indicated linkage group(s). Indicate the position of all markers and any centromeres called for by the data.

24. Diagram the origin of tetrads 1 to 9 in Table 14-3, showing the crossover events, if any, that would produce the nine tetrad genotypes.

25. Individuals from a stock of *Chlamydomonas* containing genes *a*, *b*, and *c* were mated to those from a wild-type stock. The following *unordered* tetrads were formed.

| (1) | (2) | (3) | (4) |
|---|---|---|---|
| + + + | a + + | a b + | a + + |
| + + + | a + c | a b + | + + + |
| a b c | + b + | + + c | + b c |
| a b c | + b c | + + c | a b c |
| 268 | 12 | 20 | 180 |

| (5) | (6) | (7) | (8) | (9) |
|---|---|---|---|---|
| a + + | + + c | + + + | a + + | + b c |
| a + + | + + c | a + c | + + + | a b + |
| + b c | a b + | + b + | + b c | a + c |
| + b c | a b + | a b c | a b c | + + + |
| 18 | 12 | 142 | 18 | 20 |

(a) Are the genes linked? If so, determine the order and linkage distance. (b) Is it more correct to calculate the intermediate distance between a group of genes and then add them or calculate the distance between the two outermost genes in one step? Why?

26. The following *Neurospora* cross was made

$+ sp inos \times ala + +$

where *inos* and *ala* are auxotrophic markers for inosine and alanine requirements, and *sp* is a mutant allele resulting in stunted growth. The data below were obtained from "random spores." The spores were grown in minimal media or with supplements as indicated:

| Type of growth (% total) | Medium I: Alanine and Inositol | Medium II: Alanine | Medium III: Inositol | Medium IV: Minimal |
|---|---|---|---|---|
| wild | 50.6 | 43.4 | 3.7 | 0.2 |
| stunted | 49.4 | 6.5 | 46.1 | 3.4 |

(a) Determine the gene order. (b) Determine the distances between the genes. Hint—predict the genotypes expected to grow on each of the four types of media. (c) Can the centromere be located using this data? if not, what additional types of data would be required?

27. In simple eukaryotes, an individual chromosome often comprises several meiotic linkage groups. Explain this statement. Does it suggest that simple eukaryotes have high or low chromosome numbers? High or low rates of meiotic recombination?

28. Suppose the analysis of 100 ordered tetrads from a *Neurospora* cross gave the data below (top and bottom of tetrads are not distinguished, i.e., the orders *AaAa* and *aAaA* are considered to be identical.)

| No. tetrads | 85 | 2 | 3 | 2 | 3 | 3 | 1 | 1 |
|---|---|---|---|---|---|---|---|---|
| Tetrad type | *Ab* | *Ab* | *Ab* | *AB* | *AB* | *Ab* | *Ab* | *aB* |
| | *Ab* | *AB* | *AB* | *Ab* | *Ab* | *aB* | *aB* | *Ab* |
| | *aB* | *ab* | *aB* | *ab* | *aB* | *Ab* | *aB* | *Ab* |
| | *aB* | *aB* | *ab* | *aB* | *ab* | *aB* | *Ab* | *aB* |

(a) Are the *A* and *B* loci linked or unlinked? Explain.
(b) What are the genotypes of the parents?
(c) At what stage of meiosis did the crossovers occur which produced each tetrad?

# 15 Somatic Cell Genetics

**A. Introduction**

**B. The Parasexual Cycle of**
   *Aspergillus*
   15.1 The Formation of
        Heterokaryons and Diploid
        Synkaryons
   15.2 Haploidization and Mitotic
        Crossing Over
   15.3 Genetic Analysis Using the
        Parasexual Cycle

**C. Genetic Analysis of Cultured**
   **Somatic Cells**
   15.4 Properties of Cultured
        Somatic Cells
   15.5 Intraspecific Somatic Cell
        Genetics
   15.6 Interspecific Somatic Cell
        Genetics
        Human-rodent hybrids
        Chromosome elimination and
           HAT selection

Detecting human traits in
   hybrid cells
Identifying human
   chromosomes in hybrid
   cells
Mapping by chromosomal
   mutations
Mapping by hybridization

**D. Gene Transfer or Eukaryotic**
   **Transformation**
   15.7 Gene Transfer Mediated by
        Chromosomes
   15.8 Gene Transfer Mediated by
        DNA Fragments
   15.9 Gene Transfer Mediated by
        Shuttle Vectors

**E. The Human Chromosome Map**

**F. Questions and Problems**

## Introduction

The eukaryotic genetics we have considered thus far (*Chapters 6 and 14*) has relied exclusively on sexual crosses: gametes were allowed to fuse, and the resultant zygotes or their meiotic products were analyzed. A far more uncommon event in nature is the fusion of somatic or vegetative cells. When such fusions occur between heterologous cells, however, the nuclei may

482

also fuse and undergo processes akin to segregation and recombination. It thus becomes possible to analyze the genomes of somatic cells, and the strategies employed have come to be called **somatic cell genetics.**

"Natural" somatic fusions occur with regularity only in cases of certain fungi, notably *Aspergillus,* that undergo what is known as a **parasexual cycle.** G. Pontecorvo and E. Käfer, among others, developed principles to analyze the resultant organisms, and we begin this chapter by presenting the parasexual cycle of *Aspergillus* in some detail. We then show how these same genetic principles are applied to cultured somatic cells that have undergone "artificial" cell fusions in the laboratory. Analysis of such cells is becoming one of the most active fields in genetics and has, in particular, already yielded information about the human genome that could never have been obtained from sexual crosses.

We conclude the chapter by describing techniques that are collectively known as either **gene transfer** or **eukaryotic transformation** because they are formally analogous to bacterial transformation. In this case, a somatic cell is presented not with a second cell type but with chromosomal DNA from a second cell type. Since DNA cloning technology is producing a vast array of defined eukaryotic genes (*Chapter 4*), gene transfer offers exciting possibilities for eukaryotic genetics.

# The Parasexual Cycle of Aspergillus

## 15.1 The Formation of Heterokaryons and Diploid Synkaryons

The haploid filamentous fungus *Aspergillus nidulans* has a sexual cycle quite similar to that of *Neurospora crassa* (*review Figure 5-9*). In addition, both *Aspergillus* and *Neurospora* can undergo a process known as **heterokaryosis,** which is depicted in Figure 15-1. Cytoplasmic bridges first form between the somatic hyphae of two haploid organisms (also called **homokaryons**) that differ from one another at one or more genetic loci (Figure 15-1a). Nuclei and cytoplasm migrate across the bridges as well as up and down each hyphae. The resultant organism therefore comes to attain a mix of the two parental nuclei and is called a **heterokaryon** (Figure 15-1b). If the two parents are auxotrophic, requiring, for example, adenine and para-amino-benzoic acid as in Figure 15-1, then the nuclei will complement one another and the heterokaryons can be selected for growth on minimal medium.

In a *Neurospora* heterokaryon, the nuclei always remain haploid. In *A. nidulans,* however, two haploid nuclei of differing genotype will very occasionally fuse together to form a diploid heterozygous nucleus or **synkaryon** that, after several rounds of mitotic duplication, comes to generate a **diploid sector** of the hyphae (Figure 15-1c). One way to recognize such sectors is by the judicious use of genetic markers. In our example in Figure 15-1, one parent carries a gene *y* for yellow conidia, the other a nonallelic

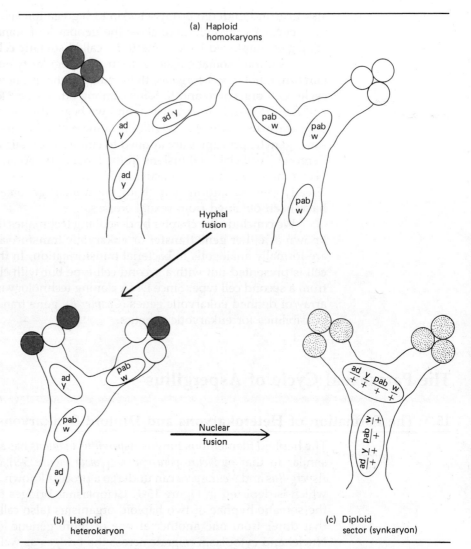

**FIGURE 15-1 Heterokaryon and diploid formation** between two *Aspergillus* homokaryons having the genotypes *ad y* and *pab w*. Yellow conidia are colored gray; green conidia are stippled. Genetic markers: *ad*, adenine-requiring; *y*, yellow conidia; *pab*, *p*-aminobenzoic acid–requiring; *w*, white conidia.

gene *w* for white conidia. The heterokaryotic conidia will continue to express one or the other color gene (Figure 15-1b), but the $\frac{+}{y}\frac{+}{w}$ diploid sectors will produce wild-type (green) conidia at their tips (Figure 15-1c). The green conidia, moreover, will be larger and will contain more DNA than do their haploid counterparts.

When diploid conidia are isolated and germinated, stable diploid organisms will form. Occasional sectors, however, will exhibit recessive traits; a hypha bearing white or yellow conidia, for example, may appear among the great majority bearing green conidia. In these sectors, **segregation** has occurred: $w$ and $y$ are no longer accompanied by their + alleles and are therefore expressed.

## 15.2 Haploidization and Mitotic Crossing Over

Segregant sectors can arise in stable diploid *Aspergillus* strains by one of two mechanisms. The first is a process known as **haploidization.** Presumably because of **mitotic nondisjunction** (Figure 15-2) or because one or more chromosomes do not become effectively included in the spindle at metaphase, aneuploid cells are produced that are either **hyperploid** ($2n + 1$, $2n + 2$, and so on) or **hypoploid** ($2n - 1$, $2n - 2$, and so on) (Figure 15-2). The latter cells turn out to be very unstable, to exhibit poor growth, and to continue to "throw off" chromosomes until they finally attain a stable haploid chromosome number. If during haploidization an eliminated chromosome carries a + gene for which the cell is heterozy-

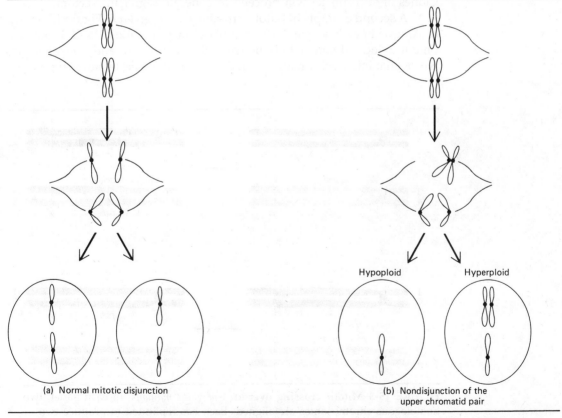

(a) Normal mitotic disjunction

(b) Nondisjunction of the upper chromatid pair

**FIGURE 15-2 Mitotic nondisjunction** caused by failure of the centromeres to divide. Only two chromosomes of the somatic genome are illustrated.

gous, then the emerging cell line will display the phenotype dictated by its mutant allele.

The second way that segregant sectors can arise in heterozygous diploids is by **mitotic crossing over.** Homologous chromosomes do not regularly undergo synapsis and crossing over during mitosis (*review Figure 2-9*), as they do during meiosis, but they become aligned sufficiently often to engage in occasional genetic exchange. In the mitotic anaphase that follows, each centromere divides and one copy of each homologue passes to each daughter nucleus. If, as in Figure 15-3a, a mitotic crossover occurs between the *pab* and *y* loci on chromatids 2 and 3, then the recombinant chromosomes drawn in Figure 15-3b will result. If a daughter nucleus receives chromatids 1' and 3', it will have the genotype $\dfrac{+ \quad +}{pab \quad +}$ and will continue to generate a normal phenotype. Normal phenotypes will also be generated by daughters that receive chromatids 2' and 3' or the nonrecombinant chromatids 1' and 4'. Any nuclei receiving chromatids 2' and 4', however, will have the genotype $\dfrac{+ \quad y}{pab \quad y}$, allowing expression of the *y* trait. Clones of such nuclei will proceed to generate segregant sectors.

A second example of mitotic crossing over is given in Figure 15-3c and d. An exchange has again occurred between chromatids 2 and 3, but this time it is located somewhere between the centromere and the *pab* locus. Nuclei that inherit chromatids 2' and 4' will in this case be homozygous for

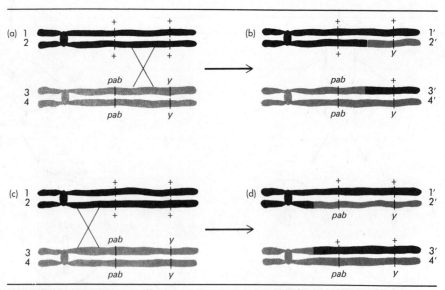

**FIGURE 15-3 Mitotic crossing over** in *Aspergillus nidulans*. In each panel, two homologous chromosomes are aligned. Each has replicated to produce two sister chromatids, labeled 1 through 4. After mitotic crossing over, the chromatids are labeled 1' through 4'.

*pab* as well as for *y* and will generate segregant sectors for both phenotypes.

The two examples given in Figure 15-3 illustrate an important general principle of mitotic crossing over. Using the terms **proximal** and **distal** to denote positions relatively close to and distant from the centromere, we can state the principle as follows. **Because the centromere divides at mitosis, any genes proximal to the site of a single genetic exchange in a given chromosome arm will always find themselves associated with the same alleles in the daughter nuclei as in the parent nucleus.** Thus in Figure 15-3a, where the parent nucleus is heterozygous for *pab* and its + allele, all possible daughter nuclei will also be heterozygous for *pab* and +. In contrast, all genes linked in coupling and distal to the site of the single exchange (in this case, *y* and its + allele) have the chance of being transmitted to a daughter nucleus in the homozygous condition. In Figure 15-3b, both *y* and *pab* and their + alleles are distal to the exchange, and segregant nuclei can therefore be homozygous at both loci.

## 15.3 Genetic Analysis Using the Parasexual Cycle

The series of events we have been describing—the fusion of two haploid nuclei in a heterokaryon, the occurrence of mitotic crossing over, and the return of the diploid nucleus to a (recombinant) haploid state via haploidization—has been called the **parasexual cycle.** The parasexual cycle has been used for genetic analysis of *A. nidulans* in two ways.

First, it is possible to assign markers to linkage groups by following their patterns of inheritance during haploidization. A haploid segregant will inherit one copy of each chromosome type, and this collection of *n* chromosomes should represent a random sampling of the two parental sets. In other words, some of the chromosomes that are thrown off during haploidization will be of "maternal" origin and some will be of "paternal" origin. It follows that markers linked in coupling on a "maternal" chromosome are usually inherited together, as are markers linked on its "paternal" homologue; the linkage is broken only in the relatively unlikely event that crossing over takes place in the appropriate region before haploidization gets underway. As an example we can consider a diploid made between a "maternal" *adenine*-requiring strain (*ad*) and a "paternal" strain carrying *y, w,* and *lys* (*lysine*-requiring). We can provisionally write the genotype of the diploid as $\frac{ad + + +}{+ \; y \; w \; lys}$. We now examine haploid segregants of this diploid and learn that they are often *ad w* or *ad lys* but almost never *ad y*. Since **markers of the same linkage group and contributed by different parents are only rarely inherited together during haploidization,** we can conclude that the *ad* and *y* loci are linked. The diploid's genotype is therefore more correctly written as $\frac{ad + \quad + \quad +}{+ \; y \quad w \quad lys}$.

The second use of the parasexual cycle is to determine **mitotic recombination frequencies** between genes along a chromosome. We can take as an example a diploid *A. nidulans* strain heterozygous for *y*, *ad*, *pab*, and *pro* (*proline-requiring*) and the + alleles of these genes. Haploidization data have indicated that the markers are all linked and the order of the markers has been determined by meiotic analysis to be *ad*, centromere, *pro*, *pab*, *y*. The four recessive alleles are introduced into the diploid in coupling (Figure 15-4). Diploid segregants that bear diploid *yellow* conidia are selected first, these being the most easily detected and *y* being the most distal

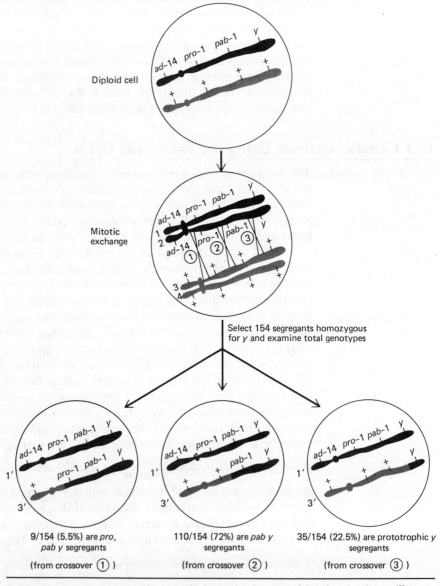

FIGURE 15-4 Genetic analysis utilizing mitotic recombination in *Aspergillus nidulans*.

marker on the chromosome. The complete genotypes of the segregant nuclei are then determined by the phenotypes they produce. It is found (Figure 15-4) that among segregants homozygous for *y*, 5.5 percent are homozygous for *pab* and *pro*, 72 percent are homozygous for *pab*, and 22.5 percent are prototrophic. Because none of the *y* segregants are adenine-requiring, one can deduce that *ad* is located on the other arm of the chromosome. From Figure 15-4 it should be apparent that **mitotic crossing over in one arm of the chromosome is without effect on the segregation of alleles in the other arm,** and the simultaneous occurrence of mitotic crossovers in both arms of the same chromosome will be very rare.

The preceding mitotic recombination frequencies cannot be converted into map distances in the same way that meiotic data are treated, for the total number of nuclei present in a hyphal mass (a number that corresponds to "total progeny" in conventional genetic analysis) is unknown. Moreover, it is not possible to recover all the products of a recombinational event, since only some will have a recognizable phenotype (in Figure 15-4, only the cells receiving chromatids 1' and 3' are scored as *y* segregants). A mitotic map therefore simply utilizes the recombination percentages we have obtained previously as map distances separating the markers. Obviously, the scale established for one chromosome arm will not necessarily relate to the scale established for any other arm.

In an organism such as *A. nidulans*, in which both a sexual and a parasexual cycle are known, it becomes possible to compare mitotic and meiotic map distances by expressing the latter on a percentage basis similar to the former. Typical results, shown in Table 15-1, demonstrate little corre-

**TABLE 15-1  A Comparison of Meiotic and Mitotic Linkage Data** in *A. nidulans*. The linkage-group I markers are those followed in the text.

Linkage group I, "left" arm (85 meiotic units total)

| Marker | *su* | | *ribo*1 | | *an*1 | | *ad*14 | |
|---|---|---|---|---|---|---|---|---|
| Interval | I | | II | | III | | IV | |
| Mitotic | 23 | | 7.4 | | 6.2 | | 63.4 | |
| Meiotic | 45.8 | | 22.4 | | 8.2 | | 23.6 | |

Linkage group I, "right" arm (44 meiotic units)

| Marker | | *pro*1 | | *pab*1 | | *y* |
|---|---|---|---|---|---|---|
| Interval | | V | | VI | | VII |
| Mitotic | | 5.5 | | 72.0 | | 22.5 |
| Meiotic | | 46.1 | | 18.0 | | 35.8 |

Linkage group II, "left" arm (46 meiotic units)

| Marker | *Acr*1 | | *w*2 | |
|---|---|---|---|---|
| Interval | | VIII | | IX |
| Mitotic | | 14.7 | | 85.3 |
| Meiotic | | 54.3 | | 45.7 |

From J. A. Roper, in *The Fungi*, Vol. 2, G. C. Ainsworth and A. S. Sussman, Eds. New York: Academic Press, 1968, p. 604.

spondence between the meiotic and mitotic distances. For example, the mitotic-recombination data indicate that the *pro1* locus should be close to its centromere, whereas the meiotic data would map it farther away. Thus there are undoubtedly fundamental differences between the processes of mitotic and meiotic crossing over.

# Genetic Analysis of Cultured Somatic Cells

## 15.4 Properties of Cultured Somatic Cells

When a cell is taken from a sexually reproducing "higher" eukaryote and is cloned under defined conditions *in vitro*, the resultant culture is said to represent a **line** of **somatic cells.** Cultured somatic cells are of two types. **Primary** somatic cells are taken directly from a plant or animal and maintained for a finite period; these are most likely to retain the chromosomal makeup of the parental organism and to retain the phenotypic traits of the parental tissue. **Established cell lines,** on the other hand, have been selected for their ability to grow indefinitely in culture. Such cells frequently derive from malignant tissues, and their chromosomal makeup is typically different from that of the other tissues of the parental organism: chromosomes may be lost, added, deleted, translocated, and so on.

Somatic cells have been subjected to two sorts of genetic analysis, which are described in the following sections. The first, which we designate **intraspecific** somatic cell genetics, refers to studies of cells from a single species; these have been subjected to mutagenesis, selection, parasexual manipulations, and complementation analysis, the goal being to augment the genetic information available from sexual crosses. The second, called **interspecific** somatic cell genetics, refers to the analysis of **hybrid cells** from two different species, most often between human and rodent. Such hybrids prove to eliminate human chromosomes preferentially, a phenomenon that has permitted the assignment of many human genes to particular chromosomes.

## 15.5 Intraspecific Somatic Cell Genetics

For a somatic cell line to be amenable to genetic analysis, it should be haploid or "functionally haploid," so that recessive mutations can be induced and expressed. This requirement is most easily met by higher plant cells, since cultures can be started from haploid pollen cells and, as noted in Section 5.15, the haploid state is more likely to be tolerated by higher plants. Haploid clones can then be mutagenized and subjected to selection procedures, much the same as those described in Chapter 7 for haploid microorganisms. Thus amino acid and vitamin auxotrophs of cultured *Nicotiana tabacum* cells have been isolated using the BuDR enrichment procedure (*Section 7.3*); streptomycin-resistant lines of *Petunia* have been selected from streptomycin-containing media; and so on. Of direct agricultural interest are attempts to isolate mutant plant cell lines resistant to fungal or bacterial pathogens or to such adverse conditions as cold or wind. Since it is possible to induce plant somatic cell lines to develop into complete plants that produce seed, such new cell lines are possible progenitors of new and more vigorous crop plants.

True haploidy may be incompatible with animal cell viability in culture since attempts to produce haploid animal cell lines have so far been unsuccessful. The Chinese Hamster Ovary **(CHO)** cell line has, however, proven to be "functionally" haploid at many loci. Although CHO cells possess a near-diploid amount of DNA and most of the original bands of the Chinese hamster karyotype remain, the chromosomes have been extensively rearranged and this appears to have resulted in the effective inactivation of many loci throughout the genome. Consequently, although CHO cells are demonstrably diploid for certain genes, they are "hemizygous" for many others, and these can be marked by recessive gene mutations. F. Kao, T. Puck, and others have, for example, isolated CHO glycine, proline, and adenine auxotrophs; temperature-sensitive conditional lethals; and cells with enzyme deficiencies.

As we have stressed throughout the text, genetic analysis begins with a collection of mutants. The next requirement is that the mutant genomes be brought together. In somatic cell genetics, this is accomplished by **somatic cell fusion,** a technique pioneered by G. Barski, B. Ephrussi, and H. Harris. Two mutant cell types are cultured together in the presence of either **Sendai virus** particles or **polyethylene glycol;** either agent causes cells to fuse together, and the resultant **heterokaryons** typically go on to fuse their nuclei and become **synkaryons,** much as we saw for *Aspergillus* (*Figure 15-1*). The heterokaryons and synkaryons are usually selected from the background of unfused and fused homokaryotic cells on the basis of complementation between two of the markers introduced. For example, two CHO glycine auxotrophs, *gly A* and *gly B,* may be grown and exposed to Sendai virus in a glycine-supplemented medium; the culture is then transferred to a glycine-free medium on which only the complementing heterokaryons can grow.

Using this system, it is possible to ask whether a set of mutant cell lines carries dominant or recessive mutations and whether these belong to the same or to different complementation groups. Suppose, for example, the *gly A* and *gly B* lines are subjected to mutagensis and adenine-requiring lines (*ad-1, ad-2,* and so on) are selected by replica plating. One first wants to know whether the *ad* mutations are dominant or recessive. This is accomplished by fusing, say, a *gly A ad-1* line with a *gly B +* line and asking whether the resultant diploids (glycine prototrophs) are auxotrophic for adenine (dominant mutation) or prototrophic for adenine (recessive mutation). All recessive lines can then be subjected to pairwise complementation tests (*Section 12.4*). Thus if *ad-1* and *ad-2* are recessive, then a *gly A ad-1* line can be fused with a *gly B ad-2* line and the selected *gly A gly B* heterokaryons transferred to an adenine-free medium. If *ad-1* and *ad-2* complement (that is, affect independent genes), then the cells will grow; there will be no growth if the two mutations affect the same gene. In this fashion, CHO cells have been shown to possess at least ten essential genes involved in adenine biosynthesis.

The final events in a fungal parasexual cycle—mitotic recombination and chromosome elimination—are as yet impossible to manipulate in somatic cells. If mitotic recombination occurs at all, its detection awaits larger collections of linked markers, and chromosome elimination is very rare in CHO/CHO fused cells. Nonetheless, L. Chassin has been able to ascertain that the loss of the chromosome bearing the *gly A* marker is not accompanied by the loss of *gly B*, and vice versa, indicating that the two markers lie in different chromosomes. In the terminology of somatic cell genetics, *gly A* and *gly B* are said not to be **syntenic.** Synteny means, literally, "located in the same strand," where "strand" designates a chromosome and not a polynucleotide chain. Thus, **synteny designates linkage between markers as analyzed in somatic cell crosses.**

## 15.6 Interspecific Somatic Cell Genetics

The experiments described in the previous section concern the fusion and interaction of cells of the same species. It is also possible to fuse cells of different species, forming **interspecific cell hybrids.** In the case of plant cells, development of this approach has great agronomic potential, since the desirable features of two crop plants might be combined. Thus far, only closely related tobacco species have generated a hybrid plant from a fused somatic cell, but an increasing number of laboratories are investigating such approaches.

For animal cells, interspecific somatic fusion is different from intraspecific fusion in that it is often followed by a very **efficient elimination of chromosomes.** Usually, although not inevitably, the chromosomes of one species are lost in preference to the other. Specifically, in human-rodent hybrids, human chromosomes are preferentially eliminated, a phenomenon that has had far-reaching consequences for human genetics since it allows the assignment of human genes to human chromosomes. Details are given in the paragraphs that follow.

**Human-Rodent Cell Hybrids**  In a typical protocol for obtaining human-rodent cell hybrids, primary human cells, usually leukocytes or fibrocytes, are obtained from donors with a known karyotype and genetic constitution. These are then mixed with CHO cells or with cells from established mouse lines that carry conditional mutations; for example, a CHO glycine auxotroph may be used. Following Sendai-mediated fusion, cells are transferred to glycine-free medium. Since the hybrid cells contain human wild-type genes for glycine prototrophy, they are able to flourish on this medium. Any unfused CHO cells, on the other hand, fail to grow without glycine, and unfused human leukocytes also fail to proliferate in culture and are gradually diluted out. Therefore, only the human-CHO hybrids survive.

**Chromosome Elimination and HAT Selection**    Although it is not yet clear how chromosomes are lost from hybrid cells, it is found experimentally that after human-rodent hybrids have undergone about 30 rounds of mitosis, they contain a full or near-full complement of rodent chromosomes but an average of only seven human chromosomes (the range extending from 1 to 20). When the karyotypes of such cells are analyzed, the human chromosomes are usually found to have been lost in a random fashion. It is possible, however, to assure that a particular human chromosome will be among those retained by including an additional selective feature in the hybrid cell growth medium.

An important example is a medium known as **HAT** (for **h**ypoxanthine, **a**minopterin, and **t**hymidine), which is widely used as a selective agent in somatic cell genetics. Figure 15-5 summarizes the biochemical pathways that form the basis of HAT selection. The important points are as follows.

1. In the *de novo* synthesis of purine nucleotides, the first nucleotide intermediate is IMP; this is then converted to AMP or to GMP (via XMP). **Synthesis of IMP is inhibited by aminopterin** present in **HAT.**

2. When *de novo* synthesis is blocked, purine nucleotides can still be produced via salvage pathways wherein free purines are condensed with phosphoribosylpyrophosphate (PRPP) in the presence of enzymes known as phosphoribosyl transferases. Specifically, if a cell is poisoned by aminopterin but possesses normal levels of hypoxanthine guanine phosphoribosyl transferase (**HGPRT**), it will be able to use the hypoxanthine in **HAT** medium to synthesize IMP, and therefore the other purine nucleotides.

3. Aminopterin also blocks the synthesis of thymidine nucleotides (Figure 15-5b). Therefore, aminopterin-poisoned cells must also be provided the thymidine in HAT medium in order to grow. The exogenous thymidine cannot be used unless the cells possess normal levels of the enzyme **thymidine kinase (TK)** (Figure 15-5b).

To illustrate **the use of HAT medium for chromosome selection,** let us assume that a parent mouse cell line lacks functional TK, whereas human primary cells have normal levels of TK. When these are fused and plated to HAT, the human-mouse hybrids are initially selected for growth because they possess the human TK. More importantly, if subsequent chromosome elimination causes any hybrid cell to lose both human chromosomes carrying the *Tk* gene, that cell will die. Therefore, all hybrids that survive for long periods on HAT will be expected to possess at least one *Tk*-bearing human chromosome.

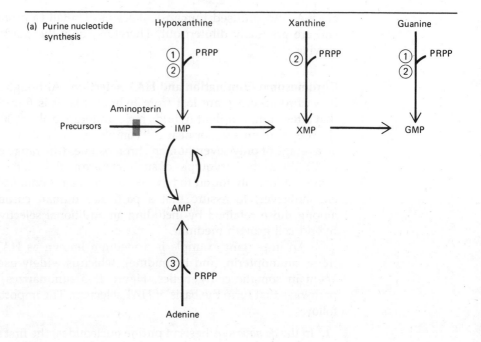

(a) Purine nucleotide synthesis

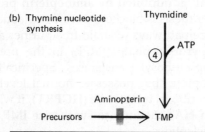

(b) Thymine nucleotide synthesis

**FIGURE 15-5 Purine and thymine nucleotide synthetic pathways in HAT selection.** (1) Reactions catalyzed by eukaryotic hypoxanthine guanine phosphoribosyl transferase (HGPRT). (2) Reactions catalyzed by the *E. coli* enzyme xanthine guanine phosphoribosyl transferase (XGPRT). (3) Reaction catalyzed by adenine phosphoribosyl transferase (APRT). (4) Reaction catalyzed by thymidine kinase (TK).

**Detecting Human Traits in Hybrid Cells**   Once stable clones of hybrid cells have been obtained, the next step is to ask which clones express a particular human trait. If it is a **selected** human trait (that is, if the parent rodent cells lack the activity, as with the TK-deficient mouse line), then it should be expressed by all the surviving clones, as in the foregoing example. If it is an **unselected** trait, then individual clones must be analyzed for the particular human gene product of interest. To give an example, the human form of lactate dehydrogenase B (LDH-B) has a different electrophoretic mobility pattern (*Box 3.2*) from the corresponding mouse or CHO enzymes and can be detected in gels by a specific staining reaction. Clones that carry the human form of the enzyme can therefore be identified by subjecting samples of numerous hybrid clones to the electrophoresis assay.

**Identifying Human Chromosomes in Hybrid Cells**  The final step in assigning a human gene to a chromosome is to **determine which human chromosome is shared in common by all the clones exhibiting the particular human trait;** this chromosome is then assumed to carry the relevant gene. Thus in the case of thymidine kinase, karyotypes were prepared from a number of hybrid clones that survived extended growth on HAT medium, and the one human chromosome consistently present was found to be chromosome 17. Here, of course, banded chromosome preparations (*Section 2.11*) are critical to assure correct identifications.

Linkage between markers can also often be deduced from segregation data. The human gene for LDH-A, for example, was first shown in chromosome elimination studies to reside on chromosome 11. The hybrids that retained LDH-A activity were then analyzed for the presence of various cell-surface antigens, and all were found to express an antigen known as $A_1$. It is concluded, therefore, that both the *ldh-a* and the $a_1$ genes are syntenic and lie on chromosome 11.

**Mapping by Chromosomal Mutations**  The ultimate goal of somatic cell genetics is, of course, to define the loci of individual genes along identified chromosomes. We give below two examples to illustrate how chromosomal mutations (*Sections 7.7 through 7.9*) are being used to attain this goal.

The first example entails **translocation mapping.** Translocation chromosomes (*Section 7.9*) can be recognized in hybrid somatic cells by their altered lengths and banding patterns. The reciprocal translocation illustrated in Figure 15-6a was found to move a large piece of the human X chromosome onto chromosome 14. When hybrid cells carrying this chromosome were induced to lose it, they simultaneously stopped expressing four human genetic markers: *Hgprt* (the enzyme HGPRT; Figure 15-4), *Pgk* (phosphoglycerate kinase), *G6pd* (glucose-6-phosphate dehydrogenase), and *Np* (nucleoside phosphorylase). The first three markers were known to reside on the X chromosome; the fourth was known to reside on chromosome 14. Since the translocation has caused all four to become syntenic, it follows that the three X-linked markers must all reside in the arm of the X chromosome distal to band q15, the site of the translocation break (Figure 15-6a).

The second example entails **deficiency mapping** (*Section 14.7*). As illustrated in Figure 15-6b, a second cell line was found to carry a human X chromosome missing all the genes distal to band q24. This cell line was found to express the human phosphoglycerate kinase gene but was lacking the *Hgprt* and *G6pd* markers. Combining this result with the information obtained from the translocation chromosome (15-6a), the *Pgk* gene can be localized between the q15 and q24 bands, and *Hgprt* and *G6pd* must lie between q24 and the X terminus.

FIGURE 15-6
**Translocation and deficiency chromosomes** used to map human genes in hybrid cells. (After F. H. Ruddle and R. P. Creagan, *Ann. Rev. Genetics* **9**:407, 1975.)

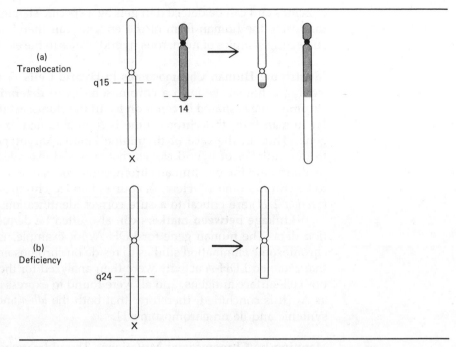

**Mapping by Hybridization** The procedures just described can only be used for genes that are expressed in cultured cells and that are therefore scorable by some phenotypic assay. This is not the case, however, for many genes of interest, an example being the genes that code for hemoglobin.

*In situ* hybridization (*Section 4.10*) can provide chromosome assignments for such genes. Thus F. Ruddle and coworkers prepared a radiolabeled probe of cloned DNA from the human β-globin gene (*see Figure 3-5*), and they applied it to somatic cell hybrids containing various combinations of human chromosomes. They found hybridization occurred only in cell lines that carried human chromosome 11, localizing the gene to that chromosome.

D. Housman, F. T. Kao, and their colleagues carried this analysis further. They began with a CHO line that had retained only human chromosome 11. They treated the cell line with clastogens (*Section 7.10*), and isolated five subclones that had lost various selectable chromosome 11 markers and that carried various deficiencies; a summary is presented in Figure 15-7a. By matching the lost traits with the lost bands, they were able to construct the chromosome 11 map shown in Figure 15-7b. They then isolated DNA from each of the five deficiency-carrying subclones, treated each DNA sample with the restriction enzyme *Eco R1* (*Section 4.11*), displayed the restriction fragments by gel electrophoresis (*Figure 4-8*), and asked which samples would form hybrids with the same radiolabeled β-globin probe used in the *in situ* hybridization experiment described in the

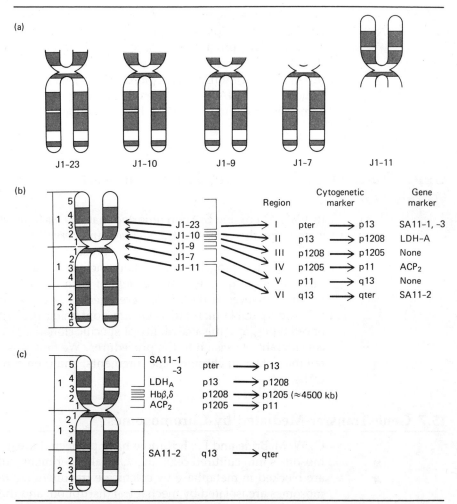

**FIGURE 15-7 Mapping the gene for human β-globin** within chromosome 11. The J1 designations refer to CHO-human hybrids carrying deficiencies at indicated bands. The numbering system to designate breakpoints is as follows: a band (e.g., p12) is visually subdivided into ten equal parts (numbered 01, 02, . . . 10), the 01 region being that closest to the centromere, so that p1205 (J1-9) is more proximal than p1208 (J1-10). The gene markers are the enzymes LDH-A and acid phosphatase 2(ACP₂) plus three cell-surface antigens coded by chromosome 11(SA11). (From J. Gusella, *et al., Proceedings of the National Academy of Science U.S.* **76**:5239, 1979.)

preceding paragraph. They found that hybrids formed when the DNA was derived from the parental hybrid line and from subclones J1-23, J1-10, and J1-11 (Figure 15-7a). No hybridization occurred, however, with the DNA derived from either J1-9 or J1-7. Thus the β-globin gene was pinpointed to the interval between the J1-10 breakpoint and the J1-9 breakpoint (Figure 15-7c).

As increasing numbers of human genes are being cloned by recombinant DNA techniques (*Chapters 4 and 10*), the chromosome mapping of these genes is progressing apace. In a typical experiment, a series of rodent-human hybrids with known compositions of human chromosomes (often called a **hybrid clone panel**) is subjected to hybridization with the radiolabeled DNA from a cloned human gene, and the labeled chromosome sector is then identified by autoradiography.

# Gene Transfer or Eukaryotic Transformation

In 1966, A. Fox and S. Yoon reported that if *D. melanogaster* mutants, homozygous for the recessive traits *brown* and *vermillion*, were mated and the fertilized eggs were incubated in the presence of DNA derived from wild-type flies, the progeny occasionally included a fly with wild-type traits. They went on to show that the yield of such "transformed" flies could be greatly enhanced if the DNA were injected into the eggs.

Comparable manipulations are now being performed with a variety of cell types, and the availability of specific cloned DNA species has added considerable precision to the procedures. We first present studies that entail the uptake of chromosome fragments, and then consider the uptake of DNA.

## 15.7 Gene Transfer Mediated by Chromosomes

O. W. McBride and F. H. Ruddle have pioneered the introduction of chromosomes into cultured cells. In a typical experiment, cultured human cells are blocked in metaphase by colchicine treatment (*Section 2.9*); their chromosomes are isolated by mechanical disruption; and the chromosomes are applied to recipient cells. If the recipient is a mouse cell line lacking thymidine kinase (*Section 15.6*) and the donor human cell line is TK$^+$, then transformed cells alone will grow on HAT medium (*Section 15.6 and Figure 15-5*).

Transformed cell lines prove to be of two types. One type, known as **unstable transformants,** carry independent human chromosomes or chromosome fragments that are rapidly lost from the genome. These can be used to establish genetic linkage in the same fashion that we saw for *Aspergillus* (*Section 15.2*) and for hybrid somatic cells (*Section 15.6*): **genes that are cotransferred during transformation and that are simultaneously lost during subsequent incubation are likely to be syntenic.** The second type, known as **stable transformants,** carry human chromosome fragments integrated into the mouse genome. Figure 15-8a illustrates a number of such hybrid chromosomes, where portions of lightly stained human chromosome (bracketed) are ligated to the dark mouse chromosomes. The integration process appears to be nonspecific: different mouse chromosomes are

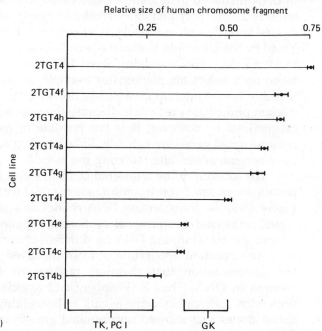

**FIGURE 15-8 Chromosome-mediated gene transfer.** (a) A mouse TK cell line transformed with human chromosome 17 containing the TK[+] gene is designated 2TGT7. Stable subclones of this cell line (e.g., 2TGT4a) carry fragments of chromosome 17 (lightly stained and bracketed) associated with various mouse chromosomes (darkly stained). Centromeres are identified by arrows. (b) Histogram displaying the sizes of the human fragments present in 2TGT4 and its eight subclones. The regions containing the genes for TK, GK, and PCI are indicated at the base of the histogram. (From L. A. Klobutcher and F. H. Ruddle, *Nature* **280**:657–660, 1979.)

involved in different cell lines, and the fragments may either fuse with centromeric regions or add onto chromosomal termini.

The stable transformants carrying the TK-containing chromosome regions displayed in Figure 15-8a were analyzed for their expression of the syntenic human genes GK (galactokinase) and PCI (type I procollagen). As summarized in Figure 15-8a and b, the three clones carrying the smallest human chromosome fragments have lost the GK gene but not the PCI gene. Banding analysis reveals, moreover, that the missing GK-containing portion is proximal to the human chromosome 17 centromere. These data establish a gene order of centromere-GK-(TK, PCI), where the order of TK and PCI remains undetermined. This approach is, of course, identical in concept to the deficiency mapping experiments summarized in Figure 15-7.

We noted in Section 2.12 that techniques are presently being developed for the purification of specific chromosome types. When available, they can clearly be used to advantage in chromosome gene transfer experiments.

## 15.8 Gene Transfer Mediated by DNA Fragments

In DNA-mediated gene transfer (also called **transfection**), the DNA of interest is usually precipitated with calcium phosphate and the precipitate is layered directly onto a monolayer of cultured cells. Uptake is quickly followed by the formation of **concatemers:** the entering DNA becomes ligated to form long pieces averaging about 1000 kilobases. Cell lines that have taken up a selectable marker (for example, a $TK^+$ gene) are commonly found to have taken up as well any nonselectable genes included in the calcium phosphate precipitate. Because the entering DNA undergoes concatamerization, however, it is not possible to conclude that groups of cotransformed genes are naturally linked in chromosomal DNA; they may have become linked after entering the cell.

The location of the transfered DNA in stable transformants has been probed by *in situ* hybridization (*Section 4.10*), with the results shown in Figure 15-9: the transforming DNA resides at a single chromosome site. Again, no unique chromosomal locations are apparent, since different lines contain the transforming DNA on different chromosomes.

An important application of DNA-mediated gene transfer has been the demonstration that **chemical carcinogens bring about heritable changes in DNA.** Thus R. Weinberg and associates treated various cell lines with such carcinogenic agents as benzo[a]pyrene (*Section 7.12*), selected clones that showed uncontrolled growth in culture, isolated their DNA, and used the DNA to transfect a stable recipient cell line. In one third of the cases, the recipient cell lines were converted into malignant variants that not only displayed uncontrolled growth in culture but also caused tumors when injected into newborn mice. DNA from untreated cells, by contrast, had no such ability to induce malignant transformation. These experiments appear to demonstrate that the **chemically induced**

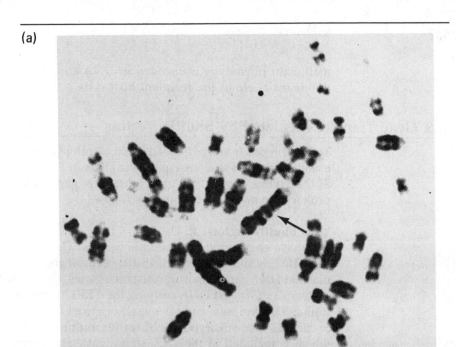

(a)

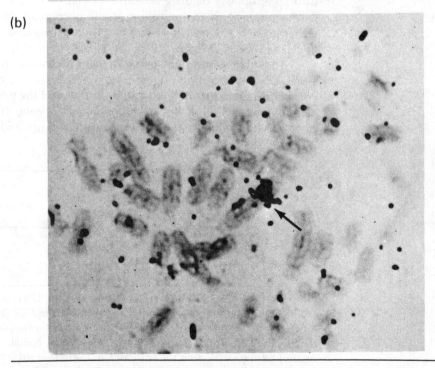

(b)

**FIGURE 15-9 DNA-mediated gene transfer.** (a) Giemsa-banded metaphase plate of a cultured TK⁻ rat liver cell line cotransformed with a TK⁺ gene from herpes virus and with cloned human DNA encoding genes for human growth hormone (HGH). (b) The same spread after annealing with radiolabeled HGH DNA. The large arrow in both photographs identifies the chromosome labeled after hybridization. (From D. M. Robins, S. Ripley, A. S. Henderson, and R. Axel, *Cell* **23:**29–39, 1981.)

malignant phenotype is encoded in DNA and that **the trait behaves as a dominant allele in the recipient host cells.**

## 15.9 Gene Transfer Mediated by Shuttle Vectors

You will recall from Section 4.13 that cloning vectors such as pBR322 and phage λ are used to clone eukaryotic DNA in *E. coli* hosts. **Shuttle vectors,** as their name implies, are used to **transfer genes between eukaryotic and prokaryotic hosts.** Two examples follow.

**Yeast Shuttle Vectors**  R. Davis, L. Clarke, and J. Carbon are among those who have engineered yeast shuttle vectors. The structure of one such vector, YRp7, is shown in Figure 15-10a. Critical are the presence of (1) *E. coli* plasmid DNA carrying drug-resistance genes; (2) a bacterial origin of DNA replication; (3) yeast DNA carrying the *TRP1*$^+$ gene that functions in tryptophan biosynthesis; and (4) a yeast origin of DNA replication (recall from Section 2.7 that eukaryotic DNA carries multiple replication origins, one of which is included in the *TRP*$^+$ fragment). The vector can be introduced into spheroplasts of drug-sensitive *E. coli* or *trp*$^-$ yeast cells; because it carries two origins of replication, it will perpetuate itself autonomously in either host. *E. coli* transformants can then be selected for their drug resistance, and yeast transformants are selected for tryptophan prototrophy.

To isolate a new gene using this shuttle vector, random fragments of yeast DNA are simply ligated into YRp7, and the vector is used to cotransform mutants defective in both trp1 and the gene of interest. As an example, we can follow the isolation of *cdc28*, a gene that encodes a protein necessary for cell division in yeast. K. A. Nasmuth and S. I. Reed prepared

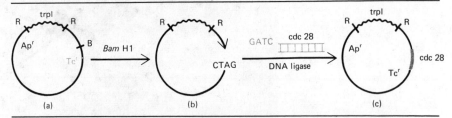

FIGURE 15-10 **Yeast Shuttle vector** YRp7, and the DNA transfer of the *cdc28* yeast gene. Solid line, DNA from *E. coli* plasmid pBR322 (Figure 4-9) containing a bacterial replication origin and genes for ampicillin resistance (Ap$^r$) and tetracycline resistance (Tc$^r$). Wavy line, yeast DNA containing a yeast replication origin and the yeast gene for anthranilate isomerase (trpl) essential for tryptophan biosynthesis. R, *Eco* R1 endonuclease cleavage site; B, *Bam* H1 cleavage site; cdc28, yeast cell division gene inserted into the vector. (After K. Struhl, D. T. Stinchcomb, S. Scherer, and R. W. Davis, *Proc. Natl. Acad. Sci. U.S.* **76**:1035–1039, 1979, and K. A. Nasmyth and S. I. Reed, *Proc. Natl. Acad. Sci. U.S.* **77**:2119–2123, 1980.)

random yeast DNA fragments with cohesive ends complementary to those present after treating YRp7 with the restriction endonuclease *Bam* H1 (Figure 15-10b) and ligated the DNA into the vector (Figure 15-10c).

The new vectors were introduced into *E. coli;* drug-resistant colonies were selected; and clones of their DNA were used to transform *trp⁻* yeast mutants carrying a temperature-sensitive (*ts*) mutation (*Section 7.5*) in the *cdc28* gene. Transformants were then selected for their ability to undergo cell division at high temperature in a medium lacking tryptophan. When plasmid DNA was isolated from such cotransformants and recloned in yeast, it could be shown to carry both $TRP^+$ and $CDC^+$ gene sequences.

Yeast cotransformants are genetically unstable. Thus in the previous experiment, when the cotransformants were returned to normal temperatures and grown on tryptophan-containing medium (permissive conditions), most of the cell lines lost their $TRP^+$ and $CDC^+$ genes: since both genes are carried on a plasmid that is not regularly included in the mitotic spindle, the plasmid is quickly lost unless selective pressure is maintained. Several cell lines, however, became **stably transformed** for both wild-type genes, and these could be shown to have **integrated a copy of the vector-borne yeast sequences into the *cdc28* locus of the host genome**. This **site-specific recombination** contrasts with the random chromosomal location of the donor fragments in the gene-transfer experiments we have described (*Figures 15-8 and 15-9*). The difference, of course, is that in the present case, donor and host DNA derive from the same species (yeast), whereas in the previous experiments they derived from different species.

**Mammalian Shuttle Vectors**   Shuttle vectors that can grow in both *E. coli* and mammalian cultured cells have been constructed using the same principles as for the yeast vector YRp7 (*Figure 15-10*). They typically include an origin of replication derived from the animal virus SV40; they also include SV40 termination sequences to insure that the spliced-in eukaryotic gene will be transcribed and translated correctly. Selection for mammalian cells carrying the vector is made by transforming auxotrophic cell lines (e.g., $Tk^-$) with vectors containing a wild-type sequence (e.g., $Tk^+$). In theory, such shuttles will allow the cloning of any mammalian gene in *E. coli* and its subsequent functional identification using complementation in a suitably mutant mammalian cell line.

# The Human Chromosome Map

Figure 15-11a summarizes the present status of the human chromosome map; Figure 15-11b presents the map from a clinical perspective. Much additional information will almost certainly be available by the time this text is in print. Most of the data for this map was generated by somatic cell genetics and by recombinant DNA hybridization probes, a fact that can be

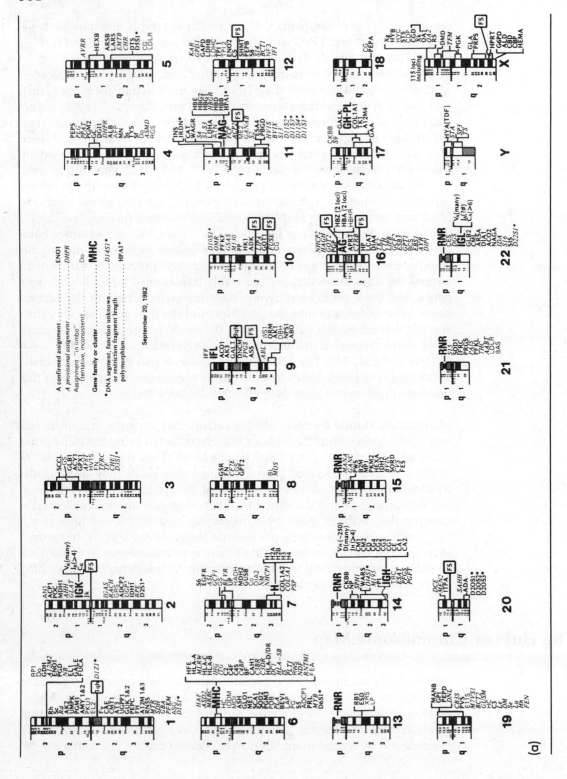

(a)

505

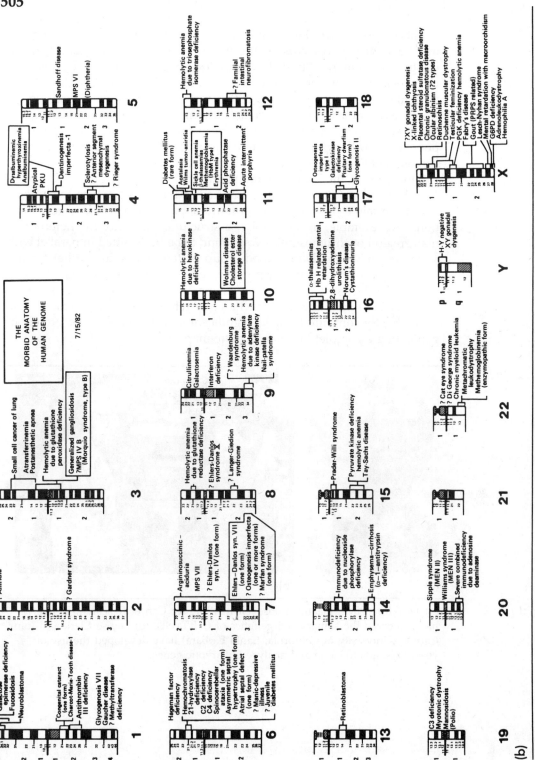

FIGURE 15-11 Genetic maps of the human chromosomes. (a) Genetic markers; (b) inherited clinical traits. Details of the banding patterns and numbering systems are found in Figure 2-18. Key to genetic symbols can be found in Cytogenet. Cell Genet. 32:1, 1982. (Courtesy of V. McKusick.)

(b)

readily appreciated when one considers the impossibility of asking humans to engage in test crosses. It is important to note, however, that familial studies of inherited traits have provided important information as well. In particular, as we stressed in Section 6.15, traits encoded in the X chromosome are effectively subjected to a test cross when male offspring are produced, and a wealth of familial information is therefore available for sex-linked genes. The *Hgprt* gene, for example, is marked in the human population by the *hgprt* mutation that causes an HGPRT deficiency in hemizygous males and leads to an affliction in purine metabolism (*see Figure 15-5*) known as the **Lesch-Nyhan syndrome.**

Familial studies have also permitted correlations between the inheritance of certain traits and the inheritance of particular chromosome aberrations. For example, in one family, a structural variant of the heterochromatic region of chromosome 1 was found to be transmitted in parallel with a marker for the *Fy* (Duffy blood group system) locus; in another family, loss of expression of acid phosphatase, red-cell type, was associated with a deletion in the short arm of chromosome 2, thus localizing the *Acp-1* gene to the 2p23 region. One can conclude, therefore, that familial studies will undoubtedly continue to make significant contributions to the project of generating a complete human genetic map.

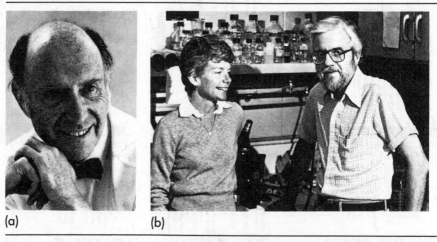

(a) Theodore Puck (Eleanor Roosevelt Institute for Cancer Research, Denver) has pioneered the somatic cell genetics of CHO cells. (b) Louise Clark and John Carbon (University of California, Santa Barbara) have developed the use of yeast plasmids to clone and analyze eukaryotic genes.

# Questions and Problems

1. Distinguish between the following: (a) homokaryon, heterokaryon, and syn-karyon; (b) primary and established cell lines; (c) intraspecific and interspecific cell fusions; (d) unstable and stable transformants.

2. Meiotic nondisjunction can occur at either anaphase I or II (*Section 5.15 and Problem 5-12*). Which, if either, is comparable to mitotic nondisjunction? Explain.

3. *Aspergillus* synkaryons are formed by the cross + + + + × a b c d. Fifty haploid sectors are selected that express allele *a*; the sectors are then scored for expression of the remaining genes. Twelve are found to be *b c d*; 16 are *b + d*; 10 are *+ c +*; and 12 are *+ + +*. Which genes are likely to be linked and which unlinked? Explain your answer.

4. Genes *f*, *g*, and *h* have been shown by haploidization studies to be linked on chromosome 7 of *Aspergillus*. Each recessive allele confers the conidia with a distinctive visible phenotype when either haploid or homozygous diploid. Syn-karyons are formed by the cross + + + × f g h, and diploid conidia that display the *f*, *g*, and *h* phenotypes are analyzed. Conidia that express *g* alone can be found, whereas no conidia are found that express *f* or *h* without also expressing *g*. Of the *g* conidia, 14 express *g* alone, 156 express *g* and *h*, and 83 express *f*, *g*, and *h*. Use this information to construct a linkage map of the chromosome.

5. (a) The mitotic crossing over data presented in Section 15.3 do not in fact establish, as drawn in Figure 15-4 and Table 15-1, that the *ad*-14 gene is located in the left arm of linkage group I. The data are also consistent with the possibility that the *ad* gene is very closely linked to the centromere in the right arm. Explain.
(b) The location of the *ad* gene was therefore determined by constructing *ad y/ + +* synkaryons and showing that they could produce *ad y/ad +* diploid sectors. Explain the reasoning behind this demonstration.

6. Describe how you would use HAT selection to identify the human chromosome carrying the HGPRT gene.

7. (a) When human fibroblasts are injected into rabbits, antisera (*Box 2.3*) are raised that react with antigens exposed on the human cell surfaces and, in the presence of proteins called complement, cause the cells to be killed. The antihuman sera have no effect on CHO cells. When CHO/human fibroblast hybrids are selected that retain only human chromosome 11, the cells are killed by the antiserum plus complement, whereas CHO/human hybrids containing only chromosome 12 are unaffected. What would you conclude from these observations?
(b) An antiserum is now prepared in rabbits using human brain cells, and this antiserum, plus complement, is very active in killing both the chromosome-11 and chromosome-12 lines. What would you conclude from this observation?
(c) You wish to use this approach to identify the human chromosome specifying the cell surface receptor for the serum protein transferrin, a receptor you have purified. Describe how you would design your experiments.

8. Human chromosome 11 is found to carry three genes for the cell surface antigens A, B, and C, genes we can designate *A*, *B*, and *C*. You now wish to determine their relative order and map distance. You treat CHO/human hybrids containing the *A B C* chromosome 11 with high doses of radiation that

produce chromosome breaks, and you probe 5000 of the resultant clones with anti-A, anti-B, and anti-C sera plus complement. Most of the clones continue to be sensitive to all three antisera; however, a few clones are sensitive to only one or two antisera, with the following frequencies (K = killed; L = lives):

| | Response to Complement Plus | | | Number Isolated |
| --- | --- | --- | --- | --- |
| Category of Clone | Anti-A | Anti-B | Anti-C | 5000 Screened |
| 1 | K | L | K | 56 |
| 2 | L | K | K | 21 |
| 3 | K | K | L | 0 |
| 4 | L | K | L | 58 |
| 5 | K | L | L | 18 |
| 6 | L | L | K | 1 |

Use these data to determine the order of genes *A*, *B*, and *C* and the distances between them. State the assumptions you are making. (Note: for purposes of this exercise, assume that all broken chromosome fragments continue to carry centromeres, an assumption clearly invalid for a real experiment of this kind.)

9. You wish to construct a recombinant DNA library (*Section 4.14*) of restriction fragments that contain genes from human chromosome 11. You generate large restriction fragments from the CHO/human hybrid containing chromosome 11, separate these on gels, and identify those bands that form hybrids with $^{32}$P-labeled human middle-repetitive DNA (*Section 4.8*). The positive bands are isolated, treated with restriction enzymes that cleave them into smaller fragments, and retested with the same $^{32}$P middle-repetitive DNA probes; this time all positive bands are discarded, whereas those that fail to hybridize with the probe are cloned. Explain the rationale of the various steps in this experiment.

10. How would you use genetics to demonstrate that yeast stable transformants for $CDC^+$ (*Section 15.9*) have integrated a copy of the wild-type gene into the correct locus on the yeast chromosome?

11. Mutations in the *CYC1* gene of yeast (*Section 14.15*) can be mapped by **X ray induced mitotic recombination:** diploid cells with a noncomplementing $\frac{cyc1 \; +}{+ \; cyc2}$ genotype, for example, are subjected to X rays to induce rare recombination events that generate $\frac{+ \quad +}{cyc1 \; cyc2}$ cells; such recombinants are detected by their ability to use lactate as a carbon source. The results obtained for various pairwise combinations are shown on page 509, where the numbers are given in X ray mapping milliunits which represent recombinants/$10^{11}$ survivors/roentgen of X ray administered. A dash means no data; a 0 means any value less than 1 milliunit.

| | 1 | 2 | 3 | 4 | 5 | 6 | 7 | 8 | 9 | 10 | 11 | 12 | 13 | 14 | 15 | 16 |
|---|---|---|---|---|---|---|---|---|---|---|---|---|---|---|---|---|
| 1 | 0 | | | | | | | | | | | | | | | |
| 2 | 0 | 2 | | | | | | | | | | | | | | |
| 3 | 0 | 87 | 0 | | | | | | | | | | | | | |
| 4 | 0 | 73 | 70 | 0 | | | | | | | | | | | | |
| 5 | 0 | 78 | 120 | 0 | 0 | | | | | | | | | | | |
| 6 | 0 | 90 | 165 | 0 | 0 | 0 | | | | | | | | | | |
| 7 | a | 11 | — | — | — | 105 | — | | | | | | | | | |
| 8 | 0 | 63 | 0 | — | — | 183 | — | — | | | | | | | | |
| 9 | a | 112 | — | — | — | 67 | — | 134 | — | | | | | | | |
| 10 | 0 | 81 | 67 | 7 | 8 | 6 | 73 | — | 42 | 0 | | | | | | |
| 11 | 0 | 2200 | a | a | a | b | — | — | — | — | | | | | | |
| 12 | 0 | 14 | — | — | — | 103 | 1 | — | — | — | — | — | | | | |
| 13 | 0 | 350 | — | — | — | 176 | — | — | 10 | 169 | — | — | — | | | |
| 14 | 0 | 35 | — | — | — | 52 | — | — | — | 16 | — | — | — | — | | |
| 15 | 0 | 134 | — | — | — | 115 | 39 | 80 | — | 200 | — | — | 390 | — | 0 | |
| 16 | 0 | 48 | — | — | — | 11 | 43 | — | — | 17 | — | — | 170 | 25 | 1 | 3 |
| | 1 | 2 | 3 | 4 | 5 | 6 | 7 | 8 | 9 | 10 | 11 | 12 | 13 | 14 | 15 | 16 |

[a] A cross for which no clear result was obtained due to high residual growth or high spontaneous recombination frequency.

[b] It is clear that the cross of $cy_{1\text{-}6}/cy_{1\text{-}11}$ gives a value of 2000 or greater but accurate values have not been obtained because of the extremely high spontaneous recombination frequency.

(a) What is unusual about mutant 1? How could you explain its properties?

(b) What explanations can you give for the 2 × 2 and 16 × 16 results?

(c) Which mutations appear to affect identical positions in the *cyc1* gene?

(d) Order markers 2, 3, 6, 9, and 13 with respect to one another. How can you explain any nonadditive distances?

(e) Which value utilizing marker 15 is unexpected? How might it be explained?

12. Gene-product mapping (*Section 14.15*) of three *cyc1* mutants indicates that each is 13 base-pairs apart:

C. Moore and F. Sherman made diploids carrying pairwise combinations of these mutations, and prototrophs were recovered that arose by X ray-induced, sunlamp-induced, uv-induced, or spontaneous mitotic recombination. The diploids were also allowed to sporulate, and meiotic prototrophic recombinants were scored. The resulting data are summarized below:

**Mean Prototrophic Frequencies**

| Cross | X ray | Sunlamp | UV | Spontaneous | Meiotic |
|---|---|---|---|---|---|
| cyc1-13 × cyc1-239 | 2200 | 2175 | 1345 | 6.1 | 78 |
| cyc1-239 × cyc1-179 | 84 | 297 | 143 | 1.7 | 40 |
| cyc1-13 × cyc1-179 | 340 | 1309 | 641 | 3.7 | 21 |

(a) Diagram the relative distances between these markers indicated by the five methods.

(b) What genetic principle(s) are illustrated by this study? Explain.

13. Two recessive mutations, $a$ and $b$, are located in the same chromosome in a diploid yeast strain and the homologous chromosome is wild type.

   Consider first that $a$ and $b$ lie on the same side of the centromere and answer the following questions. (Assume that $a$ is the more distal marker.) (a) If a single crossover occurs between $a$ and $b$ at meiosis, what will be the genotypes of the meiotic products? (b) If a single crossover occurs between $a$ and $b$ at mitosis, what will be the genotypes of the products? (c) If the centromere is between $a$ and $b$, would your answer to (a) remain valid? If not, give the correct answer. (d) If the centromere is between $a$ and $b$, would your answer to (b) remain valid? If not, give the correct answer.

# 16

# Extranuclear Genetic Systems

**A. Introduction**

**B. Molecular Studies of Mitochondrial Genetic Functions**
16.1 The Mitochondrial Genetic Apparatus
16.2 Nuclear Versus Mitochondrial Specification of Mitochondrial Components

**C. Genetic Analysis of the Yeast Mitochondrial Genome**
16.3 Petite Mutations in Yeast
16.4 Structural Gene Mutations in Yeast Mitochondrial DNA
16.5 Genetic and Physical Mapping of the Mitochondrial Genome
16.6 Introns that Code for Proteins

**D. Mitochondrial Genomes of Higher Eukaryotes**
16.7 Maternal Inheritance of Mitochondrial DNA in Higher Eukaryotes
16.8 The Human Mitochondrial Genome

16.9 Mitochondrial Genes and Male Sterility in Plants

**E. Chloroplast Genomes**
16.10 Molecular Studies of Chloroplast Genetic Functions
16.11 Maternal Inheritance of Chloroplast Mutations
16.12 Nuclear Influence Over Chloroplast Inheritance: The *Iojap* Gene

**F. Endosymbiosis and the Origins of Organelle Genetic Systems**
16.13 Modern Endosymbiotic Relationships
16.14 The Origins of Organelle Genetic Systems

**G. Inheritance of Preformed Structures**
16.15 Cortical Inheritance in Ciliates
16.16 Inheritance of Cytoplasmic Structures in Other Eukaryotes

**H. Questions and Problems**

# Introduction

This chapter considers a collection of eukaryotic genetic systems that have been individually described as **non-Mendelian, nonchromosomal, extra-chromosomal, extranuclear, cytoplasmic, uniparental,** and **maternal.** Major emphasis is given to the genetic systems present in mitochondria and chloroplasts. We also describe certain viruses, bacteria, and algae that take up residence within other cells and develop a permanent and mutually dependent relationship with their hosts. Finally, we present the concept that components of the cell surface may possess genetic information that can be modified by the environment and inherited independently of the nuclear genome.

These various extranuclear genetic systems share certain similarities. (1) Their mode of inheritance is often distinct from the inheritance of nuclear chromosomes, and their genetic material is often replicated at a different time or by a different set of enzymes. (2) The genetic information carried by the extranuclear elements may be unnecessary for the survival of the organism, although this is not true in several important cases. (3) The elements usually have a physical location outside the nucleus, but even the term "extranuclear" is not uniformly applicable, for a genetic element that resides within the nucleus but does not follow the hereditary patterns of the main chromosomes would also qualify as the kind of genetic system we are considering.

A compelling reason to consider these phenomena together is to convey the actual genetic complexity of eukaryotic cells. There is no question that the major set of chromosomes carries most of the genes and that the transmission of these genes dominates patterns of eukaryotic heredity. At the same time, the major chromosomes usually reside in cells in which extranuclear genes are also making their modest but often critical contribution to the phenotype of the organism. Once we have described these additional contributors, we shall be in a position to determine how the full phenotype is constructed and regulated, the subject of Chapters 19 to 25.

# Molecular Studies of Mitochondrial Genetic Functions

## 16.1 The Mitochondrial Genetic Apparatus

Mitochondria are cytoplasmic organelles found in all eukaryotes (*Figure 2-6*). They contain the enzymes of the Krebs cycle, carry out oxidative phosphorylation, and participate in fatty acid biosynthesis. Mitochondria also possess their own endowment of DNA and ribosomes, both of which are distinctive from their counterparts in the nucleus and cytoplasm of the cell.

TABLE 16-1   Physical Properties of Mitochondrial (mito) Compared with Cytoplasmic (cyto) Components in Three Organisms

| | Human (Hela) | | Yeast | | Neurospora | |
|---|---|---|---|---|---|---|
| | *Mito* | *Cyto* | *Mito* | *Cyto* | *Mito* | *Cyto* |
| Length of genome | 16.5 kb | — | 78 kb | — | 60 kb | — |
| Ribosomes | 60S | 74S | 75S | 80S | 73S | 77S |
|   Large subunit | 45S | — | 53S | 60S | 50S | 60S |
|   Small subunit | 35S | — | 35S | 40S | 37S | 37S |
| Ribosomal RNA | | | | | | |
|   Large subunit | 16S | 28S | 21S | 26S | 24S | 28S |
|   Small subunit | 12S | 18S | 15S | 18S | 17S | 18S |

Adapted from P. L. Altman and D. D. Katz, Eds. *Cell Biology*. Bethesda: Fed. Am. Soc. Exp. Biol., 1976, Cell Biology Federation of American Societies for Experimental Biology, Bethesda, Md.

Table 16-1 summarizes some of the physical properties of representative mitochondrial DNAs **(mtDNA)** and ribosomes. Mitochondrial genomes in all but the ciliated protozoa take the form of closed, circular duplexes (Figure 16-1a and b) that range in length from 5 μm (animals) to 25 μm (fungi) and 30 μm (higher plants). Basic histone-like proteins (*Section 3.4*) associate with this DNA and presumably help package it into the mitochondrion, but **the DNA is not organized into chromatin-like fibers.** Because each cell contains numerous mitochondria, each cell also possesses multiple copies of mtDNA (an estimated 100 copies for diploid yeast and 1000 to 10,000 copies for a mammalian cell). In all cases, mtDNA replication is carried out by unique mitochondrial DNA polymerases that function independently of their nuclear counterparts.

Mitochondrial ribosomes and ribosomal RNAs are distinctive from those present in the cytoplasm, although again there is considerable variation from one type of organism to the next (Table 16-1). Regardless of ribosomal size, protein synthesis on mitochondrial ribosomes is sensitive to the same spectrum of antibiotics as is protein synthesis on bacterial ribosomes (for example, erythromycin and chloramphenicol).

Mitochondria contain RNA polymerases that are unique in that they possess but a single polypeptide (recall from Section 9.3 that the RNA polymerases of both bacteria and the eukaryotic nucleus are high-molecular-weight complexes). Some mitochondrial RNA polymerases are sensitive to the antibiotic rifampicin (*Section 9.3*), whereas others are not.

Mitochondria contain at least one tRNA species for every amino acid, but there are not as many isoaccepting species (*Section 10.1*) as are present in prokaryotic and eukaryotic cytoplasms. We have already described in Section 11.13 a particularly intriguing feature of the translation process in mitochondria, namely, that several codons are used that differ from the "universal" genetic code.

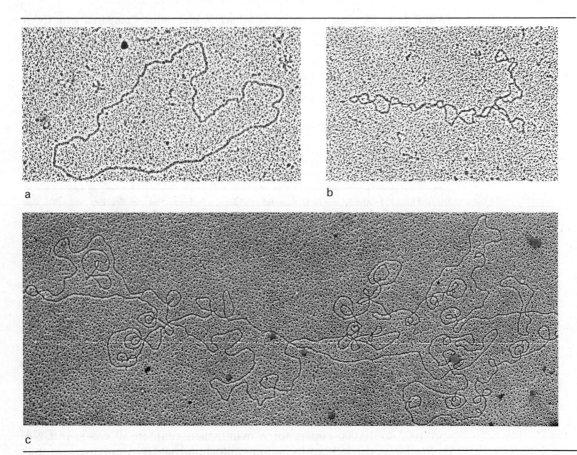

a

b

c

FIGURE 16-1 **Mitochondrial and chloroplast DNA.** Plates a and b show DNA from rat liver mitochondria at 48,000×. The chromosome in (a) has been nicked with endonuclease to display its circular form; the chromosome in (b) is not nicked and displays the native supercoiled configuration of intact circular DNA molecules (Courtesy of Dr. David Wolstenholme). Plate c shows DNA from the chloroplast of *Euglena gracilis* at 35,000×. Two molecules of phage φX174 circular DNA are also present in the field; can you find them? (Courtesy of Dr. Richard Hallick.)

## 16.2 Nuclear Versus Mitochondrial Specification of Mitochondrial Components

In the mid-1960s, as the molecular properties of the mitochondrial genetic apparatus were being analyzed, it became clear that the numerous enzymes, membrane components, and ribosomal proteins found in mitochondria could not possibly all be encoded within only 5 to 30 μm of mtDNA (an animal mitochondrial genome, for example, has enough DNA for perhaps 15 to 18 structural genes). Therefore, a number of laboratories investigated which mitochondrial components were specified by the nuclear-cytoplasmic genetic system and which by the mitochondria themselves.

One approach taken was to poison the organelle system with such antibiotics as rifampicin or chloramphenicol and then to test which components continued to be synthesized and which were inhibited. An alternate approach was to isolate RNA molecules found within mitochondria and to test whether they annealed to mtDNA in DNA-RNA hybridization studies (*Section 4.10*). A third approach was to study the mitochondrial biosynthetic capacities of a particular class of "petite" mutants of yeast (described in more detail further on) which carry major alterations in the structure of their mtDNA. The net results of such studies are summarized in Tables 16-2 and 16-3. Most mitochondrial components are indeed encoded by the nucleus and synthesized in the cytoplasm. The mitochondrial genome is devoted to the synthesis of certain RNA molecules and to certain subunits of three respiratory enzymes. Genetic evidence for these statements is given shortly.

# Genetic Analysis of the Yeast Mitochondrial Genome

Genetic mapping of a mitochondrial genome requires a large collection of different mutations, and, to date, the yeast *Saccharomyces cerevisiae* is the one organism to meet this criterion. The major advantage in studying yeast is that it is a **facultative anaerobe,** meaning that it can grow by fermenting glucose in the complete absence of mitochondrial respiratory functions. **Mutations to respiratory deficiency in yeast are therefore automatically conditional mutations:** they do not prevent normal growth by fermentation and place limitations on growth only when yeast cells are forced to utilize a carbon source such as ethanol, which requires that they respire. An obligate aerobe, by contrast, can survive most mitochondrial mutations only if they are rather leaky (and therefore unsatisfactory as genetic markers).

## 16.3 Petite Mutations in Yeast

In the late 1940s, B. Ephrussi and his colleagues in France observed that when yeast cells are plated under conditions in which either respiratory or fermentative metabolism can take place, an occasional **petite** colony appears among normal **grande** colonies. When cells of the *petite* class were tested for their ability to respire, it was found to be nil or negligible. In other words, *petite* yeast cells can derive energy only by the relatively inefficient process of fermentation so that their growth is expectedly slow, particularly on an agar plate where glucose is limiting.

Perhaps the most surprising, and still unexplained, property of the *petite* state is the frequency with which it arises. Spontaneously, 1 in 100 to 1000 cells in a yeast population exhibits the *petite* phenotype, and if a population is treated with such acridine mutagens (*Section 8.11*) as ethidium bromide, 100 percent of the cells can be converted to the *petite* state.

Some *petite* yeast cells prove to carry mutations in their nuclear genes. These *petite* markers, denoted as *pet*, are called **nuclear** or **segregational petites** (*Figure 14-16*) and, as illustrated in Figure 16-2a, they exhibit classic Mendelian inheritance: in a cross between *pet* and + cells, the diploids are +/*pet* wild type, and the meiotic products of sporulation are 2 +:2 *pet* (review the yeast life cycle in Figure 5-10). Most petite yeast, however, prove to carry mutations in their mitochondrial DNA. These mutations are designated *rho*⁻ (also symbolized **ρ**⁻) in contrast to *rho*⁺ (ρ⁺) wild-type cells.

Two classes of *rho*⁻ petites—*neutral* and *suppressive*—are known, and their properties are described separately in the following sections.

**Neutral Petites** When a neutral *petite* is crossed with a wild-type strain, the *rho*⁻ trait exhibits **non-Mendelian inheritance.** As illustrated in Figure 16-2b, the diploid progeny of the cross prove to be *rho*⁺ in phenotype, as with the recessive nuclear *petites*, but when these diploid progeny undergo sporulation, **all four meiotic products are *rho*⁺.** If nuclear gene markers (for example, mating-type alleles) are followed in the same cross, they segregate in the expected 2:2 ratio. Moreover, if the *rho*⁺ progeny are back crossed to the original neutral *petite* strain, the emergent meiotic progeny

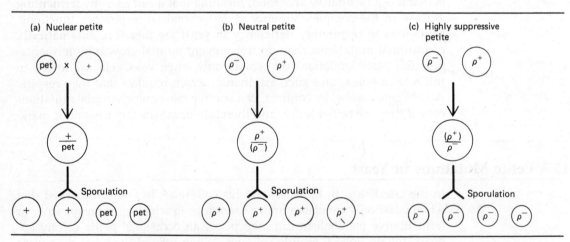

FIGURE 16-2 *Petite* **mutations in genetic crosses.** The nuclear *petites* (a) show Mendelian segregation; the neutral *petites* (b) show 4+:0 *petite* non-Mendelian segregation; the highly suppressive *petites* (c) show 0+:4 *petite* non-Mendelian segregation.

are again $rho^+$. Ephrussi interpreted these studies to mean that the **4:0 pattern of inheritance** of the $rho^-$ mutation was the consequence of its localization in some genetic element that resided in the cytoplasm rather than the nucleus. The loss of $rho^-$ in the cross could then be attributed to a "takeover" by the more functional $rho^+$ elements and the concomitant dilution of $rho^-$ in successive cell divisions.

That mtDNA represents the cytoplasmic genetic element affected by neutral *petite rho*$^-$ mutations seemed likely in view of the fact that $rho^-$ cells are respiratory deficient. A direct demonstration came when it was shown that most neutral *petites* lack mtDNA altogether (the so-called **rho$^0$** strains). In other words, it is clear in the case of neutral *petite* yeast that heritable alterations in mitochondrial DNA lead to heritable alterations in mitochondrial phenotype.

**Suppressive Petites**   Most $rho^-$ mutations convert yeast cells into suppressive *petites*. Suppressive strains are also disabled in mitochondrial function, but they differ from neutral *petites* in that their mutations are expressed in diploid cells: as illustrated in Figure 16-2c, a fully suppressive *petite* will convert all diploids arising from a $rho^+ \times rho^-$ cross to the $rho^-$ phenotype. When such diploids are sporulated, moreover, one obtains **0 wild-type**: **4 *petite*** ascospore ratios (Figure 16-2c). Partially suppressive strains have the same properties but affect only a portion of the diploids. A number of theories have been offered to explain how suppressiveness works, but the actual mechanism remains unknown.

Suppressive *petite* mitochondria usually possess about as much DNA as wild-type mitochondria, but the buoyant density of their DNA is typically shifted toward a lower GC content (*Section 4.2*). Further analysis of the DNA reveals that **the suppressive rho$^-$ mutations generally originate as deletions.** The sequences that are not deleted proceed to reiterate and sometimes rearrange themselves. If rearrangements occur, the result is often a highly scrambled version of the original genome.

**Phenotype of Non-Mendelian *Petites***   Virtually all neutral and suppressive *petites* exhibit the same phenotype regardless of the compositions of their surviving mtDNA: all are primarily affected in their ability to carry out normal mitochondrial protein synthesis. As we will see shortly, genes coding for essential components of the mitochondrial protein-synthesizing apparatus are distributed throughout mtDNA; therefore, it is hardly surprising that most deletions and rearrangments of the genome will disrupt the translational process. This disruption, in turn, blocks the synthesis of all proteins produced in the mitochondria (Table 16-2). Such drastic mutations, and such complex phenotypes, are not suitable for the mapping of individual genes. Therefore, it became essential to isolate mutant strains that carry point mutations in the mitochondrial genome.

**TABLE 16-2  Mitochondrial Components in "*Petite*" Yeast Mutants**

**Absent**

Functional respiratory chain; functional cytochrome $bc_1$ and cytochrome oxidase

Functional energy-transfer system

Functional protein-synthesizing system; ribosomes

**Present**

Outer membrane

Inner membrane (altered)

Parts of respiratory chain (cytochromes $c$ and $c_1$, some subunits of cytochrome $aa_3$)

Parts of the energy-transfer system: $F_1$-ATPase

Transport systems

Krebs cycle

DNA and RNA polymerases, ribosomal proteins, elongation factors

From P. Borst, in *International Cell Biology*, B. R. Brinkley and K. R. Porter, Eds. New York: Rockefeller University Press, 1977.

## 16.4 Structural Gene Mutations in Yeast Mitochondrial DNA

Figure 16-3 outlines the procedures developed by A. Tzagaloff to obtain structural gene mutations in yeast mtDNA.

(1) Wild-type cells are mutagenized with manganese and spread on agar plates with limiting glucose concentrations to differentiate respiratory-sufficient (large) from respiratory-deficient (small) colonies.

(2) Cells from the small colonies are mated with wild-type cells to determine whether their defects are inherited in a Mendelian or a non-Mendelian fashion.

(3) The non-Mendelian mutants are assessed for their ability to carry out protein synthesis in the presence of **cycloheximide,** an inhibitor of translation of nuclear mRNAs on cytoplasmic ribosomes.

(4) Clones that fail to respire but retain mitochondrial (cycloheximide-resistant) protein synthesis are designated *mit⁻*. These are assayed for individual mitochondrial respiratory functions, and each is typically found to be defective in one or more of the polypeptide subunits listed in the right-hand column of Table 16-3.

(5) Clones that have lost mitochondrial (cycloheximide-resistant) protein synthesis could be $\rho^-$, or they could carry point mutations in individual genes of the translational apparatus. To distinguish between the two, the clones are first checked for their ability to revert: neither neutral nor suppressive *petites* should be able to revert to wild type, since they carry physically altered genomes. The revertible clones are then subjected to complementation tests (*Section 12.4*) using deletion mutants that have lost respiratory genes but not "synthetic" tRNA or rRNA genes. If the revertible mutation can be complemented by such deletion chromosomes, then it is assumed to affect a "synthetic" gene important for protein synthesis, and the new mutation is designated *syn⁻*. If the new mutant is nonrevertible and is not complemented by the dele-

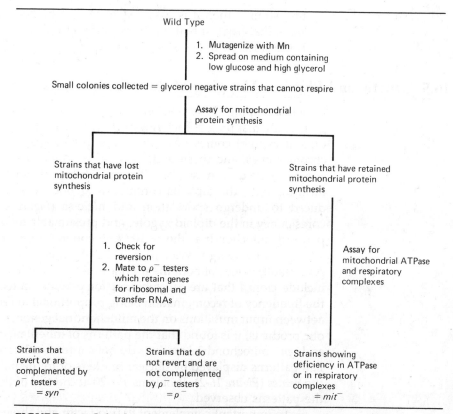

**FIGURE 16-3 Selection for structural gene mutations in yeast mtDNA.**
Manganese (Mn) is used as a mutagen because it induces point mutations and
deletions primarily in mitochondrial DNA. Cells lacking adequate mitochondrial
function are unable to grow on glycerol. (From A. Tzagaloff, in *The Genetic
Function of Mitochondrial DNA,* C. Saccone and A. M. Kroon, Eds. Amsterdam:
North-Holland, 1976.)

**TABLE 16-3  Biosynthesis of Major Mitochondrial Enzyme Complexes in Yeast**

| | Number of Polypeptide Subunits | | |
| | --- | --- | --- |
| Enzyme Complex | Total | Made in Cytoplasm | Made in Mitochondria |
| Cytochrome oxidase (cytochrome $aa_3$) | 7 | 4 | 3 |
| Cytochrome $bc_1$ complex | 7 | 6 | 1 |
| ATPase (oligomycin sensitive) | 9 | 7 | 2 |
| Large ribosomal subunits | 30 | 30 | 0 |
| Small ribosomal subunits | 22 | 21 | 1 |

From P. Borst, in *International Cell Biology,* B. R. Brinkley and K. R. Porter, Eds. New York: Rockefeller
University Press, 1977.

tion strains, then it is assumed to carry a highly rearranged chromosome that has lost both protein-synthetic and respiratory genes, and it is classified as $\rho^-$.

## 16.5 Genetic and Physical Mapping of the Mitochondrial Genome

Genetic mapping of the yeast mitochondrial genome is conceptually analogous to the mapping of bacteriophages (*Chapter 12*) in that a "host cell" (the diploid zygote) comes to contain multiple copies of genetically marked chromosomes, and these under genetic recombination. Specifically, yeast strains marked with two or more mitochondrial mutations are mated to form diploids; the diploids continue to divide by mitosis and can be induced to undergo sporulation and meiosis (*Figure 5-10*). During their coresidency in the diploid zygote, and presumably following the fusion of parental mitochondria, the ~ 100 mitochondrial genomes introduced by each parent undergo cycles of both DNA replication and recombination. As a result, clones of both the mitotic and the meiotic progeny of the cross include clones that are recombinant for parental mitochondrial markers, the frequency of recombination being proportional to the physical distance between input mutations on the mitochondrial genome. In the case of meiotic products, it is found that the patterns of inheritance displayed by *mit⁻* and *syn⁻* mitochondrial markers do not conform either to the strict Mendelian patterns displayed by nuclear markers or to the patterns displayed by the *petites* (*Figure 16-2*). Problems 18–20 at the end of this chapter describe the patterns observed.

An important complement to two-factor and three-factor cross data is provided by deletion mapping of the mitochondrial genome. As with phages (*Section 12.13*), one assembles a collection of mutants carrying deletions in known regions of the genome. These mutants are then mated with point mutants, and recombinants are sought. Wild-type function is restored in recombinant clones only when the deletion does not overlap the site of the point mutation, just as we have seen with other forms of deletion mapping.

In the past several years, such genetic approaches have been largely replaced by physical mapping techniques, this replacement being obligatory for animal mtDNA, for which genetic analysis is largely impracticable. The most widely used approach is **restriction mapping** using cloned fragments of mtDNA; this is performed as described in Section 12.16. **Hybridization mapping** has also yielded important information. As noted in Box 4.1, when small chromosomes are denatured (*Section 4.10*) and separated into their two component strands, one strand frequently carries more purines than the other and therefore has a greater buoyant density (the **heavy,** or **H,** strand) than the other (the **light,** or **L,** strand). To obtain a hybridization map of animal mtDNA, H and L strands of mtDNA are separately hybridized with a mitochondrial RNA species (for example, a transfer RNA) that has been tagged with **ferritin,** a large, electron-dense protein. The resultant ferritin–RNA-DNA hybrids are then characterized with the electron microscope. This technique not only locates the gene physically but also demonstrates that several mtDNA genes are encoded on the opposite strand from the majority of the genes (recall from Section 9.4 that this pattern is also found for a number of viral chromosomes).

The combination of these approaches has produced the map of yeast mtDNA shown in Figure 16-4, where the genes are separated by long ex-

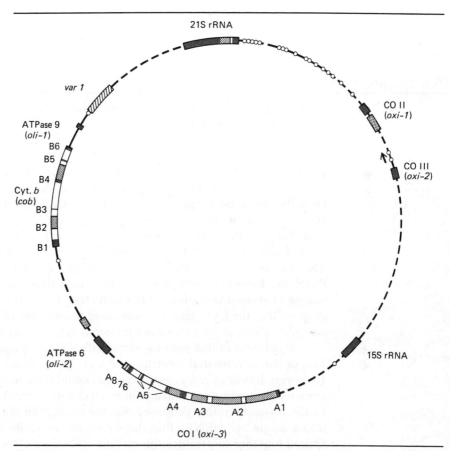

**FIGURE 16-4 Map of yeast mitochondrial DNA.** Sequenced areas are denoted by continuous lines, and unsequenced areas are shown by dashed lines. Small circles indicate tRNA genes, one of which (arrow) is encoded on the strand opposite the strand that encodes most genes. The *var1* gene (cross-hatched) specifies a mitochondrial ribosomal protein (*Table 16-3*). The CO I, II, and III genes encode subunits I, II, and III of cytochrome oxidase; their alternate genetic nomenclature is *oxi-3*, *oxi-1*, and *oxi-2*, respectively. ATPase 6 and 9 are genes for subunits 6 and 9 of the ATPase complex (*Table 16-3*); they are also designated *oli-2* and *oli-1*, respectively, for oligomycin resistance. Cyt*b* is the gene for apocytochrome b (also designated *cob*). The CO I, Cyt*b*, and 21S rRNA genes are split into exons (black bars or gray regions) and introns; introns that contain an open reading frame are stippled, and other introns are shown as open bars. (From P. Borst and L. A. Grivell, *Nature* 290:443–445, 1981.)

panses of very AT-rich DNA. One interesting feature of this map is that the two genes for rRNA (21S and 15S) are separated by a large segment of DNA containing other structural genes. Such a separation is found in several other (but not all) types of mtDNA, but it does not occur in the genomes of prokaryotes and eukaryotes (*Sections 10.6 and 10.7*).

A second notable feature of the yeast mtDNA is that introns (*Section 9.9*) are present: the 21S rRNA gene contains one, whereas the Cyt*b* and CO I genes each contain many (Figure 16-4). This discovery was of interest in that introns were initially thought to be present only in eukaryotic nuclear DNA.

# 16.6 Introns that Code for Proteins

A remarkable feature of the yeast mitochondrial genome is that **at least some of its introns appear to code for proteins.** This was first suspected from the results of genetic studies. When *mit*⁻ mutants defective in cytochrome *b* production were subjected to pairwise complementation tests in diploid yeast, they fell into two classes. Mutants in the first class failed to complement each other and mapped within a 9-kilobase sector of the mitochondrial genome at the Cyt*b* locus; they therefore appeared to define a classic complementation group or gene (*Section 12.4*), although the sector was clearly much too long to code for a single protein. Mutants of the second class were able to complement mutants of the first class and one another, as if they defined separate complementation groups or genes. Puzzling, however, was the fact that **mutations of the second class mapped between mutations of the first class.** It was therefore necessary to propose that the Cyt*b* gene is interrupted by mutable DNA sequences that also play a role in specifying a functional cytochrome b.

Resolution of this paradox came with DNA sequencing of the region. Two of the introns that interrupt the Cyt*b* gene (those stippled in Figure 16-4) were found to carry sequences that could code for protein, having an **open reading frame** without any internal chain-terminating codons (*Section 11.15*). It was therefore proposed that the entire 9 kb of DNA is transcribed into a single pre-mRNA, that the six exons (B1 to B6 in Figure 16-4) are spliced together to produce the mature mRNA for cytochrome *b*, and that the "intron RNA," which is ultimately spliced out, has genetic functions as well.

In the case of the intron lying between B2 and B3, the function appears to be the specification of a splicing enzyme **(maturase)** necessary for removing the intron itself. Thus, missense mutations in this intron lead to the accumulation of long pre-mRNA molecules containing the intron sequence; since these cannot be spliced out, the mRNAs are translated into long, bizarre polypeptides that are related in sequence to cytochrome b but contain intron-coded amino acids, and these are unable to function as respiratory proteins. Such intron mutants would be expected to be complemented by chromosomes that carry normal intron sequences specifying normal maturase enzymes; this explains how the two classes of cyto-chrome b mutants are able to complement one another.

In short, the yeast mitochondrial genome contains at least one, and probably several "genes within genes." This arrangement is not at present believed to apply to most other split genes. The real mystery is how such an arrangement evolved: what is the point of a minigene that encodes a protein necessary to splice out the gene's information from a longer tran-script?

# Mitochondrial Genomes of Higher Eukaryotes

## 16.7 Maternal Inheritance of Mitochondrial DNA in Higher Eukaryotes

As noted earlier, obligate aerobes are unlikely to carry nonleaky mutations in their mitochondrial genome. To study the pattern of inheritance of plant and animal mtDNAs, therefore, **the DNA itself has been used as a genetic marker.** Specifically, individuals within a population often carry somewhat different mtDNAs, and these have distinctive patterns of susceptibility to digestion by restriction endonucleases (*Section 4.11*). The diagnostic cleav-age patterns then become markers that are followed in genetic crosses.

In the example illustrated in Figure 16-5a, human mtDNAs were found to differ in their susceptibility to the enzyme *Hae* II, the individuals depicted by shaded symbols having a more complex pattern than the indi-viduals depicted by open symbols. The inheritance of these cleavage pat-terns was then followed in the three-generation family illustrated in Figure 16-5b (review human pedigree construction in Section 6.6), and the results

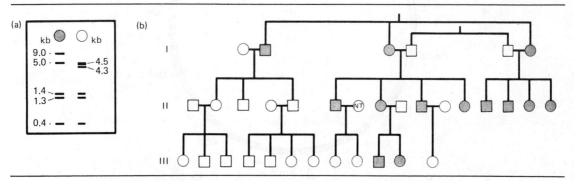

**FIGURE 16-5 Maternal inheritance of human mitochondrial DNA.** (a) Pattern of *Hae* II restriction fragments in an atypical (shaded) and typical (unshaded) digest of human mtDNA. (b) Pedigree of a three-generation family illustrating the inheritance of the two cleavage patterns. Circles are females, squares are males, and NT denotes "not tested." (From R. E. Giles, H. Blanc, H. M. Cann, and D. C. Wallace, *Proceedings of the National Academy of Science of the United States* **77**:6715–6719, 1980.)

were decidedly non-Mendelian. Thus the shaded male in the first generation fails to transmit his mitochondrial genotype to any of his progeny, whereas the generation I females do; this same pattern of maternal inheritance is repeated in subsequent generations. Such **maternal inheritance of mtDNA** has been documented for many other animals and plants as well.

Why should mitochondria be inherited only through the maternal line? The most likely answer rests with the biology of fertilization. Both animal eggs and plant embryo sacs (*Figures 5-6 and 5-7*) are large cells with numerous mitochondria, whereas spermatozoa usually contribute only nuclei to the zygote. Therefore, the **cytoplasmic inheritance of genetic traits is usually synonymous, in the higher eukaryotes, with maternal inheritance.**

## 16.8 The Human Mitochondrial Genome

Again because mutations are lacking, the mitochondrial genomes of obligate aerobes are most productively analyzed by physical mapping procedures coupled with cloning and sequencing analysis. The human mitochondrial genome, shown in Figure 16-6, has been completely sequenced

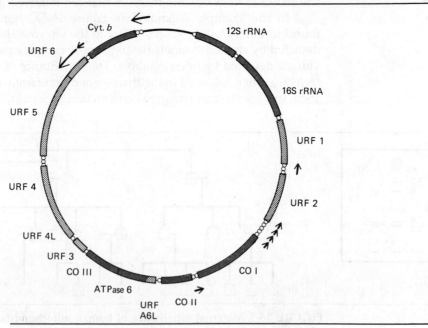

**FIGURE 16-6 Map of human mitochondrial DNA.** Small circles indicate tRNA genes, several of which (arrows) are encoded on the strand opposite the strand that encodes most genes. Sequences encoding known gene products are shown as gray bars, whereas URFs (unassigned reading frames) are indicated as stippled areas. Other abbreviations are as in Figure 16-5. The entire genome contains 16,569 nucleotide pairs, all of which have been sequenced. (From P. Borst and L. A. Grivell, *Nature* **290**:443–445, 1981.)

in the laboratory of F. Sanger. A comparison of this map with the yeast map (Figure 16-4) reveals a number of interesting contrasts.

1. The two human rRNA genes are separated only by a tRNA gene rather than by long expanses of DNA coding for other proteins.
2. None of the human genes contains introns.
3. It has been reported that there is only one promoter for each human mtDNA strand (compared with at least five in yeast), meaning that the entire sequence may be copied into two giant transcripts that are subsequently processed.
4. One or more tRNA genes lie at either end of each human coding sequence, suggesting that they serve a "punctuation" role for RNA-processing enzymes.
5. The entire human genome shows a remarkable "economy": the long expanses of AT-rich DNA found in yeast are absent, as are leader and trailer sequences (*Sections 9.6 and 9.8*). The map explains, therefore, how the mammalian mitochondrion can carry as much genetic information as the yeast mitochondrion with only one fifth the amount of DNA. Presumably, selective pressures have pushed animal mtDNA into economy and yeast into maximizing the A + T content, but the nature of these pressures is unknown.

## 16.9 Mitochondrial Genes and Male Sterility in Plants

**Cytoplasmic male sterility (CMS)** is a trait in higher plants that causes pollen to abort in the anthers. A CMS plant is therefore unable to self-fertilize. If fertilized by wild-type pollen, it will produce only male-sterile progeny, indicating that the trait is maternally inherited.

The CMS phenotype has had major importance in agriculture, since it greatly simplifies the construction of **hybrid plant lines** that combine the best traits of two inbred parental lines. To illustrate its use, we can follow the production of hybrid maize seeds (review the maize life cycle in Figure 5-7). Ordinarily, hybrid corn seeds are produced by manually removing the male flowers (tassles) from Parent A so its self-fertilization is prevented and then pollinating with Parent B. The need for the laborious detassling step is eliminated if Parent A carries the CMS trait called **Texas cytoplasm (cms-T):** such plants produce no pollen and therefore serve as the female. Parent B, with normal cytoplasm **(cms-N)**, produces pollen and serves as the male. The hybrid seed that results from such an A × B cross would ordinarily give rise to male-sterile plants, owing to maternal inheritance, and these would be unable to self-fertilize and produce corn. Therefore, as illustrated in Figure 16-7, parent B is also made homozygous for a dominant nuclear gene, called **Rf** (restores fertility), which suppresses the CMS phenotype. The hybrid, heterozygous for *Rf*, can therefore proceed to self-pollinate and yield a normal crop (Figure 16-7).

The widespread use of the cms-T trait, we should note, had a disastrous consequence. The fungal pathogen *Helmiothosporium maydis* proved

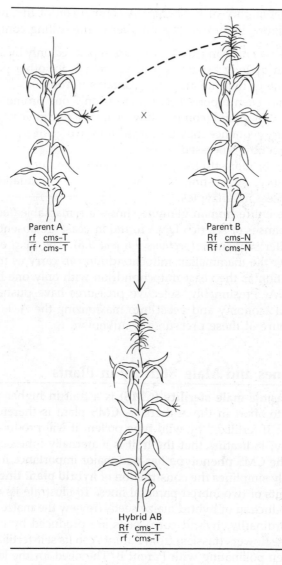

FIGURE 16-7 Cytoplasmic male sterility (cms-T) used to produce hybrid plants.

Parent A
rf   cms-T
rf ′ cms-T

Parent B
Rf   cms-N
Rf ′ cms-N

Hybrid AB
Rf   cms-T
rf ′cms-T

to be far more virulent against cms-T varieties of maize than against cms-N varieties, and the resultant Southern corn leaf blight in 1970 caused losses of over one thousand million dollars in maize production.

What does the cms-T trait do? When pollen development is studied in male-sterile anthers, mitochondria are found to degenerate very early. The toxin produced by *H. maydis*, moreover, has been shown to affect respiratory phosphorylation in cms-T mitochondria but to be without effect on cms-N mitochondria. It appears, therefore, that **the mitochondrial genome of maize carries the maternally inherited CMS trait, which affects both pollen formation and toxin susceptibility.** The genetic basis for the trait,

however, is not yet clear: some investigators report differences in the restriction-endonuclease cleavage patterns (*Figure 16-5*) of cms-N and cms-T
mtDNA and propose that the trait is controlled by an mtDNA gene product; other investigators suspect the participation of small plasmids (*Section
13.16*), about 2 kilobases in length, which reside within maize mitochondria.

# Chloroplast Genomes

## 16.10 Molecular Studies of Chloroplast Genetic Functions

Chloroplasts contain DNA **(cpDNA)** and a transcription-translation system
that are very similar to their counterparts in mitochondria (*Sections 16.1 and
16.2*). The DNA exists in covalently closed circles (Figure 16-1c), which, in
most organisms, are present in several thousand copies per cell. Each circle
is larger than mtDNA (typically 45 μm in length) but specifies similar components: rRNAs, about 30 tRNAs, and a handful of chloroplast-specific
proteins, the best characterized being the large subunit of the critical
photosynthetic enzyme, ribulose-1,5-biphosphate (RuBP) carboxylase.
Chloroplast ribosomes are small (~67S) and are sensitive to the same spectrum of antibiotics as mitochondrial ribosomes. Certain chloroplast ribosomal proteins are believed to be encoded in chloroplast DNA, since mutations to antibiotic resistance have in some cases been shown to exhibit
maternal inheritance.

As with mitochondria (Table 16-2), most chloroplast-specific proteins
are encoded in the nucleus, translated in the cytoplasm, and then imported
into the organelle. This is the case for many chloroplast ribosomal proteins.
It is also the case for the small subunit of RuBP carboxylase, meaning that
the functional enzyme is a hybrid molecule: it contains eight subunits of
the nuclear-coded polypeptide and eight subunits of the chloroplast-coded
species.

Restriction mapping of cpDNA has lagged behind the effort to map
mtDNA, in part because of its much larger size. Figure 16-8 illustrates a
recent map of corn cpDNA, where two copies of rRNA gene clusters are
seen to be encoded on opposite strands in reverse orientation and where
the gene for the large RuBP carboxylase subunit has been localized. Transfer RNA genes have also been mapped on a variety of cpDNA species, and
most are found clustered in a single region, much as in yeast mtDNA (*Figure 16-4*).

## 16.11 Maternal Inheritance of Chloroplast Mutations

Maternal inheritance, introduced in Section 16.7 for mitochondrial mutations, is also the pattern followed during the transmission of many muta-

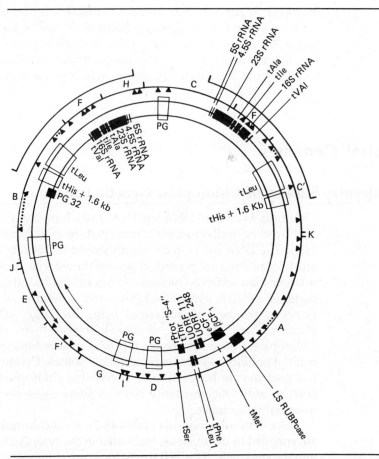

**FIGURE 16-8 Map of corn chloroplast DNA.** The two inner circles show the 2 strands of the chromosome, the innermost strand transcribed in a clockwise direction and the other in a counterclockwise direction. Filled boxes or lines show genes precisely located; open boxes show genes approximately located. The outer complete circle shows the sites of *Sal* I (lines) and *Bam* HI (arrowhead) cleavage, and the outermost pair of arcs shows the positions of two repeated DNA sequences that are inverted orientation to one another (genes present here are represented twice per chromosome). Abbreviations: tAla, etc., genes for transfer RNA (more than 25 tRNA genes are present, but only sequenced genes are shown); LS RUBPcase, large subunit of ribulose bisphosphate carboxylase; $\epsilon$ and $\beta$ $CF_1$, subunits of the chloroplast coupling factor; PG, photogene (gene whose transcription requires light); r Prot "S-4," ribosomal protein homologous to the S4 protein of *E. coli* ribosomes. (Courtesy of L. Bogorad.)

tions affecting the chloroplast. As early as 1909, C. Correns described this phenomenon for the plant *Mirabilis jalapa*, also known as the "four o'clock." The distribution of green pigment in a given *Mirabilis* leaf varies from branch to branch: the leaves on certain branches may be fully green; other branches will have patchy leaves in which green tissue is inter-

**TABLE 16-4   Chloroplast Inheritance in Variegated Four O'Clocks**

| Branch of Origin of the Male Parent | Branch of Origin of the Female Parent | Progeny |
|---|---|---|
| Green | Green | Green |
|  | Pale | Pale |
|  | Variegated | Green, pale, variegated |
| Pale | Green | Green |
|  | Pale | Pale |
|  | Variegated | Green, pale, variegated |
| Variegated | Green | Green |
|  | Pale | Pale |
|  | Variegated | Green, pale, variegated |

spersed with pale green–to–white tissue in a **variegated** pattern; still others will bear fully pale leaves. Microscopic examination of the green leaves, and the green areas of the variegated leaves, shows that the cells contain normal chloroplasts and chlorophyll pigment; by contrast, the pale leaves and the pale patches lack normal chloroplasts and pigment. These cells cannot, of course, carry out photosynthesis, but they are fed by other portions of the variegated plants.

When Correns followed the inheritance of these pigment traits, he found that **the progeny inherit the phenotype of the female parent.** For example, if ovules derive from fully green portions of the plant, then regardless of the source of the pollen, only fully green plants will result, and the variegated character will not reappear in subsequent generations. Similarly, when ovules derive from a wholly white branch, sickly white plants emerge, even when the pollen comes from a green-branch flower. These results are summarized in Table 16-4.

When the ovule instead derives from variegated branches, three types of seed are produced in variable numbers, again regardless of the male parent: some give rise to pure green, some to pure white, and the majority to variegated offspring (Table 16-4). The classic explanation for this outcome is to assume that the variegated ovule contains a mixture of normal and abnormal plastids. Should the embryo sac happen to receive only normal plastids from this collection, a green plant will result; if it receives only abnormal plastids, a white plant will result. When the embryo sac receives a mixture of both plastids, which should happen most often, the plant will be variegated, the patchiness and the occasional pure-colored branches arising as the normal and abnormal plastids **sort out** during vegetative cell divisions.

# 16.12 Nuclear Influence over Chloroplast Inheritance: The *Iojap* Gene

Maternal inheritance suggests the presence of a mutation in organelle DNA, but this is not invariably the case. To illustrate possible alternatives, we can consider the *iojap* trait in corn, first studied by M. Rhoades. The normal *Iojap* (+) gene and its recessive *iojap* allele (*ij*) map to linkage group VI of the maize nuclear genome (*Figure 14-12*). Plants that are +/+ or +/*ij* have fully green leaves and apparently normal plastids, but if +/*ij* heterozygotes are inbred, one quarter of the progeny will have the *ij*/*ij* genotype and the *iojap* phenotype. Many *iojap* plants have no green pigment and are inviable; others are green with a characteristic white striping, a pattern known as *striped*.

**Once established in an *ij*/*ij* plant, the *iojap* phenotype is transmitted in a maternal fashion, regardless of the nuclear genotype.** This maternal inheritance is illustrated in Figure 16-9: *striped* females give rise to *striped* progeny (plus some *white* and *green* progeny), even if the paternal genome carries + alleles (Figure 16-9a), whereas *green* females give rise to *green* progeny, even if the paternal genome carries *ij* alleles (Figure 16-9b). Moreover, if striped +/*ij* females are crossed to normal +/+ males, *striped* and *white* plants will appear among the progeny even though many of these now have a homozygous +/+ nuclear genotype (Figure 16-9c).

Once it was realized that chloroplasts possess an independent genome, it was postulated that the effect of the *iojap* state, or the absence of wild-type *Iojap* function, was to create mutations in plastid DNA. The mutant plastids would then be transmitted to progeny in a maternal fashion and would remain mutant even when the +/+ nuclear genotype was restored. Recent studies of *iojap* plastid DNA, however, show no detectable alterations in its size or susceptibility to restriction endonuclease digestion.

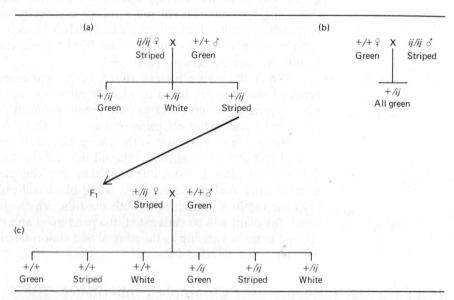

**FIGURE 16-9 *Iojap* inheritance.** (a) Cross between *iojap* (striped) and normal (green) plants. (b) Reciprocal of cross (a). (c) Cross of $F_1$ striped females to normal males.

On the other hand, *iojap* plastids are completely devoid of plastid ribosomes and, therefore, are unable to synthesize normal chlorophyll-containing membranes. Since most chloroplast ribosomal proteins are encoded in the nucleus, it is postulated that the *iojap* effect may be totally independent of the plastid genome. By this theory, ribosome-less plastids, created against an *iojap* background, are themselves transmitted in a maternal fashion and cannot be "cured" by normal nuclear function; they may not, however, carry chloroplast DNA mutations at all.

# Endosymbiosis and the Origins of Organelle Genetic Systems

## 16.13 Modern Endosymbiotic Relationships

A number of present-day eukaryotic cells or organisms are known to maintain populations of viruses or microorganisms that are transmitted *via* the cytoplasm. For example, *Paramecium bursaria* harbors a collection of small, eukaryotic algae that cannot be maintained in culture once they are removed from the *Paramecium* cytoplasm. Since the algae clearly derive some essential component or components from *Paramecium* and since they are known to provide their hosts with photosynthetic products, the interaction is regarded as mutually beneficial and is called **endosymbiosis**. In other cases the "mutant benefits" are not as apparent. For example, a maternally inherited infectious virus known as **sigma** is found in the cytoplasm of $CO_2$-sensitive strains of *D. melanogaster*. Such flies become permanently paralyzed when exposed to concentrations of $CO_2$ that do not affect normal flies, but whether more adaptive traits are conferred by the virus is unclear.

The most extensively studied case of endosymbiosis is given by *Paramecium aurelia*, which can harbor some ten different types of Gram-negative bacterial species in its cytoplasm (Figure 16-10), the best known being called **kappa particles**. Maintenance of each bacterial species requires particular nuclear genes of the host *Paramecium*. The bacteria may be transferred from one cell to the next during conjugation (*Figure 5-11*), and they often elaborate toxins. Further information on this system is given in the problems at the end of this chapter.

## 16.14 The Origins of Organelle Genetic Systems

Although a number of interesting theories have been proposed for the origin of organelle genomes, the best known has come to be called the **endosymbiosis theory**. First suggested by R. Altmann in 1890, this theory postulates that organelles are modern descendants of bacteria-like organisms within primitive, perhaps even then eukaryotic, cells. The presence of these endosymbionts was presumably advantageous to the host cells, and with time the endosymbionts evolved into present-day organelles.

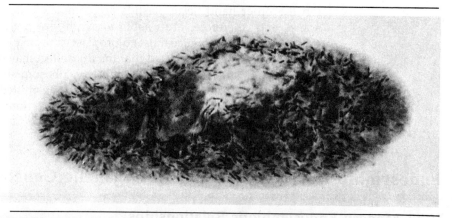

FIGURE 16-10 Endosymbiosis in *Paramecium tetraurelia*. The dark filamentous objects represent endosymbiotic bacteria known as lambda. (Courtesy of L. B. Preer, *Bact. Rev.* **38**:113, 1975.)

This evolution clearly entailed major genetic rearrangements: many genes that must have been encoded in the genomes of the original free-living invaders are no longer found in organelle genomes; some of these genes may have been lost entirely, whereas others may have been transferred into the nuclear genome of the host. The transfer process, moreover, must not have occurred in a uniform fashion: for example, the gene for subunit $F_0$-3 of the mitochondrial ATPase is located in the mitochondrial genome of yeast (Figure 16-4) but is found in the nuclear genome of another fungus, *Neurospora crassa*.

The strongest case for a prokaryotic origin of organelles is given by the chloroplast. Thus the 3'-terminal end of chloroplast 16S rRNA is homologous with *E. coli* 16S rRNA; chloroplasts utilize $tRNA_i^{fMet}$ as an initiator tRNA (*Section 11.3*); and, most strikingly, genes cloned from maize and wheat chloroplast DNA can be directly transcribed and translated in *E. coli* hosts, indicating that the transcription and translation apparatuses of *E. coli* are very similar to those in the chloroplasts of higher plants. It is therefore plausible to argue that bacteria-like algae, analogous perhaps to the modern cyanobacteria, were the ancestors of eukaryotic chloroplasts.

If mitochondrial genomes originally derived from prokaryotic chromosomes, they have undergone extensive modifications, a notable example being the presence of split genes in yeast mtDNA (*Figure 16-4*). Since split genes are absent from mammalian mtDNA (*Figure 16-6*), however, simple evolutionary pathways are not apparent.

The endosymbiotic theory has been very important in stimulating research on organelle genomes. In the future, there will clearly be equivalent interest in learning how present-day organelle genomes are related to one another.

# Inheritance of Preformed Structures

## 16.15 Cortical Inheritance in Ciliates

The transmission of endosymbionts and organelles can be attributed to extranuclear DNA found either in infectious particles or in organelles. There are, however, hereditary traits whose origin is not yet easily ascribed either to nuclear or extranuclear DNA. The best studied of these characteristics concern the cortex of ciliated protozoa such as *Paramecium*. The mouth and contractile vacuole are two prominent features of the cortex of *Paramecium*, and T. Sonneborn and his coworkers have shown that these existing, or **preformed,** structures can be transmitted from one cell generation to the next, independent of the transmission of nuclear genes and cytoplasmic genetic factors.

During normal sexual reproduction in *Paramecium* (*Figure 5-11*) two cells conjugate and then separate after they have exchanged nuclei. On occasion, however, the conjugants fuse and form a double animal (a **doublet**) with two sets of cortical structures. When such doublets reproduce asexually by mitosis, they give rise to doublets, showing that the double nature of the cortical structures has a genetic basis. Significantly, when doublet Paramecia are mated with normal (singlet) Paramecia, the progeny of the doublet exconjugant are doublets, and the progeny of the singlet exconjugant are singlets. Furthermore, when the progeny go through autogamy (*Figure 5-12*), they maintain their doublet or their singlet properties. Nuclear gene markers introduced into the crosses between doublets and singlets are transmitted and segregate according to the pattern expected for nuclear genes.

The mode of inheritance of the duplicated structures therefore appears to be independent of the mode of inheritance of nuclear genes. It also appears to be unaffected by exchange of cytoplasm, since doublet and singlet cells retain their identities and their ability to reproduce true to type even after they are allowed to conjugate under conditions in which cytoplasm as well as nuclei are exchanged.

An additional observation supporting the idea that the cortical structures of *Paramecium* are genetically autonomous concerns the natural grafting of a piece of cortex from one conjugant to the other. This grafting happens only rarely, but when it does, an animal that has duplex cortical structures is recovered. When this animal reproduces, the new structures reproduce autonomously.

Clearly, the preformed cortical structures are maintained by cell division; they appear to be essential for their own reproduction, and their inheritance is not under the control of nuclear genes or cytoplasmic genetic factors. Twenty years ago, Sonneborn suggested that different parts of the cortex might serve as sites for the "specific absorption and orientation of molecules derived from the milieu and genetic action" and that pre-existing cortical structures could act by "determining where some gene products go in the cell, how these combine and orient, and what they do." Although such phrases are carefully stated generalizations, it is not yet possible to explain the *Paramecium* phenomena in more specific terms.

## 16.16 Inheritance of Cytoplasmic Structures in Other Eukaryotes

The existence of nonnuclear, nonorganelle genetic systems is repeatedly suggested for various eukaryotes, although none of these studies have as yet established specific mechanisms. Regular reports are made of DNA associated with the plasma membrane, and in some cases this DNA is claimed to have molecular properties distinct from those of nuclear or organelle DNA. In addition, RNA has been shown to be intimately associated with **basal bodies** (structures akin to centrioles that give rise to cilia and flagella), and basal bodies are thought to have certain autonomous properties. Therefore, there is some reason to believe that such phenomena as cell symmetry and form may, at least in some cases, prove to depend on strategically localized species of extranuclear nucleic acids.

(a)                                              (b)

(a) Tracey Sonneborn has made many contributions toward an understanding of the extranuclear genetics of *Paramecium*. (b) Ruth Sager (Sidney Farber Cancer Center) discovered extranuclear inheritance in *Chlamydomonas reinhardi* and now studies its molecular basis.

# Questions and Problems

1. The trait *yellow embryo* in a mammal does not affect the viability of either sex. When *yellow embryo* females are crossed to normal males, the $F_1$ are all *yellow embryo*. When the $F_1$ is inbred, 3/4 of the progeny are *yellow embryo* and 1/4 are normal. Is the trait maternally inherited? Explain.

2. To what extent are plasmids and temperate bacteriophages like the extra-nuclear genetic systems considered in this chapter and to what extent are they different?

3. When females of a certain mutant strain of *D. melanogaster* are crossed to wild-type males (or to males of any strain), all of the viable offspring are females. This result could be the consequence of either a sex-linked lethal mutation or a maternally inherited factor that is lethal to males. What kinds of crosses would you perform in order to distinguish between these alternatives?

4. Certain plants from the $F_2$ of the cross given in Figure 16-9 are subjected to further crosses. What progeny genotypes and phenotypes do you expect from the following (assume that the white plants are viable, although in fact they are not).

   (a) $+/+$ white females (from $F_2$) $\times$ $+/+$ green males
   (b) $+/+$ white females (from $F_2$) $\times$ $+/ij$ green males
   (c) $+/+$ green females (from $F_2$) $\times$ $+/ij$ striped males
   (d) $+/+$ striped females (from $F_2$) $\times$ $ij/ij$ striped males

5. You suspect that certain species of mRNA from the nucleus are translated exclusively in organelles. Devise hypotheses to explain how it is that such organelle-destined mRNA is not translated in the cytoplasm while en route to the organelle by postulating unique organelle tRNA species.

6. It has been proposed that cortical patterns in *Paramecium* are determined by DNA located within the *Paramecium* cell membrane. How would the experiments concerning the inheritance of preformed structure be interpreted in light of this hypothesis?

7. The endosymbiotic bacterium *kappa* is maintained only by *Paramecium* cells that carry at least one dominant nuclear gene called *K*; in the absence of *K*, the bacteria are lost from cells after several rounds of mitosis. A *K/k* strain of *Paramecium* harboring *kappa* is crossed with a *k/k* strain. Diagram all possible outcomes of this cross (a) when conjugation involves nuclear exchange only and (b) when conjugation proceeds under conditions in which cytoplasm (and therefore endosymbionts) as well as nuclei are exchanged.

8. *Paramecium* cells that harbor *kappa* are called "killers" because a toxin released by the endosymbionts is lethal to "sensitive" cells that lack *kappa*, although it is not harmful to cells that possess *kappa*.

   (a) Killer strain *A* and sensitive strain *B* are allowed to conjugate without cytoplasmic exchange. The two exconjugant clones *C* and *D* are cultured separately; each is then induced to undergo autogamy (*Figure 5-12*). When the exautogamous cells deriving from clone *C* are examined several generations later, half are found to be killers and half sensitives, whereas those deriving from clone *D* prove to be all sensitives. Give the nuclear genotypes of the various strains and clones and explain the results.

(b) How would the outcome differ, if at all, if the same crosses were performed except that cytoplasmic exchange is allowed during the initial conjugation?

9. When a *Neurospora* cross is made such that the protoperithecium derives from a *poky* strain and the conidia are from a wild-type strain (see Figure 5-9 for the *Neurospora* life cycle), what mitochondrial genotype do you expect of the eight ascospores assuming maternal inheritance of the trait?

10. Mitochondria from the respiratory-deficient *abn-1* strain of *Neurospora* (*Figure 7-2*) were isolated and physically injected into wild-type *Neurospora* hyphae. What is the expected outcome if the *abn-1* phenotype is controlled by a nuclear mutation? A recessive mitochondrial mutation? A suppressive mitochondrial mutation?

11. When an erythromycin-resistant (*ery-r*) strain of *Chlamydomonas* is crossed with an erythromycin-sensitive strain, the following data are obtained:

$$ery\text{-}r\ mt^+ \times ery\text{-}s\ mt^- \qquad ery\text{-}s\ mt^+ \times ery\text{-}r\ mt^-$$
$$\downarrow \qquad\qquad\qquad\qquad \downarrow$$

All progeny are *ery-r*      All progeny are *ery-s*

(a) Devise a hypothesis to explain these results.

(b) Both mating types of *Chlamydomonas* contribute apparently equal amounts of cytoplasm to the zygote. How does this affect your answer to (a)?

(c) Explain these results according to a model which proposes that the *mt* loci encode restriction endonuclease and modification enzymes (*Section 4.11*).

12. Hybridization mapping can be performed with cloned restriction fragments of yeast DNA as well as with RNA products as described in the chapter. Outline how you would perform this experiment and how $\rho^-$ deletion mutants would serve as controls.

13. The offspring of male horses and female donkeys are known as hinnies; the offspring of reciprocal crosses are known as mules. You discover that the restriction endonuclease patterns from horse mitochondrial DNA are very different from donkey patterns. Outline experiments utilizing this information which would determine whether or not mitochondria are maternally inherited in ungulate mammals.

14. Fungal mitochondrial DNA contains long tracts of AT-rich DNA.

(a) How would you ask whether this DNA contains functional structural genes?

(b) How would you use denaturation mapping to locate them in wild-type and $\rho^-$ yeast?

(c) What results would you need to obtain from (a) and (b) to argue that these tracts account for the particularly large size of mitochondrial chromosomes in the fungi?

15. A nonsense mutation occurs at the upstream end of the intron lying between exons B2 and B3 of the cytochrome *b* gene of yeast.

(a) How, if at all, would this affect the length of the resultant pre-mRNA transcript?

(b) How, if at all, would it affect the length of any polypeptides translated from this transcript?

16. Is the Figure 16-5 pedigree compatible with an X-linked inheritance of the mtDNA phenotype? Explain.

17. Deletion strains of yeast mtDNA are described as being used in both complementation and mapping studies. Describe each use and explain how you would distinguish the two outcomes.

18. Mitochondrial genes in yeast display unique patterns of sexual inheritance that are explored in this and the next few problems. We first perform a one-factor cross (*Section 12.8*), which we can denote *a mit*$^+$ × α *mit*$^-$, and allow the resultant diploid (*a*/α *mit*$^+$/*mit*$^-$) to undergo several rounds of mitosis. Each mitotic division is accomplished by a process known as **budding**: (1) a daughter mitotic nucleus moves into a small cytoplasmic protrusion; (2) cytokinesis yields a large and a small daughter cell; and (3) the small daughter enlarges and both again divide by budding.

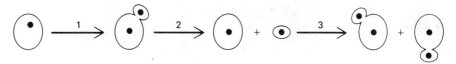

Show, with drawings, how this pattern of mitosis is the most likely explanation for the fact that when the diploid cells from the one-factor cross are cloned and the clones are scored for mitochondrial phenotype, roughly half are *mit*$^+$ and half are *mit*$^-$: none are *mit*$^+$/*mit*$^-$ heterozygotes. (Hint: the phenomenon is often called a **bud position effect**).

19. A diploid zygote usually undergoes several rounds of mitosis before the cells can be induced to enter the sporulation/meiosis stage of the life cycle (*Figure 5-10*). When individual asci are subjected to tetrad analysis, the following two tetrad classes (*Section 6.9*) are found with roughly equal frequency: *a mit*$^+$, *a mit*$^+$, α *mit*$^+$ α *mit*$^+$; and *a mit*$^-$ *a mit*$^-$ α *mit*$^-$ α *mit*$^-$. Give a probable explanation for these results.

20. A two-factor cross is now performed: *a mit*$^+$ *syn*$^+$ × α *mit*$^-$ *syn*$^-$. The diploid zygotes are sporulated, and the following four tetrad classes emerge with the following frequencies:

| Class | Number |
|---|---|
| 1. *a mit*$^+$ *syn*$^+$, *a mit*$^+$ *syn*$^+$, α *mit*$^+$ *syn*$^+$, α *mit*$^+$ *syn*$^+$ | 300 |
| 2. *a mit*$^-$ *syn*$^-$, *a mit*$^-$ *syn*$^-$, α *mit*$^-$ *syn*$^-$, α *mit*$^-$ *syn*$^-$ | 227 |
| 3. *a mit*$^+$ *syn*$^-$ *a mit*$^+$ *syn*$^-$ α *mit*$^+$ *syn*$^-$ α *mit*$^+$ *syn*$^-$ | 45 |
| 4. *a mit*$^-$ *syn*$^+$ *a mit*$^-$ *syn*$^+$ α *mit*$^-$ *syn*$^+$ α *mit*$^-$ *syn*$^+$ | 21 |
| | Total 576 |

(a) When do you think mitochondrial DNA crosses over during the yeast life cycle?

(b) How does this pattern of inheritance differ from that observed with neutral and suppressive petites?

(c) What recombination frequency between the *mit* and the *syn* loci is indicated by these data? Why might these numbers be in error, given the nature of early events in the diploid zygote?

21. The pattern of organelle gene transmission of yeast and in *Chlamydomonas* (Probem 11) is called uniparental inheritance, whereas the pattern in most complex eukaryotes is called maternal inheritance. How are the three patterns distinctive?

# 17 General Recombination Mechanisms

**A. Introduction**

**B. Models of General Recombination**
- 17.1 The Holliday Model of Reciprocal Recombination
- 17.2 The Fox Model for Nonreciprocal Recombination
- 17.3 Fate of "Outside" and "Inside" Genetic Markers in Models of General Recombination
- 17.4 Overview

**C. Enzymes that Mediate General Recombination**
- 17.5 DNA Replication and Repair Enzymes Involved with General Recombination
- 17.6 Enzymes Specifically Involved in General Recombination

**D. Formation and Segregation of Physical and Genetic DNA Hybrids During General Recombination**

- 17.7 Hybrid DNA Formation During Bacterial Transformation
- 17.8 Heteroduplex DNA Formation and Its Segregation

**E. Mismatch Repair of Heteroduplex DNA During General Recombination**
- 17.9 Mismatch Repair of Heteroduplexes Created *in Vitro*
- 17.10 Intragenic Marker Effects and Mismatch Repair
- 17.11 High Negative Interference and Mismatch Repair
- 17.12 Gene Conversion and Mismatch Repair

**F. Questions and Problems**

# Introduction

In the five previous chapters, as strategies for mapping chromosomes and genes were outlined, it was assumed at the outset that markers would recombine when physical exchanges occurred between chromosomes. Cytological evidence for the occurrence of such exchanges during meiosis was presented in Section 14.3. In this chapter we ask how chromosomal exchanges come about at a molecular level, using data derived from biochemical, structural, and genetic studies.

The chapter begins with two models for **general recombination,** the process that mediates most forms of chromosomal exchange. The models may not be correct in detail, since the actual mechanisms of general recombination are by no means firmly established, and alternate recombination models have also been proposed. They suffice, however, to organize and offer a conceptual framework for the molecular and genetic observations presented in the ensuing sections of the chapter. In the next chapter, we go on to consider alternative modes of recombination utilized by temperate-phages and by transposable elements.

# Models of General Recombination

## 17.1 The Holliday Model of Reciprocal Recombination

**Recombination is said to be reciprocal when the two interacting duplexes exchange equivalent lengths of polynucleotide chains.** Most of the recombination events we followed in Chapters 12 to 16 entailed reciprocal exchanges: in meiosis, for example, all four meiotic products receive a complete set of genetic information, some of which may retain its parental linkage and some of which may have acquired a recombinant linkage during crossing over.

In 1964, R. Holliday proposed a model for reciprocal general recombination. It has since been refined by a number of other geneticists, notably M. Meselson and C. M. Radding, so that a more accurate (but cumbersome) designation of the scheme we consider would be a "modified Holliday" model.

Figure 17-1 diagrams the three major stages of the Holliday model, giving an overview for the more detailed presentation in Figure 17-2. In the first stage of Figure 17-1, called **recognition and alignment,** two homologous duplexes position themselves in register so that the subsequent exchange does not delete or duplicate any genetic information. In the second stage, two of the strands break, invade the other helix, and are ligated to the opposite strand, forming a molecule called a **Holliday intermediate.** The position where the two "swapped" strands cross one another is known as the **branch.** In the final **cleavage-and-ligation** stage, the Holliday

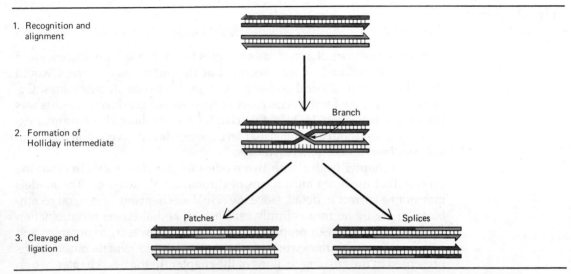

1. Recognition and
   alignment

2. Formation of
   Holliday intermediate

Branch

Patches

Splices

3. Cleavage and
   ligation

**FIGURE 17-1 Three major stages in the Holliday model of reciprocal recombination** (From F. W. Stahl, *Ann. Rev. Genet.* 73:7–24, 1979.)

intermediate is cut in the region of the branch, and the cut ends are covalently sealed (ligated) to form mature recombinant molecules. Cleavage can occur in two ways. If the "swapped" strands are cut, then the resultant duplexes each contain a **patch** of hybrid (red-black) DNA. If the "unswapped" strands are cut at the equivalent position, then the resultant duplexes appear **spliced,** with hybrid DNA in the spliced region.

In the more detailed model in Figure 17-2, the three major stages we have just described are designated 1, 2, and 3. Recognition and alignment (a) are drawn as before, but the formation of the Holliday intermediate (b through i) is shown in more detail. Breaks are first made at comparable positions in two of the polynucleotide chains having the same polarity (b). Each broken chain then invades the opposite helix (c) and establishes base pairs with complementary nucleotides on the other strand (d). Ligase enzymes (*Figure 4-11*) seal the discontinuities (e) to produce a Holliday intermediate with an internal branch point. Because the Holliday intermediate has the four-armed shape of the Greek letter chi ($\chi$), it has also been referred to as a **chi form.** Space-filling molecular models of such intermediates predict that they should be very stable structures: the nucleotides remain close enough to one another to form stable base pairs, even in the region of the branch.

Although stable, a Holliday intermediate need not be static. If both duplex chromosomes rotate in the same direction, the branch point should be free to swivel to the right (as in f) or to the left. Such shifts are called **branch migrations,** and they have the effect of creating sectors of red-black

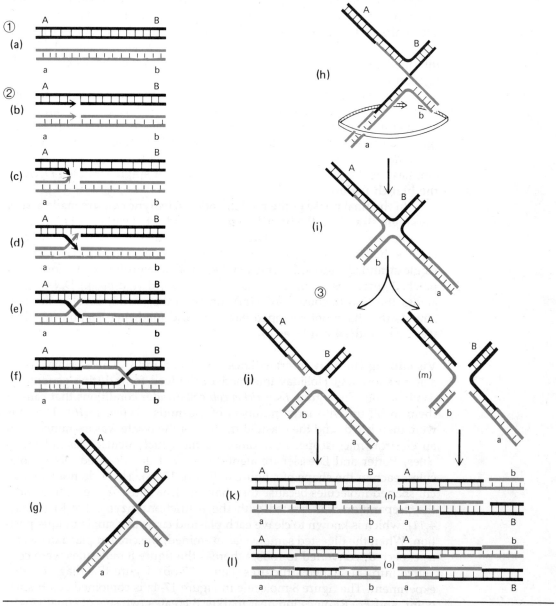

**FIGURE 17-2 Holliday model for reciprocal genetic recombination.** The three major stages drawn in Figure 17-1 are indicated at one to three. (From H. Potter and D. Dressler, *Proceedings of the National Academy of Science United States* **73**:3000, 1976, after R. Holliday, *Genetics* **78**:273, 1974, and previous publications cited therein.)

hybrid DNA in the Holliday intermediate (f). It has been calculated that under physiological conditions, a branch could easily migrate through 1000 base pairs of DNA in 20 seconds. Therefore, extensive regions of hybrid DNA could be created in a matter of seconds between the time a Holliday intermediate first forms and the time it matures in the final cleavage-and-ligation phase of the model.

The cleavage-and-ligation phase is easiest to visualize if we redraw the Holliday intermediate. If we take the molecule depicted in (f) and simply pull its four arms apart, we create the planar configuration drawn in (g). Two of the arms can then be rotated with respect to the other two, as diagrammed in (h), to generate the structure shown in (i). This drawing emphasizes the presence of **four short sectors of single-stranded DNA in the branch region.**

In the final phase of the model, **endonucleolytic cuts are made across the single strands in the branch region.** As indicated earlier in Figure 17-1, these cuts can occur in one of two ways. Cuts across the "east-west" axis (j) involve the two "swapped" strands. These generate two duplexes carrying single-stranded gaps (k), which, when sealed with ligase (l), form two **patched duplexes** with central sectors of hybrid (red-black) DNA. Cuts across the "north-south" axis (m), on the other hand, involve the two strands that have not swapped partners, and following ligation (n), they yield two **spliced duplexes** (o).

**Visualizing Holliday Intermediates**   H. Potter and D. Dressler have been able to show that Holliday intermediates indeed form during general recombination. They grew *Escherichia coli* cells under conditions that caused them to fill up with large numbers of plasmids (*Section 13.16*). The cells were then lysed, and the plasmid fraction of the lysate was examined with an electron microscope. In addition to the usual circular plasmid molecules, Potter and Dressler frequently observed double-sized DNA molecules, each in the shape of a figure 8 (Figure 17-3a). These do not look like chi-shaped molecules because they have no free ends. To create free ends, the preparation was treated with the restriction enzyme *Eco* RI (*Section 4.11*), which is known to cleave each plasmid circle at a single unique position. When the digested sample was examined, the circular plasmids were, as expected, converted to rods, whereas the figure 8 molecules were converted into chi-shaped structures (Figure 17-3b). Figure 17-4 diagrams this experiment. The figure 8 molecule in Figure 17-4a is conjoined at a branch point, and *Eco* RI digestion near marker B creates two sets of equal-length arms (Figure 17-4b).

By denaturing the DNA slightly, Potter and Dressler were able to create chi-shaped molecules of the sort shown in Figure 17-3c. The branch point has opened up sufficiently to provide a view of its organization, and it is seen to carry four short spans of single-stranded DNA, exactly as predicted by the Holliday model (*Figure 17-2i*).

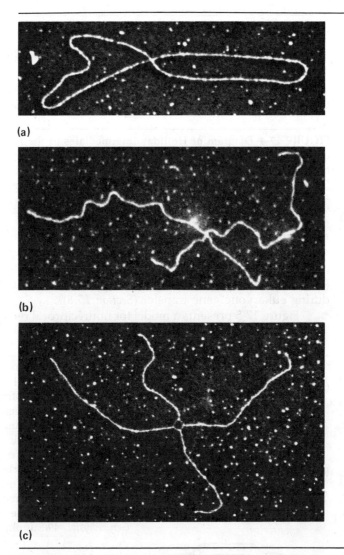

**FIGURE 17-3** (a) A double-sized pMB9 plasmid with the shape of a figure 8 isolated during the Potter and Dressler experiment. (b) A chi-shaped DNA molecule isolated after *Eco* RI treatment of the plasmid preparation. (c) A chi-shaped DNA molecule showing single strands in the crossover region. (From H. Potter and D. Dressler, *Proc. Nat. Acad. Sci. U.S.* **74**:4168, 1977.)

## 17.2 The Fox Model for Nonreciprocal Recombination

In **nonreciprocal recombination, only one of the two interacting duplexes retains its original length.** Thus if we denote the conserved chromosome as the **host** and the nonconserved chromosome as the **donor**, then during a

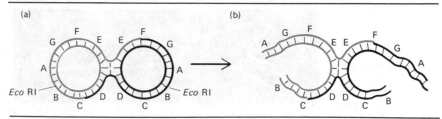

FIGURE 17-4 **Diagram of Holliday intermediates** in the Potter and Dressler experiment. (a) Figure-eight molecule, comparable to Figure 17-3a. Branch migration following exchange has created regions of red-black hybrid overlap. (b) Same molecule after *Eco* RI digestion at a single cleavage site near marker B. A chi-shaped structure is generated, comparable to Figure 17-3b.

nonreciprocal exchange, a piece of donor DNA is inserted into the host chromosome, and all pieces of "extra" donor DNA are degraded by nucleases. Nonreciprocal recombination occurs during both bacterial transformation (*Section 13.8*) and transduction (*Section 13.11*), and it may occur as well during eukaryotic gene transfer (*Section 15.8*).

    Figure 17-5 presents a model for nonreciprocal recombination as visu-

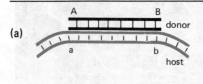

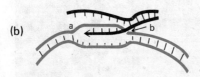

FIGURE 17-5 **The Fox model for nonreciprocal general recombination** during bacterial transformation of host by donor DNA. (a) Recognition and alignment. (b) Denaturation of host and donor DNA and pairing of host and donor strands. Endonucleases nick donor DNA at arrow. (c) Endonucleases nick host DNA at arrows. (d) Inserted donor DNA with gaps. (e) Gaps repaired by polymerase and ligase. (After M. Fox, *J. Gen. Physiol.* **49**(Suppl.): 183, 1966.)

alized for bacterial transformation by M. Fox. Recognition and alignment (a) occur as before, and host DNA is invaded by a donor strand (b). Unlike reciprocal recombination (*Figure 17-2b through d*), however, **the donor DNA is not invaded by a host strand** (and therefore the interaction is not a reciprocal one). Instead, the intermediate is resolved by making one endonucleolytic cut in the donor DNA (b) and two cuts in the host DNA (c) to create a sector of hybrid DNA (d). This is then sealed in place by the action of DNA ligase (e).

# 17.3 Fate of "Outside" and "Inside" Genetic Markers in Models of General Recombination

We can now examine the Holliday and the Fox models from a genetic point of view by following the fate of genetic markers. First, let us examine what happens to **outside markers,** defined as those lying outside regions of hybrid-overlap DNA. In Figures 17-2 and 17-5, the outside markers on the original duplexes are designated *AB* and *ab*. Examination of the final products of reciprocal recombination (Figure 17-2 l and o) reveals that they are different: **the patched duplexes retain the outside markers in their parental configurations,** *AB* and *ab*, whereas **the spliced duplexes carry the outside markers in a recombinant configuration,** *Ab* and *aB*. If we make the likely assumption that endonucleases acting in the branch region are blind about which pair of strands they are cutting, then the Holliday model predicts that a **physical exchange between homologous chromosomes should result in the genetic exchange of outside markers about 50 percent of the time.** Looking next at Figure 17-5, we see clearly that **during a nonreciprocal exchange, host chromosomes inevitably emerge as patched duplexes that retain the parental configuration of outside markers.**

We can now consider **inside markers,** defined as those lying within regions of hybrid-overlap DNA. These are designated *m* and *n* and their + alleles in Figure 17-6a (reciprocal recombination) and 17-6b (nonreciprocal recombination). It is clear, first, that **inside markers will always retain their parental linkages along a given polynucleotide strand:** hybrid DNA is thought to form either by branch migration (a) or by strand insertion (b), and neither event should break parental linkages. Second, it is evident that **sectors of DNA that are physical hybrids (red-black) are also genetic hybrids** $\frac{m\ n}{+\ +}$. If we consider the simplest case—namely, that *m* differs from its + counterpart by a single base—then if the *m* sense strand (*Section 9.5*) reads 5′ . . . ACAAT . . . 3′ and its + counterpart reads 5′ . . . ACAGT . . . 3′ (so that its antisense strand is 3′ . . . TGTCA . . . 5′), then the *m*/+ hybrid region will have the structure

$$5'\ \ldots \text{ACAAT} \ldots 3'$$
$$3'\ \ldots \text{TGTCA} \ldots 5'$$

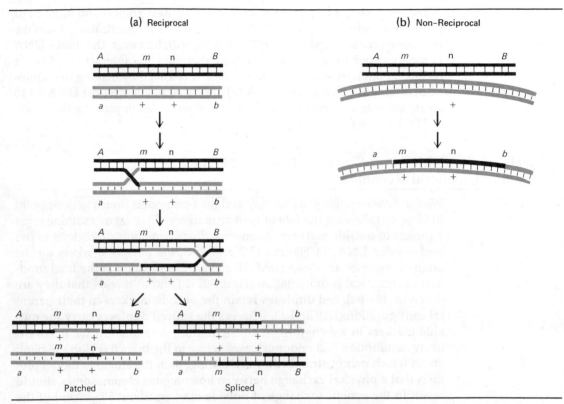

**FIGURE 17-6 Formation of heteroduplex DNA** during reciprocal (a) and nonreciprocal (b) recombination.

This hybrid DNA carries a **mismatched** A-C base pair. It is therefore said to contain **heterologous** base sequences and is called a **heteroduplex** (the term **heterozygous DNA** is also used). At least one site of mismatch will also be present at the $n/+$ site.

Figure 17-7 diagrams two of the possible "fates" that can befall a sector of such heteroduplex DNA once it has formed. In Figure 17-7a, the heteroduplex region is simply replicated, the result being two daughter **homoduplexes** that differ from one another at a single position. In such cases, the heterologous sequences are said to **segregate** from one another. Figure 17-7b illustrates an alternate possibility, known as **mismatch repair,** during which one strand of the heteroduplex is removed and replaced by a new stretch of homologous DNA.

## 17.4 Overview

In the remaining sections on general recombination, we present the molecular and genetic studies that have given rise to, and supported, the theo-

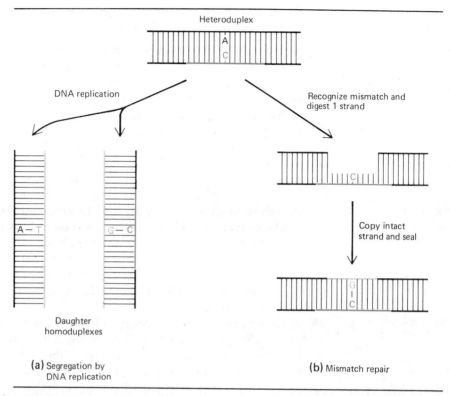

**FIGURE 17-7 Resolution of heteroduplex DNA** by (a) DNA replication and segregation or (b) mismatch repair.

retical models. We first consider how a battery of enzymes, notably a critical molecule called the recA protein, is thought to mediate key steps in general recombination. We next present experiments showing that physically hybrid DNA forms during transformation and that genetically hybrid DNA forms during both reciprocal and nonreciprocal exchanges. Finally, we examine the extensive evidence showing that this heteroduplex DNA can be "resolved" both by segregation and by mismatch repair.

# Enzymes that Mediate General Recombination

## 17.5 DNA Replication and Repair Enzymes Involved with General Recombination

Examination of the Holliday and Fox models (*Figures 17-2 and 17-5*) at once suggests that **many of the enzymes described in Chapter 8 as being involved with DNA replication and repair are also involved with general recombination.** That this is indeed the case can be shown by demonstrating that mutant strains defective in replication and/or repair are often also defective in recombination. We can give three examples.

1. Mutant strains of T4 carrying an *amber* mutation in gene 32 fails to form recombination intermediates in a nonsuppressor host. Gene 32 encodes a phage-specific version of a **single-strand binding protein (SSB),** which, as described in Section 8.2, stabilizes single-stranded regions that form within DNA duplexes. This suggests that the formation of single-stranded regions is an important early event in the recombination sequence, a concept expanded in the next section.

2. T4 mutants carrying *amber* mutations in gene 30, which encodes a phage-specific **ligase,** are only able to form recombinant molecules with the configuration drawn in Figure 17-2 k and n: long gaps persist in each strand, so that they are held together only by base pairs. These can be converted into covalently sealed recombinant chromosomes by isolating them and presenting them with wild-type T4 ligase *in vitro*. Yeast mutants carrying a *ts* mutation (*cdc9*) that inactivates the ligase involved in DNA replication are also defective in recombination. Therefore, for both T4 and yeast, DNA ligase is clearly an essential enzyme in recombination.

3. A number of mutants in yeast and *Drosophila* are simultaneously hypersensitive to mutagens and defective in meiosis. Although the affected enzymes have not yet been identified, these phenotypes suggest that enzymes involved in the repair of mutagen-induced lesions (*Section 8.10*) may also participate in recombination during meiosis.

## 17.6 Enzymes Specifically Involved in General Recombination

A number of the events depicted in the Holliday model have no obvious counterparts in DNA replication and repair. One would predict, therefore, that **certain enzymes should be specifically involved in mediating recombination.** Again, this deduction has been verified genetically: certain mutant strains, notably in *E. coli*, are selectively affected in their ability to recombine. We can focus here on two enzymes identified by mutant analysis.

The first is called **exonuclease V** or the **recBC protein.** It is a dimer of two subunits, one encoded by *recB* and the other by *recC*, and since it can act as an endonuclease and as a DNA-unwinding protein as well as an exonuclease, it represents the kind of enzyme one would expect to participate in the breakage and rejoining of chromosomes. Recombination is defective but not absent in *recBC* mutants, indicating that other enzymes are able to substitute their activities if exonuclease V is nonfunctional.

The second is the **recA protein.** This protein was already introduced in Section 8.12, where we described its role as a protease that destroys the lexA protein and thus facilitates the expression of SOS repair genes. In addition, constitutive (noninduced) levels of recA protein mediate recombination, so that *recA⁻* mutants are completely blocked in general recombination as well as unable to muster an SOS response.

The gene for the recA protein has been cloned and sequenced, and the protein, a tetramer of a 38,000 dalton polypeptide, has been purified. In addition to its protease activity, which is probably not relevant to recombination, the recA protein has two additional properties. (1) It is a **DNA-dependent ATPase,** meaning that it will hydrolyze ATP when incubated with ATP in the presence of single-stranded DNA. (2) It catalyzes both **homologous DNA pairing and strand assimilation,** the first critical events in general recombination. Specifically, a number of investigators have shown that the recA protein catalyzes the reaction diagrammed in Figure 17-8: a piece of single-stranded DNA recognizes a homologous sequence in duplex DNA; the duplex is unwound; the single strand invades the helix and pairs with its complementary strand; and the displaced strand bulges out to form a **D loop** (recall analogous R loops described in Section 9.9). This reaction is greatly facilitated by the addition of SSB, which presumably helps hold the helix apart, and it requires ATP hydrolysis by the ATPase component of the recA protein.

Even more relevant to general recombination is the demonstration, by P. Howard-Flanders and his coworkers, that in the presence of ATP, recA protein will induce two *duplex* DNA molecules to swap strands at homologous positions. This observation supports the concept, inherent in our models, that **the recognition-alignment phase of general recombination can be carried out by intact DNA duplexes** (*Figure 17-1*). The obvious question, of course, is how this works, since Watson-Crick base pairing cannot occur between bases that are already paired and tucked within helical DNA. The most logical answer is to propose that **the base pairs themselves somehow recognize one another.** Thus one can show, with molecular models, that the edges of an AT base pair that protrude into the major groove of helical DNA (*Figure 1-8*) are different in configuration from the edges that protrude from a GC pair. It is therefore possible that AT base pairs might form hydrogen bonds with other AT pairs but not with GC

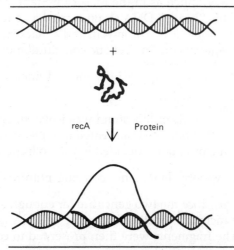

**FIGURE 17-8 Actions of recA protein** *in vitro.* A single-stranded DNA fragment (heavy line) is induced to undergo pairing and strand invasion with a homologous region of duplex DNA. The resultant bulged structure is called a D loop. (From R. P. Cunningham, T. Shibata, C. DasGupta, and C. M. Radding, *Nature* **281**:191–195, 1979.)

pairs, and vice versa, and that a mutual recognition of homologous regions would result. It is widely believed that recA protein somehow catalyzes this recognition event, but the mechanism is not yet clear.

Mutual recognition would clearly be a more probable event if homologous chromosomes were to carry special sites that specifically promote the onset of pairing. The nucleotide sequence 5'GCTGGTG3' (also called a **chi hotspot**) has recently been shown to play such a role: when present in a phage chromosome, this sequence greatly enhances the probability that recombination will occur in its local vicinity. F. Stahl has proposed, therefore, that some recombination-promoting enzyme recognizes and binds to such chi hotspot sequences in two homologous duplexes, thereby aligning them so that mutual recognition events can more easily occur in contiguous regions.

# Formation and Segregation of Physical and Genetic DNA Hybrids During General Recombination

## 17.7 Hybrid DNA Formation During Bacterial Transformation

As stressed in Section 17.3, the Holliday and Fox models predict that recombinant chromosomes should be physically hybrid. Figure 17-9 summarizes an experiment, performed by T. Gurney, Jr., and M. Fox, that demonstrates that physical hybrids form during the nonreciprocal act of bacterial transformation (*review Section 13.8*). Analogous experiments have shown that physically hybrid chromosomes are the result of reciprocal recombination in bacteriophages.

Gurney and Fox first grew *Diplococcus pneumoniae* (pneumococcus) cells in a medium containing $^{15}N$ and $^2H$ (deuterium) so that the DNA of the cells was heavy, much as in the Meselson-Stahl experiment described in Section 2.3. This "heavy" DNA strain carried a genetic marker, *ery$^r$*, for resistance to the antibiotic erythromycin and a marker *str$^s$* for sensitivity to the antibiotic streptomycin. The genetic constitution of the "heavy" strain can thus be written as $\frac{ery^r\ str^s}{ery^r\ str^s}$, where the red lines symbolize the heavy isotope label (Figure 17-9a).

The DNA from cells of this strain was isolated, and during the isolation process, the DNA became fragmented into pieces that were the appropriate size for transformation. Included in this collection were fragments of the $\frac{ery^r}{ery^r}$ and $\frac{str^s}{str^s}$ variety. Fragments carrying both *ery$^r$* and *str$^s$* were not present, however, since the two genes are far enough apart in the chromosome not to be included together in a piece of DNA in the 0.3 to $8 \times 10^6$ dalton range. The fragments were then presented to competent cells hav-

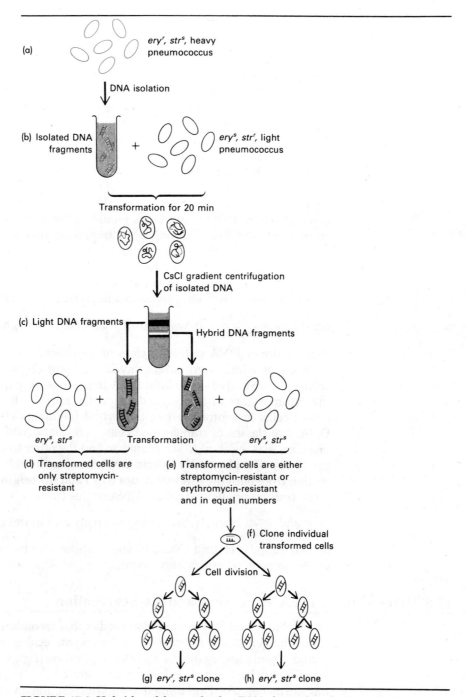

**FIGURE 17-9 Hybrid and heteroduplex DNA formation during transformation,** as demonstrated in the experiments with pneumococcus performed by T. Gurney and M. Fox. Heavy ($^{15}$N, $^{2}$H-labeled) DNA is depicted in red, and light ($^{14}$N, H-labeled) DNA is depicted in black.

ing light ($^{14}$N, $^{1}$H) chromosomes with the genetic constitution $\dfrac{ery^s\ str^r}{ery^s\ str^r}$; in other words, the recipients were erythromycin sensitive ($ery^s$) but streptomycin resistant ($str^r$) (Figure 17-9b).

After the donor fragments had been taken up and about 25 minutes was allowed for transformation, the recipient cells were themselves broken and their fragmented DNA was subjected to centrifugation in CsCl (*Box 2.1*). These fragments distributed themselves in the gradient according to their buoyant density, and two peaks were found: a large peak with the density of a fully light DNA duplex and a smaller peak with the density of a **hybrid** DNA—a duplex composed of one heavy and one light polynucleotide chain (Figure 17-9c). Almost no DNA was found in that region of the gradient where fully heavy DNA would be expected to equilibrate, suggesting that the heavy donor DNA fragments had somehow been converted into hybrid DNA after being taken up by the cells.

The fragments from the two peaks were then isolated, and each fraction was tested for its ability to transform a *third* strain of pneumococcus, a strain sensitive to *both* streptomycin and erythromycin and thus having the constitution $\dfrac{ery^s\ str^s}{ery^s\ str^s}$. It was found that when the light fragments were used as donor DNA only one class of transformed cells could be recovered, all of which were streptomycin resistant (Figure 17-9d). In other words—as expected—the light DNA fragments were derived solely from the chromosomes of the original recipients, namely, the streptomycin-resistant cells. In contrast, when the hybrid fragments were used as donor DNA, *two* classes of transformed cells were found with equal frequency: one class was resistant to streptomycin and the other to erythromycin (Figure 17-9e). **The physically hybrid DNA fraction thus clearly carried information from both the original donor ($ery^r$) and the original recipient ($str^r$);** it can thus be predicted that the DNA in this fraction had the composition $\dfrac{str^s}{str^r}$ and $\dfrac{ery^r}{ery^s}$, being hybrid both in density and in the genes it carried on its two strands. Hybrid DNA, in short, appears to be the result of a successful genetic transformation event.

## 17.8 Heteroduplex DNA Formation and Its Segregation

The Holliday and Fox models also predict that recombinant chromosomes should be genetically hybrid (heteroduplex) molecules, which, if left unrepaired, should **segregate** during DNA replication (*Figure 17-7a*). Additional experiments by Gurney and Fox demonstrated this to be the case during transformation. As diagrammed in Figure 17-9f, the transformed cell population was diluted and plated on agar to allow individual cells to give rise to discrete colonies. When the cells in each colony were tested for their drug sensitivity, two distinct clones were found to be present: one clone was found to be erythromycin resistant (and streptomycin sensitive), whereas the other was erythromycin sensitive (and streptomycin sensitive) (Figure 17-9g and h). As drawn in Figure 17-9, these results can best be explained by assuming that, as a result of transformation, the original cell came to have a heteroduplex chromosome of the $\dfrac{ery^r\ str^s}{ery^s\ str^s}$ type, whose $ery^r$ sequence was inserted into it by recombination. When this chromosome underwent a round of semiconservative DNA replication, one daughter came to have the chromosome structure $\dfrac{ery^r\ str^s}{ery^r\ str^s}$, whereas the other daughter had the structure $\dfrac{ery^s\ str^s}{ery^s\ str^s}$. The two daughters, genetically distinct with respect to erythromycin sensitivity, then each went on to generate two distinct clones. Thus the $ery^r$ and $ery^s$ sequences, once together in the same heteroduplex, subsequently segregated and found themselves in two different cells.

A second example of heteroduplex DNA formation and its segregation is given by phage T4. When mixed infections are carried out between $r^+$ and $r$ (rapid-lysis) T4 phages (*Section 7.2*) and the progeny phages are allowed to form plaques on a bacterial lawn, most of the plaques are either small and fuzzy edged (wild type) or large and clear ($r$ mutant)—as is to be expected for a one-factor cross. Occasional plaques, however, are found to have a "mottled" appearance: roughly half the plaque is fuzzy edged, and half is clear (*see Figure 7-3*). Since wild-type phage particles can be picked from the fuzzy side of the plaque and $r$ phages from the clear side, such plaques are interpreted to have been formed by phages that were **partial heterozygotes** (or **partial hets**) with heteroduplex $\dfrac{r}{r^+}$ chromosomes.

A partial het is thought to represent a phage that participated in recombination, acquired heterologous sequences in the $r$ locus, and then became packaged into a protein coat before it had time to replicate its DNA. When such a phage lands on a bacterial lawn and replicates its DNA during the course of its infection cycle, it will generate two types of phage progeny and will hence produce a mottled plaque. Support for this interpretation comes from experiments where the parental phages are constructed to carry outside markers so that the cross becomes $A\,r^+\,B \times a\,r\,b$. Progeny phages are then plated, $r^+$ and $r$ phages are picked from either side of mottled plaques (as before), and the genetic constitution of these phages is assessed. It is found that a sizable proportion are indeed recombinant for the outside markers (recall from Section 17.3 that this proportion should approach 50 percent for reciprocal recombination).

# Mismatch Repair of Heteroduplex DNA During General Recombination

That heteroduplex DNA occasionally forms, and segregates, during general recombination is well established. The alternate fate of heteroduplex DNA, its resolution by mismatch repair (Figure 17-7b), has been inferred by a variety of observations, some of which we consider in the next few sections. We first describe *in vitro* experiments with phage λ showing that heteroduplex DNA can, indeed, be repaired essentially as diagrammed in Figure 17-7b. We then describe genetic phenomena in both prokaryotes and eukaryotes that are best explained by assuming that heteroduplexes in fact form during recombination and are resolved by mismatch repair.

## 17.9 Mismatch Repair of Heteroduplexes Created *in Vitro*

R. Wagner, J. Wildenberg, and M. Meselson have demonstrated that heteroduplex chromosomes are efficiently repaired by *E. coli* cells. They first created, by the procedures outlined in the upper portion of Figure 17-10, heteroduplex λ chromosomes with the general structure $\dfrac{am1\ +}{+\ am2}$, where *am* stands for an *amber* mutation. These chromosomes were presented to *E. coli* cells under *in vitro* conditions that allow the uptake of naked DNA, and the resultant progeny were analyzed for the presence of wild-type ($am^+$) phages. Since Wagner and colleagues were interested in learning whether such $am^+$ phages could arise by mismatch repair alone, they used $recA^-$ *E. coli* cells (*Section 17.6*) and λ strains defective in phage-recombination genes to prevent any $am^+$ phages from arising as the consequence of recombination.

Wild-type phages were indeed generated from heteroduplex DNA under these conditions, and Wagner and colleagues were able to argue that the $am^+$ chromosomes had been generated by one of the three patterns of mismatch repair illustrated in Figure 17-10. You will note in Figure 17-10 that for each avenue of mismatch repair, a long tract of DNA is digested in the vicinity of the mismatch and a comparable length of DNA is regenerated by repair synthesis (gray strands). The existence of such **repair tracts** was demonstrated by repeating the Figure 17-10 experiments using λ heteroduplexes that were mismatched at a *third* locus, *cI*. These heteroduplexes were again used to generate $am^+$ progeny, but this time the progeny were also tested for their ability to produce a *clear* (*cI*) or a turbid ($cI^+$) plaque. If we write the general structure of the triply marked heteroduplexes as

$$\frac{am1 \quad cI \quad +}{+ \quad cI^+ \quad am2}$$

then if an *am1* marker was chosen that mapped close to the *cI* locus com-

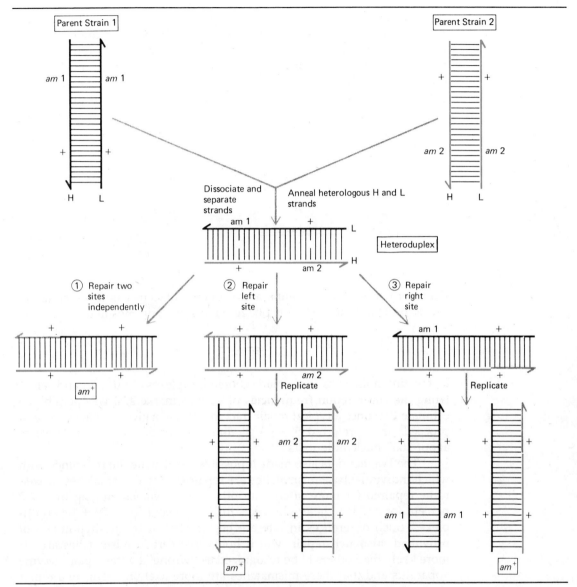

**FIGURE 17-10 Mismatch repair *in vitro*.** Heteroduplex formation *in vitro* using the *am1* and *am2* strains of phage λ, as performed by Wagner, Wildenberg, and Meselson, and the three avenues (1 to 3) by which such a heteroduplex can be repaired to produce wild-type (*am*⁺) DNA. Avenues two and three require DNA replication as well as mismatch repair. DNA produced by repair synthesis is shown in gray.

pared with *am2*, the *am*⁺ progeny were found to be far more likely to produce *turbid* than *clear* plaques. If, on the other hand, *am2* mapped close to
the *cI* locus compared with *am1*, then the *am*⁺ progeny produced *clear*
plaques far more frequently than *turbid*. Explaining these results in terms
of repair tracts, Wagner et al. proposed that in the first case, the *am*⁺ *turbid*
progeny had arisen by digesting the strand carrying *am1* and the contiguous *cI* gene

$$\frac{am1 \quad cI \quad +}{+ \quad cI^+ \quad am2}$$

followed by repair synthesis and DNA replication

$$\frac{+ \quad cI^+ \quad +}{+ \quad cI^+ \quad am2} \quad \Bigg\langle \quad \begin{array}{l} \dfrac{+ \quad cI^+ \quad +}{+ \quad cI^+ \quad +} \\[2ex] \dfrac{+ \quad cI^+ \quad am2}{+ \quad cI^+ \quad am2} \end{array}$$

Draw for yourself how a comparable sequence of steps would generate, for
the second case, an *am*⁺ *clear* phage having the chromosome

$$\frac{+ \quad cI+}{+ \quad cI+}$$

By creating numerous triply marked heteroduplexes of this sort and calculating the **cocorrection frequencies** of various markers, it was possible to
estimate that once a typical repair tract initiates at a given mismatch site, it
proceeds in a preferred 5′ $\longrightarrow$ 3′ direction for an average distance of
about 3000 nucleotide pairs.

The Wagner data also made it possible to calculate the frequency with
which individual base mismatches are repaired. Some mismatches proved
to be repaired far more often than others, with values ranging from 2.2
percent for *am53/* + heteroduplexes to 20.7 percent for *am2/* + heteroduplexes. Such differences in rate are best explained by the hypothesis of
**provoked mismatch repair,** which holds that certain mismatches are far
more likely than others to be recognized as "wrong" by the repair-enzyme
apparatus and that these mismatches are more likely to instigate a repair
tract.

The experiments of Wagner and coworkers, in short, not only demonstrate that heteroduplexes can be repaired but also predict two features of
the repair process in *E. coli*: (1) different mismatches are likely to be repaired at different rates; and (2) closely spaced mismatches (within 3000
base pairs of one another) are more often "corepaired" than distantly
spaced mismatches.

With these experiments in mind, we can turn to consider three phenomena that are encountered when closely linked markers are followed in
genetic crosses, namely, **intragenic marker effects, high negative interfer-**

ence, and **gene conversion.** All these phenomena are readily explained by proposing that short stretches of heteroduplex DNA are often created during the course of recombination and are often resolved by a mismatch-repair mechanism.

## 17.10 Intragenic Marker Effects and Mismatch Repair

**When very closely linked markers are being followed in a prokaryotic or eukaryotic cross, recombination frequencies between various markers do not necessarily reflect the physical distances between them.** This phenomenon is well illustrated by the data of L. Norkin on *lac* operon mutations in *E. coli.* Norkin found that individual *lac* nonsense mutations could be ordered with respect to one another by two techniques that do not depend on recombination frequencies, namely, deletion mapping (*Section 12.13*), using partially transducing *lac* phages (*Section 13.13*), and gene-product mapping (*Section 14.15*), using the lengths of several "nonsense fragment" polypeptides produced by the mutant alleles in his collection. These two approaches proved to yield congruent physical maps of the *lac* mutations, as shown in Figure 17-11a. The various mutations were then crossed against one another in *Hfr* × *F⁻* matings, and the frequency of *lac⁺* recombinants from these matings was scored for each cross. Figure 17-11b shows a typical set of results; in this case, recombination frequencies are plotted for crosses involving mutant 118 and other markers in the collection. From the figure, it is clear that mutant 118 recombines far more often with its neighboring marker 545 than with the more distant 498, that it recombines most often with the still more distant marker 608, and so on. Most striking was Norkin's demonstration that the substitution of one nucleotide pair for another at a mutant site could change recombination frequencies by several orders of magnitude.

Such **intragenic marker effects** on recombination frequencies are readily explained by the mismatch-repair hypothesis. As with the similar observations made by Wagner, Wildenberg, and Meselson, it is only necessary to postulate that certain mismatches provoke repair more readily than others and that such mismatch effects can overshadow or obscure true recombinational exchanges between closely linked loci.

## 17.11 High Negative Interference and Mismatch Repair

In Section 14.6 we considered a phenomenon called **interference,** which relates to the inhibitory effect exerted by one recombinational event on the probability of a second recombinational event. We further defined a parameter called the **coefficient of coincidence** (abbreviated $s$), which measures the amount of interference occurring in a particular three-factor cross, and we noted that if the number of double exchanges in the cross is less than the predicted number, $s$ will be a fractional number.

For bacteriophages, $s$ typically has a value greater than 1; for example,

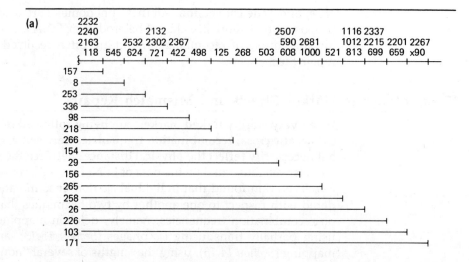

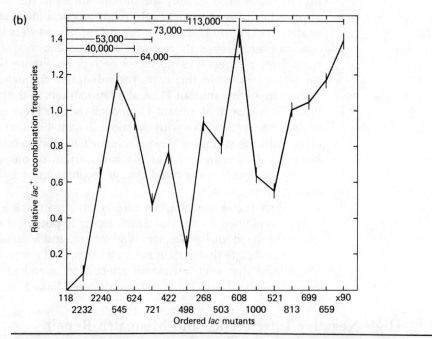

**FIGURE 17-11 Intragenic marker effects.** (a) Deletion map of the *lacZ* gene of
*E. coli* showing point mutations above the line and deletions below. (b) Relative
*lac*⁺ recombination frequencies from crosses of mutation 118 with the *lac*
mutations shown on the abcissa. The molecular weights of the nonsense
fragment polypeptides produced by some of these mutant strains are given at
the top of the graph to provide some indication of real distances. (From L. C.
Norkin, *J. Mol. Biol.* **51**:633, 1970. Copyright by Academic Press Inc. [London]
Ltd.)

in phage λ, $s = 5$ for most sets of three markers. These high coefficient of coincidence values are thought to have a mechanical explanation: exchanges between genetically distinct chromosomes will occur only if phages of one genotype find themselves undergoing replication and recombination in the same geographic portion of a bacterium as phages of a second genotype, and once two such chromosomes achieve sufficient proximity to recombine at one locus, it is probable that they will remain in the same vicinity long enough to undergo a second (and perhaps a third or fourth) exchange at nearby loci. This phenomenon has come to have the cumbersome name of **low negative interference,** a term that originated as follows: interference was first defined, as noted previously, as the *inhibitory* effect that one crossover exerted on the establishment of a second; when it was subsequently learned that there are cases in which one genetic exchange seems to *enhance* the probability of a second, the original phenomenon was renamed **positive interference** and the new phenomenon was called **negative interference,** with the double negation in fact connoting an enhancement of recombination.

In addition to demonstrating low negative interference, bacteriophages (and bacteria and eukaryotes as well) exhibit a distinct phenomenon called **localized high negative interference.** This is seen only if the markers followed in a cross lie within a short interval on the genetic map—that is, when the markers are highly localized. In such cases the value of $s$ becomes dramatically high.

A concrete example is given by P. Amati and M. Meselson's analysis of closely linked *sus* markers in the right arm of the λ chromosome. The map order of the markers involved, shown in Figure 17-12, was first deduced by two-factor crosses, and values for $R_1$ and $R_2$ were determined for the recombinational distances separating individual pairs of markers, as shown in the first two columns of Table 17-1. Three-factor crosses were then performed, and double-exchange frequencies ($R_{1,2}$) were obtained for various sets of three markers. Using these values, Amati and Meselson calculated coefficients of coincidence (*review Section 14.6*) for a number of

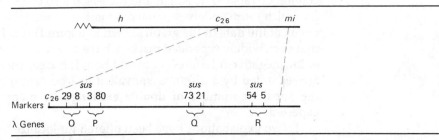

FIGURE 17-12 **The Amati-Meselson experiment.** Portion of the genetic map of phage λ showing the *sus* markers used to demonstrate high negative interference.

**TABLE 17-1 The Amati-Meselson Experiment.** Crosses Involving $c_{26}$ and Various Pairs of *sus* Markers in Phage $\lambda$

$$c_{26} \quad\vdash\quad \overset{sus}{\underset{R_1}{\vdash}} \quad \overset{sus}{\underset{R_2}{\vdash}}$$

| *sus* Markers | $R_{1'}$ % | $R_{2'}$ % | $s = \dfrac{R_{1,2}}{R_1 R_2}$ |
|---|---|---|---|
| 29,8 | 0.48 | 0.30 | 72 |
| 29,3 | 0.48 | 0.51 | 41 |
| 29,73 | 0.48 | 1.4 | 18 |
| 29,21 | 0.48 | 1.3 | 17 |
| 29,54 | 0.48 | 3.3 | 11 |
| 8,3 | 0.64 | 0.32 | 28 |
| 8,80 | 0.64 | 0.40 | 21 |
| 8,54 | 0.64 | 3.5 | 8 |
| 3,80 | 1.1 | 0.25 | 35 |
| 3,73 | 1.1 | 1.35 | 15 |
| 3,21 | 1.1 | 1.5 | 13 |
| 3,54 | 1.1 | 3.1 | 8 |
| 80,73 | 1.1 | 1.5 | 15 |
| 80,21 | 1.1 | 1.1 | 15 |
| 80,54 | 1.1 | 2.1 | 11 |
| 80,5 | 1.1 | 3.7 | 10 |
| 73,21 | 2.6 | 0.08 | 12 |
| 73,54 | 2.6 | 1.0 | 8 |
| 73,5 | 2.6 | 1.7 | 5.5 |
| 21,54 | 2.7 | 1.0 | 8 |
| 21,5 | 2.7 | 1.6 | 6 |
| 54,5 | 3.0 | 0.35 | 17 |

From P. Amati and M. Meselson, *Genetics* 51:369, 1965.

the map intervals being studied. The resultant data are shown in the third column of Table 17-1. Amati and Meselson also collected data from crosses for which more widely separated $\lambda$ markers were followed, and they plotted all of the data in the graph shown in Figure 17-13. The graph illustrates that even widely separated markers have an *s* value of about 5, an example of low negative interference. In addition, it is clear that when two markers are separated by a distance equivalent to 1 percent recombination or less, **the apparent number of double exchanges is many times higher than expected.**

Two explanations have been offered for localized high negative interference; both may well prove to be applicable to particular situations. The first proposes that physical exchanges occur in **clusters,** the idea being that once chromosomes become "effectively paired" for one break-and-rejoin

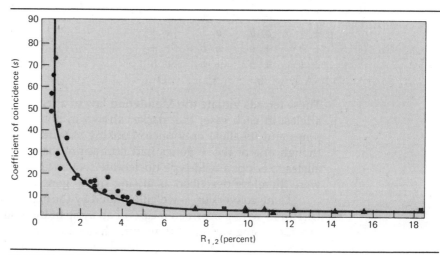

**FIGURE 17-13 High negative interference.** The relationship between $s$ (the coefficient of coincidence) and $R_{1,2}$ (the simultaneous occurrence of exchanges in both the $a$-$b$ and the $b$-$c$ intervals in $a\ b\ c \times\ +\ +\ +$ three-factor crosses). Symbols ■ and ▲ represent data from crosses using widely separated markers; the symbol ● represents data from the closely linked markers used by Amati and Meselson (*Figure 17-12*). (From P. Amati and M. Meselson, *Genetics* **51**:369, 1965.)

event, the attendant enzymes may well proceed to mediate numerous such events in one highly localized region. The alternative explanation holds that high negative interference is simply an exaggerated view of provoked mismatch repair. Supporting this argument are observations by Norkin on the *lac* operon mutations in *E. coli* (*Section 17.10*); he found that those *lac* markers that gave the highest apparent recombination frequencies in two-factor crosses (*Figure 17-11b*) also gave the highest values of high negative interference in three-factor crosses (see Problem 11 at the end of this chapter).

## 17.12 Gene Conversion and Mismatch Repair

**Gene conversion** was first observed in fungi that can be subjected to tetrad analysis (*Sections 14.12 and 14.13*). The phenomenon is best illustrated by recalling first what is expected when the markers $a$ and $b$ are linked and an $a+ \times +b$ cross is performed: $a+$ $a+$ $+b$ $+b$ parental (P) tetrads will be numerous, whereas $++$ $++$ $ab$ $ab$ nonparental ditype (NPD) tetrads will be rare, and $a+$ $++$ $ab$ $+b$ tetratype (T) tetrads will be intermediate in frequency. If the same cross is performed but $a$ and $b$ are very closely linked—if, for example, $a$ and $b$ lie in the same gene—then, in addition to P, T, and very rare NPD tetrads, four new tetrad classes are found:

| $a$ + | $a$ + | $a$ + | $a$ + |
|---|---|---|---|
| **+** + | $a$ **$b$** | $a$ + | $a$ + |
| + $b$ | + $b$ | **$a$** $b$ | + **+** |
| + $b$ | + $b$ | + $b$ | + $b$ |
| A | B | C | D |

**These tetrads violate the Mendelian law of a reciprocal 2 : 2 segregation of alleles:** in each case, the marker shown in boldface is represented three times and its allele only once. Looking at tetrad A, for example, it is as though one of the + genes had nonreciprocally "converted" one of its $a$ alleles to adopt a wild-type nucleotide sequence. These new tetrad classes were therefore described as arising from "gene conversion."

Gene conversion is well illustrated by the experiments of S. Fogel and D. Hurst, who performed a careful analysis of the *histidine-1* (*his1*) gene in yeast, crossing various *his1* auxotrophs against one another. In a cross that can be designated $his^a$ + × + $his^b$, Fogel and Hurst isolated and analyzed 1081 asci in which at least one member of the tetrad was a + + prototroph. They found that 101 of these asci were classic tetratypes, 847 were convertants for the proximal $his^a$ allele (convertant-tetrad class A above), and 133 were convertants for the distal $his^b$ allele (convertant-tetrad class D above). In other words, in crosses involving auxotrophic markers **within** the *his1* gene, prototrophic cells arise by conversion far more often than by "classic" patterns.

**Evidence that Gene Conversion Results from Mismatch Repair**   Figure 17-14 diagrams the origins of a convertant tetrad assuming mismatch repair. The parent homologues are shown synapsed in Figure 17-14a. Figure 17-14b shows an exchange producing two patched duplexes (*recall Figure 17-1*); the outside markers *EF* and *ef* retain their parental configurations, and mismatch regions are present in two of the four chromatids. By drawing the repair tracts as in Figure 17-14c, a convertant tetrad of the Class A type is created wherein the proximal locus exhibits a 3+ : 1$a$ ratio (Figure 17-14d). If different sets of repair tracts are drawn, convertant tetrads of the B, C, and D classes can readily be generated; you should convince yourself of this with diagrams.

A mismatch-repair explanation for conversion, although not yet proven, is supported by several kinds of observations.

First, it is clear from *his1* data generated by Fogel and Hurst that the $his^a$ mutation is subjected to conversion almost seven times as often as $his^b$, an asymmetry that immediately recalls the marker effects described in Section 17.10 for the *lac* mutations. A "mismatch-repair explanation" in this case would state that a heteroduplex containing $his^a$ and its + allele "provokes" repair seven times more frequently than a $his^b$/+ heteroduplex.

Second, conversion is often, but not necessarily, accompanied by the recombination of outside markers. The conversion tetrad diagrammed in Figure 17-14 happens to retain parental *EF* linkages, but similar events can

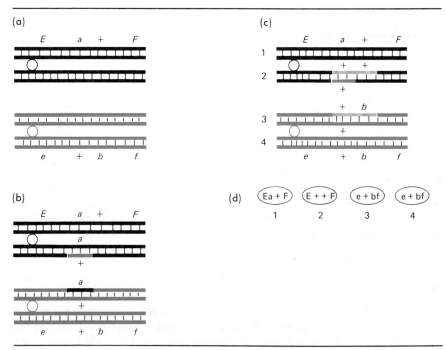

**FIGURE 17-14 Gene conversion by mismatch repair at two sites.** (a) Parent homologues in meiotic prophase. (b) A recombination event occurs between the two inner chromatids, generating a patched duplex with two mismatches between *a* and + sequences. (c) Excision and DNA synthesis (gray color) repair both mismatches. (d) The two meiotic divisions generate a tetrad with a 3:1 conversion for the + allele of *a*.

be diagrammed starting with "spliced" recombinant chromatids of the sort diagrammed in Figure 17-1. In an extensive analysis of 907 conversion events in six genes in three yeast chromosomes, Fogel and colleagues found that 49.1 percent of these events were in fact associated with reciprocal recombination of outside markers, whereas 50.9 percent retained the parental disposition of outside markers. Such a tidy 50:50 proportion of recombinant-to-nonrecombinant outside markers is predicted from our theoretical considerations (*Section 17.3*), although it is not always as closely approximated in actual crosses.

Third, a number of crosses demonstrate that two markers very close together in the same gene will often **co-convert,** meaning that both exhibit a 1:3 (or 3:1) segregation in a tetrad, whereas outside markers are segregating in a 2:2 ratio. Markers judged to be relatively far apart, on the other hand, are usually converted at one or the other site within a single tetrad, "double" events being rare in this case. The modal length of a co-converted segment has been estimated to be about 1000 nucleotide pairs, which is of the same magnitude as the estimated length of a "corepaired" excision

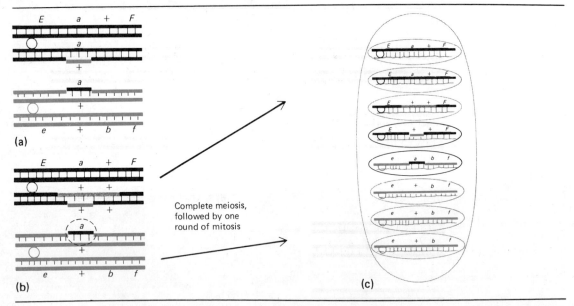

(a)

(b)

Complete meiosis,
followed by one
round of mitosis

(c)

**FIGURE 17-15 Gene conversion by mismatch repair at one site and segregation at another.** (a) Meiotic prophase chromatids with two heteroduplexes, generated as in 17-14b, at the proximal (*a*/+) locus. (b) Repair of the upper mismatch. Lower mismatch (circled) remains unrepaired. (c) Ascus of the eight products resulting from one round of postmeiotic mitosis. Newly replicated DNA daughter strands are drawn as wavy gray lines. The markers at the proximal locus are present in a 5+:3*a* ratio.

tract in the artificially created heteroduplex DNA of phage λ described in Section 17.9.

Fourth, at least two mutations have been shown to affect both repair activities and conversion rates simultaneously. In the smut fungus *Ustilago*, Holliday and colleagues isolated a strain defective in gene conversion and in DNAse I, an enzyme that specifically recognizes and excises heteroduplex DNA *in vitro*. In yeast, W. Boram and H. Roman described a mutation that prevents repair of radiation-induced single-strand chromosome breaks and enhances gene conversion; they proposed that the unrepaired single-strand ends in the mutant are prone to invade homologous chromosomes (Figure 17-2c) and create heteroduplex DNA, which is then subjected to mismatch repair.

Finally, we can return to our original concept of heteroduplex DNA (*Figure 17-7*) and recall that any mismatches *not* corrected by mismatch repair should segregate at the time of DNA replication. The same holds true for meiotic tetrads. Figure 17-15 illustrates an ordered tetrad (*Section 6.8*) in which one mismatch undergoes repair whereas a second escapes repair and is copied during the S phase of the ensuing postmeiotic mitosis.

The resultant **postmeiotic mitotic segregation** is seen to yield a 5:3 ratio of the $+:a$ markers.

In summary, therefore, the predictions of the Holliday and the Fox models for general recombination are supported by numerous observations made with both prokaryotes and eukaryotes. We should conclude, however, by noting that the field of recombination research is by no means in the tidy, self-consistent state that this chapter perhaps implies. Many observations remain unexplained, and the broad outlines of general recombination described here are sure to be filled in with a wealth of experimental detail.

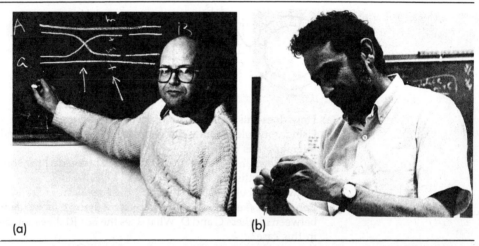

(a)                                          (b)

(a) Robin Holliday (National Institute for Medical Research, London) has formulated much of the theoretical framework for our current understanding of general recombination. (b) Maurice Fox (Massachusetts Institute of Technology) has worked out many molecular features of recombination during bacterial transformation.

# Questions and Problems

1. In the Gurney/Fox experiment outlined in Figure 17-9, the transformed cells in (e) are resistant to either streptomycin or erythromycin but not to both. Explain (*review Section 13.9*).

2. In the Gurney/Fox experiment, hybrid DNA was isolated immediately after transformation had occurred. If the DNA had instead been allowed to undergo a round of replication in light medium and hybrid DNA had then been isolated, how would its transformation properties differ when presented to the doubly sensitive third strain, as in Figure 17-9e. Explain using diagrams.

3. Three mutations of unknown map order lie within the sex-linked *vermilion* eye gene of *Drosophila*. They are designated *a*, *b*, and *c*. Females of the constitution

$$\frac{w(a\ b\ +)m}{+(+\ +\ c)+}$$

where *w* and *m* are outside markers give rise to a few males with wild-type eye color. These are nearly all wild type with respect to the two outside markers. What can you say about the map order of the mutations *a*, *b*, and *c* if you assume gene conversion does not occur in *Drosophila*? How does your answer change if you assume gene conversion can occur?

4. The figure 8 molecule drawn below was generated during recombination of the pMB9 plasmid.

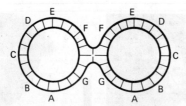

(a) How does this molecule differ from that drawn in Figure 17-4a?
(b) Using two colors, draw the two parental nucleotide strands if no branch migration has occured.
(c) Draw the strands if branch migration has proceeded the equivalent of three marker lengths.
(d) Draw and label the structure that would result from *Eco* RI digestion of (b).
(e) Repeat this exercise, this time drawing a figure 8 molecule with the branch between markers C and D. What does the *Eco* RI digestion product look like in this case?

5. A stock of female *Drosophila* have attached X ($\widehat{XX}$) chromosomes (*Figure 6-20*) and $\widehat{XX}$Y karyotypes. The genes on one X chromosome are *t* $r^a$ + *f*; the allelic genes on its homologue are + + $r^b$ +. All marked loci are distal to the centromere and all mutations are recessive. The $r^a$ and $r^b$ mutations lie within the same gene. The attachment between the homologues prevents the homologues from separating but does not prevent centromere division and chromatid separation.

(a) Diagram a meiosis-without-crossing-over in such females.
(b) What are the genotypes and karyotypes of the surviving progeny if the resultant eggs are test-crossed to males carrying recessive mutations at all four loci (refer to *Figure 6-20* to answer this question).
(c) Diagram two reciprocal exchange events that could produce surviving progeny that are wild type with respect to gene *r* function.
(d) Diagram two conversion events that would have the same result as (c).
(e) Explain why genetic analysis performed with $\widehat{XX}$ chromosomes is called **half-tetrad analysis.**

6. Two pairs of markers are used in two *Drosophila* matings. With the first pair of markers (*a,b*), the interference (observed from the progeny phenotypes) is 0.85. When a second pair of markers (*P,Q*) is used an interference of 0.15 is observed.

(a) Which of these two matings possesses the highest coefficient of coincidence?

(b) Which mating will possess the fewest numbers of double crossovers in the $F_1$?

(c) Is it necessary to have a third marker between $a$ and $b$ and between $P$ and $Q$ in these matings? Explain.

7. The *rosy* (*ry*) locus in chromosome 3 of *D. melanogaster* contains the structural gene for xanthine dehydrogenase (XDH); strains homozygous for *ry* mutations have a *rosy* eye color and are killed by high levels of exogenous purine. When flies of the genotype $\dfrac{kar\ ry^x\ l}{+\ \ \ ry^y\ +}$ are test-crossed with *kar ry l*, where *kar* and *l* represent flanking outside markers, and the progeny are reared in high-purine medium, only $ry^+$ recombinants survive. These are found to fall into two crossover classes with nonparental combinations of outside markers, plus two conversion classes. Write out the genotypes of all four classes and diagram how they arise.

8. In a cross involving $\dfrac{kar\ ry^{41}\ l}{+\ \ \ ry^{502}\ +}$ females, $1.48 \times 10^6$ zygotes were sampled and 40 $ry^+$ progeny were recovered, of which 24 were *kar ry$^+$*, 7 were $ry^{502}$ convertants, and 9 were $ry^{41}$ convertants. What are the relative positions of the two *ry* markers with respect to the flanking markers? Do the convertants yield any information in this regard? Explain.

9. Two additional markers lie between $ry^{502}$ and $ry^{41}$; these are concerned with the electrophoretic mobility of XDH (*Box 3.2*). The combination S1 S2 leads to a slow-migrating XDH; the combination F1 F2 leads to a *fast*-migrating protein. The females in the above cross can be rewritten as

| kar | + | S1 | S2 | $ry^{41}$ | $l$ |
|-----|-----|-----|-----|-----|-----|
| + | $ry^{502}$ | F1 | F2 | + | + |

The electrophoretic mobilities of the XDH produced by the 40 $ry^+$ progeny were examined and scored as S1 S2, F1 F2, S1 F2, or F1 S2, the last two recombinant types being electrophoretically distinct from one another.

(a) Of the 24 *kar ry$^+$* crossover progeny, 23 prove to be S1 F2, whereas one is S1 S2. What does this result indicate about the relative distances between the four sites?

(b) Would you expect S2 and $ry^{41}$ to co-convert? Explain.

(c) Six of the seven $ry^{502}$ convertants were co-convertants for the F1 site. What electrophoretic class of XDH do these six progeny synthesize? What does this result indicate about the relative distances between the four sites?

10. A. Ronen and Y. Salts found in 1971 that recombination frequencies between adjacent nucleotides in the *rII* genes of T4 vary 1000-fold when 12 different sites are compared. How does this observation relate to the Benzer map of these genes (*Figure 12-15*) published ten years earlier?

11. Norkin performed *E. coli* crosses of the following type: *Hfr* 721 $\times$ F$^-$ 624 *lac$_x$*, where 721, 624, and *lac$_x$* represent three mutations in the *lacZ* gene, and scored for *lac$^+$* recombinants. The results for various *lac$_x$* markers were as follows:

| $lac_x$ | $lac^+$ Observed | $lac^+$ Expected |
|---------|------------------|------------------|
| 608     | 0.08             | 0.0012           |
| 1012    | 0.003            | 0.00033          |
| 200     | 0.007            | 0.00030          |
| 90      | 0.02             | 0.00063          |
| 707     | 0.08             | 0.0012           |

(a) Calculate the coefficient of coincidence for each cross.

(b) At which site(s) in the gene would you expect to find the most pronounced marker effects? The least pronounced? Explain.

12. Conversion ratios of 6:2 are sometimes encountered in analyzing fungal asci. Show with diagrams similar to Figure 17-15 how they could arise.

13. The Holliday intermediate is sometimes referred to as a **half-chromatid chiasma**. Explain the origin of this term.

14. In their electron microscope study, Potter and Dressler noted that figure 8 molecules did not form in *recA* mutants of *E. coli*. They also found that their chi-shaped molecules were stable only at 0°C; when warmed to 37°C, they "roll apart" within a matter of minutes into two separate rods. Explain how these two observations support the argument that the molecules shown in Figure 17-3 represent recombination intermediates.

# 18

# Transposition and Mutagenesis by Temperate Viruses and Transposable Elements

A. **Introduction**

B. **Integration and Excision by Temperate Bacteriophages**
- 18.1 Phage λ Integration and Excision
- 18.2 Phage Mu Integration and Excision
- 18.3 Tumor Virus Integration and Excision

C. **Integration and Excision by Transposable Elements**
- 18.4 Insertion Sequences in Prokaryotes

18.5 Transposons in Prokaryotes

18.6 Transposable Elements in Eukaryotes
Selfish DNA

18.7 Controlling Elements in Corn: Dissociator and Dotted

18.8 Hybrid Dysgenesis in *Drosophila*

18.9 Mutagenic Effects of Transposable Elements

D. **Questions and Problems**

## Introduction

A variety of genetic elements are **transposable:** they can integrate into a chromosome or plasmid and, at a later time, either excise themselves again or make a new copy of themselves for a new integration event. In this chapter, we first consider how various temperate viruses execute the transposition process. For a phage such as **λ**, integration and excision ordinarily occur at a precise location in the *Escherichia coli* chromosome, whereas for the phage **Mu** and for certain **animal tumor viruses,** insertion is a far more random process. We then consider **transposable elements,** which, unlike viruses, cannot replicate independent of the host replicon and must therefore always insert if they are to be inherited. These include **insertion sequences** and **transposons** in bacteria and analogous elements in maize and *Drosophila*. In addition to analyzing the genetic properties of these elements and their transposition mechanisms, we also describe how their insertion and/or excision can often generate gene mutations.

# Integration and Excision by Temperate Bacteriophages

## 18.1 Phage λ Integration and Excision

F. Jacob and E. Wollman coined the term **episome** to describe a segment of DNA that can exist either as an independent replicon in a bacterial cell or as a passively replicated sector of a bacterial chromosome. We have already considered the episome properties of the *E. coli* F plasmid in Chapter 13 and noted that F can insert into the *E. coli* genome at several positions (*Figure 13-5* and *Section 13.18*).

Phage λ, although also an episome, differs from F in that it ordinarily integrates into the *E. coli* chromosome at only one position, which, as noted in Section 12.14, is flanked on one side by *gal* genes and on the other by *bio* genes. This site is known as *attB* (for bacterial attachment site), and recognition is mediated by *attP* (phage attachment site), which is flanked by the *int* and *J* genes on the λ chromosome (*Figure 12-8a*). Recognition is followed by a **site-specific recombination** event that results in prophage integration and lysogeny (*Figure 12-4*). A second site-specific recombination event, again involving the *att* sites, is responsible for prophage excision at the time of induction (*Figure 12-16*).

Integration and excision require proteins encoded by both the phage and the bacterium. Phage proteins include the products of the *int* and *xis* loci: *int⁻* phages are unable to integrate, whereas neither *int⁻* nor *xis⁻* prophages can excise. Bacterial proteins include the *himA* (*host integration mediator*) gene product, which functions in some unknown way during the exchange reaction. Significantly, site-specific integration and excision require neither the RecA nor the RecBC proteins that are essential for general recombination (*Section 17.6*); they are therefore said to be mediated by a **RecA-independent pathway.**

Mutations in either *int* or *xis* can be complemented by coinfecting wild-type λ phages (also known as **helpers**). A second class of phage mutations cannot be complemented by helpers, and these map to the *attP* region. Some of these are deletions of *attP*, whereas others are point mutations. Deletion mapping and recombination analysis of these mutations have led to the subdivision of *attP* into three segments, called *P, O,* and *P'*, as diagrammed in Figure 18-1a. The exchange event itself occurs within the short internal O segment, whereas the flanking P and P' sequences facilitate the reaction. Similarly, *attB* can be subdivided into *B, O,* and *B'* (*Figure 18-1b*), and it is proposed that the two O segments interact during the integration reaction to form the prophage diagrammed in Figure 18-1c. An analogous diagram of the excision process is given in Figure 12-16.

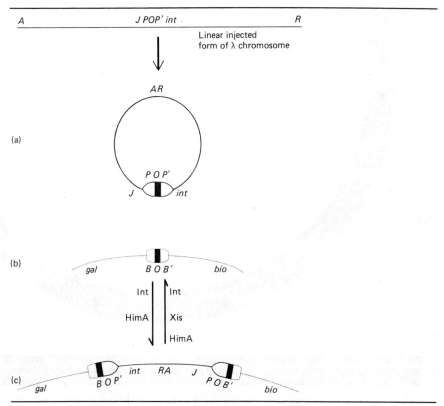

**FIGURE 18-1 Integration of the circular phage λ chromosome** (a) into the *E. coli* chromosome (b) to form a prophage (c). The reciprocal excision reaction is diagrammed in Figure 12-16. The phage and bacterial *att* sites are given distinct shapes to clarify their orientation in the prophage. Bacterial DNA is gray, and phage DNA is black. The essential features of this model were first proposed by A. Campbell (*Adv. Genet.* **11**:101–145, 1962).

**Molecular Features of Site-Specific Recombination**    The nucleotide sequences of *attP* and *attB* have been determined. The DNA sequences corresponding to the *P*, *P′*, *B*, and *B′* segments all prove to be different from one another. By contrast, both *attP* and *attB* carry an identical internal sequence, 15 base pairs long, termed the **common core** (Figure 18-2a and b). Analysis of *in vitro* integration reactions has demonstrated that integration involves exchanges within these common-core sequences. Moreover, as predicted by the Campbell model (*Figure 18-1c*), the core sequences are found at either end of the prophage at the conclusion of integration (Figure 18-2c). Therefore, the common cores can be considered synonymous with the O segments of the *att* regions (*Figure 18-1*).

To determine which of these segments is recognized by the *int* gene product, *att* sequences were allowed to interact with purified Int protein

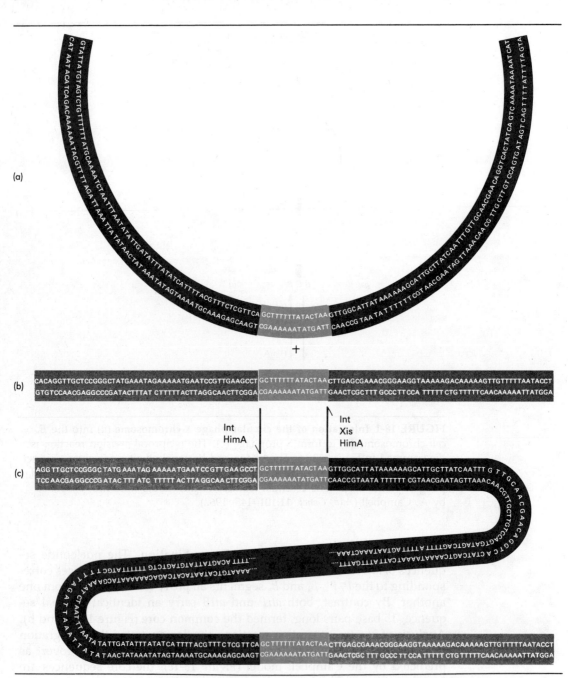

**FIGURE 18-2 Site-specific recombination** between the circular λ chromosome (a) and the *E. coli* chromosome (b) to yield an integrated λ prophage (c). The λ chromosome is shown in black, and the *E. coli* chromosome in gray except for the 15-nucleotide common-core sequences, which are shown in color. (From A. Landy and W. Ross, *Science* **197**:1147, September 16, 1977. Copyright 1977 by the American Association for the Advancement of Science.)

and then digested with nucleases to destroy all but the protected regions. Such "footprinting" experiments (*Section 9.7*) indicate that although Int protein interacts strongly with the common cores, it also reacts strongly with two sites to the left of the core and one site to the right. It appears, therefore, that Int protein mediates critical interactions at several positions within the two *att* regions.

If wild-type λ phages infect mutant *E. coli* cells in which the *gal-bio* region (and therefore *attB*) has been deleted, λ insertion occurs, with 200-fold lower frequency, into a hierarchy of secondary sites located elsewhere in the *E. coli* genome. Since secondary insertion requires Int and excision requires both Int and Xis proteins, it is believed that the secondary sites possess sufficiently similar sequences to *attB* that they can serve as substrates for the Int-mediated reaction.

**Int protein is also a topoisomerase.** We encountered an example of a topoisomerase when considering DNA replication in Section 8.2: we noted that the enzyme DNA gyrase acts to break and reseal the phosphodiester backbone of superhelical DNA (*Section 3.7*) to "relieve" the twisting. Int protein has the same effect on the phage λ chromosome, which assumes a supercoiled configuration after it enters the host cell, and it has been suggested that once Int protein mediates the alignment of common-core sequences, it proceeds to nick the juxtaposed helices, creating broken DNA strands that can go on to form Holliday intermediates (*Figure 17-2*).

## 18.2 Phage Mu Integration and Excision

Phage Mu resembles phage λ (and other temperate phages of *E. coli*) in that it can either infect and lyse the host or lysogenize by inserting its chromosome into the *E. coli* genome by a RecA-independent pathway. The integrated state is stabilized by a Mu-specified repressor, which blocks expression of Mu lytic genes, and the prophage can be later induced to excise. Two phage genes, denoted *A* and *B*, are required for Mu integration, and the protein products of these genes appear to mediate the association of the Mu chromosome with host DNA, reminiscent of the λ Int protein. Mutants in either *A* or *B* can be complemented by helper phages (*Section 18.1*). Helper phages are of no assistance, however, if Mu carries mutations in either of its chromosomal ends, indicating that the ends of the Mu chromosome play a structural role in the integration process analogous to the *attP* sequences of λ.

Four major differences exist between Mu and λ. The first is that the Mu chromosome does not appear to circularize after it infects *E. coli*. Thus circular molecules consisting entirely of Mu DNA are never found inside infected cells—in contrast to the ease with which λ DNA circles are found inside λ-infected cells (*detailed in Problem 12-24*). Moreover, the prophage genetic map of Mu is colinear with the vegetative map, whereas for λ the two are permutations (*Section 12.10*) of each other (*Figure 18-1*).

A second difference is that, unlike λ, **Mu integrates efficiently at**

**multiple sites,** if not at random, in the *E. coli* chromosome. Genetic analysis of 75 independent Mu insertions into the *lacZ* gene of *E. coli*, for example, showed that at least 50 occurred at different positions from one another. Thus Mu integration is clearly independent of any extensive sequence homology, in contrast to the *attP/attB* interaction.

The third difference is that **Mu generates a 5 bp repeat of host DNA when it integrates.** This is best visualized by example. Figure 18-3a shows a portion of the *lacZ* gene prior to Mu insertion, and Figure 18-3b shows the same sector after integration. The sequence CTGGA, which flanks each end of the inserted phage, was present only once in the original gene. When additional Mu lysogens are subjected to this sort of analysis, different 5 bp duplications are found to have been generated in each case. Therefore, although λ and Mu prophages are superficially similar in that they carry short repeated sequences at either end, the common-core sequences of λ are always the same and are contributed equally by phage and host, whereas **the flanking Mu sequences usually differ from one insertion event to the next and are contributed only by the host.**

The fourth difference is that **both Mu integration and Mu transposition require DNA synthesis,** whereas λ integration and transposition do not. Specifically, when Mu transposes from one chromosome site to another, the original prophage remains *in situ*, where it is copied by localized semiconservative DNA replication, and the daughter copy inserts elsewhere. Figure 18-4 illustrates a current model of how Mu might integrate by what is termed "roll-in replication"; details are given in the figure legend. A reversal of these steps could generate an independent copy of Mu while the original remains in position.

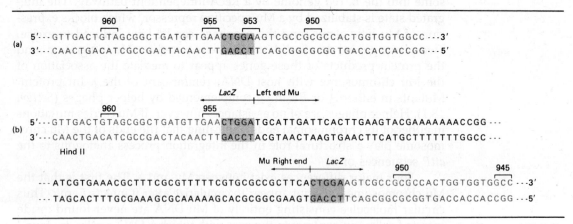

FIGURE 18-3 **Mu integration into lacZ.** (a) Sequence of the *lacZ* gene of *E. coli* where triplet codons are marked off and numbered (960, 955, and so on). (b) Sequence of the *lacZ* gene after Mu insertion. The five bases in the box in (a) have been duplicated and flank each end of the Mu prophage. (From B. Allet, *Cell* **16**:123–129, 1979.)

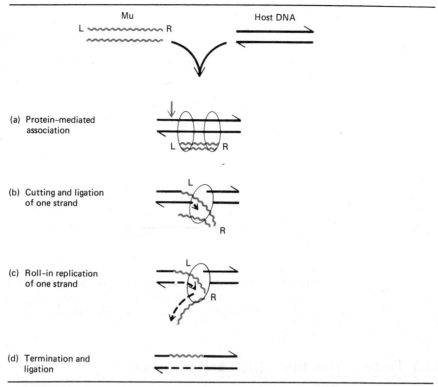

**FIGURE 18-4 Model of Mu integration.** (a) Mu and host DNA are brought together by proteins (ellipses) that recognize sequences at both ends (L and R) of Mu and sequences in the host DNA. For clarity, these proteins are omitted in subsequent panels. (b) The host DNA undergoes a slightly staggered, double-stranded cut at the arrow in (a), and one strand of the L end of Mu is ligated to the exposed 5'-P of the host strand. The exposed 3'-OH of the complementary strand is used as a primer for DNA replication (dashed lines), where the oval represents the replication complex. (c) DNA is replicated at this fixed complex as it reels through. (d) Replication terminates when the R end passes through the replication complex, and both the newly synthesized strand and the R end are ligated to host DNA. (From R. M. Hershey, and A. I. Bukhari, *Proc. Natl. Acad. Sci. U.S.* **78**:1090–1094, 1981.)

To explain the generation of the 5 bp repeat of host DNA during Mu insertion, it is proposed that the double-stranded endonucleolytic cut diagrammed in Figure 18-4a is in fact a "staggered cut," as amplified in Figure 18-5a. The 5 bp repeat of host DNA that resides within the two cleavage sites will then be copied at either end of the Mu prophage, as drawn in Figure 18-5b and c. Thus the model embodies the key features of Mu integration: (1) the mediation of Mu-encoded proteins; (2) the importance of the Mu chromosome ends in the insertion process; (3) the need for DNA replication; and (4) the generation of a 5 bp repeat of host DNA.

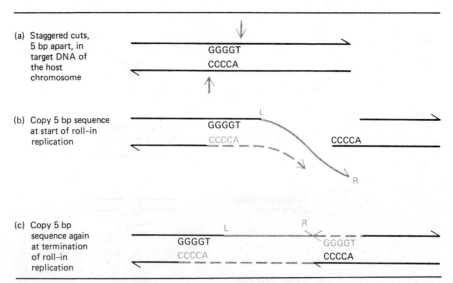

(a) Staggered cuts, 5 bp apart, in target DNA of the host chromosome

(b) Copy 5 bp sequence at start of roll-in replication

(c) Copy 5 bp sequence again at termination of roll-in replication

FIGURE 18-5 **Generation of 5 bp repeats during Mu integration.** Symbols and abbreviations are as in Figure 18-4. Final structure (c) is comparable to Figure 18-3b except that the duplicated bases are different.

## 18.3 Tumor Virus Integration and Excision

The life cycle of SV40, a eukaryotic DNA tumor virus, is described in Sections 12.3 and 12.16 and should be reviewed at this time. The **retroviruses** constitute a second class of tumor virus. They are similar to SV40 in that they also integrate into the genome of their host and can cause malignant transformation. A major difference, however, is that in the virion, the retrovirus genome is in the form of single-stranded RNA. The virion also contains a virus-coded **RNA-dependent DNA polymerase** (also called **reverse transcriptase**), which, upon infection, copies the single-stranded RNA into a duplex DNA chromosome known as a **provirus.** The provirus then proceeds to integrate into the host genome during both productive and nonproductive infections, and viral genes are transcribed. The integrated provirus is also replicated along with the host chromosome and is transmitted from parent to offspring of the host in a Mendelian fashion.

Both SV40 and the retroviruses, therefore, eventually insert double-stranded DNA versions of their genomes into host chromosomes. To what extent do these insertions resemble those of the temperate bacteriophages? Present evidence indicates that a number of familiar strategies are employed.

SV40 resembles Mu phage in that integration appears to be a random or quasirandom event, occurring at no fixed site in the host genome. Therefore, different lines of transformed cells contain different quantities of viral sequences integrated at different chromosomal locations. Unlike both λ and Mu, however, there is **no fixed site for integration in the SV40 genome.** Thus when integrated SV40 chromosomes are cloned and the viral-host DNA junctions are sequenced, the junction sometimes involves one region of the viral genome and sometimes another region. Despite this apparent haphazardness, the integrated provirus is very stable and cannot be induced to excise unless the transformed cell is fused with a permissive cell (*Section 12.3*).

The retroviruses resemble both SV40 and Mu in the sense that they are promiscuous with regard to the cellular DNA sequences into which they insert. They differ from SV40 but resemble Mu in two additional respects. First, the integrated retrovirus is always colinear with the nonintegrated viral chromosome, rather than some circular permutation, as with SV40. This suggests that **insertion is somehow mediated by the two ends of the retrovirus,** as is believed to be the case with Mu. Second, **retrovirus integration generates a flanking repeat of host DNA sequences.** Thus the murine sarcoma provirus is flanked by a 4 bp repeat of cellular DNA, the spleen necrosis virus by a 5 bp repeat, and the mouse mammary tumor virus by a 6 bp repeat. Therefore, the "staggered cut" model for Mu integration (*Figure 18-5*) may prove to be relevant for retroviruses as well.

# Integration and Excision by Transposable Elements

## 18.4 Insertion Sequences in Prokaryotes

An **insertion sequence (IS)** is a short length of DNA duplex, ranging from 768 to 5700 base pairs, that has no origin of replication and must therefore be inherited in an integrated form. Six distinct insertion sequences have been identified to date (Table 18-1), and most are components of the *E. coli* genome: thus, *E. coli* normally carries five to eight copies of IS1, five copies of IS2, and so on (Table 18-1). As we learned in Section 13.18, the F element normally carries one copy of IS2, one copy of γδ, and two copies of IS3, and

**TABLE 18-1    Representative Insertion Sequences**

|  | Normal Occurrence in *E. coli* | Length (bp) | Inverted Repeats (bp) | Repeat Generated upon Insertion |
|---|---|---|---|---|
| IS1 | 5–8 copies on chromosome | 768 | 18/23* | 9 bp |
| IS2 | 5 on chromosome, 1 on F | 1327 | 32/41 | 5 bp |
| IS3 | 5 on chromosome, 2 on F | 1400 | 32/38 | 3 or 4 bp |
| IS4 | 1–2 on chromosome | 1400 | 16/18 | 11 bp |
| IS5 | ~10 on chromosome | 1195 | 15/16 | 4 bp |
| γδ | 1 on F, none on chromosome of most *E. coli* strains | 5700 | 35 | 5 bp |

*The inverted repeats are often imperfect. This notation indicates that 18 bp on the left end of the IS can be found, in inverted order, within 23 bp on the right end. (After M. P. Calos and J. H. Miller, *Cell* **20**:579–595, 1980.)

a RecA-mediated homologous recombination between plasmid and chromosomal IS elements is thought to account for the "hot spots" of F plasmid insertion.

**IS elements can also insert into chromosomes by a RecA-independent pathway.** This insertion reaction is catalyzed by **transposase** proteins, analogous to the A and B proteins of Mu, that are specified by the IS element itself, although such proteins have not as yet been well characterized. It is also likely that insertion is mediated by the IS termini: as summarized in Table 18-1, each terminus carries an **inverted repeat (IR)** (*Section 4.9*), and the integrated sequence always carries the repeat at each IS/chromosome junction.

IS integration is identical to Mu and retrovirus integration in that **a short repeat of cellular DNA is generated** on each side of the integrated element. The length of this repeat is specific for each IS, as summarized in Table 18-1, and once again, the staggered-cut hypothesis (Figure 18-5) is the favored explanation for this phenomenon. Available sequencing data indicates that the IS elements may be somewhat selective about the sequences they seek in target DNA. Thus in one study of IS1 insertion into the *E. coli gal* operon, six independent insertion events were observed, but they involved only three sites. Thus IS elements appear to be "intermediate" in their specificity, being less conservative than λ but more conservative than Mu or the retroviruses.

**The transposition of an IS sequence to a new chromosomal position does not necessitate the loss of a copy from the old chromosomal position.** As with phage Mu, therefore, many IS transposition events appear to be mediated by DNA copies of the original sequence. IS elements can, however, also be spontaneously lost from a chromosomal site, suggesting that they may either "loop out" in the fashion of phage λ (*Figure 12-16*) or be removed from the chromosome by some deletion mechanism.

## 18.5 Transposons in Prokaryotes

Transposons were also introduced in Chapter 13 (*Section 13.19*), where we described their ability to encode, and to transpose, genes for antibiotic resistance within R plasmids. Essential features of transposon structure are diagrammed in Figure 13-13: each contains one or more resistance genes, and each gene is flanked by an inverted repeat of DNA. Table 18-2 summarizes the properties of several well-studied transposons (a total of 35 have been characterized to date), and Figure 18-6 diagrams the structure of two representative examples, Tn9 and Tn3.

As drawn in Figure 18-6a, **Tn9 carries a copy of IS1** that is repeated directly at each terminus. Since IS1 carries a short (~20 bp) inverted repeat at either end (Table 18-1), this means that the transposon itself is flanked by short inverted repeats. The favored explanation for the origin of Tn9, therefore, is to suppose that a chloramphenicol-resistance gene came to lie between two IS1 elements, and the resultant unit thereby acquired the ability

TABLE 18-2   Representative Transposons

| Marker | Transposon | Length | Inverted Repeats (bp) | Repeat Generated upon Insertion |
|---|---|---|---|---|
| Ampicillin | Tn3 | 4,957 | 38 | 5 bp |
| Ampicillin, streptomycin, sulfanilamide | Tn4 | 20,500 | short | — |
| Chloramphenicol | Tn9 | 2,638 | 18/23* | 9 bp |
| Kanamycin | Tn5 | 5,700 | 1,534 | 9 bp |
|  | Tn903 | 3,100 | 1,050 | 9 bp |
| Tetracycline | Tn10 | 9,300 | 1,400 | 9 bp |

*The inverted repeats are often imperfect. This notation indicates that 18 bp on the left end of the IS can be found, in inverted order, within 23 bp on the right end. (After J. P. Calos, and J. H. Miller, *Cell* **20**:579–595, 1980, and C. M. Berg, and D. E. Berg, *Microbiology*, D. Schlessinger, Ed. American Society of Microbiology, 1981, pp. 107–116).

to transpose not only itself but also selectively advantageous drug-resistance genes. It has been shown that transposase functions are encoded within the IS elements themselves and that the terminal inverted repeats function during integration.

Tn3 is a larger transposon than Tn9 (*Figure 18-6b*), but it is speculated to have originated in the same fashion as Tn3, namely, as an ampicillin-

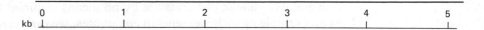

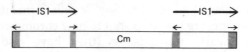

(a) Tn9

(b) Tn3

**FIGURE 18-6   Maps of transposons Tn9 and Tn3.** Cm encodes the enzyme chloramphenicol acetyl transferase; Ap encodes β-lactamase, which modifies ampicillin. The transposase gene is located on the opposite strand from the repressor and Ap genes.

resistance (Ap) gene that came to be flanked by two copies of a large insertion element, probably related to γδ. Two genes have been identified in the rearranged DNA, one encoding a transposase and the other a repressor that is thought to regulate the levels of transposase and hence the mobility of the unit.

Consistent with the theory that transposons were once flanked by *bona fide* insertion sequences is the fact that **all transposons insert in the same fashion as IS elements:** each integrates as a discrete, nonpermuted linear segment of DNA; each generates either a 9 bp or a 5 bp repeat of target DNA (Table 18-2); and each can transpose either via excision or via a new DNA copy. Some transposons are like IS elements, moreover, in that they are selective about their choice of insertion sites. N. Kleckner and associates, for example, analyzed the insertion of the large Tn10 transposon into the *hisG* gene of *Salmonella* and found that insertion occurred 53 times at one "hot spot," 4 times at a second "warm spot," and not at all anywhere else in the 900 bp gene. Transposons vary in this regard, however, since another large transposon, Tn5, shows little preference in its choice of target DNA. Thus it is safe to speculate that modern transposons, like modern insertion sequences, will prove to have evolved unique strategies for carrying out what are fundamentally similar insertion reactions.

## 18.6 Transposable Elements in Eukaryotes

Transposable genetic elements are by no means confined to prokaryotes. As an example of such elements in eukaryotes, we can consider sequences found in *Drosophila* that are known as **copia.** A typical *Drosophila* genome carries about 50 copies of *copia* in widely scattered positions. When two noninterbreeding strains of *Drosophila* are compared, however, they usually differ considerably in the chromosomal locations of their *copia* sequences. Moreover, if a line of *Drosophila* cells is maintained in tissue culture, *copia* elements are found to shift to new chromosomal loci. Thus *copia* elements are clearly transposable (they have also been designated **nomadic**).

Hybridization studies indicate that the various copies of *copia* present in a genome are closely homologous but not necessarily identical, so that one commonly speaks of the copia family of transposable sequences. Each integrated *copia* element, however, is a nonpermuted version of its relatives; therefore, *copia* insertion is not a random event in the fashion of SV40 but rather involves terminal sequences in the fashion of retroviruses and prokaryotic elements. The termini of *copia* are indeed specialized, but in a unique way: each is a 276 bp **direct repeat** of the other, rather than an inverted repeat. Therefore, if insertion of IS elements and transposons proves to depend on the presence of inverted repeats as recognition sites for transposase enzymes, as many believe, then insertion of *copia* sequences must entail a different mechanism of transposase recognition. The consequences of insertion, however, are strikingly parallel: a 5 bp repeat of cellular DNA is created on either side of the inserted *copia* element, with the sequence of cellular DNA chosen for insertion differing from one insertion event to the next.

Each *copia* element is about 5 kilobases long. In addition to its direct terminal repeats, the element carries coding sequences that are transcribed into poly-A-tailed mRNAs and translated into several proteins. Whether these proteins function as transposases or have other roles in the cell is not yet known.

Additional *copia*-like families in *Drosophila* include those designated 412 and 297, and it appears that most of the middle-repetitive DNA of *Drosophila* (*Section 4.8*) will prove to constitute nomadic families similar to *copia*. An analogous element called Ty1 has been found in yeast, and it is generally believed that **transposable genetic elements are ubiquitous in eukaryotes** and simply await discovery in other species.

**Selfish DNA**   In Section 4.1, we noted that certain eukaryotes, such as salamanders and lilies, were found to have as much as 20 times the amount of DNA (the C value) as is found in the human genome. Because it seemed unlikely that these species have 20 times as many genes as humans, the question of the origin and "purpose" of this DNA came to be called the **C-value paradox.** With continued research it became clear that in all higher eukaryotes, much DNA is "noncoding"; this includes introns, the long tracts of DNA that lie between genes, and the satellite-heterochromatic sequences. It became logical to propose, therefore, that more "extra" DNA was allowed to accumulate in the salamander and lily lineages than in the human lineage and to state **the real "paradox" as the existence of the extra DNA in the first place.**

One solution to the paradox is to propose that the extra DNA in fact does serve functional roles. We have offered W. Gilbert's hypothesis that introns play an evolutionary role in gene construction (*Section 9.11*), and we have noted that middle-repetitive DNA has been postulated to serve regulatory roles (*Section 4.8*).

The other extreme is to propose that the DNA is present only because it is able to be duplicated. The term **selfish DNA** has been coined by W. F. Doolittle and C. Sapienza and by L. E. Orgel and F. H. C. Crick to describe this species of DNA. They suggest that any transposable element, or any element that has lost its ability to transpose but retained its ability to be replicated, would simply parasitize the genome. It would not need to confer any phenotypic property, nor would the cell have any easy mechanism to get rid of it. It would, of course, create an energetic "burden" in that it would need to be replicated at every mitotic cycle, but higher eukaryotes may have minimized this problem by replicating their DNA less often than, say, prokaryotes and lower eukaryotes.

Clearly, continued research into the properties of nongenic eukaryotic DNA will help decide between these theories, and each may well prove to be applicable. Indeed, a tract of DNA with purely "selfish" properties in a twentieth century genome may well have been part of an evolutionarily important sequence during eukaryotic evolution and/or may survive to play a crucial evolutionary role in future genomes.

## 18.7 Controlling Elements in Corn: Dissociator and Dotted

Transposable elements that can cause mutations during transposition are said to have **mutator activity.** The examples we will consider from maize involve three interacting genetic units: (1) the transposable element itself; (2) a regulatory gene that determines or facilitates the transposition activities of the element; and (3) a target gene or genes that experiences insertion or excision of the element.

**Dissociator-Activator**  In our first example from maize, the transposable element is called *Dissociator (Ds)* and the regulatory gene is called *Activator (Ac).* *Ds* will enter or leave a target gene only in the presence of *Ac.* It has been shown to insert into many chromosomal positions, but since a rich array of visible markers exists on chromosome 9 of maize, these loci prove to be particularly favorable for study.

Transposition of *Ds* often causes a **chromosomal break** at the site of excision, producing a deletion of all genes distal to the break. Such defi-

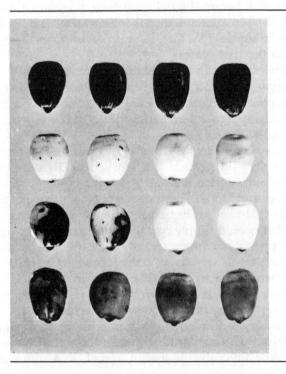

**FIGURE 18-7 Controlling element action in corn.** Top row shows fully pigmented kernels; lower rows show varying degrees of variegation. (Courtesy of Dr. Jerry L. Kermicle)

ciency-carrying chromosomes can frequently be detected cytologically (*Section 14.7*); alternatively, they can be detected because they allow expression of recessive genes in heterozygous organisms. Thus if a corn plant heterozygous for *Wx* (*waxy* kernels) and *C*(*colored* kernel coat) on chromosome 9 experiences a *Ds*-induced deletion in the chromosome carrying the dominant genes, it will display *wx c* (nonwaxy, colorless) kernels.

*Ds* insertion may instead cause an **unstable suppression of gene expression.** Thus if *Ds* is introduced into a heterozygous *C/c* plant, the resultant kernels are often patchy or **variegated,** some areas being colorless and others pigmented (Figure 18-7). As diagrammed in Figure 18-8, these effects are explained by postulating that a *C* target gene has been entered by a *Ds* element, creating a mutant gene *C\**, which cannot dictate the synthesis of normal pigment, and that this integrated *Ds* occasionally excises itself from *C\** to restore a normal *C* gene, which proceeds to dictate the synthesis of normal pigment in a patch of kernel tissues.

Both the insertion and the subsequent excision of *Ds* are dependent upon *Ac*. Specifically, if a *C\** gene is created in an *Ac*-containing organism and the gene is then crossed into an *Ac*-free background, *C\** loses its ability to "revert" to a normal gene and instead behaves as a mutant *c* allele.

**Dotted**   A second example from maize is analogous to the first, but with one exception: **the putative transposable element has a specific target gene** rather than a variety of targets. Thus the element, which we can call *T*, preferentially inserts into the locus $A_1$ responsible for anthocyanin (pigment) production in the presence but not the absence of a distinct regula-

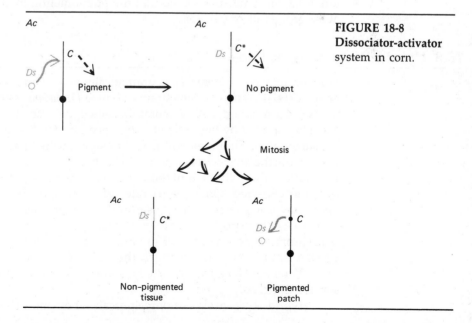

**FIGURE 18-8**
**Dissociator-activator** system in corn.

tory gene called *Dotted* (*Dt*). Insertion of *T* into the $A_1$ gene creates an $A_1$\* gene with properties analogous to the *C*\* gene in the previous example. An $A_1$\*/$a_1$ heterozygote, therefore, produces *dotted* kernels, the purple dots produced by each clone of cells that undergoes an $A_1$\* $\longrightarrow$ $A_1$ excision event. Similarly, the occurrence of $A_1$\* $\longrightarrow$ $A_1$ events during the formation of vegetative tissues causes purple streaks to appear on the stems and leaves.

The elements *Ds* and *T* have not yet been identified in maize, nor are the regulatory effects of *Ac* and *Dt* yet understood. *Ac* and *Dt* are of additional interest, however, in that they are themselves transposable, being able to move from place to place in the maize genome. They also both have cumulative effects on the behavior of the elements that they regulate. Thus as maize strains are bred to carry more and more copies of *Ac*, *Ds*-induced variegation is initiated at progressively later times in kernel development, whereas increased copy numbers of *Dt* increase the transposition frequency of *T*. Specifically, the kernel endosperm (*Figure 5-7*), which is triploid, can have the composition *Dt dt dt*, in which case an average of 7.2 dots appear on each kernel; it can be *Dt Dt dt*, in which case 22.2 dots are found; or it can be *Dt Dt Dt*, in which case 121.9 dots appear on an average kernel. All of these phenomena are certain to be the subject of intensive research in the coming years.

It is pertinent to note here that much of the variegation seen in so-called "Indian corn" is due to transposable elements, and Native Americans were probably the first collectors of transposable-element strains—both for religious and aesthetic reasons. McClintock's genetic studies of these strains, moreover, were carried out in the 1940s, some 20 years before it began to be appreciated that the phenomenon might be relevant to other groups of organisms.

## 18.8 Hybrid Dysgenesis in Drosophila

Mutator activity in *Drosophila* is dramatically displayed during **hybrid dysgenesis (HD)**. When certain strains of male *D. melanogaster*, isolated from the wild, are mated with "laboratory wild-type" females, their offspring are characterized by high sterility, enhanced mutability, and chromosome breakage. M. J. Simmons and J. K. Lin studied the effects of hybrid dysgenesis on the *zeste-white* region of the *Drosophila* X chromosome (*Section 14.8*) and found that the mutation rate in the region was enhanced tenfold over the expected spontaneous rate. Moreover, many of the mutations involved chromosome breaks, insertions, or inversions. Finally, there were mutational "hot spots" in the region, notably *zw1*, and analysis of the polytene chromosomes in such *zw1* mutants revealed that extra DNA was often present in the 3A4 band to which the *zw1* gene has been mapped (*Figure 14-8*). Thus the HD effects clearly have the properties expected of chromosomal insertions.

Hybrid dysgenesis is nonreciprocal: when the male parent derives from a laboratory stock and the female from the wild, mutator activity is usually nil. These observations have led to the proposal diagrammed in Figure 18-9. It is suggested that certain families of transposable elements are integrated into the genomes of wild populations but are absent from laboratory strains. The integrated wild elements are thought to specify repressor molecules that inhibit transposability and, hence, inhibit the potential to cause deleterious mutations (recall the transposase repressors encoded by Tn3 [*Figure 18-6*], which are believed to have an analogous function in bacterial cells). Therefore, as diagrammed in Figure 18-9a, repressor levels would be high in wild female eggs, and transposable elements introduced by the sperm of wild males would remain in their integrated positions and cause no damage. Laboratory females, on the other hand, would lack both the elements and their repressor products; there-

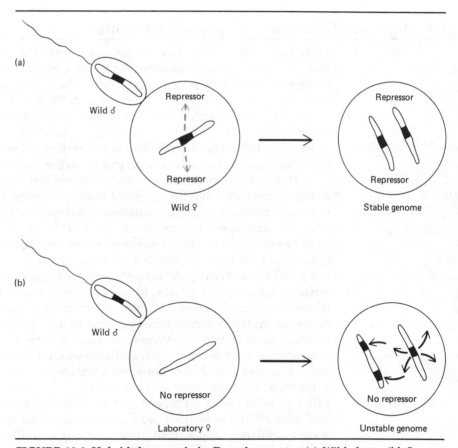

**FIGURE 18-9 Hybrid dysgenesis in *D. melanogaster*.** (a) Wild ♂ × wild ♀. (b) Wild ♂ × laboratory ♀. Dark sectors in chromosomes represent transposable elements.

fore, incoming wild chromosomes from the sperm would encounter a repressor-free environment and would proceed with an orgy of mutagenic transposition (*Figure 18-9b*) before eventually synthesizing enough repressor to settle down into new integrated positions.

We should note that hybrid dysgenesis has exciting potential for the genetic analysis of *Drosophila*. It is now possible to identify, and clone, many of the sequences responsible for HD insertions. With these clones, it should be possible to isolate any gene in *Drosophila* for which there exists a detectable mutant phenotype. If, for example, one is interested in learning which piece of genomic DNA specifies *zw1*, one can obtain DNA from dysgenic flies carrying a mutagenic insertion in the *zw1* gene; one can treat this DNA with restriction enzymes and identify the fragments that hybridize with the cloned HD element; these fragments can then be cloned and used as probes to "fish out" their nonmutated *zw1*$^+$ counterparts from a library of normal *Drosophila* genomic DNA (*Section 4.16*).

## 18.9 Mutagenic Effects of Transposable Elements

**Polar effects of IS insertion** IS elements were first recognized by their ability to induce polar mutations. As is often the case with biological discoveries, these effects were noticed during the course of experiments that had a very different goal. Thus in the mid-1960s, J. Shapiro, S. Adhya, H. Saedler, and P. Starlinger set out to map the genes involved with galactose utilization in *E. coli*. They therefore isolated *gal*$^-$ mutants and subjected them to genetic analysis. One set of mutants allowed them to accomplish their goal; a second set of mutants led to the discovery of IS elements.

The first group of mutants was obtained following treatment with chemical mutagens, and these proved to carry "conventional" point mutations and deletions. Complementation and recombination analysis generated the map illustrated in Figure 18-10. The mutations defined three linked genes, one encoding a galactose *kinase* (*galK*), one a galactose *transferase* (*galT*), and one a galactose *epimerase* (*galE*), which mapped in the order *galKTE*. Moreover, the properties of supressible *amber* and *ochre* nonsense mutations (*Section 11.14*) in these genes indicated that all three constituted an operon (*Section 9.2*). You will recall from Section 11.14 that if a nonsense mutation occurs within a gene in a bacterial operon, then the resultant mutant fails to express not only that gene but also all genes downstream from the mutation, a phenomenon termed **polarity**. Shapiro and colleagues found that nonsense mutations in *galE* also blocked expression of the transferase and kinase genes, that nonsense mutations in *galT* had polar effects on kinase activity, and that nonsense mutations in *galK* affected kinase alone. They therefore concluded that the three genes were ordinarily transcribed into a single mRNA having order *galE* $\longrightarrow$ T $\longrightarrow$ K (*Figure 18-10*).

The second group of mutants was obtained by selecting for spontaneous mutations in the region. This time, Shapiro and colleagues used a

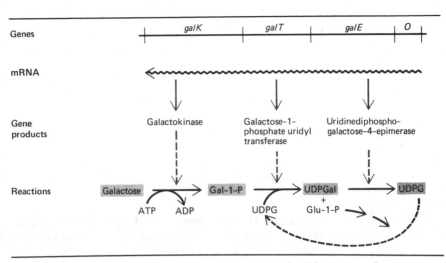

**FIGURE 18-10 The galactose (*gal*) operon of *E. coli*.** *Abbreviations:* O, operator; Gal-1-P, galactose-1-phosphate; UDPG, uridine diphosphoglucose; UDPGal, uridine diphosphogalactose; Glu-1-P, glucose-1-phosphate. (After S. Adhya and J. Shapiro, *Genetics* **62**:231, 1969.)

screening procedure that kills all cells that have normal galactokinase activity but allows kinase-defective ($k^-$) cells to grow. Most of the mutant strains proved to lack not only kinase but also transferase or both transferase and epimerase; in other words, most were either $tk^-$ or $etk^-$, suggesting that each carried a polar mutation. Mapping studies supported this interpretation: the $tk^-$ strains carried mutations in *galT* and the $etk^-$ strains carried mutations in *galE*, just as one would expect for *amber* or *ochre* mutations. However, **none of these spontaneous polar mutations was suppressible by amber or ochre suppressors alone** (*Section 11.17*). Moreover, none could be suppressed by frameshift suppressors (*Section 11.7*), none were revertible by mutagens that cause transitions or transversions (*Sections 8.3 to 8.5*), and none carried deletions, since they could undergo spontaneous reversion. The notion therefore developed that an extra piece of DNA might have inserted into either the *galT* or the *galE* genes.

Subsequent heteroduplex mapping analysis (*Section 12.15*) established that sequences homologous to IS1 were present in the *gal* regions of many spontaneous $tk^-$ strains, and N. Grindley went on to subject one of these strains to DNA sequencing analysis. The results are shown in Figure 18-11. The strain indeed carries an IS1 sequence within *galT*, generating the canonical 9 bp repetition of genomic DNA at either end (*recall Figure 18-3 and Table 18-1*). Since IS1 contains a number of internal "stop" codons, transcription and translation of this new piece of DNA results in abortive expression of *galT* and, therefore, of *galK* as well. It is thus not surprising that IS-induced mutations in operons are found to be "extremely" polar: negligible gene expression occurs downstream.

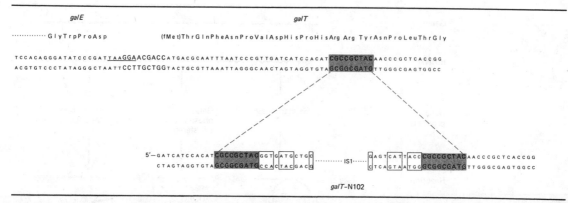

**FIGURE 18-11 IS1 insertion into the *galT* gene of *E. coli*.** Normal sequence is shown in the upper line; sequences surrounding the termini of the *galT*-N102 insertion of IS1 are shown below. 9 bp repeat of *galT* DNA shown in gray; imperfect IR sectors of IS1 are in boxes. The hexanucleotide underlined on the sense strand is a putative ribosome binding sequence (Shine-Dalgarno sequence) *(Section 9.6)*. (From N. D. F. Grindley, *Cell* **13**:419–426, 1978.)

**Chromosome Rearrangements Induced by IS Elements**    In the previous section, we noted that eukaryotic transposable elements induce not only mutations but also chromosome breaks. **Prokaryotic transposable elements are also found to generate deletions and inversions.** Some of these may well be the consequence of enzymatic errors during integration, replication, or excision, perhaps analogous to the deletions occasionally incurred by λ excision (*Figure 12-16*). In addition, however, **the presence of an integrated element causes the adjacent DNA to undergo rearrangement at a very high rate,** typically some three orders of magnitude higher than in the genome as a whole. An example is illustrated in Figure 18-12. The transpo-

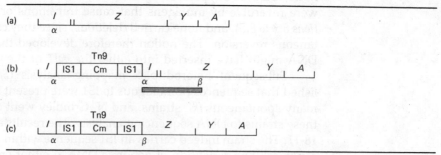

**FIGURE 18-12 Deletions generated by transposon insertion.** (a) The *lac* operon of *E. coli*. (b) Tn9 has inserted into *lacI*, generating a 9 bp duplication of the *lacI* sequence α. The gray bar denotes the region that is to undergo deletion. (c) Chromosome after deletion. The *lacZ* sequence β now adjoins the right-hand end of Tn9. The transposon itself is preserved. (From M. P. Calos, and J. H. Miller, *Cell* **20**:579–595, 1980.)

son T9 initially inserted into the *I* gene of the *lac* operon (*Figure 18-12a*), thereby inactivating it and generating a 9 bp duplication of the cellular sequence denoted α (*Figure 18-12b*). This *lacI⁻* strain was subsequently observed to undergo a high rate of mutation in the contiguous *lacZ* gene. Genetic analysis revealed that these *lacZ⁻* mutations were deletions, and sequencing studies showed that each deletion left the Tn9 transposon intact, removed the right-hand α sequence and contiguous DNA, and fused the right end of Tn9 to an internal *lacZ* sequence, denoted β in Figure 18-12c. Since β is different from α, the transposon is no longer flanked by a short direct repeat of cellular DNA, but since it can still transpose normally, the direct repeat is clearly irrelevant to the transposition process. How and why these rearrangements are induced by integrated transposable elements is still unclear, but their understanding may help explain the instability of eukaryotic DNA carrying inserted genetic elements.

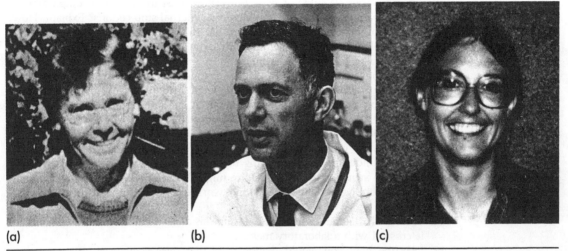

(a)                    (b)                    (c)

(a) Barbara McClintock (Cold Spring Harbor Laboratory) pioneered the genetic analysis of transposable elements in eukaryotes. (b) Allan Campbell (Stanford University) formulated the episome model of phage lambda integration based on genetic studies. (c) Nancy Kleckner (Harvard University) studies the genetic structure of transposons in bacteria.

# Questions and Problems

1. Is there any distinctive feature of the base compositions in the regions depicted in color in Figure 18-2? Do you expect that these regions would be denatured readily or with difficulty? What base composition might you expect to find in secondary *attB* sites in *E. coli*?

2. Would you postulate that the formation of a specialized transducing phage (*Figure 12-16*) requires Int, Xis, or RecA proteins? Explain.

3. Branch migration (*Figure 17-2f*) does not normally occur during phage λ integration. Explain.

4. Explain why *int* mutations can be complemented by helpers, whereas *attP* mutations cannot.

5. The Mu phage earned its name from its dramatic ability to induce mutations: 1 to 3 percent of the bacteria lysogenized by Mu carrying new auxotrophic mutations located throughout the genome. (a) Explain this property in the context of Figure 18-3. (b) You wish to prove that a mutation that arises during Mu lysogeny is due to Mu insertion. How would you show this by P1 transduction? By mating?

6. Mutator activity has been encountered in the context of DNA-related enzymes (*Section 8.13*) and transposable elements. How do they compare?

7. Plant variegation can be caused by chloroplast genes (*Section 16.11*) or by transposable elements. How would you distinguish the two?

8. IS1 contains 768 base pairs and induces a nine base-pair repeat of cellular DNA upon insertion. Would you expect insertion of IS1 to shift the reading frame of downstream sequences if it contained no internal stop codons? Explain. How would you answer this question for IS2, Tn3, and Tn9?

9. The *galT*-NIO2 mutation illustrated in Figure 18-11 is crossed into strains carrying either *amber* suppressors, *ochre* suppressors, or both. The IS1 element is known to contain both TAG and TAA codons. Would you expect expression of galactose transferase, kinase, or epimerase activity in any of these strains?

10. A spontaneous *gal* mutation was sequenced, and an IS1 element was found to reside in the region *between* the *galE* and *galT* genes, involving the nine base pairs depicted in large type in Figure 18-11. (a) Draw the DNA sequence of this region in the mutant strain, abbreviating the IS sequence as in Figure 18-11. (b) The underlined bases in this region are believed to constitute a Shine-Dalgarno sequence (*Table 9-2*). Why might insertions here cause polar mutations in both *galT* and *galK*? Would you expect *galE* to be affected as well? Explain.

11. Show with diagrams why hybrid dysgenesis does not occur when a wild female is crossed with a laboratory male.

12. Describe how you would demonstrate that *copia* sequences are present in widely scattered positions in *Drosophila* and that their positions differ between strains.

13. Explain the following statement: at least some cancers caused by tumor viruses may be due to their mutagenic activity.

# 19

# Related Genes: Alleles, Isoloci, and Gene Families

A. **Introduction**

B. **Traits Controlled by a Single Gene Locus**
   19.1 Inborn Errors of Metabolism
   19.2 The ABO Blood-Group Locus
   19.3 Dimeric Enzymes: Peptidase A
   19.4 Dominant Mutations: Penetrance and Expressivity

C. **Isozymes Specified by Isoloci**
   19.5 Unlinked Isoloci: Human Phosphoglucomutase Isozymes

19.6 Linked Isoloci: Mouse Esterase Isozymes

D. **Gene Families**
   19.7 "Homogeneous" Gene Families
   19.8 The "Heterogeneous" Hemoglobin Gene Family

E. **Distinguishing Alleles, Isoloci, and Gene Families in Genetic Crosses**

F. **Questions and Problems**

## Introduction

**Trait** is a useful term in genetics: it indicates an aspect of the phenotype, sometimes a specific and sometimes a rather generalized one. Thus the ability to utilize lactose, the possession of short fingers, and a sensitivity to loud noise can all be considered traits. In this and the next four chapters we consider how single genes and combinations of genes produce particular phenotypic traits. We have, of course, been considering such material throughout the text—whenever we have stated, for example, that gene $A$ gives rise to a particular polypeptide or that gene $A$ is dominant to its allele $a$. Now, however, we explore the construction of the phenotype more carefully, using examples from human genetics whenever feasible, since the human phenotype has an obvious inherent interest.

This chapter is particularly concerned with traits specified by genes that are similar to one another. We first consider the simplest case of such traits in diploid organisms—namely, a trait dictated by one locus having two or more alleles. In such a case, the phenotype depends on which two alleles are present in the diploid and on their relative degrees of dominance in a heterozygote. We then consider traits that are determined by two or more distinct, but related, gene loci. Since each such locus will be represented twice in a diploid, and since dominant or recessive alleles may be found at each locus, considerable variability is possible in such cases.

# Traits Controlled by a Single Gene Locus

In the sections that follow, we consider human traits that are controlled by a single gene locus that may have two, three, or multiple alleles. For any two alleles in a diploid, several relationships are possible. On the one hand, one allele may be **fully recessive,** so that its presence exerts no detectable influence on the phenotype of a heterozygote (the allele may, for example, produce a nonfunctional enzyme or a nonsense fragment of a structural polypeptide). On the other hand, an allele may be **fully dominant** so that its presence inevitably defines the phenotype, regardless of which other allele is present. An allele may be **partially** or **incompletely dominant** so that its presence in a heterozygote invariably influences the phenotype but need not necessarily define the phenotype, depending on the nature of the allele at the paired locus. Finally, two alleles may be **codominant,** each making an equivalent contribution to the phenotype.

With the preceding definitions as guidelines, we can proceed to consider the rich phenotypic detail of specific human traits.

## 19.1 Inborn Errors of Metabolism

The term "inborn error of metabolism," first coined by A. Garrod in 1902, has come to refer to a pathological human trait caused by an inherited defect in a single enzyme. Usually, although not invariably, the affected individual is homozygous for a recessive gene that codes for the enzyme, as in the following four examples.

**Alkaptonuria**  This condition was the first to be described by Garrod. Approximately one person in every million is homozygous for the alkaptonuria gene, known as $a$, and thereby suffers from the disorder. Affected individuals may develop arthritic ailments in later life but otherwise appear to be quite healthy except for the color of their urine, which becomes black shortly after exposure to air. A single biochemical reaction has been shown to be absent in alkaptonuric individuals. Thus, normal individuals catabolize the amino acids phenylalanine and tyrosine via a substance called homogentisic acid, which in turn is oxidized to fumaric and acetoacetic

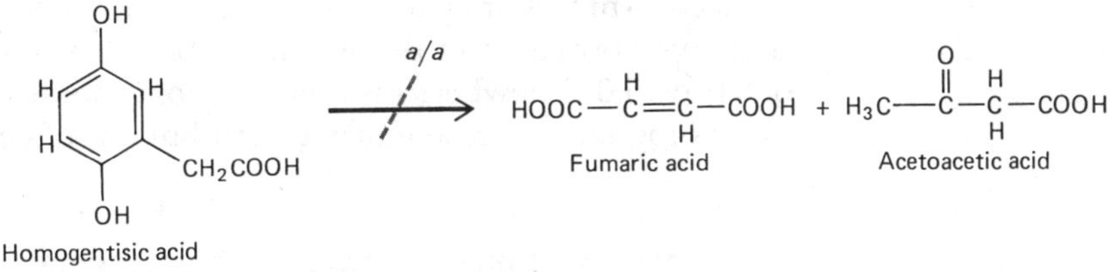

FIGURE 19-1 Alkaptonuria. The normal metabolism of homogentisic acid, a reaction blocked in *a/a* persons suffering from alkaptonuria.

acid (Figure 19-1); alkaptonurics lack an active homogentisic acid oxidase enzyme and therefore excrete homogentisic acid into their urine. It is this homogentisic acid that turns black upon oxidation by air.

**Phenylketonuria (PKU)** The most common form of phenylketonuria is produced by a defect in the enzyme phenylalanine hydroxylase. This enzyme normally converts phenylalanine to tyrosine (Figure 19-2); in phenylketonuric individuals, however, phenylalanine accumulates and is converted to other phenyl derivatives (Figure 19-2). One of these, phenyl-pyruvic acid, can be detected in the urine. If left undiagnosed and un-treated in infancy, the accumulated phenyl compounds of phenylketonuria produce irreversible and severe mental retardation. Therefore, the urine of newborn infants is routinely tested for the presence of phenylpyruvic acid,

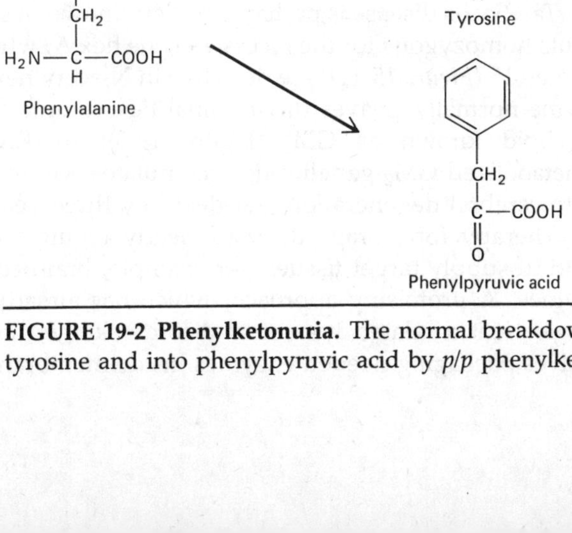

FIGURE 19-2 Phenylketonuria. The normal breakdown of phenylalanine into tyrosine and into phenylpyruvic acid by *p/p* phenylketonuric individuals.

and affected infants are placed on a diet low in phenylalanine, a treatment that allows normal brain development. Since, in Caucasian populations, about 1 in 15,000 newborns is a $p/p$ homozygote, the PKU screening and therapy programs represent major contributions of genetics to human welfare.

**Lesch-Nyhan Syndrome**  A recessive human gene in the X chromosome causes afflicted males to develop a complex spectrum of traits. Such individuals are typically subnormal in intelligence, spastic, destructive, and self-mutilating, with a particular tendency to bite their fingers and lips. These children prove to lack activity of the enzyme hypoxanthine-guanine phosphoribosyltransferase (HGPRT), an enzyme normally involved in purine metabolism. A discussion of this enzyme and its location on the human X chromosome is found in Section 15.6 and Figure 15-5.

Although it is not yet known why HGPRT inactivity produces such a bizarre pattern of behavior, the enzyme deficiency clearly exerts a **pleiotropic** effect, meaning that many facets of the phenotype are altered as a consequence of a single mutational lesion. Most inborn errors of metabolism are, in fact, highly pleiotropic, reflecting the multiple instances during fetal and infant development in which the absence of an enzyme (or the excess of a metabolite) can cause damage. Thus phenyketonuria produces not only brain defects but also a reduced pigmentation in hair and skin and a defective metabolism of the amino acid tryptophan. Pleiotropy often proves to be a major obstacle in determining the primary defect in an inherited clinical syndrome and, therefore, in devising therapy for the syndrome.

**Tay-Sachs Disease**  Homozygosity for a number of mutations in the human gene pool produces a variety of diseases known collectively as **lysosomal storage diseases.** The individual affected by such diseases lacks a specific lysosomal enzyme that normally acts to break down some type of complex macromolecule (a polysaccharide, lipid, protein, or nucleic acid), with the result that such macromolecules accumulate in the individual's tissues.

Tay-Sachs disease is perhaps the best known of such storage diseases. Infants homozygous for the recessive gene hex A, which is located in chromosome 15 (*Figure 15-11*), are defective in N-acetylhexosaminidase A. This enzyme normally cleaves the terminal hexosamine from a brain ganglioside lipid known as $GM_2$ (Figure 19-3). In Tay-Sachs infants, the unmetabolized $GM_2$ ganglioside accumulates within the brain cells, leading to cerebral degeneration and death by three years of age.

Therapy for storage diseases clearly requires that some means be found to supply target tissues (for example, brain cells) with the missing enzymes. A promising approach, which has already been used successfully to treat a related lysosomal disorder known as **Gaucher's disease,** involves packaging needed enzymes inside membrane vesicles and inject-

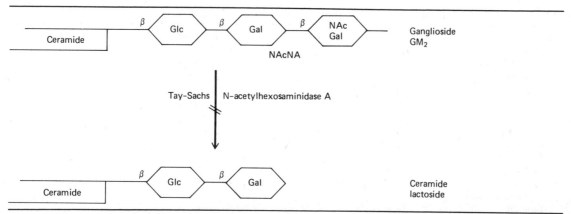

FIGURE 19-3 Tay-Sachs Disease. The normal breakdown of ganglioside GM$_2$, a reaction blocked in Tay-Sachs individuals. Abbreviations: NAcGal, N-acetylgalactosamine; NAcNA, N-acetylneuraminic acid; Gal, galactose; Glc, glucose. Ceramide is a long-chain amino alcohol linked to a long-chain fatty acid.

ing the vesicles into the patient's blood. The vesicles proceed to fuse with the target-cell membranes, introducing their enzyme contents into the cytoplasm where they can act on accumulated macromolecules.

**Detecting Heterozygous Carriers of Inborn Errors of Metabolism** Although most inborn errors of metabolism produce clinical symptoms only in homozygotes, it is possible in many instances to ascertain whether a person is a heterozygous carrier of a recessive mutant allele. For example, parents of phenylketonuric infants exhibit a slight, but significant, increase in blood phenylalanine levels, which has no effect on their health; persons heterozygous for the Tay-Sachs allele have reduced serum levels of N-acetylhexosaminidase A; and Lesch-Nyhan heterozygotes can be detected by reduced HGPRT activity in their hair follicle cells. In many cases a heterozygote will exhibit only 50 percent of the level of normal activity for a particular enzyme. This means that fully normal persons have, in fact, twice as high a level of that enzyme as needed for unimpaired health.

**Detecting Homozygotes for Inborn Errors of Metabolism in Utero** A sample of amniotic fluid can be withdrawn from a pregnant uterus (**amniocentesis**), and its constituent fetal cells can be subjected to karyotyping or biochemical assays. Fetuses with abnormal chromosome constitutions (*Sections 7.7 to 7.9*) and a number of metabolic diseases (Table 19-1) can in this way be detected. Such prenatal diagnosis, coupled with therapeutic abortion, has permitted Tay-Sachs heterozygotes, for example, to escape the anguish of bearing children who die in infancy.

With the increasing understanding of human genetic diseases and the

**TABLE 19-1  Some Inborn Errors of Metabolism Diagnosable before Birth**

### Lipidoses

Cholesterol-ester
    storage disease*
Fabry disease†
Farber disease*
Gaucher disease†
Generalized
    gangliosidosis
    (GM$_1$ gangliosidosis
    Type 1)†
Juvenile GM$_1$
    gangliosidosis
    (GM$_1$ gangliosidosis
    Type 2)†
Tay-Sachs disease
    (GM$_2$ gangliosidosis
    Type 1)†
Sandhoff's disease
    (GM$_2$ gangliosidosis
    Type 2)†

Juvenile GM$_2$
    gangliosidosis
    (GM$_2$ gangliosidosis
    type 3)*
GM$_3$ sphingolipo-
    dystrophy*
Krabbe's disease
    (globoid-cell
    leukodystrophy)†
Metachromatic
    leukodystrophy
Niemann-Pick disease
    Type A
Niemann-Pick disease
    Type B*
Niemann-Pick disease
    Type C*
Refsum disease*
Wolman disease†

### Mucopolysaccharidoses

MPS I—Hurler†
MPS I—Scheie*
MPS—Hurler/Scheie*
MPS II A—Hunter†
MPS II B—Hunter†
MPS III—Sanfilippo A†
        Sanfilippo B*

MPS IV—Morquio's
    syndrome
MPS VI A—
    Maroteaux-Lamy
    syndrome*
MPS VI B—
    Maroteaux-Lamy
    syndrome*
MPS VII—β-
    glucuronidase
    deficiency*

### Amino Acid and Related Disorders

Argininosuccinic
    aciduria†
Aspartylglucosaminuria*
Citrullinemia†
Congenital
    hyperammonemia*
Histidinemia*
Hypervalinemia*
Iminoglycinuria*
Isoleucine catabolism
    disorder*
Isovaleric acidemia*

Cystathionine synthase
    deficit
    (homocystinuria)*
Cystathioninuria*
Cystinuria†
Hartnup disease*
Methylenetetrahydro-
    folate reductase
    deficiency*
Ornithine-α-ketoacid
    transaminase
    deficiency*

Maple syrup urine
    disease:
    Severe infantile†
    Intermittent*
Methylmalonic aciduria
    Unresponsive to
    vitamin B$_{12}$*
    Responsive to vitamin
    B$_{12}$†

Propionyl CoA
    carboxylase
    deficiency (ketotic
    hyperglycinemia)*
Succinyl CoA: 3
    ketoacid
    CoA transferase
    deficiency*
Vitamin B$_{12}$ metabolic
    defect*

### Disorders of Carbohydrate Metabolism

Fucosidosis*
Galactokinase
    deficiency*
Galactosemia†
Glucose-6-phosphate
    dehydrogenase
    deficiency*
Glycogen storage
    disease
    (Type II)†
Glycogen storage
    disease
    (Type III)*

Glycogen storage
    disease
    (Type IV)†
Mannosidosis*
Phosphohexose
    isomerase
    deficiency*
Pyruvate decarboxylase
    deficiency*
Pyruvate
    dehydrogenase
    deficiency*

### Miscellaneous Hereditary Disorders

Acatalasemia*
Acute intermittent
    porphyria*
Adenosine deaminase
    deficiency†
Chediak-Higashi
    syndrome*
Congenital
    erythropoietic
    porphyria*
Congenital nephrosis†
Lysosomal acid
    phosphatase
    deficiency†
Lysyl-protocollagen
    hydroxylase
    deficiency*
Myotonic muscular
    dystrophy*
Nail-patella syndrome*
Orotic aciduria*

Cystinosis†
Familial
    hypercholesterolemia*
Glutathionuria*
Hypophosphatasia†
I-cell disease†
Leigh's
    encephalopathy*
Lesch-Nyhan
    syndrome†
Protoporphyria*
Saccharopinuria*
Sickle cell anemia*
Testicular feminization*
Thalassemia†
Xeroderma
    pigmentosum*

*Prenatal diagnosis potentially possible.
†Prenatal diagnosis made.
From A. Milunsky, *New Eng. J. Med.* **295**:377, 1976.

availability of heterozygote screening and prenatal diagnostic methods, **genetic counseling** is an increasingly important service provided by hospitals and other centers. Genetic counselors are trained not only to provide potential parents with an accurate estimate of the probabilities that they will have a child with a genetic disorder but also to help parents adjust to the birth of such a child or to reach a decision about prenatal diagnosis and possible abortion.

## 19.2 The ABO Blood-Group Locus

The ABO locus, which maps to chromosome 9 in the human genome (*Figure 15-11*), has three common alleles, two of which ($I^A$ and $I^B$) are codominant and the third of which ($I^O$) is recessive. Since the locus controls an important surface property of human erythrocytes, it is considered in some detail.

**The A, B, and H Glycolipids**  The ABO locus is concerned with the production of **glycosyltransferases**—enzymes that mediate the synthesis of polysaccharides. We are concerned here in particular with polysaccharides that attach to lipids and associate with cell membranes, forming **cell-surface glycolipids.** By influencing which glycosyltransferases are present in a cell, therefore, the ABO locus influences the types of surface glycolipids a cell can produce.

Glycosyltransferases act in a very precise fashion. If we denote a polysaccharide as R-glucose-glucose-galactose, where the R represents the end of the chain that will attach to the lipid and the galactose represents the free end of the chain, then one type of glycosyltransferase may add a fucose molecule to the chain to produce R-glucose-glucose-galactose-fucose. A second type of glycosyltransferase may recognize the same polysaccharide chain but will add a mannose, instead of a fucose, to it, producing an R-glucose-glucose-galactose-mannose chain.

The $I^A$ allele of the ABO locus codes for a glycosyltransferase known as α-N-acetylgalactosamyl transferase. This enzyme recognizes the polysaccharide chain drawn in Figure 19-4a and abbreviated as **H**. Recognition is followed by the addition of the sugar α-N-acetylgalactosamine, and, with this addition, the H polysaccharide is converted to a polysaccharide known as **A** (Figure 19-4b). The $I^B$ allele of the ABO locus, on the other hand, codes for an α-D-galactosyltransferase. This enzyme also recognizes H, but instead of adding an α-N-acetylgalactosamine, it adds a galactose residue, producing the **B** polysaccharide (Figure 19-4c). In both cases, **some H chains fail to be acted upon by either enzyme and retain their H structure,** so that all individuals produce some unmodified H.

**Expression of the $I^A$, $I^B$, and $I^O$ Alleles**  The H, A, and B polysaccharides are found as components of glycolipids (and probably glycoproteins as

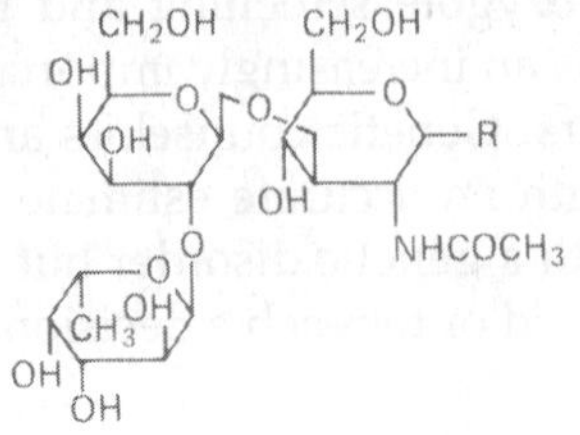

β-Gal-(1 → 3)-GNAc-R
↑ 1, 2
α-Fuc

**(a) H antigen**

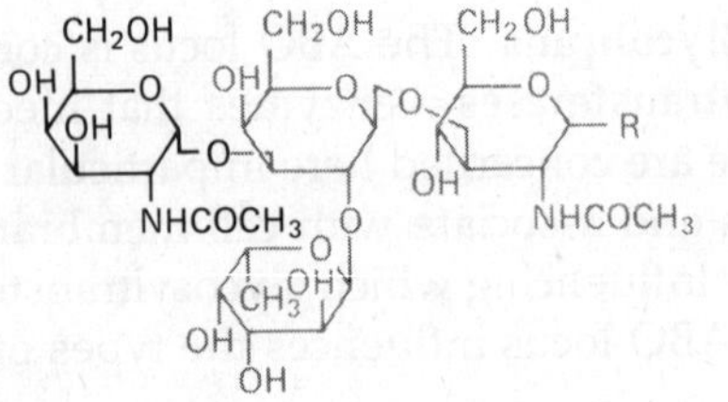

α-GalNAc-(1 → 3)-β-Gal-(1 → 3)-GNAc-R
↑ 1, 2
α-Fuc

**(b) A antigen**

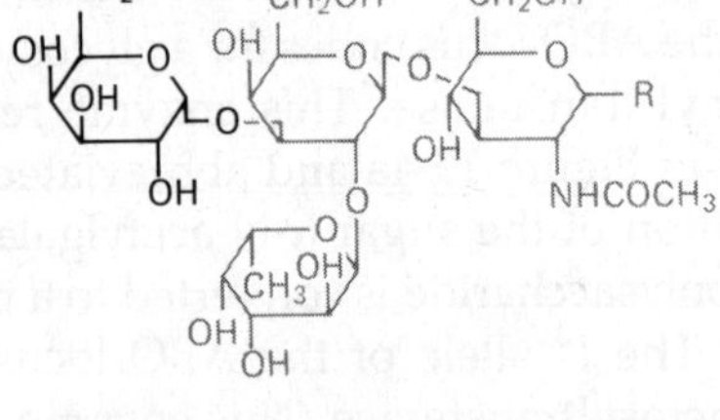

α-Gal-(1 → 3)-β-Gal-(1 → 3)-GNAc-R
↑ 1, 2
α-Fuc

**(c) B antigen**

**FIGURE 19-4 ABO Blood Groups.** Terminal sugar sequences in polysaccharide chains of glycolipids that confer A, B, and H specificity. Gal: D-galactose; Fuc: L-fucose; GNAc: N-acetyl-D-glucosamine; GalNAc: N-acetyl-D-galactosamine.

well) on the erythrocyte surface, and their distribution depends strictly on the ABO locus genotype. Thus, an $I^A/I^A$ homozygote produces erythrocytes that carry H and A chains but never B chains, whereas an $I^B/I^B$ homozygote produces erythrocytes with H and B, but never with A chains. The $I^A/I^B$ heterozygote possesses both enzymes and therefore carries H, A, and B polysaccharides on his or her red cell surfaces. The $I^A$ and $I^B$ alleles are therefore said to be **codominant,** since **each allele makes a comparable contribution to the phenotype of the heterozygote.**

The third $I^O$ allele of the ABO locus codes for neither the *A* nor the *B* type glycosyltransferase enzyme; therefore, an $I^O/I^O$ homozygote is unable to produce either A or B polysaccharides and places only "unmodified" H glycolipids on the erythrocyte surface. As might be predicted, the $I^O$ allele is fully recessive to $I^A$ and to $I^B$, so that $I^A/I^O$ heterozygotes produce H and A chains and $I^B/I^O$ heterozygotes produce H and B chains.

**Detecting the ABO Blood Type** Polysaccharides A, B, and H can act as **antigens,** meaning that they are capable of eliciting specific anti-A, anti-B, and anti-H antibodies when injected into experimental animals (*Box 2.3*). The antisera from such animals can then be used to carry out **blood typing** of human beings. If a sample of red blood cells from a person is found to **agglutinate** (stick together) only in the presence of the anti-H serum, then the cells clearly possess only the H polysaccharide, and the person being tested must be an $I^O/I^O$ homozygote. The person is then said to be of **blood type O.** If a sample of cells from a second individual agglutinates with both anti-H and anti-A antisera, then the cells possess H and A and the individual is said to be of blood **type A;** correspondingly, cells that agglutinate with anti-H and anti-B sera derive from **type B** persons; and cells that agglutinate in the presence of all three antisera are of **type AB.**

The cell-surface polysaccharides A, B, and H are not only antigenic to experimental animals. They are also potential antigens in humans—with one important proviso: **a normal individual will not produce antibodies against himself or herself.** Thus the serum of a person with type A blood will contain neither anti-A nor anti-H antibodies; however, since the B polysaccharide is not an element of that person's "self," the production of anti-B antibody is not "forbidden" and proceeds apace. Similarly, the blood sera of type B persons contain anti-A antibodies, the sera of type O persons contain both anti-A and anti-B antibodies, and the sera of type AB persons contain no antibodies against this group of cell-surface polysaccharides. These relationships are summarized in Table 19-2.

The clinical importance of the blood type can now be appreciated, for when a person is in need of transfused erythrocytes, the donor blood type must be carefully checked. The injection of type A cells into type O or type B persons, for example, will result in potentially fatal agglutination of the transfused erythrocytes by circulating anti-A antibodies in the serum of the recipient.

TABLE 19-2    **ABO Blood Typing.** The agglutination reactions observed with the A, B, AB, and O blood groups

| Blood Group | Blood Type–Specific Sugar | Type(s) of Antibody Produced | Types of Red Blood Cells Agglutinated | Transfusions Accepted from |
|---|---|---|---|---|
| A | Galactosamine | Anti-B | B, AB | Type A, Type O |
| B | Galactose | Anti-A | A, AB | Type B, Type O |
| AB | Galactosamine + galactose | None | None | Any donor |
| O | None | Anti-A and Anti-B | A, B, and AB | Type O only |

# 19.3 Dimeric Enzymes: Peptidase A

Codominance arises when each gene product is equally active and operates independently. Thus the codominance of the $I^A$ and $I^B$ blood type alleles in a heterozygote results from the fact that each glycosyltransferase gene product has an equivalent chance of modifying an H polysaccharide. A slightly more complex kind of relationship between pairs of alleles is illustrated by the human locus coding for **peptidase A,** an enzyme found in erythrocytes that acts to hydrolyze dipeptides into their individual amino acids.

When erythrocyte extracts from individual persons are subjected to gel electrophoresis (*Box 3.2*) and the gel is stained to reveal peptidase A activity, three types of enzyme patterns are found (Figure 19-5): some individuals possess only the peptidase A enzyme (called 1) that migrates relatively slowly toward the cathode; some individuals possess a more rapidly migrating peptidase A species (called 2); and some individuals possess three types of peptidase A enzymes—namely, enzyme 1, enzyme 2, and an enzyme denoted 2-1 with intermediate electrophoretic mobility (Figure 19-5). Such patterns are explained in the following way:

1. Peptidase A is a dimeric protein (*Section 3.3*) having the composition $\alpha\alpha$.
2. The locus coding for the $\alpha$ polypeptide of peptidase A exists in a number of allelic forms; here we consider two, called PEP $A^1$ and PEP $A^2$.
3. The gene product of PEP $A^1$, called $\alpha^1$, has a slower electrophoretic mobility than the $\alpha^2$ product of the PEP $A^2$ allele.
4. PEP $A^1$/PEP $A^1$ homozygotes produce $\alpha^1\alpha^1$ **homodimer** enzymes equivalent to enzyme 1 in Figure 19-5; PEP $A^2$/PEP $A^2$ homozygotes produce $\alpha^2\alpha^2$ homodimers equivalent to enzyme 2; and PEP $A^1$/PEP $A^2$ heterozygotes produce, in addition to the two types of homodimers, a

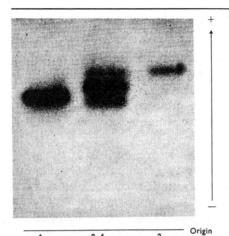

**FIGURE 19-5 Peptidase A.** Electrophoretic patterns of peptidase A types, 1, 2-1, and 2. (From H. Harris, *The Principles of Human Biochemical Genetics.* Amsterdam: North-Holland, 1975.)

1    2-1    2    Origin

Peptidase A types

**heterodimer** enzyme, $\alpha^1\alpha^2$, with an electrophoretic mobility equivalent to enzyme 2-1.

The validity of this series of postulates for peptidase A is demonstrated by the following experiment. Enzymes 1 and 2 were isolated and dissociated into their component polypeptides by incubating in mercaptoethanol and urea to reduce S-S bridges (*Section 3.3*). The polypeptides derived from dissociation of each enzyme were then mixed together, allowed to reassociate, and subjected to gel electrophoresis. The gel electrophoresis patterns then revealed not only enzymes 1 and 2 but also enzyme 2-1, which, in this case, had formed *in vitro* from enzyme 1 and enzyme 2 subunits.

In humans, numerous cases have been described that are qualitatively similar to the peptidase A situation in which **a single locus having two or more alleles codes for a dimer-forming polypeptide, with pure homodimers present in homozygotes and both homo- and heterodimers present in heterozygotes.** What are the effects of such heterozygosity on the phenotype? In many cases, homozygotes can be distinguished from heterozygotes only by electrophoretic analysis of their component enzymes: the heterodimer enzymes appear to function quite as well as the homodimer enzymes, and heterozygosity has no other measurable effect on the phenotype. In some cases, however, the heterodimer may be more active, or more stable, than either homodimer. In still other cases, one type of homodimer may be particularly unstable. For example, individuals heterozygous for another PEP A allele, called PEP A[8], possess no active $\alpha^8\alpha^8$ homodimers in their red blood cells and thus appear to have only two electrophoretic variants of the peptidase A enzyme.

## 19.4 Dominant Mutations: Penetrance and Expressivity

Dominant mutant genes are relatively uncommon compared with "recessives" or "partial dominants." In theory, dominant mutations should be of particular value for genetic analysis, since they should exert an immediate effect on the phenotype of diploid organisms. In fact, however, many dominant mutant genes prove to affect the human phenotype in a variable fashion. This variability is due to two phenomena known as **penetrance** and **expressivity.**

**Penetrance** is an all-or-nothing concept: a gene is either completely penetrant or incompletely penetrant. An incompletely penetrant gene finds expression in some individuals but not others. Figure 19-6, for example, shows two identical, or **monozygotic,** twins who derived from the same fertilized egg and can therefore be assumed to have identical genotypes. The genetically based developmental abnormality leading to the production of a harelip has clearly "penetrated" the twin on the left but not the twin on the right.

Incomplete penetrance is a property of many dominant genes. The human dominant gene *D*, for example, causes a bent and **stiff little finger.** Pedigrees are known, however, in which a woman exhibits this trait and her son does not, whereas two of his three children do. Clearly, the son was at least heterozygous for the gene, since he was able to pass it on to his children, but the gene did not affect his own finger development at all. Another example of incomplete penetrance is given by some forms of the psychotic disease **schizophrenia:** present evidence indicates that in some cases the illness may be controlled by two incompletely penetrant dominant genes.

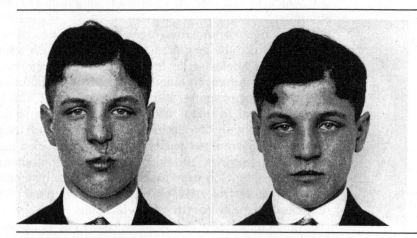

**FIGURE 19-6 Incomplete penetrance.** A pair of identical twins: *left:* penetrance of harelip; *right:* lack of penetrance. (From F. Claussen, *Zeitschr. Abstgs. und Vererbgsl.* **76:**30, 1939.)

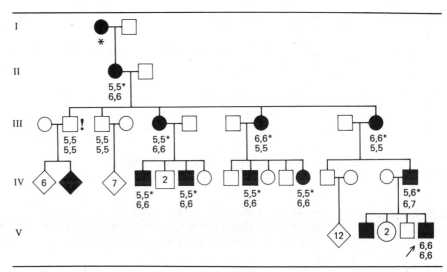

**FIGURE 19-7 Variable expressivity.** A pedigree of polydactyly, showing only selected individuals. The groups of four numbers below affected individuals represent the number of fingers (above) and toes (below) on the left and the right. An asterisk indicates that the nature of the polydactyly was not reported unequivocally. Arrow points to the propositus (affected individual who attracts the geneticist's attention). (From C. Stern, *Human Genetics*, 2nd ed. San Francisco: Freeman, 1960, after Lucas, *Guy's Hosp. Rpts.*, 3rd Ser. **25,** 1881.)

Once a gene is penetrant (finds expression), it will frequently exhibit a **variable expressivity,** meaning that the same gene will produce a range of phenotypes in various "penetrated" individuals. There are many examples of genes with variable expressivity in *Drosophila*, higher plants, and animals. Again using a human example, we can cite **polydactyly,** a condition leading to an excessive number of fingers and toes. It is clear from the pedigree shown in Figure 19-7 that this trait is determined by a dominant gene and that the expressivity of this gene varies widely: some persons have normal hands but six-toed feet; others have six-fingered hands and normal feet; one individual has more digits on his right extremities than on his left; and one has six digits on all four extremities. The pedigree for polydactyly also illustrates the incomplete penetrance of the gene: the male marked (!) does not exhibit the trait and yet passes it on to a child. As a general rule, genes that are incompletely penetrant are found to show variable expressivity as well, and indeed "no penetrance" may often represent one extreme in the observed range of variation, with "highly abnormal" perhaps representing the other.

**Factors Influencing Penetrance and Expressivity**  The variation in the degree of expression attained by the genes we are now discussing is thought to be due to two factors: **genetic background** and **environmental influ-**

ence. **Genetic background** refers to the fact that an individual inherits a genotype and not a particular gene. A particular gene product will therefore normally be operating in the presence of countless different combinations of other gene products. When a particular gene product is especially sensitive to the nature of other gene products in a cell, its expression may well be variable from one individual to the next and, in some cases, inhibited altogether.

The effect of genetic background can be reduced or eliminated by studying highly inbred strains or monozygotic twins, and it is here that the effects of **environmental influence** can best be appreciated. A straightforward example is given by the dominant trait **Curly** of *D. melanogaster*. The *Curly (Cy)* gene gives rise to flies with curled-up wings when pupae are maintained at 25°C; when the pupae are maintained at 19°C, however, many individuals have apparently normal wings (incomplete penetrance), whereas the rest show a range of abnormalities (variable expressivity). It appears that the *Curly* gene produces wings that are unusually sensitive to heat at the time the adult fly emerges from the moist pupal case; at 19°C the wings dry more or less normally, whereas at 25° the upper and lower portions of the wings dry at different rates and a *Curly* wing results.

The environmental influence in the case of *Curly* wing is easy to analyze, but this is not always the case. The intrauterine environment of mammals, even for two monozygotic twins, is believed to be quite variable—twins will occupy different positions in the uterus, placental connections may differ, and so forth, and the array of effects that may influence gene expression after birth is so vast that it would be quite impossible to specify them all. In other words, when a gene exhibits incomplete penetrance and variable expressivity, particularly in humans, it is usually difficult to determine which environmental factors are involved.

# Isozymes Specified by Isoloci

The remainder of this chapter describes traits that are specified by two or more distinct, but related, gene products. We first consider traits that are controlled by **isoloci** and carried out by **isozymes,** terms that are best defined by examples.

## 19.5 Unlinked Isoloci: Human Phosphoglucomutase Isozymes

Suppose the phenotypic trait being followed is the ability to transfer phosphate groups from glucose-1-P to glucose-6-P during carbohydrate metabolism, a reaction catalyzed by an enzyme activity called **phosphoglucomutase.** Phosphoglucomutase enzymes can be detected in cell extracts subjected to gel electrophoresis, much as with peptidase A (*Figure 19-5*), and these enzymes are known to function as monomeric proteins. Therefore, if human phosphoglucomutase activity were dictated by a single gene locus, as with peptidase A, one would expect either one band (for homozygotes) or two bands (for heterozygotes) when erythrocytes from a particular individual are subjected to electrophoresis and stained for the enzyme. Instead, H. Harris and coworkers have found multiple phosphoglucomutase bands in every human extract tested. Pedigree and somatic cell hybridization studies (*Sections 6.6 and 15.6*) reveal that one group of these bands is specified by the $PGM_1$ locus, located in human chromosome 1. The $PGM_2$ locus, in chromosome 4, is found to specify a second group of bands; and the $PGM_3$ locus in chromosome 6 specifies a third group. In other words, the human genome carries **three unlinked loci that specify similar enzymes.** The various phosphoglucomutase monomers present in the electrophoresis gels are collectively known as **isozymes**—enzymes that have similar substrates but distinctive electrophoretic or biochemical properties—and the $PGM_1$, $PGM_2$, and $PGM_3$ loci can be considered as a set of **isoloci.**

Consider next the individual PGM loci. At the $PGM_1$ locus, two alleles are common, one specifying a phosphoglucomutase known as $PGM_1$-1 and the other specifying the electrophoretically distinct $PGM_1$-2 enzyme. Since $PGM_1$-1 and $PGM_1$-2 are **alle**les, the two enzymes are commonly referred to as **allozymes.** Similarly, the human gene pool contains two alleles of the $PGM_2$ gene that specify the allozymes $PGM_2$-1 and $PGM_2$-3.

How are allozymes distinguished from isozymes and alleles from distinct isoloci? The usual demonstration utilizes Mendelian genetics: alleles are expected to segregate at anaphase I of meiosis, whereas unlinked isoloci should assort independently. Thus a $PGM_1$-1/$PGM_1$-2 heterozygote will transmit either $PGM_1$-1 or $PGM_1$-2 to individual offspring, never both. If this same person is also a $PGM_2$-1/$PGM_2$-3 heterozygote, again only one or the other $PGM_2$ allele will be transmitted to individual children. Each child will, however, receive both a $PGM_1$ allele and a $PGM_2$ allele from this parent, the identity of each being the chance consequence of independent assortment.

## 19.6 Linked Isoloci: Mouse Esterase Isozymes

As a second example of isozymes, we can consider the **esterases.** Esterase enzymes resemble phosphoglucomutases in that they function as monomers and exist as electrophoretic variants detectable by subjecting electrophoresis gels to specific staining reactions. In contrast to the unlinked isoloci responsible for the phosphoglucomutases, however, several mouse esterases are specified by isoloci that are closely linked. Specifically, F. Ruddle and colleagues have determined that the mouse esterase loci *Es-1*, *Es-2*, and *Es-5* are all found within about ten map units of one another in linkage group XVIII of the mouse. At each of the three loci, at least two allelic genes are known that specify identifiable allozymes; thus the *Es-1a* and *Es-1b* alleles give rise, respectively, to *fast* and *slow* electrophoretic variants of the *Es-1* esterase. In this case, of course, **the genes at linked isoloci will not assort independently.** Instead, a chromosome carrying, for example, genes *Es-1a*, *Es-2b*, and *Es-5a* will usually be inherited as a unit, with occasional recombinational events breaking up the general linkage.

The esterase isozymes specified by *Es-1*, *Es-2*, and *Es-5* are in many functional respects similar to one another: all are found in the serum and all are resistant to the inhibitor eserine. A fourth esterase isolocus in the mouse, *Es-3*, is not linked to the others and encodes a distinct kind of esterase: it is largely restricted to kidney cells, and it is sensitive to eserine. Two general principles are illustrated by this example. First, **isozymes that catalyze the same enzymatic reaction are not necessarily structurally similar** (indeed, the three classes of human phosphoglucomutase enzymes described previously prove to differ in size, in tissue distribution, in substrate specificity, and in stability). Second, **clustered isoloci are more likely to be related than are unlinked isoloci** (although this rule has its exceptions) and are indeed likely to represent a **gene family.**

# Gene Families

Most genes in both prokaryotes and eukaryotes are believed to be present in only a single copy per haploid genome, but there are a number of important cases in which **each haploid genome contains two or more genes that are either identical to, or very similar to, one another in their nucleotide sequence.** Such gene sets are usually denoted as **gene families.** It is generally believed that **the genes in such families originated as duplications of some common ancestral sequence.** The subsequent evolutionary history of the duplicated sequences varies considerably from one example to the next: in some cases the reiterated genes remain clustered together; in other cases their linkage has been broken and they may even reside on different chromosomes; finally, there are cases in which some members of a gene family are clustered and others are dispersed throughout the genome.

Two classes of gene families can be recognized: in a "homogeneous" family, all the genes are somehow kept identical to one another, whereas in a "heterogenous" family, the genes have been free to accumulate differences in nucleotide sequence. The homogeneous families, exemplified by the rRNA and histone gene clusters, usually encode gene products that the cell requires in large quantities. The heterogeneous families, exemplified by globin, actin, and tubulin, typically specify similar proteins that are utilized by different cell types and/or cells at different stages of development. Thus one form of actin is localized in skeletal muscle, whereas other forms are synthesized in various types of nonmuscle cell, and different globin genes are expressed at different times during embryonic development. The special case of the immunoglobulin gene families is the topic of the next chapter.

## 19.7 "Homogeneous" Gene Families

The "homogeneous" rRNA and histone gene families were described in detail in Chapter 10. As detailed in Sections 10.7 through 10.9, the rDNA genes of *X. laevis* are repeated, in tandem array, hundreds of times along a single length of DNA (*Figure 10-11*). The clusters are arranged in a . . . gene-spacer-gene-spacer . . . array (*Figure 10-10*), with the genes virtually identical and the spacers showing some heterogeneity in the number of times that their short internal repeats are reiterated. Similarly, in our description of the histone family (*Section 10.11*), we noted that a given cluster of histone genes is tandemly repeated from 10 to 100 times.

The identity, or virtual identity, of the many rRNA genes and the many histone genes within a given species poses an evolutionary riddle previously raised in Section 10.11: **How are the nucleotide sequences of identical gene families preserved?** Over the millions of years of a species' existence, spontaneous mutations must occur in these genes. How are they eliminated? One explanation, the **selection theory,** states that such mutations are immediately lethal. Thus it is argued for rRNA that a single mutant rDNA gene in a cluster of 400 would cause 1 of every 400 ribosomes to be faulty; many essential mRNAs bound to these defective ribosomes would be improperly translated, and the growth and development of the organism would be fatally impaired. The alternate **correction theory** holds that any mutated rDNA or histone sequences are somehow repaired before they have a chance to be transmitted through the germ line. There is at present insufficient data to decide between these two theories, but the question is being actively studied.

The rDNA and histone gene families are permanently amplified. **Temporary amplification** can also occur in response to environmental pressures. We have already noted (*Section 7.7*) that tumor cells treated with the drug methotrexate often respond by reversibly amplifying their genes encoding the enzyme that is inhibited by methotrexate. Similarly, pesticide-resistant insects have been found to have duplicated those genes that specify insecticide-degrading enzymes.

## 19.8 The "Heterogeneous" Hemoglobin Gene Family

Hemoglobin (*Figure 3-5*) is a tetramer with the general structure $x_2y_2$, where x and y are two of the six known globin chains—$\alpha$, $\beta$, $\delta$, $\gamma$, $\epsilon$, and $\zeta$. The most common form of adult hemoglobin has the structure $\alpha_2\beta_2$ and is known as **HbA:** a minor form of adult hemoglobin, $\alpha_2\delta_2$, is called **HbA$_2$.** The most prominent fetal hemoglobin, **HbF,** is $\alpha_2\gamma_2$, whereas hemoglobins composed of the $\epsilon$ and $\zeta$ chains are present only in the first few weeks of embryonic life. All six globin chains are structurally similar to one another: the $\alpha$-chain has 141 amino acids, whereas the others have 146 amino acids. Thus it has long been suspected that all are encoded by similar genes. In the following sections, we first describe genetic studies indicating the existence in the mammals of two globin gene clusters, one containing the **α-like** genes and the other the **β-like** genes. We then present the results of sequencing studies that confirm and expand this conclusion. The control over expression of these genes is the subject of Sections 25.17 and 25.18.

**Evolutionary Relationships Between the α-like and β-like Genes**  The cyclostomes (lamprey and hagfish) have primitive hemoglobins containing only a single type of globin chain, whereas carp hemoglobins have both α- and β-chains. Thus it is assumed that the vertebrate line originally carried a single globin gene and that this gene duplicated some time during the evolution of the bony fish, about 500 million years ago. The two duplicates, it is proposed, went on to evolve into the modern α- and β-gene families. That the modern α- and β-chains are quite distantly related is apparent from the fact that they now differ by 80 amino acids. By contrast, the β- and γ-chains differ by only 30 amino acids and β- and δ-chains by only 10. These data indicate that the β, δ, and γ-chains are closely related, earning them the designation ''β-like.''

By comparing both the amino acid and the nucleotide sequences of the various hemoglobins and cloned hemoglobin genes, it has been possible to construct the **evolutionary tree** drawn in Figure 19-8. The original α-β split at 500 million years (MY) is drawn at the base. Subsequent splits at 200, 100, and 40 MY created the present-day β-like family. The β family is thought to have arisen in the canonical fashion of other gene families: a given gene duplicates; the duplicates go on to accumulate sufficient mutations to encode distinct, but related, polypeptides; one or both of the duplicates duplicate again; and so on.

**Nonlinkage of α- and β-like Genes**  The α- and β-like genes are sufficiently distant that they by now are encoded in separate chromosomes. This conclusion has been most directly demonstrated for humans by somatic cell genetics. In probing mouse-human cell hybrids (*Section 15.6*) for their retention of human chromosomes that hybridize with α-globin or β-globin mRNA, it was found that some cell lines retained human α-chain sequences but lacked β-sequences, whereas others lost α but retained β. Subsequent *in situ* hybridization studies localized the α-globin sequences to human chromosome 16 and β-sequences to human chromosome 11; the experiments pinpointing the precise location of the β-gene sequences on chromosome 11 have already been described in Section 15.6 (*Figure 15-7*).

**Duplicate Copies of the α-Gene**  Genetic studies on mice by R. Popp established that mice carry two closely linked copies of the α-gene per haploid genome, a conclusion that has since been extended to humans. Popp discovered that a mouse strain called SEC produces two types of α-chain, one carrying a serine at position 68 and the other a threonine. Since SEC is highly inbred and presumably homozygous at most of its gene loci, the persistence of these two forms of the α-chain suggested the presence of

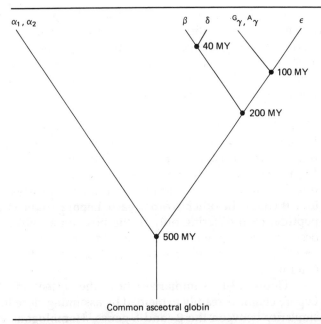

**FIGURE 19-8 Hemoglobin evolution.** Distances along the branches are additive and proportional to evolutionary time. Branch points represent the times in millions of years (MY) from the present at which genes or gene lineages began to diverge. These branch points do not necessarily represent gene duplication events: the duplication could have occurred prior to the onset of divergence. (From A. Efstratiadis, et al., *Cell* **21**:653–668, 1980.)

two copies of the α-gene in the mouse genome. That these two copies are closely linked was demonstrated by crossing SEC mice with C57BL mice, a second inbred strain that produces only one class of α-chain, having an asparagine at position 68. When the heterozygous $F_1$ of the cross were backcrossed to either parent (Table 19-3), the ability to synthesize α-68Ser and α-68Thr was inevitably inherited together: no recombinants were

**TABLE 19-3   Linkage of α-chain Genes.** Progeny of (C57BL × SEC) $F_1$ mice backcrossed to either C57BL or SEC mice

|  | Matings | |
|---|---|---|
| | (C57BL × SEC)$F_1$ × C57BL | (C57BL × SEC)$F_1$ × SEC |
| **Alpha-Chain Genotype** | | |
| $\alpha^{68asn}/\alpha^{68asn}$ | 63 | — |
| $\alpha^{68asn}/\alpha^{68ser,thr}$ | 73 | 216 |
| $\alpha^{68ser,thr}/\alpha^{68ser,thr}$ | — | 199 |
| Possible recombinants | 0 | 0 |
| Totals | 136 | 415 |

From R. A. Popp, *J. Hered.* **60**:131, 1969, Table VII.

found in 551 progeny examined, indicating a recombination frequency of less than 0.005 for the two loci (Table 19-3). It therefore appears that SEC carries an α-gene duplication and that the two genes have come to differ by a single codon. A similar duplication is probably carried by C57BL mice as well, but this fact cannot be detected genetically as long as the two genes are identical to one another in sequence.

**Linkage of the β-and δ-Genes**   The close linkage of the human β- and δ-genes was first inferred from the properties of two variant types of non–α-globin chains called **Lepore** and **anti-Lepore.** As summarized in Figure 19-9a to c, the N-terminal amino acids of Lepore chains are identical to those in a δ-chain, whereas the C-terminal sequences are identical to those in a β-chain. In other words, **each Lepore chain is a δ-β hybrid polypeptide,** each differing only in the position at which the δ→β transition occurs. As their name implies, anti-Lepore chains have the opposite construction: these chains are β-δ hybrids, with β-like N-termini and δ-like C-termini.

   Figure 19-10 summarizes how the formation of Lepore and anti-Lepore chains is readily explained by assuming close linkage and sequence similarity between the β- and δ-genes. **Homologous chromosomes carrying similar genes in tandem linkage are thought to pair out of register occasionally** so that, as illustrated in Figure 19-10a, a β-sequence pairs with a δ-sequence. If a crossover then occurs in the mispaired region, one chromosome will emerge with a hybrid δ-β Lepore gene (Figure 19-10b), whereas the other will acquire a β-δ anti-Lepore gene (Figure 19-10c). This phenomenon, termed **unequal crossing over,** is invoked to explain the genetic behavior of many other duplicated genes as well, as we will see in subsequent sections of this chapter.

**Duplicate Copies of the γ-Gene**   When normal human fetal erythrocytes are analyzed, they are found to contain equal amounts of two types of fetal hemoglobin (HbF): one, called $\alpha_2{}^A\gamma_2$, contains γ-chains with alanine at position 136; the other, called $\alpha_2{}^G\gamma_2$, contains γ-chains with glycine at position 136. The presence of the two forms of γ could be explained by widespread heterozygosity for a pair of alleles ($^A\gamma$ and $^G\gamma$); it could also be explained by widespread homozygosity at the loci of a duplicate set of γ-genes ($^A\gamma/^A\gamma$ and $^G\gamma/^G\gamma$).

   To test these alternatives, blood samples were taken from adults with a condition called **hereditary persistence of fetal hemoglobin (HPFH).** Some of the donors proved to synthesize variant γ-chains marked with amino acid substitutions at positions other than at 136: for example, a $^G\gamma$-chain might carry a valine rather than lysine at position 9, a polypeptide we can denote $^G_{Val}\gamma$. When a person homozygous for this variant $^G_{Val}\gamma$-chain condition marries a normal spouse, their newborn children are found to produce *three* types of HbF: $\alpha_2{}^A\gamma_2$, $\alpha_2{}^G\gamma_2$, and $\alpha_2{}^G_{Val}\gamma_2$. Convince yourself with diagrams that this outcome indicates that the human haploid genome carries two gene loci for γ-globin.

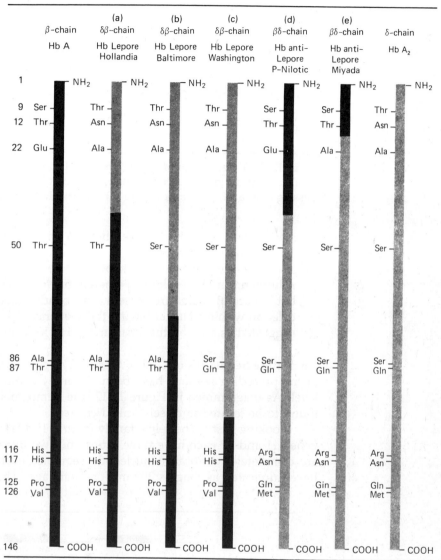

**FIGURE 19-9 Lepore and anti-Lepore hemoglobins.** Relationships of the amino acid sequences of the non–α–chains of three Lepore (δ-β) (a–c) and two anti-Lepore (β-δ) (d,e) hemoglobins to those of the normal β- and δ-hemoglobin chains. Only the amino acids that differ in the normal β- and δ-polypeptide sequences are shown. (From H. Harris, *The Principles of Human Biochemical Genetics*. Amsterdam: North Holland, 1975.)

**Linkage of γ-, β-, and δ-Genes**  Two kinds of genetic evidence indicate that the γ-genes are closely linked to the β-δ cluster.

1. A variant hemoglobin known as **Kenya** carries N-terminal $^A$γ-sequences fused to C-terminal β-sequences. As diagrammed in Figure 19-11, this variant is readily explained by postulating a misaligned pair-

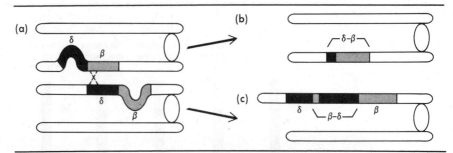

**FIGURE 19-10 Generation of Lepore and anti-Lepore hemoglobins.** Misalignment during the meiotic synapsis of homologues (a), followed by crossing over at X, generates Lepore (b) and anti-Lepore (c) hybrid genes.

ing between $^A\gamma$ and $\beta$ followed by unequal crossing over (*recall Figure 19-10*). The diagram predicts that "anti-Kenya" hemoglobins should also exist, but these have yet to be identified in the human population.

2. A human condition is known wherein no $^A\gamma$- or $\delta$-chains are produced but $^G\gamma$- and $\beta$-chains are present. This condition is best explained as a deletion within a cluster having the sequence $^G\gamma$-$^A\gamma$-$\delta$-$\beta$, a gene order suggested as well by the structure of Hb Kenya (Figure 19-11).

**Nucleotide Sequences of Globin Genes**  The genetic predictions outlined in the preceding sections have been strikingly confirmed at a molecular level. As diagrammed in Figure 19-12, the hemoglobin genes are indeed found to be located in closely linked clusters.

Looking first at the $\alpha$-like family (Figure 19-12a), we see that human genes $\alpha1$ and $\alpha2$ each have three exons and two introns, and both genes are completely identical in nucleotide sequence even though they may have duplicated as long as 300 million years ago. Therefore, as with the

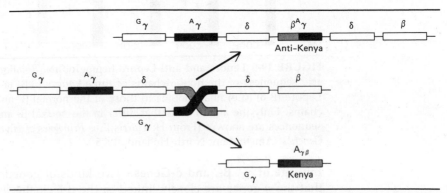

**FIGURE 19-11 Generation of hemoglobin Kenya** by an unequal crossover involving $\beta$ and $^A\gamma$. An anti-Kenya chain should also be produced. (From D. J. Weatherall, and J. B. Clegg, *Cell* **16**:467–479, 1979.)

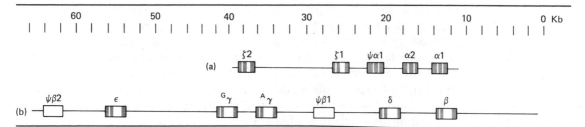

**FIGURE 19-12 Maps of globin gene families.** Gene sequences are denoted by the solid (exons) and open (introns) boxes. The 5'⟶3' direction of transcription is from left to right. (a) Human α-like cluster. (b) Human β-like cluster. (From A. Efstratiadis, et al., *Cell* **21**:653–668, 1980.)

homogeneous rRNA genes (*Section 19.7*), some mechanism must exist to prevent the accumulation of mutations in this gene pair. Both α1 and α2 are expressed equivalently in eythroid cells. In addition, the cluster contains an α-**pseudogene** (ψα1), so called because although its sequence is α-like, it does not encode a functional globin polypeptide and is thought to represent a vestigial duplicate gene. Finally, the α-like cluster carries two copies of the related ζ-globin sequences expressed during early fetal development.

The β-like cluster (Figure 19-12b) contains two pseudogenes and the gene for fetal ε-globin as well as the $^{G}γ$-$^{A}γ$-δ-β cluster in the order predicted from genetic studies. Many of the long sectors of DNA between each coding sequence are also known, and it is within these sequences that investigators are now searching for clues regarding how it is that the ε- and γ-genes are expressed exclusively during fetal life and the δ- and β-genes exclusively during adult life. These regulatory aspects of hemoglobin gene expression are considered in Chapter 25 (*Sections 25.17 and 25.18*).

# Distinguishing Alleles, Isoloci, and Gene Families in Genetic Crosses

We conclude this chapter by reviewing the often subtle but important distinctions between (multiple) alleles at one locus and a set of isoloci (or a family of genes), each of which may, in turn, be mutiply allelic. The formal distinction can be visualized by recalling our definition of a gene locus as a physical location in a particular chromosome. Thus, if gene locus 3 is defined as that group of nucleotides transcribed from promoter number 3 when counting from one end of human chromosome 7, then all extant nucleotide sequences that occupy this particular locus are alleles of one another. The nucleotides following promoter number 9 on this same chromosome may well code for a similar polypeptide, and multiple versions of this gene may also exist in the human gene pool. Nevertheless, a gene occupying locus 9 is never considered allelic to a gene occupying locus 3. Instead, genes 3 and 9 are considered either to occupy isoloci or to belong to a gene family, depending on their degree of relatedness.

The experimental distinction between alleles at one locus and genes marking isoloci is readily made when isoloci are loosely linked or unlinked. In this case, markers for isoloci will recombine freely or assort independently at meiosis, whereas alleles of a single locus will almost always segregate at anaphase I. When isoloci are closely linked, on the other hand, the distinction is more difficult. It is especially important to guard against the erroneous concept that the distinction can be made by recombination frequencies alone. The fact that markers $a$ and $b$, for instance, are "never observed to recombine," whereas both recombine at measurable frequency with marker $c$, **cannot** be used to argue that $a$ and $b$ are allelic and are not alleles of $c$. We noted in Chapter 17 how marker effects and related phenomena can cause recombination frequencies to be quite unreliable for closely linked markers. Moreover, if two isoloci lie next to one another in a chromosome, markers in adjacent genes may well recombine with one another less often than will widely spaced markers in the same gene. Therefore, a test of independent function [for example, some form of a complementation test (*Section 12.4*)] must ultimately be performed to show that two genes are not alleles of one another.

(a) Harry Harris (University of Pennsylvania) studies the biochemical basis of human inborn errors of metabolism. (b) D. J. Weatherall (Oxford University, England) has focused his research on the hemoglobin gene families.

# Questions and Problems

1. A schizophrenic marries a nonschizophrenic person. Assuming the disease to be determined by two incompletely penetrant dominant genes, $P$ and $Q$, list all possible genotypes of the two parents.

2. How would you determine whether a phenotype that differs in the degree of its expression is determined by a gene with variable expressivity versus a series of multiple alleles?

3. Huntington's chorea is a fatal disease leading to the degeneration of the nervous system. It is found in persons heterozygous for the dominant allele $Ht$, but has variable expressivity in the sense that a person may develop symptoms as early as childhood or as late as 60 years old, with the mean age of onset at about 40 years.

   (a) A man died of Huntington's chorea. The disease is not known in his widow's family. What are the chances that his son carries the gene?

   (b) In certain families known to carry the gene, the disease is found to "skip" several generations before reappearing. How might this be explained?

4. A woman having blood type A had a type O and a type B child. What is the genotype and phenotype of the father? What are the genotypes of the woman and her children? Could the legitimacy of children with blood type AB or blood type A have been questioned?

5. Persons known as *secretors* have the same AB blood group antigens in their saliva as are found on their blood cell surfaces, whereas *nonsecretors* do not have antigen in their saliva. The gene *Se* (*secretor*) is dominant to its allele *se* (*nonsecretor*), and the *secretor* locus is not linked to the ABO locus. In the following matings, what are the expected proportions of A *secretors*, B *secretors*, AB *secretors*, and *nonsecretors* among the progeny?

   (a) AB *secretor* (*Se/se*) × O (*Se/se*)

   (b) AB *secretor* (*Se/se*) × O (*se/se*)

6. The antigens M and N are located on human red cells, along with antigens A and B. All persons are M, N, or MN, and the trait is determined by a pair of codominant alleles, $L^M$ and $L^N$. Persons do not make antibody against M or N antigens, whereas rabbits do. Therefore persons are typed as M, N, or MN by mixing a sample of their blood with anti-M or anti-N serum from rabbits: anti-M will agglutinate M or MN cells, whereas anti-N will agglutinate N or NM cells.

   In an O,M × AB,MN mating, the progeny are A,M, B,M, A,MN, and B,MN in equal numbers.

   (a) Is the $L$ locus linked to the $I$ locus?

   (b) What are the genotypes of the parents in this mating?

   (c) List all possible genotypes and phenotypes in the progeny of an A,M × B,N mating.

7. A child has the blood type O,N and is a *secretor*. Which of the following matings could have produced the child and which could not? Explain in each case.

   (a) O,N *secretor* × O,N *secretor*

   (b) A,N *secretor* × A,N *secretor*

   (c) A,N *nonsecretor* × A,N *nonsecretor*

   (d) AB,N *secretor* × O,N *nonsecretor*

   (e) A,N *nonsecretor* × B,N *secretor*

8. A highly simplified view of the human Rh blood groups divides persons into two categories: those whose blood cells agglutinate with the antiserum anti-Rh$_0$ (also called anti-D) and are therefore **Rh positive** and those whose cells do not react with this serum and are therefore **Rh negative**. An Rh-positive person can be designated as $R/R$ or $R/r$ and an Rh-negative person as $r/r$. Knowing that this locus is not linked to other blood group loci, give all possible genotypes of the parents of persons with the following phenotypes:
   - (a) O,MN, Rh negative
   - (b) A,M, Rh negative
   - (c) AB,MN, Rh positive

9. Discuss the kinds of dominance relationships described in this chapter in terms of the concept of complementation (*Section 12.4*).

10. In inbred mice the enzyme isocitric dehydrogenase is made up of two identical subunits. The electrophoretic band patterns of this subunit in two inbred strains of mouse (BALB/c and C57/b) and in their F$_1$ hybrid are as follows:

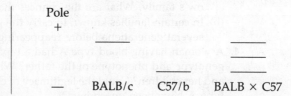

Explain these results. What other pattern might you have expected in the F$_1$ gel?

11. (a) If two alleles at a particular locus are codominant, how might a monosomic or trisomic condition aid in the chromosomal location of the gene locus?
   - (b) If two alleles are dominant and recessive, which of the following aberrations would aid in locating them to a specific chromosome: duplication, deficiency, monosomy, trisomy? Explain.

12. Four babies were born in a hospital on the same night, and their blood groups were later found to be O, A, B, and AB. The four parents were:

   O and O
   AB and O
   A and B
   B and B

   Assign the four babies to their correct parents.

13. Two *Drosophila* isoloci specify two esterases. Each has two codominant alleles, and all four products can be distinguished electrophoretically. Each enzyme is active as a dimer; the products of one locus do not associate with the products of the other locus. Show with diagrams the expected progeny phenotypes and their relative proportions when flies heterozygous at both isoloci are inbred and the isoloci are (a) unlinked or (b) closely linked.

14. (a) Persons homozygous for the Lepore hemoglobin gene develop moderate to severe hemolytic anemia. Which hemoglobin or hemoglobins would you expect to be abnormal in these persons and which normal? Explain.
   - (b) Would you expect persons homozygous for anti-Lepore hemoglobin to be as afflicted as Lepore homozygotes? Examine Figure 19-9 carefully in making your answer.

15. In a technique known as **peptide mapping,** polypeptides are isolated and digested with an enzyme such as trypsin; the resultant **tryptic peptides** (*Section 3.2*) are then separated by chromatography to generate a **peptide map.** If you prepared peptide maps of the non–α-hemoglobin chains of normal persons and Lepore homozygotes, what would you expect to find?

16. Type O persons are sometimes called universal blood donors, and type AB persons are called universal recipients. Explain.

17. The sex-linked *Bar* (*B*) mutation results from a duplication in the X chromosome (*Figure 6-16*). Females homozygous for *Bar* and heterozygous for the closely linked outside markers *f* (*forked* bristles) and *fu* (*fused* wings) were crossed to *forked fused* males, and rare flies recombinant for the outside markers were recovered. Some of these proved to have normal eyes; others exhibited an extreme eye phenotype called **ultrabar.** Cytological examination of polytene chromosomes revealed that the normal progeny carried one dose of the 16A segment, whereas the *ultrabar* progeny carried three. Explain these results, using diagrams.

18. In a Hungarian family, two α-chain variants are known that are called Buda and Pest. One individual in this family was found to have 25 percent Buda, 25 percent Pest, and 50 percent normal hemoglobin A. What does this suggest about the number of genes specifying the α chain? How would you write this person's genotype?

19. One model for the conservation of homogenous gene families proposes the existence of a single "master gene" against which all duplicate copies are compared and, if necessary, corrected by gene conversion. What unlikely property does this model require of the "master gene"?

20. Diagram the genetic demonstration, using persons with HPFH, that the human haploid genome carries two gene loci for γ-globin.

# 20 Immunogenetics

A. **Introduction**

B. **Properties of the Immunoglobulins**
  20.1 General Features of the Immune Response
  20.2 General Properties of IgG Antibodies
  20.3 Variable and Constant Portions of IgG Molecules
  20.4 Constant and Variable Domains of the Light Chain Allotypes and idiotypes
  20.5 Constant and Variable Domains of the Heavy Chain
  20.6 Overview of the Genetic Specification of Immunoglobulins

C. **Construction and Expression of Light-Chain Genes**
  20.7 Construction of the $V_L$ Domain of MOPC-41 Light Chains
  20.8 Construction of $V_L$ Domains in Other Light Chains
  20.9 Creating Light-Chain Diversity During V-J Joining
  20.10 Creating Functional Light-Chain Genes During V-J Joining
  20.11 Allelic and Isotypic Exclusion

D. **Construction and Expression of Heavy-Chain Genes**
  20.12 Construction of $V_H$ Domains: V-D-J Joining
  20.13 DNA Sequences Encoding the $C_H$ Domains of IgG
  20.14 DNA Sequences Encoding the $C_H$ Domains of Other Immunoglobulins
  20.15 Chain Alternations and Class Switching

E. **Somatic Mutation of Antibody Genes, and a Summary of Antibody Diversity Mechanisms**
  20.16 Evidence for Somatic Mutation of Antibody Genes
  20.17 Summary of Antibody Diversity Mechanisms

F. **Questions and Problems**

# Introduction

The genes that specify **antibody** or **immunoglobulin** molecules belong to several gene families, and, in previous editions of this textbook, these genes were considered in the "Gene Family" chapter. They are accorded an independent chapter in this edition because they have been found to have evolved unique patterns of interacting with one another. These patterns, although complex, appear to utilize many of the genetic mechanisms we have been considering in previous chapters, including transposition, site-specific recombination, and somatic mutation. As we consider the intricacies of the immunoglobin genes, therefore, we will have the opportunity to review these phenomena and apply them to what is perhaps the most fascinating genetic system of all.

Although we call this chapter "Immunogenetics," the reader should be aware that the field of immunogenetics easily fills a textbook of its own, and many topics ordinarily considered under immunogenetics, including those with strictly medical implications, are omitted here. The genetics of histocompatibility, which is usually also included, is presented in this text in Sections 21.4, 26.6, and 27.2.

# Properties of the Immunoglobulins

## 20.1 General Features of the Immune Response

It is estimated that **a mouse can synthesize somewhere between $10^7$ and $10^8$ different kinds of antibodies,** and humans probably synthesize even more. Each antibody is an oligomeric protein whose structure is described shortly, and each has the ability to complex with and thus remove from the tissues or blood of the body a particular **antigen** or family of antigens (*recall Box 2.3*). The immunoglobulins are synthesized by blood cells called **B lymphocytes** and by differentiated forms of these lymphocytes known as **plasma cells.** Each lymphocyte is highly specialized to synthesize antibodies directed against a single type of antigen. Therefore, an organism produces millions of unique lymphocyte clones, each committed to the production of a unique antibody.

Full details of the immune response are beyond the scope of this text. For our purposes, we can give a vastly simplified account of what occurs, using as an example a human lymphocyte committed to synthesize an antibody against a protein found in the coat of the smallpox virus. We can call this antibody antipox. The lymphocyte and its clonal descendents synthesize antipox throughout their lifetimes, whether or not they ever actually encounter the smallpox protein. In other words, **the synthesis of the antibody does not require the presence of the antigen** as some sort of stimulant. When, however, the antigen is introduced, a series of events occurs. The antigen first becomes complexed with molecules of antipox that are located on the surface of the antipox lymphocytes. This interaction stimulates the lymphocytes to undergo many rounds of mitosis and to differentiate into specialized, antipox-secreting plasma cells. As a result, an organism that has been exposed to the smallpox virus soon acquires high blood-serum titers of the antipox antibody. This is the basis for immunizing persons by giving inoculations of attenuated forms of pathogenic viruses.

## 20.2 General Properties of IgG Antibodies

Plasma cell cancers, called **plasmacytomas** or **multiple myelomas,** result in the proliferation of a single type of plasma cell, and a number of antibody-secreting mouse myeloma cell lines are now established in culture. In addition, humans with such tumors often overproduce single types of immunoglobins that can be recovered from their blood or urine. Analysis of the chemical properties of these "purified" proteins has provided considerable information about the structure of immunoglobulins; this must be mastered before we consider the genes that specify them.

Figure 20-1 shows a model of an immunoglobulin G (**IgG**) antibody,

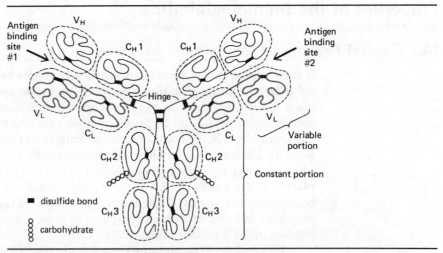

**FIGURE 20-1 Immunoglobulin structure.** A model of a human immunoglobulin G (IgG) molecule. Light polypeptide chains are shown in gray and heavy are in black. The variable regions of heavy and light chains ($V_H$ and $V_L$), the constant region of the light chain ($C_L$), and the homology regions in the constant region of heavy chain ($C_H1$, $C_H2$, and $C_H3$) are thought to fold into compact domains (delineated by dotted lines), but the exact conformation of the polypeptide chains has not been determined. (From J. A. Gally and G. M. Edelman, *Ann. Rev. Genetics* **6**:1, 1972.)

**TABLE 20-1    Classes of Human Immunoglobulins**

|  | IgG | IgA | IgM | IgD | IgE |
|---|---|---|---|---|---|
| **Heavy chains:** | | | | | |
| Class | $\gamma$ | $\alpha$ | $\mu$ | $\delta$ | $\epsilon$ |
| Subclass | $\gamma_1\gamma_{2a}\gamma_{2b}\gamma_3$ | $\alpha_1\alpha_2$ | — | — | — |
| Mol wt $\times 10^{-4}$ | 5.5 | ~6 | 6–7 | ~6 | ~7.5 |
| **Light chains:** | | | | | |
| Class | $\kappa, \lambda$ | $\kappa, \lambda$ | $\kappa, \lambda$ | $\kappa, \lambda$ | $\kappa, \lambda$ |
| Mol wt $\times 10^{-4}$ | 2.2–2.3 | 2.2–2.3 | 2.2–2.3 | 2.2–2.3 | 2.2–2.3 |
| **Whole molecule:** | | | | | |
| Formula | $\kappa_2\gamma_2$ or $\lambda_2\gamma_2$ | $(\kappa_2\alpha_2)_n$ or $(\lambda_2\alpha_2)_n$ $n = 1, 2, 3 \ldots$ | $(\kappa_2\mu_2)_5$ or $(\lambda_2\mu_2)_5$ | $\kappa_2\delta_2$ or $\lambda_2\delta_2$ | $\kappa_2\epsilon_2$ or $\lambda_2\epsilon_2$ |
| Mol wt $\times 10^{-5}$ | ~1.5 | 1.6 | 9–10 | 1.7–1.8 | ~2 |
| Carbohydrate, % | 2.9 | 7.5 | 7.7–10.7 | 12 | 10.7 |

After H. Metzger, *Ann. Rev. Biochem.* **39**:889, 1970, Table I.

the most prominent of the five immunoglobulin classes produced by humans (Table 20-1). In overall topology, an IgG molecule is seen to resemble a two-pronged fork. Four polypeptide chains, interconnected by disulfide bridges, are found in each molecule. Two of these chains are longer than the other two, so that each IgG is said to have two **heavy chains** (drawn in black in Figure 20-1) and two **light chains** (drawn in gray). **In any one molecule, the two heavy chains are identical and the two light chains are identical.**

In considering immunoglobulins, the units of interest are not only the 4 polypeptide chains but also specific **domains** within each chain. The domains in an IgG molecule are delineated in Figure 20-1 by dashed lines. Domains labeled $V_H$ and $C_H1$ of the heavy chain, and $V_L$ and $C_L$ of the light chain, are located in each "prong" of the fork, whereas domains $C_H2$ and $C_H3$ of the heavy chain are found in the "handle." All six domain types are generally homologous to one another and probably derive from some common ancestral sequence. As we shall see, however, each has undergone distinct evolutionary modifications, and each performs unique functions.

## 20.3 Variable and Constant Portions of IgG Molecules

The six domains identified in the previous section were designated either **V** ($V_H$ and $V_L$) or **C** ($C_H1$, $C_H2$, $C_H3$, and $C_L$). The V domains are so called because they reside in the variable portion of an IgG located at the tip of each prong in the fork, whereas the C domains make up what is called the constant portion of the molecule in the center and handle sections.

As its name suggests, **the variable portion of an antibody molecule constitutes the antigen-combining site.** Each antigen-binding site must be structurally different to account for the millions of different antigens recognized by the immune system. Since there are two combining sites per antibody (Figure 20-1) and each contains an identical amino acid sequence, it follows that **an antibody is bivalent,** being able to bind to two antigens at once. As diagrammed in Figure 20-1, an antigen-binding site is created by the complex three-dimensional interactions of both the $V_H$ and the $V_L$ domains of a heavy and a light chain.

**The constant portion of an antibody molecule is responsible for eliciting immune responses.** Specifically, once the variable portions of the protein have recognized and bound to antigens, the constant portion somehow changes its configuration in such a way that macrophages and T lymphocytes, cellular mediators of the immune response, are induced to migrate to the site of the antigen and eliminate it from the body. Proteins known collectively as complement are also induced to interact with the constant region once antigen binding has occurred, and the presence of complement further stimulates the killing activities of the macrophages and the T cells. The constant portion of an antibody also binds to the surface of B lymphocytes during the initial phase of antibody display described in Section 20.1. Thus the constant portion is a critical component of an antibody molecule, and four of its six domains are found in the constant sector.

## 20.4 Constant and Variable Domains of the Light Chain

We are now in a position to explore at a molecular level what is meant by "constant" and "variable," focusing first on the light chains. As drawn in Figure 20-2a, the variable domain of the light chain ($V_L$) constitutes amino acids 1 to 107, whereas the $C_L$ domain is composed of amino acids 108 to

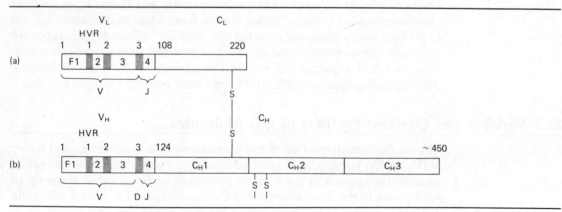

**FIGURE 20-2 Light chain (a) and heavy chain (b) of an antibody molecule.** Numbers indicate amino acid positions in the chain, where 1 is the N-terminal amino acid. HVR, hypervariable region; F, framework region. V, D, and J indicate sectors of the variable regions encoded by V, D, and J sequences in the germline DNA.

220. The constant domains fall into two subclasses, called **κ** and **λ**. Thus if 20 different human myeloma IgG antibodies are analyzed, 12 of the 20 might contain $C_L$ domains of the κ variety (also called $C_κ$), whereas 8 of the 20 might contain $C_L$ domains of the λ variety ($C_λ$). Thus to say that C domains are "constant" is misleading in the sense that they are by no means invariant; they are, however, relatively constant by comparison with the large amount of variability found in the V domain.

As drawn in Figure 20-2a, the $V_L$ domain is a mosaic of three **hypervariable regions** (HVR) and four **framework regions** (F). In the folded immunoglobulin molecule, the three hypervariable regions come together to form the antigen-binding site, and the four framework regions support the structure. As their names imply, the sequences of amino acids in the hypervariable regions are generally very different from one antibody type to the next, whereas the sequences in the framework regions are less variable.

**Allotypes and Idiotypes** In the preceding paragraphs, we made a number of statements about how antibodies are found to be relatively similar or dissimilar from one another in their structure. How is this in fact determined? The two methods of choice, which will certainly be used primarily in the years to come, are either to sequence the antibody directly or to isolate the gene encoding that antibody and deduce its amino acid sequence from the codons found in the gene. Until recently, however, neither of these techniques was practical for large-scale studies, and immunogeneticists therefore characterized particular antibodies according to their allotype and their idiotype.

Details of allotyping and idiotyping fall beyond the scope of this text, but the general approach can be illustrated by example. A single IgG species from a mouse myeloma is injected into a goat (or some animal other than mouse). The goat responds by producing an array of "anti-mouse-IgG" antibodies. Some of the goat antibodies will be directed against constant regions of the molecule: thus if the mouse light-chain constant region is of the κ variety, "anti-mouse-kappa" antibodies will be elicited in the goat's immune system. Other goat antibodies will recognize particular sections of the variable regions, some interacting with antigenic framework regions and others finding antigenic hypervariable regions. The goat antiserum is now fractionated such that antibodies with distinct specifications are separated from one another. The four general classes of antibodies present will be anti-$V_L$, anti-$C_L$, anti-$V_H$, and anti-$C_H$. If we focus on the anti-$V_L$ fraction, continued purification will eventually yield a species that recognizes only a particular sector of the $V_L$ region, perhaps a framework 3 (F3) region. If this "purified" anti-F3 antiserum is now presented to 25 additional kinds of mouse myeloma IgG, it may ignore 23 of them but immunoprecipitate the other 2. The two IgGs that are recognized do not necessarily have F3 regions that are identical to the original IgG species, but their F3 regions are at least sufficiently similar to the original that the probing antiserum recognizes all three. The three IgGs are in this case said to share a common **idiotype,** idiotype connoting "variable-region pheno-type as probed by heterologous antisera." Other idiotypes, of course, will be defined by other F regions or by hypervariable regions.

If we focus instead on the anti-$C_L$ fraction, continued fractionation should yield a goat antiserum that specifically immunoprecipitates mouse IgG molecules bearing kappa domains. These are said to share a common **allotype,** allotype being synonymous with "constant-region phenotype." Idiotypes and allotypes also exist, of course, for the variable and constant regions of heavy chains.

With this background, we can now offer more formal definitions. **Allotypes** refer to constant-region sequences that express limited genetic polymorphism. **Idiotypes** refer to variable-region sequences. Both allotypes and idiotypes are defined serologically (that is, using heterologous antiserum) and therefore are less precisely defined than an amino acid sequence of an immunoglobulin or the nucleotide sequence of its gene.

## 20.5 Constant and Variable Domains of the Heavy Chain

Figure 20-2b diagrams the structure of the heavy chain, where the $V_H$ domain again occupies the N-terminus (amino acids 1 to 123) and the 3 $C_H$ domains make up the remainder of the polypeptide. As indicated earlier and summarized in Table 20-1, the $C_H$ domains belong to one of five different classes known as $\gamma$, $\alpha$, $\mu$, $\delta$, and $\epsilon$. Confining ourselves for the present to mouse IgG molecules, the $\gamma$ domains belong to one of 4 **subclasses** called $\gamma_1$, $\gamma_{2a}$, $\gamma_{2b}$, and $\gamma_3$. These are analogous to the $\kappa$ and $\lambda$ allotypes in the $C_L$ domains; thus if the 20 IgG molecules classified as $\kappa$ or $\lambda$ were instead examined for their $C_H$ subclass, 5 might prove to have a $C_H$ domain falling into the $\gamma_1$ subclass, 3 might carry $\gamma_{2a}$, 8 might carry $\gamma_{2b}$, and 4 might carry $\gamma_3$. Again, therefore, constant domains are only relatively "constant."

The $V_H$ domain is very similar in overall structure to the $V_L$ domain (Figure 20-2b): three hypervariable regions, which interact with the $V_L$ hypervariable regions to form the antigen binding site, are flanked by four framework regions. Subclasses of these $V_H$ domains can be categorized by the idiotyping technology.

## 20.6 Overview of the Genetic Specification of Immunoglobulins

We are now in a position to appreciate three remarkable features of immunoglobulin genetics, each of which is summarized here and detailed in the following sections.

The first relates to **the mechanism of antibody gene construction.** In 1965, W. Dryer and J. Bennett proposed an astonishing concept: they suggested that in the germline and embryonic DNA, multiple $V_L$ and $C_L$ sequences exist separately from one another; that $V_H$ and $C_H$ sequences are similarly separate; and that during lymphocyte differentiation, $V_L$ sequences recombine with $C_L$ sequences to produce an array of functional light-chain genes, whereas $V_H$ and $C_H$ sequences recombine to create heavy-chain genes. The Dryer-Bennett proposal is correct in concept, but the details of "V-C joining" prove to be even more extraordinary than they imagined. In fact, **it is the V domain itself that is separated in the germline genome.** Specifically, looking at Figure 20-2a, the portion of the $V_L$ domain denoted V is located in one sector of DNA, whereas the portion of the $V_L$ domain denoted J is contiguous to a $C_L$ sequence in a second sector of DNA. During lymphocyte differentiation, a copy of V, is joined to a copy of the J-$C_L$ complex via recombination, and a functional light-chain gene results. Analogous but even more complex events bring *three* subregions of the $V_H$ domain (denoted V, D, and J in Figure 20-2b) into proximity to form a functional heavy-chain gene. Thus in the case of the immunoglobulins, the adage "one gene $\longrightarrow$ one polypeptide" is replaced by the more cumbersome **"several DNA sequences recombined $\longrightarrow$ one gene $\longrightarrow$ one polypeptide."** As you may have already guessed, this strategy of creating genes *de novo* is a major mechanism for providing the organism with different antibody types.

A second remarkable feature of antibody genetics relates exclusively to heavy chains and is known as **class switching.** The newly created $V_H$ portion of an immunoglobin gene is at first associated with a $C_H$ sequence that specifies $\mu$ chains; therefore, the first immunoglobulins produced by a B lymphocyte line are of the IgM class (*Table 20-1*). As the clone continues to differentiate, however, **this same $V_H$ sequence becomes coupled instead with one of the $\gamma$ sequences,** so that the same cell line no longer carries a functional IgM gene but instead transcribes and expresses a new IgG gene whose antibody product recognizes the same antigen as its predecessor. Similar "class switches" are able to generate the genes for IgE and IgA heavy chains.

The third remarkable feature of antibody genes, and one that is currently least understood, is that they appear to undergo **somatic mutations** after they are created, the result being the creation of still more diversity in the antibody repertoire.

In the sections that follow, we scrutinize what is known about antibody gene construction, class switching, and somatic mutation of antibody genes. Although many scientists have contributed to elucidating these mechanisms, the laboratories of L. Hood, P. Leder, and S. Tonegawa have been particularly productive during the past few years.

# Construction and Expression of Light-Chain Genes

## 20.7 Construction of the $V_L$ Domain of MOPC-41 Light Chains

As our first example of light-chain gene construction, we describe a series of experiments from P. Leder's laboratory that are focused on the gene for a particular light chain, a κ-containing polypeptide synthesized and secreted by a mouse myeloma cell line known as MOPC-41. The amino acid sequence of this light chain was known, and the abundant myeloma mRNA specifying it was isolated for use as a probe to identify homologous genomic DNA sequences (*Section 4.16*). The DNA to be probed was isolated from two sources: from the myeloma genome itself and from the mouse embryo. Each was treated with restriction enzymes that yielded very long DNA fragments, and fragments homologous to the mRNA probe were identified, cloned, and subjected to DNA sequencing (*review these techniques in Chapter 4*).

The myeloma genomic DNA yielded a large fragment homologous to the mRNA probe that proved to contain the entire gene for the MOPC-41 light-chain polypeptide. Its structure is diagrammed in Figure 20-3a. The $V_L$ domain of the gene is located at the 5' end of the fragment and the $C_L$ domain at the 3' end. The $V_L$ coding region (bracketed) begins with a short hydrophobic "signal sequence" (S) which specifies that the message is to be translated on rough endoplasmic reticulum (*Section 10.9*). This is followed by a short intervening sequence (IVS 1) and then the sequence encoding amino acids 1 to 108 of the $V_L$ domain. There follows a very long intervening sequence (IVS 2) and then the $C_L$ domain, beginning with amino acid 109 of the κ chain. Thus the gene for the light-chain polypeptide has the canonical structure we have come to expect for a eukaryotic gene: three exons are interrupted by two introns (IVS 1 and 2), and these are spliced out during mRNA processing (*Section 9.10*) to yield the mature mRNA with the structure $S$-$V_L$-$C_L$. The 18-amino-acid signal sequence is

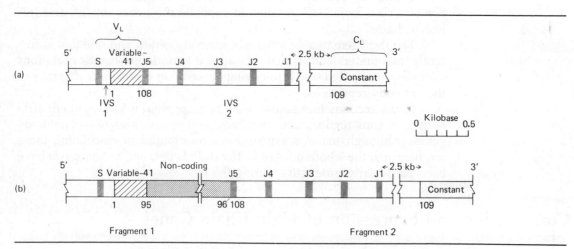

FIGURE 20-3 Somatic vs. germline light-chain sequences. MOPC-41 myeloma (a) and mouse embryo (b) restriction fragments homologous to $C_K$ and V-41 probes. Abbreviations are defined in the text. (From J. G. Seidman, E. E. Max, and P. Leder, *Nature* 280:370–375, 1979.)

cleaved from the polypeptide after translation begins, yielding the final $V_L C_L$ structure of the light chain.

The unusual features of immunoglobulin gene construction become apparent, therefore, only when this myeloma sequence is compared with the embryonic or **germline sequence,** diagrammed in Figure 20-3b. The first surprising finding is that when large germline restriction fragments carrying the $C_L$ DNA sequence (fragment 2 in Figure 20-3b) are probed for $V_L$ sequences, none are found. Continued searching reveals that DNA homologous to $V_L$ DNA probes is carried in a different restriction fragment, namely, fragment 1 of Figure 20-3. Thus **the $V_L$ sequences and the $C_L$ sequences are located in different restriction fragments in the germline genome,** whereas, as we saw, they are located in the same fragment in the lymphocyte genome.

The second surprising finding is that **the $V_L$ sequence itself is split between the two restriction fragments in germline DNA.** As drawn in Figure 20-3b, fragment 1 carries the S sequence and a sector, denoted *variable-41* or *V41*, whose DNA sequence encodes only the first 95 amino acids of the $V_L$ domain of the MOPC-41 light chain. The contiguous (stippled) DNA does *not* specify amino acids 96 to 108 and is therefore designated "noncoding." The sequence encoding these last 13 amino acids of the $V_L$ domain proves to reside in fragment 2, in the sector designated J5. The stippled DNA to the 5' side of J5, like the stippled DNA to the 3' side of V41, has nothing in common with the light-chain sequence and is also designated "noncoding." It appears, therefore, that at some point during differentiation of the MOPC-41 lymphocyte line, the variable-41 sequence became joined to the J5 sequence located far downstream, and the noncoding DNA between them was discarded.

Clues about how this **joining reaction** might occur were soon found in the "noncoding" DNA. Adjacent to the 3' end of the germline V41 sector is an **inverted repeat** sequence (*Section 4.9*), which is also found adjacent to the 5' end of the J5 sector. Leder and his associates have therefore postulated the occurrence of the reaction drawn in Figure 20-4. The two inverted repeats are thought to denature and then reanneal with one another, bringing V41 in direct proximity to J5 and causing all the DNA between them to loop out in a giant circle. The DNA in the resultant **stem** then becomes the substrate for recombination enzymes that break the V41 and J5 sequences and join them into a continuous strand. The DNA in the circle, meanwhile, is sacrificed during the recombination reaction and is discarded.

We should pause here to remind readers of Chapter 18 that there are obvious analogies between the DNA arrangements described here and those associated with transposable elements. Specifically, the "noncoding DNA" that separates V41 and J5 consists of a long sequence of DNA flanked at either end by two inverted repeats and thus is able to undergo a precise excision event. The same description applies to transposons, retroviruses, and eukaryotic elements such as the *copia* family in *Drosophila*. It is

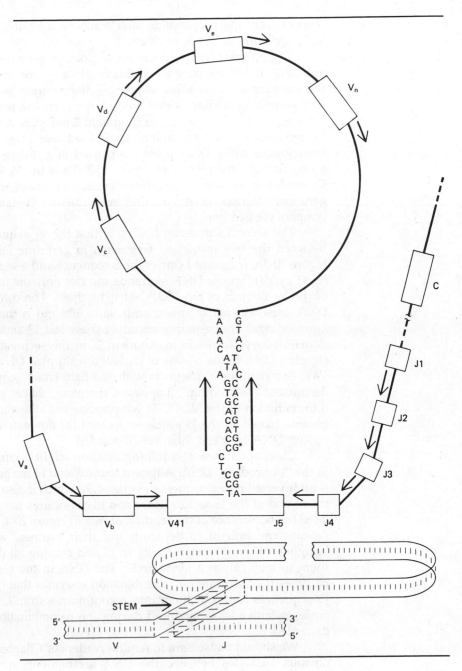

therefore exciting to speculate that the enzymes responsible for creating $V_L$ domains in the lymphocyte line are analogous to the enzymes responsible for the transposition of transposable elements. A search for the lymphocyte enzymes may be very difficult, however, for they probably operate in some ill-defined "pre-B" stem cells during a very brief period of neonatal development.

**FIGURE 20-4 V-J joining:** hypothetical stem structure formed between inverted repeats located next to germline V- and J-region genes. Above is a diagram of hypothetical intermediate in the chromosomal rearrangement of V-, J-, and C-region sequences. The distance between V- and J/C-region sequences is not known, and this is indicated by the broken line. Each V- and J-region gene is bordered by a palindrome that is also an inverted repeat sequence (boldface arrow) located on the 3' side of the V genes and on the 5' side of the J genes. Each of these sequences can be written as a complementary stem structure, as shown in the figure in which the actual sequence of MOPC-41 recombinant is used as an example. The bases marked with asterisks are those actually joined to form the recombinant. Below is a three-dimensional representation of the hypothetical stem intermediate. V- and J-coding sequences interact with their opposite strands in a normal DNA duplex, but the inverted repeat adjacent to the V gene is drawn as a stem interacting with its complement located on the same strand, presumably many thousands of bases away, adjacent to a J sequence. (From E. F. Max, J. G. Seidman, and P. Leder. *Proceedings of the National Academy of Science of United States.* **76**:3450–3454, 1979.)

## 20.8 Construction of $V_L$ Domains in Other Light Chains

The preceding example focused on the MOPC-41 immunoglobulin. Our knowledge of the immune system tells us, however, that each different lymphocyte clone should produce a different type of light chain than the MOPC-41 version. The source of some of this variation becomes apparent upon re-examination of Figures 20-3 and 20-4. Figure 20-3 shows that a total of **5 J sequences,** J1 to J5, lie in tandem array along the IVS 2 of MOPC-41 DNA, each separated by 300 bp. Figure 20-4 illustrates that the germline DNA contains not only these 5 J sequences but also **multiple copies of V sequences,** also in tandem array, one member of which is the V41 sector we have thus far been studying. When other myeloma lines are subjected to cloning/sequencing analyses and their V and J sectors are compared with their germline arrangements, the same story emerges in all cases: V and J regions are separated in the germline and recombined in the lymphocyte. **For each myeloma, however, a unique V-J combination has occurred.** Thus for MOPC-21, a V region specifying the first 96 amino acids of the MOPC-21 light chain is joined with a **J4** sequence specifying the last 12; for MOPC-149, a V149 sequence has joined with a **J2;** and so on. Each germline V sequence and each germline J sequence is flanked by the same inverted repeat, at its respective 5' and 3' ends, as the repeats flanking V41 and J5. Therefore, to explain one source of antibody diversity, it is only necessary to look at Figure 20-4 and imagine that in different lymphocyte lines the stem will bring together different V and J combinations. If, for example, J2 is "chosen," then the DNA encoding J5 to J3 will become part of the large circle and will be discarded along with the rest of the V regions; in this case, the length of the IVS 2 in the resultant gene (*Figure 20-3a*) will be shorter than in the MOPC-41 gene.

## 20.9  Creating Light-Chain Diversity During V-J Joining

The number of different light-chain genes that can be produced by different V-J combinations obviously depends on the number of V and J regions present in the genome. We have seen that five J regions are present, but it turns out that the J3 sequence is nonfunctional, for reasons detailed below, leaving four functional J sequences per haploid genome. The number of V regions is less certain. Estimates are often based on the number of light-chain idiotypes (*Section 20.4*) detected in a particular species, the notion being that each idiotype corresponds to a distinct V region that undergoes recruitment. If we take a typical value of 200 idiotypes, and therefore 200 V regions, a total of $200 \times 4 = 800$ different light chains can be created by making all functional combinations of V and J sequences.

A second, probable source of antibody diversity is detected when the recombination event itself is scrutinized more closely. Thus the crossover that joins VK41 with J5 occurs between the second and third nucleotides of codon 95 (CCG):

Recombined

| V41 | | J5 | | Pro | Trp |
|-----|---|-----|---|-----|-----|
| CC*T CCC | $\times$ | TGG* TGG | $\longrightarrow$ | CCG | TGG |
| | | | | 95 | 96 |

The starred nucleotides are those that join together to form the recombinant; these are also starred in Figure 20-4. **If the recombination frame is now shifted, a new amino acid will appear** at position 96. Thus we can imagine that in another lymphocyte line, recombination instead occurs between the first and second nucleotides of codon 96:

Recombined

| V41 | | J5 | Pro | Arg |
|-----|---|-----|-----|-----|
| CCT CCC | $\times$ | TGG TGG | CCT | CGG |
| | | | 95 | 96 |

This lymphocyte line will carry an arginine rather than a tryptophan at position 96. By making all possible shifts in the recombination frame between V41 and the four functional J sequences, it is possible to generate Trp, Arg, Pro, Tyr, His, Phe, and Leu at position 96, and all but His and Leu have in fact been found in the light chains that have been sequenced to date. Further diversity is possible if additional V sequences prove to carry terminal nucleotides other than CCT CCC.

That amino acid 96 of the light chain is particularly prone to variation has biological relevance, for this amino acid is believed to lie in or near the antigen binding site; it is also thought to be important for contacts between heavy and light chains. Therefore, a lymphocyte line producing an antibody with an arginine at position 96 might have very different antigen-recognition properties from the MOPC-41 line with a tryptophan at this position.

## 20.10 Creating Functional Light-Chain Genes During V-J Joining

If V-J joining indeed occurs as drawn in Figure 20-4, then it is not difficult to imagine that the joining reaction may not always work: **some V-J joining reactions may create "dead" genes.** We can consider two examples of such nonproductive arrangements.

The first, alluded to previously, occurs whenever a V segment joins with a J3 segment. Inspection of the sequence of J3 reveals that it ends with a C-T-3′, whereas the other four J sequences end with a G-T-3′. Looking again at Figure 20-3a, it is evident that the 3′ end of the recombinant J element must also serve as the 5′ end for the splicing reaction that removes the huge IVS2 from the gene transcript. As we learned in Section 9.10, the "consensus sequence" for intron splicing obeys the "GT-AG Rule": a GT must be present at the 5′ end of the splice site. Therefore, V-J3 recombinants, which lack the critical 5′-GT signal, would be expected to produce RNA transcripts that cannot be correctly processed and are therefore not translated into protein.

A second chance for error occurs during the splicing reaction itself. If, as proposed in the preceding section, there is indeed variability in the recombination frame, then some combinations are likely to create translational stop codons, other combinations may create GT splice junctions *within* the genes, and so on. In addition, recombination could erroneously occur at totally inappropriate positions.

These are not mere possibilities. Studies of several lymphocytic cell lines have demonstrated that such errors occur rather frequently. The mouse plasmacytoma cell line MOPC-173 can serve as an example. Cloning experiments analogous to those described for MOPC-41 have shown that *two* rearranged κ-containing light-chain genes are present in the MOPC-173 genomic DNA, even though the cells are known to synthesize only a single species of kappa polypeptide. Further study revealed that one of these genes, MOPC-173A, is the functional gene encoding the secreted antibody species, whereas the second gene, MOPC-173B, is transcribed into RNA but not translated into a functional protein. DNA sequencing of the nonexpressed MOPC-173B gene showed that it had undergone a defective V-J4 joining reaction: recombination had resulted in the deletion of a single nucleotide within codon 96, thereby throwing out of phase the reading frame for the rest of the gene and creating a UAA nonsense codon within the J4 sector. Since translation of the resultant mRNA would terminate before any of the C region was synthesized, the truncated polypeptide product would clearly be unable to function as an immunoglobulin chain.

## 20.11 Allelic and Isotypic Exclusion

The creation of dead genes during V-J joining is the apparent explanation for two longstanding observations made by immunogeneticists. The first, called **allelic exclusion,** refers to the fact that even though a diploid mammal carries two alleleic copies of the $C_\kappa$ sequence, a given lymphocyte line that synthesizes kappa-containing light chains will express only one of the two $C_\kappa$ alleles. Specifically, the locus for the human $\kappa$ sequence, known as *Inv*, exists in two allelic forms: *Inv-1* specifies a leucine at position 191, whereas *Inv-3* specifies a valine at position 191. When kappa-producing lymphocytes are cloned from *Inv-1/Inv-3* heterozygotes, roughly half the clones secrete kappa-191$^{\text{Leu}}$ chains, whereas the other half secrete kappa-191$^{\text{Val}}$ chains: **no clones express both types of allelic genes,** even though there is no obvious basis for postulating a "dominance" of one type over the other. It is as though expression of one allele automatically excludes expression of the other.

The second observation, called **isotypic exclusion,** refers to the fact that a given lymphocyte clone not only expresses only one light-chain allele but also synthesizes only either $\kappa$ or $\lambda$ light chains. You will recall from the start of the chapter that the $C_L$ domain of a light chain may be either of the $\kappa$ or $\lambda$ variety and that one third of all human lymphocytes secrete antibodies with $\lambda$ light chains rather than $\kappa$. The human genome carries at least two allelic copies of $\lambda$ sequences and, as noted earlier, two alleles of $\kappa$. Therefore, a lymphocyte appears to make two "choices": first, it makes the **isotypic choice** of whether to express $C_\kappa$ or $C_\lambda$; second, it makes the **allelic choice** of which $\kappa$ allele or $\lambda$ allele is expressed.

We can now examine the choice process in the context of the V-J joining reaction. Figure 20-5 and its legend summarize the experimental findings; we present here a somewhat anecdotal account of what is thought to occur. A neonatal pre-B cell, carrying two copies of $\kappa$ and two copies of $\lambda$, reaches the stage at which it is stimulated to create a functional light-chain gene from its embryonic DNA sequences. The cell therefore attempts to recombine a V region with a J region contiguous to one of its two $C_\kappa$ sequences. If the joining reaction is successful so that a functional light-chain gene is transcribed and translated, then some unknown "feedback" signal prevents the initiation of any further recombination reactions. In this case, **the remaining three $C_L$ regions in the genome remain in their embryonic configuration throughout the clonal life of the lymphocyte,** and the lymphocyte synthesizes a single type of $\kappa$ light chain. If, however, the joining reaction fails to produce a functional gene, as in the foregoing example with the MOPC-173B gene, then the lymphocyte does not receive the feedback signal, and it proceeds to "try again" with its other $\kappa$ allele. Should the second attempt also fail, it will make two more attempts, this time with its two $\lambda$ sequences. **Thus the cell has four chances to come up with a functional light-chain gene,** meaning that even if the V-J joining reaction is often imprecise, the chances of producing a "valid gene" are, on balance, very good. Since the creation of one valid gene excludes the creation of any other gene, both isotypic and allelic exclusions are explained.

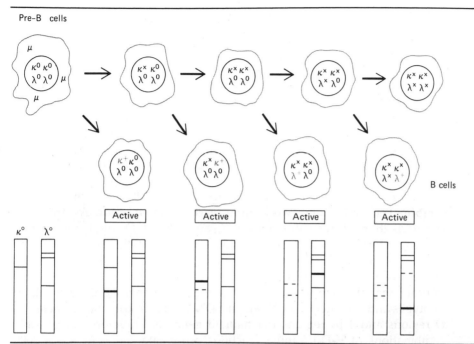

**FIGURE 20-5 Allelic and isotypic exclusion.** Developmental scheme of ordered light-chain gene rearrangements in pre-B cells. The genotype of each intermediate cell is indicated within the nucleus, and the predicted arrangement of *Bam*H1($\kappa$) or *Eco*R1($\lambda$) fragments is shown diagrammatically below each active cell. The overall scheme of lymphocyte development is not shown in order to focus on those stages concerned with light-chain gene activation. The initial cell in this scheme represents a pre-B cell already committed to the expression of cytoplasmic $\mu$ heavy chain. The expected germline configuration of the $\kappa$ ($\kappa^0$) and $\lambda$ ($\lambda^0$) light-chain gene fragments is shown schematically below. Light-chain gene rearrangements then take place (either productively or aberrantly) in a sequence such that rearrangements progress generally from the $\kappa$ to the $\lambda$ light-chain genes. Each rearrangement creates a new class of restriction fragment, detected for $\kappa$ in the *Bam*H1 digests and for $\lambda$ in the *Eco*R1 digests. Gene rearrangements cease to occur as a result of the formation of an active light-chain gene and synthesis of a viable light-chain product. The genotypes and representative light-chain gene fragment configurations for each of the four light-chain–producing cell types are shown. Bold-faced lines represent the active allele; a dotted line represents either a deleted or an aberrantly rearranged allele. Genotype superscripts signify the following: 0, germline configuration; +, viable V-J joining event; X, aberrant rearrangement or deletion. The scheme predicts a population of B-cell precursors with nonrearranged or only one aberrantly rearranged gene. It also predicts a population of cells incapable of forming valid light chains, having 'spent' their germline genes on invalid recombinations. (From P. A. Heiter, S. J. Korsmeyer, T. A. Waldmann, and P. Leder, *Nature* **290**:368–372, 1981.)

# Construction and Expression of Heavy-Chain Genes

## 20.12 Construction of $V_H$ Domains: V-D-J Joining

The creation of $V_H$ domains is sufficiently similar to the creation of $V_L$ domains that we need only present the process in a summary fashion. Returning to Figure 20-2a, we can first review the $V_L$ domain of the light chain. The V sequence is seen to denote information for frameworks 1 to 3 and for hypervariable regions 1 and 2. About half of hypervariable region 3 is also specified by the V sequence. The other half, plus all of framework 4, is supplied by the J sequence.

In the case of the $V_H$ domain, **three separate DNA regions must be combined,** meaning that **two joining reactions must occur.** As drawn in Figure 20-2b, the three subregions of $V_H$ are called V, D, and $J_H$; V extends to the terminus of F3 and specifies amino acids 1 to 101; D encodes a portion of HVR3, specifying amino acids 102 to 106; and $J_H$ completes HVR3 and encodes F4, which includes amino acids 107 to 123. Current estimates indicate that the mammalian genome expresses about 200 heavy-chain idiotypes and, therefore, 200 V regions; DNA hybridization has detected 10 D regions and 4 $J_H$ regions per haploid genome. Therefore, all possible combinations of V-D-J joining reactions can create $200 \times 10 \times 4 = 8000$ different $V_H$ genes. The V, D, and J sequences in the embryonic genome are flanked by inverted repeat sequences that are very similar to those surrounding their light-chain counterparts, and recombination mechanisms analogous to those suggested in Figure 20-4 have been proposed for the joining reactions.

If the light-chain and heavy-chain genes utilize similar joining mechanisms, you may wonder what prevents a light-chain V sequence from combining with a heavy-chain J sequence. The answer is not known, but it may reside in the fact that all the light-chain sequences are located on one chromosome (at least in mice), whereas all the heavy-chain sequences are located on a different chromosome. The recombination mechanism proposed in Figure 20-4 can work only within a continuous DNA duplex. Therefore, if such a mechanism is correct, interchromosome exchanges would be disallowed.

## 20.13 DNA Sequences Encoding the $C_H$ Domains of IgG

You will recall from Figures 20-1 and 20-2 that the $C_H$ region of an IgG heavy chain carries three homologous domains called $C_H1$, $C_H2$, and $C_H3$, plus a **hinge region** at the prong-handle junction. The DNA segment that specifies these portions of the polypeptide has been cloned and is found to have the following structure:

$$5'\text{-}C_H1\text{-IVS1-Hinge-IVS2-}C_H2\text{-IVS3-}C_H3\text{-untranslated trailer-}3'$$

Thus the four exons in the gene correspond to the four functional domains in the polypeptide, whereas the three intervening sequences separate these and are spliced out of the final product. It is pertinent to recall here our discussion of introns in Section 9.11. We noted that such a pattern has been detected in a number of other genes, leading to the generalization that **functional domains in a protein may often be specified by discrete exons in a split gene.**

Four distinct γ-chain subclasses are found in both mice and humans, suggesting the presence of four distinct γ-chain sequences per genome. Alleles exist at some of these loci, moreover, so that humans, for example, can produce two types of $γ_1$-chains, one carrying lysine at position 141 and the other carrying arginine. Mice heterozygous for alleles at various γ loci have been crossed to one another, and all four sequences are found to be inherited as a unit, with no recombinants found among over 2000 tested mouse progeny.

The close genetic linkage of the 4 γ subclass sequences has been confirmed by DNA cloning: contiguous restriction fragments of mouse germ-line DNA display the sequence

$$5'\text{-}C_{γ3}\text{-}C_{γ1}\text{-}C_{γ2b}\text{-}C_{γ2a}\text{-}3'.$$

Each $C_γ$ gene sequence is split into its four functional domains, as illustrated above, and each is separated from its neighbor by 15 to 34 kb of "noncoding" DNA, which may, of course, serve important regulatory functions. The $γ_{2a}$ and $γ_{2b}$ sequences are very similar and probably represent a recent duplication, but all four genes clearly constitute a tandem gene family analogous to the "β-like" globin genes (*Section 19.8*).

## 20.14 DNA Sequences Encoding the $C_H$ Domains of Other Immunoglobulins

That all $C_H$ sequences constitute a single giant tandem gene family was suggested by genetic crosses but established definitively by DNA cloning, where contiguous restriction fragments demonstrate the cluster to have the following order:

$$5'\text{-}C_μ\text{-}C_δ\text{-}C_{γ3}\text{-}C_{γ1}\text{-}C_{γ2b}\text{-}C_{γ2a}\text{-}C_ε\text{-}C_α\text{-}3'$$

Each $C_H$ sequence is subdivided into exonic functional domains in the same fashion as the γ-sequences, and the entire family is believed to have evolved from a single ancestral sequence that underwent tandem duplication.

## 20.15 Chain Alternations and Class Switching

The 4 $J_H$ sequences of the variable domain (*Section 20.12*) lie next to the 5' end of the $C_μ$ sequence. Therefore, **the consequence of the V-D-$J_H$ joining reaction is to bring a recombined $V_H$ region in proximity to a $C_μ$ se-**quence, just as V-J joining brings a recombinant $V_L$ sequence in proximity to a $C_\kappa$ or $C_\lambda$ sequence during the construction of light-chain genes (*Figure 20-3*). This means that **during lymphocyte differentiation, the first antibodies produced by a clone are always of the IgM class.**

The newly created $C_\mu$ gene first dictates the synthesis of IgM molecules that carry a hydrophobic sequence of 41 amino acids at their C-termini. These amino acids interact with the membrane bilayer of the lymphocyte surface so that these early lymphocytes carry a "display" of membane-bound IgM molecules that are called **IgM$_m$.** At a subsequent state in lymphocyte development, production of IgM$_m$ gives way to the synthesis of **IgM$_s$** molecules; these are identical to their IgM$_m$ predecessors except that their C-termini carry a 20-residue tail with very different, hydrophilic amino acids that allow the IgM to be secreted. Exhaustive probing indicates that only one $C_\mu$ sequence is present per haploid genome, meaning that the **IgM chain alternation** does not involve the successive activation of two distinct $C_\mu$ genes. Instead, the alternation appears to occur at the level of mRNA splicing: in the early lymphocyte, RNA sequences transcribed at the 3' end of the $C_\mu$ sequence are processed to yield the 41 amino acid coding sequence, whereas different splicing patterns produce an mRNA that dictates the shorter, more soluble C-terminus of the IgM$_s$ product.

As lymphocyte clones continue to develop, a second kind of differentiation occurs that is known as **class switching.** As noted earlier in the chapter, an IgM-secreting clone converts to secreting an IgG or an IgA species with the same $V_H$ domain as was present in the IgM product. Switches also occur to IgD and IgE production, but these mechanisms are less well understood at a genetic level and will not concern us here.

Figure 20-6 diagrams one model of how such class switches might

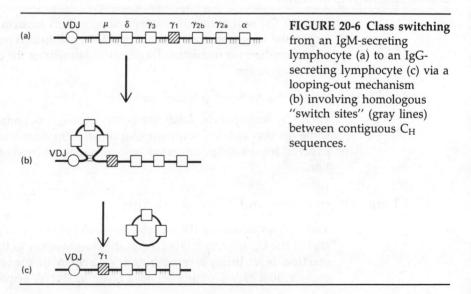

**FIGURE 20-6 Class switching** from an IgM-secreting lymphocyte (a) to an IgG-secreting lymphocyte (c) via a looping-out mechanism (b) involving homologous "switch sites" (gray lines) between contiguous $C_H$ sequences.

occur. In Figure 20-6a, the original IgM-secreting lymphocyte genome carries its newly created V-D-J region adjacent to the μ sequence, forming a functional μ-chain gene. Drawn between each of the contiguous $C_H$ regions are gray lines that represent **switching sites:** regions of homologous DNA that are likely to be involved in the switching process. In Figure 20-6b, two of these switching sites have interacted, causing a looping out of the DNA lying between them. Recombination excises the looped-out sequences, bringing the VDJ segment adjacent to the $\gamma_1$ sequence so that the lymphocyte now secretes $IgG_1$. Another lymphocyte might "choose" to join up with an α-sequence and thus generate a clone of IgA-secreting cells, and so on.

# Somatic Mutation of Antibody Genes, and a Summary of Antibody Diversity Mechanisms

## 20.16 Evidence for Somatic Mutation of Antibody Genes

The preceding sections have documented that the large amount of amino acid sequence diversity found in hypervariable region 3 of both light and heavy chains can be explained, at least in part, by the occurrence of different recombinational frames in this region of the gene. Hypervariable regions 1 and 2, however, are not sites of junction formation, and the ~ 400 different germline sequences for the V sectors of heavy and light chains are inadequate to account for all the HVR1 and 2 sequences found in the immunoglobulin repertoire. Moreover, antibodies of the same idiotype are often found to carry different amino acids in their framework sequences. The source of this additional diversity in antibody genes is believed to be **somatic mutation.**

Strong evidence for the occurrence of antibody gene mutations comes from studies by L. Hood and his associates. Hood chose to analyze the complete repertoire of antibodies elicited in mice by a small antigen called phosphorylcholine. A total of 19 independently arising immunoglobulins that recognize phosphorylcholine were isolated from various lymphocytic clones. These were first subjected to amino acid sequencing. A cDNA probe was then prepared that was complementary to the mRNA specifying the heavy chain of 1 of these 19 immunoglobulins, called T15, and homologous DNA sequences were located in mouse germline DNA. The mouse germline proved to have four V sequences related to the T15 sequence, but Hood was able to demonstrate that the T15-encoding V sequence itself had been used to create all 19 antiphosphorylcholine genes; the remaining three sequences were, for some reason, never "chosen" for V-J joining. Hood was further able to show that all 19 heavy-chain genes had been created using the $J_{H1}$ segment, whereas 3 to 6 different D segments were employed in the various combinatorial reactions. Therefore, he was able to account for amino acid sequence diversity in the V-D and the $D-J_H$ boundary regions by alternative recombination points and by the use of different D elements. Because the same V segment had been utilized to encode all 19 heavy chains, however, Hood could not use junctional mechanisms to explain amino acid differences occurring in the first 100 positions of the heavy chains. As shown in Figure 20-7, these amino acid differences concentrate in the HVR1 and 2 regions but occur as well in flanking framework regions. The only remaining alternative, therefore, is to postulate that they arise as the consequence of somatic mutation.

An interesting feature of the somatic mutation process becomes apparent when the 19 polypeptides depicted in Figure 20-7 are categorized as IgM, IgG, or IgA. It is seen that the first 100 amino acids of the 5 antiphosphorylcholine IgM species are all identical to one another, even though they derive from independent isolates. Four of the IgA antibodies are also identical to the IgM species. However, the remaining IgA polypeptides and all of the IgG polypeptides carry one or more amino acid substitutions. This result has led Hood to postulate that the newly created μ-encoding genes are not subject to somatic mutation—perhaps because the lympho-

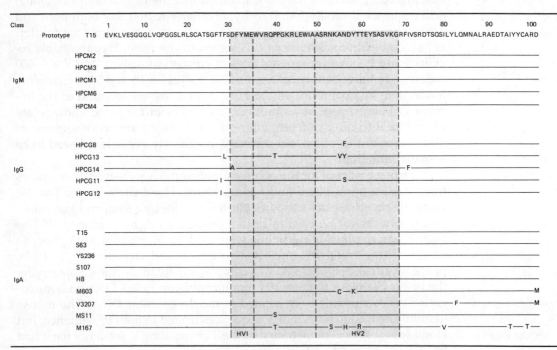

**FIGURE 20-7 Somatic mutation of antibody genes.** The $V_H$ portions of 19 heavy chains form antibodies that bind phosphorylcholine. Amino acids are depicted by a single-letter code; their absolute identification does not concern us here. The two hypervariable regions are demarcated by vertical dashed lines. (From S. Crews, J. Griffen, H. Huang, K. Calame, and L. Hood, *Cell* **25**:59–66, 1981.)

cytes are still "young" in clonal age, perhaps because they have not yet responded to some developmental signals—whereas **the genes that have undergone class switching are more prone to undergo somatic diversification.**

## 20.17 Summary of Antibody Diversity Mechanisms

We conclude by summarizing the six distinct ways that antibody genes and gene products can create the enormous diversity found in the mammalian immune system.

1. **Germline diversity.** Each haploid genome carries ~200 V sequences for light chains, ~ 200 V sequences for heavy chains, 4 $J_L$ segments, 4 $J_H$ segments, and at least 10 D segments.

2. **Combinatorial joining.** The joining of the separated germline sequences in all possible combinations is predicted to create ~800 different V-$J_L$ units and ~8000 different V-D-$J_H$ units.

3. **Junctional diversity.** Shifts in the frame of the combinatorial joining reaction are expected to create diverse amino acid codons at the sites of V-$J_L$ and V-D-$J_H$ recombination.

4. **Class switching.** Heavy-chain $V_H$ regions can switch from one $C_H$ class to another. Although this has no effect on the total number of antigen recognition sites, the different antibody classes elicit different types of immune responses; moreover, the more "mature" classes appear to be more prone to somatic mutation.

5. **Somatic Mutation.** V sequences that are identical in the germline come to carry mutations in the lymphocyte genome such that they specify $V_L$ and $V_H$ domains carrying different amino acid sequences and, hence, the potential for different antigen recognition properties.

6. **Combinatorial association.** This mechanism of antibody diversification, not specifically emphasized in preceding sections, refers to the fact that any set of light chains is presumably able to form a double-pronged antibody molecule with any set of heavy chains. Therefore, since a given lymphocyte clone creates its light-chain genes independently of its heavy-chain genes, 800L × 8000H = $6.4 \times 10^6$ different antibody proteins are possible by combinatorial joining alone. This number becomes much higher when junctional diversity and somatic mutation are factored in. Since antibody binding sites are created jointly by $V_L$ and $V_H$ domains (*Figure 20-1*), each of these proteins is likely to have distinct antigen recognition properties: some may well bind the same antigen, as with the 19 antiphosphorylcholine species, but each presumably does so in a slightly different fashion, recognizing different topological features of the antigen shape.

Therefore, the six means of creating antibody diversity that we have described can easily account for the $10^7$ to $10^8$ different antibody species produced by a mammal.

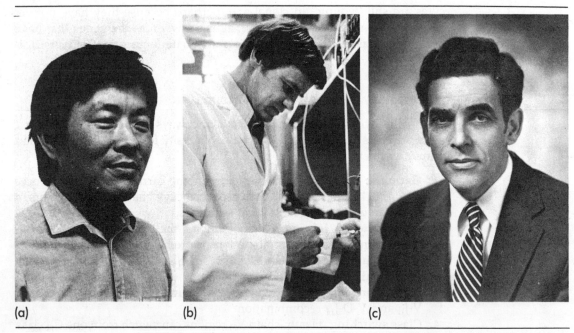

(a)                    (b)                                (c)

(a) Susumu Tonegawa (Massachusetts Institute of Technology), (b) Leroy Hood (California Institute of Technology), and (c) Philip Leder (Harvard University) have utilized recombinant DNA technology to make major contributions towards our current understanding of the immunoglobulin genes.

# Questions and Problems

1. The following questions are designed to help you review antibody structure.
   (a) How many domains are in an IgG molecule? What are their names?
   (b) How many antigen binding sites are in an IgG?
   (c) How many domains are there per antigen binding site? What are their names?
   (d) How many constant portions are there per IgG?
   (e) How many domains are there per constant portion? What are their names?
   (f) How many classes of human constant light-chain domains are there? What are their names?
   (g) How many classes of human constant heavy-chain domains are there? What are their names?
   (h) How many IgG subclasses are there? What are their names?
   (i) How many subregions of the variable domain of light chains are there? Of heavy chains? What are their names?
   (j) How many DNA sequences are brought together to create a light chain gene? A heavy chain gene? What are their names?
2. The following terms are encountered in considering antibody genes. Describe where each applies, and give an example where each term is encountered in some other genetic context.

(a) Tandem repeats

(b) Inverted repeats

(c) Stem-and-loop structures

(d) GT-AG Rule

(e) Frameshift mutation

(f) Somatic mutation

3. Define the following terms.

(a) Allelic exclusion

(b) Allotype

(c) Isotypic exclusion

(d) Hypervariable region

(e) IgM chain alternation

(f) Class switching

(g) Combinatorial joining

(h) Combinatorial association

4. Draw the chromosome sector containing $C_\gamma$ information, including in your diagram all subclasses, exons, introns, and "noncoding" DNA.

5. Redraw Figure 20-2, indicating where variation is thought to be produced primarily during joining and where it is thought to be produced primarily by somatic mutation.

6. Until recently, many immunogeneticists espoused the **germline theory of antibody diversity,** which proposed that each immunoglobulin species had a corresponding gene in the germline DNA. How many genes would this theory require? How could hybridization studies be used to test its validity?

7. Are the following statements true or false? Explain in each case.

(a) All immunoglobulin molecules in an individual have a common polypeptide chain (i.e., one with an identical amino acid sequence).

(b) All IgG immunoglobulin molecules in an individual have the same heavy chain and differ from each other only by the type of light chain they contain.

(c) Variable light regions ($V_L$) within a given immunoglobulin molecule associate with either a kappa or a lambda $C_L$ subgroup.

(d) An IgG immunoglobulin molecule consists of two polypeptide chains linked by a disulfide bridge.

(e) $\gamma_2$- and $\alpha_1$ $C_H$-chains can be found in an immunoglobulin molecule.

(f) A given $C_H$ gamma-chain is specified by multiple gene loci at which there are multiple alleles.

8. You prepare a radiolabeled cDNA probe from a κ light-chain mRNA isolated from a human plasmacytoma. You now prepare genomic DNA from the plasmacytoma line and from human fetal placenta tissue, subject the DNA to a restriction endonuclease that generates large fragments, prepare a Southern Blot (*Box 12.1*), and look for fragments that hybridize with your radiolabeled probe. The fetal DNA displays one radiolabeled band, whereas the plasmacytoma DNA has both the fetal band and an additional radiolabeled band. Explain this result.

9. Explain what is meant by the following statement: each functional J segment must contain a DNA-DNA splice signal and an RNA splice signal.

10. Draw a model wherein an IgM ⟶ IgG class switch is effected by unequal crossing over between sister chromatids during mitosis (*recall Figures 19-10 and 19-11*).

# 21

# Genes that Cooperate to Produce Complex Phenotypes and Quantitative Traits

**A. Introduction**

**B. Clustered Genes Specifying One Trait**
- 21.1 Operons
- 21.2 "Gene Clusters" in Fungi: The *HIS4* Locus
- 21.3 Complex Loci in *Drosophila:* The Rudimentary Locus
- 21.4 The Major Histocompatibility Loci

**C. Dispersed Genes Specifying One Trait**
- 21.5 Dispersed Genes Specifying One Enzyme: Human Lactate Dehydrogenase
- 21.6 Dispersed Genes Specifying a Multienzyme Complex

**D. Biochemical Genetics**
- 21.7 Biochemical Genetics of Haploid Organisms

- 21.8 Biochemical Genetics of Diploid Organisms: Modified Dihybrid Ratios and Epistasis
- 21.9 Genetic Dissection of Complex Pathways
- 21.10 Dispersed Genes Controlling Composite Traits

**E. Polygenes and Continuous Variation**
- 21.11 Examples of Polygenic Inheritance
- 21.12 General Principles of Polygenic Inheritance
- 21.13 Statistical Analysis of Polygenic Traits
- 21.14 Factors Contributing to Phenotypic Variance
- 21.15 Heritability

**F. Questions and Problems**

## Introduction

This chapter explores the properties of genes that are apparently not structurally related to one another—that is, are not alleles or members of gene families—but nonetheless cooperate to generate a particular phenotypic trait. In some cases such genes prove to be grouped together in one region of the chromosome, as with **operons** in bacteria, **gene clusters** in fungi, and

**642**

complex loci in *Drosophila*. In other cases the genes are not linked but nonetheless function in a cooperative, regulated way. And finally, there are numerous traits such as egg yield, height, and kernel size that are influenced by an unknown but sizeable number of genes that have not been individually characterized. The inheritance of such traits is the subject of a discipline known as **quantitative genetics,** to which we give an introduction at the end of this chapter.

# Clustered Genes Specifying One Trait

## 21.1 Operons

In bacteria and bacteriophages different genes concerned with the same trait are often (but by no means always) found clustered together in a group known as an **operon.** In the operon, the component genes are transcribed as a unit and are thereby under joint control. We have made reference to operons on several occasions throughout this text. In this section we examine operons from a functional point of view. Regulation of operon transcription is covered in Chapters 22 and 23.

The genes in a bacterial operon most frequently control sequential metabolic steps. A straightforward example of an operon of this type is the *gal* operon of *Escherichia coli*, which we have already encountered in Section 18.9. This is illustrated in Figure 18-10, where the sequence of genes in the operon directly mirrors the sequence followed by the gene products as they carry out the degradation of galactose into glucose-1-phosphate and UDP-glucose. A slightly more complex example is given by the tryptophan (*trp*) operon of *E. coli* (Figure 21-1), where the sequence of genes again reflects the sequence of reactions of the gene products, but where various gene products form enzyme complexes that act in concert; specifically, the *trpE* and *trpD* gene products form an enzyme complex, as do the *trpB* and *trpA* products. We should note that the organization of the *trp* operon in *Salmonella typhimurium* is identical to that shown for *E. coli*.

Exceptions to these simple types of operons are well known. In the *leu* operon of *S. typhimurium* (Figure 21-2), for example, the *leuA* gene is in phase with the reaction sequence of the gene products, whereas the *leuB* gene is out of order: its gene product, a dehydrogenase, acts *after* the combined gene products of *leuC* and *leuD* catalyze an isomerization reaction. Far greater disparity is found when the gene order of the *his* operon in *S. typhimurium* is compared with the sequence of reactions in the pathway of histidine biosynthesis. As seen in Figure 21-3, there is no apparent relation between the sequence of *his* genes in the operon and the order in which the gene products act.

A somewhat tangential, but nonetheless important, point about bacterial operons should be made at this point: **the clustering together of cer-**

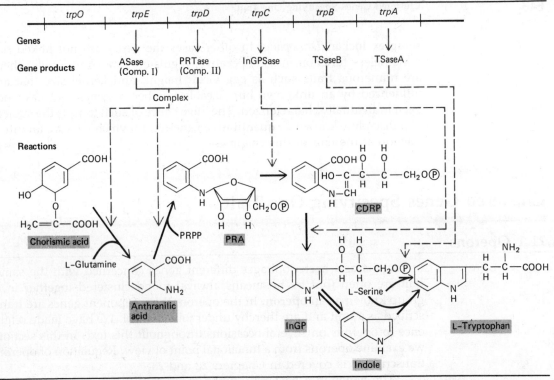

**FIGURE 21-1 The tryptophan (*trp*) operon of *E. coli*.** In *S. typhimurium*, the comparable genes are called trpA, B, E, D, and C. *Abbreviations:* O, operator; ASase, anthranilate synthetase; PRTase, phosphoribosyl transferase; InGPSase, indole glycerol phosphate synthetase; TSase, tryptophan synthetase; PRA, phosphoribosyl-anthranilic acid; CDRP, carboxyphenylamino-1-deoxyribulose-5-phosphate; InGP, indole-3-glycerol-phosphate. (From G. Wuesthoff and R. H. Bauerle, *J. Mol. Biol.* **49**:171, 1970.)

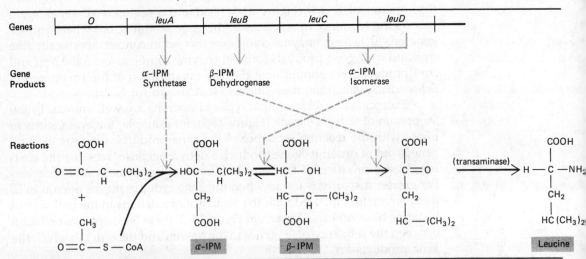

**FIGURE 21-2 The leucine (*leu*) operon** of *S. typhimurium. Abbreviations:* O, operator; α-IPM, α-isopropylmalonate; β-IPM, β-isopropylmalonate. (After J. Calvo and H. Worden, *Genetics* **64**:199, 1970.)

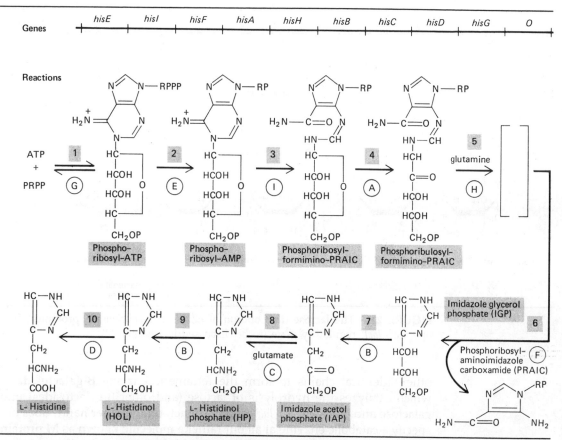

**FIGURE 21-3 The histidine (*his*) operon** of *S. typhimurium*. Circled letters in the biosynthetic pathway relate enzymes to the lettered *his* operon genes from which they derive; boxed numbers refer to the ten sequential steps in the pathway. The structure of the compound in brackets is uncertain. *Abbreviations:* R, ribosyl; P, phosphate; PRAIC, phosphoribosylaminoimidazole carboxamide. (From M. Brenner and B. N. Ames, in *Metabolic Regulation*, H. J. Vogel, Ed. New York: Academic, 1971.)

**tain functionally related genes in a bacterial operon does not preclude there being other related genetic loci elsewhere in the genome.** For example, in addition to the *his* operon, at least five other *his* loci that relate to histidine biosynthesis are found in the *S. typhimurium* genome. None of these appears to specify an enzyme involved in the major histidine biosynthetic pathway, but all relate to normal histidine production.

The *lac* operon of *E. coli* (Figure 21-4) represents a very different sort of operon from those we have been describing: among its three structural genes (*lacZ*, *lacY*, and *lacA*), only the *lacZ* gene specifies an enzyme known to be directly involved in a metabolic pathway. The *lacZ* gene product is a single, long polypeptide chain (134,000 daltons); this combines with three

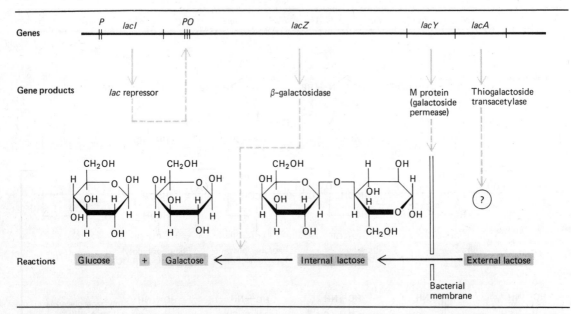

**FIGURE 21-4 The lactose (*lac*) operon** of *E. coli. Abbreviations: P,* promoter; *O,* operator; *lacI,* regulatory gene (to be discussed in Chapter 22).

other, identical chains to form the tetrameric enzyme **β-galactosidase,** which catalyzes the hydrolysis of lactose (and other β-galactosides) into galactose and glucose. The *lacY* gene product, on the other hand, does not specify a catabolic enzyme at all but rather a molecule known as **M protein** or **galactoside permease,** which is located in the bacterial cell membrane and acts to facilitate the specific uptake of lactose (and similar molecules) from the external medium. The last gene in the operon, *lacA,* specifies the enzyme **thiogalactoside transacetylase,** which can, *in vitro,* transfer the acetyl group from acetyl coenzyme A to isopropyl-D-thiogalactoside **(IPTG),** a synthetic galactoside derivative. The function of the enzyme *in vivo* is unknown, but since strains carrying deletions of the *lacA* gene behave normally in all respects, *lacA* appears not to specify an enzyme essential for lactose utilization. The *lac* operon, in summary, specifies three proteins that all interact with β-galactoside but do not do so in a series of related catabolic reactions.

This survey of representative operons has stressed their dissimilarities, but all operons appear to be eminently logical ways of organizing functionally related genes. The evolutionary origin of an operon is, however, difficult to visualize. If we assume that the different genes in ancient organisms arose through more or less random events, it is not readily apparent how several quite different genes concerned with the same trait came to find themselves next to one another in the same segment of a chromosome. One can only suppose that when such an arrangement did, perhaps by chance, come into being, it endowed an organism with such superior adaptive value that the arrangement was immediately conserved.

## 21.2 "Gene Clusters" in Fungi: The *HIS4* Locus

Fungi such as yeast and *Neurospora* possess a number of so-called **gene clusters,** each of which contains genetic information for several steps in a biosynthetic or degradative pathway. Table 21-1 summarizes the known gene clusters in yeast; in addition, M. Case and N. Giles have performed extensive genetic analyses of the *arom* gene cluster (encoding five enzymatic activities in the biosynthetic pathway of aromatic amino acids) and the *qa* gene cluster (encoding three enzymatic activities in the catabolism of aromatic amino acids) in *Neurospora*.

As a specific example of a gene cluster, we can consider the *HIS4* locus in yeast, studied extensively by G. Fink and his colleagues. This cluster specifies three of the ten enzymatic activities involved in histidine biosynthesis that are summarized in Figure 21-3. A combined genetic and complementation analysis of this locus defined three regions, *HIS4A, B,* and *C* (Figure 21-5) that behave as three distinct genes. Thus missense mutations defective in the same biochemical reaction were found by Fink and his coworkers to map to the same region of the locus: mutants defective in hydrolase activity (step 3 in the histidine biosynthetic pathway,

**TABLE 21-1   Gene Clusters in Yeast**

| Gene Cluster | Enzyme Activities Encoded |
|---|---|
| 1. *HIS4* | 1. Dehydrogenase<br>2. Cyclohydrolase<br>3. Pyrophosphorylase |
| 2. *ARO2* | 1. Dehydroquinase<br>2. DHQ synthetase<br>3. DHS reductase<br>4. Shikimic acid kinase<br>5. EPSP synthetase |
| 3. *ADE3* | 1. Tetrahydrofolate ligase<br>2. Methenyl tetrahydrofolate cyclohydrolase<br>3. Methenyl tetrahydrofolate reductase |
| 4. *TRP5* | 1. InGP $\longrightarrow$ Indole<br>2. Indole + serine $\longrightarrow$ tryptophan |
| 5. *FAS1* | 1. Keto reductase<br>2. Condensing enzyme<br>3. Acyl carrier protein |
| 6. *FAS2* | 1. Dehydratase<br>2. Enoyl reductase |
| 7. *URA3* | 1. Carbamyl phosphate synthetase<br>2. Aspartate transcarbamylase |

FIGURE 21-5 The
HIS4 locus of yeast
and the steps of
histidine
biosynthesis
controlled by each
region of the gene.

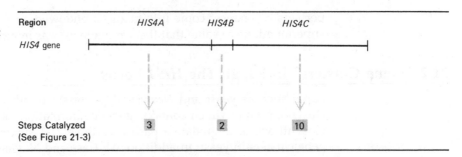

Figure 21-3) all map to *HIS4A;* mutants defective in a pyrophosphohydrolase activity (step 2) all map to *HIS4B;* and mutants defective in a dehydrogenase (step 10) all map to *HIS4C.* Moreover, missense mutations in any one region complement missense mutations in either of the other two regions.

Additional experiments in the Fink laboratory revealed that the *HIS4* locus functions as a single unit of transcription. Thus nonsense mutations (*Section 11.14*) in the *HIS4A* region were found to abolish *HIS4A, B,* and *C* activities; nonsense codons in *HIS4B* affected *HIS4B* and *HIS4C* expression, and so forth. Such polarity of a proximal nonsense mutation on the expression of distal genes is, of course, observed also for bacterial operons (*Sections 11.14 and 18.9*). In many respects, therefore, *HIS4* resembles an operon.

A major distinction exists between *HIS4* and a bacterial operon, however, for **the *HIS4* transcript proves to specify a single polypeptide chain,** with a molecular weight of 95,000, **that carries all three enzymatic activities.** In other words, *HIS4* is in fact a single gene, as we have defined a gene in Section 9.2, whereas a bacterial operon, although transcribed into a single mRNA, contains internal AUG "start" codons and is translated into distinct, separate polypeptide chains.

Are the other known "gene clusters" in fungi (Table 21-1) similar to the *HIS4* locus in being single genes specifying multifunctional polypeptides? Although the answer to this question is not yet known, the tendency of the enzymatic activities specified by fungal gene clusters to remain associated through extensive purification procedures argues that this may be the case. Indeed, Fink speculates that a major difference between prokaryotes and eukaryotes may be that eukaryotes are incapable of reinitiating translation in response to the sequence AUG . . . UAAAUG, perhaps because eukaryotic ribosomes actually dissociate from the message when they encounter UAA termination signals, and can "reenter" an mRNA only at the 5' end. Granted such a difference between the structure of bacterial operons and fungal "gene clusters," the fact remains that the fungi have often found it advantageous to encode several related enzymatic activities in a common transcriptional unit. Such an arrangement clearly allows the expression of these activities to be regulated as a unit.

In yeast, three steps in histidine biosynthesis are carried out by enzymatic activities that are specified by *HIS4*, whereas the remaining seven steps are specified by genes unlinked to *HIS4*. In bacteria, on the other hand, we have seen that all 10 *his*-related enzymes are encoded in a single giant operon under the control of a single operator region (Figure 21-3). We may well ask how and why these two divergent arrangements of similar genetic information have evolved in prokaryotes and eukaryotes.

## 21.3 Complex Loci in *Drosophila:* The Rudimentary Locus

The rudimentary (*r*) locus can serve to exemplify a number of so-called **complex loci** in *Drosophila* (other examples include *white, Notch, dumpy, miniature,* and *lozenge*), which appear, at many levels of analysis, to contain a number of genes concerned with a single trait. Mutations in the *r* locus all produce similar phenotypes, namely, truncated wings and female sterility, but a number of independent *r* mutations have been shown to fall into seven complementation groups, I to VII, based on their ability to yield normal wings when brought together in diploid flies. S. Nørby was able to show that larvae homozygous for a given *r* allele are in fact auxotrophic for pyrimidines or pyrimidine precursors. Since pyrimidines are normally supplied in the *Drosophila* diet, wing development and female fertility are usually the only traits affected in these homozygotes. When enzyme assays are performed, however, flies homozygous for *r* alleles in complementation group VI are found to lack carbamyl phosphate synthetase (CPSase), the first enzyme in the pyrimidine biosynthetic pathway (Figure 21-6); flies homozygous for *r* alleles in complementation groups I or III lack the second enzyme in the pathway (aspartate transcarbamylase or ACTase); and complementation class VII mutants lack the third enzyme in the sequence (dihydroorotase or DHOase). A complementation map of the *r* locus region can therefore be drawn as in Figure 21-7, where the boundaries of the

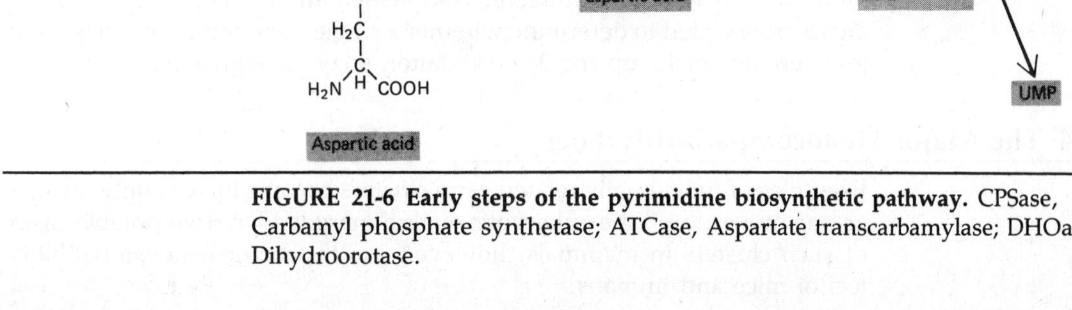

**FIGURE 21-6 Early steps of the pyrimidine biosynthetic pathway.** CPSase, Carbamyl phosphate synthetase; ATCase, Aspartate transcarbamylase; DHOase, Dihydroorotase.

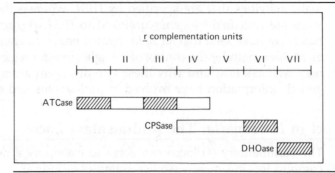

FIGURE 21-7 Complementation map of the *rudimentary* locus of
*D. melangaster*. Complementation units are designated as I to VII. Horizontal
bars indicate those units for which control of ATCase, CPSase, or DHOase
levels is demonstrated (cross-hatched) or suspected (open). (From J. M. Rawls
and J. W. Fristrom, *Nature* **255**:739, 1975.)

regions specifying ACTase and CPSase are seen to be as yet unclear. The
fourth enzyme of the pyrimidine biosynthetic pathway, dihydroorotate
dehydrogenase (DHOdehase), is unaffected by *r* mutations and its gene
presumably maps outside the *r* locus.

The complementation map of the *r* locus is in fact more complex than
that shown in Figure 21-7: a mutation in complementation group I, for
example, may be able to restore normal wing morphology when combined
with one group VI mutant but will do so only poorly when combined with
another group VI mutant. Such patterns allow two interpretations. CPSase
and ACTase activities are known to reside in large multimeric proteins with
a molecular weight of 340,000; therefore, one possibility is that CPSase
activity resides in one polypeptide and ACTase in a second polypeptide
and that final levels of enzyme activity are dependent, in part, on the way
the two polypeptides associate with one another. By this interpretation, a
given mutant group I polypeptide might associate in a stable fashion with
one mutant form of group VI polypeptide but poorly with a second mutant
form, and overall levels of enzymatic activity would fluctuate accordingly.

The alternate possibility, of course, is that the product of the *r* locus is
a single, multifunctional polypeptide—much like the yeast *HIS4* product—that associates with itself to form multimeric complexes and that exhibits various levels of intragenic complementation (*Section 12.5*). Studies
are clearly needed to determine whether a single class or multiple classes of
polypeptides make up the 340,000-dalton enzyme aggregate.

## 21.4 The Major Histocompatibility Loci

Examples of functionally related gene clusters become increasingly meager
as one moves "up" the eukaryotic evolutionary ladder. Two notable cases
of such clusters in mammals, however, are the major histocompatibility
loci of mice and humans.

In mammals, the fate of a tissue graft is determined by the **histocompatibility system** of the donor and host. Two organisms are said to be histocompatible if they can accept solid tissue transplants from each other and incompatible if they cannot. Compatible organisms have the same spectrum of **histocompatibility antigens** associated with their cell surfaces and therefore lack the same spectrum of antibodies (recall from Section 19.2 that antibodies are not made against "self" antigens). In addition to defining "self" versus "nonself," the histocompatibility antigens play a major role in eliciting immune responses from a class of cells known as T lymphocytes.

In all species that have been studied, an organism possesses a single **major histocompatibility locus** or **complex (MHC)** and multiple minor loci. (The difference between major and minor is operational: the major locus gives rise to antigens that elicit a more vigorous immune response in an incompatible host than the antigens specified by the minor loci.) Thus at least 30 loci have been identified in the mouse, including the major **H-2** locus, and many are present in humans in addition to the major **HLA** locus. Many of these loci are not linked, nor is there any reason to expect them to be linked, since the products they specify undoubtedly mediate quite different cellular functions.

Most intensively studied of the MHCs have been the mouse H-2 locus in chromosome 17 and the human HLA locus in chromosome 6. Both loci occupy more than one recombination unit (one cM) of the chromosome, which represents enough DNA to code for several thousand polypeptides. Whether or not the full coding potentials of these loci have been realized, there is no question that numerous genes map to each region.

A map of the mouse H-2 locus is shown in Figure 21-8a, and the human HLA locus is mapped in Figure 21-8b. Of the H-2 locus genes, those denoted K, D, and L are responsible for producing the major cell surface histocompatibility antigens; the corresponding human genes are called A, B, and C. Each set of three genes is closely related: amino acid analysis of their respective glycopolypeptide gene products has established extensive sequence homologies. Moreover, the mouse family is related in sequence to the human family. Finally, DNA cloning and sequencing has established the presence of multiple genes homologous to the major antigen genes, and although it is not yet clear how many of these are expressed and how many are "pseudogenes," it is clear that the loci bear many similarities to the hemoglobin gene families (*Section 19.8*).

The MHC loci differ from a typical gene family cluster in that **they also contain a number of genes that are unrelated in structure, but related in function, to the MHC genes.** As detailed in the legend to Figure 21-8, some of these genes specify proteins that affect the nature of lymphocyte interactions, others control features of the immune response, and others specify or control a number of the serum complement proteins that function in many immunologic reactions. Their presence in the MHC is presumed to reflect some functional requirement that the genes governing these related activities be inherited as a closely linked unit along with the

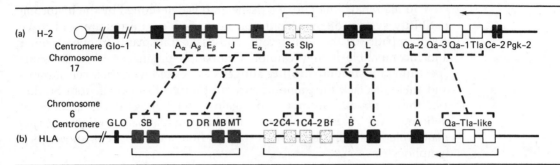

FIGURE 21-8 Major histocompatibility loci. A map of the mouse H-2 locus is drawn in (a); the human HLA locus is drawn in (b). The genes for the three major surface antigens (K, D, L and A, B, C) are depicted in black; they are separated in the mouse and contiguous in humans. Genes depicted in gray or open gray specify lymphocyte cell-membrane components that play key roles in the various functions carried out by B and T lymphocytes. Genes depicted in gray stipple specify components of the complement system. Genes depicted as open black squares specify additional transplantation antigens, similar in structure to the major antigens. Glo, Ce-2, and Pgk-2 code for various unrelated enzymes. Brackets and arrows indicate that the order of loci is not known. Dotted lines indicate presumed homologies between human and mouse loci. (From *Immunology Today* 2:192a, 1981.)

major histocompatibility genes. The MHC genes prove to be of additional interest from the standpoint of population genetics and will be considered in this context in Sections 26.6 and 27.2.

# Dispersed Genes Specifying One Trait

**The overwhelmingly common finding in the eukaryotes is that genes concerned with the same trait are dispersed throughout the genome.** Thus a glance at the linkage map of *C. reinhardi* (*Figure 14-15*) shows that genes involved with thiamin synthesis (*thi*) map to several different chromosomes, as do the *arg* genes involved with arginine synthesis. The same is found in yeast (*Figure 14-16*), *Neurospora* (*Figure 14-17*), *Drosophila* (*Figure 14-11*) and the mouse (*Figure 14-13*). Even in *E. coli* related genes are not necessarily clustered. The genes that have been identified for the different aminoacyl-tRNA synthetases in *E. coli*, for example, map to widely dispersed regions of the chromosome, as do the genes for arginine biosynthesis.

The dispersed loci that specify a common trait do so by dictating the structures of polypeptides that act in concert to create a certain phenotype. Various types of such interactions have been recognized, depending on whether the trait being monitored is dependent on a single protein, a biochemical or developmental sequence of reactions, or a composite of two, several, or many proteins. These possibilities are considered in the following sections.

## 21.5 Dispersed Genes Specifying One Enzyme: Human Lactate Dehydrogenase

Because proteins are often oligomeric, dispersed genes will often specify the same trait because they specify subunits of the same oligomeric protein. An example is given by the genes specifying human lactate dehydrogenase (LDH). LDH catalyzes the formation of lactic acid from pyruvate. When human cell extracts are subjected to electrophoresis and stained for lactate dehydrogenase activity, five isozymes (*Section 19.5*) are typically found, the relative proportions of each varying from one tissue to the next. Each isozyme proves to be a tetramer (composed of four polypeptide chains), and *in vitro* analysis reveals that two different kinds of polypeptides, known as A and B, give rise to the five isozymes. The two polypeptides do this in the following manner: LDH 1 is composed of four B chains (B4); LDH 2 contains one A and three B chains (AB$_3$); LDH 3 is A$_2$B$_2$; LDH 4 is A$_3$B$_1$; and LDH 5 is A$_4$. Somatic cell hybridization has demonstrated that the A polypeptide chain is specified by the *A* locus in human chromosome 11 (*Section 15.6*), whereas the B chain is coded by the *B* locus in human chromosome 12. Allelic forms of both genes are known, generating more complex isozyme patterns in heterozygotes, but most humans are homozygous at both loci.

## 21.6 Dispersed Genes Specifying a Multienzyme Complex

Multienzyme complexes in fungi can be specified by a single locus, as we saw for the *HIS4* product in yeast (*Figure 21-5*). Multienzyme complexes may also be composed of polypeptides from unlinked genes, and the interactions between them may become very complex, as exemplified by the *trp-1* and *trp-2* genes of *Neurospora crassa*. These genes are not linked and each controls the synthesis of a polypeptide chain. Four of the *trp-1* polypeptides interact with two of the *trp-2* polypeptides to form a hexameric protein that exercises three enzymatic activities in the pathway of tryptophan biosynthesis (*Figure 21-1*): anthranilate synthetase, phosphoribosylanthranilic acid (PRA) isomerase, and indole-3-glycerol-phosphate (InGP) synthetase activity. Mutations in the *trp-2* gene lead to the loss of anthranilate synthetase activity, and mutant *trp-2* cells produce a tetrameric protein that represents an aggregate of the *trp-1* product. Mutations in *trp-1* lead to a variety of phenotypes: various *trp-1* mutant strains exhibit alterations in any or all of the activities normally catalyzed by the enzyme complex. Such pleiotropy suggests not only that the *trp-1* polypeptides are essential to PRA isomerase and InGP synthetase activity but also that they create the appropriate macromolecular configuration for anthranilate synthetase activity, presumably by their interaction with the *trp*-2 polypeptides.

Some generalized principles should be stated at this point regarding the genetic analysis of multimeric proteins. Most eukaryotic proteins, and particularly enzymes, function *in vivo* as multimers that can become extremely complex, having many polypeptides, catalyzing several distinct reactions, and encoded by several genes. This is also the case for proteins that perform membrane-associated activities, as exemplified by the components of the mitochondrial electron transport chain (Section 16.2). Analysis of the genes that specify such multimeric protein polypeptides requires considerable care, for the kinds of pleiotropic effects described previously for the *trp*-1 mutations can give erroneous impressions regarding the function of the gene product being analyzed. It is essential for such studies that many mutant alleles of the gene in question be obtained and that their phenotypic effects be monitored both alone and in combination with other alleles.

# Biochemical Genetics

Mutations in genes that control sequential biological steps, be the genes clustered or dispersed, have proved to be invaluable for "dissecting" the number and order of reactions in a given metabolic pathway. In particular, a field of study known as **biochemical genetics** has combined genetics and biochemistry to elucidate the nature of metabolic pathways, most notably in haploid organisms whose growth requirements are known and whose gene expression is not complicated by allelic interactions.

The approach followed in biochemical genetics is to assemble a collection of mutant strains that cannot synthesize a particular metabolic component and thus require it for their growth. These strains are then subjected to complementation tests to estimate how many separate genes are involved in the synthesis of the component, and the genes are mapped to determine their linkage relationships. Strains in a given complementation group are then tested for their ability to grow when supplied with known metabolic precursors of the final component. The following general rule is then applied: if they **can** grow in the presence of a certain precursor, they must suffer from a genetic lesion affecting a step **before** the synthesis of that precursor; if instead they **cannot** grow when a certain precursor is supplied, the genetic lesion must affect a step that **follows** the synthesis of the precursor.

## 21.7 Biochemical Genetics of Haploid Organisms

As a first example of a biochemical genetic analysis, we can cite mutant strains of *E. coli* that require arginine for growth. Complementation analy-

**FIGURE 21-9
Pathway of
arginine
biosynthesis** and a
spectrum of mutant
strains of *E. coli*
blocked at the
indicated positions.

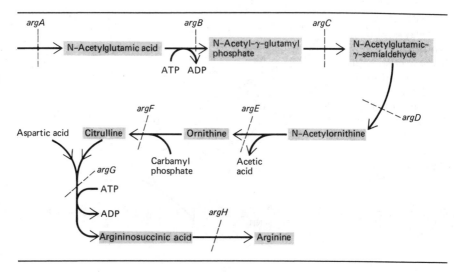

sis of partial diploids (*Section 13.15*) places these strains into eight complementation groups, and crosses show that the strains carry mutations in
eight genes, *argA* through *argH*. The *argH* strains can grow only if they are
given arginine; no other precursors will do. The *argG* strains, on the other
hand, will grow either in the presence of arginine or of argininosuccinic
acid. The *argF* strains can grow only if supplied with arginine, argininosuccinic acid, or citrulline, and so on. By comparing the growth requirements
of the different strains, it is possible to determine the sequence of precursors in the pathway leading to arginine biosynthesis. These results are
illustrated in Figure 21-9.

Biochemical genetics can also determine whether a biosynthetic pathway is linear (as is the arginine pathway in *E. coli*) or branched. For example, three unlinked complementary genes that we can call *thi*-1, *thi*-2, and
*thi*-3 are known to be involved in thiamin biosynthesis in *Neurospora*. The
growth requirements of strains carrying these mutations are summarized in
Table 21-2. When an attempt is made to order these precursors into a linear
pathway, the data argue otherwise: a thiazole $\longrightarrow$ pyrimidine $\longrightarrow$ thia-.

**TABLE 21-2   Growth Requirements of
*Neurospora* Strains Requiring Thiamin**

|  | Precursor | | |
| --- | --- | --- | --- |
| **Strain** | *Thiazole* | *Pyrimidine* | *Thiamin* |
| *thi*-1 | + | − | + |
| *thi*-2 | − | + | + |
| *thi*-3 | − | − | + |

A plus sign indicates growth.
A minus sign indicates no growth.

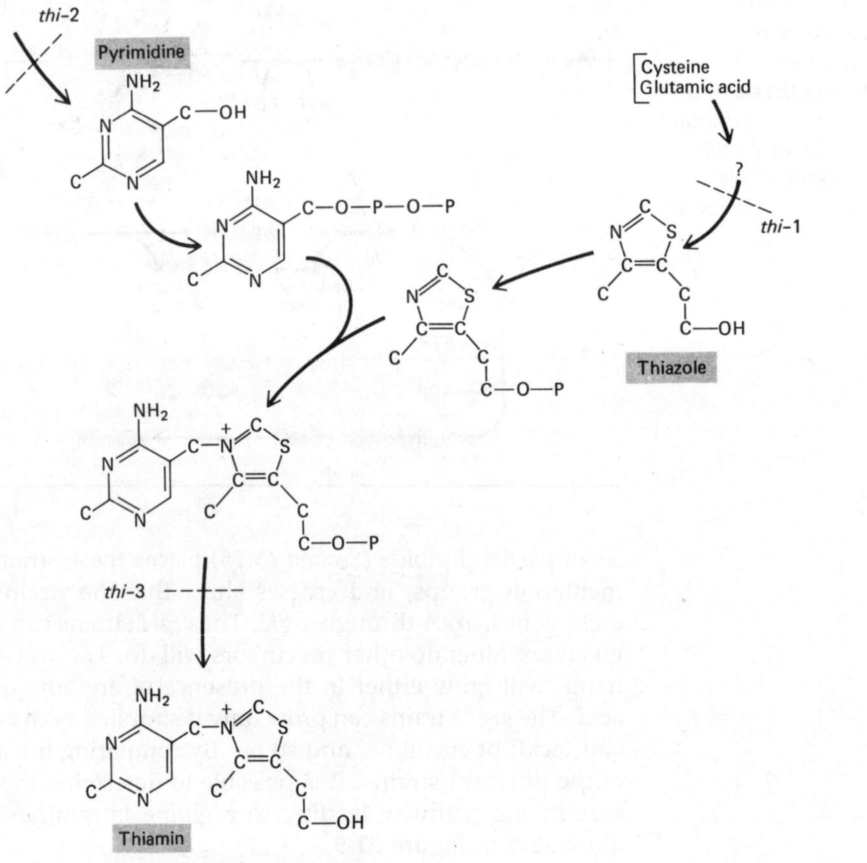

**FIGURE 21-10
Pathway of thiamin
synthesis** in
*Neurospora* showing
sites blocked in the
mutant strains *thi*-1,
*thi*-2, and *thi*-3.
(From R. P. Wagner
and H. K. Mitchell,
*Genetics and
Metabolism*, 2nd ed.
New York: John
Wiley, 1964.)

min sequence suggests that *thi*-1 should grow when supplemented with pyrimidines, which is not observed, and a pyrimidine ⟶ thiazole ⟶ thiamin sequence suggests that *thi*-2 should grow when supplemented with thiazole, which is also not observed. The correct pathway, therefore, must include both thiazole and pyrimidine as common precurosrs of thiamin synthesis, as drawn in Figure 21-10.

As a final example, we can consider the observation that single-gene mutations in *Neurospora* block the biosynthesis of two quite distinct amino acids—methionine and threonine. A study of the accumulated precursors of these two amino acids reveals that the mutations prevent the biosynthesis of homoserine from aspartic acid (Figure 21-11, site 1) and that homoserine serves as a precursor for the biosynthesis of both methionine and threonine along independent pathways. Other sites in which mutations are known in these pathways are also indicated in Figure 21-11.

**FIGURE 21-11 Biosynthesis of the amino acids methionine and threonine** in *Neurospora* showing reactions blocked in various mutant strains (① to ⑤). (From R. P. Wagner and H. K. Mitchell, *Genetics and Metabolism*, 2nd ed. New York: John Wiley, 1964.)

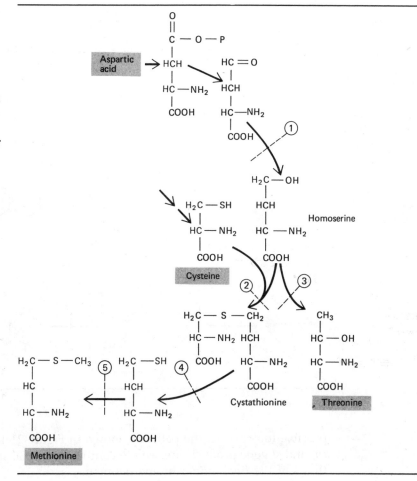

## 21.8 Biochemical Genetics of Diploid Organisms: Modified Dihybrid Ratios and Epistasis

**Eye Color in *Drosophila*** Biochemical genetic analysis in diploid organisms is best described by giving specific examples. We can first consider the genes participating in the synthesis of the brown ommochrome pigments that are prominent in the wild-type eye of *D. melanogaster*. Three loci, one sex linked (*vermilion, v*), one in chromosome II (*cinnabar, cn*), and one in chromosome III (*scarlet, st*), are implicated as being involved in ommochrome biosynthesis, since flies homozygous or hemizygous for mutations in *v, cn,* or *st* have red eyes and synthesize only the red pterin eye pigments. When the ommochrome pigment precursors accumulated by the homozygous mutant strains are isolated and analyzed, it becomes

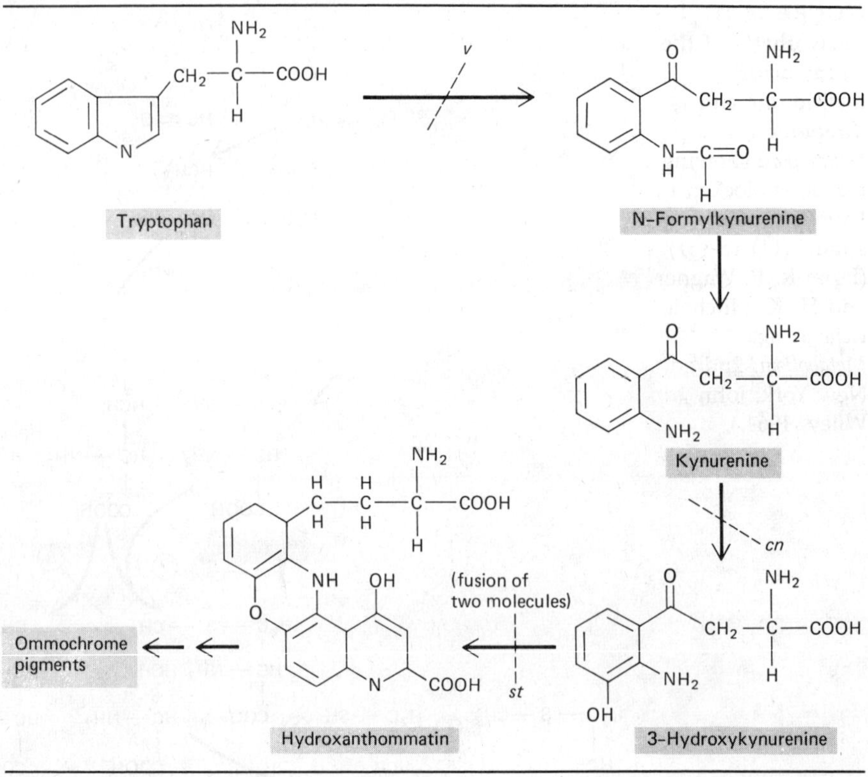

FIGURE 21-12
Synthesis of the
ommochrome
pigments in
*Drosophila*
*melanogaster*
showing sites
blocked in mutant
strains homozygous
for *vermilion* (*v*),
*cinnabar* (*cn*), and
*scarlet* (*st*).

possible to construct the pathway shown in Figure 21-12, in which the *v*, *cn*, and *st* gene products are seen to control sequential steps in the biosynthesis of the final ommochrome pigment.

The fact that the *v*, *cn*, and *st* loci are involved in specifying a common trait is also evident when the various eye color mutants are crossed to one another. If, for example, *scarlet* females are mated with *vermilion* males and the $F_1$ is inbred, then a 9:3:3:1 ratio of phenotypes (*Section 6.13*) is not observed among the $F_2$, even though the alleles can be shown to assort independently. Instead, as diagrammed in Figure 21-13, the $F_2$ progeny exhibit a **9:7 phenotypic ratio.** More generally, if a 9:3:3:1 is considered the "standard" obtained when **dihybrid** $\left( \dfrac{+}{a} \dfrac{+}{b} \right)$ organisms are inbred, then variations such as 9:7 are termed **modified dihybrid ratios,** and they signal that **the two loci are interacting to produce the trait.**

**Coat Color in Mice** A second and more complex example can be given in which the biochemical basis for the phenotype is not yet fully understood but in which interrelated pathways also appear to be involved. This concerns two independently assorting loci, *c* and *a*, that control coat color in

FIGURE 21-13
**Modified dihybrid
ratios.** Result of
inbreeding the $F_1$
obtained from a
*scarlet (st)* ×
*vermilion (v)* cross
of *D. melanogaster.*
The 16 genotypic
classes of $F_2$
progeny are shown
in the
checkerboard; the
seven shaded
classes will have a
mutant (*red eye*)
phenotype and the
nine unshaded
classes will have a
wild-type (*purple-
brown eye*)
phenotype.

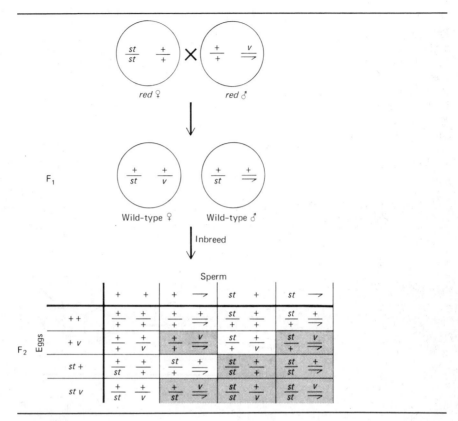

mice. Mice homozygous for the recessive gene *c* cannot synthesize pigment anywhere in their bodies and have white hair (*albino*). Mice that are
homozygous for *a* produce completely black hair. When *white* and *black*
mice are crossed $\dfrac{+}{+}\dfrac{c}{c} \times \dfrac{a}{a}\dfrac{+}{+}$, the $\dfrac{+}{a}\dfrac{+}{c}$ mice in the $F_1$ all have the
grayish coat known as *agouti.* Individual agouti hairs are black with a yel-
low band near the tip. When the $F_1$ mice are inbred, 9/16 of the $F_2$ are
*agouti,* 3/16 are *black,* and 4/16 are *white.* To explain this result, it is proposed that the *c* locus constitutes the structural gene for tyrosine oxidase,
an enzyme that acts early in the biosynthesis of melanin (Figure 21-14). The
*a* locus, we can then imagine, is involved with the placement of the melanin pigment in the hair. The melanin is dispersed throughout the hair
when the *a* gene alone is present, whereas its wild-type allele specifies the
unusual pigment arrangement of agouti hair. You should work this out for
yourself using the steps outlined in Figure 21-13 as a model.

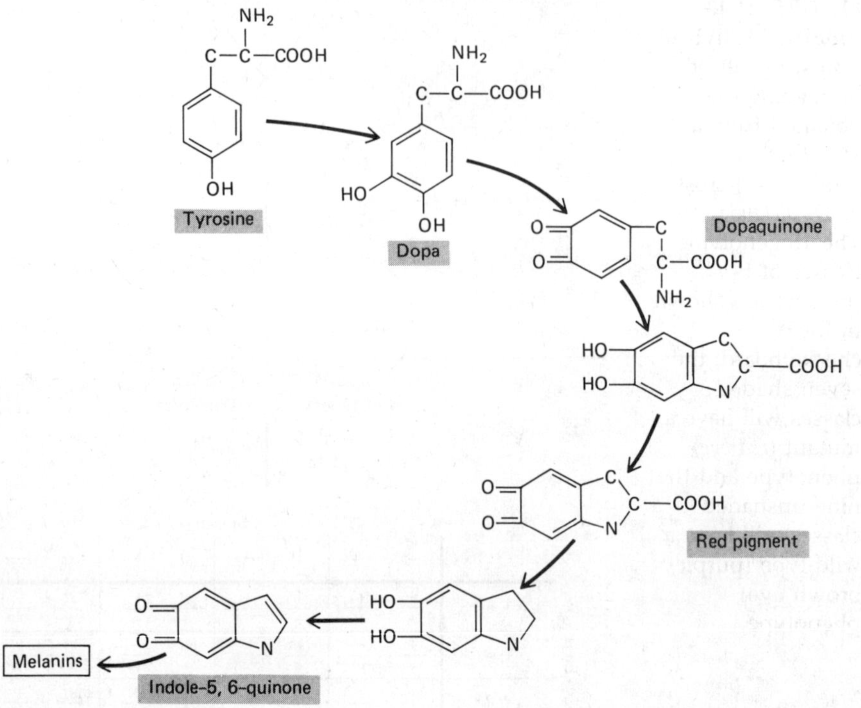

FIGURE 21-14
**Pathway of
melanin
biosynthesis** from
tyrosine. (From
R. P. Wagner and
H. K. Mitchell,
*Genetics and
Metabolism*, 2nd ed.
New York: John
Wiley, 1964.)

FIGURE 21-14 **Pathway of melanin biosynthesis** from tyrosine. (From R. P. Wagner and H. K. Mitchell, *Genetics and Metabolism*, 2nd ed. New York: John Wiley, 1964.)

**Epistasis**   A term that is frequently encountered when complex phenotypic traits are being analyzed is **epistasis,** which means literally "standing above." A gene or gene pair at one locus is said to be epistatic to a gene or gene pair at a second locus when **the gene product of the first locus, or the presence of nonfunctional genes at the first locus, masks or prevents the expression of the second.** Thus the homozygous *albino* gene pair in the mouse is said to be epistatic to the *agouti* and *black* genes. Additional examples of epistasis are given in the problems at the end of the chapter.

## 21.9 Genetic Dissection of Complex Pathways

The approach of biochemical genetics can be used to dissect very complex metabolic pathways. Perhaps the most elegant of these applications, initiated by R. Edgar and W. Wood, concerns morphogenesis of the phage T4, a process that involves the elaborate and sequential self-assembly of numerous different kinds of proteins.

   The analysis begins by isolating conditional lethal strains of T4 that carry *amber* or temperature-sensitive mutations in genes essential to the assembly of mature phages. Each strain is allowed to infect *E. coli* cells under nonpermissive conditions, whereupon phage assembly proceeds up to the point at which an essential protein is missing or inactive. Depending on the protein that is absent, different precusor phage structures will accumulate in the infected cells. If the cells are now lysed artificially, the precursor structures can often be isolated and observed with the electron microscope; they can also be subjected to *in vitro* complementation tests (*see Problem 12-25*) with precursor structures derived from other lysates. The complementation patterns of the various T4 strains can also be studied *in vivo*. In this way, more than 50 genes have been shown to be involved in T4 morphogenetic processes, either by directly specifying some structural protein of the phage coat or by specifying a protein required in smaller amounts for some delicate assembly process. The sequence of events and some of the relevant genes are shown in Figure 21-15. The mutant T4 strains have clearly permitted an understanding of phage assembly in exquisitely intricate detail.

In principle, the study of any morphogenic or biochemical sequence can be analyzed with the use of mutant strains, and many laboratories are currently isolating mutant strains blocked in pathways leading to the development of the brain or of particular behavior patterns. Thus, two single-gene learning mutants have been characterized in *Drosophila*: the *"dunce"* mutant is defective in learning, whereas the *"amnesiac"* mutant learns well but forgets abnormally rapidly.

In practice, as the trait to be analyzed becomes more complex, it becomes increasingly difficult to recognize and maintain mutant strains that are *directly* affected in the pathway of interest. Many strains that have, for example, abnormal brain development may prove to be suffering from a metabolic lesion that affects the brain only indirectly (for example, phenylketonuria); other mutations exhibit considerable pleiotropy. This does not mean that complex sequences will not eventually be understandable in terms of the genes that control them, but only that achievement of this goal appears distant at the present time.

## 21.10 Dispersed Genes Controlling Composite Traits

There are many examples in which the phenotype of interest in a diploid eukaryote is best categorized as a **composite trait.** The relevant gene products do not appear to interact to produce a single protein, nor do they appear to be acting in sequence in a biosynthetic pathway. Instead, **the gene products seem to interact by masking, modifying, or enhancing the effects of one another.**

In some cases the components of a composite trait can be distinguished; in such cases the genetic analysis becomes relatively straightforward. For example, the purple eye color of *D. melanogaster* is a composite of two classes of pigments—the brown ommochromes and the red pterins. We have already analyzed mutations such as *st* and *v*, which produce red-eyed flies when homozygous and affect the ommochrome pathway (*Figure 21-12*). A second class of eye color mutation in *D. melanogaster*, typified by

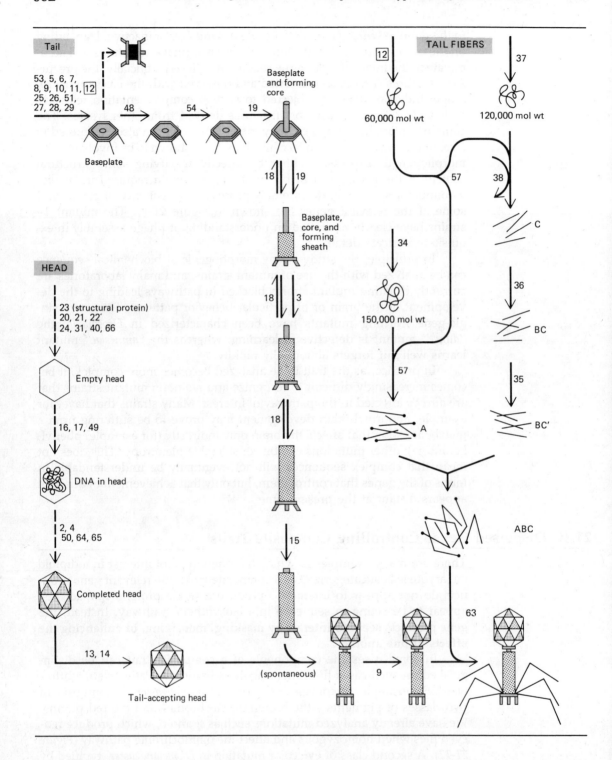

◀ **FIGURE 21-15 Morphogenetic pathway of phage T4** showing individual pathways for head, tail, and tail fiber assembly. Dashed arrows indicate steps that had not been demonstrated *in vitro* at the time this diagram was drawn. Number(s) associated with each arrow indicates the gene controlling that step or protein in the assembly process. Gene 12 is seen to participate both in the tail and in the tail fiber sequence. In the tail fiber sequence, molecular weights of the intermediates are given, and capital letters denote antigenic compositions. (From M. Levine, *Ann. Rev. Genetics* **3**:323, 1969; F. R. Frankel, H. L. Batcheler, and C. K. Clark, *J. Mol. Biol.* **62**:439, 1971; and J. King, *J. Mol. Biol.* **58**:693, 1971.)

*brown* (*bw*), gives rise to *brown*-eyed flies in homozygotes. To demonstrate that the synthesis of red pigments and brown pigments proceeds by independent, additive pathways, a cross can be made between *bw/bw* and *st/st* flies and the $F_1$ can be inbred. In the $F_2$, we find that 9/16 have normal purple eyes, 3/16 have *scarlet* eyes, 3/16 have *brown* eyes, and 1/16 have *white* eyes. In the last, doubly homozygous recessive organisms, neither pigment pathway is operative and neither trait is present. Otherwise, the phenotypic ratio is that expected for two kinds of gene products that do not interact, a statement you should verify for yourself with diagrams.

    Genetic analysis of a composite trait becomes increasingly difficult as the biochemical basis of the trait becomes less well known and as the number of loci controlling the trait becomes greater than two. The loci concerned with the red kernel color in wheat can serve as an example. In a study published in 1909, H. Nilsson-Ehle identified three kernel color loci, $r_1$, $r_2$, and $r_3$. A monohybrid cross ($R/r \times R/r$) for any gene pair results in the expected phenotype ratio of 3 *red*:1 *white* and a genotypic ratio of 1 *RR*:2 *Rr*:1 *rr*. Dihybrid crosses $\left(\text{for example, } \dfrac{R_1 R_2}{r_1 r_2} \times \dfrac{R_1 R_2}{r_1 r_2}\right)$ give an overall phenotypic ratio of 15 *red*:1 *white*, with only the doubly homozygous recessive kernels producing no pigment. Among the red progeny, however, a **discontinuous gradation** of intensity is observed between dark red and light red, with each category corresponding to a particular combination of *R* and *r* alleles. Thus five phenotypic classes can be identified, with the ratio 1 dark red:4 medium dark:6 medium:4 light red:1 white (Table 21-3).

    When trihybrid crosses are performed $\left(\dfrac{R_1}{r_1} \dfrac{R_2}{r_2} \dfrac{R_3}{r_3} \times \dfrac{R_1}{r_1} \dfrac{R_2}{r_2} \dfrac{R_3}{r_3}\right)$, 64 combinations are possible (Figure 21-16), again only one of which is *white*. The 63 *red* classes again show a gradation of intensity and can in this case be grouped into 7 phenotypic classes having a 1:6:15:20:15:6:1 ratio (Figure 21-16). Here, however, the discontinuities between one class and the next are less distinct, so that had one simply inbred two red strains and obtained the progeny shown in Figure 21-16, it would be no mean task to classify them into their phenotypic groupings and conclude that three independently assorting pairs of alleles were involved in determining red

**TABLE 21-3**   Dihybrid Cross for a Composite Trait:

$$\frac{R_1}{r_1}\frac{R_2}{r_2} \times \frac{R_1}{r_1}\frac{R_2}{r_2} \text{ in Wheat}$$

| | | Gametes | | |
|---|---|---|---|---|
| | | $R_1R_2$ | $r_1R_2$ | $R_1r_2$ | $r_1r_2$ |

| Gametes | | $R_1R_2$ | $r_1R_2$ | $R_1r_2$ | $r_1r_2$ |
|---|---|---|---|---|---|
| | $R_1R_2$ | **Dark** $\frac{R_1\;\;R_2}{R_1\;\;R_2}$ | **Medium Dark** $\frac{R_1\;\;R_2}{r_1\;\;R_2}$ | **Medium Dark** $\frac{R_1\;\;R_2}{R_1\;\;r_2}$ | **Medium** $\frac{R_1\;\;R_2}{r_1\;\;r_2}$ |
| | $r_1R_2$ | **Medium Dark** $\frac{R_1\;\;R_2}{r_1\;\;R_2}$ | **Medium** $\frac{r_1\;\;R_2}{r_1\;\;R_2}$ | **Medium** $\frac{R_1\;\;R_2}{r_1\;\;r_2}$ | **Light** $\frac{r_1\;\;r_2}{r_1\;\;R_2}$ |
| | $R_1r_2$ | **Medium Dark** $\frac{R_1\;\;R_2}{R_1\;\;r_2}$ | **Medium** $\frac{R_1\;\;R_2}{r_1\;\;r_2}$ | **Medium** $\frac{R_1\;\;r_2}{R_1\;\;r_2}$ | **Light** $\frac{R_1\;\;r_2}{r_1\;\;r_2}$ |
| | $r_1r_2$ | **Medium** $\frac{R_1\;\;R_2}{r_1\;\;r_2}$ | **Light** $\frac{r_1\;\;R_2}{r_1\;\;r_2}$ | **Light** $\frac{r_1\;\;r_2}{R_1\;\;r_2}$ | **White** $\frac{r_1\;\;r_2}{r_1\;\;r_2}$ |

color. Instead, it would appear that a **continuous variation** in red coloring was present among the progeny. If four or five loci were each contributing to kernel color, then phenotypic classes would be even less distinct, and the impression of continuous variation even more pronounced.

# Polygenes and Continuous Variation

When several gene loci contribute to the expression of a trait, as in the foregoing example of kernel color in wheat, the trait is said to be **polygenic.** The term **quantitative trait** is often used as well, particularly when such parameters as height, weight, or length are being analyzed, and the study of such traits is usually called **quantitative genetics.**

·   We will begin by considering three additional examples of polygenic inheritance and will then formulate some general principles and develop some straightforward mathematics to describe what occurs.

## 21.11 Examples of Polygenic Inheritance

**Ear Length in Corn**   In 1913, R. Emerson and E. East published the results of experiments on the inheritance of ear length in corn. They developed

**FIGURE 21-16 Trihybrid Cross for a Composite Trait** (the Nilsson-Ehle experiment). Kernel color in wheat is determined by three loci marked by gene pairs $R_1/r_1, R_2/r_2,$ and $R_3/r_3$. Homozygous parental lines are crossed, and the $F_1$ is self-crossed to give the $F_2$ distribution of color shown in the histogram. (Reprinted with permission of Macmillan Publishing Co., Inc., from *Genetics* by M. Strickberger. Copyright © 1968 by M. W. Strickberger.)

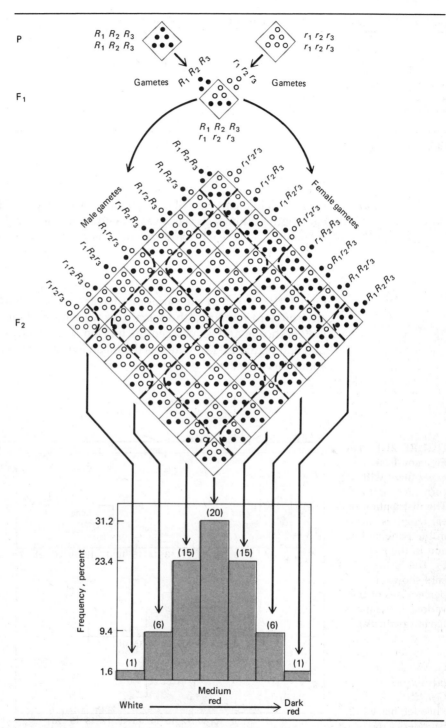

**TABLE 21-4 Frequency Distributions of Ear Length in Maize**

| | Length of Ear, cm | | | | | | | | | | | | | | | | | N | $\bar{x}$ | $\sigma$ | $\sigma^2$ | S.E. |
|---|---|---|---|---|---|---|---|---|---|---|---|---|---|---|---|---|---|---|---|---|---|---|
| | 5 | 6 | 7 | 8 | 9 | 10 | 11 | 12 | 13 | 14 | 15 | 16 | 17 | 18 | 19 | 20 | 21 | | | | | |
| Short Parent | 4 | 21 | 24 | 8 | | | | | | | | | | | | | | 57 | 6.632 | 0.816 | 0.666 | 0.108 |
| Long Parent | | | | | | | | | 3 | 11 | 12 | 15 | 26 | 15 | 10 | 7 | 2 | 101 | 16.802 | 1.887 | 3.561 | 0.188 |
| F$_1$ (60 × 54) | | | | 1 | 12 | 12 | 14 | 17 | 9 | 4 | | | | | | | | 69 | 12.116 | 1.519 | 2.307 | 0.183 |
| F$_2$ | | | 1 | 10 | 19 | 26 | 47 | 73 | 68 | 68 | 39 | 25 | 15 | 9 | 1 | | | 401 | 12.888 | 2.252 | 5.07 | 0.112 |

After R. Emerson and E. East, *Nebraska Research Bull.* **2**, 1913.

two strains of maize, one giving rise to long (13 to 21 cm) ears and the other to short (5 to 8 cm) ears of corn (Table 21-4 and Figure 21-17). The range in ear length within each class was judged to be due largely to environmental influences, since each parental line was highly inbred and therefore presumably homozygous. When the two strains were crossed, the mean F$_1$ ear length was intermediate between the two parents, with a similar range on either side of the mean (Table 21-4 and Figure 21-17). When the F$_1$ was inbred, **the spread of the variation increased significantly even though the environment remained the same:** ears were found that were several centimeters shorter and several centimeters longer than any of the F$_1$ ears (Table 21-4 and Figure 21-17), whereas the mean length for the F$_1$ and F$_2$ remained about the same.

**FIGURE 21-17 The Emerson-East experiment with corn ear length.** The distribution of ear length is shown in the parental lines and in the F$_1$ and F$_2$. The histogram bars show percentages of the various generations having particular ear lengths. (From A. Sturtevant and G. Beadle, *An Introduction to Genetics.* Philadelphia: W. B. Saunders Co., 1940.)

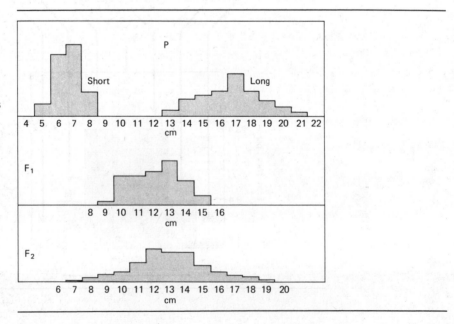

**Flower Length in Tobacco**    A second study by E. East, published in 1916, illustrates the same patterns of inheritance. Again, as illustrated in Figure 21-18, the two inbred parental strains had very different mean flower lengths, and each displayed some variation in length that was attributed to environmental vagaries. Again, the $F_1$ hybrid between the two parents was intermediate in average length and showed a similar "spread" on either

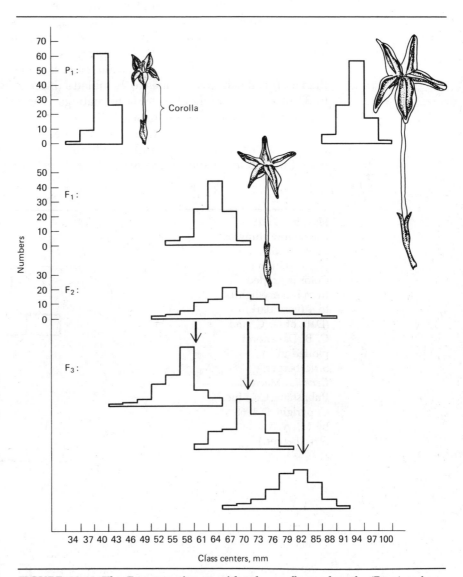

**FIGURE 21-18 The East experiment with tobacco flower length.** (Reprinted with permission of Macmillan Publishing Co., Inc., from *Genetics*, by M. Strickberger. Copyright © 1968 by M. W. Strickberger.)

side of this mean value. And again, the $F_2$ hybrid was similar to the $F_1$ in its mean flower length but showed an increased spread in variability.

In this experiment, East went on to breed three sets of individuals from the $F_2$, producing the three sets of $F_3$ plants shown in Figure 21-18. Here it is clear that the short $F_2$ plants give rise to progeny with distinctly shorter mean flower lengths than do the long $F_2$ plants.

**Skin Color in Humans**    A familiar example of a polygenic trait concerns skin pigmentation in humans. The trait is determined by gene loci that govern the distribution and pigment production of melanocytes (pigmented cells beneath the skin). In an early study, G. and C. Davenport classified skin color on a scale of 0 (white parents) to 4 (black parents). They then assigned skin color values to 29 mulatto (black × white) persons and to 32 progeny of mulatto × mulatto matings. The results are plotted in Figure 21-19. Although these data are different from those on plants in that the individuals in each generation are not necessarily related, the same patterns obtain. More recent studies indicate that the Davenports oversimplified the situation and that human skin color in fact varies continuously

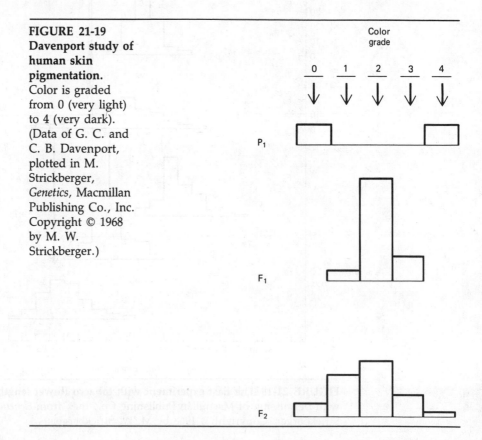

**FIGURE 21-19 Davenport study of human skin pigmentation.** Color is graded from 0 (very light) to 4 (very dark). (Data of G. C. and C. B. Davenport, plotted in M. Strickberger, *Genetics*, Macmillan Publishing Co., Inc. Copyright © 1968 by M. W. Strickberger.)

rather than in five incremental jumps, but the Davenport experiment stands as one of the first inquiries into human polygenic inheritance.

## 21.12 General Principles of Polygenic Inheritance

From our four examples of polygenic traits, we can formulate some general principles.

1. **A polygenic trait typically shows a difference in degree** among related individuals. Thus some kernels are redder than others, some skin is darker than others, and so on.

2. **The greater the number of loci determining the trait, the more continuous the variation.** Thus phenotypic differences between kernel colors were much sharper when two loci were segregating than when three were segregating.

3. **As the number of loci determining a trait increases, the proportion of extreme phenotypes among the progeny decreases** precipitously. Thus in the Nilsson-Ehle experiment with wheat, 1/4 of the $F_2$ were white and 1/4 were dark red in the monohybrid cross; in the dihybrid cross, 1/16 were white and 1/16 were dark red; in the trihybrid cross, only 1/64 were white and 1/64 were dark red. Stated generally, when $F_1$ individuals are heterozygous at $n$ loci that govern a polygenic trait, then $(1/4)^n$ of the $F_2$ offspring will have the phenotype (and genotype) of one of the original homozygous parents. As the value of $n$ increases, this number becomes vanishingly small, as does the number of some of the less extreme phenotypes in the $F_2$.

   An example of the use of this principle can be given for the East experiments with flower length in tobacco. East examined 444 $F_2$ plants and found that none achieved the extremes in flower length exhibited by either parent (Figure 21-18). If three loci were governing the trait, one would have predicted that $(1/4)^3$, or 1 of every 64, $F_2$ plants would be as extreme as one of the parents. Were four loci involved, then $(1/4)^4$, or 1 of every 256, plants should display an extreme flower length. Since none of 444 plants were extreme, the trait is estimated to be controlled by more than four gene loci.

4. **When highly inbred lines continue to exhibit variation in a polygenic trait, the variation is attributed to environmental rather than genetic causes.** This rule-of-thumb applies both to inbred parental lines and to their $F_1$ offspring, which should be genetically identical to one another and uniformly heterozygous for the input parental polygenes.

5. **An $F_1$ and an $F_2$ population usually display the same average value for a polygenic trait,** a value that is usually intermediate between the two parents. Thus the mean ear length in the Emerson-East experiment is about 12 cm for both the $F_1$ and $F_2$ (*Figure 21-16*), an intermediate value between the average of 6.6 cm for one parent and 17 cm for the other.

6. **The range of phenotypes in the $F_2$ is usually greater than in the $F_1$.** This is due to the fact that all possible combinations of alleles are likely to turn up in the $F_2$, whereas a much smaller group of combinations is possible in the homogeneous $F_1$. The limit, of course, is set by the phenotypic extremes of the two parents. Thus all possible combinations of the alleles determining wheat kernel color can never create parents that are more extreme than dark red $\left(\dfrac{R_1}{R_1} \dfrac{R_2}{R_2} \dfrac{R_3}{R_3}\right)$ and white $\left(\dfrac{r_1}{r_1} \dfrac{r_2}{r_2} \dfrac{r_3}{r_3}\right)$.

## 21.13 Statistical Analysis of Polygenic Traits

The distribution of individuals displaying various degrees of a polygenic trait can usually be approximated by a **bell-shaped curve.** Such a curve is drawn in Figure 21-20. If you re-examine the $F_2$ phenotypes plotted for wheat kernel color (*Figure 21-16*), corn ear length (*Figure 21-17*), tobacco flower length (*Figure 21-18*), and human skin color (*Figure 21-19*), you will see that they indeed follow a bell-shaped distribution, with the largest proportion of individuals displaying the average or mean phenotype and the smallest proportion of individuals at either extreme.

A symmetrical distribution of this type is called a **normal distribution.** As you may well have learned at some previous time, two parameters define a normal distribution, the mean and the standard deviation. The **mean,** expressed as $\bar{x}$ or $\mu$, is obtained simply by summing all the values observed ($\Sigma x$) and dividing by the number of values ($\Sigma x/n$). Thus in the Emerson-East experiment (Table 21-4), the means were 6.6 and 16.8 cm for the short- and long-eared parents, respectively, 12.1 cm for the $F_1$, and 12.9 cm for the $F_2$. The **standard deviation,** which expresses the amount of variability in the sample, is calculated as follows:

**FIGURE 21-20 A normal distribution,** showing standard deviations ($\sigma$) from the mean ($\mu$).

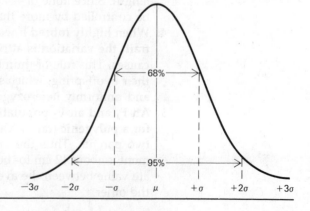

1. Subtract the mean ($\bar{x}$) from each individual measurement ($x$) to obtain a deviation ($\bar{x} - x$).
2. Square each deviation ($\bar{x} - x$)$^2$.
3. Sum all the squared deviations ($\Sigma(\bar{x} - x)^2$) and divide the sum by $N - 1$, where $N$ is the sample size.
4. The square root of this value yields the standard deviation $\sigma$:

$$\sigma = \sqrt{\frac{\Sigma(\bar{x} - x)^2}{N - 1}}$$

For every normal distribution, approximately 2/3 of the measurements (68 percent) lie within one standard deviation from the mean ($\bar{x} \pm \sigma$), whereas approximately 19/20 of the measurements (95 percent) lie within two standard deviations ($\bar{x} \pm 2\sigma$), and more than 99 percent are within three standard deviations ($\bar{x} \pm 3\sigma$). As drawn in Figure 21-20, the position of the first standard deviation can be identified as the point where the curve first begins to inflect outward.

Quantitative traits are usually described by a parameter called **variance,** which is simply the square of the standard deviation ($\sigma$). Therefore,

$$\sigma^2 = \Sigma \frac{(\bar{x} - x)^2}{N - 1}$$

Variance describes the way that values are distributed about the mean: the larger the variance, the greater the spread. Thus in the Emerson-East experiment (Table 21-4), the variance in the $F_2$ is greater than in the $F_1$, even though both generations exhibit about the same mean value for ear length.

## 21.14 Factors Contributing to Phenotypic Variance

We noted in our description of the Emerson-East experiment that there was a considerable range of sizes (in fact, a normal distribution of sizes) in the initial long- and short-ear corn samples, even though each sample derived from highly inbred lines. Most of this variability was attributed to environmental influences, an attribute that can be expressed more quantitatively by the following expression:

$$V_T = V_G + V_E + V_{GE} + 2\,C_{GE}$$

$V_T$ = **Total Phenotypic Variance**   In the Emerson-East experiment (Table 21-4), for example, $V_T = \sigma^2$ for each parental and offspring sample.

$V_G$ = **Genetic Variance**   This variance would be equal to zero if the individuals in the sample were identical in genotype, as in a sample of monozygotic twins. It describes the genetic contribution of all loci that contribute to the trait in an additive manner (such as the red kernel genes in wheat), although it ignores genes that exhibit dominance or loci that are epistatic to one another.

$V_E$ = **Environmental Variance**  This variance would be equal to zero if the individuals in the sample were all reared in an identical environment.

$V_{GE}$ = **Variance Due to Genetic and Environmental Interactions**  If two genotypes are raised in the same environment and one genotype responds differently to that environment than the other, then variance is created above and beyond the $V_G$ and $V_E$ parameters and is termed $V_{GE}$.

$C_{GE}$ = **Covariance of Genotype and Environment**  This component arises if two genotypes are selectively raised in two distinct environments so that genotype A is correlated with environment A and genotype B with environment B. To assess the contributions of genotype and environment to artistic ability, for example, we would want to examine children in random environments. If all children perceived by their parents as talented are sent to art class, this produces a covariance that skews the analysis.

For most purposes, the $V_{GE}$ and the $C_{GE}$ terms are ignored because they are difficult to measure, so that

$$V_T \cong V_G + V_E$$

## 21.15 Heritability

The preceding equation is often rewritten to define a parameter known as **heritability (H):**

$$H = \frac{V_G}{V_T} = \frac{V_G}{V_G + V_E}$$

**Heritability expresses the proportion of the total phenotypic variance in a sample that is contributed by genetic variance.** Thus a trait with a heritability of 1 is said to be expressed without any environmental influence, whereas a trait with a heritability of 0.5 would have half its variability (from individual to individual) determined by environmental factors (for example, diet and incidence of disease) and half by genotypic factors. Table 21-5 summarizes some representative heritability values, where all are expressed as percentages.

We can illustrate how heritability is estimated by returning to the data in Table 21-4 from the Emerson-East experiment. We make the standard assumption that the inbred parents and the inbred $F_1$ are each genetically homogeneous so that all the variance observed in these samples is due to environmental influence, or $V_E$. This assumption is supported by the fact that the average $V_E$ for the two parental lines, $\dfrac{0.666 + 3.561}{2} = 2.113$, is equivalent to the $V_E$ of the $F_1$ (2.307). We can obtain an overall value for $V_E$ by averaging these two numbers $\left( \dfrac{2.113 + 2.307}{2} = 2.21 \right)$. If we now define $V_T$ as the variance displayed by the $F_2$ ($V_T = 5.07$), then the genetic

**TABLE 21-5   Estimates of Heritability (in Percent) for Various Characters in Different Varieties of Farm and Laboratory Animals**

|  | Percent Heritability |
|---|---|
| Cattle | |
| Birth weight (Angus) | 49 |
| Gestation length (Angus) | 35 |
| Calving interval (Angus) | 4 |
| Milk yield (Ayrshire) | 43 |
| Conception rate (Holstein) | 3 |
| White spotting (Friesians) | 95 |
| Sheep | |
| Birth weight (Shropshire) | 33 |
| Weight of clean fleece (Merino, 22 months) | 47 |
| Length of wool fiber (Rambouillet, 14 months) | 36 |
| Multiple birth (Shropshire) | 4 |
| Chickens | |
| Body weight (Plymouth Rock, 8 weeks) | 31 |
| Shank length (New Hampshire) | 50 |
| Egg production (White Leghorn) | 21 |
| Egg weight (White Leghorn) | 60 |
| Hatchability (composite) | 16 |
| Mice | |
| Tail length (6 weeks) | 60 |
| Body weight (6 weeks) | 35 |
| Litter size | 15 |
| *Drosophila melanogaster* | |
| Abdominal bristle number | 52 |
| Thorax length | 47 |
| Wing length | 45 |
| Egg production | 18 |

From M. Strickberger, *Genetics*. New York: Macmillan, 1968.

variance revealed in this sample becomes, from $V_T = V_G + V_E$

$$V_G = V_T - V_E$$
$$= 5.07 - 2.21$$
$$= 2.86$$

Therefore,

$$H = \frac{2.86}{5.07}$$
$$= 0.56$$

so that about 56 percent of the variance in corn ear length is contributed by genetic factors in the Emerson-East experiment, the rest being contributed by the environment.

A subsequent study yielded a heritability estimate for corn ear length of only 0.17. Does this mean that one or the other study must be wrong? The answer is an emphatic *no*. The environmental conditions in the cornfield studied by Emerson and East may have been far more favorable for expressing genetic differences in ear length than the environment in the later study, allowing genetic differences to assume a far larger share of the total phenotypic variance. Alternatively, the corn strains may have undergone different levels of inbreeding, or the genetic backgrounds of the corn strains in the two experiments might have been very different, with the strains in the later study carrying fewer genes affecting ear length.

The general principle here is an important one: **the heritability of a trait cannot be extrapolated from one population and set of environmental conditions to another.** Such a fallacy is often committed in human studies, where selective breeding is not possible and the temptation to compare two different populations is great. Thus one might establish that height has a high heritability in a given sample: the height of children may be well correlated with the height of their parents, and environmental factors may be definitively ruled out as an important contributory cause. If a second sample of humans displays short stature, however, this does *not* necessarily mean that this group possesses "fewer height genes" or any similar conclusion. The people in this sample may well suffer from poor diet or endemic disease. Stated more formally, **within-group heritability** cannot automatically be converted to **without-group heritability.**

The practice of confusing the applicability of a human heritability measurement has created damaging misconceptions. For example, some studies have claimed that identical twins, even when reared apart, have very similar IQ scores, and the heritability of IQ has been assigned a value of 0.8 on this basis. The most extensive of these studies has been shown to have been fabricated from falsified data, and all such studies have ignored the fact that adoption agencies usually place twins in similar home environments so that environmental influence is hardly randomized in such analyses. Even if we assume that the 0.8 value is correct, however, the estimate is applicable only to the authentic samples of identical twins reared apart. In particular, it is not valid to examine a second population of humans, observe that they have below average IQ scores, and conclude that they are inferior in "intelligence genes" (if intelligence is indeed what an IQ test measures). There is much evidence to support the view that the development of intelligence is highly dependent on a supportive environment, and an adverse environment can greatly hinder the expression of those "intelligence genes" a human being inherits.

(a) George Snell (The Jackson Laboratory, Bar Harbor) has made fundamental contributions to the histocompatability genetics of the mouse. (b) I. Michael Lerner established much of the theoretical and experimental framework for quantitative genetics.

(a)

(b)

# Questions and Problems

1. Spore color in *Neurospora* is autonomous, meaning that it is determined by the genotype of the haploid spore nucleus. The wild-type genotype gives a *black* spore color. Two mutations singly result in *tan* or *gray* spores, but when the two mutations are carried in the same spore it is *colorless*. Tetrad analysis of a cross yields the following results:

|  | | |
|---|---|---|
| *gray* | *black* | *black* |
| *gray* | *black* | *gray* |
| *tan* | *colorless* | *tan* |
| *tan* | *colorless* | *colorless* |
| Percentage 93 | 3 | 4 |

   (a) Classify each type of tetrad as PD, NPD, or T.
   (b) What are the genotypes of the parental strains?
   (c) Are the *gray* and *tan* loci linked? If so, how many map units separate them?

2. As shown in Figure 21-11, mutant strains of *Neurospora* can be blocked at several positions in the pathways leading to methionine and threonine synthesis. Locate the sites where the following strains carry genetic lesions based on their growth requirements (+ means growth, − no growth).

| Mutant Strain | Aspartic Acid | Homoserine | Cysteine | Threonine | Cystathionine | Methionine |
|---|---|---|---|---|---|---|
| A | − | − | − | − | + | + |
| B | − | + | + | + | + | + |
| C | − | − | − | − | − | + |
| D | − | − | − | + | − | − |

3. Two new arginine-requiring strains of *C. reinhardi*, *arg*-8 and *arg*-9, are crossed. The resulting tetrads are found by replica plating to be of three types: Type I: all four tetrad products require arginine for growth; Type II: 3/4 of the products require arginine for growth; Type III: 1/2 of the products require arginine for growth. Type I tetrads represent 80 percent of the total, Type II represent 18 percent, and Type III represent 2 percent.
   (a) Are the mutations allelic or in different genes? Explain.
   (b) If in different genes, is there linkage between them? By how many map units?
   (c) Give the genotypes of the various tetrad products.

4. Three *trp* mutant strains of *E. coli* (*trp*-1 through *trp*-3) are isolated, all requiring tryptophan for growth. Each is tested to determine whether it will grow in the presence of tryptophan precursors, with the following results (see Figure 21-1 for a key to abbreviations):

| Mutant Strain | Chorismic Acid | Anthranilic Acid | PRA | CDRP | InGP | Tryptophan |
|---|---|---|---|---|---|---|
| *trp*-1 | − | − | − | + | + | + |
| *trp*-2 | − | − | − | − | − | + |
| *trp*-3 | − | − | + | + | + | + |

   (a) In which gene (or genes) of the *trp* operon (Figure 21-1) could each mutation lie? Explain ambiguous cases.
   (b) Which is the most likely candidate as a polar mutation? Where would such a mutation be most likely to map in the *trp* operon?

5. A *vermilion* female is crossed to a *scarlet* male and the resulting $F_1$ is inbred. Give the phenotypes and genotypes of the $F_1$ and $F_2$ flies, each classified according to sex.

6. A hen with a *rose*-shaped comb is crossed with a rooster having a *pea*-shaped comb. The $F_1$ all have *walnut*-shaped combs. When the $F_1$ is inbred, 9/16 of the $F_2$ have *walnut* combs, 3/16 have *pea* combs, 3/16 have *rose* combs, and 1/16 have *single*-shaped combs.
   (a) How would you explain these results?
   (b) What progeny would you expect from a cross *rose* × *pea*? What would be the phenotypes in the $F_2$ when this $F_1$ was inbred?

7. An *agouti* mouse is crossed with a *white* mouse. Their progeny are 3/8 *agouti*, 1/2 *white*, and 1/8 *black*. What are the genotypes of the parents and of the progeny?

8. In corn, the three dominant genes *A*, *C*, and *R* are all necessary to produce a colored seed. Plants with the genotype $\dfrac{A}{-}\ \dfrac{C}{-}\ \dfrac{R}{-}$ have colored seeds; all others are colorless. A colored plant is subjected to crosses with two tester plants with the following results:

(a) A cross with $\dfrac{a}{a}\ \dfrac{c}{c}\ \dfrac{R}{R}$ plants yields 50 percent colored seeds.

(b) A cross with $\dfrac{a}{a}\ \dfrac{c}{c}\ \dfrac{r}{r}$ plants yields 25 percent colored seeds.

What is the plant's genotype?

9. Two different strains of summer squash produce a *spherical*-shaped fruit, and both breed true. When they are crossed, the $F_1$ all have a *disc*-shaped fruit. Inbreeding the $F_1$ yields the following ratio among the $F_2$: 9/16 *disc*, 6/16 *spherical*, and 1/16 *long*. Explain these results.

10. *Barred* feathers in the chicken are produced in the presence of the dominant sex-linked gene *B* (*Note:* review the chromosomal basis of sex determination in birds in Section 6.19); a chicken hemizygous or homozygous for its *b* allele has *nonbarred* feathers. Chickens also possess the non–sex-linked *C/c* alleles for pigmentation: *c/c* chickens have solid white feathers regardless of the presence of the *B* allele, whereas birds heterozygous or homozygous for *C* have colored feathers.

 Predict the outcome of the following cross in terms of sex, genotype, and phenotype of progeny chicks:

$$\frac{B}{b}\ \frac{C}{c}\ \male\ \times\ \frac{b}{W}\ \frac{C}{c}\ \female$$

11. The plant *Bursa* (shepherd's purse) can have seed capsules in two shapes: *triangular* and *oval*. The $F_1$ progeny of a *triangular* × *oval* cross is all *triangular*; when the $F_1$ is inbred, the $F_2$ are in the ratio 15 *triangular*: 1 *oval*. How might this result be explained?

12. Four adjacent genes code for enzymes required in the tryptophan biosynthetic pathway in an organism. Complementation tests were made by crossing the haploid *trp⁻* strains 1 through 5 in different pairwise combinations, and the resulting diploid cells were tested for their ability to grow on minimal medium. Results were:

|   | 1 | 2 | 3 | 4 | 5 |
|---|---|---|---|---|---|
| 1 | 0 |   |   |   |   |
| 2 | + | 0 |   |   |   |
| 3 | + | + | 0 |   |   |
| 4 | + | + | + | 0 |   |
| 5 | 0 | 0 | 0 | 0 | 0 |

+ = growth on minimal
0 = no growth on minimal

Another haploid *trp⁻* strain (6) was discovered which could complement strains 1, 3, and 4. When this new auxotroph was crossed with strain 5, a very few cells of the resulting diploid strain were found to have the ability to grow on minimal medium. These rare cells probably arose as a result of one of the following:

(a) complementation.

(b) reversion to wild type of the strain 6 mutation in the diploid cells.

(c) rare events in which a triple recombination exchange occurred between three different loci along the chromosome carrying the mutant genes.

(d) a four-strand double crossover between the mutant genes in strains 5 and 6.

Explain your choice of answer.

13. Four mutant strains of *E. coli* that required serotonin for growth were tested for their growth responses to substances thought to be intermediates in the pathway of serotonin synthesis. The following results were obtained (+ = growth when substance was added to minimal medium; − = no growth). What is the order of the intermediate substances in the serotonin pathway and the position of the mutant blocks?

| Mutant | Serine | Tryptophan | Anthranillic Acid | Serotonin |
|--------|--------|------------|-------------------|-----------|
| a      | −      | +          | −                 | +         |
| b      | −      | −          | −                 | +         |
| c      | +      | +          | −                 | +         |
| d      | −      | +          | +                 | +         |

14. Three *arg* mutant strains of *E. coli* (*see Figure 21-9*) are streaked close together (but not touching) on a petri dish of minimal medium in the following fashion:

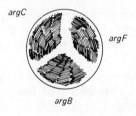

Each mutation completely blocks the activity of its respective enzyme. Assuming that cross-feeding can take place, what growth response might you expect to see several days after streaking?

(a) No growth of any of the strains on the dish.

(b) Cells cf *arg* C nearest those of *arg* F growing; and cells of *arg* B nearest those of *arg* C and *arg* F growing.

(c) Cells of *arg* F nearest those of *arg* C growing.

(d) Cells of *arg* C nearest those of *arg* B growing.

Explain your choice of answer(s).

15. An essential multimeric enzyme in rabbits is composed of two different protein subunits. Gene locus 1 (allele A dominant to allele a) codes for one subunit, whereas the unlinked gene locus 2 codes for the other (allele B dominant to b). The homozygous recessive condition at locus 2 (b/b) results in the inability to produce a functional subunit and is lethal at a very early embryonic stage. Homozygous recessives for locus 1 (a/a) produce a marginally effective enzyme subunit, which leads to metabolic disturbances invariably resulting in dwarfism. All other genotypes are phenotyically normal.

If rabbits heterozygous at both loci are crossed, what is the proportion of *viable* progeny that would have a *normal* phenotype? Explain your answer.

16. A scientist was studying mutants of *E. coli* and isolated a number that required compound G to grow. A number of compounds in the pathway to G were known, and each was tested for the ability to support the growth of the mutant. A+ indicates that growth occurred when the gene product was added.

**Gene Products**

| Mutant | A | B | C | D | G |
|--------|---|---|---|---|---|
| 1 | − | − | − | + | + |
| 2 | + | + | − | + | + |
| 3 | − | − | − | − | + |
| 4 | − | + | + | + | + |
| 5 | − | + | − | + | + |

(will grow if both A + C are added together)

(a) What is the pathway?

(b) At which step is each mutant blocked?

17. Two different true-breeding strains of sweet peas having white flowers were crossed. The flower color of the $F_1$ was purple. When the $F_1$ was inbred, 9/16 of the progeny had purple flowers and 7/16 had white flowers. Explain these results in terms of epistatic relationships between two gene loci concerned with the biosynthesis of the purple pigment anthocyanin.

18. Two albinos have five children, all normally pigmented. Assuming illegitimacy is ruled out, how would you explain this phenomenon?

19. Listed below are the HLA phenotypes of a family.

(a) Identify the genotypes of parents and offspring and write out their haplotypes.

(b) Child number 2 requires a kidney transplantation. Which, if any, of the parents and/or sib(s) would be fully histocompatible donors? Explain.

    Mother: A2; B7, B8
    Father: A1, A3; B7, B9
    Child number 1: A1, A2; B8, B9
    Child number 2: A2, A3; B7, B8
    Child number 3: A1, A2; B7, B8

20. Two varieties of beans have mean leaf lengths of 9.6 cm and 6.3 cm, respectively. The average leaf length of the $F_1$ and $F_2$ obtained by crossing these two varieties is 8.0 cm. About 6 percent of the $F_2$ have a leaf length of 9.6 cm, and 6 percent have a leaf length of 6.3 cm. How many gene loci are involved in determining leaf length in beans?

21. Twenty calves in an inbred herd are found to have the following birth weights (in pounds). What is the expectant mean ± standard deviation birth weight for 95 percent of the calves in this herd?

| | | | | |
|---|---|---|---|---|
| 81 | 81 | 83 | 101 | 86 |
| 65 | 68 | 77 | 66 | 92 |
| 94 | 85 | 105 | 60 | 90 |
| 94 | 90 | 81 | 63 | 58 |

22. The data on the $F_3$ plants provided by the East experiment on tobacco (*Figure 21-18*) indicate that part of the variance observed in the $F_2$ is genetic. Explain.

23. The variance shown by the two parental lines and the $F_1$ in the East experiment on tobacco (*Figure 21-18*) averages at 8.76. The total variance of the $F_2$ plants is 40.96. Calculate the heritability of flower length in tobacco.

24. In a more sophisticated study of human skin pigmentation than that performed by the Davenports, Harrison and Owen obtained the following data on the reflectance of the skin exposed to different wavelengths of light, where "European" denotes "true-breeding" Caucasoid, "African" denotes "true-breeding" blacks, and $F_1$ hybrid denotes the mulatto progeny of European × African matings.

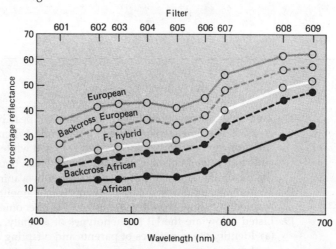

(a) Do the data indicate that skin pigmentation is a quantitative trait? Explain.

(b) When the $F_1$ mulattos "backcross" with either Europeans or Africans, the data shown in the graph are obtained for their offspring. Are the data compatible with the notion that pigmentation is controlled by two loci? More than two loci? Explain. (Assume simple dominance/recessive relationships at each locus, with pigmentation dominant.)

# 22

# Control of Gene Expression in Bacteria

A. **Introduction**

B. **General Features of Gene Regulation**
   - 22.1 Constitutive Versus Regulated Genes
   - 22.2 General Properties of Genes That Are Regulated
   - 22.3 General Properties of Molecules Involved in Regulation
   - 22.4 Inducible Vs. Repressible Operons

C. **Regulation of Lactose Utilization**
   - 22.5 General Features of *lac* Regulation
   - 22.6 Allolactose as the True Effector
   - 22.7 The *lacI* Regulatory Gene: Genetic Studies
   - 22.8 The *lacI* Regulatory Gene: Molecular Studies
   - 22.9 The *lac* Operator and Operator Mutations

22.10 How Repressor Binding Inhibits *lac* Operon Transcription

22.11 The CAP Protein and Catabolite Repression

22.12 Overlap Between *lacZ* and the Operator

D. **Regulation of Tryptophan Biosynthesis**
   - 22.13 General Features of Tryptophan Regulation
   - 22.14 Feedback Inhibition
   - 22.15 Operator-Repressor Regulation of the *trp* Operon
   - 22.16 Attenuator Regulation

E. **Translational Control**
   - 22.17 Representative Translational Control Mechanisms

F. **Questions and Problems**

681

# Introduction

In thinking of an animal or a higher plant, it is perhaps not difficult to conceive of its growth and differentiation as an intricate chain of events in which some genes function at all times, others are switched on or off as the cellular or extracellular environment changes, and still others function only at specific places and times. Such a concept may be less obvious for an organism like *Escherichia coli*, whose chromosome appears to be in a continuous state of replication, whose life cycle can take as little as 20 minutes and whose capacity to differentiate in the eukaryotic sense appears negligible. The concept might also, at first sight, seem inapplicable to a bacteriophage whose genes are limited in number, whose life cycle is very short, and whose phenotypic capacities also appear to be limited.

In fact, however, the genes of bacteria and of their phages are under exquisitely sophisticated kinds of controls, and our deep understanding of these controls stands in sharp contrast to our present ignorance of how gene expression is regulated in eukaryotes (*Chapters 24 and 25*). This chapter focuses on the different kinds of control mechanisms that have evolved to regulate gene expression in bacteria; the following chapter considers similar phenomena in viruses. We begin with some general definitions to give an overview of what is involved in gene regulation and then turn to several specific examples that give experimental evidence for our introductory statements.

# General Features of Gene Regulation

## 22.1 Constitutive Versus Regulated Genes

Basic distinctions can be made between genes whose expression is regulated and genes whose expression is not. The latter can be called **constitutive genes.** A constitutive gene is expressed continuously, so that its product (for example, an enzyme) is invariably found in the cell at roughly the same concentration, regardless of growth conditions and regardless of whether the substrate for the enzyme is ever presented to the cell or not. The expression of a **regulated gene,** on the other hand, is subject to modification by chemical stimuli that appear in and disappear from the gene's environment.

It should be stressed at once that a constitutive gene by no means experiences unregulated expression in the sense, say, that cancerous growth is unregulated. There are many levels at which control is exerted over the expression of *all* genes, including constitutive genes. We can list three ways that such control may be exercised:

1. The structure of a promoter may influence the frequency with which transcription of a gene is initiated by a DNA-dependent RNA polymer-

ase, and this feature can govern how much of the gene's product is present in the cell.

2. The structure of a gene's mRNA transcript may determine how rapidly the message is digested by cellular ribonucleases after being created.

3. The base sequence associated with the AUG start signal on a transcript may influence how readily its translation is initiated.

A more precise distinction between constitutive and regulated genes, therefore, might be that **the controls governing expression of a constitutive gene are invariant,** whereas **the controls governing regulated gene expression are subject to modification.**

There seems to be some logic regarding which enzymes of a bacterium are synthesized constitutively and which are synthesized "on call." The constitutive enzymes are generally those that are needed at all times. Thus the enzymes involved with glucose metabolism—enzymes of glycolysis and the hexose monophosphate pathway—are constitutive, and one can imagine that the metabolism of glucose is sufficiently central to the physiology of a bacterium that these pathways should be continuously functional. The metabolism of less common sugars such as lactose, arabinose, and galactose, on the other hand, is effected by regulated enzymes, and it is clearly economical to elaborate such enzymes only on those occasions when their substrates are present. Regulated genes also specify the enzymes of most biosynthetic pathways—pathways for the synthesis of amino acids, purines, pyrimidines, and so on—and it again seems logical that these enzymes would be synthesized only as there was a demand for their biosynthetic products.

## 22.2 General Properties of Genes that are Regulated

Two general mechanisms can be imagined for limiting the expression of regulated genes to specific occasions. Control could be exerted at the **transcriptional level,** meaning that a regulated gene would be transcribed only when conditions are appropriate. Alternatively, control could be exerted at a **translational level** so that an mRNA from a regulated gene is translated only under appropriate conditions. A few examples of translational control have been found in bacteria, as described later in the chapter. In general, however, **bacterial genes are regulated at a transcriptional level.**

If a gene's transcription is to be regulated, we can imagine that there must be some genetic elements associated with the gene that respond to environmental signals and either allow or disallow mRNA to be synthesized from that gene. Two major classes of such **controlling elements** have been identified in bacteria and in DNA-containing phages. Both are portions of the chromosome, and both lie very close to the gene(s) they govern. One is the promoter and the other the operator.

A **promoter,** as we learned in Section 9.7, represents a sequence of bases that is recognized by a DNA-dependent RNA polymerase, with promoter-polymerase binding initiating transcription of the neighboring gene. In many senses, the promoter of a regulated gene is likely to be similar to the promoter of a constitutive gene. Presumably, the major difference between the two types of promoters is that promoters of constitutive genes have relatively invariant initiation properties, whereas promoters of regulated genes are more or less active in promoting transcription of contiguous genes, depending on the presence or absence of specific proteins known as **regulator proteins.** Examples of regulator proteins and their effects on transcription will be presented shortly.

The second main class of controlling element is the **operator.** An operator also represents a sequence of bases that interacts with a protein. The protein is also called a regulator molecule, and operator-regulator interaction will either prevent or promote transcription of the regulated gene, depending on which operator is being considered. The major distinction to be made between an operator and a promoter element is that the operator is not, in and of itself, an RNA polymerase binding site. Instead, **the operator acts to influence whether or not the polymerase can bind to its promoter** and from there, progress to transcribe the gene.

Let us pause here to clarify some terminology. The operator was first postulated by F. Jacob and J. Monod in 1961 as a controlling element for the genes involved in lactose utilization. These genes constitute the *lac* operon, an **operon** (*Sections 9.2 and 21.1*) being a group of related genes that is transcribed into a single molecule of mRNA. Therefore the initial, and very simple, definition of an operator was that it controlled the transcription of an operon. More recently, however, single genes have been found that are regulated, and these possess operator loci in the same way that operons do. These single genes are occasionally called operons, but it seems preferable to conserve the term operon for *groups* of genes that are co-transcribed. Therefore, in this text **an operator refers to any locus that interacts with a regulator molecule and lies between a promoter and a gene,** with the understanding that an operator locus may lie next to a single regulated gene or next to a regulated operon.

## 22.3 General Properties of Molecules Involved in Regulation

The preceding section has focused on a regulated gene. We now describe the features of the diffusible molecules that do the regulating. As already noted, the **regulator proteins** (also called regulators, affector proteins, or affectors) are key molecules in any regulation process. They have the potential to bind specifically to controlling elements, the specificity residing in their three-dimensional configurations.

If gene transcription is to be regulated by the binding and dissociation of regulator proteins to or from controlling elements, then there must obviously exist some additional class of molecules that governs whether such binding in fact takes place. These additional molecules are known as **effector molecules.** An effector molecule is classically a small molecule—a

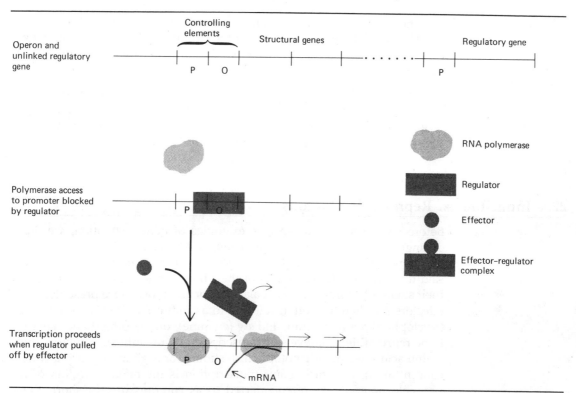

**FIGURE 22-1 Relationships between promoter (P), operator (O), and two regulated structural genes** in an operon, showing RNA polymerase attachment to a promoter, a regulator protein binding to the operator, and, in the presence of effector, the effector-regulator binding, which removes the regulator from the operon. This diagram presents the essence of the 1961 Jacob-Monod model for *lac* operon regulation, although many of the molecular details have since been added.

sugar, an amino acid, or a nucleotide—that can bind to a regulator protein and thereby change its ability to interact with a controlling element. Perhaps the most important concept to come from the work of Jacob and Monod in the late 1950s was the idea that the regulator protein of the *lac* operon could bind to the operator and inhibit operon expression as long as the effector was *not* present. Once the effector molecule appeared in the intracellular environment, it would bind to the regulator and "pull" the regulator off the operator, thus allowing operon expression to proceed. These relationships are diagrammed in Figure 22-1.

Regulator molecules, being proteins, must themselves be specified by genes; these are called **regulatory genes.** The reader should fix clearly in mind the distinction between a gene that is regulated and a regulatory gene, since this distinction is basic to an understanding of this chapter.

Four key elements thus emerge as essential to genetic regulation in bacteria and phages:

1. The regulated structural gene (or gene cluster) itself.
2. The controlling elements (including the promoter and the operator).
3. The regulator protein (and its structural gene, the regulatory gene).
4. The effector.

All are diagrammed in Figure 22-1. As we go on to consider actual examples of genetic control, it will become apparent that highly complex kinds of gene regulation can be understood in terms of the interplay between these four kinds of molecules or regions of DNA.

## 22.4 Inducible vs. Repressible Operons

Before we proceed with specific examples of gene regulation, we must distinguish between inducible and repressible operons.

Operons such as *lac* (*lac*tose), *gal* (*gal*actose), and *ara* (*ara*binose) are said to be **inducible;** full levels of their transcription do not occur unless their sugar effectors—lactose, galactose, or arabinose—are presented. The effectors are therefore often referred to as **inducers.** Operons such as *trp* (*tryp*tophan), *his* (*his*tidine), and *arg* (*arg*inine), on the other hand, are said to be **repressible:** full levels of transcription occur until such time as their amino acid effectors—tryptophan, histidine, or arginine—reach a critical concentration, at which point transcription is inhibited. Effectors of repressible operons are often referred to as **corepressors.** In short, then, **inducible operons are "off" unless the effector is present, whereas repressible operons are "on" until the effector is present.** Both kinds of control can clearly accomplish the same goal, but by reciprocal means.

As a rule of thumb, inducible operons in bacteria typically carry genes for "gratuitous" enzymes (that is, enzymes not required for survival), and the enzymes are usually involved in the catabolism of a substrate not always present in the bacterial medium. Thus it "makes sense" for the enzymes of arabinose utilization to be inducible by arabinose: arabinose may be only infrequently taken up by a bacterial cell, and it is energetically most efficient to produce enzymes for arabinose degradation only "on call."

Repressible operons, in contrast, typically carry genes for "necessary" enzymes that participate in biosynthetic processes. Thus histidine is in constant demand by a growing cell, and a steady output of *his*-encoded gene products is generally occurring; such a cell is said to be **derepressed.** Should free histidine levels become too high, however (as, for example, when rates of protein synthesis abate or when histidine is supplied in the growth medium), then the expression of these genes is repressed and histidine biosynthesis soon abates as well.

With this much background we can turn to specific descriptions of two representative bacterial operons, the inducible *lac* operon and the repressible *trp* operon of *E. coli*. We follow with a brief description of translational control mechanisms.

# Regulation of Lactose Utilization

## 22.5 General Features of *lac* Regulation

As described in detail in Section 21.1, lactose utilization in *E. coli* involves the products of the three structural genes that constitute the *lac* operon; these products are **β-galactosidase**, a lactose-splitting enzyme specified by gene *lacZ*; **M protein**, a molecule necessary for lactose uptake and specified by gene *lacY*; and **transacetylase**, a poorly understood enzyme specified by gene *lacA*. A map of the *lac* operon is found in Figure 21-4 and should be reviewed at this time.

Analysis of *lac* operon regulation has reached a high degree of sophistication, in part because pure *lac* DNA has for some time been available from specialized transducing phages (*Section 13.14 and Figure 13-11*). The nucleotide sequence of most of the operon has been determined, and the relevant regulatory molecules have been isolated and analyzed. The molecular basis for *lac* regulation can therefore be described in some detail. Briefly stated, *lac* regulation exhibits the following features.

1. The operon is under **negative control,** meaning that when it interacts with a regulator protein known as the *lac* repressor, operon transcription is inhibited. The repressor is specified by the *lacI* regulatory gene, which maps just outside the operon (*Figure 21-4*).
2. The *lac* repressor binds to the *lac* operator, thereby preventing transcription, unless an effector (lactose or a lactose analogue) is presented to the cell; in the presence of the effector, the repressor leaves the operator, permitting *lac* transcription. Thus the *lac* operon is **inducible.**
3. The operon is also under **positive control,** meaning that when it interacts with a regulator protein known as **CAP,** operon transcription is enhanced. The regulatory gene for CAP is known as *crp* and maps elsewhere in the *E. coli* genome.
4. The CAP protein binds to a controlling element near the *lac* promoter, thereby enhancing transcription, as long as glucose levels are low. A rise in glucose causes CAP to leave its controlling element, and *lac* transcription soon drops to very low levels.

In the sections that follow, we present the genetic and molecular evidence for this picture of *lac* regulation.

## 22.6 Allolactose as the True Effector

When wild-type *E. coli* cells are grown on a carbon source other than lactose, the intracellular levels of β-galactosidase, M protein, and transacetylase are only about 1/1000 of the levels these substances attain when lactose is present (it is important, however, to realize that low levels of all three proteins are present at all times). Introduction of lactose (Figure 22-2a)

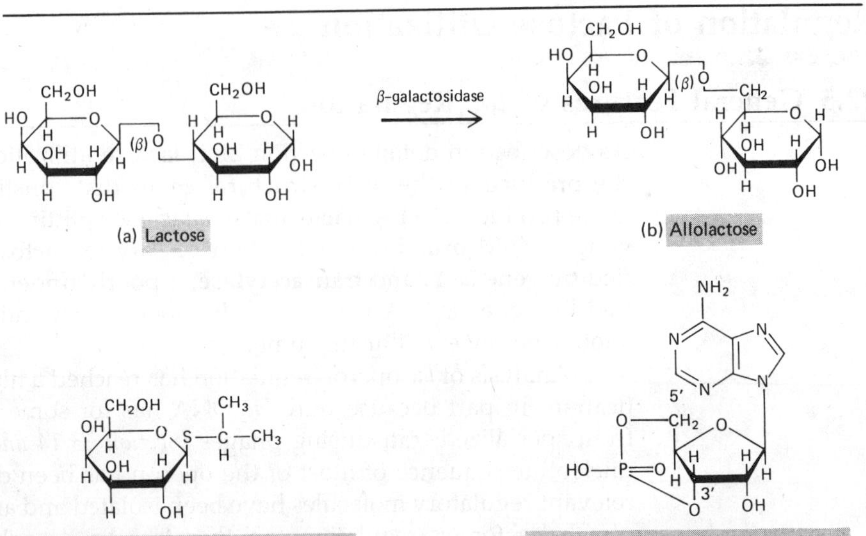

FIGURE 22-2
Small molecules
active in *lac* operon
regulation.

(a) Lactose

(b) Allolactose

(c) IPTG (Isopropyl β-D-thiogalactoside)

(d) Cyclic AMP (3', 5'-cyclic adenylic acid)

FIGURE 22-2
Small molecules active in *lac* operon regulation.

stimulates the transcription of the *lac* operon almost immediately, and *de novo* synthesis of the three proteins commences within three to four minutes. Early observers of such phenomena therefore concluded that lactose was able to induce the *lac* operon, and lactose was termed an inducer.

Recently, however, it has been shown that once lactose enters the cytoplasm, it is acted upon by one of the few molecules of β-galactosidase that manage to be synthesized in the uninduced cell. The enzyme transforms lactose into **allolactose** (Figure 22-2b), and it is **allolactose that acts as the true effector molecule,** interacting directly with the repressor protein to allow *lac* transcription.

The fact that allolactose and not lactose is the natural effector of *lac* transcription explains why **mutations in the *lacZ* and *lacY* genes of the *lac* operon cause cells to be uninducible by lactose.** A *lacZ⁻* cell cannot produce a functional β-galactosidase and therefore cannot transform entering lactose into allolactose, whereas a *lacY⁻* cell cannot produce even a few molecules of functional M protein and therefore cannot transport lactose molecules into the cytoplasm for conversion to allolactose. To minimize these effects, it is customary in experimental studies to induce the *lac* operon with the lactose analogue isopropylthiogalactoside **(IPTG)** (Figure 22-2c). This molecule can be taken up by *lacY⁻* cells; moreover, it does not require β-galactosidase modifications, since it can interact directly with the *lac* repressor.

## 22.7 The *lacI* Regulatory Gene: Genetic Studies

The "target" of both IPTG and allolactose is the *lac* repressor protein. Important information on the nature of the repressor was obtained from genetic studies of its gene, *lacI*. Proximal nonsense mutations in *lacI*, which we will call **lacI⁻ mutations,** were found to lead to the **constitutive synthesis** of the *lac* operon products, meaning that the cells synthesize both β-galactosidase and M protein **whether or not inducer is present in the medium.** It was concluded, therefore, that the product of the normal *lacI* gene functioned to **repress** *lac* operon expression unless effector was present and that elimination of this repressor product by a nonsense mutation allowed the operon to be expressed independent of the effector.

To learn whether these *lacI⁻* mutants were dominant or recessive, geneticists performed analyses of **partial diploids.** You will recall from Section 13.15 that it is possible to construct F′ elements carrying defined groups of genes and to introduce these elements into *E. coli* hosts for functional studies. When a wild-type *E. coli* strain was infected with F′ *lacI⁻ lacZ⁺* plasmids, the resultant partial diploids were indistinguishable from wild-type cells, being fully inducible with both IPTG and allolactose. It was therefore concluded that the **lacI⁻ mutations are recessive.**

Further studies revealed that it was possible to isolate a second class of mutations in this gene, called **lacI⁻ˢ**, where **s** stands for **super-repressed.** Cells carrying such mutations are unable to express any of the *lac* structural genes unless effector concentrations are extremely high. To learn whether these mutations are dominant or recessive, wild-type *E. coli* cells were infected with F′ *lacI⁻ˢ lacZ⁺* plasmids. This time, the resultant partial diploids had the phenotype of the mutant: no β-galactosidase was synthesized unless effector concentrations were very high. Thus the **lacI⁻ˢ mutations are dominant** to the wild-type *lacI* gene. Moreover, dominance is observed whether *lac⁻ˢ* is on the plasmid or on the main chromosome. In other words, $\frac{lacI^{-s}\ lacZ^{+}}{F'\ lacI^{+}\ lacZ^{+}}$ and $\frac{lacI^{+}\ lacZ^{+}}{F'\ lacI^{-s}\ lacZ^{+}}$ partial diploids both exhibit the super-repressed phenotype. This observation was given an important interpretation, namely, that **the lacI gene product is a diffusible protein** that can regulate *lac* operon transcription *in trans* (on a different chromosome) and *in cis* (on the same chromosome).

Continued induction of mutations in the *lacI* gene lead to the indentification of a third class of mutation, called **lacI⁻ᵈ.** The *lacI⁻ᵈ* mutants resemble the *lacI⁻* mutants in that they lead to constitutive β-galactosidase synthesis in haploid cells, and they resemble the *lacI⁻ˢ* mutants in that they are dominant in partial diploids (the −*d* superscript derives from this **d**ominance).

We discuss the *lacI⁻ˢ* and *lacI⁻ᵈ* mutations at a molecular level in the next section. Before reading it, try to figure out yourself what each mutation might be doing.

## 22.8 The *lacI* Regulatory Gene: Molecular Studies

To understand the nature of the *lacI⁻ˢ* and *lacI⁻ᵈ* mutations, B. Müller-Hill and J. Miller induced large numbers of *lacI⁻ˢ* and *lacI⁻ᵈ* mutants, starting with mutant strains that overproduce repressor so that enough mutant

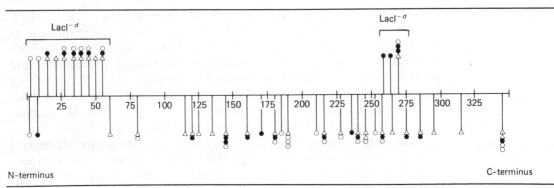

**FIGURE 22-3 Genetic map of the *lacI* gene,** with each vertical line representing missense mutations that fall within a particular deletion that extends into the gene. The circles and triangles represent mutants isolated by different conditions of mutagenesis. All mutants above the line have partial or full IPTG binding activity *in vitro;* mutants below the line have no detectable activity. Distances are given in terms of amino acids of *lac* repressor protein (at the time the map was drawn, this number was thought to be 347). (From J. Miller et. al., in *Protein-Ligand Interactions,* H. Sund and G. Blante, Eds. Berlin: Walter de Gruyter, 1975.)

protein could be isolated for *in vitro* analysis. The mutations were ordered into a linear sequence by deletion mapping (*Section 13.14*), producing the fine-structure map of the *lacI* gene shown in Figure 22-3. The repressor proteins produced by these mutants were also analyzed biochemically. The results of these studies are as follows.

Most of the *lacI*$^{-d}$ mutations were found to affect amino acids that cluster in the N-terminal region of the polypeptide, with a second cluster nearer the C-terminal end (Figure 22-3). The repressor proteins produced by such mutants were found to bind IPTG normally, but they were **defective in their ability to bind to *lac* DNA.** This property explains the constitutive phenotype of *lacI*$^{-d}$ mutants: their *lac* operons are free to be transcribed at all times, since the repressor cannot bind to them.

To understand why *lacI*$^{-d}$ mutations are dominant in partial diploids, it is necessary to know that **the *lac* repressor is a tetramer,** each of its four polypeptides being a *lacI* gene product. This means that in an F' *lacI*$^{-d}$/*lacI*$^{+}$ partial diploid, both mutant and wild-type polypeptides are produced and, as diagrammed in Figure 22-4a, these can combine with one another to form **mixed-hybrid tetramers.** It turns out that even if only one of the four polypeptides in the hybrid protein is mutant, DNA binding is impaired and constitutive transcription occurs. Therefore, the mutant phenotype is dominant in virtually all the cells in a culture.

The *lacI*$^{-s}$ mutations, by contrast, were found to map throughout the *lacI* gene, although they cluster in the central portion (Figure 22-3). The repressor proteins produced by these mutants were found to be **defective**

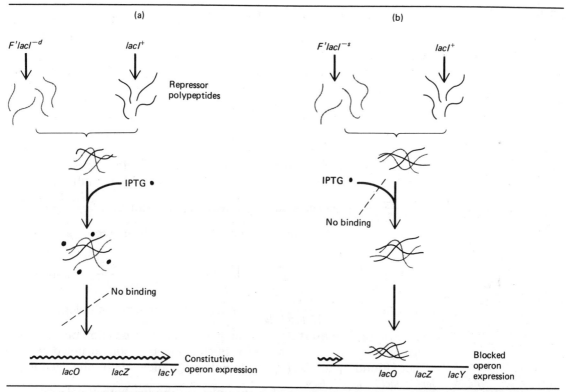

**FIGURE 22-4 Two classes of dominant mutations in the *lacI* gene** and their behavior in partial diploids. (a) The *lacI*$^{-d}$ mutation affecting binding to *lac* DNA. (b) The *lacI*$^{-s}$ mutation affecting binding to inducer (IPTG). Wavy arrow denotes transcription.

in their ability to bind to IPTG or other *lac* inducers, although their ability to bind to *lac* DNA is normal. This inducer-binding defect explains the super-repressed phenotype of *lacI*$^{-s}$ mutants. The dominance of *lacI*$^{-s}$ mutations in partial diploids is also readily explained: mutant or mixed-hybrid repressors, unable to bind normal levels of inducer, presumably come to "monopolize" all *lac* operator sites in the cell and thereby block induction. This is illustrated in Figure 22-4b.

Returning to Figure 22-3, the genetic data suggest that the *lac* repressor polypeptide has distinct functional domains. The N-terminus and C-terminus are implicated as being important for binding to the operator, whereas the central portion of the polypeptide appears crucial for binding to the inducer. Further studies on repressor function became possible with the analysis of the *lac* operator itself, as described in the next section.

## 22.9 The *lac* Operator and Operator Mutations

The *lac* operator is currently defined as the DNA sequence to which the *lac* repressor binds in such a way to prohibit transcription of the *lacZ*, *lacY*, and *lacA* genes. In the original study by Jacob and Monod, the *lac* operator was defined as the site of *lacOᶜ* (**operator-constitutive**) mutations, which lead to a constitutive synthesis of *lac*-encoded proteins. Unlike the constitutive *lacI⁻ᵈ* mutations described in the preceding section, however, the *lacOᶜ* mutations map outside the *lacI* gene, and very close to *lacZ*.

The *lacOᶜ* mutations also differ from *lacI⁻ᵈ* mutations in partial diploids: **lacOᶜ confers constitutiveness only on the *lac* genes located in the same chromosome as itself.** Thus in partial diploids with the genotype $\frac{lacO^+\ lacZ^-}{F'lacO^c\ lacZ^+}$, β-galactosidase synthesis is constitutive; in $\frac{lacO^+\ lacZ^+}{F'\ lacO^c\ lacZ^-}$ diploids, on the other hand, β-galactosidase synthesis requires the presence of inducer. This effect extends to other genes in the operon as well: thus M protein is synthesized constitutively in $\frac{lacO^+\ lacY^-}{F'\ lacO^c\ lacY^+}$ diploids but requires inducer in $\frac{lacO^+\ lacY^+}{F'\ lacO^c\ lacY^-}$ strains. In genetic terminology, the *lac* operator is seen to exert its effect on genes in the *cis* position only, a phenomenon known as ***cis*-dominance.** Such a phenomenon, as noted earlier, is not found with *lacI⁻ᵈ* mutations. Diploids with the genotype $\frac{lacI^{-d}\ lacO^+\ lacZ^+\ lacY^-}{F'\ lacI^+\ lacO^+\ lacZ^-\ lacY^+}$, for example, synthesize both β-galactosidase and M protein in a constitutive fashion, demonstrating that the dominant effect of *lacI⁻ᵈ* extends to genes located in *trans* as well as in *cis*.

The *cis*-dominance of *lacOᶜ* mutations was correctly interpreted by Jacob and Monod to signify that the operator locus is not a gene specifying a diffusible gene product that can act on all chromosomes of the cell, but instead **a controlling element that can influence only the genes directly contiguous to it.** This effect is perhaps best appreciated by considering *cis*-dominant *lacOᶜ* mutations that confer only partial constitutiveness. In *lacI⁺ lacOᶜ lacZ⁺* cells of this type, intermediate levels of β-galactosidase are synthesized at all times, with full levels produced when inducer is added. Such *lacOᶜ* operators are visualized as binding repressor protein less effectively than wild-type operators, hence allowing RNA polymerase molecules an "intermediate" access to the adjacent structural genes in the absence of inducer. Mutant operators of this sort would not be expected to allow increased access to structural genes that are located in a different chromosome.

Studies *in vitro* have demonstrated that *lacOᶜ* mutations indeed mark the site for *lac* repressor recognition. For example, *lac* DNA isolated from wild-type strains binds *lac* repressor much more effectively than DNA from other regions of the genome; by contrast, isolated *lacO^c* DNA binds repressor only at "background" levels. It was therefore not surprising when nucleotide-sequencing studies localized *lacO^c* mutations within the center of the operator region. Figure 22-5 shows a partial nucleotide sequence of the *lac* operon, and on the left side of the figure is indicated that portion of the *lac* DNA that is protected from DNAse digestion by repressor binding. Within the repressor-protected segment, eight *lacO^c* mutations have been identified (Figure 22-5). Therefore, it seems clearly established that the *lac* repressor recognizes and singles out this sequence of nucleotides (Figure 22-5, bracket) from among the millions present in the *E. coli* cell.

How does the repressor in fact recognize this operator sequence? Since *lac* repressor fails to recognize single-stranded operator DNA, we are left with the following possibilities: (1) repressor interacts with exposed bases in partially or fully denatured operator DNA; (2) repressor recognizes some feature of the sugar-phosphate backbone of operator DNA that is modified as a consequence of the specific base sequence in the interior; or (3) repressor interacts with the edges of the bases exposed in the major or the minor grooves of the operator DNA helix (*Figure 1-8*).

The third possibility is at present considered the most likely, since there is no evidence of DNA denaturation during repressor binding to the *lac* operator and since repressor binds specifically to *lac* operator DNA synthesized chemically in the laboratory without any backbone modifications. Recent studies indicate that several tyrosines located near the amino terminus of the repressor are critical for *lac* operator recognition, and they appear to interact with the N7 regions of several guanine residues present in the *lac* operator DNA. As is evident in Figures 1-6 and 1-8, the N7 region of guanine does not participate in base pairing but would be expected to be exposed in the major DNA groove. In general, the concept that regulatory proteins recognize the edges of bases is becoming increasingly prevalent, as we have already seen in our discussion of the RecA protein (*Section 17.6*) and as we shall see again in Section 23.7.

## 22.10 How Repressor Binding Inhibits *lac* Operon Transcription

The *lac* repressor appears to recognize a very small group of nucleotide pairs but, if only by its sheer bulk, it covers a much larger region of the *lac* DNA once it binds, as illustrated in Figure 22-5. On the right-hand side of Figure 22-5 is indicated the portion of *lac* DNA that is protected from nuclease digestion by RNA polymerase binding. Considerable overlap is seen to exist between the DNA occupied by polymerase binding and the DNA occupied by repressor binding. Therefore, as might be expected, binding of repressor to the operator and binding of the RNA polymerase to the promoter are mutually exclusive events *in vitro* and, one assumes, *in vivo* as well, thus explaining how repressor binding inhibits transcription of the *lac* operon.

**FIGURE 22-5 The *lac* operator.** Sequence of nucleotides in the *lac* control region and in the initial sector of the *lacZ* gene. (Courtesy of W. Gilbert.)

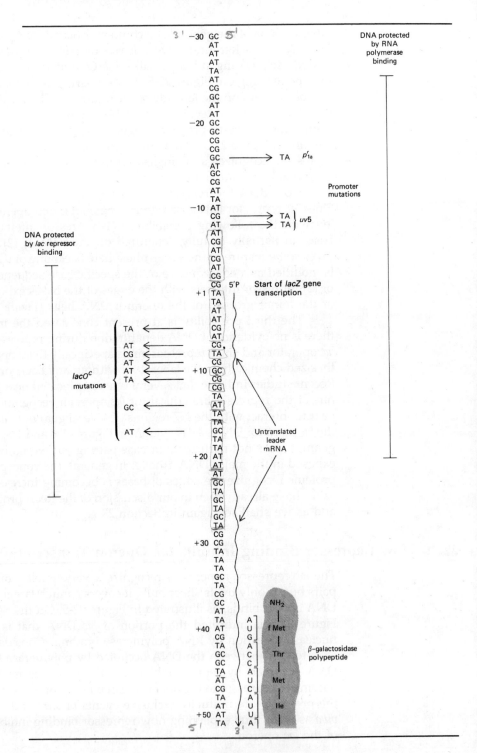

## 22.11 The CAP Protein and Catabolite Repression

For many years, regulation of *lac* operon transcription was thought to reside solely in a competition between repressor and RNA polymerase for access to the same controlling element, much as in our initial general diagram of negative control (*Figure 22-1*). Recently, however, it has become apparent that the *lac* operon is also under a positive form of control exerted by a dimeric regulator molecule generally known as **CAP** (Catabolite Activator Protein). The term **CRP** (Cyclic AMP Receptor Protein) is also used. Both names include essential information about this protein's function, as we now describe.

It has been known for some time from the work of B. Magasanik and others that the presence of glucose inhibits the synthesis of enzymes involved with utilization of such sugars as lactose and arabinose. This phenomenon is called **catabolite repression** or the **glucose effect,** and it occurs in many microorganisms, including *E. coli* and yeast. More recent experiments with *E. coli* have shown that the glucose effect is brought about by two molecular intermediaries, namely, the CAP protein and a small nucleotide known as **cyclic AMP (cAMP)** (Figure 22-2d). CAP is sensitive to intracellular levels of cAMP. As long as levels are high, CAP stimulates the expression of such operons as *lac* and *ara.* Introduction of glucose causes, by some unknown mechanism, a fall in intracellular cAMP levels and, therefore, a reduction in CAP activity. As a result, *lac* (and *ara*) transcription drops to low levels. The "purpose" of this control system is presumably to allow the preferential catabolism of glucose whenever it becomes available, since, energetically, glucose is the optimum carbon source for microbial growth.

Strains of *E. coli* that carry mutations in *crp* (the structural gene for CAP) and *cya* (the gene specifying adenylate cyclase, the enzyme necessary for cAMP production) are capable of only 2 percent of the maximal level of *lac* operon expression. Moreover, when isolated *lac* operon DNA is presented with RNA polymerase and ribonucleotides, active transcription will occur only when an extract of *crp*+ (but not *crp*−) cells is included and cAMP is provided. Such observations indicate that **CAP functions as an activator,** stimulating *lac* transcription by binding to *lac* DNA, and that **cAMP acts as an effector molecule,** interacting with CAP and thereby enhancing its activity.

How does CAP-plus-cAMP enhance the transcription of the *lac* operon? A clue comes from experiments that identify the sector of *lac* DNA protected from nuclease digestion by CAP binding. This sector proves to be "upstream" from the polymerase-binding sector identified in Figure 22-5 and centers instead around the −60 position, as drawn in Figure 22-6. Several mutations that prevent CAP binding to *lac* DNA *in vitro* are found to map to the −60 region (Figure 22-6).

Taken together, therefore, the various observations on *lac* transcription suggest that the *lac* operon promoter itself is a relatively **low-level**

**FIGURE 22-6 CAP binding site.** Numbers extend upstream from those in Figure 22-5. The L8, L29, and L1 mutations all affect CAP-stimulated transcription. (From W. Gilbert, in *RNA Polymerase*, R. Losick and M. Chamberlain, Eds. Cold Spring Harbor, N.Y.: Cold Spring Harbor Laboratories, 1976, p. 199.)

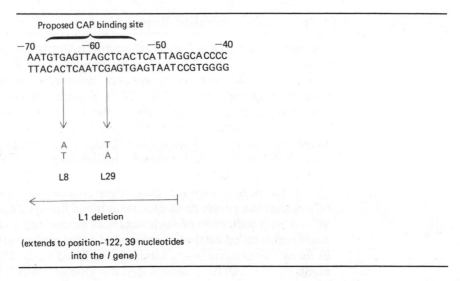

promoter, meaning that even when repressor is "pulled off" the *lac* DNA, RNA polymerases initiate transcription only relatively infrequently. When CAP interacts with *lac* DNA at its CAP binding site, on the other hand, the "strength" of the promoter is increased, and initiation frequencies are greatly enhanced. Figure 22-7 suggests one possible way that CAP binding might stimulate *lac* transcription: CAP may interact directly with RNA polymerase as well as with DNA, thereby increasing polymerase binding to

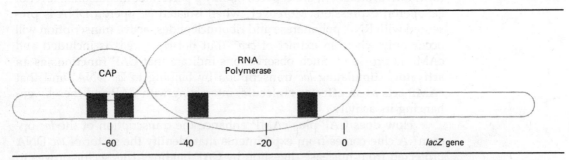

**FIGURE 22-7 CAP regulation model,** showing a hypothetical arrangement of the CAP protein and the RNA polymerase protein on the *lac* operon. If the two proteins are laid out along the DNA, assuming that each has a 2:1 axial ratio, then they can contact each other as well as the DNA recognition regions (gray boxes). (After W. Gilbert, in *RNA Polymerase*, R. Losick and M. Chamberlain, Eds. Cold Spring Harbor, N.Y.: Cold Spring Harbor Laboratories, 1976, p. 199.)

its promoter. A second possibility is that CAP binding may alter the conformation of the downstream DNA so that the promoter sequences are more easily recognized and/or entered by polymerase enzymes.

## 22.12 Overlap Between *lacZ* and the Operator

Before leaving the *lac* operon, we should consider Figure 22-5 one more time and note an interesting feature of its organization, namely, that *lac* operator DNA serves a double role. Nucleotides 1 to 21 of the operator interact with repressor in the uninduced cell and then, when the cell is induced and repressor leaves the operator, this same sequence is transcribed into the nontranslated "leader sequence" of *lac* operon mRNA. It should be stressed that operator sequences are not always transcribed, as we shall see when we consider the *trp* operon.

# Regulation of Tryptophan Biosynthesis

## 22.13 General Features of Tryptophan Regulation

The biosynthesis of each amino acid in a bacterial cell is performed by a series of enzymes. The tryptophan pathway, presented in Figure 21-1, involves five distinct enzymatic activities, all of which are specified by a single operon called *trp*. Regulation of these enzymes in *E. coli* proves to be carried out at three distinct levels; these are described in the following sections. Since similar mechanisms are believed to govern many other biosynthetic pathways, tryptophan regulation in *E. coli* can serve as our model for biosynthetic control in bacteria.

## 22.14 Feedback Inhibition

The first level of control for tryptophan biosynthesis, known as **feedback inhibition,** does not involve genes and is therefore not of direct relevance to this text. It is, however, an important form of control for most biosynthetic pathways and must be mentioned for completeness. During feedback regulation, a product of a biosynthetic pathway, usually the end product, inhibits the activity of an early enzyme in the pathway. Specifically, in the presence of increasingly high concentrations of tryptophan, the enzyme anthranilate synthetase (*Figure 21-1,* ASase) is found to increasingly reduce its affinity for its two substrates, glutamine and chorismate. Such a control system can clearly function at the phenotypic level to assure relatively constant rates of tryptophan biosynthesis.

## 22.15 Operator-Repressor Regulation of the *trp* Operon

Regulation of tryptophan biosynthesis at the gene level occurs at two separate sites in the *trp* operon, the first of these being the *trp* operator locus **trpO** (*Figure 21-1*). As with the *lac* operon, the sequence of nucleotides in the *trp* operator is recognized by a repressor protein. The **trp repressor** is encoded by the *trpR* gene, a gene that maps far away from the operon (in contrast to the contiguous *lacI* gene of the *lac* operon). Each repressor polypeptide is relatively small (108 amino acids) and, like the *lac* repressor, the *trp* repressor protein is a tetramer of four of these polypeptides.

The *trp* operon is repressible, a term we defined earlier in this chapter: transcription of its structural genes proceeds as long as tryptophan concentrations are low and falls when tryptophan concentrations rise. This effect is due, in part, to the fact that **the trp repressor alone is incapable of binding to the trp operator** (for this reason, it is more correctly referred to as an **aporepressor**). Binding can occur only when tryptophan, acting as an effector or **corepressor,** associates to form an **aporepressor-corepressor complex.** This complex then proceeds to bind to the operator and inhibit *trp* transcription in wild-type strains. In *cis*-dominant *trpO^c* mutants or in *trpR^−* mutants, binding fails to occur and transcription becomes constitutive. Studies *in vitro* have shown that the bound repressor-tryptophan complex excludes RNA polymerase from its promoter binding site; conversely, once RNA polymerase has bound to the *trp* operon, transcription cannot be inhibited by subsequent addition of repressor-tryptophan. Therefore, as with the *lac* operon (*Figure 22-5*), the two proteins appear to compete for binding to the same sector of the genome.

The *trp* operon has been transduced to both λ and ϕ80 phages (*Section 13.13*), allowing analysis of *trp* DNA and *trp* mRNA *in vitro*. Such studies, most notably those of G. Zubay, C. Yanofsky, and their coworkers, have generated the information summarized in Figure 22-8. The operator is contained in the region to the left of the purine number 1 (adenine) that initiates *trp* mRNA transcription, and it appears not to be transcribed. The hyphens and bars drawn in Figure 22-8 indicate two inverted repeats (*Section 4.9*) in the *trp* operator and in the first few nucleotides of the gene. If, for example, one starts at the dot in Figure 22-8 and reads to the right on the upper strand, the sequence encountered is the same as the sequence found on the opposite strand reading to the left of the dot. Inverted repeats are also present in the *lac* operator (*Figure 22-5*), in the CAP binding site (*Figure 22-6*), and in a number of other control regions that have been sequenced. The significance of this **dyad symmetry** is believed to reside in the fact that regulator proteins are usually active as dimers or tetramers. Thus two binding sites on the regulator protein are thought to make contact with the two symmetrical sites in the controlling element, allowing a far more stable and discriminatory interaction than if only one point of contact were involved. This concept is considered again in Section 23.7.

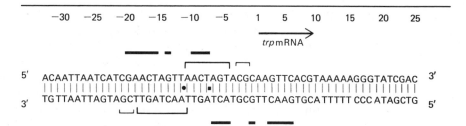

**FIGURE 22-8** *Trp* **Operator.** The initiation site for *trp* mRNA synthesis is numbered position 1. The centers of the hyphenated symmetries are indicated by a dot and a square between the two DNA strands. The nucleotide pairs involved in each symmetry are denoted by the presence of similar bars above and below the sequence. (From C. N. Bennett, M. E. Schweingruber, K. D. Brown, C. Squires, and C. Yanofsky, *Proc. Natl. Acad. Sci. U.S.* **73**:2351, 1976.)

**Multiple Functions of the *trp* Repressor**    The *trp* operon, located at 27 min on the *E. coli* map, is not the only target of the aporepressor encoded by the *trpR* gene. A second locus, known as *aroH* and mapping at 37 min, is also controlled by the same aporepressor-tryptophan complex. The first step in aromatic amino acid biosynthesis is catalyzed by the *aroH* gene product, an enzyme known as DAHP synthase, and the *aroH* gene is preceded by an operator locus that responds to the same fluctuations in tryptophan/repressor levels as does the *trp* operon. The *aroH* gene therefore behaves as an operon (as noted in Section 22.2, operators can control the expression of single genes as well as groups of genes), and the *trpR* repressor protein exerts control over two separate genetic units that function in tryptophan biosynthesis.

The third site of the *trp* repressor action is particularly interesting: the *trp* repressor binds to an operator that precedes the *trpR* gene itself. In other words, **transcription of the trpR gene is regulated by its own protein product,** a phenomenon called **autogenous regulation** (several other cases of autogenous regulation are described in the next chapter). The "purpose" of autogenous regulation is easy to imagine: levels of *trp* repressor must be kept very low (at about 20 to 30 molecules per cell) if the system is to be quickly responsive to fluctuations in free tryptophan concentration, and one means to achieve this end is to have the repressor repress its own synthesis as long as tryptophan levels are adequate.

## 22.16 Attenuator Regulation

The *trp* operon (and repressible operons generally) does not respond to catabolite repression and appears to have no regulation at the promoter level comparable to that exerted by CAP in the *lac* operon. Transcription of *trp* is, on the other hand, affected at the level of a controlling element known as the **attenuator region.** Since attenuator regions with similar properties have been discovered in the *phe, his, thr,* and *leu* operons as well, **attenuator control** is likely to represent a general regulatory mechanism for repressible operons.

The attenuator is so called because its net effect is to **attenuate transcription of the operon.** As diagrammed in Figure 22-9, transcription of the *trp* operon can occur in two ways. One mode of transcription (Figure 22-9a) is unremarkable: the transcript initiates at a promoter, copies a long (162-nucleotide) leader sequence (*Section 9.6*), and then copies the *TrpE* gene, which is translated into the TrpE protein. Under the influence of the attenuator, however, transcription follows a very different course (Figure 22-9b): after the first 140 nucleotides of the leader are copied, transcription terminates so that the structural genes are neither copied nor expressed. **The choice of transcriptional mode is dictated by intracellular levels of tryptophan:** when cells are starved of tryptophan, transcription proceeds past the attenuator and into the structural genes, as in Figure 22-9a, whereas when cells are grown in excess tryptophan, the attenuator dominates the process and foreshortened transcripts are produced (Figure 22-9b).

The existence of the attenuator was first suspected from the phenotype of *E. coli* mutants carrying *trp* deletions that map close to, but not within, *trpE*. These mutants carried out a constitutive expression of the *trp* operon structural genes, and, since the mutations were *cis*-dominant, it appeared that a controlling element had been deleted that ordinarily had the effect of attenuating transcription in the presence of tryptophan.

An understanding of how the attenuator might work came with the elucidation of its DNA sequence: the region deleted by the *cis*-dominant mutations was found to contain a string of T's in an AT-rich region, preceded upstream by an inverted repeat lying in a GC-rich region. This description will hopefully sound familiar: it is identical to our description of terminator sequences normally found at the **ends** of genes (*Section 9.8*). You will recall from reviewing Figure 9-6 that termination is believed to entail the formation of a stem-and-loop structure between the inverted repeats in the RNA transcript, followed by transcriptional run-off at the thymine-rich sector. Returning to the *trp* operon, it appears that soon after transcription

**FIGURE 22-9
Attenuator control
of *trp* operon
transcription.**
Transcription
pattern in the
absence of
tryptophan (a) and
the presence of
tryptophan (b).

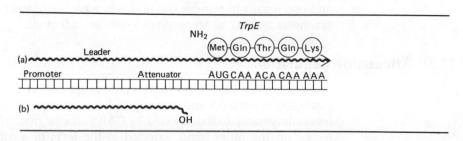

initiates, and in the presence of tryptophan, the polymerase encounters a functional termination sequence at the *trp* attenuator site. Somehow, however, the polymerase reads past this sequence if tryptophan is absent.

**How does tryptophan influence whether the attenuator sequence is active or inactive as a terminator of transcription?** The answer is again found in the DNA sequence. Careful scrutiny of a cloned *trp* operon shows that the "leader DNA" in fact encodes an AUG "start" at nucleotide number 27 and a UGA "stop" at nucleotide number 69; in other words, the so-called leader in fact encodes a short (14 amino acid) peptide in the following fashion:

| Met | Lys | Ala | Ile | . . . . . | **Trp** | **Trp** | Arg | Thr | Ser |
|-----|-----|-----|-----|-----------|---------|---------|-----|-----|-----|
| . . . . AUG | AAA | GCA | AUU . . . . . | | **UGG** | **UGG** | CGC | ACU | UGA . . . . |
| 27 | | | | | 54 | | | UCC | 69 |

Prominent in this sequence are two tandem UGG codons specifying tryptophan. This data led Yanofsky and coworkers to propose the **arrested ribosome model of attenuation** diagrammed in Figure 22-10. They suggest that soon after the leader DNA sequence starts being transcribed, a ribosome attaches to the AUG "start" and proceeds to begin translating the message into protein (*recall Figure 9-11*). If the cell is starved for tryptophan and, therefore, lacks aminoacyl tRNA$^{Trp}$, the ribosome will "stall" when it reaches the two UGG codons (Figure 22-10a). For reasons detailed further on, this stall prevents the attenuator portion of the transcript, located downstream, from forming its stem-and-loop structure, the result being that the polymerase proceeds into the *trp* structural genes (Figure 22-10b). If, on the other hand, ample aminoacyl tRNA$^{Trp}$ is available (Figure 22-10c), then the ribosome stall will not occur, the stem-and-loop structure is free to form, and transcription halts downstream at the attenuator (Figure 22-10d).

This leaves us with the following final question: why does the stem-and-loop structure form at the attenuator site when the ribosome is freely translating but fail to form when the ribosome is stalled? The proposed answer is illustrated in Figure 22-11. Figure 22-11a shows that the RNA transcript of the region contains four inverted repeats, labeled 1 to 4. The terminator stem-and-loop structure that we have been considering up to now involves repeats numbers 3 and 4, as diagrammed in Figure 22-11b. The second set of repeats, numbers 1 and 2, proves to be homologous to numbers 3 and 4. Thus the RNA can potentially form three different stem-and-loop structures, which we can call 3 · 4 (Figure 22-11b), 1 · 2 (Figure 22-11c) and 2 · 3 (Figure 22-11d). When the ribosome is freely translating the proximal end of the message in the presence of excess tryptophan, regions 1 and 2 are masked by the ribosome (Figure 22-11e) so that when 3 and 4 are transcribed they are free to form a 3 · 4 termination structure (Figure 22-11e). When, on the other hand, the ribosome stalls so that only region 1 is masked by the ribosome (Figure 22-11f), then a 2 · 3 loop will form,

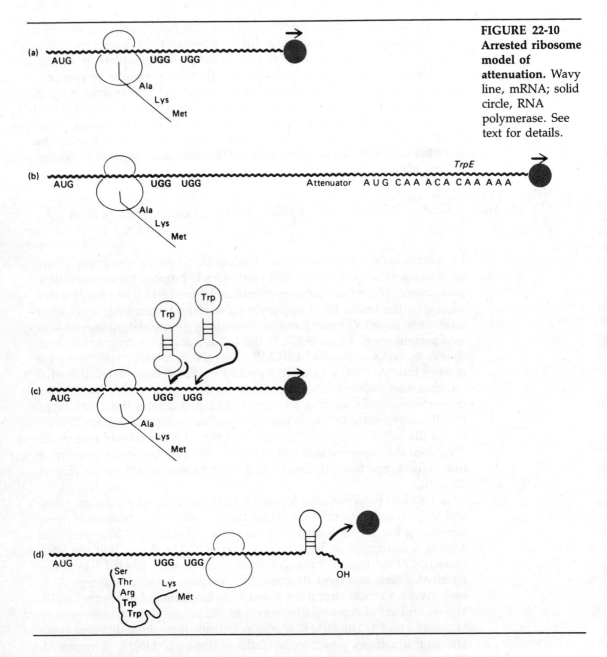

**FIGURE 22-10**
**Arrested ribosome model of attenuation.** Wavy line, mRNA; solid circle, RNA polymerase. See text for details.

leaving region 4 without a partner (Figure 22-11f). Thus the formation of the inhibitory 3 · 4 structure cannot occur, and RNA polymerase can continue transcription into the structural genes.

Before leaving the *trp* operon, we can pause to ask why it might be regulated in two such independent ways. The Yanofsky group has proposed that the **repressor-operator mechanism and the attenuator mecha-**

**FIGURE 22-11 Role of inverted repeats in attenuation.** See text for details. (After D. L. Oxender, G. Zurawski, and C. Yanofsky, *Proc. Natl. Acad. Sci. U.S.* **76:**5524–5528, 1979.)

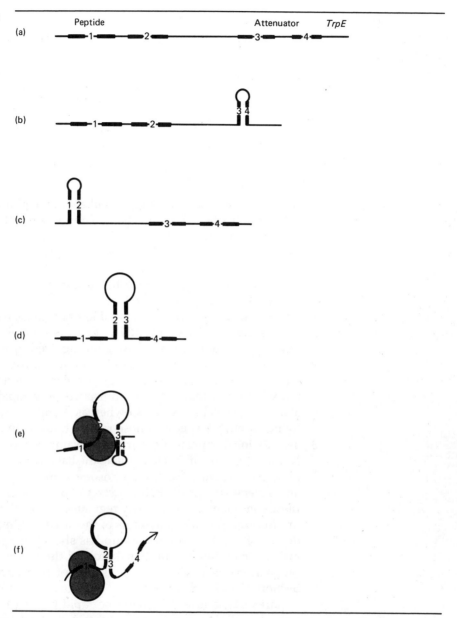

**nism may effectively back each other up:** as tryptophan levels drop, a two-stage response is launched during which repression at the operator is first lifted, followed by a relaxation of transcriptional attenuation. This perhaps enables the cell to regulate the expression of *trp* over a greater range of tryptophan concentrations than would be possible by the repression system alone.

# Translational Control

## 22.17 Representative Translational Control Mechanisms

The attenuation control mechanism described previously involves a finely tuned interaction between translation and transcription, the net effect of which is to regulate the transcription of particular sets of genes. Strictly defined, **translational control mechanisms operate after a gene transcript has left the template:** they may govern the rate at which ribosomes initiate translation of the message, the rate at which the message is read, the stability of the mRNA, and so on. We can cite three representative examples.

1. **Ribosome binding efficiency** is a likely mode of translational control. For example, the mRNA for the *trpR* repressor protein has been found to lack any recognizable Shine-Dalgarno sequence for ribosomal binding (*Section 9.6*). You will recall that very few *trpR* repressor proteins are present per cell. It has therefore been suggested that **efficient ribosome-binding sequences may be absent from mRNA species whose translation must be restricted.**

2. **Codon usage** is a second potential kind of translational control. Some codons are recognized by tRNAs that are very abundant in the cell; other synonymous codons may be recognized by isoaccepting tRNAs (*Sections 10.1 and 11.2*) that are present at relatively low levels. Therefore, it has been proposed that "popular" codons will be found in mRNA species that must be translated very rapidly, thus ensuring abundant protein products, whereas "unpopular" codons will be found in mRNA species whose rate of translation must be restricted.

3. **Translational repression,** our third example, has been found by M. Nomura and coworkers to regulate the balanced synthesis of ribosomal proteins. You will recall that a ribosome is made up of some 50 proteins and several species of rRNA (*Section 11.3*). How is the synthesis of all these components coordinately regulated so that the cell does not wind up, for example, with an excess of the large-subunit protein L11 and a dearth of L5? The Nomura group has shown that **certain free ribosomal proteins inhibit the translation of their own mRNAs** (an effect reminiscent of autogenous regulation of transcription, described in Section 22.15 for the *trpR* gene). Thus, as long as ribosome assembly requires L11, and L11 is therefore rapidly being removed from solution, the mRNA specifying this protein escapes inhibition and continues to dictate the synthesis of more L11. If L11 is overproduced, however, it binds to a particular target site on its mRNA and translation is blocked.

(a) Suzanne Bourgeois (Salk Institute), (b) Benno Müller-Hill (University of Cologne, Germany), and (c) Walter Gilbert (Harvard University) are all students of repressor-operator interactions in the *lac* operon of *E. coli*.

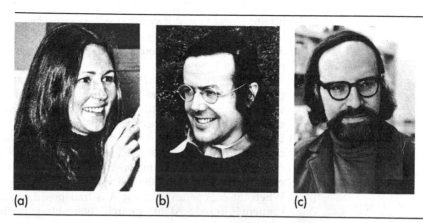

(a)    (b)    (c)

# Questions and Problems

1. The synthesis of enzyme E is inducible and dependent on three loci: a regulatory gene (*R*), a structural gene (*S*), and an operator (*o*). Haploid bacterial strains are found with the following properties:
   (a) Enzyme synthesis occurs only when inducer is present.
   (b) Enzyme synthesis never occurs.
   (c) Enzyme synthesis occurs in the presence or absence of inducer.
   (1) List all possible genotypes producing each of the phenotypes listed.
   (2) When the F′ element F′ $R^+$ $o^+$ $S^-$ is introduced into these haploid cells, which of the resultant partial diploids can synthesize E and which cannot?
   (3) Of those partial diploids that can synthesize E, which require the presence of an inducer?

2. Give the phenotypes of *E. coli* cells with the following genotypes:
   (a) $lacI^{-s}$ $lacO^c$ $lacZ^+$ $lacY^+$
   (b) $lacI^+$ $lacO^+$ $lacZ^-$ $lacY^+$ where *lacZ* carries a proximal nonsense mutation
   (c) $lacI^{-s}$ $lacO^+$ $lacZ^+$ $lacY^-$
   (d) $lacI^{-d}$ $lacO^c$ $lacZ^-$ $lacY^+$
   (e) $\dfrac{L1\ lacO^c\ lacZ^+\ lacY^+}{\text{F}'\ lacI^+\ lacO^+\ lacZ^-\ lacY^-}$ (see Figure 22-6 for location of *L1*)

3. In what respect is a repressor protein similar to an aminoacyl-tRNA synthetase molecule?

4. Four haploid strains of *E. coli* denoted A, B, C, and D carry mutations affecting the expression of the *lac* operon (none of the mutations lie in genes *lacY* or *lacA* or in promoter regions). The following observations are made:

(a) Strains B and C do not synthesize β-galactosidase, even in the presence of lactose.

(b) Strains A and D continuously synthesize β-galactosidase regardless of the presence of lactose.

(c) When the F' element F'$lacI^+$ $lacZ^-$ is introduced into strain A, the resultant partial diploid synthesizes β-galactosidase when lactose is present but not when it is absent. When the same F' element is introduced into strain D, no effect on strain D's phenotype is observed.

(d) Strain C can be induced to synthesize a material that is immunologically cross-reactive with β-galactosidase in a precipitin test (*see Box 2.3*). Strain B makes no cross-reactive material, but it will do so if an *amber* suppressor gene is introduced into the cell via an F' element.

(e) When the *lac* operon of strain A is transposed to an F' element and this element is introduced into strain B, the resulting partial diploid is inducible with lactose.

What are the genotypes of the four strains with regard to their *lacI, lacO,* and *lacZ* loci? Give your reasoning in each case. Indicate any cases where uncertainty remains.

5. Three mutations, *a, b,* and *c,* affect the expression of a repressible operon of *E. coli* where *abc* is the correct map order as determined by transduction. Strains carrying various combinations of these markers are tested to see whether they form the regulated enzyme in the presence or absence of corepressor. The following results are obtained (P means the enzyme is present in large amount; A means it is absent or present in only trace amounts):

| | Corepressor Present | Corepressor Absent |
|---|---|---|
| *a* + + | A | A |
| + *b* + | P | P |
| + + *c* | P | P |
| + + + /F' *a b c* | A | P |
| + + *c* /F' *a b* + | A | P |
| + *b* + /F' *a* + *c* | P | P |
| *a* + + /F' + *b c* | P | P |

Which mutation lies in the regulatory gene for the operon? Which lies in the structural gene for the regulated enzyme? Which lies in the operator locus of the operon? Give your reasoning in each case.

6. The feedback inhibition of enzyme activity occurs immediately, whereas the loss of enzyme activity by genetic repression requires several generations during which existent enzymes are diluted out. The following table summarizes the effects that two products of biosynthesis, A and B, exert on the activities of three biosynthetic enzymes (Immed. = immediate effect; Delay = effect after several generations; + means enzymatic activity; − means no enzymatic activity)

| | Presence of A | | Presence of B | | Presence of A + B | |
|---|---|---|---|---|---|---|
| | Immed. | Delay | Immed. | Delay | Immed. | Delay |
| Enzyme 1 | + | + | + | + | − | − |
| Enzyme 2 | + | − | + | + | + | − |
| Enzyme 3 | + | + | + | − | + | − |

Which enzymes are inhibited and which are repressed by the biosynthetic products? On the basis of these results, suggest a reasonable metabolic pathway leading to the production of A and B.

7. Would you expect *amber* mutations in *lacO*? Explain.

8. Cells carrying a deletion covering the *lac* repressor binding site and the *lacZ* gene continue to express the *lacA* gene.
   (a) Would you expect this expression to be regulated or constitutive? Explain.
   (b) Would you expect this mutation to be *cis*-dominant or also *trans*-dominant? Explain.
   (c) What does this mutation reveal about the relative locations of the *lac* promoter and *lac* operator?

9. How would you distinguish between *trpR⁻* and *trpOᶜ* mutations in *E. coli*?

10. A mutation at the *hisT* locus affects the expression of the repressible *his* operon of *E. coli* (*see Figure 21-3*). It is found that the mutation changes one nucleotide in the tRNA^His anticodon such that the tRNA^His fails to recognize the *his* codon in mRNA. Would you expect cells carrying this mutation to have high or low levels of *his* operon expression? Explain your answer.

11. When *lac* repressor proteins are briefly exposed to trypsin, 50 to 60 amino acids are clipped off one end; the resultant "core" tetramer continues to bind IPTG normally but no longer binds to operator DNA. Does trypsin cleave at the N-terminus or the C-terminus? Explain.

12. The behavior of *lacI⁻ᵈ* mutations in partial diploids has been termed **negative complementation.** Explain the meaning of this term.

13. The seqence of the proximal region of the *lacI* gene is shown below.

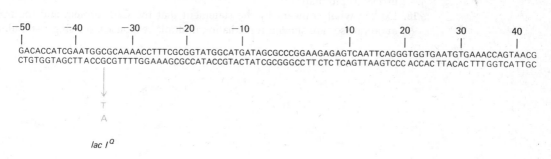

(a) The *lacI^Q* mutation occurs at a critical site. Explain.

(b) The *lacI^Q* mutation is called an "up promoter mutation," and Q stands for quantity. What is its effect on *lacI* expression?

(c) Partial diploids of *E. coli* that are *lacI⁻ˢ/lacI⁺* are noninducible, whereas *lacI⁻ˢ/lacI^Q* partial diploids can be induced to 25 percent of the normal level. Explain. (Hint: the *lac* repressor is a tetrameric protein.)

14. The **stringent response** occurs when auxotrophic *E. coli* are starved for their required amino acid. Thus for a starved valine auxotroph, the response will be triggered when a valine codon moves into the aminoacyl site on the ribosome

(*Figure 11-3*), and a charged valyl-tRNA$^{Val}$ is not available. Under these circumstances, an uncharged tRNA$^{Val}$ attaches to the ribosome, and this binding triggers the association of a protein, the **stringent factor,** with the ribosome. The cell responds by inhibiting the transcription of all tRNA and rRNA genes. Mutants of *E. coli* known as **relaxed** (*relA⁻*) cannot produce a normal stringent factor and therefore do not show this response to amino acid depletion.

(a) Explain why the terms "stringent" and "relaxed" were chosen to describe this control system.

(b) Would you classify the stringent response as an example of transcriptional or translational control? Explain.

(c) In what ways are the attenuator mechanism and the stringent response similar? How are they different? How might you test these conclusions using *relA⁻* and *trp⁻* strains of *E. coli*?

15. The *phe, his, thr,* and *leu* operons are regulated, in part, by an attenuator mechanism similar in many respects to that described for *trp*. The most important difference lies in the tandem codons present in the leader sequence. What do you think these codons would be in each case? Explain.

16. The *trpL29* mutant carries an AUA instead of an AUG at position number 27 of the *trp* leader sequence and therefore cannot initiate synthesis of the 14 amino acid peptide. As a result, stem-and-loop 1 · 2 (*Figure 22-11c*) is able to form.

(a) Does this prevent or allow the formation of 2 · 3 structures? Of 3 · 4 structures? Explain with diagrams.

(b) Would you expect *trpL29* to be able to relieve transcription termination in response to tryptophan starvation? Explain.

17. The *trpL75* mutant carries a mutation that destabilizes the 2 · 3 stem-and-loop structure (*Figure 22-11d*). Do you expect this strain to be more or less effective than wild type at relieving transcription termination in response to tryptophan starvation? Explain. How would you explain the observation that *trpL75* is more efficient than wild type in terminating transcription in the presence of excess tryptophan?

18. Explain what is meant by the statement that the CAP protein and the *trpR* aporepressor are similar in mediating multiple responses to a single stimulus.

# Control of Gene Expression in Bacteriophages and Eukaryotic Viruses

A. **Introduction**

B. **Regulation of Gene Expression by Lytic Bacteriophages**
    23.1  Phage T7 Regulation
    23.2  Phage SPO1 Regulation

C. **Regulation of Gene Expression During Phage λ Infection**
    23.3  General Features of λ Infection
    23.4  Regulation During the λ Lytic Cycle
    23.5  Regulation During the λ Lysogenic Cycle

23.6  Repressor-Operator Interactions in Phage λ
23.7  The Lysis-Lysogeny "Race"
23.8  The Lysogenic State
23.9  Prophage Induction

D. **Gene Regulation During SV40 Infection**
    23.10 SV40 Regulation in "Permissive" Hosts
    23.11 SV40 Regulation in "Nonpermissive" Hosts

E. **Questions and Problems**

## Introduction

Bacteriophages and eukaryotic viruses face unique challenges when they infect their hosts. Since the coding potential of the small viral chromosome is necessarily limited, it is advantageous to develop strategies for subverting host-directed DNA and protein synthesis into phage-directed DNA and protein synthesis. Moreover, it becomes important to keep these vital host functions operative during the course of the infection process, meaning that the host cell must be kept alive throughout at least most of the cycle. Indeed, many phages and viruses avoid killing the host cell altogether and instead set up an endosymbiotic relationship called lysogeny.

The enormous number of phages and viruses that inhabit the biosphere have developed an equally enormous array of specific mechanisms for responding to these challenges; we limit ourselves to three representative phages and one animal virus in this chapter. It is useful to realize, however, that all of the strategies described to date appear to be mediated by the same four classes of virus-specified macromolecules:

1. **Novel RNA polymerases** may selectively transcribe the viral chromosome.
2. **Novel RNA polymerase subunits** may associate with the host polymerase and divert its activities to the transcription of the viral chromosome.
3. **Repressor proteins** may inhibit host cell functions and/or modulate the sequence of viral gene expression.
4. **Activator proteins** may elicit necessary host-cell functions and/or modulate the sequence of viral gene expression.

In the examples of virus regulation presented in this chapter, we illustrate how these four kinds of proteins are utilized to carry out quite different kinds of infection cycles.

# Regulation of Gene Expression by Lytic Bacteriophages

## 23.1 Phage T7 Regulation

The infection cycle of T7, described in Section 12.1, is diagrammed in Figure 12-3, which should be reviewed at this time. Figure 23-1 presents a map of the T7 chromosome, which displays the kinds of genes encoded by the phage. Transcription of this genome proves to be controlled in an elegantly simple fashion. At the extreme "left end" of the T7 chromosome (the "top end" of the Figure 23-1 map) lie three sequences recognized as promoter sites by *Escherichia coli* RNA polymerase (*recall Figure 9-4*). When a T7 chromosome enters an *E. coli* host cell, therefore, a host RNA polymerase proceeds to initiate transcription from one of these sites and produces a single long transcript covering genes 0.3, 0.7, 1, 1.1, and 1.3, which are known as Class I genes (Figure 23-1). This transcript is next cleaved by host RNAse III molecules into five mRNA molecules, each encoding one of the five gene products, and the mRNAs are translated on host ribosomes. The gene 0.3 product acts to overcome the restriction/modification system of the *E. coli* host (*Section 4.11*) and is the first to act. Translation of a functional gene 1 product is, however, the pivotal event, for **gene 1 encodes a new T7 RNA polymerase,** which alone is able to recognize the promoters governing the remaining genes in the T7 chromosome.

The appearance of the T7 polymerase allows the remaining 80 percent of the genome to be expressed (Figure 23-1). This encodes the Class II gene products involved in phage DNA replication (along with the gene 1.3 product, a DNA ligase) and the Class III gene products involved with phage morphogenesis, DNA packaging, and host lysis. At the same time, the gene 0.7 product, a protein kinase, is thought to modify host RNA polym-

FIGURE 23-1
**Genetic map of T7**
showing essential
genes 1 to 19,
defined by
conditional lethal
mutations, and
unessential genes
mapping in
between. Classes I
to III represent the
three transcriptional
gene classes
described in the
text, and
percentages indicate
the proportion of
the chromosome
occupied by each
class. Numbers in
parentheses give
the molecular
weights of the
various gene
products. (After
F. W. Studier,
"Bacteriophage T7,"
*Science* **176:**367–376,
1972.)

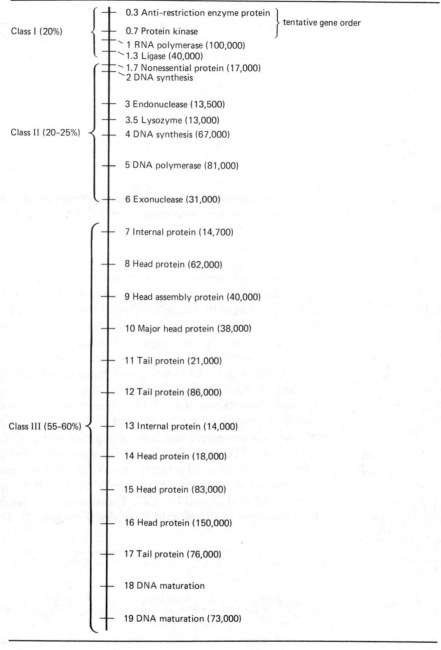

Class I (20%)
- 0.3 Anti-restriction enzyme protein ⎫
- 0.7 Protein kinase       ⎬ tentative gene order
- 1 RNA polymerase (100,000) ⎭
- 1.3 Ligase (40,000)

Class II (20–25%)
- 1.7 Nonessential protein (17,000)
- 2 DNA synthesis
- 3 Endonuclease (13,500)
- 3.5 Lysozyme (13,000)
- 4 DNA synthesis (67,000)
- 5 DNA polymerase (81,000)
- 6 Exonuclease (31,000)

Class III (55–60%)
- 7 Internal protein (14,700)
- 8 Head protein (62,000)
- 9 Head assembly protein (40,000)
- 10 Major head protein (38,000)
- 11 Tail protein (21,000)
- 12 Tail protein (86,000)
- 13 Internal protein (14,000)
- 14 Head protein (18,000)
- 15 Head protein (83,000)
- 16 Head protein (150,000)
- 17 Tail protein (76,000)
- 18 DNA maturation
- 19 DNA maturation (73,000)

erases in such a way that they no longer transcribe either the Class I genes
of the phage or the bacterial genes of the host. As a result, all nucleotide
precursors for RNA synthesis become available for the transcription of the
Class II and Class III genes; this assures the rapid and efficient production
of phage progeny.

Looking at the overall pattern of T7 infection, therefore, we find that the phage genes fall into two categories: the **early genes,** clustered together as Class I, are the first to be expressed; the **late genes,** clustered together as Classes II and III, **are not expressed until products of early-gene transcription become available.** Thus a temporal sequence of gene expression is assured by requiring that early-gene products be synthesized before late-gene transcription can begin.

## 23.2 Phage SPO1 Regulation

Phage SPO1 is a large virus that carries out a virulent infection of *Bacillus subtilis.* Many features of its lytic cycle are similar to those of T4, the *E. coli* bacteriophage that has been the more intensively studied. But whereas the molecular basis for gene control in T4 appears to be fairly complex, the analogous control system of SPO1, as elucidated by J. Pero and colleagues, is elegantly straightforward.

Gene expression during the SPO1 cycle occurs in the temporally defined sequence summarized schematically in Figure 23-2. **Early genes** are transcribed by *B. subtilis* RNA polymerases almost immediately after infection, exactly as with phage T7. One of the products of early transcription is the polypeptide, specified by phage gene *28,* which is required for subsequent events in the cycle, again as in the case of T7. Gene *28,* however, does not specify a new phage-specific polymerase, but rather a polypeptide known as **gene *28* protein,** which associates with the host polymerase in place of the host sigma (σ) factor. You will recall (*Section 9.3 and Figure 9-2*) that the σ factor of an RNA polymerase is responsible for recognizing critical sequences at the "−35" and "−10" regions of a promoter. In both the host chromosome and the early genes of the SPO1 phage, these sequences are TTGACA and TATAAT, respectively. Replacement of the host σ factor by the gene *28* σ factor causes the RNA polymerase to recognize a different set of sequences at the "−35" and "−10" regions of a promoter, namely, AGGAGA and TTTTTT. **These sequences are found exclusively in the promoters of phage middle genes** (Figure 23-2). Therefore, the transi-

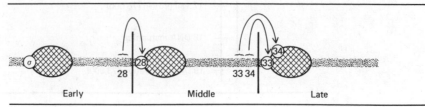

**FIGURE 23-2 Phage SPO1 gene regulation.** The figure illustrates schematically the three major categories of SPO1 gene transcription and is not intended to represent the actual physical arrangement of early, middle, and late genes on the phage genome or the number and arrangement of promoter sites. The cross-hatched ellipses represent the host subunits of RNA polymerase other than sigma factor. (From T. D. Fox, *Nature* **262:**748, 1976.)

tion from "early" to "middle" involves **a modification of the host RNA polymerases,** rather than a substitution for them, such that they can recognize and transcribe only phage middle genes.

Two of the phage middle genes are genes *33* and *34*, which specify gene *33* protein and gene *34* protein. These polypeptides interact to form yet another σ factor, which replaces the gene *28* protein and directs the RNA polymerase to transcribe the **late genes,** which, it is presumed, have unique "−35" and "−10" sequences of their own. The SPO1 control mechanism presumably eliminates the need for an equivalent of the 0.7 gene product of T7, which shuts off host RNA synthesis: here, host transcription abates as polymerases become modified in phage-specific directions.

# Regulation of Gene Expression During Phage λ Infection

More is known about the control of gene expression in phage λ than in any other phage. An understanding of λ regulation is therefore immensely rewarding, but it is not easily acquired. A number of different operons, regulatory proteins, and "feedback loops" are involved that cannot be accurately described in any simplified way. The rather detailed account of the lytic and lysogenic cycles of λ presented here will allow a full appreciation of their complexity of design.

## 23.3 General Features of λ Infection

As noted at several points in previous chapters, phage λ can follow one of two pathways after it infects an *E. coli* cell: it can direct a lytic cycle that results in the ultimate lysis of the host cell, releasing a burst of about 100 phage progeny; or it can lysogenize the host, during which the infecting phage chromosome inserts into the *E. coli* chromosome (*see Figure 12-4*). The decision of whether λ pursues a lytic or lysogenic course depends on the relative levels of a number of λ-coded regulatory proteins, some of which are lysis-promoting and others of which are lysogeny-promoting. To simplify the presentation, a description is first given of a λ infection that results in lysis, where the lysogeny-promoting genes are ignored. A λ infection is then described that results in lysogeny; this time the lysis-promoting genes are ignored. Finally, consideration is given to the competition between these opposing sets of gene products during an actual λ infection.

## 23.4 Regulation During the λ Lytic Cycle

The λ lytic cycle occurs in three phases, much as with the SPO1 cycle, but for λ these have come to be called the immediate-early, delayed-early, and late phases.

The **immediate-early phase** begins immediately after a λ chromosome enters the host cell, and, like the early phases of T7 and SPO1, transcription is directed solely by unmodified host RNA polymerases. The polymerases initiate transcription at two promoter sites. As diagrammed in Figure 23-3, one of these **immediate-early promoters** is adjacent to a sense sequence that is transcribed in a rightward direction; this sequence is therefore said to reside in the **R** (right) DNA strand, and the promoter is called $P_R$. The second immediate-early promoter, $P_L$, is located in the left (L) DNA strand and is contiguous to genes transcribed in a leftward direction. Transcription initiated at $P_R$ leads to the expression of a gene called *cro*, which we consider in a later section. Transcription initiated at $P_L$ leads to the expression of **gene N** and the production of **N protein**. Since N protein is required for the immediate-early $\longrightarrow$ delayed-early transition, its role in the lytic cycle can be analogized to the gene *28* protein of SPO1 (*Section 23.2*).

The **delayed-early phase** begins when N protein stimulates transcription of delayed-early genes. As diagrammed in Figure 23-3, these include genes *cII* and *cIII*, which are relevant only to lysogeny; genes *O* and *P*, which are required for replication of the λ chromosome; and **gene Q**. The product of gene *Q* is required for the delayed-early $\longrightarrow$ late transition, and its role in the lytic cycle is therefore analogous to the gene *33* and gene *34* proteins of SPO1.

The **late phase** begins when Q protein stimulates transcription of late genes. These include genes that encode head and tail proteins and genes that encode lytic factors. Meanwhile, λ DNA replication is proceeding apace under the influence of genes *O* and *P*; the new chromosomes are packaged into fully assembled phages; and the cell is lysed to release the phage progeny.

**FIGURE 23-3**
**Immediate-early transcription** of genes *N* and *cro* from the $P_L$ and $P_R$ promoters by *E. coli* RNA polymerase (gray object) in the circular chromosome of phage λ. Strand L is "sense" for leftward transcription; strand R is "sense" for rightward transcription.

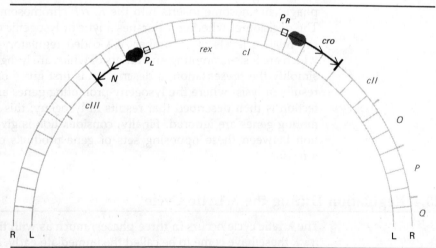

**N and Q Proteins**   We can now consider the N and Q proteins in more detail. The role of the N protein was first inferred by studies of $N^-$ mutant phages. When these phages infect *E. coli*, immediate-early transcription begins normally, but the host polymerases remain "frozen" at the ends of the *N* and *cro* genes (Figure 23-3) and soon fall off the template. By contrast, when phages are able to dictate a normal N protein, the polymerases are able to move past these stopping points and transcribe the downstream genes (*cIII* for the leftward unit and *cII* for the rightward unit, as illustrated in Figure 23-4). Thus **the N protein functions as an antiterminator.** If we focus on the rightward unit, a terminator sequence appears to reside at the *cro/cII* gene boundary (Figure 23-3, black bar), which causes unmodified RNA polymerases to stop transcription; this sequence is ignored, however, by RNA polymerases that have associated with N protein. A similar event occurs at the terminator sequence located at the *N/cIII* gene boundary in the leftward unit (Figure 23-3).

That the N protein exerts its antiterminator effect by combining with RNA polymerases was first inferred by studies of mutant strains. In this case, C. Georgopoulos began by isolating mutants of *E. coli* that could not support a λ lytic infection. Among these he singled out strains carrying mutations that map to the *rpoB* locus, which encodes the β subunit of *E. coli* RNA polymerase (*Section 9.3*), and determined that these strains indeed synthesized mutant forms of the polymerase enzyme. He then proceeded to select for λ mutants that were able to lyse the *rpoB* mutants and found that some of the λ strains carried mutations in gene N. These results suggested that the mutant *E. coli* polymerases were unable to interact with wild-type N protein and therefore could not read past the immediate-early termination signals when infected by wild-type phages, whereas they were quite capable of interacting with the mutant N proteins in such a way that read-through could occur and "lysis genes" could be expressed. To make

FIGURE 23-4
**Delayed-early
transcription** of the
λ chromosome
stimulated by N
protein.

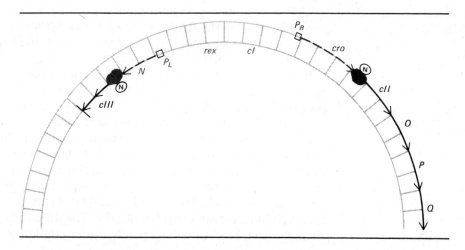

sure you understand these studies, try drawing diagrams of the two wild type and the two mutant strains and their interactions during infection.

Studies *in vitro* have established that N protein will exert its antiterminator influence on *E. coli* RNA polymerase only if it combines with the polymerase at a site near the $P_L$ or $P_R$ promoters. In other words, once an N-free polymerase transcribes an immediate-early gene and freezes at the termination signal, the addition of N cannot relieve the blockage: N must combine with the polymerase before or at the time transcription is initiated. Presumably, therefore, N protein either modifies the polymerase in some way at the promoter site or else travels with it to the site of N action.

Phages that are $Q^-$ express late genes only at a very low level. Although less is known about the Q protein, it is believed to act similarly to the N protein, allowing read-through of termination signals at the boundaries of delayed-early/late genes.

## 23.5 Regulation During the λ Lysogenic Cycle

Lysogeny requires two critical events: the λ **repressor protein** must be synthesized to turn off lytic-cycle functions, and the λ **Int protein** must be synthesized to mediate integration of the prophage into the *E. coli* chromosome. The λ repressor is encoded by the gene *cI*, whose transcription is governed by the promoter $P_{RE}$ (promoter for repressor establishment) (Figure 23-5). The Int protein is encoded by the gene *int*, whose transcription is governed by the promoter $P_{INT}$ (Figure 23-5). Both promoters are similar in sequence, and **both promoters are recognized by RNA polymerase only in the presence of the cII/cIII proteins.** Therefore, lysogeny first necessitates the N-dependent transcription of genes *cII* and *cIII* (Figure 23-4). It then depends on the cII/cIII stimulation of *cI* and *int* transcription (Figure 23-5). Finally, it depends on the actions of the repressor and Int proteins themselves.

The cII protein proves to act as a positive regulator or activator protein, analogous to the CAP protein of *E. coli* (*Section 22.11*). M. Rosenberg has shown that cII binds near the "−35" region of the $P_{RE}$ and $P_{INT}$ promoters, thereby stimulating RNA polymerase binding and the transcription of both operons. The cIII protein plays a crucial but more indirect role: it appears to inactivate a protease that would otherwise rapidly cleave cII protein. Its net effect, therefore, is to stabilize the cII protein.

The action of the λ repressor is a familiar one: it binds to operator sites and prevents transcription of neighboring genes (*recall Figure 22-1*). Specifically, it binds to two operator sites, called $O_L$ and $O_R$. Since these overlap the $P_L$ and $P_R$ promoters, λ **repressor binding prevents transcription of the leftward and rightward operons.** As a result, transcription of genes N and *cro* (*Figure 23-3*) is abolished, and, since the N and Cro proteins are unstable, their intracellular levels fall rapidly. The drop in N protein concentrations, in turn, blocks the N-dependent transcription of genes O, P, and Q

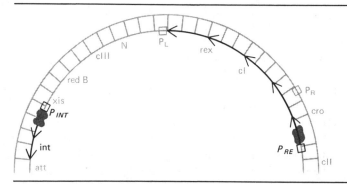

**FIGURE 23-5 Transcription stimulated by cII/cIII proteins to establish phage λ lysogeny.** The *cI* gene is read from the $P_{RE}$ and the *int* gene from the $P_{INT}$ promoter. A gene called *rex* is also transcribed from $P_{RE}$; its protein product, in some unknown fashion, prevents λ lysogens from being lysed by infecting T4 phages carrying a mutation in the *rII* gene (*Section 12.13*).

(*Figure 23-4*) and therefore prevents continuation of the delayed-early and late phases of the lytic cycle. Therefore, **if sufficient repressor becomes bound to $O_L$ and $O_R$, the lysogenic stage is set.**

The action of the Int protein has been considered in detail in Section 18.1. Briefly, it is required to mediate the site-specific recombination between *attP* and *attB*, which results in λ prophage integration.

At this point, therefore, we find that phage λ uses three kinds of regulatory strategies: (1) it can produce the N and Q proteins, which interact with RNA polymerase and influence its reading of termination sequences; (2) it can produce the cII activator protein, which binds to upstream sequences in promoters and facilitates RNA polymerase/promoter binding; and (3) it can produce a classic repressor protein, which binds to operator loci in DNA.

## 23.6 Repressor-Operator Interactions in Phage λ

Mutations in the *cI* gene that encodes the λ repressor cause a plaque phenotype (*Section 7.2*) known as *clear*: whereas wild-type λ will lyse some host cells and lysogenize others, producing a *turbid* plaque on a bacterial lawn, *cI⁻* phages can only lyse their hosts and therefore leave a clear patch in the lawn. Since the expression of the *cI* gene is dependent on the *cII* and *cIII* gene products, *cII⁻* and *cIII⁻* mutants also produce *clear* plaques.

Wild-type and mutant forms of λ repressor have been isolated and studied by M. Ptashne and collaborators. It proves to be a stable protein, with a monomer molecular weight of 26,000, which interacts with its operator targets as a dimer.

The two target operators, $O_L$ and $O_R$, have also been probed by genetic analysis. Mutant strains of λ, known as **virulent**, direct the synthesis of normal repressor but cannot respond to it; they are therefore unable to lysogenize, and they produce *clear* plaques. These strains prove to carry mutations that are the equivalent of *lacO^c* mutations (*Section 22.9*): their **operator DNA is unable to bind repressor.**

Ptashne and coworkers have performed studies to determine which regions of DNA are protected from DNAse digestion by repressor binding. The results, diagrammed in Figure 23-6, are somewhat surprising. **Each operator is found to contain not one but three repressor binding sites:** these are denoted $O_L1$, 2, and 3 for the $O_L$ locus and $O_R1$, 2, and 3 for the $O_R$ locus. The various *virulent* mutations in the $O_R$ region, denoted by arrows in Figure 23-6, are seen to map within the three binding sites, whereas two promoter mutations map to the regions betwen the binding sites. Since λ repressor will prevent the binding of *E. coli* RNA polymerase to operator DNA *in vitro*, there is clearly considerable overlap between operators and promoters, just as we saw for bacterial operons (*Section 22.10*). The regions are therefore usually referred to as $P_LO_L$ and $P_RO_R$.

The DNA sequence of the $P_RO_R$ region is given in Figure 23-7. Each repressor binding site is 17 base pairs long, and each is separated from the next by short AT-rich regions, which may aid in signaling polymerase binding. Each sequence is approximately symmetrical, carrying an imperfect inverted repeat (*Section 4.9*), just as we saw for the operator locus of the *trp* operon of *E. coli* (*Section 22.15 and Figure 22-8*). As with bacterial operons, this dyad symmetry is believed to be critical for repressor recognitions, a concept we develop more fully in the next section.

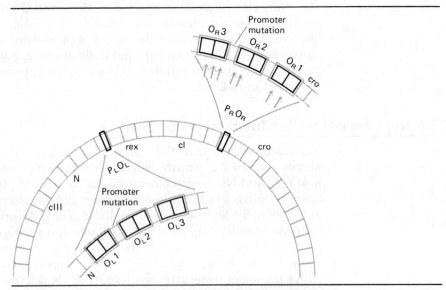

**FIGURE 23-6 The two operator regions of the λ chromosome,** showing their three operator sites. Positions of *virulent* mutations in the $O_R$ region are indicated by arrows.

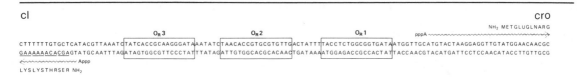

FIGURE 23-7 DNA sequence of the $O_R$ region of phage λ. (From M. Ptashne, A. Jeffrey, A. D. Johnson, R. Maurer, B. J. Meyer, C. O. Pabo, T. M. Roberts, and R. T. Sauer, *Cell* 19:1–11, 1980.)

## 23.7 The Lysis-Lysogeny "Race"

We can now consider an actual λ infection and describe the competition between lytic and lysogenic pathways. Although all six regulator proteins of λ are involved in the competition—namely, the N, Q, cII, cIII, cI, and *cro* gene products—the action of the **Cro protein** is perhaps the most pivotal.

The *cro* gene derives its name from control of repressor and other things. As described in Section 23.4, *cro* is expressed immediately after a λ infection and does not require N function (*Figure 23-3*). The Cro protein therefore comes to be present in the cell somewhat before the cII and cIII products appear and therefore before *cI* expression and repressor synthesis are activated. **Cro is also a repressor,** and it interacts with the same binding sites in the $O_L$ and $O_R$ operators as does the *cI*-specified repressor; its binding, for example, is sensitive to the same *virulent* mutations. **The Cro protein does not, however, bind nearly as efficiently as the *cI* repressor, nor is it nearly as stable.** Therefore, as it binds to $O_L$ and $O_R$, it slows down but does not completely inhibit expression of genes N and *cIII* in the leftward operon and genes *cro*, cII, O, P, and Q in the rightward operon.

We now come to the critical point in the overall process. You will recall that *cI* gene expression from the $P_{RE}$ promoter will occur only if the cII and cIII proteins are present, whereas expression of lysis-promoting genes requires the antiterminator activities of the N and Q proteins. It turns out that $P_{RE}$**-mediated transcription of *cI* is more sensitive to a decline in cII/cIII levels than the expression of lysis-promoting genes is sensitive to a decline in N and Q levels.** Therefore, if enough Cro protein binds to $O_L$ and $O_R$ so that cII and cIII levels are low, then the *cI* gene is never adequately activated. As a result, no stable inhibition of phage DNA replication and head and tail synthesis is achieved, and a lytic cycle ensues. If, on the other hand, insufficient Cro is produced or its binding to the two operators is not effective enough to abate *cII* and *cIII* expression, then abundant cI repressor will be synthesized and lysogeny will ensue. Such factors as temperature, the metabolic state of the host, and the genotypes of the host and the infecting phage all appear to influence the level of Cro production so that, in any given infection, the final balance between Cro-mediated "pseudo-repression" and *cI*-mediated "true" repression may tip one way or the other.

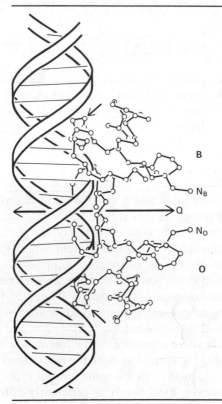

**FIGURE 23-8 Model of Cro repressor interacting with DNA.** Two monomers of Cro (labeled B and O), related by the twofold symmetry axis Q, interact with the DNA so that axis Q coincides with the twofold (dyad) symmetry axis of the operator DNA. The respective amino termini of the two Cro molecules are labeled $N_O$ and $N_B$. A pair of twofold-related $\alpha$-helices (arrows) are exactly 34 Å apart, meaning that they could occupy successive major grooves of the DNA. Closer to the symmetry axis, two extended polypeptide strands run parallel to the backbone of the DNA. (From W. F. Anderson, D. H. Ohlendorf, Y. Takeda, and B. W. Matthews, *Nature* 290:754–758, 1981.)

**Cro Binding to the λ Operator**   A model of the interaction between Cro protein and λ operator DNA is shown in Figure 23-8. Details of the model are given in the legend; the key points are as follows.

1. Cro is seen to fit very snugly into two adjacent major grooves of the DNA helix. As with models of RNA polymerase binding (*Section 9.7*), Cro is thought to bind to only one side of the helix, where it is presumably able to recognize the "edges" of the base pairs exposed in the two major grooves (*recall Figure 1-8 and Sections 17.6 and 22.9*).
2. The protein has the structure of a mirror-symmetrical dimer: if you imagine the line labeled Q in Figure 23-8 as a mirror, then the upper monomer is a "reflection" of the lower monomer.

3. As we have already noted in Figure 23-7, the λ operator binding sites possess dyad symmetry, meaning that they are also mirror-symmetrical. Since a 17-base-pair tract of DNA would occupy a full turn of a double helix (*Section 1.2*), one half of an operator would be expected to be exposed in one major groove of the helix, whereas its inverted repeat would appear in the adjacent major groove.

**Thus the twofold symmetry axis of the Cro repressor coincides exactly with the twofold symmetry axis of the operator DNA.** This type of interaction is expected to have been widely adopted by proteins that recognize symmetrical, sequence-specific regions of the genome, additional examples being other repressors and activators, polymerases, and transposase enzymes involved in site-specific recombination (*Section 18.4*).

## 23.8 The Lysogenic State

A diagnostic property of bacteria lysogenic for λ [*E. coli*(λ)] is that they are **immune to superinfection** by additional λ phages. In other words, when a culture of *E. coli*(λ) is presented with λ particles, the phage DNA is able to enter the bacterial cells, but no lysis of the bacteria occurs. This immunity results from the important fact that **the prophage dictates the synthesis of a low level of cI repressor.** The repressor serves to block transcription of the $P_LO_L$ and $P_RO_R$ operons in the prophage, thereby sustaining the lysogenic condition. It also represses the $P_LO_L$ and $P_RO_R$ operons of any superinfecting λ chromosomes that may happen to enter the cell. As a result, the infecting chromosomes cannot replicate, nor can they ordinarily insert into the host chromosome, since its *attB* site (*Section 18.1*) is already occupied. They therefore simply remain in the host cytoplasm and are diluted out by successive host cell divisions.

The low levels of repressor found in an *E. coli*(λ) lysogen prove to be transcribed not from the $P_{RE}$ promoter we have described earlier, but from a second promoter called **$P_{RM}$** (Promoter for Repressor Maintenance). As diagrammed in Figure 23-9, the $P_{RM}$ is directly adjacent to the *cI* gene, in contrast to the more distant $P_{RE}$. Unlike the $P_{RE}$, moreover, **the $P_{RM}$ does**

**FIGURE 23-9**
**Transcription from the $P_{RM}$ promoter** once lysogeny is established.

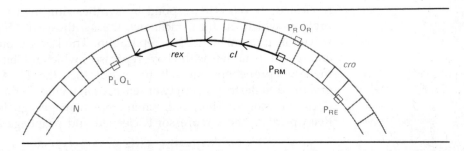

not require cII or cIII activation—indeed, neither *cII* nor *cIII* is expressed in a lysogen, since both the $P_LO_L$ and $P_RO_R$ operons are repressed. **Control over $P_{RM}$ transcription is instead exerted by the cI protein itself:** when repressor levels become low, repressor protein stimulates transcription from $P_{RM}$; when repressor levels "overshoot," transcription from $P_{RM}$ is turned off by repressor. How these effects are exerted is not yet clear, but such **autogenous regulation** (*recall Section 22.15*) is clearly designed to ensure that repressor synthesis continues at a slow, steady rate in the lysogenic bacterium.

## 23.9 Prophage Induction

The spontaneous loss of a λ prophage from an *E. coli* chromosome occurs only rarely. Three experimental means have therefore been developed to bring about the induction of a lytic cycle in lysogenic cells. All share a common feature: **the cI repressor protein is somehow prevented from maintaining repression of the λ prophage operons.**

The first approach to lysis induction is known as **zygotic induction.** Matings are induced between lysogenic *Hfr* donor cells and nonlysogenic recipient cells, as described in Section 13.2. When the chromosomal segment containing the prophage enters the recipient cell, it encounters a cytoplasm free of repressor protein. Transcription of the $P_LO_L$ and $P_RO_R$ operons therefore commences, and a lytic cycle frequently ensues. No such events occur when the recipient cells are also lysogenic for λ.

The second method of prophage induction utilizes a *cI* mutation known as *cI* 857. This mutation results in the synthesis of a thermolabile repressor protein, one that is stable at 30°C but becomes inactive at 42°C. Induction in this case simply requires raising the temperature of a lysogenic *E. coli* culture. Since the outcome of this procedure is far more reliable than other methods of induction, the *cI* 857 strain is now routinely used in studies of the lysogenic process.

The third method involves treating a culture of lysogenic bacteria with any of a variety of agents that damage DNA, including UV irradiation, mitomycin, or nitrogen mustard. The mechanism here proves to be quite interesting, for it involves the **SOS repair pathway.** You may recall from Section 8.12 that when DNA is damaged in some way and cannot be replicated, *E. coli* cells respond by synthesizing high levels of the RecA protein, which, in turn, facilitates the synthesis of "SOS enzymes" involved in untemplated DNA replication. The RecA protein is a protease, and one of its targets is the lexA repressor, which inhibits the synthesis of the error-prone enzymes. It turns out that **the RecA protein is programmed to destroy a variety of repressors,** including the λ repressor in an *E. coli*(λ) lysogen. Therefore, when DNA damage signals the synthesis of RecA protein, the λ repressor is cleaved and prophage is induced.

# Gene Regulation During SV40 Infection

We conclude our consideration of virus infection cycles with SV40, which, although very much a virus like λ in its infection pattern, can grow only in mammalian cells and therefore conducts its infection in a eukaryotic environment.

We have described SV40 in various contexts during the course of this text: its life cycle is presented in Section 12.3, its genetic map in Sections 12.16 and 12.18, and its integration into the host chromosome in Section 18.3. This material should be reviewed as background for the sections that follow.

## 23.10 SV40 Regulation in "Permissive" Hosts

The critical regulatory molecule in a productive infection of a monkey cell line is the T antigen, encoded by a DNA sequence known as gene A. During the eight to ten hours that follow viral infection, this "early" gene (*Figure 12-24*) is transcribed by host RNA polymerase II (*Section 9.3*) and translated on host ribosomes. Mutant strains that produce a thermolabile T antigen (*tsA* mutants) cannot direct either a lytic or a transforming infection at nonpermissive temperatures. Specifically, they cannot initiate replication of the viral chromosome, nor can they initiate the transcription of "late" genes for viral capsid proteins (*Figure 12-24*). Moreover, the thermolabile T antigen fails to bind to the SV40 chromosome in the region of the origin of replication, in contrast to its wild-type counterpart. Therefore, the T antigen is formally analogous to an essential "early" regulatory protein, such as the N protein of λ, in a bacteriophage cycle.

Figure 23-10 diagrams the critical control sector of the SV40 chromosome in the replication-origin (ORI) region, where the early genes are drawn lying to the right of ORI. The region contains two early promoters, called EI and EII, where EI lies within the ORI sequence itself. When the viral chromosome first enters the host, RNA polymerases select EI and generate T antigen transcripts (Figure 23-10a). As T accumulates, it forms tetramers that have a strong binding affinity for a DNA sequence called site I. Binding of T to site I (Figure 23-10b) blocks initiation from EI, but the polymerase remains capable of initiating at EII and producing more T antigen transcripts. Once all site I's are filled, the next wave of T antigen proceeds to bind to site II, located upstream from site I (Figure 23-10c). Site II binding disallows any further initiation of early gene transcription and simultaneously permits the onset of SV40 DNA replication and the transcription of late genes. Although the early $\longrightarrow$ late switch is not well understood, it is thought that the EI and EII promoters are much "stronger" than the late promoter, and that they therefore monopolize available polymerases until they are repressed by T antigen binding.

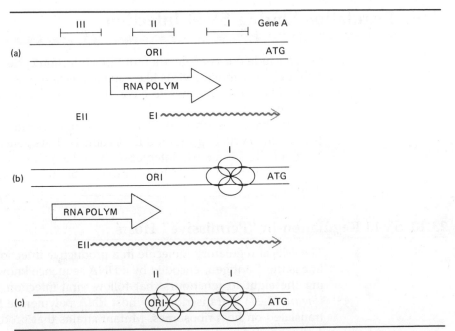

**FIGURE 23-10 T antigen regulation of SV40 early gene transcription.** (a) Early promoter region in the vicinity of the origin of replication (ORI) contiguous to early gene A. Three binding sites for T antigen (I, II, and III) are unoccupied at the onset of infection, allowing the RNA polymerase to initiate transcription (wavy lines) from site EI. (b) Binding of T antigen (tetramer) to site I causes transcription to initiate from site EII. (c) Binding of T antigen to sites I and II blocks further transcription of gene A. (After U. Hansen, et al., *Cell* **27**:603–612, 1981.)

It should be apparent from the previous paragraph that T antigen exerts **autogenous regulation** over its own synthesis, in much the same fashion as the λ repressor and the *trp* repressor. This is readily demonstrated with the *tsA* mutant. When shifted to restrictive temperature shortly after infection, existent thermolabile T antigen is degraded. This causes a 15-fold stimulation in early gene transcription, followed by a similar overproduction of mutant T protein (which proceeds to undergo degradation), exactly as one would predict, from Figure 23-10, if T were prevented from binding to sites I and II and thereby unable to modulate its own synthesis.

The final similarity we can cite between the lytic cycles of phage and animal viruses is that both employ **strand switching.** Thus in SV40, the capsid protein genes are encoded in the late (L) strand of the chromosome, whereas the early genes are encoded in the opposite early (E) strand and are therefore transcribed in the opposite direction (*Figure 12-24*), just as we saw with phage λ (Figure 23-3). The arrangement presumably facilitates expression of discrete segments of the genome at discrete phases of an infection cycle.

## 23.11 SV40 Regulation in "Nonpermissive" Hosts

In a mouse cell, expression of gene A sets the stage either for an abortive infection or for a transformation event, which, as we saw in Section 18.3, involves integration of the SV40 genome into the mouse genome. That expression of gene A is vital for transformation can again be demonstrated with *tsA* mutants: at high temperatures, such mutants are unable to initiate a transformation event. Even more dramatically, if cells are transformed by *tsA* mutants at permissive temperature and are then shifted to a restrictive temperature, the transformed state is in most cases aborted, and the cells revert to their normal state. Such events are, of course, formally analogous to λ prophage induction in the *cI* 857 strain that produces temperature-sensitive repressor proteins. It perhaps goes without saying that intensive research is presently focused on determining the molecular basis for SV40 transformation and the role of T and t antigens in this process.

(a)                            (b)

(a) Mark Ptashne (Harvard University) and (b) Dale Kaiser (Stanford University) have each made important contributions to our understanding of phage λ genetics and the process of lysogeny. Dr. Ptashne continues to analyze the molecular basis of λ gene expression; Dr. Kaiser now studies the genetics of cell–cell recognition in bacteria.

# Questions and Problems

1. What phenotypes would you expect of phage λ carrying the following mutations? Specifically, can they carry out lysogeny or a lytic cycle? What proteins can they synthesize?
   (a) A deletion of $cI$
   (b) A deletion of Q
   (c) A deletion of N
   (d) A deletion of $cro$
   (e) A deletion of $P_{RE}$
   (f) A deletion of $P_R$
   (g) A deletion of $P_L$
   (h) A deletion of $P_{RM}$

2. The **immunity region** of a temperate-phage chromosome contains its repressor-encoding gene and the operators controlled by the repressor. The temperate phage 434 is sufficiently similar to λ that the two chromosome types will undergo recombination in a mixed infection, yet phage 434 will readily superinfect and lyse an E. coli(λ) lysogen.
   (a) Explain why this information suggests that the immunity regions of λ and 434 are different.
   (b) A hybrid phage is constructed by recombination between λ and 434. Describe two ways you might test whether the hybrid contains a λ or a 434 immunity region.

3. Hybrid λ phages are constructed in which the $lacZ$ gene is fused next to the $λP_{RM}$ promoter. Levels of β-galactosidase expression were then observed in E. coli cells lysogenic for the hybrid phages. When the hybrid phages were $cI^+$, 2600 units of enzyme activity were detected per cell. Would you expect, in the following two cases, enzyme levels to be higher, lower, or the same as for the $cI^+$ hybrid? Explain.
   (a) The hybrid is $cI^-$.
   (b) The hybrid is $cI^+$ and the E. coli lysogens also contain a recombinant plasmid carrying a $cI$ gene that greatly overproduces $cI$ protein.

4. A phage T7 strain carries an *amber* mutation in gene 1. Compare the course of an infection by this strain of permissive and nonpermissive E. coli (*Section 11.17*).

5. The mRNA transcribed from $P_{RE}$ (*Figure 23.5*) has a much longer leader (*Section 9.6*) than the mRNA initiated at $P_{RM}$. Explain how this observation might help account for the different levels of λ repressor in a lysogen compared with levels found after the first few minutes of a λ infection.

6. A mutation in the $cI$ gene of λ, called $ind^-$, renders the repressor insensitive to uv induction; a second mutation, $cI^s$, renders the repressor unusually sensitive to uv induction. How would you explain these mutations by the two theories for uv-induction mechanisms offered in Section 23.9.

7. A mutation in the rightward operon of λ allows $N^-$ phages to carry out a normal lytic cycle. Explain the effect of this mutation and its probable location in the λ chromosome.

8. The $trp$ repressor of E. coli regulates its own synthesis (*Section 22.15*); an analogous regulation is described in this chapter for the λ repressor. How are the two similar and how are they different?

9. In an early experiment by R. Jayaraman and E. Goldberg demonstrating that T4 early genes are transcribed from one strand of the T4 chromosome and that T4 late genes are transcribed from the opposite strand, mRNA was isolated from *E. coli* cells shortly after T4 infection ("early") and again late in the cycle ("late"). T4 chromosomes were then dissociated into their heavy (H) and light (L) strands (*Box 4.1*), the separate strands were immobilized on filters, and hybridization was allowed with either early or late mRNA. The DNA was then digested with an endonuclease that destroyed all nonhybridized DNA, and the protected fragments were eluted from the filter. The four sets of fragments, which we can designate H/early mRNA, H/late mRNA, L/early mRNA, L/late mRNA, were now tested for their ability to restore the function of two mutant phages, one defective in early gene *rIIB* and other defective in late gene 21. In the first experiment, four cultures of bacteria were induced to take up *rIIB* phages along with one of the four sets of protected fragments; in the second experiment, four cultures were induced to take up *gene 21⁻* phages along with one of the four sets of fragments.

(a) Diagram this experiment.

(b) In which bacterial culture or cultures would you expect to observe normal phage lysis if early genes are encoded in the L strand? If late genes are encoded in the H strand?

10. Describe the expected effects of *amber* mutations in genes *28*, *33*, and *34* on the lytic cycle of phage SPO1.

11. When the bacterium *Bacillus subtilis* is starved for nutrients, it enters a developmental pathway that culminates in the formation of a dormant cell type known as the endospore. This **sporulation** program involves at least five temporally defined stages of gene expression. How would you test whether the strategies employed by the *B. subtilis* phage are also employed during sporulation of the host?

12. The order of expression of the three proteins encoded by such RNA phages as Qβ and R17 is determined as follows: the AUG codon of the coat-protein sequence is the only one exposed when a phage enters *E. coli*; translation of the coat sequence brings about a major conformational change in the structure of the phage RNA, with the result that the AUG codons for the replicase and A-protein sequences are unmasked.

(a) Is this an example of transcriptional or translational control (*Section 22.17*)? Explain.

(b) Would you expect proximal *amber* mutations in the coat-protein gene to be polar (*Section 11.14*)? Explain.

# 24

# Control of Gene Expression in Eukaryotes: Short-Term Regulation

A. **Introduction**

B. **Short-Term Regulation in Fungi**
   24.1 The *qa* Gene Cluster in Neurospora
   24.2 The *GAL* Gene Cluster in Yeast
   24.3 Regulation of Nitrogen Metabolism in *Aspergillus*

C. **Short-Term Regulation in Higher Eukaryotes**
   24.4 The Heat-Shock Response
   24.5 Hormonal Regulation in Higher Eukaryotes: General Features
   24.6 Ecdysone Stimulation of

Gene Transcription in Diptera
   24.7 Steroid-Hormone Stimulation of Gene Transcription in Mammals

D. **Mechanisms of Short-Term Regulation in Eukaryotes**
   24.8 Binding of Regulatory Molecules to Chromatin
   24.9 Controlling Elements Associated with Regulated Structural Genes
   24.10 Physical Conformation of Chromatin and DNA

E. **Questions and Problems**

## Introduction

Two different classes of regulatory phenomena can be recognized in eukaryotes. The first, which we can call **short-term** or **reversible regulation**, corresponds to the kind of regulation we have studied in bacteria (*Chapter 22*). It represents a cell's response to fluctuations in the environment; specifically, it involves changes in activities or concentrations of enzymes as particular substrates or hormone levels rise and fall and as the cell cycle (*Section 2.6*) is traversed. The second kind of eukaryotic regulation can be termed **long-term** or (usually) **irreversible** ("usually" in that there are occasions, such as cancerous growth, when the process appears to reverse itself). Long-term regulation includes the phenomena associated with

728

determination, differentiation, or, more generally, development: it is involved in the numerous steps by which a fertilized egg becomes an organism of perhaps trillions of cells that have diverse and, ultimately, quite permanent roles to play in the maintenance of the whole.

Regulation in eukaryotes is the subject of this and the next chapter. In this chapter we consider examples of short-term regulation in fungi, in insects, and in mammals. In the next chapter we examine a number of long-term differentiation phenomena that occur during eukaryotic embryology, focusing on studies that use genetic approaches. At the conclusion of each chapter, we consider how each kind of regulation might be brought about at the level of chromatin and DNA.

# Short-Term Regulation in Fungi

Many studies have been made of short-term control systems in fungi, most being concerned with either biosynthetic or catabolic pathways. The three examples presented here illustrate relatively straightforward, but representative, modes of control. Emphasized are similarities and differences between these and analogous bacterial control systems.

## 24.1 The *qa* Gene Cluster in Neurospora

A *Neurospora* cell contains all the enzymes required for the synthesis of unsaturated ring structures known as aromatic molecules, of which the amino acids tyrosine and phenylalanine (*Figure 3-1*) are familiar examples. When levels of such aromatic compounds in the cell become too high, several "scavenging enzymes" are induced that break down the compounds into nonaromatic forms. The induction of three of these enzymes is accomplished by elevated levels of an aromatic metabolite known as quinic acid, and the three induced enzymes—a dehydroquinase, a dehydrogenase, and a dehydrase—are encoded by the linked genes *qa-2*, *qa-3*, and *qa-4*. (Figure 24-1). Once quinic acid is catabolized, enzyme synthesis is turned off. The *qa-2*, *qa-3*, and *qa-4* genes in the cluster can, therefore, be classified as inducible genes, much as the inducible catabolism-related genes in *Escherichia coli* (*Section 22.4*). Each enzyme, moreover, functions as an independent protein, so that we are not dealing with a multifunctional protein like the *HIS4* gene product (*Section 21.2*).

We saw in Chapters 22 and 23 that the induction of gene expression in prokaryotes proceeds under the influence of positive or negative control mechanisms. In a positive-control system such as that exerted by CAP or N protein, gene transcription is stimulated directly; in a negative-control system such as that involving the *lac* or λ repressors, induction involves the lifting of transcriptional repression. The *qa* genes of *Neurospora* appear to be under positive control: the inducer (quinic acid) is thought to combine

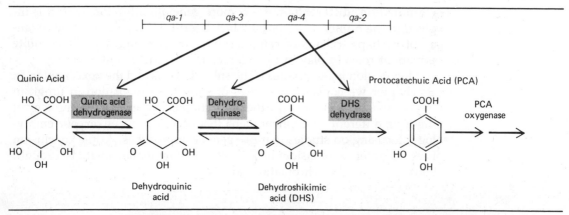

**FIGURE 24-1** *Qa* **gene cluster in** *Neurospora crassa* and the enzymes they specify in the catabolic quinate-shikimate pathway.

with a regulator (the product of the closely linked gene *qa-1*) to turn on *qa* gene expression.

Evidence for this model comes from the properties of *qa-1* mutations. A number of *qa-1⁻* mutant strains have been isolated by M. Case and N. Giles. All of these produce what is called a **pleiotropic negative** phenotype, meaning that quinic acid cannot induce any of the *qa*-coded enzymes. Several of these strains, moreover, prove to carry deletions in the *qa-1* gene. Since deletions in a repressor-protein gene would be expected to lead to constitutive enzyme synthesis, whereas deletions in a positive-regulator gene would be expected to generate a pleiotropic-negative phenotype, it is argued that the *qa-1* gene product somehow acts in a positive manner.

Since *qa-1* maps close to the regulated genes (Figure 24-1), an alternate possibility is that it might be a controlling element analogous to a bacterial or phage operator. Two observations, however, indicate that *qa-1* is a gene and not a controlling element.

1. Heterokaryons (*Section 15.1*) have been constructed that carry two (nondeletion) *qa-1⁻* alleles, and certain of these are found to complement one another; that is, they exhibit intragenic complementation (*Section 12.5*). This result is consistent with the idea that the *qa-1* gene product is a polypeptide that forms a multimeric protein; on the other hand, it is inconsistent with the postulate that *qa-1* is a controlling element, since controlling elements should not be able to complement each others' defects.

2. Certain *qa-1⁻* mutations are found to be temperature sensitive. Since almost all temperature-sensitive gene products are polypeptides, it again appears that *qa-1* encodes a protein having a regulatory function.

Although many *qa-1* mutations are pleiotropic negative, the *qa-1* gene can also mutate to a constitutive form. In such *qa-1ᶜ* strains, all three *qa* enzymes are synthesized even when quinic acid is absent. Case and Giles have interpreted this observation in the following way. They propose that the active *qa-1* gene product carries two functional domains: an **initiator region** and an **inducer-binding region.** Mutations in the initiator region would prevent activation of transcription (for example, by preventing binding to promoter sequences in *qa* DNA) and would thereby create the pleiotropic-negative phenotype. Mutations in the inducer-binding site, on the other hand, could be of two types. Those that simply prevent inducer binding should also be pleiotropic negative, whereas mutations that abolish the need for inducer binding to exert positive control should generate a constitutive phenotype: the *qa* gene would be "ON" all the time.

Expression of the *qa* gene cluster, in summary, appears to be regulated by the *qa-1* gene product, and certain of its genetic properties suggest that it acts as a positive regulatory molecule. Prokaryotic cells display two types of positive regulators: molecules that interact directly with DNA to stimulate transcription (e.g., the CAP protein [*Section 22.11*]) and molecules that interact with RNA polymerases to modify their initiation or termination specificities (e.g., the cII and the N proteins of phage λ [*Sections 23.4 and 23.5*]). Unfortunately, there is as yet insufficient understanding of *Neurospora* gene expression to distinguish between these possibilities. Indeed, despite extensive searches, no *cis*-dominant mutations (*Section 22.9*) have been found adjacent to the three structural *qa* genes that might identify controlling elements for the cluster. Moreover, no nonsense mutations in the proximal *qa-3* gene have been found that exert polar effects (*Section 11.14*) on the expression of the two distal genes. In other words, there is no evidence that the *qa* genes are coordinately transcribed in the fashion of an operon, even though it is clear that all three enzymes are coordinately synthesized in response to quinic acid induction. Therefore, prokaryotic models may or may not prove relevant to understanding how the *qa-1* product and quinic acid stimulate expression of these catabolic enzymes. We explore this concept more fully in the final sections of this chapter.

## 24.2 The *GAL* Gene Cluster in Yeast

Galactose fermentation in yeast is carried out by three enzymes—a kinase, a transferase, and an epimerase—in the same fashion as for *E. coli* (*Figure 18-10*). These enzymes are specified by three closely linked loci known as *GAL1*, *GAL7*, and *GAL10*. Whether these *GAL* loci are in fact three genes or encode three portions of a multifunctional polypeptide (*Section 21.2*) is not known; for present purposes we can consider them as three genes. The three enzyme activities are coordinately induced by galactose. In many respects, therefore, the system is analogous to the *qa* cluster in *Neurospora* described above.

The yeast *GAL* cluster differs from *qa*, however, in that the *GAL* cluster is controlled by the product of an unlinked gene. Moreover, since most mutations in this gene lead to a constitutive synthesis, and not an inhibi-

FIGURE 24-2
Regulation of
galactose
fermentation in
yeast.

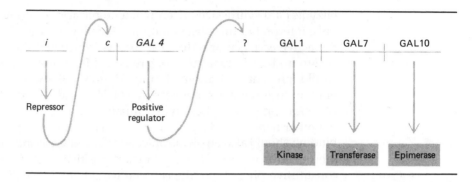

tion, of the galactose-catabolizing enzymes, this unlinked gene appears to encode a repressor-like protein. Additionally, mutations in the regulatory gene prove to be recessive to wild type in diploids and thus behave very much like certain *lacI⁻* mutations in the *lac* operon (*Section 22.7*). For this reason, the gene regulating the *GAL* gene cluster has been called *i*.

Further study has shown that the analogy between the *i* gene of yeast and the *lacI* gene of *E. coli* goes only so far. It turns out that **the *i* repressor of yeast acts not on the *GAL* gene cluster itself but on yet another unlinked gene,** *GAL4*. Mutations in the *GAL4* gene (denoted as *gal4*) generate a pleiotropic-negative phenotype for the kinase, transferase, and epimerase enzymes. Moreover, *gal4 i⁺* single mutants and *gal4 i⁻* double mutants are both pleiotropic negative. Therefore, the *GAL4* locus appears to encode a positive regulator protein, much like the *qa-1* protein, which directly stimulates transcription in the *GAL* gene cluster; this *GAL4* locus, in turn, is negatively controlled by the *i* gene product. These relationships are diagrammed in Figure 24-2.

Associated with *GAL4* is a region called *c*, which behaves formally like an operator locus in that it can undergo mutations to an *oᶜ*-like condition. Such mutations, known as *C*, map right next to *GAL4* (Figure 24-2) and can confer constitutive enzyme synthesis on the *GAL* gene cluster only if they are *cis* to a wild-type *GAL4* allele. In this condition, moreover, they are dominant to the wild-type *c* locus, so that a $\dfrac{CGAL4}{cGAL4}$ diploid produces enzymes constitutively. Presumably, therefore, the *c* locus is the usual site for *i*-repressor binding (Figure 24-2), and when this property is lost by mutation, unregulated gene expression ensues.

The *i* locus in yeast can experience *iˢ* mutations that are formally analogous to *lacI⁻ˢ* mutations in *E. coli* (*Section 22.7*): a repressor is produced that is no longer responsive to inducer action, and the resultant cell is noninducible by galactose. Such "super-repressed" *iˢ* mutations are dominant to *i* in diploids but are not expressed in haploid strains carrying *C* operator mutations. Work out for yourself how these outcomes are predicted, given analogies with *lac* operon mutations.

## 24.3 Regulation of Nitrogen Metabolism in *Aspergillus*

The *qa* and *GAL* control systems are analogous to *lac* and *trp* in *E. coli* in that regulator molecules are apparently specialized to control a single gene or gene cluster. We have also encountered, in Chapter 22, a second class of regulator molecule that acts not at one but at several unlinked sites in the genome. A prominent example is the CAP protein (*Section 22.11*), which responds to fluctuations in glucose/cAMP levels by altering the expression of several discrete operons in *E. coli*, all of which are related to sugar catabolism.

In eukaryotes, regulatory mechanisms of this kind would appear to be very common, since, as noted in Chapter 21, functionally related genes in eukaryotes are usually not linked in operon-like clusters yet are often co-induced by the same effector molecules or environmental stimuli. One can therefore predict that regulatory genes like *crp*, which specifies the CAP protein in *E. coli*, should exist in eukaryotes and that mutations in these genes should disallow the induction of a variety of related, but unlinked, genes.

The *intA* (for *int*egrator) locus in linkage group II of the fungus *Aspergillus nidulans* appears to contain such a gene. Mutations in *intA* affect the expression of three inducible structural genes that specify enzymes involved in amino acid metabolism: *lamA* (in linkage group VIII), *gatA* (LG VII), and *gabB* (LG VI). It also regulates a fourth gene, *amdS* (LG III), whose enzyme product is not obviously related to nitrogen metabolism. Thus *intA^c* mutations lead to high constitutive levels of all four enzymes, whereas *intA^-* mutations reduce the levels of the four enzyme activities.

The *intA* product responds to at least two effector molecules, namely, the amino acid alanine and GABA ($\gamma$-amino-n-butyric acid). All four structural genes, moreover, are sensitive to both effectors. With molecular details still unavailable, readers of Chapter 22 are again presented with familiar forms of control circuits, and it is tempting to speculate that since the fungi lie at the "primitive" end of the eukaryotic spectrum, they will prove to have control mechanisms similar to those of bacteria.

# Short-Term Regulation in Higher Eukaryotes

## 24.4 The Heat-Shock Response

Certainly the most ubiquitous example of short-term regulation in higher eukaryotes is known as the **heat-shock response**. When *Drosophila* or cultured *Drosophila* cells are insulted by raising the temperature or by exposing them to a variety of metabolic inhibitors, a rapid and fully reversible sequence of events takes place: the cells stop transcribing most of their active genes and instead transcribe a small number of **heat-shock genes;** the cells also stop translating extant mRNAs in the cytoplasm and instead translate the newly synthesized heat-shock mRNAs into **heat-shock proteins.** The response is a massive one: total RNA synthesis drops by 70 percent, and, at its peak, at least 40 percent of the total protein synthetic capacity of the cell is devoted to producing heat-shock proteins. Although we will focus on *Drosophila*, analogous responses to heat have been described in yeast, *Dicteostelium*, *Tetrahymena*, corn, and cultured avian and mammalian cells, indicating that the reaction is of fundamental importance to cell metabolism and survival.

In *Drosophila*, a temperature rise from 25 to 37°C elicits the same heat-shock sequence in a variety of tissues and at all developmental stages, but the response was first described in larval tissues as a **heat-induced change in the puffing patterns** of salivary gland chromosomes. As detailed in Section 9.12 and Figure 9-10, puffing of a band in a polytene chromosome indicates intense transcriptional activity. F. Ritossa noticed that within one minute after the temperature was raised, most pre-existing puffs began to regress and nine new puffs were detected. Figure 24-3 illustrates this response for two bands, 87A and 87C, which reside in the right arm of chromosome 3; other responsive bands are located elsewhere in the *Drosophila* genome. Thus heat causes the **coordinate induction of nine scattered gene loci;** this is followed, several minutes later, by the preferential translation of their gene transcripts.

A variety of experimental studies have confirmed what this sequence of events suggests, namely, that **the stimulation of gene transcription is the primary response to heat shock** and that the synthesis of heat-shock

**FIGURE 24-3 The heat-shock response.**
(a) Control polytene chromosome showing flat bands at 87C1 and 87A7. (b) Heat-shocked chromosome (40 min at 37°) showing puffs at these two loci. Note that pre-existing puffs have diminished (dashed lines). (From M. Ashburner and J. J. Bonner, *Cell* 17:241–254, 1979.)

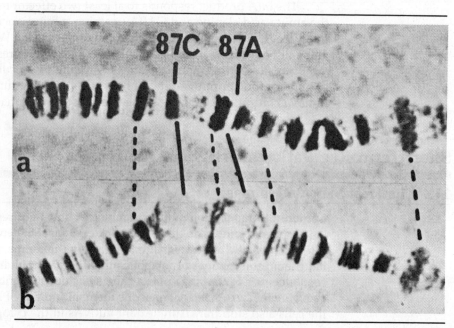

proteins and their subsequent (and poorly understood) activities in protecting the cell from damage represent a series of secondary responses. How, then, do heat and other forms of stress induce the specific transcription of nine gene loci in *Drosophila?* The answer is unknown, and since no regulatory mutations have yet been described that modify the heat-shock response, strictly genetic clues are lacking. The system may soon yield some of its secrets to molecular geneticists, however, who have cloned several of the heat-shock genes using recombinant DNA technology (*Section 4.13*). The initial results of analyzing these clones are presented at the end of this chapter.

## 24.5 Hormonal Regulation in Higher Eukaryotes: General Features

The heat-shock response in higher eukaryotes is unusual in that all the cells in an organism respond in an identical fashion. The usual situation in higher eukaryotes is very different: a given stimulus or environmental change is perceived and acted upon in a particular fashion by only a subset of cells in the body. A fundamental feature of higher eukaryotes is their exploitation of **cell specialization,** wherein a given cell type becomes able to perform particular functions and concomitantly unable to perform other functions. In Chapter 25 we will consider how such cell specialization is achieved during development. Here we assume that it has occurred, and we ask how an organism composed of numerous specialized cell types can bring about short-term regulation of gene expression.

Cell specialization is made possible by the fact that all cells in a complex eukaryote are presented with a relatively homogeneous external environment, namely, blood, hemolymph, or extracellular sap. Consequently, most cells never experience the drastic fluctuations in the quality of the environment that simple eukaryotes must be prepared to face, and they are therefore free to focus on particular metabolic activities. Critical, therefore, are those cell types that function to maintain the constancy of the extracellular fluids. Some cell types have a direct effect: in a vertebrate, these include intestine, liver, and kidney cells, which add the appropriate concentrations of ions and substrates and remove any excesses or waste products. Other cell types have an indirect effect: they sample the extracellular fluid, detect a particular imbalance, and send signals to the "direct-acting" tissues, which ultimately alter their processing activities.

The "signaling cells" of interest here are **endocrine cells,** which secrete **hormones.** An important property of hormones is that they typically have multiple **target cells;** this is best illustrated by example. When blood calcium levels fall below optimum, endocrine cells in the parathyroid gland detect the imbalance and secrete parathyroid hormone. The hormone has at least three target tissues, and it stimulates a different response in each: thus, it promotes calcium resorption from the bone, increases calcium permeability in the intestine, and reduces calcium excretion by the kidney. In other words, the "direct-acting" kidney, liver, and bone tissues do not respond to blood calcium levels *per se* but rather to hormonal signals that dictate what the level of calcium should be.

Hormones may well have evolved to coordinate the activities of specialized cells that act to maintain a constant extracellular environment. However, a second class of hormones has come to govern more "insulated" functions such as sexual reproduction and behavior. A familiar example is the delicate balance of secretions by the pituitary and the ovary, which collectively orchestrate the complex changes in a woman's uterus during a menstrual cycle. The course of the menstrual cycle is not completely free of dietary or other external influences, but it is relatively free of such influences when compared with the endocrine response to a low-calcium diet.

From this brief discussion, then, we can conclude that **short-term regulation in complex eukaryotes is largely mediated by hormones.** Many of these, most notably those classified as polypeptide hormones, act at the cytoplasmic level to stimulate or repress levels of specific enzymatic activities. We focus here on examples of hormones that appear to have a direct effect on gene expression.

## 24.6 Ecdysone Stimulation of Gene Transcription in Diptera

The first report of a specific hormonal effect at the chromosomal level in any organism was made in 1960 by U. Clever and his associates for the dipteran *Chironomus tentans.* Clever was studying the patterns of polytene chromosome puffing at various stages of *Chironomus* larval development, and he established that these patterns followed a fixed sequence, with certain bands puffing at one stage and regressing at a later stage. He then took nonmolting **(intermolt)** larvae, in which puffing of band I-19A was known to be particularly active, and injected them with **ecdysone,** an insect steroid hormone that was known to trigger molting. Within 10 to 15 minutes, puff I-19A regressed, band I-18C began to puff, and band IV-2B followed suit. This was exactly the puffing sequence known to take place just before molting, a time at which the insect spontaneously releases ecdysone into the hemolymph. Clever's experiments therefore indicated that **ecdysone turns on the transcription of particular genes.**

More recent studies have been made by M. Ashburner using ecdysone on *Drosophila* salivary glands isolated from third-instar larvae. These were cultured *in vitro* and presented with ecdysone at particular intervals. Approximately six bands responded very rapidly to ecdysone—bands 23E and 74E of the larval chromosomes, for example, showed puffing within five minutes of ecdysone presentation—and all reached a maximum size and then regressed within the first four hours. These so-called **early puffs** were unaffected by inhibitors of protein synthesis, and they regressed prematurely if ecdysone was washed from the culture medium during the four-hour period. Between three and ten hours after exposure to the hormone, a group of abut 100 **late puffs** were visualized. Since late puffing does not occur in the presence of protein synthesis inhibitors, Ashburner concluded that the late puffs are triggered, at least in part, by some ecdysone-stimulated early-puff gene products rather than by ecdysone alone. In other words, the rapid, and transitory, early puffing qualifies as the primary response to the hormone, and the more delayed, and massive, late puffing as the secondary response. The parallels between this and the heat-shock response (*Section 24.5*) should be obvious, the major differences being that the stimulus is a defined hormone rather than "stress," and the response is confined to particular tissues in the larva.

Recent studies by O. Pongs and colleagues have demonstrated that ecdysone stimulates *Drosophila* gene expression at the level of the chromosome. The hormone was presented to early third-instar larva for a short period of time, and its location was then determined by incubating the salivary gland chromosomes with a labeled antibody to ecdysone (*Box 2.3*). The label was found to localize specifically over the same set of early puffs defined by the Ashburner studies. Still unknown is whether ecdysone interacts directly with target DNA or with chromosomal proteins associated with this DNA.

Ecdysone is also secreted by *Drosophila* at the end of the third larval instar, and this time it has a very different effect: the larval tissues are induced to disintegrate, whereas cells destined to give rise to adult tissue, known as **imaginal disc cells** (*Section 25.8*), are induced to initiate their transformation. J. J. Bonner and M. L. Pardue designed experiments to ascertain whether the same group of genes is stimulated by ecdysone in the third-instar imaginal discs, as is stimulated in the larval polytene cells. Since imaginal disc nuclei do not undergo polytenization, Bonner and Pardue had to make their inquiry somewhat indirectly. They excised imaginal wing discs, cultured them *in vitro*, and exposed them to ecdysone and $^3$H-uridine for various lengths of time. The cultured cells were then broken open, their $^3$H-labeled RNA was extracted, and the poly A–containing mRNA fraction of this RNA was prepared. Finally, the $^3$H-mRNA was exposed to salivary gland chromosome preparations under *in situ* hybridization conditions (*Section 4.10*). The resultant tritium grains indeed localized over specific polytene bands, but none of these proved to correspond to the six bands that undergo early puffing in larval tissues.

These results demonstrate clearly that a single hormone activates one set of genes in one cell type and a second set of genes in a second cell type. Since the hormone is the same, these different responses might be explained by tissue-specific receptor proteins, which, when complexed with ecdysone, recognize distinct subsets of controlling elements. Alternatively, each cell type might so modify its target controlling elements that the identical hormone or receptor-hormone complex recognizes a different subset of genes. It is at present not possible to state which, if either, of these models is correct.

## 24.7 Steroid Hormone Stimulation of Gene Transcription in Mammals

That ecdysone receptor proteins exist in insect tissues has not yet been clearly demonstrated, but such proteins appear to mediate the response to steroid hormones (e.g., the **estrogens, androgens,** and **glucocorticoids**) in mammalian and avian tissues. Steroids first enter their target cells and then rapidly accumulate in the nucleus, as can be shown using $^3$H-steroid hormones and autoradiography (*Box 2.3*). Transport of the hormone to the nucleus is brought about by specific steroid receptors that reside in the cytoplasm of an uninduced cell. These proteins (perhaps 10,000 per cell) bind particular steroids with very high affinity. Binding is thought to cause a change in the shape of the receptor and is quickly followed by nuclear migration and a primary response to the hormone. Thus, in prokaryotic terminology (*Chapter 22*), **steroids act as effectors and their receptor proteins act as regulator molecules, becoming active only when bound to the effector.**

Genetic evidence for the importance of steroid-receptor binding and nuclear migration in launching a primary steroid hormone response has been provided by G. Tomkins, K. Yamamoto, C. Sidley, and associates. These investigators began with the mouse lymphoma cell line S49, which is rapidly killed by physiological concentrations of glucocorticoids, and proceeded to isolate mutant cell lines resistant to such cytotoxic effects of the steroid. These resistant mutants, they reasoned, might be expected to be blocked at various stages of steroid utilization. Analysis of several mutant lines confirmed this prediction. Most of the mutants were found to be receptor deficient ($r^-$), meaning that labeled steroid enters the cells but is not bound to any cytoplasmic or nuclear protein. Certain mutant lines proved to transport steroid to the nucleus with reduced efficiency ($nt^-$), whereas others showed an excessively large proportion of steroid binding by the nucleus and were called $nt^i$ (increased nuclear transfer). When receptor proteins were isolated from $nt^-$ and $nt^i$ mutants, they exhibited altered molecular properties. Therefore, it was proposed that at least some $nt^-$ and $nt^i$ mutations mark the structural gene or genes for steroid receptor proteins.

Several investigators have demonstrated that once the steroid-receptor complexes enter the nucleus, they stimulate the transcription of specific genes. Thus in a representative series of experiments from R. Palmiter's laboratory, presentation of the steroid hormone estrogen to immature chicken oviduct cells was shown to elicit the specific synthesis of pre-mRNA transcripts for the egg-specific protein ovalbumin. Such experiments indicate that the hormone-receptor complexes do not act in some indirect way, such as stimulating pre-mRNA $\longrightarrow$ mRNA processing or transport of ovalbumin mRNA to ribosomes; rather, they appear to have a direct effect on gene expression.

# Mechanisms of Short-Term Regulation in Eukaryotes

We arrive, therefore, at the heart of the matter: how is the transcription of specific genes stimulated by steroid-receptor complexes, by heat shock, and by the *qa-1*, *GAL4*, and *IntA* gene products? The question may obviously have a different answer for each regulatory event; the problem is that at present no definitive answer can be given for any of the eukaryotic systems that have been amenable to genetic analysis. We therefore present experiments and suggestive models derived from a variety of other experimental systems, the expectation being that these will soon be subjected to genetic scrutiny.

## 24.8 Binding of Regulatory Molecules to Chromatin

As we learned in Chapters 22 and 23, prokaryotes and their phages carry out short-term regulation with two kinds of proteins: proteins that bind directly to controlling elements in target genes and proteins that bind to RNA polymerases such that their specificity of initiation or termination is altered. There are as yet no definitive reports of polymerase-associated factors in eukaryotes; there are, however, several examples of proteins that bind to eukaryotic DNA.

A well-studied example relates to the **5S genes** of *Xenopus laevis*. Readers of Section 10.8 will recall that these genes possess an internal promoter and that they are transcribed by RNA polymerase III. R. Roeder and colleagues were able to show that 5S gene transcription is dependent on the presence of a 40,000-dalton protein factor that binds to the internal promoter and facilitates the binding of the polymerase. Moreover, the factor also binds to 5S RNA itself and, when bound, is no longer able to stimulate transcription of the gene. In other words, **the 5S genes appear to be subject to autogenous regulation.** This is analogous to the autogenous regulation of the $trp^R$ *cI*, and A genes (Sections 22.15, 23.8, and 23.10) except that **an RNA transcript is acting to inhibit its own synthesis** rather than a protein product of the gene. The following model has been proposed for this system. Cells are thought to contain a pool of factor-bound 5S RNA that is available for incorporation into ribosomes. As ribosomes are assembled, 5S-free factor is released, allowing it to rebind to the 5S gene and promote the synthesis of more 5S RNA. The new 5S RNA proceeds to bind factor so that transcription is turned off; the factor-bound 5S RNA then migrates to the site of ribosome assembly to complete the cycle.

We have already described experiments showing that the steroid hormone ecdysone, or perhaps an ecdysone-receptor complex, binds to the loci of specific target genes in *Drosophila*. Is there any evidence that steroid-receptor complexes bind to chromatin in vertebrate nuclei? The answer is yes, and the affirmative is considerably strengthened by the demonstration that glucocorticoid-receptor proteins from $nt^-$ steroid-resistant variants of the lymphoma line S49 (*Section 24.7*) have a lower binding affinity for chromatin than do controls. The answer must at once be qualified, however, for it appears that both normal and mutant steroid-receptor complexes have a low affinity for a very large number of sites in the eukaryotic genome, many more than they presumably activate. Therefore, in most binding studies, massive nonspecific binding creates a "noise level" above which any specific binding is very difficult to detect. Similar "noise," we should quickly note, is present in prokaryotic studies as well: CAP protein binds nonspecifically to many sites in the *E. coli* chromosome, and its specific affinity for the *lac* promoter was only demonstrated once purified *lac* DNA became available. Indeed, the factor required for 5S RNA transcription was detected only after the 5S genes were isolated as recombinant DNA clones. Since several genes that are regulated by steroid hormones have now been cloned, reports on their affinity for steroid-receptor complexes will undoubtedly be appearing in the near future.

## 24.9 Controlling Elements Associated with Regulated Structural Genes

The 5S genes of *Xenopus* have unique promoters and are transcribed by a unique class of RNA polymerase. If transcriptional control mechanism are to explain other kinds of short-term fluctuations in structural gene expression, then these genes must also carry controlling elements that are uniquely responsive to particular regulatory molecules. This prediction has been tested with the heat-shock genes of *D. melanogaster* (*Section 24.4*). M. Meselson and colleagues subjected four different heat-shock genes to cloning and DNA sequencing. They then ordered the sequences such that their "TATA boxes," the sites of polymerase recognition (*Section 9.7*), were lined up, as shown in Figure 24-4, and they asked whether any sequences upstream from the TATA box might be shared in common by all four genes. The sequences underlined in Figure 24-4 were singled out as being of interest in that they possess imperfect dyad symmetry, a property of many controlling elements (*Sections 22.15 and 23.7*). Moreover, they are present at comparable locations in all four genes and are not found in a comparable location in non–heat-shock insect genes. It is therefore possible to imagine that the unidentified regulatory molecules activated by heat shock may recognize sequences of this sort at all nine heat-shock loci.

A different approach, taken by W. Gehring and his associates, has been to incubate cloned heat-shock DNA in the presence of *Drosophila* nuclear extracts and look for proteins that bind specifically to heat-shock genes. The one binding activity detected to date associates with a sequence 800 to 1000 nucleotides upstream from the TATA box, which is clearly a very different location from the dyads identified in the Meselson studies.

```
    -30         -20         -10         0          10          20          30  ↓       40          50          60
    |           |           |          |          |           |           |           |           |           |
AGAAGTTTCTAGAGACTTCCAGTTCGGGTCGGGTTTTTC⌐TATAAAA⌐GCAGACGCGCGGCGTTTGCCGGTTCGAGTCTTGAAAAAAATTTCGTACGGTGTG
CGCAGGGAAATCTCGAATTTTCCCCTCCCGGCGACAGAG⌐TATAAAT⌐ACGGGCGCAAATTTCCCAGACGCTACATTTGAAATCAAACAGTCAAAGTGAAAA
GAGCGCGGCCTGCGAATGTTCGCGAAAAGAGCGCCGGAG⌐TATAAAT⌐AGAGGCGCTTCGTCGACGGAGCGTCAATTCAATTCAAACAAGCAAAGTGAACAC
CTGTCACTTTCCGGACTCTTCTAGAAAAGCTCCAGCGGG⌐TATAAAAG⌐CAGCGTCGCTTGACGAACAGAGCACAGATCGAATTCAAAAATCGAGCAGTGAA
```

FIGURE 24-4 **Four heat-shock gene sequences** derived from clones of genomic DNA coding for heat-shock proteins (hsp) 83, 68, 70, and 26, showing the probable site for initiation of transcription (arrow), the "TATA Boxes" (boxed), which occur about 30 bp upstream from the initiation site, and regions of imperfect dyad symmetry (underlined) upstream from the TATA boxes. (From R. Holmgren, V. Corres, R. Morimoto, R. Blackman, and M. Meselson, *Proc. Natl. Acad. Sci. U.S.* **78**:3775–3778, 1981.)

We present these two experiments, even though they are preliminary, to illustrate both the power and the difficulty with "*in vitro* genetics" as practiced with recombinant DNA. Detection of a putative controlling element or a putative binding site does not, in itself, provide much insight into gene regulation. Fortunately, however, it is now possible to subject DNA sequences of potential interest to *in vitro* mutagenesis (*Section 8.7*) such that particular bases are modified or deleted. Therefore, it should be possible to mutate the two candidate heat-shock controlling elements and ask whether transcription of the contiguous genes is modified; if it is, then the sequences are likely to be important in controlling gene expression.

## 24.10  Physical Conformation of Chromatin and DNA

The eukaryotic genome possesses many more levels of organization than the prokaryotic genome: its DNA wraps around nucleosomes and the nucleosomes are packed into ordered fibers (*Chapter 3*). In addition, the nonhistone proteins that associate with eukaryotic DNA are known to differ considerably from one region of the genome to the next, and the physical conformation of the DNA may change accordingly. There has therefore been considerable interest in the possibility that physical conformation is a eukaryotic DNA parameter that is subject to regulation. The basic notion is that DNA cannot be transcribed unless its bases are exposed to polymerase enzymes; therefore, certain physical conformations may prevent gene expression, whereas others may promote it. Experimental tests of this notion have thus far focused on long-term differentiation systems, and these are therefore presented in the next chapter. There is every reason to extend the idea to short-term regulation, however, and to imagine that at least some short-term regulatory molecules will act to change the physical state of their target genes.

(a)                                (b)

(a) Norman Giles and Mary Case (University of Georgia, Athens) study the genetics of regulated gene clusters in *Neurospora*. (b) Gordon Tomkins pioneered a combined genetic and biochemical approach to the analysis of mammalian steriod receptors.

# Questions and Problems

1. Do you expect *qa-1* mutations to be recessive, codominant, or dominant in *Neurospora* heterokaryons if they are of the pleiotropic-negative class? The constitutive class? Explain with diagrams.

2. Heterokaryons were constructed with the following genotypes: heterokaryon A carried *qa-1$^c$* and *qa-3$^+$* in one nucleus and *qa-1$^+$* and *qa-3$^-$* in the other; heterokaryon B carried *qa-1$^c$* and *qa-3$^-$* in one nucleus and *qa-1$^+$* and *qa-3$^+$* in the other. In the absence of quinic acid, heterokaryon A was much more effective in promoting the synthesis of dehydroquinic acid than was heterokaryon B. Explain how this observation supports the contention that regulation by *qa-1* is at the transcriptional and not the translational level.

3. Give the phenotypes of each of the following yeast mutants. Gene loci not designated in the genotype are wild type.
   (a) *C gal4*
   (b) *C gal4/c GAL4*
   (c) *c gal4/C GAL4*
   (d) *i$^-$ C GAL4*
   (e) *i$^s$ C GAL4*

(f) $i^s$ c GAL4/$i^+$ c gal4

(g) $i^s$ c gal4/$i^+$ c GAL4

4. Give the phenotypes of each of the following *A. nidulans* mutants. Gene loci not designated in the genotype are wild type.

(a) *intA* + alanine

(b) *intA$^c$* + alanine

(c) *intA$^-$* + alanine

(d) *intA$^-$/intA$^c$* without alanine or GABA

(e) *intA$^-$/intA$^c$* + alanine

(f) *intA/intA$^c$* without alanine or GABA

(g) *intA/intA$^-$* without alanine or GABA

5. How would you ask whether alterations in puffing patterns represent the primary response to heat shock in *Drosophila* larvae or whether heat shock induces the translation of certain proteins, which, in turn, go on to induce novel puffing patterns as a secondary effect?

6. *Tetrahymena* cells have recently been shown to undergo a response to heat that includes repression of most gene transcription and selective transcription of a few novel genes.

(a) How would you screen for mutants unable to launch this response?

(b) How would you go about identifying and cloning the relevant heat-shock genes?

(c) You postulate that heat might convert some pre-existing hormone-like molecule from an inactive to an active form and that the active species moderates the subsequent heat-shock response. How might you test this theory (Hint: *Tetrahymena* cells readily take substances up from their media)?

7. How would you design experiments to test whether (a) ecdysone receptor proteins exist and, if so, whether (b) different ecdysone-receptor complexes exist in different tissues or whether the same ecdysone-receptor complex recognizes modified controlling elements in the chromosomes of different tissues?

8. Are steroid receptor proteins analogous to the *qa-1* gene product of *Neurospora* or the *i* gene product of yeast?

9. Compare examples of autogenous regulation in phage λ, *E. coli*, SV40, and *Xenopus laevis*.

10. Control over eukaryotic gene expression may well be exerted in some cases by post-transcriptional mechanisms. Explain how regulation over the following systems could bring about short-term fluctuations in levels of a particular gene product.

(a) Intron-encoded splicing enzymes (*Section 16.6*)

(b) Ribonucleoprotein particles (*Section 9.13*)

(c) Ribosome binding (*Section 11.3*)

(d) Post-translational processing (*Section 11.5*)

11. The salivary gland and the liver of the mouse synthesize an identical α-amylase enzyme, dictated by the same gene (*Amy 1$^A$*), but whereas α-amylase mRNA accounts for 2 percent of the mRNA in the salivary gland, it is only 0.02 percent in the liver. Both mRNA species have been isolated and partially sequenced, and they prove to differ only at their 5′ ends: the 48 nucleotides immediately upstream from the initiator AUG are identical, but whereas the salivary gland leader possesses an additional 47 nucleotides, the liver leader contains an additional 158 nucleotides, all different from the salivary sequence.

(a) Speculate on how this difference in mRNAs might occur.

    (b) Speculate on how this difference might account for the tissue-specific levels of salivary amylase.

12. Two related Hawaiian species of *Drosophila* synthesize homologous versions of the enzyme alcohol dehydrogenase (ADH), which are encoded by equivalent structural genes; however, the two enzyme species have diverged sufficiently to migrate differently in electrophoresis gels (*Sections 19.3 through 19.6*). The larval midgut tissue of *D. grimshawi* synthesizes at least 64-fold more enzyme than the midgut tissue of *D. orthofascia*, whereas enzyme levels are equivalent in the fat body. Rare hybrid larvae will emerge from forced mass matings between the two species.

    (a) Diagram the patterns of ADH from the fat body and larval midgut tissue of *D. grimshawi* and *D. orthofascia* (assign arbitrary electrophoretic mobilities to each species).

    (b) Diagram the ADH pattern you would expect from midgut tissue of the hybrid larvae if control is exerted in *cis,* and the two patterns you might expect if control is exerted in *trans.*

    (c) Gels supporting the *cis* interpretation were in fact obtained in this study. How would you proceed to analyze this system using recombinant DNA technology?

# 25

# Control of Gene Expression in Eukaryotes: Long-Term Regulation

A. Introduction

B. General Features of Long-Term Differentiation
- 25.1 The Constancy of Nuclear DNA
- 25.2 Selective DNA Transcription
- 25.3 Regulatory Molecules in the Cytoplasm

C. The Differentiation of the Egg and Maternal Influences on Development
- 25.4 Maternal Ribosomal RNA Synthesis
- 25.5 Maternal Messenger RNA Synthesis
- 25.6 Maternal Patterning of the Egg Cytoplasm
- 25.7 Maternal-Effect Mutations that Affect Egg Patterning

D. Developmental Genetics of *Drosophila*
- 25.8 Normal Development in *Drosophilia*
- 25.9 Fate Maps and Sexual Mosaics

25.10 Founder Cells for Imaginal Discs

25.11 Compartmentalization

25.12 Homoeotic Mutations

E. Developmental Genetics of Vertebrates
- 25.13 Mutations Affecting Early Development: The Mouse *T* Locus
- 25.14 Gonadal Sex Determination in Mammals
- 25.15 Genital-Duct Sex Determination in Mammals
- 25.16 Facultative Heterochromatization

F. Differential Expression of Hemoglobin Genes
- 25.17 Switching Hemoglobin Genes On and Off
- 25.18 Activation of Globin Genes During Chicken Development

G. Questions and Problems

# Introduction

Long-term differentiation in higher eukaryotes commences with a fertilized egg, itself a highly specialized cell that performs certain functions that no mature cells ever perform. The zygote then undergoes a series of mitotic divisions. At various times during this mitotic period, individual cells become **determined**—a process that no one understands—such that they and their clones become committed to forming particular cell types—skin, muscle, leaf, or whatever. Following the determination event—and often many cell generations later—individual cells undergo **differentiation** into their specialized forms. Differentiation may involve mitosis, cell fusion, cell migration, or intercellular interactions. These events occur in an interdependent fashion, interdependent in the sense that event A must be completed before event B can possibly begin, the onset of event B triggers a biosynthetic event or cellular migration necessary for event C, and so forth.

The process of differentiation lies outside the scope of genetics, since it represents the acting out of particular developmental programs. The geneticist is concerned with the nature of the developmental programs themselves and how different programs are imposed on particular cells at particular times during embryogenesis.

In this chapter we first present experiments that illustrate general features of long-term differentiation in eukaryotes, using examples from a variety of developmental systems. We then examine more closely the determinative events that accompany egg maturation and early embryology, and we conclude with an analysis of erythrocyte differentiation, where it has been possible to study the differential expression of genes at a molecular level.

# General Features of Long-Term Differentiation

Three axioms appear to be generally applicable to long-term differentiation in eukaryotes.

1. Differentiation does not involve massive, permanent changes in nuclear DNA.
2. Differentiation involves selective gene transcription.
3. Differentiation involves self-reinforcing changes in the cytoplasm.

Experiments that support these conclusions are presented in the following three sections.

## 25.1 The Constancy of Nuclear DNA

A conceivable way to visualize determination is to propose that a cell destined to give rise to a liver cell simply degrades all of its DNA except the

DNA relevant to liver-specific and general, cellular "housekeeping" activities; a cell determined to give rise to a bone cell would similarly discard all but its bone-making and housekeeping information; and so forth. Alternatively, one could propose that a series of directed mutational events might change the informational content of liver DNA so that it would proceed in one direction while bone DNA would proceed in another. Numerous experiments performed over the past five decades have indicated that neither of these proposals is correct. Instead, the nuclei of differentiated cells have been shown to carry most, if not all, of the genetic information present in the zygote nucleus. We can cite three lines of experimentation that demonstrate this important principle; additional approaches can be found in texts on developmental biology.

**Nuclear Transplantation Experiments**    In experiments performed in J. Gurdon's laboratory, individual nuclei were isolated from adult *Xenopus laevis* skin cells that had been cultured *in vitro;* they were then injected into fertilized *Xenopus* eggs whose nuclei had been surgically removed. Most of these hybrids did not survive, presumably because of the trauma incurred during their preparation, but the few that did went on to develop into normal swimming larvae, although none was able to mature into an adult toad. As proof that the survivors did indeed inherit their genomes from the transplanted nuclei, Gurdon used $+/0$-*nu* heterozygotes as his donor strain and $+/+$ toads as his recipient strain. As noted in Section 10.7, $+/0$-*nu* nuclei contain only one nucleolus, whereas $+/+$ nuclei contain two nucleoli. Gurdon showed that in numerous tissue samples from various survivors only uninucleolate cells were present. Therefore, **nuclei from the adult toad skin must continue to carry all the DNA necessary to support early embryonic development.**

K. Illmensee and P. Hoppe have performed similar experiments with mice. In this case, nuclei were isolated from **blastocysts,** the mammalian embryonic stage that directly precedes implantation. The zygote has at this point cleaved many times and the cells have differentiated into two major tissues: the **trophectoderm (TE),** which participates in uterine implantation, and the **inner cell mass (ICM),** which will become the embryo. As diagrammed in Figure 25-1 and detailed in the figure legend, donor nuclei were genetically marked with gray coat color genes; they were transplanted into enucleated eggs from a black female; and the transplanted eggs were allowed to develop in culture to the blastocyst stage and were then placed in the uteri of white female mice. When the donor nuclei derived from ICM cells, three of the transplanted eggs went on to produce normal gray mice. Moreover, the isozyme patterns (*Section 19.5*) and the karyotypes of the nuclear transplant offspring were those of the nuclear donor parent and, in subsequent test crosses, two of the nuclear transplant offspring were able to give birth to several normal progeny, all of which displayed the nuclear donor phenotypes. Thus **the ICM cells,** which had probably already differentiated into primitive ectoderm and endoderm,

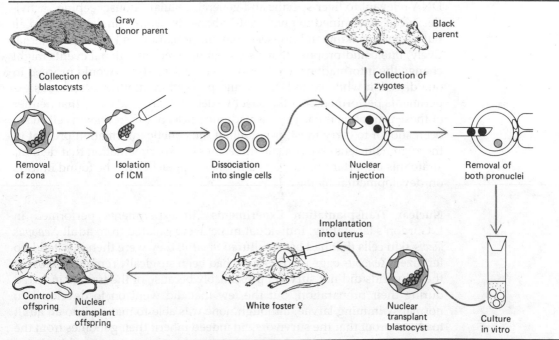

**FIGURE 25-1 Nuclear transplantation in the mouse.** Donor blastocysts were collected from a gray (CBA/HT6) strain and kept in culture medium. After surgical removal of the zona pellucida, the blastocysts were dissected manually into inner cell mass (ICM) and trophectoderm (TE), which were then dissociated enzymatically into single cells. An ICM or TE cell was mechanically disrupted by sucking it into a small glass pipet, and the cell nucleus with surrounding cytoplasm was subsequently injected into a fertilized egg from a black (C57BL/6) strain. Following nuclear injection, the genome of the recipient egg was removed by sucking the male and female pronuclei into the micropipet. The nuclear transplant embryos were cultured *in vitro* to the blastocyst state and then transferred, together with some white control embryos, into the uterus of a pseudopregnant white (ICR/Swiss) foster female in order to allow development to term. Live-born mice were analyzed chromosomally and biochemically for the genetic markers of the transplanted nuclei. In breeding tests, the nuclear transplant mice were examined for the functional germ-line transmission of the nuclear transplanted genome. (From K. Illmensee and P. C. Hoppe, *Cell* **23**:9–18, 1981.)

clearly **retain nuclei that remain developmentally equivalent to the toti-potent zygote nucleus.**

An important caveat in these nuclear transplantation experiments should now be stressed. Illmensee and Hoppe found that when the same experiment was performed with trophectoderm nuclei, none of the transplanted eggs went on to develop normally. Trophectoderm cells, in contrast to ICM cells, are already programmed for a particular cell lineage and normally cease to divide and become polyploid at the time of implantation. Similarly, it has not yet been possible to produce an *adult* toad derived from the transplanted nucleus of a differentiated adult cell. Although it is possible that the more differentiated cells are more prone to trauma during surgery, the present experiments in fact tell us only that **nuclear totipotency is retained only by the still quite undifferentiated ICM cells and that adult amphibian nuclei can program development only to the larval stage.**

A second potential difficulty with nuclear transplantation experiments is that they may not detect "semipermanent" changes in the genome. If, for example, DNA undergoes selective recombination events during differentiation so that new genetic units are created, as indeed occurs with the immunoglobulin genes during the differentiation of lymphoid cells (*Section 20.9*), then these rearrangements might possibly be capable of reversing themselves, either after the differentiation event takes place or when the nucleus is transplanted to an oocyte cytoplasm.

**Restriction Enzyme Experiments**   To test whether genomic rearrangements occur during *Drosophila* development without introducing the possible ambiguities of nuclear transplantation assays, A. Garen and his colleagues performed a series of experiments using recombinant DNA techniques. As detailed in Figure 25-2a, they first identified recombinant clones carrying *Drosophila* genes that are transcriptionally active in fat body tissues of late third-instar larvae but are inactive in other larval tissues and in gastrulating embryos. They then isolated total DNA from larval fat bodies, from other larval tissues, and from embryos and digested each with various restriction endonucleases (*Section 4.11*). Finally, the fragments from each digestion were fractionated by electrophoresis in agarose gels (*recall Figure 4-8*) and tested for hybridization to the labeled cloned DNA, as diagrammed in Figure 25-2b. In all cases the results were the same: the genomic samples from different embryonic stages produced identical restriction maps when probed with the "fat-body-specific" clones. These results indicate that intense expression of the genes in the larval fat bodies is not associated with DNA rearrangements that alter any of the tested endonuclease recognition sites. Since 13 different endonucleases were used and since this same endonuclease approach can readily detect the rearrangements undergone by immunoglobulin genes (*Chapter 20*), the experiments appear to demonstrate that major rearrangements of genomic DNA do not occur in regions of structural genes that are stringently regulated during *Drosophila* development. Although the evidence is by no means in for all developmentally regulated genes in all organisms, we can safely conclude that **embryonic development need not be accompanied by a permanent loss or reorganization of genetic information,** and organisms are likely to utilize this strategy only in limited situations (e.g., during lymphoid differentiation). Therefore, other mechanisms must be identified.

(a)

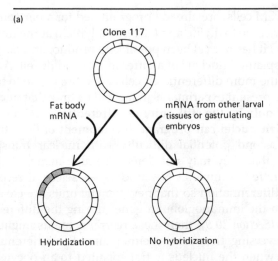

Clone 117

Fat body mRNA

mRNA from other larval tissues or gastrulating embryos

Hybridization

No hybridization

(b)

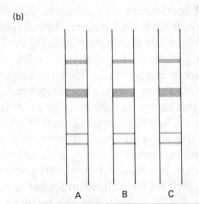

A     B     C

Genomic DNA from embryos (A), fat body (B), and other larval tissues (C), cleaved with the restriction enzyme *Bgl* II and probed with labeled DNA from clone 117. The patterns are identical. Had rearrangements occurred during development, then the patterns would be different (e.g., a new *Bgl* II recognition site might have been created in the vicinity of the gene, generating a new band recognized by the probing clone 117 DNA).

FIGURE 25-2 **Restriction enzyme analysis of developmentally regulated genes in *Drosophila*.** The drawing illustrates the use of one clone and one restriction enzyme; 6 clones and 13 restriction enzymes were used in the complete series of experiments. (After M. Levine, A. Garen, J. Lepresant, and J. Lepresant-Kejzlarova, *Proc. Natl. Acad. Sci. U.S.* **78:**2417–2421, 1981.)

## 25.2 Selective DNA Transcription

Since the wholesale elimination or alteration of DNA does not appear to accompany most determination or differentiation events, it is logical to propose next that differentiated cells transcribe only a small fraction of their genome and that cells with different specializations transcribe different portions (transcribing in common, perhaps, certain "housekeeping genes.") Numerous demonstrations of this principle have been made, many of which have already been considered in Chapter 24 and in earlier chapters. Thus we have already seen that the lymphocyte line becomes committed to antibody production, the erythroid line to hemoglobin production, and so on; we have also described the specific puffing patterns of polytene chromosomes at particular developmental stages and, in the preceding section, the expression of a unique set of genes in *Drosophila* larval fat body tissues.

Granted that selective gene transcription is a fundamental concomitant of differentiation in multicellular organisms, several investigators have asked a more general question, namely, what fraction of the genome is expressed during development and what fraction is expressed by fully differentiated cells? An early series of experiments designed to answer this question was performed by E. H. Davidson and colleagues for sea urchins. As summarized in Figure 25-3, they performed molecular hybridization studies between the single-copy DNA of the sea urchin genome and the RNA synthesized at various developmental stages or by various differentiated tissues. They found that maximal transcription occurs in the oocyte, a cell that is particularly active in synthesizing "maternal mRNAs" to be used during subsequent stages of early embryonic development (*Section 25.5*). Transcriptional activity continues to be high during the blastula, gastrula, and pluteus stages of sea urchin development and then drops to

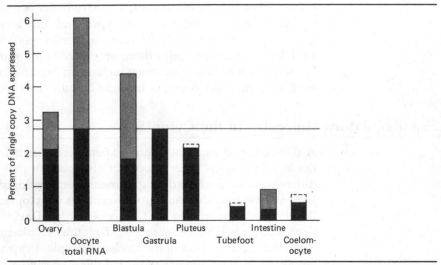

FIGURE 25-3 Selective DNA transcription during development, as measured by DNA-RNA hybridization in sea urchin embryos and adult tissues. The black portion of each bar indicates the amount of single-copy sequence shared between gastrula mRNA and the RNA preparations listed along the abscissa. The gray portions show the amount of single-copy sequence present in the various RNAs studied but absent from gastrula mRNA. Dashed lines indicate the maximum amount of reaction that could have been present and escaped detection. (From G. A. Galau, W. H. Klein, M. M. Davis, B. J. Wold, R. J. Britten, and E. H. Davidson, *Cell* **7**:487–505, 1976. Copyright © MIT.)

relatively low levels in such differentiated tissues as tubefoot or intestine. In other words, it would appear that **most of the genes in the sea urchin genome encode information required for early events in embryology.**

How many genes are we talking about? Most estimates on the number of genes per eukaryotic nucleus are in the range of 50,000 (there is enough single-copy DNA per nucleus to code for about a million genes, but most of this is thought to take the form of intergenic DNA and intervening sequences). Taking 50,000 genes as a working number, the Davidson study suggests that about 20,000 to 30,000 of these find expression in the oocyte and no more than one quarter that number in differentiated cells. Similar numbers have emerged from a comparison, by R. Axel and colleagues, of the RNA sequences present in chicken liver *versus* chicken oviduct cells. Each tissue proved to express on the order of 15,000 genes. Of these, 85 percent appear to be "shared"; that is, about 13,000 of the single-copy genes transcribed in oviduct nuclei are also transcribed in liver nuclei. These presumably encode proteins that serve housekeeping functions common to both, and perhaps all, tissues in the chicken. The remaining 15 percent of the transcripts appear to dictate the tissue-specific proteins that determine the very different properties of oviduct and liver cells.

In summary, therefore, a small proportion of the total eukaryotic genome contains protein-encoding genes; a large percentage of these genes is expressed early in development; a much smaller percentage of these genes is expressed by differentiated cell types; and only a very small number of genes experience unique (or at least major) expression in specific differentiated cell types. Ultimately, then, an understanding of the differentiation "program" of an organism must include an explanation of how these patterns of gene expression are brought about.

## 25.3 Regulatory Molecules in the Cytoplasm

A differentiated nucleus clearly experiences constraints on its transcriptional activities, but the cytoplasm also plays a major role in long-term differentiation. Such early experimental embryologists as T. Boveri and E. B. Wilson provided numerous examples of cytoplasmic influences over development (many are well described in E. Davidson's excellent book *Gene Activity in Early Development*). That the cytoplasm can induce nuclear changes is also obvious in the nuclear transplant experiments described in Section 25.1. The influence of the cytoplasm becomes particularly dramatic in some of the toad experiments: Gurdon has found that when a small, nondividing brain nucleus is injected into an enucleate fertilized egg, the nucleus swells to the size characteristic of a fertilized egg nucleus. It also modulates its rate of RNA synthesis and increases its rate of DNA synthesis in an apparent effort to mimic the biosynthetic properties of a rapidly dividing fertilized egg nucleus. When a brain nucleus is instead injected into an egg undergoing meiosis, its chromosomes condense and abortive spindle formation occurs.

A cornerstone of developmental genetics is therefore the concept that the cytoplasm is of critical importance in maintaining the differentiated state. It is imagined that when certain determined genes turn on, certain proteins are synthesized; some act in the nucleus to sustain the transcription of particular genes and others act in the cytoplasm. The cytoplasm of a differentiated cell is thought to acquire in this manner a collection of regulatory molecules that reinforce the differentiated condition. When a foreign nucleus is introduced into this cytoplasm by transplantation, the nucleus is expected to come rapidly under the regulatory influence of the host cytoplasm.

# The Differentiation of the Egg and Maternal Influences on Development

The transformation of an immature oocyte into a mature egg involves a developmental sequence that is of immediate interest in understanding how cells undergo differentiation. Egg differentiation has a second level of interest, moreover, for the patterns of early embryonic development are critically influenced by cytoplasmic determinants that are organized during oogenesis. Therefore, an understanding of early differentiation depends on an understanding of how egg cells develop.

## 25.4 Maternal Ribosomal RNA Synthesis

A number of egg cells, notably those of fish and amphibia, store enormous numbers of **maternal ribosomes** for the use of the developing zygote. This means that oocytes must produce, and store, enormous numbers of 5S, 18S, and 25S rRNA molecules (*Sections 10.7 and 10.8*), far more than are ever generated during a "normal" cell cycle. In the toad *X. laevis*, this selective expression of the oocyte genome is accomplished by two different mechanisms.

The stimulation of 5S rRNA gene expression during oocyte maturation is accomplished by **transcriptional control.** We noted in Section 24.8 that 5S rDNA transcription in somatic cells is regulated by a protein factor that binds to the internal promoters of the tandemly repeated 5S rRNA genes and stimulates the binding of RNA polymerase III. D. Brown and coworkers have found that oocytes produce a similar, but slightly smaller, 5S rRNA binding factor that is present in enormous concentrations—some seven orders of magnitude higher than in somatic nuclei. This **oocyte transcription factor** is able to stimulate massive transcription of the 5S rDNA genes; it also binds to the 5S rRNA product so that autogenous inhibition of transcription (*Section 24.8*) is prevented.

The stimulation of 18S and 25S rRNA synthesis is accomplished by a distinct process known as **gene amplification.** As diagrammed in Figure

10-11, the nucleolar organizer of *Xenopus laevis* contains hundreds of tandemly repeated copies of 18S and 28S rDNA genes interspersed with spacer DNA. During oogenesis, this DNA undergoes repeated rounds of replication, using the same type of rolling-circle replication mechanism that generates copies of *E. coli* plasmids (*Figure 13-2*). As a result, 1000 to 1500 additional copies of nucleolar organizer DNA are produced during the pachytene stage of prophase I and, at diplonema, the nucleus contains 1000 to 1500 tiny nucleoli, each independently engaged in the rapid synthesis of rRNA. At the conclusion of meiosis I, when maternal ribosomes are assembled, the extra nucleolar DNA is discarded into the cytoplasm.

Gene amplification is a mechanism that in some ways violates our general rule that the genome is not altered during differentiation, although the change is obviously a temporary one. It proves to be more than an rRNA curiosity, moreover: *Drosophila* follicle cells, for example, amplify structural genes that code for proteins of the egg shell, proteins that must also be produced in abundance at a particular developmental stage.

## 25.5 Maternal Messenger RNA Synthesis

The fertilized amphibian (and sea urchin) egg also contains a large amount of mRNA, which is synthesized during oogenesis and which encodes much of the informational program for early embryogenesis (*Figure 25-3*). Thus, inhibitors of gene transcription have little, if any, effect on protein synthesis or normal development until the amphibian mid–blastula stage, at which time the embryo consists of from 6000 to 15,000 cells; moreover, embryonic nuclei seem virtually inactive in RNA synthesis during the early cleavage stage. The amphibian egg therefore appears to accumulate well in advance the informational RNA it will need to take it through all the early stages of development.

Until recently it was believed that maternal mRNA stores were produced by the "lampbrush chromosomes" (*Figure 9-9*) that appear in diplotene oocyte nuclei and are conspicuously active in transcription. M. Robash and his colleagues have shown, however, that lampbrush chromosomes are active well *after* all the stable mRNA of the *Xenopus* oocyte has been transcribed and accumulated. Therefore, lampbrush chromosome transcripts appear to play as yet undefined roles in egg physiology.

If stored mRNA is indeed to direct the early stages of embryogenesis, then some form of **translational control** (*Section 22.17*) must regulate which of these transcripts is expressed in different tissues. J. V. Ruderman and her colleagues have indeed shown that one subset of maternal mRNA is associated with oocyte ribosomes, whereas a second set is associated with early embryo ribosomes. Discrimination appears to be exerted by factors present in egg and embryo cytoplasms, but these have yet to be identified, and nothing is known about how they are sequentially expressed.

The storage of maternal mRNA for use in the zygote is most conspicuous in amphibia, echinoderms, and insects. Relatively little mRNA appears to be present in the mammalian egg, and the mammalian zygote nucleus commences mRNA synthesis by the two- to four-cell stage. It is not known why mammals are distinctive in this feature of development.

## 25.6 Maternal Patterning of the Egg Cytoplasm

For most organisms, the first two or four cells that result from early cleavage of the fertilized egg can give rise, when dissected apart, to small but complete embryos. This capacity is progressively lost, however, as cleavage proceeds. Instead, each cell undergoes a **primary determination** such that it becomes committed to form only particular portions of the embryo. Primary determination events are relatively crude; they seem to direct a cell along general rather than specific lines of development. Once a cell has become primarily determined as an endodermal cell, however, its descendants experience the more finely tuned determinative events that dictate whether the cell will become liver, gut, or pancreas. In other words, the primary events appear to set the stage for the rest of the developmental process.

In many organisms, primary determination events appear to be dictated by what have come to be called **maternal influences.** It seems that during differentiation of the egg, the maternal genome directs a specific kind of patterning within the oocyte cytoplasm so that "determinative molecules"—possibly proteins, possibly RNAs—are differentially placed in various parts of the egg. Thus the eggs of many lower vertebrates appear to contain "endodermal determinants" in one region, "ectodermal determinants" in another, and so on. Then, as the egg cleaves and individual daughter cells arise that contain only the endodermal factors, these cells experience primary determination in the endodermal direction.

Some of these factors may be synonymous with the determinants that govern which subset of maternal mRNA is translated on a particular set of ribosomes (*Section 25.5*). Others have been identified in association with visible cytoplasmic inclusions. Although nothing is yet known about how they function, we can present three examples that illustrate their properties.

**Shell Coiling in the Snail**    A classic example of maternal influence is given by the direction of shell coiling in the snail *Limnaea paregra*. The shell may coil to the left or to the right, and the internal organs of the snail adopt a corresponding handedness. A single genetic locus determines the trait, with the *dextral* (D) allele dominant to the *sinistral* (d) allele. The expression of this locus is unusual, however, in that it occurs during oogenesis. As shown in Figure 25-4, a *d/d sinistral* mother will give rise only to *sinistral* offspring, even when the father provides D-bearing sperm and the offspring are all D/d heterozygotes. Cytological studies indicate that when eggs derive from *sinistral* mothers, the plane of the first zygotic cleavage is at right angles to the plane adopted when eggs derive from *dextral* moth-

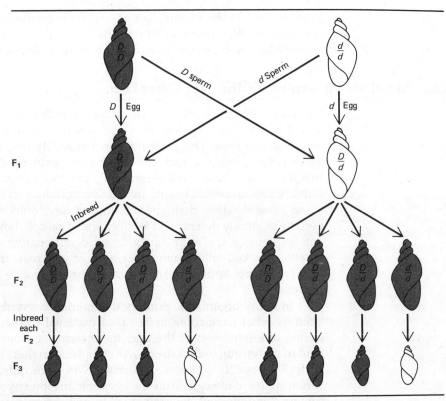

**FIGURE 25-4 Maternal effects:** the inheritance of coiling direction in the shells of the snail *Limnaea*. Gray snails are dextral; white snails are sinistral. (After E. Sinnot, L. C. Dunn, and T. Dobzhansky, *Principles of Genetics*, 5th ed. New York: McGraw-Hill Book Company. Used with permission of McGraw-Hill Book Company © 1958.)

ers. Therefore, at some time before the second meiotic division of snail oogenesis, the *D* and *d* alleles in the maternal genome specify some property of the cytoplasm that ultimately determines the orientation of the first cleavage furrow of the zygote.

**The Germ Plasm in Amphibia**    Certainly the best studied examples of cytoplasmic determinants are those that localize in one sector of amphibian and insect eggs to form what embryologists have labeled the **germ plasm.** This cytoplasm is readily identified because it contains conspicuous granules.

In *Xenopus*, the fate of the germ plasm granules has been followed during embryogenesis, and they are found to become segregated into only a few cells by the time of gastrulation. If these cells are removed from the gastrula, the resultant toad is sterile; therefore, the granules mark cells that are the progenitors of the gamete-producing **germ line.** If the germ plasm is instead surgically removed from a fertilized egg, a sterile toad also results. Therefore, the maternal genome somehow patterns a specific sector of the egg cytoplasm that will, later in embryogenesis, exert a determinative influence.

**Polar Plasm in Drosophila**    The germ plasm in *Drosophila* is localized at the posterior end of the oval-shaped *Drosophila* egg and is therefore known as the **polar plasm,** with its granular inclusions being known as **polar granules.** At the conclusion of the early cleavage stage of the zygote, the polar granules become localized in about ten **pole cells,** which divide several times and then form a ridge of tissue from which are generated the gametes of an adult fly.

Experiments by K. Illmensee and A. P. Mahowald have provided a definitive demonstration of the determinative properties of polar plasm in *D. melanogaster*. We will first describe the overall scheme of their experiments and then examine the details.

Illmensee and Mahowald first took cytoplasm containing polar plasm from **donor** zygotes, injected it into the anterior ends of young **recipient** zygotes, allowed development to proceed for about two hours, and tested whether the anterior cells exposed to the polar plasm had been influenced to become presumptive germ cells. To answer this question, it was necessary to transplant the anterior cells to the posterior region of **host** embryos where they would be assured of being included in adult gonadal tissue. It was also necessary to use genetic markers so that donor, recipient and host cell types could be distinguished, as described in detail below. The result was unambiguous: the anterior region of the recipient zygote, which normally differentiates into somatic structures, acquired the ability to differentiate into germ-line cells when exposed to the polar-granule–containing cytoplasm.

The specifics of the Illmensee-Mahowald experiments are diagrammed in Figure 25-5. Three sets of genetically marked flies were used. Donor cytoplasm was derived from young zygotes carrying the *Bar* mutation (Figure 25-5a) so that if any nuclei were inadvertently included in the extract, dominant *Bar* genes would be detectable later in the experiment. Recipient embryos were homozygous for the chromosome-3 markers *mwh* (*multiple wing hairs*) and *e* (*ebony* body) (Figure 25-5b). Host embryos were wild type for both of these chromosome-3 loci but were homozygous or hemizygous for the X chromosome markers *y* (*yellow* body), *w* (*white* eyes), and *sn* (*singed* bristles) (Figure 25-5c). Recipient embryos were separated into two groups, the "experimentals" and the "controls." Experimentals received anterior injections of the donor cytoplasm, whereas controls received mock injections. Host embryos were similarly divided into two groups, one receiving posterior-region implantations from experimental embryos (Figure 25-5c) and the other receiving implantations from control embryos.

Experimental and control host embryos were then allowed to mature

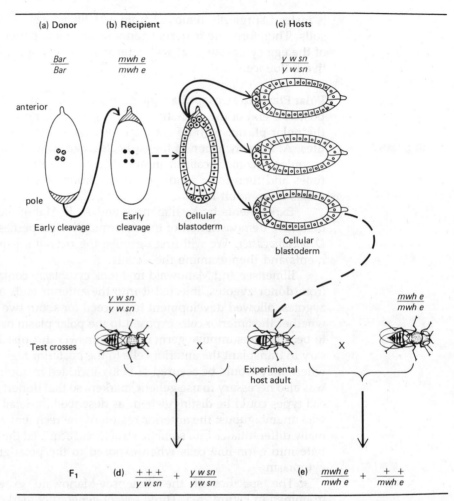

(a) Donor    (b) Recipient    (c) Hosts

$\dfrac{Bar}{Bar}$    $\dfrac{mwh\ e}{mwh\ e}$    $\dfrac{y\ w\ sn}{y\ w\ sn}$

anterior

pole

Early cleavage    Early cleavage    Cellular blastoderm

Cellular blastoderm

Test crosses    $\dfrac{y\ w\ sn}{y\ w\ sn}$    X    Experimental host adult    X    $\dfrac{mwh\ e}{mwh\ e}$

$F_1$    (d) $\dfrac{+\ +\ +}{y\ w\ sn}$  +  $\dfrac{y\ w\ sn}{y\ w\ sn}$    (e) $\dfrac{mwh\ e}{mwh\ e}$  +  $\dfrac{+\ +}{mwh\ e}$

**FIGURE 25-5 Polar plasm transplantation experiments.** (From K. Illmensee and A. P. Mahowald, *Proc. Natl. Acad. Sci. U.S.* **71**:1016, 1974.)

into adult flies, and each was mated with *y w sn* homozygotes (Figure 25-5d). The progeny from the control matings were, as expected, all *yellow white singed*. Four of the experimental flies, however, were found to produce wild type as well as *yellow white singed* offspring (Figure 25-5d), meaning that these four flies produced gametes carrying the + alleles of *y*, *w*, and *sn*. If these gametes derived from the implanted cells, reasoned Illmensee and Mahowald, then their full haploid genotype should be + + + *mwh e*, meaning that the wild-type progeny flies should be

$\dfrac{+\ \ +}{mwh\ e}$ heterozygotes. To prove this, the wild-type flies were test-crossed to *mwh e* homozygotes (Figure 25-5e), and about half of the test progeny indeed proved to express both *multiple wing hairs* and *ebony* body color.

The Illmensee-Mahowald experiments demonstrate, therefore, that **the anterior region of a fertilized *Drosophila* egg, which normally gives rise to various somatic tissues, can be converted to germ-line cells simply by exposure to a polar cytoplasm extract.** This cytoplasm, therefore, can cause a cell type to differentiate in a particular direction.

**General Considerations**  If, as seems clearly the case, cytoplasmic factors serve to influence different portions of the early embryo to differentiate in particular directions, then many of the important questions to be asked in developmental genetics are simply pushed back to the oocyte maturation stage. What kinds of genetic programs "tell" an egg cell to put certain molecules in one sector of the cytoplasm and other molecules in another? Is this all accomplished by gene products of prophase I egg nuclei, when genes inherited from both parents are present, or are subsequent stages also important? How do the cytoplasmic factors work? Some of these questions can be productively approached by studying "maternal-effect" mutants, as described in the next section.

## 25.7  Maternal-Effect Mutations that Affect Egg Patterning

If maternal genes are responsible for the patterning of the egg, then it should be possible to isolate mutations that interfere with the patterning process and thus with normal embryogenesis. In *Drosophila*, mutations of this class are known as **maternal-effect mutants: the altered embryonic pattern depends on the genotype of the mother who laid the egg and not on the genotype of the embryo.** Thus if a female is homozygous for a recessive maternal-effect mutation *mat*, then her eggs will fail to undergo a normal zygote development even if they are subsequently fertilized by a wild-type sperm and the zygote comes to have a +/*mat* genotype. By contrast, a +/*mat* heterozygous female will produce eggs that develop into normal flies even if their genotype, following fertilization, is *mat/mat*. This is in fact how *mat/mat* females are generated in the first place: +/*mat* females are mated with *mat/mat* males, and half the female progeny are viable, but sterile, *mat/mat* flies. In short, the + alleles of *mat* loci are somehow essential for producing an egg that can give rise to a normal embryo.

The course of abortive development exhibited by *Drosophila* embryos derived from *mat/mat* eggs is consistent with the postulate that **maternal-effect genes play a role in egg patterning.** Thus the maternal-effect mutation known as *bicaudal* induces a range of abnormalities that appear to affect the anteroposterior ("head-to-tail") polarity of the embryo. Some of the embryos have two posterior ends arranged in mirror-image symmetry; others have far more posterior tissue than normal, with a concomitant reduction in anterior tissues. A second maternal-effect mutation, called *dorsal,* causes structures that are usually confined to dorsal areas to appear on all regions of the embryo; in other words, the dorsoventral axis of the embryo is disrupted. A third example, called *grandchildless*, has been described in *D. subobscura*. Flies homozygous for this mutation produce eggs that, when fertilized, go on to generate normal-looking adult flies. Both the males and the females are sterile, however, so that the original mutant has children but no grandchildren. Examination of the sterile adults reveals that they have no gametes in their gonads, and examination of the original *grandchildless* eggs shows that they fail to form pole cells, just as one would expect from the Illmensee-Mahowald experiments described in the preceding section.

# Developmental Genetics of *Drosophila*

## 25.8 Normal Development in *Drosophila*

The previous sections allow us to come away with a picture of a highly organized *Drosophila* egg, carrying determinants that somehow define the anteroposterior and the dorsoventral axes of embryo, plus determinants for such specific cell types as the gametes. In the next sections of this chapter, we will follow the fate of this egg after it is fertilized, stressing the key role played by genetic analysis in understanding the course of dipteran embryogenesis.

First we must describe the overall course of *Drosophila* development. The nucleus of the fertilized egg (Figure 25-6a) undergoes a rapid and synchronous series of mitoses. Each of the resultant nuclei migrates to the periphery of the zygote, and most divide a few more times so that about 6500 nuclei are present. Each is then surrounded by a cell membrane so that the egg is converted to a **cellular blastoderm**, one cell thick, as illustrated in Figure 25-6b.

The cells of the blastoderm proceed to differentiate in one of two directions. Most of the cells gastrulate, undergo organogenesis, and give rise to the body parts of the larva, which hatches from the egg shell 22 hours after fertilization. Both the late embryo and the larva have a simple segmented body plan (Figure 25-6c): five segments compose the head region, three the thoracic region, and eight the abdominal region. Following pupation and metamorphosis, the same overall segmentation pattern can be recognized in the adult fly, but the thorax is now disproportionately large and carries the wings and the six legs of the insect (Figure 25-6d).

Some of the cells of the blastoderm experience a very different fate. Instead of forming larval tissue, they form invaginations in the epidermis and grow extensively during the three larval instars to form discrete flattened sacs of essentially undifferentiated cells, called **imaginal discs**. During metamorphosis, most of the adult epidermis derives from these disc cells. Moreover, each disc gives rise to a discrete adult structure: as diagrammed in Figure 25-7, a disc exists for each wing, for each leg, for genital structures, and so on.

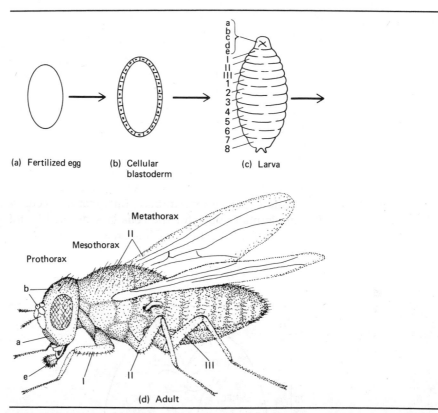

(a) Fertilized egg   (b) Cellular blastoderm   (c) Larva

(d) Adult

**FIGURE 25-6 Stages in *Drosophila* development.** The larva, whose segmentation pattern has been exaggerated for emphasis, bears three thoracic segments (I–III), eight abdominal segments (1–8), and five regions of the head that correspond, in the adult, to the clypeolabrum (a), eye-antenna (b), mandibles (c), maxilla (d), and proboscis (e).

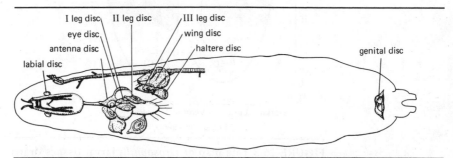

**FIGURE 25-7 Imaginal discs in a mature larva** of *D. melanogaster*. A ventral view of the larva is shown, and the brain and pharynx are depicted. Each disc except the genital disc is represented twice, producing a total of 17. (From D. Bodenstein, in *Biology of Drosophila*, M. Demerec, Ed. New York: John Wiley, 1950.)

## 25.9 Fate Maps and Sexual Mosaics

Because *Drosophila* embryogenesis is such an ordered process, it is possible to ask **which cells of the blastoderm are destined to form larval body parts and which will form imaginal discs,** the resultant distribution being known as a **fate map.**

A fate map of presumptive larval tissues, drawn in Figure 25-8, was constructed by tracking the developmental course of individual marked cells. As detailed in the legend, the precursor cells for the **hypoderm,** which will secrete the external tissue of the larvae, are located on the two sides of the blastoderm. At gastrulation, they come to cover the embryo, whereas the rest of the cells invaginate to form the internal organs.

A fate map of presumptive imaginal disc cells cannot be constructed by simple inspection, for the cells lack defined histological markings and become inaccessible to view. Such maps have therefore required the analysis of **sexual mosaics,** or **gynandromorphs,** which carry cells marked by visible gene mutations.

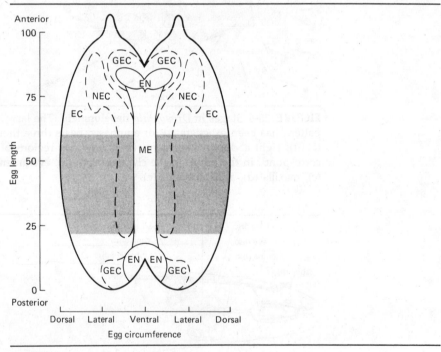

FIGURE 25-8 Fate map of *Drosophila* larval tissues drawn on the cellular blastoderm. The blastoderm is cut open along the dorsal midline and spread out to approximate a flat sheet of cells. Me, mesoderm anlage (precursor cells); En, endoderm anlage; Ec, ectoderm; Gec, gut ectoderm; Nec, neurogenic ectoderm; shaded area, hypoderm anlage. (Courtesy of C. Nüsslein-Valhard.)

The most widely used method for creating sexual mosaics is diagrammed in Figure 25-9. A female zygote is constructed to carry two marked X chromosomes. The X chromosome contributed by the father bears recessive mutations that affect adult cuticular structures, namely, *yellow* body color, *white* eye, and *forked* bristles. The X chromosome contributed by the mother carries the wild-type alleles of these mutations. It is also an unusual chromosome: it is ring shaped, and, for reasons not understood, it is frequently lost from the mitotic spindle during the first nuclear division but very rarely lost in subsequent mitoses. Loss of the ring X generates two nuclear clones: one is XO and can therefore express the *yellow, white, forked* mutations; the second is XX and expresses only their wild-type alleles (Figure 25-9).

The fate of the two nuclear clones is illustrated in Figure 25-10. Since the cleaving nuclei in an early *Drosophila* embryo tend to "stay put" and not intermingle, clonal descendants of the XX nucleus generally localize in one sector of the blastoderm and the XO descendants in the other sector (Figure 25-10d). Since the spindle of the first zygotic mitosis can be oriented in any direction, however, the two sectors can find themselves in a variety of orientations with respect to the zygote axis. Therefore, all sorts of planes of mosaicism are possible, including a bilateral pattern. In the adult fly, each epidermal cell is responsible for secreting the cuticle that lies directly above it; therefore, visual inspection of mosaic adults will reveal the boundaries between the *white forked yellow* nuclei and the wild-type nuclei (Figure 25-10g).

A fate map is obtained by collecting a number of such mosaic flies and determining **the frequency with which two adult body parts lie within the same mosaic region.** Thus it might be found that leg I and leg II are in the same mosaic sector 90 percent of the time, meaning that if a fly exhibits a mutant version of leg I cuticle, it has a 90 percent chance of exhibiting a mutant version of leg II and only a 10 percent chance of exhibiting a wild-type version. In the same collection, on the other hand, leg I and antenna might be of the same type only 75 percent of the time, meaning that they have a 25 percent chance of deriving from different mosaic tissues. **These frequencies are interpreted to reflect the distances that separate progenitor nuclei in the cellular blastoderm,** the principle being that the farther apart two nuclei are, the less likely it is that they will be included in the same mosaic sector of the blastoderm and, therefore, the less likely it is that they will share the same cuticular phenotype in the adult. The frequency that two body structures are **not** of the same phenotype is expressed in percentage map units known as **sturts** (after A. H. Sturtevant): since legs I and II are of different phenotypes 10 percent of the time, they are said to be separated by 10 sturts; similarly, leg I and antenna are 25 sturts apart. These distances are then plotted out on a hypothetical map of the blastoderm surface, as drawn in Figure 25-11a. A sketch of the blastoderm surface based on the fate map is shown in Figure 25-11b.

**FIGURE 25-9 Loss of the ring-X chromosome in sexual mosaics.** Details of the loss process are in fact more complex than diagrammed here, but the net effect is the same.

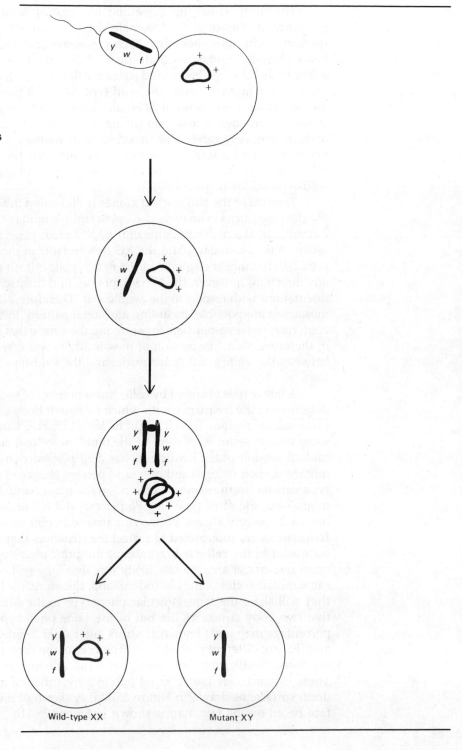

Wild-type XX                     Mutant XY

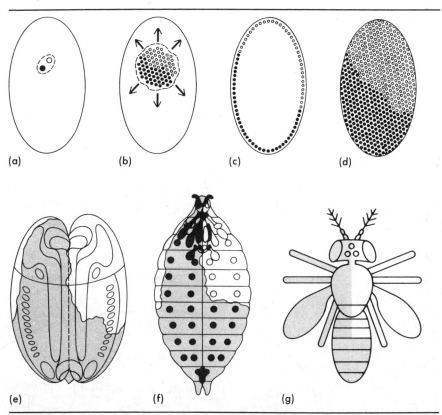

**FIGURE 25-10 Sexual mosaic formation in *Drosophila*.** (a) Fertilized female egg having undergone one nuclear division. Solid circle represents nucleus containing two X chromosomes, one a ring chromosome and the other carrying recessive genes. Open circle represents nucleus from which the ring X chromosome has been lost during mitosis; its composition is therefore XO, and the recessive genes on the remaining X can be expressed. (b) Division of the nuclei. Arrows indicate direction of nuclear migration. (c) Nuclei beneath the egg cortex. (d) Surface view of the multinucleate blastoderm showing the mosaic boundary separating the two areas. (e) Embryo map. (f) Larva map. (g) Adult fly. (From Y. Hotta and S. Benzer, *Nature* **240**:527, 1972.)

## 25.10 Founder Cells for Imaginal Discs

Sexual mosaics can be put to a second use: they permit a determination of the number of **founder cells** that give rise to a particular type of imaginal disc. For example, suppose you wish to determine the number of founder cells that serve as progenitors for the scutum (the dorsal part of the thorax). You first acquire a collection of flies that have mosaic scutal regions. For each mosaic scutum, you then estimate the relative size of a patch of mutant tissue in a predominantly wild-type scutum or a patch of wild-type

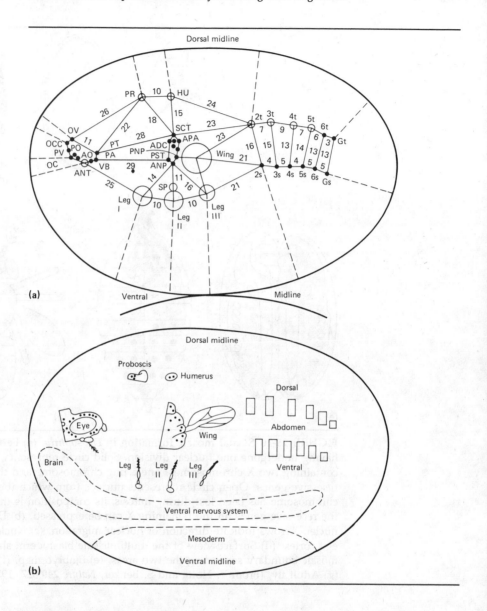

(a)

(b)

tissue in a predominantly mutant scutum, either patch being known as a **minority area.** You find that the smallest minority area occupies about one eighth of the total area of the adult scutum. This outcome is interpreted to mean that, at a minimum, one of every eight nuclei in the scutum derives from a different nucleus than the other seven, so that **at least eight nuclei had to have been determined as scutal disc precursors.** Similar estimates yield values of ten founder cells for a leg disc, six founder cells for a wing disc, and so on.

FIGURE 25-11 Fate map of *Drosophila* adult tissues drawn on the cellular blastoderm (only one side is shown, in contrast to the bilateral view presented in Figure 25-8). (a) Fate map constructed from mosaic mapping showing sites of cells that will eventually develop into the indicated external body parts of the adult fly. Distances are given in sturts. Dotted lines indicate distances to nearest midline. The size of the circle used to represent a site is proportional to the frequency with which the corresponding structure is itself split by the mosaic boundary. ADC, anterior dorsocentral bristle; AO, anterior orbital bristle; ANP, anterior notopleural bristle; ANT, antenna, APA, anterior postalar bristle; HU, humeral bristles; OC, ocellar bristle, OCC, occiput; OV, outer vertical bristle; PA, palp; PNP, posterior notopleural bristle; PO, posterior orbital bristle; PR, proboscis; PST, presutural bristle; PT, postorbital bristles; PV, post-vertical bristle; SP, sternopleural bristles; VB, vibrissae; 1t, first abdominal tergite; 1s, first abdominal sternite. (b) External parts of adult depicted on blastoderm surface. Dotted lines indicate areas that give rise to the nervous system and the mesoderm according to embryological studies. (From Y. Hotta and S. Benzer, *Nature* **240**:527, 1972.)

If a scutal imaginal disc was created by eight founder cells that are potentially marked by different sets of genes, then the cells in a scutal disc cannot be spoken of as a clone, even though all have heeded the same determinative instruction ("become a member of a scutal disc"). To emphasize both their relatedness and their distinctive nature, therefore, the cells in a given disc are referred to as a **polyclone.**

## 25.11 Compartmentalization

As detailed in Section 24.6, the hormone **ecdysone** causes the undifferentiated cells in an imaginal disc to undergo differentiation during pupation. It is possible to dissect an individual imaginal disc from a mature larva, transplant it into the abdomen of a pupa, and retrieve it from the abdomen of the resultant fly. When this is done for, say, a leg II disc, metamorphosis transforms it into the expected leg tissues. More impressive are experiments in which only small *pieces* of a leg II disc are transplanted: in this case, one piece might be found to yield exclusively femurs, another piece exclusively tibias, and so on. In other words, such **surgical fate mapping** reveals that **each cell in a mature imaginal disc is determined:** each knows precisely which cuticular structure it will form, and even which type of cuticular bristle it will elaborate. Ecdysone then simply permits the acting out of this determined state.

The transplantation studies tell us, then, that **a mature leg disc is a mosaic of cells with highly restricted developmental capacities.** Our previous analysis of sexual mosaics told us that **a presumptive leg disc initiates as a polyclone of only eight founder nuclei.** What happens between the presumptive stage and the mature stage? Are the eight founder nuclei already committed in tibia versus femur directions at their inception so that they simply expand themselves mitotically during the course of embryonic and larval development? Or do presumptive imaginative disc cells experience additional determinative events?

A series of elegant genetic experiments strongly support the second alternative, and A. Garcia-Bellido has proposed a **compartment hypothesis** to describe the pattern of determination that is followed. The data, which we consider below, suggest that **imaginal disc nuclei experience a sequence of determinative events,** each committing them to a more restricted set of possibilities than the last, with cells that have experienced a determination event in common forming a **polyclonal compartment.** In other words, nuclei are segregated into progressively more confining compartments until, at the time of pupation, each is a member of a polyclone programmed for only one pathway of differentiation.

**Clonal Analysis** The data that stimulated the compartment hypothesis were produced in part by a technique called **clonal analysis.** To perform clonal analysis, fertilized eggs are first constructed to be heterozygous for mutations affecting adult cuticular structures; as an example, we can consider $+/y$ eggs. These are allowed to develop to a selected stage, at which point the embryos or larvae are exposed to X rays. X irradiation induces **mitotic crossing over** in *Drosophila*, and, as detailed in Section 15.2 and Figure 15-3, mitotic crossing over can convert heterozygous cells into cells homozygous for particular genes. Thus in our example, a $+/y$ cell would give rise to a $y/y$ cell (and a $+/+$ cell as well). The X ray dose selected is sufficiently low that, at most, only one cell per presumptive disc region is transformed into a $y/y$ homozygote. This cell will proceed to generate a clone, which, following metamorphosis, creates a small patch of *yellow* tissue in an otherwise wild-type fly. The fly is therefore a mosaic but, unlike a gynandromorph (*Section 25.9*), **the sector represents a clone of cells created at a selected stage in development** rather than a clone generated at the early cleavage stage. Thus it becomes possible to ask whether a clone generated at one stage has a different developmental repertoire from a clone generated at a later stage.

As a specific example we can consider development of the adult mesothorax, a region that includes the wings and the second pair of legs (*Figure 25-6d*). If X irradiation is performed during the blastoderm stage, and a clone of *yellow* cells is detected in the resultant adult, the *yellow* tissue occurs either in the anterior portions or in the posterior portions of wings and legs. **Marked clones are never found to populate both anterior and posterior sectors.** Therefore, the earliest choice presented to a cell is either "be anterior" or "be posterior." Once a cell makes one or the other choice (e.g., to "anteriorness"), then all its mitotic progeny are similarly committed. Groups of cells that make the "anterior commitment" together are said to form an **anterior compartment,** whereas those that make a "posterior commitment" form a **posterior compartment** (Figure 25-12). Separating these is an anteroposterior **compartment boundary.**

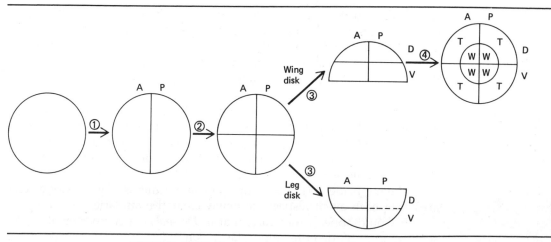

FIGURE 25-12 **Compartment formation** during the determination of mesothoracic cuticular structures. After each subdivision (represented by a straight line) no clone can include territories on both sides of that line. The first step (1) is the segregation of anterior (A) and posterior (P) polyclones, and this occurs, probably, at the blastoderm stage. The second (2) consists of the separation of the dorsal part of the segment (wing disc) and the ventral part (leg disc). This event takes place before ten hours of development. These two discs later become physically separated. During the larval period the wing disc becomes subdivided (3) into wing dorsal (D) and wing ventral (V) and about the same time (4) into thorax (T) and wing blade (W) polyclones. For the leg disc there is some evidence of the subdivision of the anterior compartment into dorsal and ventral. (From G. Morata and P. A. Lawrence, *Nature* **265**:211, 1977.)

A compartment boundary, once established, is never crossed. Thus, it is possible to construct flies whose cells grow slowly unless they have experienced mitotic crossing over, in which case they grow rapidly. When such cells also acquire the *yellow* phenotype during mitotic crossing over, it becomes possible to follow the fate of the recombinant clone. Even though the *yellow* cells could easily divide to fill the entire wing disc, and therefore the entire wing, they never do so. At most, only half the wing (or leg II) is *yellow*, meaning that the *y/y* cells divide until they reach the compartment boundary, and then stop.

The commitment to anteriorness or posteriorness occurs prior to disc determination. Thus a fast-growing "anterior" clone created at the blastoderm stage can fill the anterior half of the second leg as well as the anterior half of the wing: the only restriction at this stage is that no cells can cross into the posterior halves of either appendage.

The second determinative event in the ontogeny of the mesothorax disc, which occurs about ten hours after development begins, creates a **dorsoventral compartment boundary,** which separates cells committed to form dorsal (wing) structures from cells committed to form ventral (leg) structures (Figure 25-12). A *yellow* clone created at this stage will therefore exhibit *two* restrictions: not only does it respect the anteroposterior boundary as before; it also respects the new dorsoventral boundary. Specifically, one fast-growing *yellow* clone might be restricted to the anterior tissues of the leg; a second clone created at this time might be confined to the posterior tissues of the wing; and so on.

Once the dorsoventral commitments have been made, the two groups of cells separate physically to form an embryonic imaginal wing disc and a leg II disc. The wing disc proceeds to undergo two or more determinative events in the larva (steps 3 and 4 of Figure 25-12): the first creates wing-dorsal (D) and wing-ventral (V) compartments; the second creates wing-blade (W) and thorax (T) compartments. Again, clones created after each event respect the newly created boundaries.

It would appear, therefore, that in *Drosophila* a compartment represents a unit of determination. Each compartment is populated by a polyclone, and each founding polyclone represents from 5 to 50 cells. Determination events, in other words, affect groups of cells during *Drosophila* development rather than simply affecting single cells. Garcia-Bellido has postulated that for each determination event there is activated a **selector gene** that selects a particular developmental pathway for a particular group of cells.

Although the primary products of selector genes have yet to be identified, recent experiments suggest that they may ultimately act to specify molecules that appear on the cell surface. A cell-surface antigen, detected by its reaction with a specific antibody (*Box 2.3*), has been detected on all wing imaginal disc cells of early third-instar larvae. Midway through this instar, however, the antigen begins to disappear from the presumptive wing-ventral compartment (*Figure 25-12*), and by late third instar, it is found only on the presumptive wing-dorsal cells so that its presence defines the dorsal-ventral compartment boundary. Such **position-specific antigens,** it is proposed, may mediate subsequent stages in morphogenesis.

**Relating Maternal Influences to Compartments**  How does the compartmentalization of the embryo relate to the maternal patterning of the egg? The answer is not known, but one useful model suggests that **the maternal genome sets up the coordinates of the future animal, whereas the zygotic genome interprets and acts upon these coordinates.** For example, we learned in Section 25.7 that the anteroposterior polarity of the embryo is apparently set up by one or more maternal-effect genes. If we imagine that this **positional information** takes the form of cytoplasmic determinants placed in a particular pattern in the eggs, then the maternal genome can be said to define a set of anteroposterior **coordinates.** The zygotic genome must now interpret these coordinates: at the appropriate stage in blastoderm development, selector genes must somehow "read" the coordinates present in their immediate vicinity and interpret them as dictating either "be anterior" or "be posterior." The nuclei must then act upon this information by committing themselves, and their clonal offspring, to one or the other state. At the next determinative junction, a different set of selector genes is asked to "read" the dorsoventral coordinates in its immediate vicinity and "decide" whether they dictate "be a wing" or "be a leg," and so on. Once founder polyclones have been created under the influence of maternal determinants, they may well go on to respond to different sets of cues in making more finely tuned determinative decisions. There is at present no reason to believe that the entire course of development is maternally programmed, only the critical initial steps.

We have seen that the establishment of normal coordinates is affected by maternal-effect mutations (*Section 25.7*). If the above model is valid, then there should also be *Drosophila* mutations that act only after fertilization and that affect the interpretation process. Such mutations indeed exist.

## 25.12  Homoeotic Mutations

Mutations that interfere with the correct interpretation of positional information are known as **homoeotic**. We will describe here two examples.

**Engrailed**  The first group of homoeotic mutations, called *engrailed* (*en*), has been actively studied by P. Lawrence, G. Morata, and T. Kornberg. The *engrailed* locus maps to band 48A on the second chromosome of *D. melanogaster*. Two cuticular markings, *straw* and *pawn* (affecting the color and shape of cuticular bristles), map close to *en*. Therefore, by constructing *straw pawn en* heterozygotes and subjecting them to X irradiation at various developmental stages, it is possible to identify clones homozygous for *en* by their simultaneous expression of the *straw* and *pawn* traits. Analysis of such clones shows that **if *en/en* cells occupy any anterior compartment, development proceeds normally, whereas *en/en* cells in all posterior compartments develop abnormally.** For example, *engrailed* clones present in the posterior margin of the wing blade produce a type of bristle normally found only on the anterior wing margin, and large clones fail to respect anteroposterior boundaries and cross into anterior territory. Thus it appears that the normal product of the *engrailed* gene is required to recognize "posterior coordinates"; *en/en* cells either fail to interpret these coordinates correctly or fail to sustain determination in a posterior direction once they start, the result being that they instead adopt an anterior course. The *engrailed* locus, in short, appears to contain a selector gene that perceives anteroposterior boundaries.

Most *engrailed* alleles are embryonic lethal in homozygotes. Analysis of aborted *en/en* embryos reveals that **the engrailed gene product is necessary for forming anteroposterior borders in the embryo and larva as well as in the imaginal discs.** You will recall that *Drosophila* larvae have a segmented body pattern that reflects the organization of the adult (*Figure 25-*

*6c*). This pattern is manifest in the normal embryo as rows of **denticle belts** (Figure 25-13, left). In *en/en* embryos (Figure 25-13, right), these rows appear to fuse together, some embryos having as many as four belts fused into a single array. Thus the mutant embryo cannot recognize, or maintain, segmental boundaries any better than mutant clones can recognize, or maintain, compartmental boundaries in the imaginal disc.

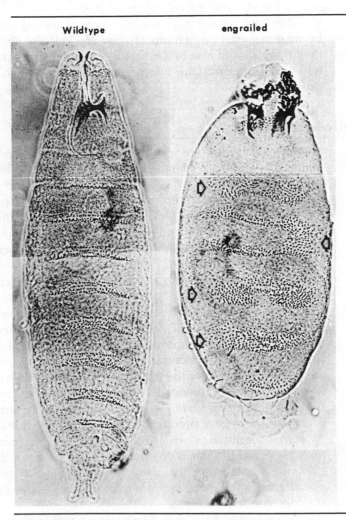

Wildtype    engrailed

**FIGURE 25-13 Engrailed mutation** affecting embryonic segmentation patterns. The wild-type 18-hour embryo (left) displays a normal series of denticle belts; the *en*^{LA4}/*en*^{LA4} 18-hour embryo (right) has disordered belts that are fused in regions marked by the arrows. (From T. Kornberg, *Proc. Natl. Acad. Sci. U.S.* **78**:1095–1099, 1981.)

**Bithorax**  The *bithorax* (*bx*) locus, at band 89E on the third chromosome of *D. melanogaster*, carries a group of at least eight genes that are important for determination of thoracic cuticular structures and have the properties of selector genes. Their genetics, which has been extensively studied by E. B. Lewis, can be illustrated by describing the determination of posterior leg III tissue.

All cells destined to give rise to thoracic structures first experience the anterior vs. posterior determination events described previously. They are next acted upon by genes that reside in the *bithorax* complex. For "posterior" polyclones destined to generate leg III structures in the metathorax (*Figure 25-6d*), the first instruction is given by the *postprothorax* (*ppx*$^+$) gene. Its gene product somehow blocks **pro**thoracic development in these cells and steers them instead toward a **meso**thoracic course. This course is prevented, several hours later, by the action of the *postbithorax* (*pbx*$^+$) gene, which somehow blocks mesothoracic development and allows determination to proceed in the desired leg III (**meta**thoracic) direction. In other words, **a temporal sequence is involved:** the "posterior" cells in this region of the blastoderm have a basal tendency to develop in a prothoracic direction. This tendency is first blocked by the *ppx*$^+$ gene, which substitutes instead a mesothoracic tendency. The mesothoracic tendency is in turn blocked by the *pbx*$^+$ gene, which permits metathoracic determination.

Evidence for this **pro⟶meso⟶meta hierarchy** comes from clonal analysis of $+/ppx$ and $+/pbx$ flies. As in the *engrailed* experiments, each *bithorax* mutation was linked to recessive cuticular markers so that homozygous clones were visibly obvious in the adult. The results were as follows.

1.  When $+/ppx$ individuals were X-irradiated at the early blastoderm stage, *ppx*/*ppx* clones in posterior leg III regions of the resultant adults were found to bear bristles characteristic of the leg **I**, a prothoracic structure. When $+/ppx$ individuals were instead irradiated as 17-hour embryos, mosaic tissues in leg III were found to bear bristles characteristic of leg **II**, a mesothoracic structure. Thus the instruction "be mesothoracic" is conveyed by the *ppx*$^+$ allele in the heterozygote sometime prior to 17 hours of development; after the instruction has been transmitted, *ppx*/*ppx* segregant clones continue to "know" that they must develop in a mesothoracic direction, even though they no longer carry the normal *ppx*$^+$ gene.

2.  When $+/pbx$ individuals are irradiated at any stage of development up to the late larval period, *pbx*/*pbx* clones in the leg III regions display bristles that should appear only on leg **II**. Thus the "be metathoracic" instruction, dictated by the normal *pbx*$^+$ gene product, must be a very late event in the sequence.

**Map Location of Selector Genes**  If we accept the hypothesis that **homoeotic mutations mark selector genes,** then the map location of these mutations becomes of interest. A group of homoeotic mutations that affect

FIGURE 25-14 *Antennapedia* **mutant** of *Drosophila*. Between the large compound eyes, where the antennae are usually located, are two legs. For comparison, normal legs appear at the bottom of the micrograph. (Courtesy of Dr. Thomas Kaufman.)

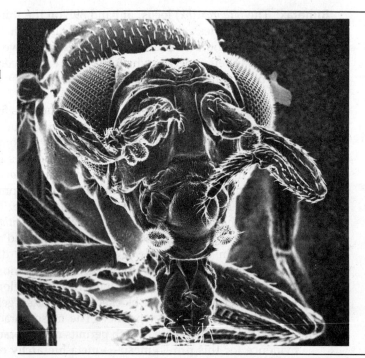

the determination of labial and maxillary structures of the head are clustered in a region called the **antennapedia complex** (*ANT-C*) (Figure 25-14), which maps to band 84B1,2 in the third chromosome. The *bithorax* complex (*BX-C*), as noted earlier, maps to the 89E band of this same chromosome, as do several other homoeotics (although not *engrailed*). It has therefore been speculated that much of the program of *Drosophila* development might be encoded in contiguous sectors of the right arm of chromosome 3.

# Developmental Genetics of Vertebrates

## 25.13 Mutations Affecting Early Development: The Mouse T Locus

The mouse **T locus** is reminiscent of the selector-gene complexes of *Drosophila* in several respects. First, many recessive *t* mutations at this locus are embryonic lethals in the homozygous condition. Second, different *t* alleles have different lethal effects: as illustrated in Figure 25-15, $t^{12}$ homozygotes fail to form blastocysts, $t^{w37}$ homozygotes fail to implant properly, and so on, as if each allele were involved in a distinct embryonic stage. Third, certain *t* alleles appear to play a role in specifying cell-surface antigens; specifically, the locus marked by the $t^{12}$ mutation has been implicated in

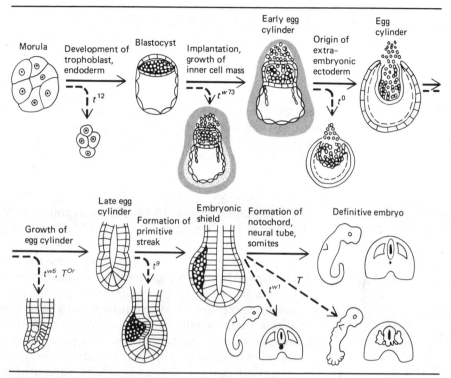

**FIGURE 25-15 Early development in the mouse and the defects seen in embryos homozygous for *T*-locus mutations.** (From D. Bennett, *Cell* **6**:441, 1975. Copyright © MIT Press.)

the production of F9, an antigen present on mouse morula cells (Figure 25-15), whereas the gene locus known as *Tcp-1* appears to code directly for a protein found on the surface of developing sperm cells. This is of course reminiscent of the position-specific antigens present on the surfaces of wing-disc cells in *Drosophila* (*Section 25.11*).

The final feature shared by selector genes and *t* alleles is that both are encoded within a large chromosomal segment. As diagrammed in Figure 25-16, the **T complex** maps to a long region of chromosome 17 that includes unrelated genetic markers as well. Interestingly, it is directly adjacent to the H-2 histocompatibility locus, which, as we learned in Section 21.4, is also a multigene complex concerned with the cell surface and the immune response.

Despite these analogies, the *T* locus has a sufficient number of unusual properties to temper facile interpretations of its role in mouse development.

1. **The *T*-interaction effect.** Wild-type mice are readily induced to mutate to a dominant form known as *T*, an allele that maps to the left side of

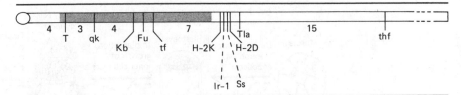

**FIGURE 25-16 Mouse chromosome 17 showing the boundaries of the *T* locus** (shaded). Other genes within the region include *qk* (quaking); *Kb* (knobbly); *Fu* (fused); and *tf* (tufted). (From M. F. Lyon, E. P. Evans, S. E. Jarvis, and I. Sayers, *Nature* **279**:38–42, 1979.)

the *T* complex (Figure 25-16). Heterozygous +/*T* mice have **short tails,** whereas *T*/*T* homozygotes inevitably die *in utero.* Recessive *t* alleles, on the other hand, cannot be generated by laboratory mutagenesis; they must therefore be isolated from heterozygous natural mouse populations. Mice carrying recessive *t* alleles are easily recognized in crosses with +/*T* heterozygotes: one quarter of the progeny, the *T*/*t* heterozygotes, prove to be **tail-less.** Thus some property of all natural *t* alleles interacts with *T* to amplify the severity of the tail-length defect.

2. **Suppression of recombination.** Whereas *T* and *tf* (Figure 25-16) are found to be separated by 7 to 12 centimorgans when marked laboratory strains are crossed, they appear to be separated by only 0.1 to 0.5 cM when *t*-bearing chromosomes are involved. In other words, *t*-bearing chromosomes severely suppress crossing over in this region of chromosome 17, a suppression that spreads to the neighboring H-2 locus as well. The suppression is thought to be due to the fact that **each *t*-bearing chromosome carries extensive rearrangements** that disallow normal synapsis with their wild-type partners and, usually, with one another. Because each *t* chromosome may carry a unique arrangement of chromatin, the term *t* allele has been replaced by the term *t* **haplotype,** which conveys its potential length and complexity.

3. **Male fertility effects.** That the *t* haplotypes are prevalent in natural mouse populations despite their recessive lethal effects is attributed to the fact that **sperm bearing *t* haplotypes are longer lived** than others and therefore more likely to reach and fertilize an egg. The basis for this success is as yet unknown. Also unknown is the basis for the **male sterility** of some $t^x/t^y$ heterozygotes (where $t^x$ and $t^y$ are different *t* haplotypes).

These disparate observations by no means rule out a role for the *T* locus in controlling development. They do, however, emphasize the caution that must be exercised in interpreting Figure 25-15; we are in fact a long way from assigning particular genes to particular developmental transitions in vertebrate organisms.

## 25.14 Gonadal Sex Determination in Mammals

Apart from the *T* locus, genetic studies pertinent to mammalian embryology are most abundant for **sex determination and differentiation,** the subject of the next few sections.

The expression of a mammal's **gonadal sex**—whether it will form testes or ovaries—necessitates the expression of the zygote genome and occurs relatively late in development. The gonad primordium region in the human embryo, for example, remains in an undifferentiated "intersex" state for about five weeks, during which time most of the other organ primordia have already begun to give evidence of performing differentiated functions. During this period, moreover, the embryo develops both **Müllerian ducts** (the precursors of oviducts) and **Wolffian ducts** (the precursors of sperm ducts), as if it were indeed uncommitted to the sex it is to become. Then, during the sixth week, Y chromosome-bearing embryos develop seminiferous tubules in their gonadal primordia, whereas the gonads of XX embryos retain their undifferentiated gonadal state, and it is only during the following week that changes indicative of follicle development begin to appear in the embryonic ovary. The development of the testis before the ovary suggests that at a critical point in development, the Y chromosome directly triggers testis determination in the primordial gonad cells. This concept is supported by the genetic data summarized in Sections 6.14 and 6.15, which indicate that the presence or absence of a Y chromosome is the critical sex-determining factor in mammals.

Testis differentiation is apparently triggered by a substance known as the histocompatability-Y **(H-Y) antigen.** If ovarian tissues are treated with H-Y antigen, they bind it to their cell surfaces and proceed to differentiate into testosterone-secreting testicular tissues. Receptors for H-Y antigen have also been identified on the surfaces of undifferentiated gonads. These and similar observations have led to the proposal that, at a critical state in embryogenesis, H-Y antigen is released, bound by gonadal receptors, and elicits testicular organogenesis. The remaining cells of the male are also induced to synthesize and display their own complement of H-Y so that H-Y comes to be a male-specific histocompatibility antigen on all tissues.

Many studies indicate that the H-Y antigen is encoded in the Y chromosome itself, which obviously explains the male-determining role of the Y most tidily. However, other studies indicate that the structural gene resides on another chromosome and that the Y somehow acts to facilitate or regulate its expression.

## 25.15 Genital-Duct Sex Determination in Mammals

The **genital-duct sex** of a mammal normally includes a penis and sperm duct in the male and a vagina, uterus, and oviducts in the female. The Wolffian and Müllerian primordia for these organs exist in the embryo

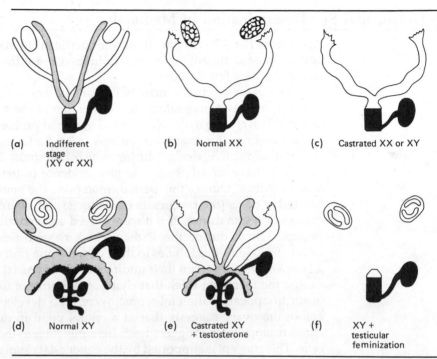

(a) Indifferent stage (XY or XX)    (b) Normal XX    (c) Castrated XX or XY

(d) Normal XY    (e) Castrated XY + testosterone    (f) XY + testicular feminization

FIGURE 25-17 Secondary sexual differentiation in mammals and the role of testosterone and the effect of the *Tfm* mutation in the embryonic development of the Wolffian duct (gray), Müllerian duct (outlined), and urogenital sinus (black), which includes bladder and urethra. The pair of bean-shaped bodies at the top of each square signify gonads: indifferent gonad (outlined), ovary (filled with small circles), and testes (containing two tubules). (From S. Ohno, *Nature* 234:134, 1971.)

before its gonadal sex is determined (Figure 25-17a). After the gonadal sex is determined, two events must occur: the "appropriate" primordia must go on to develop into fully formed structures and the "inappropriate" primordia must degenerate.

The normal female genital-duct organs are diagrammed in Figure 25-17b. These organs develop in female embryos whose ovaries have been surgically removed (Figure 25-17c), suggesting that Müllerian ducts will go on to develop into oviducts and a uterus in the absence of any signal from the gonads.

Genital-duct differentiation is more complex in the male embryo, for the normal complement of male organs (Figure 25-17d) will develop only if testes are present. If male rabbit embryos are castrated after testis differentiation takes place but before genital-duct sex determination begins and are then returned to the uterus to complete development, they are found at birth to contain oviducts, a uterus, and degenerate Wolffian ducts (Figure

25-17c). In other words, **the male genital-duct sex is apparently induced by some activity of the testis;** when the testis is removed, the female course is followed instead.

At least two secretions from the testis appear necessary to induce male genital-duct development. The first is the hormone **testosterone.** If testosterone is administered to either male or female castrated rabbit embryos during the time they are *in utero,* the rabbits are born with a normal set of male genital ducts (Figure 25-17e). They are also born with oviducts (Figure 25-17e), indicating that a second, **anti-Müllerian factor** is also required from the testis to ensure degeneration of the Müllerian ducts.

The requirement for activation by testosterone in male genital-duct determination is dramatized by the **testicular feminization syndrome,** a syndrome that has been described in humans, mice, cattle, and rats. Affected organisms show no evidence of any male genital-duct development, even though they are XY and possess normal testosterone-secreting testes (Figure 25-17f). S. Ohno has shown that the syndrome is produced in male mice when a recessive mutant gene, called *Tfm,* is present in their X chromosomes. Mice that are *Tfm*/Y are unable to respond to testosterone or testosterone derivatives. This can be demonstrated *in vitro* by taking testosterone target tissues from normal and mutant mouse embryos and presenting them with a radioactively labeled androgen. The normal tissues readily bind the male hormone, whereas the mutant tissues do not, and the difference can be shown to be caused by a deficiency in levels of androgen-binding receptor protein in the *Tfm* males. Thus the *Tfm* mutation is formally analogous to $lacI^{-5}$ mutations in the *lac* repressor gene (*Section 22.7*) and to the glucocorticoid-resistant mutations described in Section 24.7 for the mouse lymphoma line S49.

It is useful at this point to compare the effects of steroids on short-term and long-term regulation. Specific receptor proteins and differential gene transcription appear to be involved in both cases. The short-term responses are, however, reversible—androgen administration causes hypertrophy, while androgen withdrawal causes atrophy to adult male target tissues—whereas the long-term responses are essentially unidirectional; once a Wolffian duct differentiates into a sperm duct, castration will cause the duct to atrophy but not to return to its embryonic state.

## 25.16  Facultative Heterochromatization

A phenomenon more indirectly related to sexual differentiation is known as **facultative heterochromatization.** In this phenomenon a chromosome or a set of chromosomes becomes heterochromatic (*Section 3.13*) in the cells of one sex while remaining euchromatic in the cells of the opposite sex. The heterochromatized chromosomes are turned off, and the euchromatic chromosomes are expressed.

Three widely divergent cases are known in which facultative heterochromatization of this sort occurs. The first is found in such coccid insects as the mealybug: both maternal and paternal chromosome sets are active during all stages of development in the female, but the entire paternal set is selectively inactivated during the early development of the male embryo so that only the female set is active in the adult male somatic tissues and only the female set is included in spermatozoa. The second case occurs in the kangaroo and other marsupials: the paternally derived X chromosome is selectively turned off in female cells. The third case, and the one on which we shall focus, occurs in female placental mammals: either the maternally derived or the paternally derived X chromosome is turned off. This process is commonly known as **Lyonization,** after M. Lyon, an early and still active investigator of the phenomenon as it occurs in mice, and the turned off heterochromatic X chromosome is called a **Barr body,** after M. Barr.

The term facultative heterochromatization derives from the fact that the turned off chromosome is heterochromatic under the microscope, inactive in transcription by autoradiography, and late replicating like constitutive heterochromatin (*Section 3.13*). Unlike constitutive heterochromatin, however, it is not enriched in repetitive DNA sequences (*Section 4.10*) and the heterochromatic state is adopted (facultative) rather than constitutive. Thus at the onset of an XX embryo's development both X chromosomes are euchromatic and both are believed to be active in RNA synthesis; it is not until implantation that one of these chromosomes becomes heterochromatic and dysfunctional. Moreover, both X chromosomes remain euchromatic in the female germ line so that the change is not obligatory.

Numerous experiments have demonstrated that at a particular time in mammalian embryological development, each XX cell will turn off either the paternal or the maternal X chromosome. The choice is entirely random—50 percent of the time the maternal X is turned off, and 50 percent of the time the paternal X is turned off. Once the decision has been made by a given cell, however, the same choice is made by all the cell's descendants. This can be shown with cultured fibroblasts derived from human females heterozygous for sex-linked genes. For example, when fibroblasts are cloned from a woman heterozygous for a variant allele of the glucose-6-phosphate dehydrogenase (G6PD) gene, each clone is found to produce either the variant or the wild-type form of the enzyme but never a mixture. In a random sampling of cells, moreover, about 50 percent produce wild-type and 50 percent yield variant clones.

Once an X chromosome is Lyonized, the transformation is apparently irreversible. This has been shown in DNA-mediated cell transformation experiments (*Section 15.8*): DNA derived from the Lyonized chromosome of wild-type mice is unable to transform HGPRT$^-$ cell lines (*Section 15.6*) to the HGPRT$^+$ condition. DNA from their euchromatic homologous, by contrast, readily elicits transformation for HGPRT in the same hamster cell recipient.

The Lyonization phenomenon means that **all female mammals, including women, are mosaics.** Dramatic visual examples are tortoise-shell and calico cats, which are heterozygous for *black* and *orange* alleles of an

X-linked coat color gene. Tortoise-shell cats have a finely mottled black and orange coat, whereas calico cats, which also carry a "spotting" gene, exhibit an irregular patchwork of black and orange sectors all over their bodies, each sector representing a clone of cells derived from a Lyonized hair cell precursor.

Calico cats are almost always female, but an occasional male calico has been reported. These prove to be rare XXY cats and are invariably sterile, as would be expected of human males with the XXY Klinefelter syndrome described in Section 6.14 and Table 6-8. The fact that a Y chromosome does not interfere with Lyonization argues that the process is not triggered by sex hormones.

It has been proposed that the function of Lyonization is to reduce the effective X chromosome dosage of a cell. The strongest argument for this hypothesis comes from studies of mutant fetuses carrying more than two X chromosomes. Fetuses that are XXX exhibit two Barr bodies in every cell; fetuses that are XXXX exhibit three Barr bodies; and so on. In other words, **the mammalian cell seems to be programmed so that only one X chromosome per cell is allowed to remain euchromatic** and all others are inactivated. It is argued that since male mammalian cells function with only one X chromosome, Lyonization has evolved to effect the same genic balance in the female. This argument is obviously teleological and the true reason for the evolution of Lyonization is not known, nor is there any idea how it is initiated, effected, or maintained.

In *Drosophila* there is no clear counterpart of facultative heterochromatization of the X chromosome. Large portions of the X chromosome remain euchromatic in both males and females, and the remaining portions seem to be a mixture of constitutive and perhaps facultative heterochromatin. A physiological effect analogous to facultative heterochromatization, however, occurs in *Drosophila* and is known as **dosage compensation**. Male cells exhibit the same activities of most sex-linked enzymes as female cells, and this is apparently accomplished by a stimulated rate of gene transcription from the single male X chromosome.

**Position Effect Variegation**    The DNA-mediated cell transformation experiments described in the previous section suggest that **facultative heterochromatization inactivates gene expression at the level of DNA.** This leads to an obvious question: is the morphological state defined as heterochromatin the *cause* of gene inactivation or the *effect*? Thus DNA might be induced to undergo the physical transformations we associate with the heterochromatic state and, as a consequence, experience modifications (e.g., methylation or base substitution). Alternatively, modified DNA might be prone to adopt the heterochromatic physical condition.

The answer to this question is not known, but the complete answer must include an explanation for a phenomenon known as **position effect variegation.** It has been shown in the mouse that when translocations occur between an X chromosome and an autosome, the translocated seg-

ment of the X persists in undergoing Lyonization. As it does so, **the turn-
ing off process often extends into the adjacent autosomal chromatin.** The
effect is best seen by example. A well-studied translocation, drawn in Figure 25-18a, brings a piece of mouse chromosome 7 in contact with the X
chromosome such that the X is adjacent to the wild-type allele of gene *c*,
the mouse locus involved with pigment synthesis (*Section 21.8*). When
mice heterozygous for this translocation are also homozygous for the *c*
allele on chromosome 7, they display a mosaic or **variegated** coat color:
areas of pigmented fur are interspersed with areas of white (albino) fur. As
diagrammed in Figure 25-18b, the pigmented patches are generated by
those cells whose $c^+$ gene remains "on," whereas the white patches are
produced by those cells in which the Lyonization of the X has extended
into the translocated chromosome 7 and inactivated the $c^+$ gene, allowing
expression of the recessive *c* (albino) alleles at their "regular" positions on
chromosome 7.

Such variegated or **V-type position effects** display what is known as a
**spreading effect:** if translocated genes proximal to the broken heterochromatin are turned off, then distal genes may be turned off as well, with the
inactivation often extending for long distances.

Particularly striking is the fact that **position effect variegation is elicited by constitutive heterochromatin as well as by facultative heterochromatin.** This effect is well documented in *D. melanogaster*, whose genome

**FIGURE 25-18
V-type position
effects** in the
mouse. (a) The
T(X;7) Caltanach
chromosome in
which a piece of
chromosome 7 is
translocated to an
internal position in
the X chromosome;
$c^+$ designates wild-
type allele of the
*albino* locus.
(b) Effect on coat
color pigmentation
when the $c^+$ is not
Lyonized (left) or
Lyonized (right).

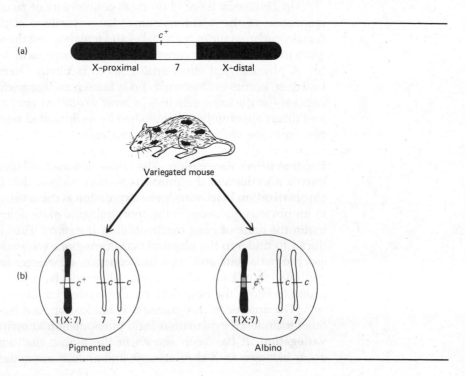

contains abundant constitutive heterochromatin (*Figure 3-14*). When a rearrangement (R) brings a piece of this heterochromatin next to the wild-type allele of the *white* eye color locus, and flies carrying this R(+) chromosome are crossed with *white* flies, the progeny with the genotype R(+)/w have red-white mosaic eyes, the red patches generated by those cells whose + gene is "on," the *white* patches produced by cells whose + gene has been turned "off." A spreading effect is also observed in these cases. Thus if the translocation heterozygote has the genotype R(+ +)/rst w, where *rst* (*roughest*) is a locus concerned with the arrangement of ommatidia in the eye, then the *white* patches in mosaic eyes invariably display the *roughest* phenotype, whereas the red patches may contain either *roughest* or wild-type ommatidia. In other words, the inhibitory effects of R may fail to reach either $rst^+$ or $w^+$ in some cells; they may spread to $rst^+$ but spare $w^+$ in other cells, and they may include both $rst^+$ and $w^+$ in others. Broken heterochromatin has been shown capable of inactivating six map units (~50 polytene bands) of euchromatin in *D. melanogaster*, which involves enormous lengths of DNA. No one yet has any idea how this comes about, nor how the properties of broken heterochromatin relate to the regulatory properties of intact heterochromatin.

# Differential Expression of Hemoglobin Genes

## 25.17 Switching Hemoglobin Genes On and Off

The hemoglobin gene family is described extensively in Section 19.8, which should be reviewed at this time. Hemoglobin genes experience two kinds of regulation. The first operates on the entire family: it ensures that the transcription of globin genes occurs only in the erythroid cell line and not in other cell types. The second operates on individual genes within the family: it ensures that synthesis of the appropriate globins occurs at the appropriate developmental stage.

**Restriction of Globin Gene Expression to Erythroid Cells**  Nonerythroid cells fail to express hemoglobin genes, but the following experiment demonstrates that this is not a permanent restriction. Human fibroblasts, which normally never synthesize hemoglobin, were fused with mouse erythroleukemia cells (*Section 15.6*), and the hybrids were found to express the *human* forms of both the α and the β hemoglobin genes. Thus the nonexpression of globin genes by nonerythroid cells is accomplished by a very different mechanism than Lyonization, for as we learned in the previous section, Lyonized chromosomes fail to be activated even when transferred into "permissive" hosts. When the experiment is modified so that human fibroblasts are fused with **non**erythroid cells, moreover, no hemoglobin synthesis is induced. Therefore, "factors" present in the erythroleukemia

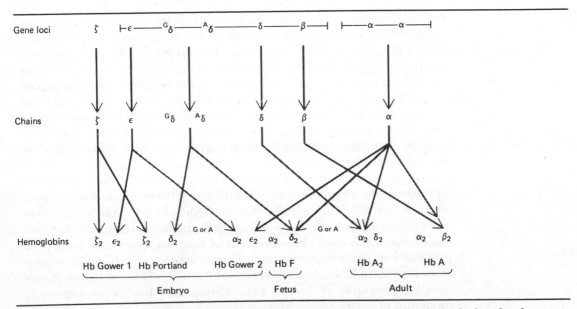

**FIGURE 25-19 Hemoglobins synthesized at various times during development.** The linkage between the ξ- and α-genes is not included in the scheme. (From D. J. Weatherall, and J. B. Clegg, *Cell* **16**:467–479, 1979.)

cell are apparently necessary and sufficient to elicit globin gene expression, factors that have yet to be identified.

**Differential Expression of Globin Genes** Figure 25-19 summarizes the pattern of globin gene expression at different developmental stages. Two genes are uniquely transcribed by the **embryo;** the **fetus** produces only α and γ chains for hemoglobin F synthesis; and the **newborn** simultaneously phases out the production of HbF and stimulates expression of the β and δ genes for the first time in development, genes that continue to be expressed throughout adult life. That these switches are qualitatively different from the overall induction of globin gene expression could be shown in the cell fusion experiment described previously: the mouse erythroleukemia cells were able to elicit expression of α and β genes from adult human fibroblasts, but they were unable to elicit expression of the γ genes, even though the γ and β genes are closely linked. Thus it appears that either the erythroleukemia cell line lacks "factors" necessary to turn on γ genes, or adult fibroblasts have heeded the instruction "turn off γ genes" even though they are not members of the erythroid lineage.

**Hereditary Persistence of Fetal Hemoglobin** The fetal →adult switch in human hemoglobin synthesis is altered in genetic disorders known collectively as hereditary persistence of fetal hemoglobin **(HPFH).** Whereas the $^G\gamma$ and $^A\gamma$ genes are normally turned down at 32 to 34 weeks of gestation and the β and δ genes are activated, persons with HPFH typically exhibit two modified patterns. First, the β- and δ-globins are not synthesized in adult HPFH erythroid cells, often because they have been (partially) deleted from the chromosome. Second, one or both of the γ-globin genes continues to be expressed postnatally, compensating for the β-globin deficiency.

A controlling element appears to be involved in HPFH phenotype. As noted in Section 19.8, the $^G\gamma$-gene exists in two allelic forms that we have called $^G_{Val}\gamma$ and $^G_{Lys}\gamma$. A person heterozygous for these alleles and for HPFH, designated $^G_{Val}\gamma$ = normal/$^G_{Lys}\gamma$ = HPFH, will synthesize both forms of polypeptide in utero, but only the $^G_{Lys}\gamma$ polypeptide postnatally. If the HPFH defect were dependent on a diffusable regulatory molecule that could act in *trans,* then either the normal chromosome should produce repressor to block expression of *all* γ-genes in the adult, or the HPFH chromosome should produce an activator to elicit expression of *both* the $^G_{Val}\gamma$ and the $^G_{Lys}\gamma$ genes in the adult. Since neither occurs, it appears that **the HPFH disorder affects the γ-gene expression in cis,** a hallmark of a controlling-element mutation (Section 22.9).

## 25.18 Activation of Globin Genes During Chicken Development

M. Groudine and H. Weintraub have performed a detailed study of globin synthesis in chick embryos. Since this study brings together the prominent current theories on how differential gene expression might be accomplished at a molecular level, we consider it in some detail.

Groudine and Weintraub isolated cells from the **area opaca,** a region known to give rise to most of the avian erythroid cells at 35 hours of development, from chick embryos 20 to 23 hours old, and grew these precursor cells in culture. The cells did not initially synthesize hemoglobin, but differentiated into hemoglobin-producing cells (erythroblasts) after several days. Groudine and Weintraub then isolated globin-gene DNA from precursor cells and from cultured erythroblasts, and were able to document a number of differences between them. The nonexpressed "precursor genes," they found, differed from the expressed "erythroblast genes" in three ways: in their degree of methylation, in their enzyme sensitivity, and in their associated nonhistone proteins. Each of these parameters is described in the following sections. It is important to note at the outset that these same three parameters are being found to characterize nonexpressed versus expressed genes in many other higher eukaryotic systems, although not all appear to apply to the lower eukaryotes. In other words, **the differential expression of hemoglobin genes appears to be accomplished by mechanisms utilized throughout the genome.**

**DNA Methylation**  In addition to the four normal bases, the DNA of most higher eukaryotes contains **5-methylcytosine ($^mC$),** found mainly at the 5' side of G residues. Methylation occurs after cytosine has been polymerized into DNA during replication, and it is accomplished by ubiquitous nuclear enzymes known as **methylases.** To ask whether genes isolated from undifferentiated cells are more or less methylated than genes isolated from differentiated cells, one compares the effects of two restriction enzymes: the enzyme **Msp I** cleaves both at CCGG and at C<sup>m</sup>CGG, whereas the enzyme **Hpa II** will not cut at this sequence if the second C is methylated. Therefore, a gene that yields a similar pattern of restriction fragments (*Section 4.10 and Figure 4-8*) with both enzymes is relatively **undermethylated** compared with a gene that yields two very different patterns.

When this test is applied to the globin genes, the following results are obtained.

1. The α-globin and β-globin genes are methylated in precursor cells but undermethylated in erythroblasts.
2. In erythroblasts, where the coding regions of the α- and β-genes are entirely undermethylated, the adjacent noncoding regions are heavily methylated.

In other words, **undermethylation of DNA is correlated with active gene expression,** whereas **methylation is correlated with nonexpression.**

**DNAse I Hypersensitivity** As first described by C. Wu and S. Elgin, **the DNA associated with the 5' end of actively transcribed eukaryotic genes is extremely sensitive to digestion by DNAse I.** The enzyme introduces double-stranded cuts at particular sequences in the 5' regions, usually located 500 to 1000 base pairs upstream from the start of the gene; these cuts produce characteristic electrophoresis patterns that need not concern us here.

Groudine and Weintraub were able to show that such hypersensitive sites are readily detectable in the globin genes of 5-day (embryonic) and 14-day (adult) red cell lineages but that no such sites are present in the globin genes of precursor cells from chick embryos 20 to 23 hours old. Since the precursor cells are not yet engaged in globin gene transcription, this result suggests that DNAse I hypersensitivity is acquired at the time of gene activation.

**DNAse I Sensitivity and the HMG Proteins** Active genes are not only **hyper**sensitive to DNAse I digestion in very local regions at their 5' ends. They are also more generally sensitive to single-stranded nicking by the enzyme than are their inactive counterparts, a parameter called **DNAse I sensitivity.** In the case of the globin genes, the γ-genes are more DNAse-sensitive than the β-gene in fetal erythroid cells, whereas the β-gene is more sensitive than the γ-genes in adult erythroid cells. That this is not simply due to transcriptional activity is dramatically demonstrated by the fact that in the mature chick erythrocyte, where hemoglobin synthesis is no longer occurring, the β-globin gene continues to be DNAse sensitive even though it is no longer transcriptionally active. In other words, the change correlates with the state of nuclear differentiation and not with gene expression *per se.*

DNAse I sensitivity appears to depend on the association of active genes with two nonhistone proteins of the HMG class (*Section 3.5*) known as **HMG-14** and **HMG-17.** If active globin genes are extracted with 0.35 M NaCl, which removes HMG-14 and HMG-17, they are no longer sensitive to DNAse I digestion; sensitivity is restored when the two proteins are added back to the DNA. The HMG proteins cannot *confer* DNAse sensitivity, however: if they are added to inactive genes (the ovalbumin gene isolated from erythroid cells, for example), the genes remain DNAse insensitive. Thus **active genes appear to undergo a transformation that allows them to bind HMG-14 and HMG-17;** binding, in turn, renders the genes DNAse I sensitive.

**Significance of the Physical and Chemical Changes in Active Genes**  It is not yet possible to specify cause-and-effect relationships between methylation, enzyme sensitivity, and HMG association. For example, it cannot yet be said whether HMGs bind preferentially to DNA having undermethylated DNA or whether HMG binding precludes methylation and perhaps stimulates the activity of demethylation enzymes. Groudine and Weintraub note that because the DNAse-hypersensitive sites occur at specific sequences in putative controlling regions, these are most likely to be established first, thereby determining the course of subsequent modifications in the downstream genes. The extreme sensitivity of these DNA sequences to double-stranded cuts suggests that whereas most eukaryotic DNA is compacted into nucleosomes (*Section 3.6*), these sequences may be nucleosome-free.

This brings us to the theory of **nucleosome phasing.** The idea is that DNA may form nucleosomes in a nonrandom fashion, leaving critical controlling sequences nucleosome free and exposed to the nucleoplasm, where regulatory molecules can find them. The fact that the origin of replication in the SV40 viral chromosome (*Section 12.16*) is nucleosome free supports the notion that such regions may be important controlling elements in eukaryotes. It is intriguing to imagine that the critical event in cell determination may be the creation, and maintenance, of specific nucleosome-free sectors upstream from genes that are to be transcriptionally active in a particular cell type.

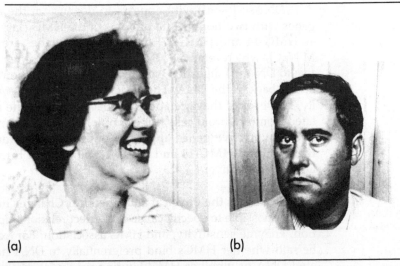

(a)                             (b)

(a) Mary F. Lyons (MRC Radiobiology Unit, England) studies the mechanism and effects of facultative heterochromatization of the X chromosome. (b) A. García-Bellido (University of Madrid, Spain) studies the genetic determination of imaginal discs in *Drosophila*.

# Questions and Problems

1. Human XO zygotes do not develop normal ovaries and are born sterile. What does this fact suggest about the facultative heterochromatization process in presumptive ovaries?

2. What can you conclude about human sexual development from the fact that XXY males are sterile and XYY males are fertile?

3. It has been reported that a burst of cell division occurs during the fifth week of human embryological development in the presumptive testes but not in the presumptive ovaries. How might this relate to other observations regarding gonadal sex determination in humans?

4. What would be the effect, if any, on the sex and fertility of humans if there were
   (a) Mitotic nondisjunction during the first cleavage of a fertilized XY egg to yield one daughter XXY cell and one daughter Y cell, with subsequent separation of the two cells to form twins;
   (b) Mitotic nondisjunction during the first cleavage of a fertilized XY egg to yield one daughter XYY and one daughter X cell, with subsequent separation of the two cells to form twins;
   (c) Injection of testosterone during and after the sixth week of life in an XX embryo;
   (d) Transplantation of a testis to an XX embryo during the sixth week of life;
   (e) Castration (removal of testes) at birth of an XY individual.

5. A 25-year-old woman has never menstruated. An analysis of her karyotype reveals that she is XY. What do you think is wrong? Would hormone therapy be helpful?

6. Two genes in the human X chromosome determine the enzymes A and B, respectively. One gene exists in the allelic forms $A_1$ and $A_2$; the other in the forms $B_1$ and $B_2$. A woman's father was $A_1B_1$ and her mother was homozygous for $A_2B_2$. Individual cells from this woman were tested for their enzyme phenotypes. Which of the following, mutually exclusive results would you expect from this study according to the Lyon Hypothesis? Explain.
   (a) There are two types of cells in roughly equal numbers: those that produce only enzymes $A_1$ and $B_1$ and those that produce only enzymes $A_2$ and $B_2$.
   (b) There are four types of cells in roughly equal numbers: those producing enzymes $A_1$ and $B_1$; those producing $A_1$ and $B_2$; those producing $A_2$ and $B_1$; and those producing $A_2$ and $B_2$.

7. Why do you think human females are not known to exhibit a patchy hair coloring in the manner of a calico cat?

8. An enzyme, BVDase, is expressed in human erythrocytes and is dictated by a sex-linked gene. Two alleles of the gene are known, $B$ and $b$, where the $b$ gene product is inactive. It is found that of women who are known from pedigree studies to be heterozygotes at this locus, about 3 percent exhibit no BVDase activity in any of their erythrocytes, whereas most have a mixed population of erythrocytes, some with and some without BVDase activity.
   (a) Explain this result in terms of the Lyon Hypothesis.
   (b) Do you predict that there are many or few erythrocyte precursors in a human embryo at the time X inactivation takes place?
   (c) Predict the frequency of heterozygous women in whom all erythrocytes contain BVDase.

9. Mosaic *Drosophila* can arise in two ways.
   (a) Describe each avenue.
   (b) How would you determine which avenue had been followed in a given sample of mosaic flies?

10. An inversion in chromosome III of *D. melanogaster* produces V-type position effects. When this R (+ + +) chromosome is coupled with a chromosome III carrying the markers *se* (*sepia* eyes), *ca* (*claret* eyes), and *r* (*rough* eyes), the variegated flies have the following phenotypes: eyes are always *rough*, may or may not be *claret*, and are never *sepia*. Relate the positions of these three markers to the position of the inversion.

11. **Sex-limited genes** produce traits that are expressed in only one of the sexes, presumably because some hormonal attribute of femaleness or maleness must be present if the genes are transcribed and/or the gene products expressed.
   The sex-limited human trait known as **pattern baldness** is expressed as follows: men heterozygous or homozygous for the autosomal gene $H^B$ will lose the hair on top of their heads as they age; women lose their hair only if they are $H^B/H^B$ homozygotes. A man exhibits pattern baldness, but his parents do not. His wife is not balding and baldness is not found in her lineage. Give the genotypes of the husband, wife, and parents. What proportion, if any, of the man's sons would you expect to become bald? His daughters? Would the phenotypes of the offspring differ if the $H^B$ gene were a sex-linked dominant gene but not sex limited? Would they differ if it were a sex-linked and sex-limited dominant gene?

12. A man who considers himself "undersexed" is given testosterone treatment, whereupon he begins losing his hair. Offer a genetic explanation for this event.

13. If a nucleus had been inadvertently included in the cytoplasmic extracts utilized in the Illmensee-Mahowald experiment (*Figure 25-5*), how would the outcome of the experiment have been different?

14. Malignant cells are sometimes found to synthesize surface molecules known as **tumor-specific antigens,** and in some cases these are reported to be immunologically similar or identical to antigens synthesized by fetal cells. Devise a theory for the origin of such cancers using one or more of the principles presented in this chapter.

15. The *O-nu* mutation of *Xenopus* can be defined as a rescuable maternal-effect mutation. Explain this statement, using the experiments summarized in Figure 10-9 to support your argument.

16. A fetus is diagnosed by amniocentesis as trisomic for chromosome 21, but the parents decide against an abortion. The child proves to have the facial characteristics of a Down's syndrome child but has normal intelligence. How might you explain this?

17. Does it make any difference in terms of progeny phenotype whether a *grandchildless* (*gs/gs*) female fly is mated with a $+/+$, $+/gs$, or *gs/gs* male? Explain.

18. In *Chlamydomonas reinhardi*, the temperature-sensitive mutation *gam-1* prevents normal sexual agglutination (*Figure 5-8*) at restrictive temperature. The mutation is not linked to the mating-type (*mt*) locus but is expressed only in $mt^-$ cells; $mt^+$ *gam-1* cells agglutinate normally and transmit the trait to their $mt^-$ progeny.

    (a) Give the genotypes, phenotypes, and expected proportions for the tetrads (*Section 6.16*) emerging from an $mt^-$ *gam-1* $\times$ $mt^+$ $+$ cross, where the mating is performed at permissive temperature and the phenotypes are assessed at restrictive temperature.

    (b) Why might a "sex-limited" gene such as *gam-1*$^+$ exist in a single-celled haploid organism like *C. reinhardi*?

19. B. Mintz and K. Illmensee have been able to inject genetically marked cells from a malignant mouse tumor cell line into blastocysts (young embryos) of a mouse embryo, implant these blastocysts into the uteri of female mice, and examine the phenotypes and genotypes of the resultant mouse progeny. A number of the progeny mice prove to be mosaics, exhibiting, for example, tumor-derived coat-pattern genes and, in some cases, forming gonads from tumor-derived cells and transmitting tumor-derived genes to their offspring. Discuss these results (a) in terms of the totipotency of differentiated cells and (b) in terms of the concept that at least some malignancies derive from developmental aberrations of gene expression rather than mutational alterations in gene structure.

20. Mutations known as **zygotic lethals** are exemplified by the *rudimentary* mutations of *Drosophila*. Females homozygous for *r* mutations produce eggs that cannot undergo normal development if fertilized by *r*-bearing sperm but that develop normally if fertilized by $+$ sperm. Eggs produced by $+/r$ females can, on the other hand, develop normally when fertilized by *r* sperm even if, at the final meiotic division, they come to carry only the female *r* allele.

    (a) Is the following statement true or false: At least one $+$ allele of the *rudimentary* gene must be present during either oogenesis or embryogenesis to ensure normal development. Explain your choice.

(b) How does the zygotic-lethal pattern differ from the pattern of a maternal-effect lethal.

(c) Review Section 21.3 and offer an explanation of the zygotic-lethal pattern for the *rudimentary* mutations.

21. The *Notch* locus of *D. melanogaster* resides in band 3C7 of the X chromosome. Point mutations and deficiencies that map to this band are all partially dominant: females heterozygous for *Notch* mutations exhibit such ectodermal abnormalities as notches and thickened veins in the wings and alterations in body bristle morphology. Embryos hemizygous or homozygous for *Notch* mutations, on the other hand, experience gross developmental abnormalities: a large portion of the ectodermal cells, which would normally give rise to such larval structures as skin, salivary gland, or pharynx, instead give rise to neural cells. As a result, the central nervous system of the embryo becomes about three times its normal size, an enormous structure that destroys the orientation of most other developing organs, and death ultimately results.

Discuss the *Notch* locus as the potential site of a selector gene. Which cytoplasmic coordinates would it most likely be "interpreting"?

22. When the genes in the 16A segment of the *D. melanogaster* X chromosome are present in a single copy, the chromosome is considered wild type ($+$), whereas a duplication of the 16A segment, known as the *Bar* mutation, creates an X chromosome denoted $B$ (*Figure 6-16*). Flies with a $+/+$ or a $+/\rightarrow$ genotype have normally sized eyes; the eye size of $B/+$ females is somewhat reduced, and the eye size of $B/B$ females and $B/\rightarrow$ males is extremely reduced. Unequal crossing over (*Figure 19-10*) in $B/B$ flies will occasionally generate **ultrabar** ($\widehat{BB}$) chromosomes that carry three copies of the 16A segment. Such chromosomes prove to have an even more drastic effect on eye size than a $B$ chromosome. Whereas a $B/B$ fly will have an average of 68 ommatidia per eye, a $\widehat{BB}/+$ fly will have an average of only 45 ommatidia per eye.

(a) Diagram the generation of an *ultrabar* chromosome. What does the other chromosome look like?

(b) Is eye size in *Drosophila* dependent on the total number of 16A segments per nucleus or upon their chromosomal position? Explain.

(c) Segment 16A contains six bands, so the region can be designated 15F-123456-16B. How would you designate the *Bar* and *ultrabar* chromosomes? Devise a hypothesis to explain the deleterious effect on ommatidia development exerted by such duplications.

23. The 5' regions of *Drosophila* heat-shock genes carry DNAase I–hypersensitive regions in all tissues, even without exposure to heat. Is this finding at odds with the theory that such regions are likely sites for control of differentiated gene expression (*review Section 24.4*)?

# 26
# Population Genetics I: General Principles and Mendelian Populations

**A. Introduction**

**B. General Principles of Population Genetics**
  26.1 Familiar Features of Human Population Genetics
  26.2 Formal Objectives of Population Genetics

**C. Genetic Variability in Populations**
  26.3 Mutations as the Source of Genetic Variability
    Mutation pressure
  26.4 Estimating Genetic Variability in a Population
    Electrophoretic variants

**D. Mendelian Populations**
  26.5 Describing Mendelian

Populations: The Hardy-Weinberg Principle
  26.6 Applications of the Hardy-Weinberg Principle
  26.7 Deviations from the Hardy-Weinberg Principle: Assortative Mating, Inbreeding, and Selfing
  26.8 Establishing a Mendelian Population and the Founder Effect
  26.9 Stable Polymorphisms in Mendelian Populations

**E. Questions and Problems**

## Introduction

Many of the principles of population genetics are very familiar to us all, for many are beautifully illustrated by the human populations that we have scrutinized daily since we were born. The chapter begins, therefore, by pulling these principles from our everyday observations and defining them in the terminology of a population geneticist. In the second section these principles are re-enunciated in a more formal, generalized context so that the reader can readily apply them to the nonhuman examples that often best illustrate a particular concept.

With the overall dynamics of population **equilibrium** and **evolution** thus defined, the remaining sections of this chapter focus on the properties of a population at equilibrium, often described as a **Mendelian population.** The next chapter explores the agents responsible for evolutionary change, and the final chapter considers how populations have evolved into species and how genes have changed over the course of evolutionary time.

# General Principles of Population Genetics

## 26.1 Familiar Features of Human Population Genetics

All humans belong to a single subspecies, *Homo sapiens sapiens*, meaning that all are genetically capable of interbreeding and producing fertile offspring. However, humans display enormous **phenotypic variability.** Some of this variability is specific to persons who live in or descend from particular geographic areas, and these are categorized as **racial** differences: the African, Oriental, and Caucasian variations in skin color and facial bone structure require no elaboration. Within a race, moreover, variability persists: Caucasians display a wide range in the color and distribution of their eye pigments, the curliness of their hair, and so on.

Human phenotypes are ultimately the product of human genotypes, although the environment will often influence how certain genes are expressed. The enormous phenotypic variability in the human species is the consequence of an **enormous genotypic variability,** meaning that if all the human gametes on the earth today (the human **gene pool**) could be scrutinized, we would find a constant number of chromosomes and gene loci in each nucleus (barring the occasional chromosome aberration and of course excluding the special case of the sex chromosomes), but we would find an enormous **diversity of alleles.**

It is possible to estimate the frequency of alleles at particular gene loci without resorting to a gamete-by-gamete inspection; the way this is done will be described later in the chapter. Table 26-1 summarizes some representative results. This table will be considered in various contexts during the course of the next two chapters; for present purposes, two features are important. The first is that it documents what we have just predicted intuitively, namely, that the human species as a whole is **multiply allelic** at many gene loci. Thus if we disregard for the moment the racial and geographic categories in Table 26-1, then the Rh blood-group locus is seen to have four widespread alleles in the human species (and many minor alleles besides), the ABO blood group has three widespread alleles, and so on. Since the presence of alleles can often, although by no means always, be detected at some level of the visible phenotype, multiply allelic loci are commonly said to display **polymorphism** (literally, "multiple morphologies").

The second feature of Table 26-1 is apparent when we compare allelic frequencies in two distinct human **populations,** a population defined as a

**TABLE 26-1   Allele Frequency Estimates** for African Blacks and for Blacks and Whites Living in Georgia and California

| Locus | Allele | Black Africa | Black California | Black Georgia | White California | White Georgia |
|---|---|---|---|---|---|---|
| Rh blood group | $R_0$ | .617 | .486 | .533 | .023 | .022 |
| | $R_1$ | .066 | .161 | .109 | .413 | .429 |
| | $R_2$ | .061 | .071 | .109 | .140 | .137 |
| | $r + R_0{}''$ | .245 | .253 | .230 | .405 | .374 |
| ABO blood group | A | .156 | .175 | .145 | .247 | .241 |
| | B | .136 | .125 | .113 | .068 | .038 |
| | O | .708 | .700 | .742 | .685 | .721 |
| Duffy blood group | $Fy^a$ | .000 | .094 | .045 | .429 | .422 |
| β-globin | $Hb^s$ (Sickle cell) | .092 | No estimate | .043 | No estimate | .000 |
| Glucose-6-P dehydrogenase | G6PD deficiency | .176 | No estimate | .118 | No estimate | .000 |

From J. Adams and R. H. Ward, *Science* 180:1137, 1973. Copyright 1973 by the American Association for the Advancement of Science.

**stable group of randomly interbreeding persons.** We can consider modern Black African and White Georgian people to compose two separate populations, since their ancestors had virtually no opportunity to interact with one another. When the allelic frequencies of these two populations are compared, striking differences are apparent. Thus the $Fy^a$ (Duffy blood group) allele is absent from the Black African population but occupies almost half (0.422) of the $Fy$ loci of the White Georgian population. On the other hand, the $Hb^s$ allele of the locus encoding β-globin and the G6PD-deficiency allele are both absent from the White Georgian population but present in the Black African. Less extreme differences characterize most of the other alleles, whereas several are present at roughly the same frequency in both populations.

In short, therefore, **allelic polymorphism exists at many gene loci when the entire species is considered. When a single population is considered, moreover, a particular allele is often present at a significantly higher or lower level than in the species as a whole.**

The source of allelic variation is easily identified as gene mutation, a fact we have amply illustrated throughout this text. The factors that act to maintain different allelic frequencies in different populations, on the other hand, are more complex, and a major goal of population genetics is to identify and analyze these mechanisms. Perhaps the most intuitively obvious mechanism is **environmental selection,** wherein a particular allele promotes survival in one environment and is less advantageous or even deleterious in another. Certainly the best known example of this mechanism is given by the $Hb^s$ allele, whose frequency is influenced by the presence in the environment of the malarial parasite.

The malarial parasite (*Plasmodium*) is carried by a genus of mosquito that thrives in certain regions of Africa. The life cycle of the parasite requires a developmental sojourn in human erythrocytes; therefore, a susceptible person who is bitten by a *Plasmodium*-infested mosquito acquires the parasite in his or her erythrocytes and contracts the disease malaria. Humans heterozygous for the $Hb^s$ allele have erythrocytes that are quite resistant to malarial infestation. These humans are thus likely to survive malaria, produce offspring, and pass the $Hb^s$ allele along to their descendants. Humans heterozygous for the $Hb^s$ allele in nonmalarial environments (such as Georgia) have obviously no such **selective advantage.** On the contrary, they have a selective disadvantage: if they marry another $Hb^s$ heterozygote, one quarter of their progeny run the risk of being $Hb^s/Hb^s$ homozygotes whose erythrocytes have an abnormal sickle shape and a severely impaired ability to transport oxygen. Homozygotes are therefore said to have **sickle-cell anemia,** and they often die unless blood transfusions are available. It follows that possession of the $Hb^s$ allele would be strongly selected against in Georgia, explaining its 0.000 frequency in the White Georgian gene pool (Table 26-1).

Clearly, $Hb^s/Hb^s$ homozygotes are also born in African populations, where the consequences of homozygosity are equally severe. Therefore, the 0.092 frequency of the $Hb^s$ allele in the African population represents a trade-off: the allele is lost from the gene pool every time an $Hb^s/Hb^s$ homozygote dies before giving birth to offspring, whereas the allele is selectively preserved in the gene pool every time a $+/Hb^s$ heterozygote survives malaria to a reproductive age while a $+/+$ individual succumbs to the disease before reproducing. Eradication of malaria from the African environment, which is rapidly taking place, will clearly change the situation: homozygotes will continue to die, but heterozygotes will no longer selectively survive. Therefore, a change in environmental selective pressures should result in a shift in the frequency of the $Hb^s$ allele until, at a new equilibrium, the allele is present at a 0.000 frequency, present in the gene pool only to the extent that it is introduced by new mutations in the β-globin gene. Similarly, if the malarial parasite were to establish itself in Georgia, the equilibrium frequency of the $Hb^s$ allele would be predicted to shift toward the 0.092 equilibrium found in Africa.

In population genetics, such shifts in allelic frequencies are called **evolutionary changes. Evolution** is a loaded word, for we are accustomed to its connotation of change in a "positive," more "advanced" direction. This is particularly true for human examples, where it is easy to label the consequences of malaria eradication as "evolution" but difficult to regard the double-lethal consequences of malarial introduction as "evolutionary." For the population geneticist, however, **evolution is a value-free term that designates a shift in allelic frequencies in a population.**

The example of sickle-cell anemia is so often repeated that many readers may assume that most polymorphisms are maintained in a similar fashion, that is, by having the **heterozygote be superior in fitness** to either form of homozygote. In fact, heterozygote superiority has only rarely been documented as an explanation of polymorphism. Other patterns of selection have therefore been sought and described, and some polymorphisms appear to be the consequence of chance and not selection at all. These various alternatives are described in subsequent sections. First, however, we must define populations in more formal terminology.

## 26.2 Formal Objectives of Population Genetics

Population genetics has as its reference point the concept of a **stable population.** Each stable population will possess a **stable gene pool** that is shuffled about from one generation to the next by interbreeding. The pool is not fixed: new alleles are constantly being introduced by mutation, whereas existing alleles are constantly being favored or eliminated by selection. Therefore, the term "stable" should not be confused with the term "static." As long as mutation rates and environmental pressures remain constant, however, the gene pool should attain some sort of **stable equilibrium,** with alleles being added and subtracted at constant rates. At equilibrium, therefore, each gene locus in the pool will display a locus-specific level of heterozygosity. For some loci, almost all the alleles in the pool will be identical (save some recently introduced mutant varieties, perhaps), and these loci can be considered homozygous. For other loci, two or more alleles will populate the pool at significant levels, and the locus is said to be polymorphic. Many population geneticists seek to understand the factors responsible for maintaining polymorphism in such a stable gene pool.

The preceding description applies to alleles that have a **selective value:** an equilibrium will eventually be reached between the pressures of mutation, positive selection, and negative selection. Some alleles, however, may be **selectively neutral,** having no influence on the survival of either heterozygotes or homozygotes. A neutral allele should experience a very different fate after its mutational entrance into the gene pool: it is expected to **drift** so that, by simple "luck of the gametic draw," one neutral allele may find itself represented more and more often in the gene pool until it eventually becomes quite frequent, whereas another neutral allele may be "drawn" less and less often until it eventually drifts to extinction. The relative contribution of neutral and selected alleles to the polymorphism of a population is at present the subject of vigorous debate, as we shall document in later sections.

Given a stable population as a starting point, population genetics next seeks to understand how such a population can evolve with time, **evolution** being defined in the "narrow sense" (*Section 26.1*) as a **net shift in allele frequencies within a population from one generation to the next.** Such a shift cannot occur unless the original population is polymorphic: a totally homozygous gene pool is obviously unable to experience shifts in its allele frequencies. Therefore, **the first prerequisite for evolution is that there be genetic variability in a population.**

**The second prerequisite for evolution is that there be some shift in the factors that were responsible for the original equilibrium.** For example, environmental pressures might change; random drift may chance to favor a particular allele over another for a considerable period of time; or mutation rates might change so that new alleles are introduced at different rates. Just as there is vigorous debate over the factors that maintain polymorphism in a population, so is there vigorous debate over the relative importance of the factors that change allelic frequencies and thus promote evolution.

We can now state an important generalization. **Populations evolve; individuals do not, and genes do not.** Genes mutate in individuals, individuals contribute new alleles to the population, and selection and drift alter allelic frequencies so that new populations, and ultimately new species, evolve from the old. In population genetics, then, **the unit of interest is ultimately the composition of the gene pool** and not the individual gene.

# Genetic Variability in Populations

## 26.3 Mutations as the Source of Genetic Variability

Mutation provides the ultimate raw material for polymorphism and, hence, for evolution. Table 26-2 presents calculations of **spontaneous mutation rates** at various loci in various organisms. For higher eukaryotes, the average value is about $10^{-5}$, meaning that about one gamete in $10^5$ will carry a new mutation at a given gene locus. Since a mammalian genome is estimated to carry about $5 \times 10^4$ functional genes, this means that about half the gametes produced by an adult mammal will carry at least one new allele. In other words, mutation provides the population with an abundant supply of new alleles at each generation.

Each new allele that is transmitted to the next generation will have unique effects on the phenotype. One may be selectively neutral, because it either is totally recessive or has a "neutral" effect (e.g., creates a "synonymous" codon). Another may be immediately lethal (e.g., a dominant embryonic lethal) or eventually deleterious, as would be the case with a recessive lethal that becomes homozygous in subsequent generations. Only very rarely is the new allele immediately, or eventually, beneficial to the population. Because many mutations are recessive lethals, the resultant accumulation of recessive deleterious mutations—often referred to as the **mutational "load"**—has been calculated to reduce the overall viability of a species by as much as 50 percent. Although many assumptions are inherent in this estimate and the true "load" may be less, it is clear that recurrent mutation must exert a large deleterious effect on natural populations. As stated by J. F. Crow: "The species pays a high price for the privilege of mutation and the evolutionary possibilities that derive from it."

**TABLE 26-2  Spontaneous Mutation Rates in Different Organisms**

| Organism | Character | Rate | Units |
|---|---|---|---|
| Bacteriophage T2 | Lysis inhibition, $r \longrightarrow r^+$ | $1 \times 10^{-8}$ | Per gene[a] |
| | Host range, $h^+ \longrightarrow h$ | $3 \times 10^{-9}$ | per replication |
| Bacterium, | Lactose fermentation, $lac^- \longrightarrow lac^+$ | $2 \times 10^{-7}$ | |
| E. coli | Phage $T_1$ sensitivity, $T_1\text{-}s \longrightarrow T_1\text{-}r$ | $2 \times 10^{-8}$ | |
| | Histidine requirement, $his^- \longrightarrow his^+$ | $4 \times 10^{-8}$ | |
| | $his^+ \longrightarrow his^-$ | $2 \times 10^{-6}$ | Per cell |
| | Streptomycin sensitivity, | | per division |
| | $str\text{-}s \longrightarrow str\text{-}d$ | $1 \times 10^{-9}$ | |
| | $str\text{-}d \longrightarrow str\text{-}s$ | $1 \times 10^{-8}$ | |
| Alga, | Streptomycin sensitivity, | | |
| C. reinhardi | $str\text{-}s \longrightarrow str\text{-}r$ | $1 \times 10^{-6}$ | |
| Fungus, | Inositol requirement, $inos^- \longrightarrow inos^+$ | $8 \times 10^{-8}$ | Mutation |
| N. crassa | Adenine requirement, $ade^- \longrightarrow ade^+$ | $4 \times 10^{-8}$ | frequency among asexual spores |
| Corn, Z. mays | Shrunken seeds, $Sh \longrightarrow sh$ | $1 \times 10^{-5}$ | |
| | Purple, $P \longrightarrow p$ | $1 \times 10^{-6}$ | |
| Fruit fly, D. | Yellow body, $Y \longrightarrow y$, in males | $1 \times 10^{-4}$ | Mutation |
| melanogaster | $Y \longrightarrow y$, in females | $1 \times 10^{-5}$ | frequency |
| | White eye, $W \longrightarrow w$ | $4 \times 10^{-5}$ | per |
| | Brown eye, $Bw \longrightarrow bw$ | $3 \times 10^{-5}$ | gamete |
| Mouse, | Piebald coat color, $S \longrightarrow s$ | $3 \times 10^{-5}$ | per sexual |
| M. musculus | Dilute coat color, $D \longrightarrow d$ | $3 \times 10^{-5}$ | generation |
| Human, Homo | Normal $\longrightarrow$ hemophilic | $3 \times 10^{-5}$ | |
| sapiens | Normal $\longrightarrow$ albino | $3 \times 10^{-5}$ | |
| Human bone | Normal $\longrightarrow$ 8-azoguanine resistant | $7 \times 10^{-4}$ | |
| marrow cells in | Normal $\longrightarrow$ 8-azoguanosine resistant | $1 \times 10^{-6}$ | Per cell |
| tissue culture | | | per division |

[a]Correction of the other mutation rates in this table to a per gene basis would not change their order of magnitude.
From R. Sager and F. J. Ryan, *Cell Heredity*. New York: Wiley, 1961.

**Mutation Pressure**  We should pause here to evaluate a common impression, namely, that repeated mutation at a given locus, from $a_1 \longrightarrow a_2$, could so increase the frequency of $a_2$ that evolutionary changes would result. If the frequency of $a_1$ in the population is defined as $p$, and if $a_1$ is mutating to $a_2$ at some constant **mutation rate μ**, then the change in $p$ per number of generations ($t$) becomes

$$\frac{dp}{dt} = -\mu p$$

This equation is solved as follows:

$$\frac{dp}{p} = -\mu dt$$

$$\int \frac{dp}{p} = -\int \mu dt$$

$$\ln p = -\mu t + \ln c$$
$$p = ce^{-\mu t}$$

If, we consider the initial frequency of $a_1$ to be $p_o$ at a time when $t = 0$, then the constant $c$ becomes equal to $p_o$ and

$$p = p_o e^{-\mu t}$$

where $e = 2.72$.

We can now turn to actual mutation rates, summarized in Table 26-2. With an average value of $10^{-5}$ for $\mu$, it becomes clear that values of $t$ must approach the reciprocal of $\mu$ if any significant change is to occur in the value of $p$. Specifically, if the value of $p$ is to be reduced by about one third of its original value ($p_o/2.72$) then, when $\mu = 10^{-5}$, $t$ must equal $10^5$, or 100,000, generations. This means 100,000 years in an organism with a 1-year generation time, and perhaps 20 times as long for humans if we assume a human population time of 20 years. These time estimates are almost certainly much too small, for the equation ignores the occurrence of $a_2 \longrightarrow a_1$ reversions. Obviously, if the reversion rate ($\nu$) is comparable to the forward mutation rate $\mu$—and there is no a priori reason to assume otherwise for point mutations—the value of $p$ will change slowly indeed. It must therefore be concluded that mutation, while critically important to evolution in providing genetic variability, does not in and of itself bring about significant changes in allelic frequencies.

## 26.4 Estimating Genetic Variability in a Population

When we speak of genetic variability in a diploid individual, we denote a given locus as either homozygous or heterozygous. When this same locus is instead considered in the population, the analogous terms are **monomorphic** and **polymorphic**: a monomorphic locus is represented by only one allele at significant frequency, whereas a polymorphic locus, as noted earlier, is represented by two or more alleles at significant frequency. The proviso "at significant frequency" is important, for as we have just seen, high mutation rates predict that at least some individuals in a large population will carry new mutant alleles at an otherwise monomorphic locus. An arbitrary cut-off has therefore been established: **if the most common allele at a particular locus is present in more than 99 percent of the population, the locus is termed monomorphic;** conversely, **if one or more additional alleles are present in more than 1 percent of the population, the locus is termed polymorphic.**

**Electrophoretic Variants**  Most current estimates of polymorphism in natural populations are based on the approach introduced by R. Lewontin and J. Hubby in 1966. They captured *Drosophila pseudoobscura* flies in their natural habitats, allowed them to breed in the laboratory, and prepared extracts of individual flies for starch gel electrophoresis (*Box 3.2*). Each gel was then subjected to enzyme assays in the fashion we described for isozymes in Sections 19.3 to 19.6. Finally, as in the isozyme studies, the gels were analyzed for the presence of "fast-slow" electrophoretic variants. By this criterion, 15 of the 28 loci under test proved to be monomorphic: only a single enzyme activity was detected in greater than 99 percent of the samples. The remaining 13 of the tested loci were polymorphic, each specifying a number of **allozymes** (*recall Section 19.5*). The most variable locus, called *esterase-5 (Est-5)*, was responsible for 11 allozymes in the population; the *malic dehydrogenase (MDH)* locus was represented by 5 allozymes; and the rest displayed 4, 3, or 2 allozymes at significant frequency. Overall, then, 13/28 of the tested gene loci, or 46 percent, were found to be polymorphic. By calculating the frequency of the various alleles at the tested loci, including the monomorphic loci, Lewontin and Hubby calculated an **average heterozygosity** value of 12 percent, meaning that **12 percent of the gene loci in a typical fly are heterozygous.**

The high levels of natural polymorphism in *D. pseudoobscura* discovered by Lewontin and Hubby were both unexpected and provocative, and scores of similar studies were soon initiated for other populations. In addition, new electrophoresis procedures were applied, and different sets of gene loci were examined. Follow-up conclusions include the following.

1. Polymorphism is a universal attribute of biological species (Figure 26-1), including haploid organisms such as *E. coli*.
2. Starch and acrylamide gel electrophoresis detects variants that differ in net charge and would not be expected to be sensitive to many classes of amino acid substitution (e.g., lysine $\longrightarrow$ arginine). Therefore, the Lewontin-Hubby technique probably *underestimates* natural levels of polymorphism.
3. Post-translational modification of proteins (e.g., glycosylation or phosphorylation) could alter their net charge, and this activity might, in some cases, be carried out on many gene products by only a few "modifier loci" that are polymorphic in the population. In such cases, electrophoretic variants would not be true allozymes at all, but simply substrates for polymorphic modifier loci, and the electrophoresis technique might *overestimate* total levels of polymorphism in a population.
4. The enzymes selected for assay in gels may be more prone to allelic variation than the bulk of the gene products. When a different set of *Drosophila* gene products was scrutinized by A. Leigh Brown and C. Langly, levels of heterozygosity were found to be 4 percent rather than 12 percent. Similarly, human heterozygosity at allozyme loci is

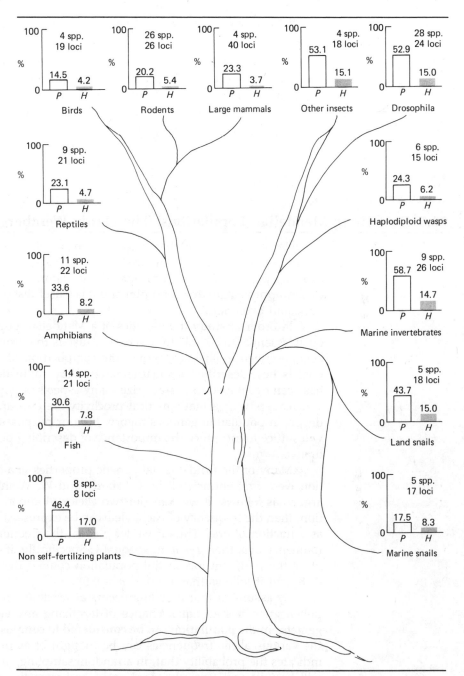

**FIGURE 26-1 Polymorphism.** Percentages of polymorphic loci (P) and average heterozygosity (H) in various animal and plant groups as estimated by electrophoretic surveys. The lengths of the branches on the evolutionary tree are arbitrary. (From D. Hartl, *Principles of Population Genetics.* Sunderland MA: Sinauer, 1980, p. 80.)

estimated at 6.3 percent, whereas it is less than 1 percent when other loci are studied.

With the debate over the actual numbers still in progress, it nonetheless appears safe to conclude that polymorphism is extensive in natural populations and that some loci are much more variable than others. Theories regarding how and why these multiple alleles are maintained in populations will be explored in the next chapter.

# Mendelian Populations

## 26.5 Describing Mendelian Populations: The Hardy-Weinberg Principle

A stable population, one that is not evolving at all, is usually called a Mendelian population for reasons that will become apparent shortly. A non-evolving population may well not exist in nature, but many populations closely approximate this condition and it is a useful starting point in thinking about evolution.

In order to study the genetics of a Mendelian population, it is obviously essential to be able to describe it. Just as the properties of genes are best described in chemical terms, **the composition of a population gene pool is best described by mathematics.** Very sophisticated mathematics has been developed to characterize many aspects of population dynamics, and readers with a mathematical proclivity are encouraged to read more deeply in population genetics theory. For our purposes, simple equations will suffice to introduce the importance of describing populations in quantitative terms.

Many of the fundamental genetic properties of a Mendelian population were first enunciated by G. Hardy and W. Weinberg in 1908, who argued as follows. If we consider two alleles, $A$ and $a$, in a stable population, then the frequency of each allele can be expressed as a percentage, or as a fraction of one. Thus if we let $p$ equal the frequency of $A$ and $q$ the frequency of $a$, then $p + q$ must equal 1. Specifically, if 48 percent, or 0.48, of all the pertinent loci in the population contain the $A$ allele, then $p = 0.48$, and it follows that $q = 1 - p = 0.52$.

By assuming that breeding occurs at random, so that any sperm or pollen grain has an equal chance of fertilizing any egg, then **all of the gametes in the population can be considered to compose a single pool** and the various allelic frequencies can be thought of as probabilities. Thus $p$ indicates the probability that, in a random sampling, an $A$-bearing gamete will be "drawn" from the pool and $q$ indicates the probability that an $a$-bearing gamete will be drawn. When we now consider the probability that two particular gametes will be drawn simultaneously—i.e., the probability that two particular gametes will unite to form a zygote—we simply multiply the probabilities of these independent events, much as we did in

Section 12.10 when we calculated the expected frequency of double cross-overs. Thus if $p$ is the probability of drawing an $A$ gamete, the probability that a particular fertilization will produce an $A/A$ homozygote is simply $p \times p = p^2$. Similarly, the probability of producing an $a/a$ homozygote is $q^2$. An $A/a$ heterozygote can be formed in two ways: an $A$ sperm can fertilize an $a$ egg, or an $a$ sperm can fertilize an $A$ egg. Therefore, the probability of producing a heterozygote is $p \times q + q \times p = 2pq$. These relationships are diagrammed in Figure 26-2. The numbers of homozygous and heterozygous individuals in an idealized population are therefore predicted as

$$p^2 = A/A \text{ homozygotes}$$
$$2pq = A/a \text{ heterozygotes}$$
$$q^2 = a/a \text{ homozygotes}$$

We noted earlier that since $p$ and $q$ are frequencies, their sum is equal to 1. The frequencies of A/A, A/a, and a/a individuals in a population must also add up to unity. Therefore, we can write the following sum:

$$p^2 + 2pq + q^2 = 1$$

The above sum can also be factored and expressed as

$$(p + q)^2 = 1$$

Indeed, we now recognize that if $p + q = 1$, then multiplying these frequencies by themselves is simply the same operation as $(p + q)^2$, which must also equal 1 ($1^2 = 1$).

The equation $p^2 + 2pq + q^2 = 1$ is a mathematical statement of the **Hardy-Weinberg Principle** (also called the **Hardy-Weinberg Law** or **Equilibrium**). Readers whose mathematics is fresh in their minds will recognize that it is also the expression for any binomial distribution. A comparison of

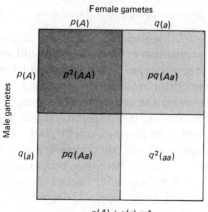

FIGURE 26-2 Hardy-Weinberg Principle for two alleles. Gamete frequencies ($p_{(A)}$ and $q_{(a)}$) are multiplied together to predict genotypic frequencies in the next generation, the sum of which must equal 1.

Figure 26-2 and Figure 6-6 will also make clear why a stable population is called a **Mendelian population:** both in a stable population and in a Mendelian cross, **the genotypic proportions of the progeny generation are determined by making all possible combinations of gametes from the parental generation.** The major difference, of course, is that when a heterozygous $F_1$ strain is inbred, $p$ is always equal to $q$ and both are equal to 0.5, whereas in the case of Mendelian populations, $p$ and $q$ need not be equal and each can vary from near 0 to near 1.

## 26.6 Applications of the Hardy-Weinberg Principle

**Calculating Allelic and Genotypic Frequencies**    The Hardy-Weinberg Principle allows one to calculate both the **frequency of alleles** and the **frequency of genotypes** in a Mendelian population. Suppose you learn that in a stable human population, 36 percent of the persons cannot taste phenylthiocarbamide (PTC), a substance that tastes very bitter to the rest of the population. Knowing that the "tasting" phenotype is controlled by a single locus and that the *nontaster* allele (*a*) is recessive to the *taster* allele (*A*), you can reason as follows:

1. **Determine the frequency of the *a/a* genotype.** Since nontasters must be homozygous for the recessive allele *a*, the *a/a* genotype frequency must be 36 percent = 0.36.
2. **Determine the frequency of the *a* allele.** This frequency is denoted as $q$. Since the frequency of *a/a* is equivalent to $q^2$, $q$ is simply the square root of the *a/a* genotypic frequency. Therefore, $q = \sqrt{0.36} = 0.6$.
3. **Determine the frequency of the *A* allele.** This frequency, $p$, is defined as $1 - q$. Therefore, p = 1 - 0.6 = 0.4.
4. **Determine the frequencies of the *A/A* and *A/a* genotypes.**

   $A/A = p^2 = 0.16 = 16$ percent homozygous tasters
   $A/a = 2pq = 2 \times 0.4 \times 0.6 = 0.48$
   $\qquad = 48$ percent heterozygous tasters

A key feature of a Mendelian population can now be appreciated: **Regardless of the proportions of particular diploid genotypes that establish a population, the genotypes assume a Hardy-Weinberg distribution in the following generation and in all succeeding generations, provided that allelic frequencies are the same in both sexes.** If the starting generation, for example, consists of 7000 *A/A* individuals and 3000 *a/a* individuals, the starting genotypic frequencies will be *A/A* = 0.7 and *a/a* = 0.3 and the starting allele frequencies will be, similarly, $p = 0.7$ and $q = 0.3$. At the next generation, when random breeding occurs and a common gamete pool is established, $p$ will still equal 0.7 and $q$ will equal 0.3, but the Hardy-Weinberg formulation will now apply. Therefore we can expect that

$p^2 = 0.49 = $ frequency of *A/A*
$q^2 = 0.09 = $ frequency of *a/a*

and

$$2pq = 2 \times 0.7 \times 0.3 = 0.42 = \text{frequency of } A/a$$

Since half the gametes produced by the $A/a$ heterozygotes will carry $A$ and all the gametes produced by $A/A$ homozygotes will carry $A$, it follows that the frequency of $A$ in the gamete pool will be

$$1/2(0.42) + 0.49 = 0.70$$

which was our original starting frequency of $A$. By similar calculations, the value of $q$ is found again to be 0.3. In other words, as long as the sampling of the gamete pool is done in a random manner, allelic and genotypic frequencies should neither rise nor fall from one generation to the next. Thus the population is aptly described as "stable."

**Considering More Than Two Alleles at One Locus**   As stressed in Section 26.4, many gene loci in natural populations are represented by three or more alleles. In such cases, the Hardy-Weinberg Principle is simply expanded.

1. For three alleles $A_1$, $A_2$, and $A_3$, let $p$, $q$, and $r$ represent their frequencies (where $p + q + r = 1$).
2. All possible gamete combinations then become $(p + q + r)^2$, which, when displayed in a Punnett Square, gives the results shown in Figure 26-3 (compare with Figure 26-2).
3. These relationships can then be used to calculate allele or genotype

|  | Female gametes | | |
|---|---|---|---|
|  | $p(A_1)$ | $q(A_2)$ | $r(A_3)$ |
| $p(A_1)$ | $p^2(A_1A_1)$ | $pq(A_1A_2)$ | $pr(A_1A_3)$ |
| $q(A_2)$ | $pq(A_1A_2)$ | $q^2(A_2A_2)$ | $qr(A_2A_3)$ |
| $r(A_3)$ | $pr(A_1A_3)$ | $qr(A_2A_3)$ | $r^2(A_3A_3)$ |

Male gametes

$$A_1A_1 = p^2$$
$$A_1A_2 = pq + pq = 2pq$$
$$A_1A_3 = pr + pr = 2pr$$
$$A_2A_2 = q^2$$
$$A_2A_3 = qr + qr = 2qr$$
$$A_3A_3 = r^2$$

**FIGURE 26-3 Hardy-Weinberg Principle for three alleles.** Gamete frequencies ($p$, $q$, and $r$) are multiplied together to predict genotype frequencies in the next generation.

frequencies. For example, given the frequencies of the $I^A$, $I^B$, and $I^O$ blood group alleles in African populations provided in Table 26-1, we can calculate the genotypic and phenotypic frequencies as follows:

$p$ = frequency of $I^A$ = 0.156
$q$ = frequency of $I^B$ = 0.136
$r$ = frequency of $I^O$ = 0.708

AA = $p^2$ = 0.024
AB = $2pq$ = 0.042
AO = $2pr$ = 0.221
BB = $q^2$ = 0.018
BO = $2qr$ = 0.193
OO = $r^2$ = 0.501

Type A persons = 0.024 + 0.221 = 0.245 = 24.5%
Type B persons = 0.018 + 0.193 = 0.211 = 21.1%
Type AB persons = 0.042 = 4.2%
Type O persons = 0.501 = 50.1%

**Considering More Than One Locus and Recognizing Linkage Disequilibrium**   When two loci are not linked, or are linked but recombine freely, the Law of Independent Assortment, defined for unlinked genes in Mendelian crosses (*Section 6.12*), applies as well to their distribution within a Mendelian population. Specifically, each locus is analyzed independently of the other, and the probability that an organism will inherit any two alleles (e.g., $A_1$ and $B_3$) is simply the product of their individual frequencies.

If, however, two loci are very closely linked, and particularly if recombination between them is suppressed, then they will tend to be transmitted as a block rather than assorting independently and are said to be in **linkage disequilibrium.** As a specific example of linkage disequilibrium, we can consider the two major histocompatibility (MHC) loci in humans, denoted $A$ and $B$ (*Section 21.4*). Each locus is highly polymorphic, there being 20 reported $A$ alleles and 42 $B$ alleles. Because $A$ and $B$ are closely linked, combinations of $A$ and $B$ alleles tend to be inherited as a unit from parent to offspring. Thus a person with an $\dfrac{A1B8}{A3B7}$ genotype produces gametes that are usually either *A1B8* or *A3B7*; only very infrequently do recombinant *A1B7* or *A3B8* gametes arise. Given enough time, recombination *should* provide a Mendelian population with equivalent numbers of *A1B8*, *A3B7*, *A1B7*, and *A3B8* chromosomes; if this were the case, the locus would be said to be in linkage equilibrium. When this is *not* the case—for example, when *A1B8* and *A3B7* chromosomes are found to be in significant excess of *A1B7* and *A3B8*—then the locus is said to be in linkage disequilibrium. Selection is usually suspected as the agent that operates to keep certain gene combinations together, as we will detail in Section 27.2.

# 26.7 Deviations from the Hardy-Weinberg Principle: Assortative Mating, Inbreeding, and Selfing

We now return to consideration of a single gene locus. In developing the Hardy-Weinberg Principle that describes allelic frequencies at that locus (*Section 26.5*), we assumed that breeding occurs at random, so that any sperm or pollen grain has an equal chance of fertilizing any egg. Any significant departure from such **random mating,** or **panmyxia,** will lead to changes in the frequencies of genotypes in a population.

**Assortative Mating**  A nonrandom mating system wherein mating choice is dictated by phenotype is called **assortative mating.** The most familiar human example is the tendency for tall women to select tall men as mates and the tendency of short men to select short women; such like-to-like choices are denoted **positive assortative mating.** When mates are chosen that are phenotypically dissimilar to oneself ("opposites attract"), then **negative assortative mating** is said to occur. "Choice" is not necessarily an activity of the organism itself; thus, if an insect vector tends to choose blue flowers over white during pollination so that blue × blue fertilizations occur more often than blue × white or white × white, then the plant population is said to show positive assortative mating. It should be intuitively obvious that positive assortative mating will produce a reduction in the heterozygosity of a population, whereas negative assortative mating will tend to increase it, relative to what is predicted for a Mendelian population.

A special but important case of positive assortative mating occurs when the "preferred phenotype" is either a close relative (**inbreeding**) or, in the case of hermaphrodites or dioecious plants, oneself (**selfing**). Again thinking intuitively, these mating systems will also act to promote homozygosity. Their effects on the population, however, depend heavily on whether the population is normally outcrossing or whether inbreeding or selfing is a common occurrence.

**Inbreeding Depression in Outcrossing Populations**  Humans constitute the most familiar example of a normally outcrossing population: selfing is impossible, parent-child or sib-sib matings are taboo, and first-cousin matings are infrequent and often taboo as well. Let us now consider a recessive deleterious allele $a_1$, newly created by mutation. In an outcrossing population, the chances are very small that this $a_1$ allele will combine with a second $a_1$ allele to create a homozygous inviable zygote whose death would rid the population of two copies of $a_1$. Instead, copies of $a_1$ would very probably be harbored by the population as rare recessives, masked by normal, dominant counterparts in $A/a_1$ heterozygotes. Should $A/a_1$ members of this outcrossing population now inbreed, however, the chances of an $a_1/a_1$ homozygous combination increase enormously. The resultant increase in debilitated or inviable offspring is known as **inbreeding depression.** This principle was in fact set forth in Section 6.7, when we noted how rare recessive human traits could become common in pedigrees involving first-cousin marriages (*Figure 6-6*). Indeed, most pedigree data on inherited human disease syndromes derive from inbred families or communities.

For a population that commonly inbreeds as well as outcrosses, on the other hand, rare recessive deleterious alleles are not given the chance to accumulate: each inbreeding or selfing event is likely to "purge" two copies of the allele from the gene pool so that only neutral or "positive" alleles remain. Many plant species both outcross and inbreed, and these are particularly suitable organisms for studying the effects of inbreeding per se.

**Self-Fertilization**  Using a plant population that inbreeds sufficiently often that it is not subject to inbreeding depression, let us calculate the effect of inbreeding on levels of heterozygosity. We can start with the simplest case, that of selfing. For a heterozygote *Aa*, the first-generation progeny that result from selfing should be in the familiar genotypic proportions

$$1/4 \; AA : 1/2 \; Aa : 1/4 \; aa$$

In the next and ensuing generations, however, selfing of the *AA* and the *aa* offspring will produce exclusively homozygous progeny; heterozygotes will only appear as half the progeny of the heterozygotes in the preceding generation. Therefore,

| Generation | *A/A* | *A/a* | *a/a* |
|:---:|:---:|:---:|:---:|
| 2 | 3/8 | 1/4 | 3/8 |
| 3 | 7/16 | 1/8 | 7/16 |
| 4 | 15/32 | 1/16 | 15/32 |

The application of these principles to a population is perhaps easiest to appreciate with actual numbers. In Figure 26-4 we start with a population of 1600 heterozygotes, all self-fertilizing, and show how obligate self-fertilization causes the proportion of heterozygotes to fall from 100 percent to a mere 6.25 percent in just four generations. Note that the input allele frequencies ($p = q - 0.5$) are not changing, only the proportions of the genotypes. It can be calculated that eight generations of obligate self-fertilization would reduce the proportion of heterozygotes to near zero.

**Inbreeding Coefficient**  More common than self-fertilization is inbreeding. To estimate the effect of inbreeding on levels of heterozygosity, one calculates the **inbreeding coefficient,** symbolized as $F$, which can be thought of as **the probability that any two gametes forming a zygote will carry identical alleles at a given locus that derive from a common ances-**

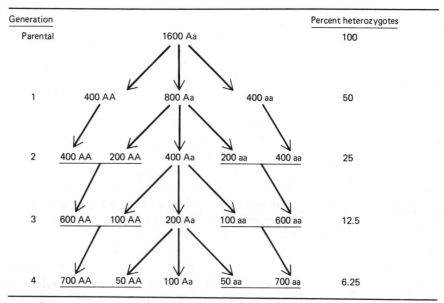

FIGURE 26-4 **Self-fertilization effects on the proportion of genotypes** in a population.

**tor.** This probability is, of course, greater as the two gametes derive from more closely related individuals, since each gamete is increasingly likely to carry a common allele. Thus, if two parents are $A_1A_2$ and $A_3A_4$, then their possible offspring are $A_1A_3$, $A_1A_4$, $A_2A_3$, and $A_2A_4$. To estimate the inbreeding coefficient for matings between these sibs, we can score each gamete combination for its likelihood of producing homozygous offspring. In $A_1A_3 \times A_1A_3$ matings, for example, half the offspring will be homozygous ($A_1A_1 + A_3A_3$) and half will be heterozygous ($A_1A_3$); in an $A_1A_3 \times A_2A_4$ mating, none of the offspring will be homozygous; and so forth. Using a Punnett Square to keep track of all possible matings (*recall Figure 26-3*), we find that 16 of the 64 possible gametic unions will produce homozygotes (either $A_1A_1$, $A_2A_2$, $A_3A_3$, or $A_4A_4$), a conclusion you can demonstrate to yourself. Therefore, $F = 16/64 = 1/4 = 0.25$ for brother-sister matings. In other words, the chance that the child of a brother-sister mating will be homozygous at a given locus is 25 percent.

These calculations become increasingly complex as more distant relatives are considered, and inbreeding coefficients are in these cases usually estimated by **path analysis.** Again considering first the offspring of a sib-sib mating, we reason as follows:

1. Denote the child of interest as Z, his brother-sister parents as X and Y, and his grandparents as V and W, and write his pedigree as

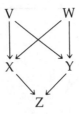

where the arrows designate the transmission of gametes and the crossed arrows indicate those ancestors that have contributed gametes to more than one relative.

2. Identify all the direct, independent transmission paths by which a common ancestor could have transmitted the same allele to Z via two different routes. Thus Z could have become an $A_1A_1$ homozygote had his grandparent V contributed an $A_1$ allele to both X and Y; this path can be written Z $\longrightarrow$ X $\longrightarrow$ V $\longrightarrow$ Y $\longrightarrow$ Z. Z could have also become an $A_1A_1$ homozygote via a Z $\longrightarrow$ X $\longrightarrow$ W $\longrightarrow$ Y $\longrightarrow$ Z pathway. (Note that Z $\longrightarrow$ Y $\longrightarrow$ V $\longrightarrow$ X $\longrightarrow$ Z and Z $\longrightarrow$ Y $\longrightarrow$ W $\longrightarrow$ X $\longrightarrow$ Z are not independent of those already scored and are therefore not counted.)

3. Apply the relationship

$$F_Z = \Sigma(1/2)^n$$

where $F_Z$ is the inbreeding coefficient for individual Z and $n$ is the number of individuals, excluding Z, in each path. Thus there are three individuals in the XVY pathway and three in the XWY pathway. Therefore,

$$F_Z = (1/2)^3 + (1/2)^3 = 1/8 + 1/8 = 1/4$$

which is the same value we obtained above by Punnett Square reasoning. Consider next the pedigree for a first-cousin marriage:

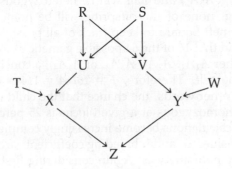

In this case Z again has two common ancestors, R and S. Again, there are only two direct, independent pathways that lead from Z to these ancestors and back. More individuals are involved, however, with one path being Z $\longrightarrow$ X $\longrightarrow$ U $\longrightarrow$ R $\longrightarrow$ V $\longrightarrow$ Y $\longrightarrow$ Z and the other

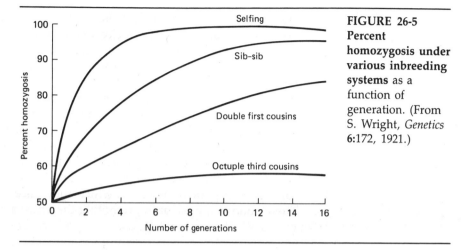

**FIGURE 26-5 Percent homozygosis under various inbreeding systems** as a function of generation. (From S. Wright, *Genetics* 6:172, 1921.)

Z $\longrightarrow$ X $\longrightarrow$ U $\longrightarrow$ **S** $\longrightarrow$ V $\longrightarrow$ Y $\longrightarrow$ Z. Therefore, $n = 5$ for both paths, so that

$$F_Z = \Sigma(1/2)^n$$
$$= (1/2)^5 + (1/2)^5 = 2/32 = 1/16$$

In other words, the chance that the child of a first-cousin marriage will be homozygous at a given locus is about 6 percent, in contrast to 25 percent for sib-sib marriages and 50 percent for the selfing of a heterozygote. The graph in Figure 26-5 displays graphically this dependency of the rate of approach to homozygosis on the closeness of relationship in an inbreeding system.

## 26.8 Establishing a Mendelian Population and the Founder Effect

We can now return to a Mendelian population and consider the two ways that it can originate: it can evolve from within a pre-existing population, or it can separate from a pre-existing population. The first mode will be described in detail in the next chapter; here we focus on the second mode.

Separation from a pre-existing population can take place in a variety of ways. Individuals from a pre-existing population may **migrate** into a new environment that is selectively identical to the original. Alternatively, some **physical barrier** may come to isolate members of a pre-existing population; a large tree, for example, may fall in the midst of an interbreeding population, dividing them into two populations with selectively identical environments. In population genetics terminology, both migration and physical barriers have the same effect: a population that was once **sympatric** becomes **allopatric**.

In some cases, the two new allopatric populations will have different gene pools from the original. This is particularly true if one population is "founded" by only a few individuals. Thus if five migratory birds are blown off course from their large flock and they proceed to establish a new flock, the genotypes of these five founder birds come to define the gene pool of the new population. If a male and a female founder bird happen to be heterozygotes for gene $a$, then even though this allele may be extremely rare in the parent population, it will be carried by 2 of every 10 gametes in the gene pool generated by the five founders. Thus $q$ will be 0.2, and the frequency of $a/a$ homozygotes in the next and subsequent generations will be expected to be $q^2 = 0.04 = 4$ percent of the population. It is this **founder effect** that is believed to be responsible, in part, for the unusual allelic frequencies (and phenotypes) of many isolated colonies compared with those of the large populations from which they once derived.

The second major factor in promoting the unusual phenotypes of isolated colonies is the tendency (and often necessity) of small populations to inbreed. As stressed in the preceding section, inbreeding does not alter allelic frequencies. It will, however, push genotypic proportions in the direction of homozygosis so that recessive alleles are more likely to be expressed than in large panmyctic populations.

## 26.9 Stable Polymorphisms in Mendelian Populations

We can conclude and summarize our consideration of Mendelian populations by stating that they display **stable polymorphism:** the frequency of alleles, and hence the overall levels of polymorphism, do not change from one generation to the next. In developing the concept of a Mendelian population, we assumed that no evolutionary agents (e.g., mutation and selection) were acting on the gene pool so that stable polymorphism was achieved via random mating alone. In fact, of course, evolutionary agents are omnipresent in all natural populations. Does this mean that Mendelian populations have no relevance to natural populations? The answer is no. Many sets of alleles in natural populations display stable polymorphisms, showing no significant fluctuations in frequency over numerous generations. How is this achieved?

There are two general sorts of answers to this question. One possibility is that the population is exposed to a number of different kinds of evolutionary agents whose effects cancel each other out, a condition usually denoted **balanced polymorphism.** We can give two examples. The first, called **mutational equilibrium,** describes a locus that is being mutated to allele $a$ with one frequency and reverted back to $A$ with a somewhat lower frequency. As long as the forward mutation and reversion rates both remain constant, a balanced level of heterozygosity at the $A/a$ locus will be sustained. In fact, the existence of the mutational events can be ignored altogether, and the population can be simply described by the Hardy-Weinberg relationship.

The second example of countervailing evolutionary forces is frequently referred to as **stabilizing selection:** in this case, selection alone acts to generate a stable balance of alleles. We have already encountered one form of stabilizing selection in considering the $Hb^s$ alleles in natural populations (*Section 26.1*), where selection was both eliminating and favoring the presence of the allele in the gene pool. Other modes of stabilizing selection (also called **balancing selection**) will be considered in the next chapter.

The alternate way that polymorphism can arise in natural populations pertains to alleles that are selectively neutral. As noted in Section 26.2, such alleles are expected to "drift" in the population, conferring no selective value to their possessors. Although drift may sometimes lead to evolutionary change if an allele is particularly "lucky" or "unlucky," the usual predicted effect of drift is to create a stable polymorphism: on the average, an established neutral allele will happen to be transmitted to fertile offspring as often as it happens not to be transmitted, and its frequency in the large gene pool will appear constant.

The preceding paragraphs have used the terms "selection" and "drift" quite generally. In the next chapter we consider these agents much more carefully. The preceding discussion should allow you to approach the next chapter with the understanding that **selection and drift can have as their net effect either evolution or balanced polymorphism:** a population is said to be evolving if the net effects of selection and drift bring about changes in the allelic frequencies of the gene pool and nonevolving if the net effects are to stabilize allelic frequencies.

(a)                                                      (b)

(a) J.B.S. Haldane made fundamental contributions to population genetics theory as it was being developed, and (b) James F. Crow (University of Wisconsin, Madison) continues to do so.

# Questions and Problems

1. The MN blood type is determined by a pair of codominant alleles, $L^M$ and $L^N$. A population of 600 persons was blood typed with the following results: 300 were type M, 240 were MN, and 60 were N. Does this represent a Mendelian population in a Hardy-Weinberg equilibrium? If not, what should the numbers of persons with each genotype have been?

2. The population described in Question 1 reproduced until it reached a size of 5000 persons. Assuming that this mating was random and that evolutionary pressures were nonexistent, how many M, MN, and N persons do you expect in the expanded population?

3. In a Mendelian population of 500 persons, five are found to be of blood type N. This group now intermarries at random with a second Mendelian population of 500 persons in which 20 are of blood group N.
   (a) What is the frequency of the $L^N$ gene in the combined population?
   (b) What proportion of the offspring of the combined population will be blood type N? Is this a lower or higher incidence than in the two populations considered separately?

4. In a Mendelian population, the expected number of $A/A \times a/a$ matings is $2(p^2 \times q^2)$, where the expression is multiplied by 2 to account for both $A/A \, ♀ \times a/a \, ♂$ matings and the reciprocal. Similarly, the number of $A/a \times a/a$ matings is expected to be $2(2pq \times q^2)$.
   (a) If $A$ shows complete dominance to $a$, then what are the proportions of the offspring expected to exhibit the dominant and recessive phenotypes in each of these matings?
   (b) You now wish to determine the proportion of recessive offspring produced by all the $A/A \times a/a$ and $A/a \times a/a$ matings in the population. Show how this proportion is most simply expressed as $\dfrac{q}{1 + q}$ (recall that $p = 1 - q$).
   (c) One person in 10,000 expresses a particular recessive trait. If these persons all marry persons who do not exhibit the trait, what percentage of their offspring will be expected to exhibit the trait?

5. The ABO blood groups are controlled by three allelic genes, $I^A$, $I^B$, and $I^O$ (*Table 19-2*).
   (a) In a certain population, the frequency of gene $I^A$ was estimated to be 0.20; $I^B$ was estimated at 0.08; and $I^O$ at 0.72. What are the expected proportions of blood-group phenotypes in this population, assuming random mating?
   (b) When the blood-group phenotypes of the persons in this population were actually tested, 1340 were found to be of type O, 895 of type A, 305 of type B, and 70 of type AB. Do the gene frequencies agree with those expected using the chi-square test (*Section 6.14*)? (*Note:* there is one degree of freedom in this situation.) Is mating in fact random?
   (c) Do you expect this population to have been predominantly white or black (*consult Table 26-1*)?

6. One person of 2500 in a Mendelian population is sterile because she or he is homozygous for the recessive gene $b$. What will be the proportion of such sterile persons in the next generation?

7. Gene loci in Mendelian populations are expected to reach a **mutational equi-**

**librium** where the change in the frequency of $a_1$ ($p$) per generation ($\Delta p$) is the gain caused by $a_2 \longrightarrow a_1$ reversions ($vq$) minus the loss incurred by $a_1 \longrightarrow a_2$ mutations ($\mu p$), whereas $\Delta q$ (the change in frequency of $a_2$) will be, reciprocally, $\mu p - vq$. At equilibrium, $\Delta p$ and $\Delta q$ will, by definition, both equal zero. Therefore, the **equilibrium allelic frequencies** of $p$ and $q$, symbolized $\hat{p}$ and $\hat{q}$, are

$$\hat{p} = \frac{v}{\mu + v}$$

$$\hat{q} = \frac{\mu}{\mu + v}$$

(a) Show how these equations are derived (you will need to do some simple algebra and factoring).
(b) Utilizing the information in Table 26-2, predict the equilibrium frequencies of the str-s and str-d alleles in E. coli populations.

8. The following data were collected by F. Ayala and J. Powell (*Proc. Natl. Açad. Sci. U.S.* **69:**109, 1972) for six gene loci in four species of *Drosophila*, where the allele numbers refer to relative rates of migration in electrophoresis studies. A dash indicates that the allele has not been found in the species.

| Gene Locus | Alleles | D. willistoni | D. tropicalis | D. equinoxialis | D. paulistorum |
|---|---|---|---|---|---|
| Lap-5 | 0.98 | 0.09 | 0.02 | — | — |
| (Leucine | 1.00 | 0.29 | 0.19 | — | — |
| aminopeptidase) | 1.03 | 0.50 | 0.63 | 0.004 | 0.004 |
| | 1.05 | 0.09 | 0.15 | 0.21 | 0.08 |
| | 1.07 | 0.007 | 0.01 | 0.71 | 0.86 |
| | 1.09 | — | — | 0.07 | 0.04 |
| Hk-1 | 0.96 | 0.04 | 0.02 | 0.08 | — |
| (Hexokinase) | 1.00 | 0.95 | 0.96 | 0.91 | 0.01 |
| | 1.04 | 0.006 | 0.02 | 0.005 | 0.97 |
| | 1.08 | — | 0.001 | 0.002 | 0.02 |
| Hk-3 | 1.00 | 0.98 | 0.97 | 0.95 | 0.07 |
| (Hexokinase) | 1.04 | 0.006 | 0.01 | 0.04 | 0.92 |
| Mdh-2 | 0.86 | 0.001 | 0.994 | 0.003 | 0.001 |
| (Malic | 0.94 | 0.02 | 0.005 | 0.994 | 0.993 |
| dehydrogenase) | 1.00 | 0.97 | — | 0.004 | 0.006 |
| Aph-1 | 0.98 | 0.02 | — | — | — |
| (Alkaline | 1.00 | 0.84 | 0.05 | 0.02 | 0.01 |
| phosphatase) | 1.02 | 0.08 | 0.90 | 0.92 | 0.93 |
| | 1.04 | 0.06 | 0.04 | 0.06 | 0.03 |
| Acph-1 | 0.94 | 0.05 | 0.95 | 0.01 | — |
| (Acid | 1.00 | 0.92 | 0.03 | 0.17 | — |
| phosphatase) | 1.04 | 0.02 | 0.006 | 0.81 | 0.16 |
| | 1.06 | — | — | — | 0.21 |
| | 1.08 | — | — | 0.004 | 0.62 |

(a) Which locus is represented by the largest number of alleles? The smallest number?

(b) Which, if any, loci would be defined as monomorphic in a given species?

(c) Which locus displays the most extensive polymorphism in a given species?

(d) Calculate the frequency of homozygotes and heterozygotes at the *Hk-3* locus in a Mendelian population of *D. paulistorum*.

9. Give two reasons why you would expect a deficiency of heterozygotes in a small population.

10. The following data were obtained for corn (N. P. Neal, *J. Amer. Soc. Agron.* **27:**666, 1935), where the inbreeding coefficient was calculated as in Section 26.7.

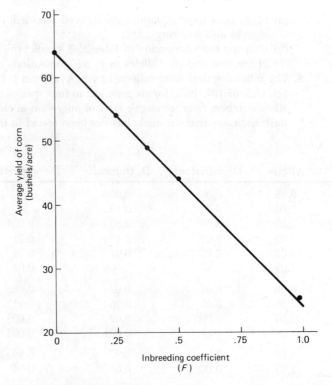

Would you predict that the cultivated corn in this study engages regularly in self-pollination or whether it usually outcrosses? Explain your answer.

11. Deduce a general expression for the proportion of heterozygotes expected after a population of *Aa* heterozygotes has been allowed to self-pollinate for *n* generations.

12. Sheep were bred to produce individual Z in the following fashion:

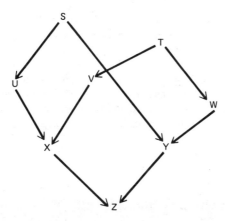

Calculate an inbreeding coefficient for Z.

13. If any of an individual's common ancestors (e.g., ancestor A) is itself inbred, its inbreeding coefficient ($F_A$) is included in the calculation by the following relationship:

$$F = \Sigma[(1/2)^n(1 + F_A)]$$

where the $1 + F_A$ term is included only in those pathways where ancestor A is involved. For Problem 12, calculate the inbreeding coefficient for Z, knowing that T has an inbreeding coefficient of 1/4.

14. Pigs were bred to produce individual Z in the following fashion:

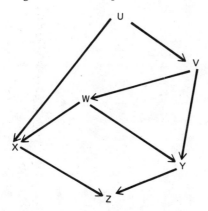

(a) Identify the three common ancestors of Z and their relationship to Z.
(b) Calculate an inbreeding coefficient for Z.
(c) Calculate an inbreeding coefficient for Z if the inbreeding coefficient of V is 1/8 (see Problem 13).

15. A survey finds that 0.016 percent of North American Jewish children are born with the autosomal recessive affliction **Tay-Sachs disease** (*Section 19.1*) and 0.00017 percent of non-Jewish children are born with the disease.

(a) What is the frequency of the Tay-Sachs allele (hex A) in the two populations? What is the frequency of heterozygous carriers? (Assume Mendelian populations).

(b) The annual number of births in the North American Jewish population is 107,567, and 3,747,844 in the non-Jewish population. What is the total number of Tay-Sachs cases in each population per year?

(c) Explain why Jews, particularly those of Ashkenazi descent, frequently choose to be screened for hex A heterozygosity, whereas non-Jews do not.

# 27 Population Genetics II: Evolutionary Agents

A. **Introduction**

B. **Fitness**
  27.1  Relative Fitness at a Single Locus
  27.2  Relative Fitness of a Genome
        Linkage disequilibrium
        Fitness modifiers and epistasis
        Polygenic traits

C. **Selection**
  27.3  Directional Selection: Theory
  27.4  Directional Selection: Observations
        Industrial melanism
        Inversions in *Drosophila*
        Phototaxis in *Drosophila*
  27.5  Plant and Animal Breeding
        General principles of a breeding program
        Effect of inbreeding and the use of hybrids
        Hybrid vigor or heterosis
  27.6  Diversifying Selection
  27.7  Stabilizing Selection
        Sickle-cell hemoglobin

Theoretical basis of stabilizing selection

D. **Migration**
  27.8  Dynamics of Migration Pressure
  27.9  Gene Flow

E. **Random Drift in Small Populations**
  27.10  Dynamics of Small Populations
  27.11  Random Drift of Alleles
  27.12  Drift Combined with Selection

F. **The Contributions of Selection and Drift to Polymorphism**
  27.13  The Neutral Mutation-Random Drift Hypothesis
  27.14  Do Neutral Mutations Occur?
  27.15  Are Patterns of Allelism Present in Parallel Populations?

G. **Questions and Problems**

819

# Introduction

The fundamental event in evolution is the change in allelic frequencies within a population. In this chapter we first describe how the "fitness" of a genotype and a phenotype are defined in genetic terms. We then describe how the most conspicuous evolutionary agent, **natural selection,** acts to convert differences in fitness into differences in allelic frequencies. In the course of this discussion, we also consider the related principle of **artificial selection** used by animal and plant breeders. Finally, we consider the dynamics of two additional evolutionary agents, **migration pressure** and **genetic drift,** and we evaluate the relative contribution of these agents to the observed polymorphism in natural populations.

# Fitness

The **fitness** of an organism is **a measure of its ability to pass its genes to the next generation.** Thus the larger the number of viable offspring, the fitter the individual. This is obviously a different use of the word than we commonly practice: we would describe a man who lived to be 100 and could still run a marathon as being extremely fit, but if he happened to be sterile, the population geneticist would assign him a fitness of 0. Fitness will clearly be a complex parameter, encompassing not only survival and fecundity but also the ability to attract a mate, care for young offspring, and so on.

Fitness is usually expressed as a relative parameter. In a population of robins with ample food supply, the fittest bird might produce 25 offspring during its reproductive life; in another population with sparser resources, the fittest bird might produce only 10 offspring. Therefore, fitness is usually defined as **relative fitness,** abbreviated **W**: the fittest organism (or genotype) in a population is assigned a fitness of 1.00, and less fit organisms (or genotypes) are assigned some fraction of 1 (or 0, in the case of non-reproduction).

The simplest way to think about fitness is to consider the relative fitness of a particular genotype at a single locus; thus an $A/A$ individual might have a fitness of 1, an $A/a$ individual a fitness of 0.9, and an $a/a$ individual a fitness of 0.2. In most situations, of course, the fitness of an organism will be determined by the collective contribution of many gene loci to its overall phenotype. These possibilities are considered in turn.

## 27.1 Relative Fitness at a Single Locus

Relative fitness is estimated in the following fashion. We determine genotype frequencies for two successive generations in a population. We know

TABLE 27-1   Relative Fitness and Selection Coefficients Calculated from Progeny Size

| Data | A/A | A/a | a/a | Total |
|---|---|---|---|---|
| (a) Number of mating individuals in one generation | 45 | 55 | 10 | 100 |
| (b) Number of progeny produced by each genotype | 65 | 95 | 15 | 175 |
| **Calculations** | | | | |
| (c) Average number of progeny per individual of a given genotype | 65/45 = 1.4 | 95/55 = 1.7 | 15/10 = 1.5 | |
| (d) Relative fitness of each genotype | 1.4/1.7 ≅ 0.8 | 1.7/1.7 = 1.0 | 1.5/1.7 ≅ 0.9 | |
| (e) Selection coefficient of each genotype | 0.2 | 0 | 0.1 | |

from Section 26.5 that if no selection is at work, the allele frequencies should be the same from one generation to the next. We find that they are not the same, indicating that selection is occurring and that some alleles are being preferentially transmitted. By comparing the observed genotype frequencies with the value expected if the population were in a Hardy-Weinberg equilibrium, we are able to determine the relative fitness of each genotype faced with the new environmental conditions.

Calculating fitnesses from observed changes in genotypic frequencies is in fact complex, requiring assumptions about the mode of selection involved. The general principles of calculating fitness, however, can be gleaned from the example provided in Table 27-1. We first ascertain the number of $A/A$, $A/a$, and $a/a$ individuals in a population (line a). We then determine the number of offspring produced by each genotype (line b). This allows us to calculate the average number of progeny produced by each individual of a given genotype (dividing line b by line a) and to identify $A/a$ as the most fit (having the largest average number of progeny). The relative fitness of the other genotypes is then calculated with reference to the $A/a$ maximum (line d).

Once the relative fitness of a genotype is calculated, its **selection coefficient (s)** can be computed as $1-W$ (Table 27-1, line e). Just as $W$ expresses the chance of an organism's reproductive success, so does $s$ reflect the chance of its reproductive failure. $W$ and $s$ are two sides of the same coin.

## 27.2 Relative Fitness of a Genome

Selection rarely acts exclusively at a single locus, although we will consider several examples in which this is a useful approximation. Usually, the environment is complex, and the phenotypes that survive and reproduce possess numerous adaptive traits that collectively add up to "optimal fitness."

Even when one is focusing on a single phenotype trait, it is unlikely that a single locus is involved. We can give several examples.

**Linkage Disequilibrium**   We noted in Section 26.6 that groups of non-alleles are sometimes found to be transmitted together much more often than predicted by chance. The resultant constellations are often called **supergenes,** and the loci are said to be in linkage disequilibrium.

One example of such a linkage disequilibrium is given by the human major histocompatibility gene cluster (*review Section 26.6*). In North European populations, the frequency of the *A1* allele at the *A* locus is 17 percent, whereas the frequency of the *B8* allele is 11 percent. Thus one would predict that the frequency of the *A1B8* chromosome (also called the *A1B8* **haplotype** after "haploid genotype") would be $0.17 \times 0.11 = 0.0187$, or ~2 percent. Instead, the *A1B8* haplotype is present at a 9 percent frequency, giving a **linkage disequilibrium (D)** value of 0.07 (observed frequency minus expected frequency). Similar disequilibria for other *AB* combinations are found in other human population groups. The selective agents maintaining these equilibria are unknown, but speculations include proposals that *A1B8* haplotypes might confer Caucasians with increased resistance to disease or with enhanced sexual attractiveness compared with *A1B_* or *A_B8* haplotypes.

The possibility of linkage disequilibrium must constantly be considered when estimating fitness. Thus if a human population were found to possess the *A1* histocompatibility allele at a 17 percent frequency and the *A2* allele at only a 1 percent frequency, one might be tempted to conclude that the *A1* allele conferred greater fitness, when, in fact, it is the *A1B8* supergene that confers a selective advantage. This problem becomes acute when populations are observed over short periods of time. Thus if we return to the example in Table 27-1 and imagine that a second locus, *B/b*, is closely linked to the *A/a* locus under test, then if the *B* allele happens to be in linkage disequilibrium with *A*, the fitness being measured may have nothing to do with possession of *A*; it may depend on the possession of *AB* or, in the extreme case, with possession of *B* alone, *A* simply "going along for the ride." In general, therefore, the fitness conferred by a given allele cannot be meaningfully estimated until it is established that the allele is not in linkage disequilibrium with an allele at another, selected locus, a proof that is obviously very difficult to make.

**Fitness Modifiers and Epistasis**   The expression of an allele at a given locus is often influenced as well by unlinked genes in the genome, and these may also be under selective pressures and contribute to the overall fitness of an individual. Thus **modifier** genes may alter the expression of a gene, whereas **epistatic** loci (Section 21.8) may interfere with or even inhibit its expression. In such cases it may appear that fitness is being determined by the product of gene $A$, whereas in fact it may (also) be determined by levels of modifier ($M$) and epistatic ($E$) genes that influence the expression of gene $A$. Thus an individual with an $A/A$ $M/M$ $e/e$ genotype may be most fit; an $A/A$ $M/m$ $e/e$ genotype less fit; and so on. Clearly, the mathematics becomes increasingly complex in such situations and untenable when the additional loci are unidentified.

**Polygenic Traits**   The ultimate extension of the previously cited cases involves phenotypic traits that are conferred by many loci, namely, the "quantitative" or "polygenic" traits described in Sections 21.11 to 21.14. It turns out that most of the traits that come to mind in describing fitness—height, weight, fecundity, and longevity—are determined by polygenic systems for which single-locus models are clearly inadequate. In other words, relative fitness $W$ is a useful concept in understanding the relationship between selection and genotype, but it is a very difficult value to calculate for natural populations.

# Selection

The **Darwin-Wallace Theory of Natural Selection,** proposed by C. Darwin and A. Wallace in 1858, is unquestionably the cornerstone of modern evolutionary theory. Darwin envisioned selection as a two-step process. He realized, first, that variation had to exist within a population. Second, he proposed that the "fittest" members of the population would be at a selective advantage and would be most likely to transmit their genes to the next generation.

Figure 27-1 illustrates the three basic patterns of Darwinian selection that will be considered in the next few sections. The curves define the distribution of phenotypes in a population, and the shading indicates those phenotypes favored by natural selection. When selection favors individuals at one extreme of the distribution, as in Figure 27-1a, selection is said to be **directional.** Selection is **diversifying** when two extremes are favored at the expense of the intermediate (Figure 27-1b), whereas selection is **stabilizing** when intermediate phenotypes are favored (Figure 27-1c and d).

## 27.3  Directional Selection: Theory

**Elimination of Recessives at One Locus**   Deleterious recessives are commonly eliminated by selection and rarely achieve significant frequencies in a population. Let us suppose, however, that the recessive allele $a$ is not deleterious under one set of environmental conditions so that the population becomes polymorphic for $a$ and its allele $A$. We now imagine a dra-

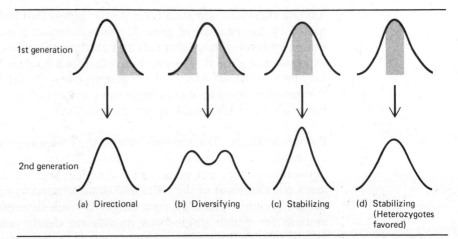

1st generation

2nd generation

(a) Directional          (b) Diversifying          (c) Stabilizing          (d) Stabilizing
                                                                            (Heterozygotes
                                                                            favored)

**FIGURE 27-1 Patterns of selection.** Shading represents the phenotypes selected as parents from the phenotypic range in the population. Arrows indicate the mean. Lower curves represent the expected progeny responding to the selection applied. (From E. B. Spiess, *Genes in Populations*. New York: John Wiley & Sons, 1977, p. 404.)

matic change in the environment such that $a/a$ individuals are totally inviable. In this case there will clearly be selection for individuals with an $A/A$ or an $A/a$ genotype. What effect will this form of directional selection have on allelic frequencies? In other words, in what fashion, and at what rate, will the population evolve?

In such cases the relative fitness of $a/a$ is set at 0, so that the selection coefficient for this set of alleles becomes $1 - 0 = 1$. If we also assume, for simplicity, that $A/a$ and $A/A$ are identically fit, that no other alleles are involved at this locus, and that no other loci contribute to the fitness of the phenotype, then by definition $A/A$ and $A/a$ each have fitnesses of 1 and selection coefficients of 0.

We next define $q_o$ as the starting frequency of $a$, and we assume that the ideal population we are dealing with is in a Hardy-Weinberg equilibrium before selection. This means that the frequency of $a/a$ individuals will be $q_o{}^2$, the frequency of $A/A$ individuals will be $p_o{}^2$, and the frequency of $A/a$ individuals will be $2p_oq_o$. The total population can then be represented as the sum

$$p_o{}^2 + 2p_oq_o + q_o{}^2 = 1$$

Once selective pressure has been imposed, each genotypic frequency is multiplied by its relative fitness. The values of $p_o{}^2$ and $2p_oq_o$ remain unchanged, since each is multiplied by 1, but the value of $q_o{}^2$ is reduced to zero, since it is multiplied by zero. The total population can therefore be expressed after selection as the sum $p_o{}^2 + 2p_oq_o$.

When this population engages in reproduction, all the $a$ alleles in the

Selection

gamete pool will be contributed by the $A/a$ heterozygotes, which constitute $2p_o q_o$ of the total. Since only half the gametes produced by the $A/a$ organisms will carry the $a$ allele, the frequency of $a$ in the next generation (expressed as $q_1$) will be

$$q_1 = \frac{1/2(2p_o q_o)}{p_o{}^2 + 2p_o q_o}$$

$$= \frac{p_o q_o}{p_o(p_o + 2q_o)} \qquad p_o = 1 - q_o$$

$$= \frac{q_o}{1 - q_o + 2q_o}$$

$$= \frac{q_o}{1 + q_o} \qquad\qquad\qquad\qquad (1)$$

By performing the same operations for successive generations we get

$$q_2 = \frac{q_1}{1 + q_1} \qquad q_3 = \frac{q_2}{1 + q_2} \qquad q_4 = \frac{q_3}{1 + q_3}$$

and, in general,

$$q_n = \frac{q_o}{1 + nq_o} \qquad\qquad\qquad\qquad (2)$$

where $q_n$ is the frequency of $a$ in the population after $n$ generations.

The elimination of "undesirable" homozygotes from a breeding population by directional selection is sometimes suggested as a way to reduce the frequency of "undesirable" alleles in the gene pool of the population. It can be seen from Equation 2, however, that the frequency of the "undesirable" allele would fall slowly indeed and that the allele would still be present after many thousands of years. Let us suppose, for example, that all albino ($a/a$) humans are prohibited from having children on the grounds that albino homozygotes can experience certain health difficulties. If the population contains 1 albino among every 10,000 persons (approximately the $a/a$ frequency in human populations), how many generations would elapse before the $a/a$ frequency became one in every million persons? The desired genotype reduction, in this case, is from $10^{-4}$ ($q_o{}^2$) to $10^{-6}$ ($q_n{}^2$). By taking the square roots of these values, we obtain the corresponding reduction in gene frequency, namely, $10^{-2}$ ($q_o$) to $10^{-3}$ ($q_n$). We can rewrite Equation 2, solving for $n$, as

$$n = \frac{q_o - q_n}{q_o q_n}$$

$$= \frac{1}{q_n} - \frac{1}{q_o}$$

The number of generations required to go from $q_o$ to $q_n$ then becomes

$$n = \frac{1}{10^{-3}} - \frac{1}{10^{-2}} = 1000 - 100 = 900 \text{ generations}$$

or about 18,000 years. Considering that the recorded history of humans began only 5000 years ago and that mutation is constantly providing the human gene pool with new $a$ alleles, a prohibition on albino reproduction would have a negligible positive effect, particularly when considered in the context of its negative effects on the human beings involved. Whereas "eugenics" programs such as these were often discussed in the first half of this century, the concept is now rarely invoked as a solution to the problem of inherited recessive genes.

**Partial Selection Against Recessives at One Locus**  We can next consider a less complete form of directional selection against homozygous recessive individuals. Here the selection coefficient $s$ is less than 1 and the relative fitness $W$ of the homozygous recessive individuals is $1 - s$, a number greater than zero. We once again begin with a Mendelian population and, following selection, multiply each original genotypic frequency by its relative fitness. The proportion of $A/A$ then remains $p_o^2$, the proportion of $A/a$ remains $2p_oq_o$, and the proportion of $a/a$ becomes $q_o^2(1 - s)$. The *total* population after selection can be expressed as $1 - sq_o^2$, where $sq_o^2$ represents the $a/a$ alleles that are eliminated by selection.

In the next generation $a$-bearing gametes derive from two sources: the $A/a$ heterozygotes as before, whose contribution is expressed as $1/2(2p_oq_o)$, and the surviving $a/a$ homozygotes, whose contribution is $q_o^2(1 - s)$. The frequency of $a$ alleles in this next generation is therefore

$$q_1 = \frac{p_oq_o + q_o^2(1 - s)}{1 - sq_o^2}$$

$$= \frac{q_o(1 - sq_o)}{1 - sq_o^2} \tag{3}$$

and, in general,

$$q_n = \frac{q_{n-1}(1 - sq_{n-1})}{1 - sq_{n-1}^2} \tag{4}$$

The only case in which this formula can be directly solved occurs when $s = 1$, in which case we have Equation 2 again. When $s$ does not equal 1, the best that can be done is to express $\Delta q$, the change in $q$ in one generation, as

$$\Delta q = q_1 - q_o$$

$$= \frac{q_o(1 - sq_o)}{1 - sq_o^2} - q_o$$

$$= \frac{-sq_o^2(1 - q_o)}{1 - sq_o^2} \tag{5}$$

At this point we can drop the subscript ($_o$), since we are actually interested in a general value for $\Delta q$ and not simply the value after the first generation. We can also simplify Equation 5 by making the assumption that $s$ has a small value such that $1 - sq_o^2$ is approximately equal to 1. Equation 5 thereby becomes

$$\Delta q \simeq -sq^2(1 - q) \tag{6}$$

When this equation is plotted for values of $q$ between 0 and 1 (Figure 27-2), it is clear that the rate of change in $q$ becomes very small as $q$ approaches 0 or 1 and that the rate attains a maximum when $q = 0.66$. This fact is of importance, for it means that when a recessive allele is prevalent in a population ($q$ near 1) and is placed under mild selective pressure it will initially be eliminated very slowly. Similarly, if a new and selectively favorable recessive allele arises, its initial rate of increase in the population will also be very slow.

**Test of Directional Selection Theory** . It sometimes becomes possible to determine a selection coefficient for a particular genotype under laboratory conditions and thus to test the applicability of relationships such as Equations 5 and 6. One study, for example, focused on the sex-linked *ras* (*raspberry*-eye) allele of *Drosophila melanogaster* affecting eye color (*recall Figure 21-12*). It was first established that under laboratory conditions *ras/ras* and *ras/→* flies had about the same viability as *+/+* or *+/ras* flies but that *ras/→* males mated with only half the efficiency of their *+/→* counterparts. This led to the prediction that the *ras/→* genotype should have a relative fitness of 0.5. A modified version of Equation 5, which takes sex linkage into account, was then developed, and a predicted rate of disappearance of the *ras* allele from a *Drosophila* population was calculated. This rate is plotted in Figure 27-3 (gray curve). Against this rate is plotted the observed rate of decline in the *ras* allele frequency (black curve) with data derived from experiments in which *+/ras* flies were introduced into population cages

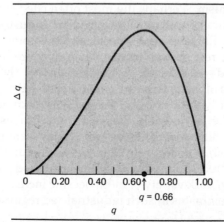

**FIGURE 27-2 Directional selection.** Graphical representation of $\Delta q \cong -sq^2 (1 - q)$ where $s$ is a constant. (From C. C. Li, *Population Genetics*, University of Chicago Press, Chicago, 1955, Figure 64.) (Reprinted from *Population Genetics* by C. C. Li by permission of The University of Chicago Press. © 1955 by The University of Chicago. All rights reserved.)

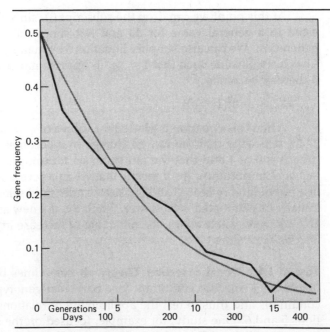

**FIGURE 27-3 Directional selection under laboratory conditions** showing decline in frequency of the *ras* allele of *D. melanogaster* (black curve) compared with the predicted (gray) curve using a modified form of Equation 5 (After D. J. Merrell, *Evolution* 7:287, 1953).

(Figure 27-4) and allowed to propagate freely over a period of about 18 generations. The similarity between the predicted and observed curves argues strongly for the relevance of the mathematical equations to such situations.

## 27.4 Directional Selection: Observations

**Industrial Melanism**   Perhaps the most often cited example of directional selection in natural populations is given by **industrial melanism.** Most moth species in England, and specifically the common Peppered Moth (*Biston betularia*), rest on tree trunks when they are not in flight. In rural parts of England—and in all of England before the Industrial Revolution—the trunks of many trees are covered with light gray lichens, and the moths are similarly light in color. As industry became prominent during the middle of the last century, however, the lichens in industrial areas were killed by falling soot, and the tree trunks were blackened. At about this time (*ca.* 1850), a collector in Manchester captured an apparently new variety of Peppered Moth with dark (melanic) body and wings. As time passed, this new, dark variety of moth became increasingly more common, and it now predominates in such industrialized regions as Manchester and Birmingham.

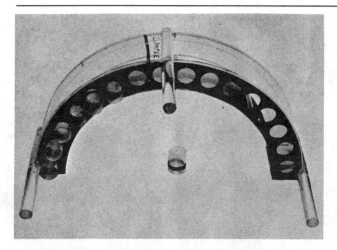

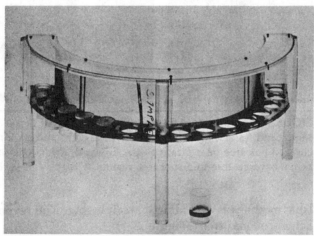

**FIGURE 27-4 Population cage.** Parent flies are introduced into the cage and the holes are filled with vials that either are empty or contain a medium used both as food and as a site for egg deposition. Food vials can be replenished at appropriate intervals, and adult flies can be sampled by examining those flies that happen to rest in the empty vial marked "sample." (Courtesy of B. Wallace.)

The inference from such observations, therefore, is that some sort of selective agents act against light moths in industrial areas and against black moths in rural areas. H. Kettlewell demonstrated the nature of these agents by observing, first, that Peppered Moths are preyed on by various bird species. He then released known numbers of dark moths into rural woods and light moths into industrial woods and noted their likelihood of becoming prey compared with the native variety of Peppered Moth. The results were clear: the soot-laden trees afforded excellent camouflage for the melanic moths relative to the light moths, and the survival of the me-

**FIGURE 27-5 Industrial melanism.** Photograph of *Biston betularia*, the Peppered Moth, and its black form, *carbonaria. Left:* Light and dark forms on a soot-covered oak tree near Birmingham, England. *Right:* The same two forms on a lichen-covered tree in a soot-free region. (Courtesy of H. B. D. Kettlewell.)

lanic moths was accordingly much higher. The reverse was true with the lichen-covered trees (Figure 27-5).

The selective agents in this case turned out to be bird predators and industrial soot. This combination has produced dramatic results in a relatively short period of time, in part because the agents of selection are themselves strong and in part for a genetic reason: dark moths are produced when a dominant allele ($B$)—which is lacking in light ($b/b$) moths—is present in the genome. Because both $B/b$ and $B/B$ individuals are thus protected in industrial areas, the $B$ allele has been able to spread much more rapidly than a recessive gene under similar pressures (recall Equation 2).

The melanic variety of the black Peppered Moth is far darker today than specimens collected 100 years ago, which suggests that other pigment-producing alleles, in addition to $B$, are now in the process of being preserved by natural selection. The fact remains that the original selective events were most probably directed at a single genetic locus.

**Inversions in *Drosophila*** Industrial melanism illustrates a common pattern of directional selection: individuals with particular genotypes at a defined locus are *prevented* from contributing their gametes to the next generation by selective pressures inherent in the environment. We now turn to examples in which selective pressures *favor* particular genotypes. Obviously, the two patterns are opposite sides of the same coin—as one group is being favored, another is being disfavored—but the focus of attention is shifted.

A classic example involves chromosomal inversions in *Drosophila*. As we noted in Section 7.8, an inversion suppresses meiotic recombination; therefore, a block of genes contained within an inversion will be in linkage disequilibrium (*Section 27.2*) and will usually be transmitted as a unit to the next generation. Because inversion heterozygotes can be readily detected by examining the polytene chromosomes of *Drosophila* (*Figure 7-8*), Th. Dobzhansky and his colleagues were able to score natural populations of flies for their endowment of inversions and to correlate inversion frequency with environmental circumstance. They concluded that inversions were most common in species or subpopulations that exploit a highly diversified environment, territories with varied food sources, changing climates and so on, whereas inversions were relatively uncommon in populations that inhabited more uniform environments.

This observation is consistent with the hypothesis that inversions, by maintaining a group of alleles in a "supergene" configuration (*Section 27.2*), can conserve a set of alleles adaptive for a particular environmental niche. By this hypothesis, species that populate diverse environments would be well served to harbor many inversions, each adaptive for a different niche. On the other hand, a population living in a uniform environment would have little need for such flexibility, and possession of a limited number of adaptive supergene clusters would suffice. Support for this hypothesis has come from laboratory studies in which inversion-rich flies were taken from their luxuriant tropical environments and maintained in uniform population cages; as one would predict, the frequency of inversions fell rapidly as the flies bred in the uniform environment.

**Phototaxis in *Drosophila***    Directional selection for the acquisition of a trait can also be simulated in the laboratory, an example being experiments by Th. Dobzhansky and B. Spassky with *D. pseudoobscura*. Dobzhansky and Spassky constructed a maze wherein flies had to make a series of light-dark choices before they could emerge at the other end. Of 300 pairs of flies tested in each generation, 25 with the strongest tendency to fly toward light (positive phototaxis) were selected, and 25 with the strongest tendency to fly toward the dark (negative phototaxis) were selected. Each collection was inbred; the offspring were challenged with the same maze; and again 25 pairs were selected with the most extreme positive and negative phototaxis. The results were striking. As plotted in Figure 27-6, selection led, in a few generations, to populations of flies with significantly higher or lower phototactic scores than the starting population. Moreover, when se-

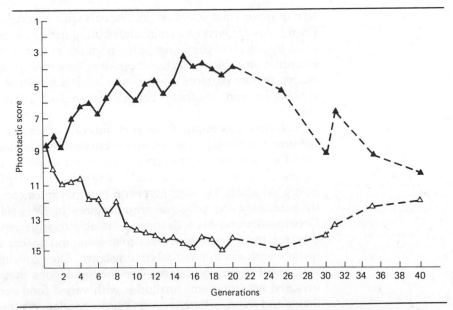

**FIGURE 27-6 Selection for phototaxis.** Flies were placed in a maze with 16 light-dark choices. If a fly chose the dark direction 15 times in a row, it wound up in vial no. 1; if a fly made 15 light choices, it wound up in vial no. 16. A mean score of 8.5 therefore connotes nonresponsiveness to light. The open triangles show mean scores after selection for positive phototaxis was imposed; the solid triangles denote negative phototaxis. Dashed line indicates that selection was relaxed. (From T. Dobzhansky and B. Spassky, *Proc. Natl. Acad. Sci. U.S.* **62**:75–80, 1969.)

lection was relaxed after 20 generations so that flies were no longer selected for their phototactic skills before breeding, the divergence achieved by the selection regimen was rapidly lost (Figure 27-6, dashed lines).

Two types of selection are illustrated in this experiment. The increased phototaxis was achieved by **artificial selection** in that the trait desired was imposed upon the flies by human manipulation. The return to a "mean" phototactic score once artificial selection was relaxed was the consequence of natural selection (albeit in the laboratory). Thus the phenotypes of the phototactically skilled flies conferred on them a high **survival value** under the artificial selection regimen, but once this "selective edge" was removed, these same phenotypes had no selective advantage and were possibly even at a disadvantage when in free competition with "normal" flies. We might speculate, for example, that flies with strong negative phototaxis had light-sensitive retinas, which, in a population cage, made them less responsive to the presence of food or to mating cues.

# 27.5 Plant and Animal Breeding

Artificial selection practiced by plant and animal breeders has made enormous progress in increasing the yield and viability of domestic animals and crop plants. Although details of their programs fall outside the scope of this text, we can outline some general principles.

**General Principles of a Breeding Program**   Plant and animal breeders are usually interested in quantitative traits (*Section 21.11*), such as corn kernel size or sweetness, whose contributing loci are numerous and undefined. Therefore, as with selection for phototactic behavior in *Drosophila* (*Section 27.4*), the individuals selected for breeding are usually chosen by empirical criteria rather than by rigorous genetic principles. Sometimes the phenotype of the individual serves as the selection criterion; in other cases, the mean phenotype of the individual's family (full sibs and half sibs) is taken into consideration. For quantitative traits, the heritability of the trait (*Section 21.15*) is obviously also relevant in designing an appropriate selection scheme.

An important problem encountered during breeding experiments, known as **correlated response,** refers to the possibility of selecting for undesired traits as well as the desired traits. For example, Leghorn chickens were selected for 12 generations for increased leg length, only to find that the eggs laid by the "leggy" birds hatched only half as well as in the starting population. Thus the products of any selection program must be carefully tested along the way for their fitness in all aspects of growth and reproduction.

Selection must obviously begin with a genetically heterogeneous population, since a homogeneous population will be unable to respond to a selection challenge. For quantitative traits with high heritability, one usually begins with a population whose variance for the trait shows a normal distribution, as in the example illustrated in Figure 27-7a, where the mean phenotype is given the value $\mu$. A **truncation point ($T$)** is then chosen such that individuals with phenotypes above this value are selected for breeding; in the case illustrated in Figure 27-7a, the truncation point is defined as a seed weight of 650 mg. In the next generation, then, the mean phenotype, $\mu'$, is found to be shifted in the direction of the mean of the selected parents ($\mu_s$), as illustrated in Figure 27-7b, the shift ($\mu' - \mu$) being defined as the **response to selection.** For obvious reasons, this mode of directional selection is often called **truncation selection.**

Figure 27-8 illustrates how the heritability ($H$) (*Section 21.15*) of a trait will influence its response to a truncation selection program. For a trait with a heritability of 1, with no environmental influence on its expression, the mean of the progeny ($\mu'$) is expected to equal the mean of the selected parents ($\mu_s$). If, at the other extreme, all of the variance in the parental population is due to environmental vagaries and the trait has a heritability

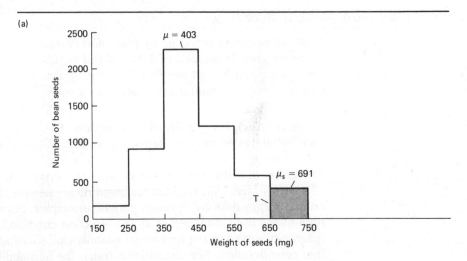

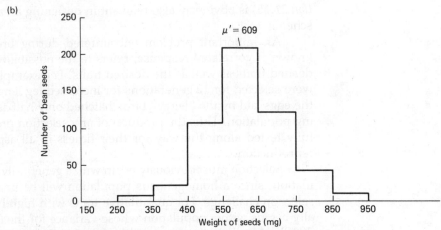

**FIGURE 27-7 Artificial selection for seed weight** in the bean, *Phaseolus*. (a) Before selection, seed weight shows a normal distribution with a mean ($\mu$) of 403 mg. Seeds were selected whose weight exceeded the truncation point ($T$) of 650 mg; the mean weight of the selected parents ($\mu_s$) is 691 mg. (b) After selection, seed weight of the progeny plants again shows a normal distribution but is shifted so that the new mean ($\mu'$) is 609 mg. (Data from W. Johannsen, 1903, plotted in D. Hartl, *Principles of Population Genetics*. Sunderland, MA: Sinauer, 1980, p. 251.)

of 0, then $\mu'$ will equal $\mu$, the mean of the parental population, and selection will have had no effect whatsoever. For a trait with a moderately low heritability (0.25), $\mu'$ will be shifted toward $\mu_s$; and so on. In general, these variables follow the relationship

$$\mu' = \mu + H(\mu_s - \mu) \tag{7}$$

Solving for $H$, we get

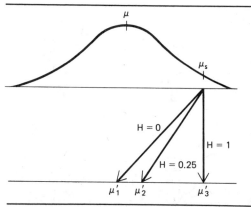

**FIGURE 27-8 Influence of heritability on artificial selection.** In a population with a mean value μ for a polygenic trait, truncation selection for parents with a mean value $\mu_s$ will produce offspring with three different mean values, $\mu_1'-\mu_3'$, depending on the heritability of the trait. (After J. F. Crow, *Genetics Notes*, 7th ed. Minneapolis: Burgess Publishing Co., 1976, p. 198.)

$$H = \frac{\mu' - \mu}{\mu_s - \mu} \tag{8}$$

Substituting the values in the selection experiment illustrated in Figure 27-7, we learn that seed weight in the bean has a high heritability (0.71), which explains its good response to the selection process.

**The Effect of Inbreeding and the Use of Hybrids**  An additional constraint on plant and animal breeders can now be appreciated. The obvious way to develop uniform strains with desirable traits is to find individuals with those traits and inbreed them or, in the case of monoecious plants, mate them with themselves ("selfing"). The price of this practice, of course, is the increased chance for homozygous lethality, which we termed "inbreeding depression" in Section 26.7.

The solution to this problem, first suggested by G. H. Shull in 1909, is to select for desirable traits by inbreeding and then cross inbred lines to produce **hybrids** for use in the field. Since each inbred line can be maintained under controlled (and expensive) environmental conditions at the breeding station, its reduced vigor is not a problem and lines with desirable traits can be developed. In theory, the breeder can then cross inbred line A with inbred line B, producing AB hybrid seed. Since it is unlikely that line A will be homozygous recessive at the same loci as line B, most loci in the AB hybrid are expected to be heterozygous, and the plants that grow from this seed are expected to display the vigor associated with outbred lines.

The major problem with this scheme is that it requires raising large numbers of the expensive inbred lines to produce enough hybrid seed for sale. Therefore, in practice, plant breeders usually adopt the **double-cross method** illustrated in Figure 27-9. In addition to the AB hybrid, they also create CD hybrid seed from inbred lines C and D; they then grow large numbers of both AB hybrids and CD hybrids under inexpensive field conditions; finally, they perform AB × CD crosses and sell the resultant four-way hybrid seed. The four-way hybrid plants are somewhat less uniform

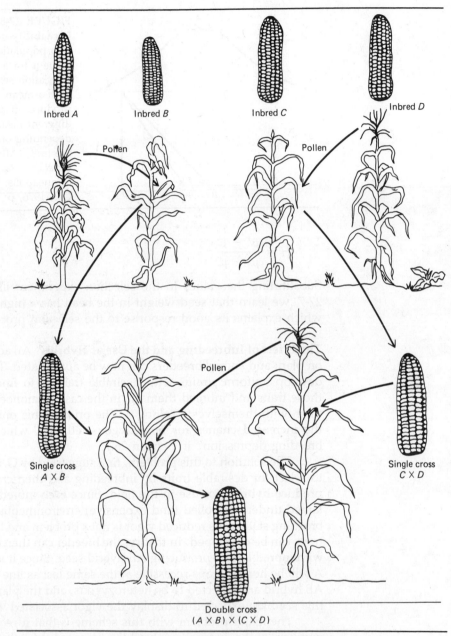

**FIGURE 27-9 The double-cross method** of creating hybrids for the improvement of yield in corn. (From Th. Dobzhansky, *Genetics, Evolution, and Man.* New York: John Wiley & Sons, 1955.)

than the two-way hybrids, but they are adequate for most agricultural purposes.

**Hybrid Vigor or Heterosis**   Figure 27-10 illustrates the phenomenon we have just described, namely, that when two inbred individuals are crossed, the resultant hybrid progeny are more vigorous (Shull described them as more "luxuriant") than their parents. If the hybrid is then inbred, luxuriance is progressively lost as inbreeding depression again takes its toll. Hybrid vigor is also known by the term **heterosis.**

Two theories have been advanced to explain heterosis. The **dominance hypothesis,** intimated in the previous section, assumes that in the course of selecting for desirable traits, the breeder has created strains that are also homozygous for somewhat deleterious recessive genes elsewhere in the genome. A hybrid formed between two inbred strains would be expected to acquire normal, dominant alleles at many of these loci, and the resultant plant might well exhibit "hybrid vigor" compared with its parents.

The second theory, known as the **overdominance hypothesis,** proposes that the hybrid is more vigorous because it is more heterozygous.

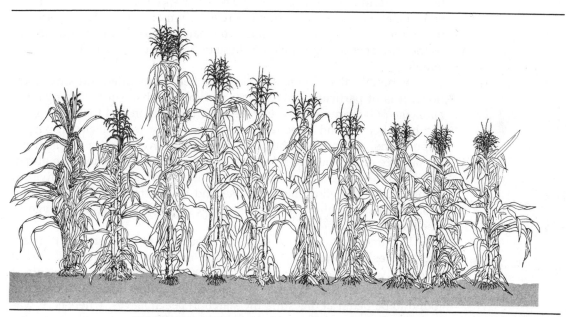

**FIGURE 27-10 Heterosis in corn.** Representative plants from the two inbred parental strains are shown at left. A representative of their $F_1$ hybrid is shown third from left, and representatives of successive inbreedings ($F_2$ through $F_8$) are shown to the right of the $F_1$ plant. (After D. F. Jones, *Genetics* **9**:405, 1924.)

How might the simple fact of heterozygosity provide a more hardy organism? One plausible, although unproved, suggestion points out that since different enzymes are used by different cell types at different developmental stages, the presence of two different versions of many different enzymes might significantly increase the probability that particular stages of an organism's development unfold with particularly optimal rates or yields. In other words, heterozygotes may simply be more flexible, and therefore more likely to come up with favorable sequences of developmental events, than homozygotes. In like fashion, the mature heterozygote may be better equipped to survive fluctuations in the environment than the homozygote.

## 27.6 Diversifying Selection

Directional selection, the topic of the preceding sections, is so named because phenotypes (and hence genotypes) are "pushed" in a single direction: seeds increase in weight; inversions increase in number; black moths replace white moths; and so on. As depicted in Figure 27-1b, **diversifying selection** (also called **disruptive selection**) has a very different outcome: *two* distinct phenotypes are selected for in the population. In all cases described, this form of selection occurs when a population inhabits a nonuniform environment. If one phenotype is successful in a certain environmental niche and a second phenotype is relatively more successful in a second niche, then both phenotypes should be maintained. As a consequence, of course, the alleles responsible for these two phenotypes will also be maintained.

Although diversifying selection has been implicated in many situations, it is of particular interest as a possible mechanism for maintaining allozyme polymorphisms. You will recall from Section 26.4 that many gene loci have been found to be represented by two or more alleles in natural populations, and many population geneticists have sought to explain the persistence of this polymorphism in terms of natural selection. An attractive hypothesis holds that if allozyme $A_1$ is advantageous in one environmental niche and allozyme $A_2$ in a second niche, then it will be advantageous to the population as a whole to have $A_1/A_1$ and $A_2/A_2$ homozygotes in equivalent numbers.

Documentation for this mode of selection has been obtained in several studies. In one case, a population of isopod crustaceans, *Asellus aquaticus*, was found to be polymorphic for two electrophoretic variants of amylase, the enzyme that acts to break down starch. The $A_1$ variant was found to be better at digesting the starch in beech leaves, whereas $A_2$ was better at digesting the starch in willow leaves. The pond inhabited by the isopods had a stand of beech trees at one end and a stand of willows at the other, and the frequency of the beech-digesting allozyme was indeed found to be higher in animals captured at the "beech end," whereas the frequency of the willow-digesting allozyme was higher in collections made at the "willow end." The data supporting this conclusion are found in Problem 27-8.

In a second case, *D. melanogaster* were collected from two locations in a Spanish wine-growing community—from the wine cellars themselves and from the compost heaps where grape skins are dumped. Each group was then tested for polymorphism at the alcohol dehydrogenase (*Adh*) locus. The wine cellar group included a significantly larger number of individuals carrying the $Adh^F$ (*fast*-migrating) allele, whereas the compost group had a higher frequency of $Adh^S$ (*slow*-migrating). Studies *in vitro* established that the $Adh^F$ allozyme is far the more efficient at metabolizing ethanol, which of course abounds in the environment inhabited by the wine cellar group. Data supporting this conclusion are presented in Problem 27-9.

Two important points must be established before such studies can be used to argue that diversifying selection plays a role in maintaining polymorphisms. First, it must be shown that the two subpopulations actually interbreed and are not simply separate populations with distinct allelic frequencies. Second, and more difficult, it must be shown that the locus under test is not in linkage disequilibrium with one or more other loci that are in fact undergoing some other form of selection (*Section 27.2*). The present consensus, however, is that diversifying selection is likely to be responsible for at least some of the balanced polymorphisms (*Section 26.9*) observed in natural populations.

## 27.7 Stabilizing Selection

The final mode of selection we will consider, called **stabilizing selection,** takes the two forms illustrated in Figure 27-1. In Figure 27-1c, some intermediate phenotype is favored over the extremes. In Figure 27-1d, an intermediate phenotype is again favored, but greatest fitness is possessed by heterozygotes for this trait, the result being that the variance (*Section 21.13*) is greater than when heterozygosity is not at a premium. When heterozygotes are favored, polymorphism will, by definition, also be promoted; therefore, this form of selection will also produce a condition of balanced polymorphism (*Section 26.9*).

**Sickle-Cell Hemoglobin**  Certainly the best documented case of stabilizing selection in human populations is the advantage accorded to $+/Hb^s$ heterozygotes in malaria-ridden environments. We have already outlined the story in Section 26.1, which should be reviewed at this time. Here we present a more complete account.

The β-chain of human hemoglobin specified by the sickle-cell $Hb^s$ allele differs from the normal by one amino acid: whereas the normal chain has the N-terminal sequence Val-His-Leu-Thr-Pro-**Glu**-Glu-Lys . . . , the mutant chain reads Val-His-Leu-Thr-Pro-**Val**-Glu-Lys. . . . When such a mutant β-chain is included in an $\alpha_2\beta_2$ hemoglobin molecule, the protein (known as hemoglobin S) is less soluble than normal A hemoglobin at low oxygen tensions, presumably because the neutral valine residue has replaced the charged glutamic acid. As a consequence the conformation of the erythrocyte at low oxygen tensions has a sickle shape rather than the normal disc shape (Figure 27-11), and these deformed cells are defective in oxygen transport.

Homozygous recessive ($Hb^s/Hb^s$) individuals synthesize only hemoglobin S (and fetal hemoglobin F) and, under primitive living conditions, rarely reach maturity (although in modern societies they can survive if given blood transfusions). Such individuals are said to suffer from **sickle-cell anemia.** Heterozygotes, on the other hand, are largely asymptomatic, even though their erythrocytes contain almost as much hemoglobin S as hemoglobin A, and such individuals are said to exhibit the **sickle-cell trait.**

Because the relative fitness of the $Hb^s/Hb^s$ genotype is near zero (so that $s \simeq 1$), directional selection models would predict that the frequency of the $Hb^s$ allele in a population should be zero or approaching zero. This prediction is confirmed in a number of human populations in which the allele is virtually absent (Table 26-1). In many African and Asian populations, however, the frequency of the $Hb^s_\beta$ allele is as high as 0.1 to 0.2, and 10 percent of Afro-Americans are believed to be heterozygous for the gene (Table 26-1). Thus some factor must prevent the extinction of the *a* allele in particular populations.

The factor, as we have already stated, is the falciparum malarial parasite, which abounds in tropical regions of Africa; it kills an estimated 15 to 20 young children per thousand, and its direct toll on the population as a

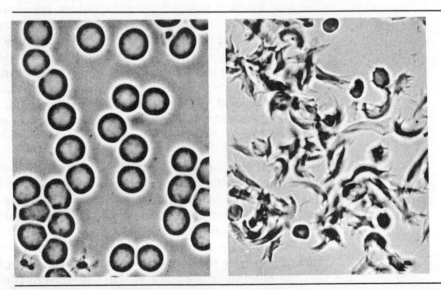

**FIGURE 27-11 Sickle-cell anemia.** Normal erythrocytes (left) and sickle-shaped erythrocytes (right). (Courtesy of A. C. Allison.)

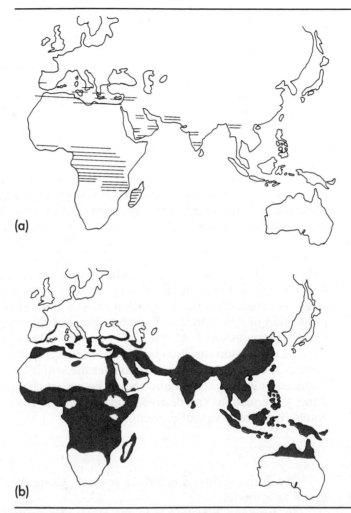

(a)

(b)

**FIGURE 27-12 Sickle-cell hemoglobin and malaria.** (a) Distribution of sickle-cell hemoglobin (bars on map). (b) Distribution of falciparum malaria (shaded areas). (From A. Motulsky, *Human Biology* **32**:43, 45, 1960. By permission of the Wayne University Press, © 1960.)

whole is greater still. Moreover, many persons who inhabit these regions and who die of other causes are actually so weakened by malaria that their susceptibility to all diseases is greatly enhanced. When the distribution of the sickle-cell gene in Africa and Asia (Figure 27-12a) is compared with the distribution of high endemic malaria (Figure 27-12b), the correlation is striking, suggesting that malaria serves to balance the deleterious effects of sickle-cell anemia so as to maintain the $Hb^s$ allele.

Malaria serves as a selective agent in that $+/Hb^s$ heterozygotes are significantly more resistant to the disease than $+/+$ homozygotes (the relative resistance of $Hb^s/Hb^s$ homozygotes is immaterial, since such persons usually die in infancy in any case). The actual basis for heterozygote resistance is thought to be as follows: the falciparum parasite develops within human erythrocytes; the erythrocytes of a heterozygote tend to adopt a sickle shape when they enter the capillaries, in which the oxygen tension is particularly low; this sickling process appears to interrupt some stage in parasite multiplication and the infection cycle is broken. Since normal erythrocytes cannot sickle, $+/+$ individuals do not enjoy this protection from infection and are fully susceptible to malaria.

**Theoretical Basis of Stabilizing Selection**  The mathematical expression of balanced polymorphism by heterozygote advantage is quite simple. We let $p^2$, $2pq$, and $q^2$ represent the proportions of $A/A$, $A/a$, and $a/a$ individuals before selection occurs, and we let $s$ and $t$ represent, respectively, the selection coefficients against the two homozygotes. The fitness of $A/A$ individuals is then $1 - s$ and the fitness of $a/a$ individuals is $1 - t$, the fitness of $A/a$ individuals being equal to 1. From here we can develop the mathematics much as we did for partial selection against recessives in Section 27.3. Thus, after selection the proportion of $A/A$ individuals is $p^2(1 - s)$, the proportion of $A/a$ individuals remains $2pq$, and the proportion of $a/a$ individuals becomes $q^2(1 - t)$, the total population being equal to $1 - p^2s - q^2t$. The expression can then be written in terms of $\Delta q$, as in Equation 5 of Section 27.3, but at this point we can let the whole be equal to 0, since, in balanced polymorphism, there should be no net change in $q$. When this is done, the equations can be greatly simplified and the expression for $\hat{q}$, the equilibrium frequency of $q$, becomes

$$\hat{q} = \frac{s}{s + t} \tag{9}$$

Try deriving this equation for yourself, using the derivation of Equation 5 as a model.

In applying Equation 9 to the sickle-cell example, where $A$ corresponds to $+$ and $a$ to the $Hb^s$ allele, we can assume that $t$, the selection coefficient against $Hb^s/Hb^s$ homozygotes, is equal to 1. This means that if the selection coefficient $s$ against $+/+$ homozygotes is equal to 0.25, $q$ will equal 0.2, whereas if $s$ is equal to 0.11, $q$ will equal 0.1. Since the frequency of the sickle-cell gene in African populations is presently estimated at from 0.1 to 0.2, values of $s$ can be assumed to fall in the 0.11 to 0.25 range.

# Migration

## 27.8 Dynamics of Migration Pressure

Although selection is the most familiar evolutionary agent, it is by no means the only one. **Migration pressure** represents an evolutionary force

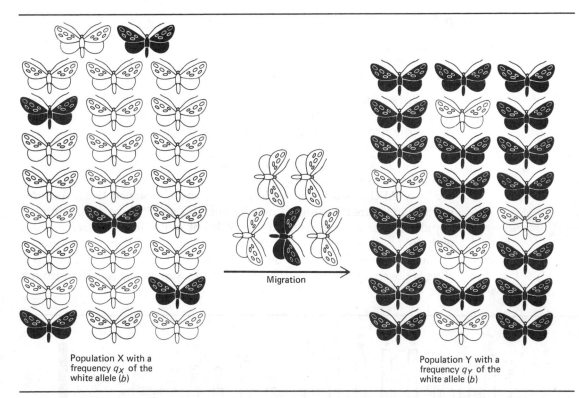

Population X with a frequency $q_X$ of the white allele ($b$)

Migration

Population Y with a frequency $q_Y$ of the white allele ($b$)

**FIGURE 27-13 Migration pressure** as an evolutionary agent. The Y population of moths receives some immigrants from the X population, an event that changes its frequency of the white allele. (After E. O. Wilson and W. H. Bossert, *A Primer of Population Biology*. Sunderland, MA: Sinauer 1971.)

that can influence allelic frequencies in a particular direction. As its name implies, migration pressure describes the effect of introducing new individuals with new genotypes into a Mendelian population. Figure 27-13 diagrams the process. Illustrated are two populations of moths, X and Y, population Y having a far higher frequency of the black allele ($B$) than population X. A random sampling of moths from population X now migrates to, and mixes with, population Y. The immigrants thus come to represent some fraction $m$ of the total number of individuals present in the now expanded population of Y. If we let $q_X$ and $q_Y$ represent the original frequencies of the $b$ (white) allele in the original populations, then a new frequency, $q'_Y$, is established in the new population. The value of $q'_Y$ will be equal to the contribution made by the immigrants ($q_X m$) plus the contribution made by the original Y population [$q_Y(1 - m)$], where $1 - m$ represents the proportion of nonmigrants. Therefore

$$q'_Y = q_X m + q_Y(1 - m)$$

and the change in $q$ after one generation in such a population becomes

$$\Delta q = q'_Y - q_Y$$
$$= q_Y - mq_Y + mq_X - q_Y$$
$$= -m(q_Y - q_X) \tag{10}$$

When numerical values are substituted into this equation, it becomes clear that significant changes in the value of $q_Y$ can result in one generation even if the two populations differ only slightly in the frequency of a given allele and if a moderate degree of migration occurs between them.

That migrations have been effective in spreading alleles from one population to the next is suggested by data on the relative frequencies of the three ABO blood group alleles, $L^A$, $L^B$, and $L^O$ (Section 19.2). As summarized in Figure 27-14 and as noted earlier in Table 26-1, particular human populations or races possess these alleles at quite different frequencies, with the $L^B$ allele being particularly common in Asia. When the frequency of the $B$ allele in European countries is assessed, a gradient known as a **cline** is observed (Figure 27-15), with $B$ alleles apparently radiating from a

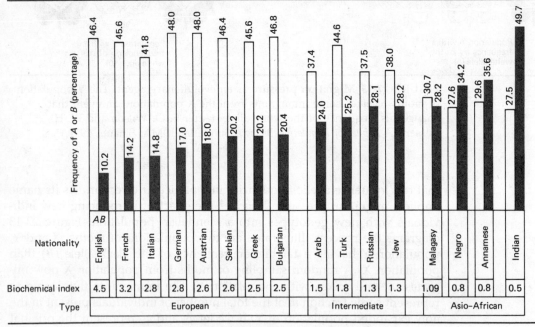

**FIGURE 27-14 Ethnic differences in ABO gene frequencies.** The figures given are percentages of positive reactions with anti-A and anti-B reagents. The frequency of the $L^O$ allele can therefore be calculated by difference. More recent data on AB gene frequencies are found in Table 26-1. The "Biochemical Index" is the ratio of A to B. (From L. Hirschfeld and H. Hirschfeld, *Anthropologie*, vol. 29, pp. 505–537, 1919, and W. F. Bodmer and L. L. Cavalli-Sforza, *Genetics, Evolution, and Man.* San Francisco, W. H. Freeman and Co., 1976.)

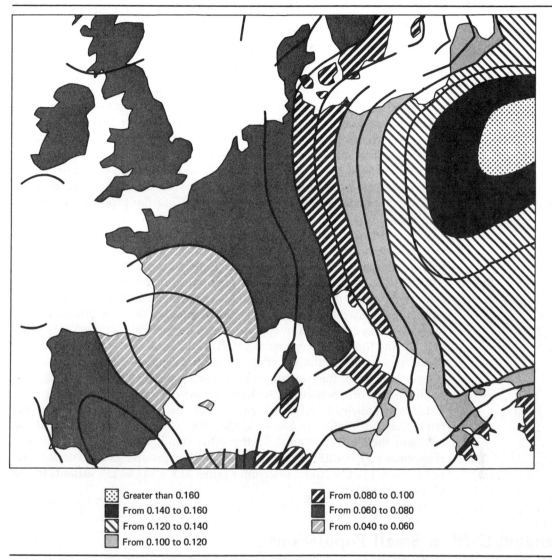

| | |
|---|---|
| Greater than 0.160 | From 0.080 to 0.100 |
| From 0.140 to 0.160 | From 0.060 to 0.080 |
| From 0.120 to 0.140 | From 0.040 to 0.060 |
| From 0.100 to 0.120 | |

**FIGURE 27-15 A cline,** or gradient of gene frequency. The computer-generated map shows the frequency of the *B* allele of the ABO blood-group system in Europe. Note the gradual change from high frequencies near central Asia toward low frequencies in western Europe. The average "slope" of the cline is about 1 percent per 400 km. The cline is probably a remnant of early population migrations and may be gradually disappearing, but it might also be maintained by selective pressures. (Courtesy of D. E. Schreiber, IBM Research Laboratory, San Jose, and of R. Matessi; taken from W. F. Bodmer and L. L. Cavalli-Sforza, *Genetics, Evolution, and Man.* San Francisco, W. H. Freeman and Co., 1976.)

very "high-*B*" focus in western Asia. Although such clines can be interpreted to result from Mongolian invasions in the twelfth century, such an argument is understandably difficult to prove.

## 27.9 Gene Flow

An influx of individuals from one population into a second generates a phenomenon known as **gene flow.** Gene flow occurs when two populations with distinct allelic frequencies become contiguous and individuals "cross the boundary" and interbreed. If individuals from population X contribute alleles to population Y more often than the reverse, then genes are said to flow from X to Y. Eventually, of course, the allelic frequencies in the two populations will become equivalent, but the process may take a long time.

A particularly dramatic example of gene flow in human populations resulted when black Africans and white Americans were brought together several centuries ago as a consequence of the slave trade. Since the interactions between the two groups were such that the "boundary" was usually crossed by white males in extramarital encounters, alleles from the white population tended to flow into the black population far more often than the reverse. Table 26-1 lists the consequences for several loci, and it is clear that, during the ten or so generations that such gene flow has occurred, the allelic frequencies in the American black populations have come to differ considerably from the African. Indeed, the data in this table can be used to calculate the proportions of African and white-American (Georgian and Californian) alleles in Afro-Americans. For the Rh, ABO, and Duffy systems, these estimates are 0.85, 0.88, and 0.84 African, respectively. For the $Hb^s$ and the G6PD systems, on the other hand, they are 0.47 and 0.67, respectively. The differences between the two groupings of gene loci can be attributed to the strong selection against $Hb^s$ and $G6PD^-$ variants in the absence of malaria.

# Random Drift in Small Populations

## 27.10 Dynamics of Small Populations

Mendelian populations are considered "infinite" in the sense that the gamete pool is very large and mixes randomly at each generation. In fact, natural populations rarely conform to this dictum. Ecological niches, limited mobility, assortative mating tendencies, and so on all conspire to create **subpopulations,** or **demes,** wherein mating is preferred. Therefore, the **effective population size** is often much smaller than the head count would suggest.

Whenever small populations are considered, either as sub-populations with limited outcrossing or as isolated **"splinter populations"** created by insurmountable boundaries, most of the assumptions about Mendelian populations no longer apply. Allelic frequencies may be very different from the population at large owing to the **founder effect,** as we have detailed in Section 26.8. Moreover, as also noted in Section 26.8, these subgroups will tend to inbreed, often uncovering recessive alleles that may have been masked in the population at large. Finally, when populations are small, alleles are subject to random drift, which, in these situations, can act as a potent evolutionary agent.

## 27.11 Random Drift of Alleles

**Random drift** is most easily visualized in the context of a coin-flipping analogy. If a coin is flipped 1000 times, the number of "heads" is expected to be roughly 500 and the number of "tails" to be roughly 500. We would not be surprised if the numbers came out 507 and 493, but we would suspect the balance of the coin if the result were, for example, 700 and 300. If, on the other hand, a coin were flipped ten times and it came out 7 "heads" and 3 "tails," we would not be startled; in such a small sampling of coin flips, we would say, such a wide fluctuation from the expected 50:50 outcome is not unexpected.

The same principle applies to what can be called gametic sampling. If we return to Figure 27-13, and imagine that the five moths diagrammed as leaving population X instead fly to an island where no other moths of that species reside, then each becomes a founder moth. If the one black moth in the group is a $B/b$ heterozygous male, then half of his sperm will carry $B$ and half $b$. Assuming that ten of his gametes manage to fertilize eggs that develop to maturity, it is not unexpected that, say, seven of these sperm will happen to carry $b$ and only three will happen to carry $B$. In a large Mendelian population, this sampling effect would presumably be canceled by a second $B/b$ moth in which seven $B$ and three $b$ sperm were involved in fertilizations. In our example, however, there is no second $B/b$ moth. Therefore, this chance sampling bias has the effect of causing the frequency of $B$ in the second-generation gene pool to be significantly less than it was in the founder gene pool.

When gametes are selected at random from the second generation of moths, two effects come into play: not only is the relative frequency of $B$-bearing gametes in the total pool reduced, but also the total gamete pool is still relatively small so that sampling errors may again occur. Therefore, once a gene such as $B$ begins to experience attrition in a small population, it is likely to move rapidly toward **extinction** ($p = 0$) while its allele ($b$) moves toward **fixation** ($q = 1$). A computer simulation of these events is shown in Figure 27-16, and an actual experiment illustrating both elimination and fixation is shown in Figure 27-17.

An often cited case of the founder effect and drift in human populations concerns the inhabitants of Pingelap, a small cluster of islands in the Pacific. In the late 18th century, all but about 30 Pingelapese were killed by

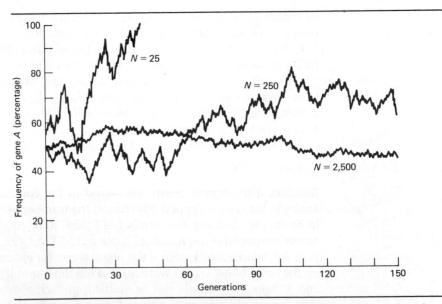

**FIGURE 27-16 Chance and population size.** The graph shows the results of computer experiments to simulate chance effects in three populations of different size, each starting with a gene frequency of 50 percent. The smallest population ($N = 25$ individuals) shows fixation of allele $A$ after 42 generations. The medium-size population ($N = 250$) shows less important fluctuations and has not reached fixation of either allele after 150 generations. Note that the frequency of allele $A$ did become greater than 80 percent shortly before generation 110 but that chance events in succeeding generations happened to carry the frequency back toward 50 percent. In the largest population ($N = 2500$), fluctuations are quite small, and fixation is very unlikely in any particular generation. However, fixation will eventually occur if the experiment is continued long enough. (From *Genetics, Evolution, and Man* by W. F. Bodmer and L. L. Cavalli-Sforza. W. H. Freeman and Company. Copyright © 1976.)

either typhoon or famine. Among the present population of 1600 that descended from these 30 founders, about 5 percent are homozygous for a rare allele (which we can call $a$) that causes a form of partial blindness known as achromatopsia. Converting this information into allelic frequences we find for the present population that

$$a/a = q^2 = 0.05$$
$$q = \sqrt{0.05} \cong 0.23$$

Among the 30 founders, on the other hand, it is likely that only one individual was heterozygous for the rare allele $a$, meaning that $q = \dfrac{1}{60} \cong 0.014$ in the original group. One is confronted, in other words, with an increase in the frequency of the $a$ allele from an estimated 1.4 percent to the present

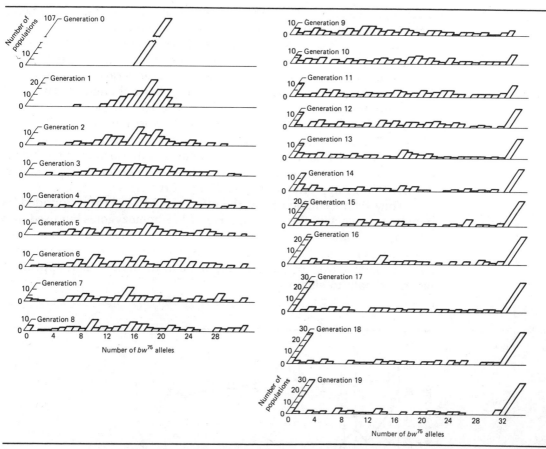

**FIGURE 27-17 Random genetic drift** in 107 actual populations of *Drosophila melanogaster*. Each of the initial 107 populations consisted of 16 *bw⁷⁵/bw* heterozygotes ($N = 16$; *bw* = *brown* eyes). From among the progeny in each generation, eight males and eight females were chosen at random to be the parents of the next generation. The abscissa of each histogram gives the number of *bw⁷⁵* alleles in the population, and the ordinate gives the corresponding number of populations. (Data from P. Buri, *Evolution* 10:367, 1956; plotted in D. Hartl, *Principles of Population Genetics*. Sunderland, MA: Sinauer, 1980, p. 146.)

23 percent, with no evidence of selection for the trait. The most plausible explanation, therefore, is to invoke random drift, although the possibility of linkage disequilibrium (*Section 27.2*) must always be borne in mind.

## 27.12 Drift Combined with Selection

We noted in Section 27.2 that fitness is commonly determined by a region of a chromosome or even a whole chromosome rather than by a single gene locus and that selection may act on these larger gene constellations. This effect is particularly dramatic in small populations, as seen in experiments that were performed by J. Powell and R. Richmond.

Powell and Richmond focused attention on the sex-linked *To* locus of *Drosophila paulistorum*, which codes for the enzyme tetrazolium oxidase. They established that wild populations of *D. paulistorum* in the Brazilian Andes carried alleles for fast (*F*) and slow (*S*) electrophoretic variants, and they found that neither allele was under noticeable selection pressure in a population-cage environment. Specifically, they demonstrated that if they "founded" a population cage with *F/F* homozygotes whose ancestors included a large number of female flies independently isolated in the Brazilian wild, then the frequency of the *F* allele after 45 generations (900 days) did not change appreciably (Figure 27-18, curve A).

They then examined what the effect would be if they "founded" a population cage with the same number of *F/F* homozygotes but used flies that had descended from only two independently isolated progenitors. As is evident in Figure 27-18, curve B, the frequency of *F* is now seen to fluctuate considerably from one generation to the next, and the allele appears to be on the path to elimination.

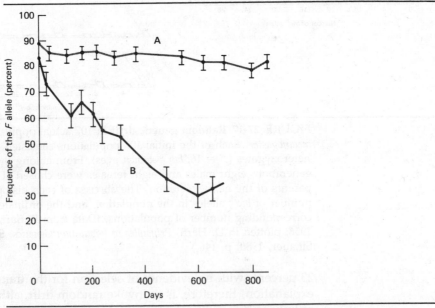

**FIGURE 27-18 Drift combined with selection.** Frequency of the *F* allele at the tetrazolium oxidase (*To*) locus in two laboratory populations of *D. paulistorum*, the "diverse" population (A) founded with over 100 different kinds of X chromosomes carrying *F*, the "limited" population (B) founded with only about 6 different X chromosomes carrying *F*. The remaining flies in the population were *S/S* homozygotes. (From J. R. Powell and R. C. Richmond. *Proc. Natl. Acad. Sci. U.S.* **71**:1663–1665, 1974.)

To best appreciate these results, visualize a collection of *F*-bearing X-chromosomes from various *D. paulistorum* flies isolated in the wild. One X-chromosome might carry alleles $a_1\ b_2\ c_1\ F\ d_2$ . . . , another might carry $a_2\ b_2\ c_1\ F\ d_1$ . . . and, in general, most of the chromosomes would be different from one another. If we now imagine that some of the gene loci closely linked to *To* carry alleles with positive or negative effects on survival in population cages, the Powell-Richardson experiments are readily explained. In the case of cage A, where many different X chromosomes "founded" the population, any deleterious combination of linked alleles would presumably be present in a relatively small proportion of the total chromosomes; moreover, X chromosomes having a positive effect on survival might also be present so that a canceling-out effect would take place. The net outcome, therefore, would be the maintenance of the *F* allele at its introduced frequency (*Figure 27-17*). In the case of cage B, on the other hand, where only a few X chromosomes are chosen to "found" the colony, one of these is likely to carry a combination of linked genes that affects population-cage fitness. Because this chromosome represents a very large proportion of the total, a marked effect on the frequency of *F* will accompany selection at the linked loci. Fitness modifier genes (*Section 27-2*) in this and other chromosomes may also be disproportionately represented.

In summary, therefore, we can say that random drift and selection pressure allow a "splinter" population to acquire very different allelic frequencies from its parent Mendelian population within a very few generations.

As noted in Section 26.8, island species are frequently found to have singularly bizarre aspects to their phenotypes, as though their ancestors were indeed forced to come up with makeshift but viable combinations of traits within short periods of time. This feature of island populations is of major historical importance, for it was the singularity of the species of animals (and particularly finches) on the Galapagos Islands that attracted the attention of Charles Darwin.

# The Contributions of Selection and Drift to Polymorphism

We can now return to the feature of natural populations that first attracted our attention, namely, their genetic diversity (*Sections 26.1 and 26.4*), and consider the factors that contribute to the maintenance of this polymorphism.

There is no question that selection can and does promote polymorphism: the *Adh* polymorphism in *Drosophila*, the amylase polymorphism in isopods, and the *Hb^s* polymorphism in humans are but three of many examples that can be given. In the past decade, however, a controversy has arisen between those who argue that virtually *all* polymorphisms are maintained through selection and those who argue that some or many polymorphisms are virtually "neutral" to selection and are established in a population by genetic drift.

## 27.13 The Neutral Mutation—Random Drift Hypothesis

The idea of drifting neutral mutations, which was first suggested by S. Wright and has been extensively developed by M. Kimura and his associates, is now known as the **Neutral Mutation—Random Drift Hypothesis,** or the **Neutral Theory** for short. The hypothesis makes two assumptions. The first is that structural genes can undergo effectively **neutral mutations** that confer no selective advantage or disadvantage to their bearers. The idea is that one acidic amino acid might replace another acidic amino acid, or a substitution might occur in an "unimportant" region of the protein, with the result that the emergent mutant protein is identical to the original in all functional respects but may differ, for example, in its electrophoretic mobility. The mutant gene is often referred to as a **neutral allele,** which is, of course, a relative term: it indicates that a given allele is selectively neutral in comparison with other alleles that have positive or negative selective value.

The second assumption of the Neutral Theory states that neutral alleles will drift in the gene pool. Thus, if a neutral mutation arises in a woman's germ cell and this germ cell gives rise to a female child, the probability is 0.5 that the mutant allele will be transmitted to a grandchild and 0.5 that it will not. If it is not, then $q$ becomes equal to zero and the allele is lost. If the allele reaches the third generation (perhaps via two or three individuals if the family is large), the chances are a little better that it will appear somewhere in the fourth generation, but the probability still remains great that its frequency will "backslide" and that $q$ will finally fall to zero. In other words, as long as the allele provides its bearer with no selective advantages or disadvantages, its fate in the gene pool is indeterminate. It may disappear as soon as it arises (elimination) or it may proceed toward $q = 1$ (fixation) in a seemingly haphazard pattern that is aptly described as a **random walk:** the course taken by the allele depends entirely on how the gamete pool happens to be sampled. This description is, of course, very similar to that given for the random drift of alleles in small populations (*Section 27.11*).

The potential importance of drift in large populations is enhanced when we recall that effective population sizes are usually much smaller than the measured size. Moreover, fluctuations in climate, predation, and similar pressures frequently cause small populations to become virtually extinct, whereupon their habitats are typically recolonized by a few founders who may import a "drifting" allele and enhance its frequency in the new subpopulation.

The Neutral Theory makes a number of predictions about the course of events leading to the ultimate fixation of a neutral allele. One is that the rate of its fixation should be approximately equal to the rate ($\mu$) at which the allele arises by mutation—that is, that the average time interval expected to elapse between the origin of a successful neutral mutation and its ultimate fixation is $1/\mu$ generations. With an average mutation rate of $1 \times 10^{-5}$ at most loci (*Table 26-2*), this means that neutral mutations should enjoy long intervals of random drift in a population and could therefore, in theory, be strong contributors to heterozygosity.

## 27.14  Do Neutral Mutations Occur?

In evaluating the contribution of neutral mutations, the first question is whether they occur at all. We have already cited several examples where two isoelectric variants of an enzyme are in fact different in substrate specificity (amylase) or effectiveness (alcohol dehydrogenase). The question is, are any polymorphisms contributed by neutral alleles?

The question is probably impossible to answer, since it is impossible to "prove" the neutrality of any one allele. For example, if two electrophoretic variants display no differences in enzymatic activity under laboratory conditions, it can simply be argued that the experiment did not simulate the appropriate *in vivo* conditions for displaying such differences. Until recently, it was believed that at least one type of mutation would surely be neutral, namely, a "synonymous" mutation in which one codon is replaced by a second dictating the same amino acid (*Section 11.10*). It is now clear, however, that most species exhibit codon preferences (*Section 11.11*), utilizing certain synonymous codons much more often than others. Thus for an insect that uses GAA much more often that GAG to specify glutamic acid and has a parallel imbalance in its levels of isoaccepting tRNAs (*Section 11.2*), a GAA $\longrightarrow$ GAG mutation might possibly modify the rate at which a critical mRNA product were translated into protein.

The "neutralist" argument, therefore, is usually made from hindsight. For example, genetic information for the synthesis of ascorbic acid (vitamin C) is clearly essential as long as the compound is limiting in the diet. When it is abundant, however, an ascorbic acid$^-$ mutation would become effectively neutral and would be expected to drift toward elimination or fixation. The fact that two distantly related mammals, guinea pigs and humans, are "fixed" in the ascorbic acid$^-$ state, whereas their close relatives are not, is cited as evidence for the drift of a neutral mutation. Just as it is impossible to prove that a mutation is neutral, however, so does it seem impossible to prove that the ascorbic acid$^+$ state was lost in pre-humans and pre–guinea pigs by some unknown form of selection. In other words, the neutralist-selectionist controversy is unlikely to be settled by attempts to demonstrate or refute the existence of neutral alleles.

## 27.15  Are Patterns of Allelism Present in Parallel Populations?

A second test of the Neutral Theory is based on the following reasoning. If polymorphism arises because various neutral alleles are drifting toward elimination or fixation in a random manner, there should be a randomness in the allelism observed from one population to the next; thus in one population, allele $a_1$ might be enjoying considerable "luck" and might be present at relatively high frequencies, whereas in another population it might be proceeding toward elimination and be present in low frequencies. If, on the other hand, polymorphic alleles are not neutral and selection is instead maintaining the polymorphism, allelic frequencies should be similar from one population to the next (assuming, of course, that the populations are experiencing similar kinds of selective pressures).

Tests of this thesis have been made in the field, and opposite results have been obtained. On the one hand, F. Ayala and his associates found patterns of allelism that support the selectionist position. They examined ten natural populations of *D. willistonii* taken from diverse locations in the western hemisphere. They looked for electrophoretic variants of 28 different kinds of enzymes and defined a locus as polymorphic when the frequency of its most common allele was no greater than 0.95. From this they estimated that about 58 percent of the *D. willistonii* enzyme loci were polymorphic, a result we have by now come to expect. What was of particular interest was the *pattern* of the allelic frequencies for these loci. Ayala found that when an allele was prevalent in one population it was, with high correlation, prevalent in the other nine populations as well; similarly, a rare allele was rare in all ten populations.

A difficulty with the Ayala study is that low-to-moderate gene flow may still be occurring between the geographically separated populations of *D. willistonii* and/or that not enough time has elapsed for allelic frequencies to change by drift since the time the *Drosophila* populations became geographically separated. A similar study by N. Aspinwall was designed to circumvent these problems, and the patterns of allelism found in his study support the neutralist position.

Aspinwall studied Pacific salmon that have a unique life cycle: at the exact age of two years (± ten days), a salmon leaves the ocean, swims up the stream in which it was born, spawns, and then dies. Therefore, a salmon born in 1982 will return to spawn in its home stream in 1984 only, and its descendants will return to spawn in every future *even*-numbered year. Similarly, a parallel geneology will spawn in this stream during *odd*-numbered years. Extensive studies have demonstrated no gene flow between odd- and even-year populations. Moreover, for any given stream, it can be assumed that odd- and even-year salmon experience nearly identical environmental circumstances and thus selection pressures. When patterns of protein polymorphisms were examined at three gene loci, however, significant differences were found in all cases between odd- and even-year fish. In other words, near-identical selection pressures have *not* produced correlated patterns in allozyme patterns, a result consistent with the tenets of the Neutral Theory.

Most nonpartisan followers of the neutralist-selectionist debate are convinced that each mechanism will turn out to contribute to some, but not all, of the polymorphisms observed in nature. Meanwhile, the debate has already served a vital function in population genetics, having stimulated both vigorous experimentation and theoretical formulations.

(a)    (b)

(a) Motoo Kimura and (b) Tomoko Ohta (National Institute of Genetics, Misima, Japan) have made important theoretical contributions to the development of the Neutral Mutation-Random Drift Hypothesis.

# Questions and Problems

1. A population of mice is "founded" by two mice, one $D/D$ (normal coat color) and the other $d/d$ (*dilute* coat color). Their offspring and succeeding offspring are allowed to mate randomly for a number of generations until a large population of mice is established. Complete selection against *dilute* mice is now imposed. What will be the change in gene frequencies after ten generations under the selection program?

2. Because of antipollution laws, trees in certain industrial areas of England have lost their sooty coating and black moths are now experiencing predation. Would you expect black moths to disappear at the same rate, more rapidly, or less rapidly than did the white moths in the nineteenth century (assume that birds can recognize black moths against a white background as readily as white moths against a black background)? Explain.

3. A population of 5000 persons contains 1500 that are of blood type M, 3000 of blood type MN, and 500 of blood type N. A representative sample of 30 persons from this population moves to join a second population whose size is 70 persons prior to the migration and where the frequency of $L^M = 0.8$ and $L^N = 0.2$ prior to the migration. What is the change in frequency of the two alleles in the expanded population after one generation?

4. Use the data in Table 26-1 to calculate the proportion of black African and black Georgian individuals born with sickle-cell anemia and with sickle-cell trait.

5. The allele $g$ exhibits a stable frequency of 15 percent in a mouse population, even though the newborn survival of $g/g$ homozygotes is half that of $G/g$ heterozygotes. Assuming the allele to be maintained by stabilizing selection, calculate the relative fitness of $g/g$, $G/g$, and $G/G$ mice. What is the selection coefficient for each genotype?

6. Mice of the $H/h$ genotype have twice the litter size of $H/H$ mice and four times the litter size of $h/h$ mice. What is the predicted equilibrium frequency of the $h$ gene, assuming stabilizing selection?

7. Derive Equation 9, using the derivation of Equation 5 as a model.

8. The frequency of amylase electrophoretic variants $A_1$ and $A_2$ for isopods living near beech and willow trees (*Section 27.6*) is as follows (from A. Christensen, *Hereditas* **87**:21, 1977).

| Collection Site | $A_1A_1$ | $A_1A_2$ | $A_2A_2$ |
|---|---|---|---|
| Beech | 165 | 129 | 72 |
| Willow | 186 | 154 | 143 |

(a) What proportion of the individuals possess the alleles for each allozyme in the two subpopulations?

(b) Are the two subpopulations, considered separately, Mendelian populations? Explain.

(c) Is the population as a whole a Mendelian population? Explain.

9. The frequency of alcohol dehydrogenase electrophoretic variants F (Fast) and S (Slow) for *Drosophila* living in wine cellars and compost heaps (*Section 27.6*) is as follows (from A. Briscoe et al., *Nature* **255**:148, 1975).

| Collection Site | F/F | F/S | S/S |
|---|---|---|---|
| Wine cellar | 177 | 15 | 0 |
| Compost heap | 152 | 37 | 3 |

(a) What proportion of the individuals possess the alleles for each allozyme in the two subpopulations?

(b) Flies are fed laboratory food mixed with 12 percent ethanol, and the allozyme composition of the flies that die quickly is determined. What result would you predict?

(c) What is the predicted result if wine cellar flies are raised in the laboratory on normal food?

(d) How do you suspect the $Adh^s$ allele is maintained at all in wine cellar populations?

10. Ayala and associates defined a polymorphic locus (*Section 27.15*) differently than we defined it in Chapter 26 (*Section 26.4*).

(a) What are the two definitions?

(b) Would the Ayala definition result in more or fewer loci being defined as polymorphic? Explain.

11. The grandparents of children homozygous for the recessive gene responsible for cystic fibrosis of the pancreas were found to have an average of 4.34 off-spring, whereas control couples had an average of 3.43, a significant difference.
    (a) How might this statistic explain the existent polymorphism for the allele?
    (b) Why were grandparents and not parents chosen for this study?

12. The maintenance of the recessive allele for Tay-Sachs disease in Jewish populations (*Problem 26-15*) has been speculated to be due to resistance to some childhood disease afforded to heterozygotes. With 0.016 percent of the population born with the disease, what would the selective advantage of heterozygosity have to be to maintain the allele at its present frequency?

13. The average rate of weight gain in a population of cattle is 2.1 lb/day. Individuals selected from this population to be parents of the next generation had an average weight gain of 3.2 lb/day.
    (a) Given a heritability of weight gain of 0.65, what is the expected average daily gain of the cattle in the next generation?
    (b) You find the average daily gain of the next generation cattle to be 3 lb/day. Calculate a revised heritability estimate.

14. In a self-fertilizing population of beans, the weight of seeds in a representative sampling was found to be (in centigrams)

| | | | | | | | | |
|----|----|----|----|----|----|----|----|----|
| 19 | 31 | 18 | 24 | 27 | 28 | 25 | 30 | 29 |
| 22 | 29 | 26 | 23 | 20 | 24 | 21 | 25 | 29 |

   (a) Calculate the mean and standard deviation for bean weight in this sample (*Section 21.13*).
   (b) What is the environmental variance ($V_E$) (*Section 21.14*)?
   (c) What is the heritability of bean weight in this population?
   (d) If the average bean weight of plants selected from this population is 300 mg, what is the predicted average bean weight of the next generation?

15. The relative fitness of a phenotype is often dependent on how frequently it occurs in a population. An example involves **mimicry in butterflies.** The palatable viceroy butterfly has evolved coloration patterns that resemble the distasteful monarch butterfly, but as the frequency of these mimics in a population increases, the relative fitness of this phenotype decreases. Explain.

# Population Genetics III: Speciation and Molecular Evolution

**28**

A. Introduction

B. Speciation
  28.1 Reproductive Isolation
      Mechanisms
  28.2 Hybridization and
      Allopolyploidy
  28.3 Allopatric Speciation
      Sexual selection by female
        mating preference
  28.4 Speciation via the Founder
      Principle

C. Molecular Evolution
  28.5 Karyotype Evolution
  28.6 Polypeptide Evolution
  28.7 Gene Evolution
  28.8 Evolution of the Primates

D. Questions and Problems

## Introduction

The previous chapter considered how selection and drift allow populations of organisms to adapt to their environment. This process is frequently termed **microevolution** and is contrasted with **macroevolution,** which leads to the formation of new species (and, then, to new genera, families, and so on). As Dobzhansky first noted, the phenomena in microevolution are repeatable and reversible, whereas those in macroevolution are creative and unique. In the first part of this chapter, we will consider these unique events and ask to what extent they are dependent on microevolutionary underpinnings.

The second part of this chapter considers macroevolution on a grand scale and at a molecular level. We describe the approaches being taken to establish evolutionary relationships between modern organisms by analyzing their genetic material and their polypeptide gene products, a field that has come to be called **molecular evolution.** Although the field is young,

858

many investigators believe that it will be able to determine evolutionary relationships and the times of species divergence more accurately than any other discipline.

# Speciation

## 28.1 Reproductive Isolation Mechanisms

If two individuals of opposite sex are brought together and are able to mate and produce fertile offspring, they are said to belong to the same **species;** if they cannot, then they belong to two separate species. Although this definition clearly cannot be used for asexual organisms such as bacteria (whose species status is commonly ascertained by immunological criteria), it holds for most eukaryotes of interest. The critical event in species divergence, therefore, is the establishment of **reproductive isolation mechanisms.**

Table 28-1 summarizes the major reproductive isolation mechanisms that have been recognized. They are subdivided into two categories, **prezygotic** and **postzygotic,** and their description in the table is sufficiently complete that it need not be repeated in the text.

Habitat and mechanical isolation prove to be the most common prezygotic mechanisms in plants. For example, two species of oak have evolved with different soil preferences, the scarlet oak found in swampy terrains and the black oak in drier soils, and many plant species have evolved critical differences in flower color or anatomy that either attract

---

**TABLE 28-1   Summary of the Most Important Isolating Mechanisms That Separate Species of Organisms**

A. *Prezygotic mechanisms.* Prevent fertilization and zygote formation.
  1. *Habitat.* The populations live in the same regions but occupy different habitats.
  2. *Seasonal or temporal.* The populations exist in the same regions but are sexually mature at different times.
  3. *Ethological* (only in animals). The populations are isolated by different and incompatible behavior before mating.
  4. *Mechanical.* Cross pollination is prevented or restricted by differences in structure of reproductive structures (genitalia in animals, flowers in plants).

B. *Postzygotic mechanisms.* Fertilization takes place and hybrid zygotes are formed, but these are inviable or give rise to weak or sterile hybrids.
  1. *Hybrid inviability or weakness.*
  2. *Developmental hybrid sterility.* Hybrids are sterile because gonads develop abnormally, or meiosis breaks down before it is completed.
  3. *Segregational hybrid sterility.* Hybrids are sterile because of abnormal segregation to the gametes of whole chromosomes, chromosome segments, or combinations of genes.
  4. *$F_2$ Breakdown.* $F_1$ hybrids are normal, vigorous, and fertile, but $F_2$ contains many weak or sterile individuals.

---

(From G. L. Stebbins, *Processes of Organic Evolution.* Englewood Cliffs, NJ: Prentice-Hall, 1966, p. 97.)

different groups of insect pollinators or reduce the chances of cross-pollination should the same insect visit two different species.

In animals the most important prezygotic mechanisms prove to involve courtship behavior. These are very familiar to us in the distinctive plumage of birds, the flashing of fireflies, the calling of crickets, and the songs of frogs. In *Drosophila*, males of each species perform a distinctive courtship "dance," which includes orienting themselves in a particular position with respect to the female, extending their wings, and licking the female with their tongues. Figure 28-1 shows the sequence of these activities for *D. melanogaster* (a) and *D. simulans* (b) males; the two patterns are seen to be very different. If a *D. melanogaster* female witnesses the dance of a *D. simulans* male, she does not become receptive to copulation, and vice versa. Female receptivity, it turns out, requires the removal of inhibitions: a female fly has repressed sexual interest unless she sees the appropriate dance.

Postzygotic mechanisms all have a similar underlying basis that can be called **genic disharmony:** the embryo fails to develop in a coordinated fashion or the adult fails to undergo normal gametogenesis or the F2 is developmentally abnormal. All of these events can be assumed to result from some form of incompatibility between the two genomes such that they can no longer cooperate to direct normal ontogenesis.

From a genetic viewpoint, the critical question is the following: how do these reproductive isolation mechanisms originate? Do they require large or small changes in the genomes of a subspecies, and how are these changes reinforced? Such questions have no single answer; it is likely that

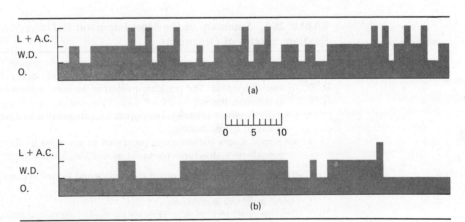

FIGURE 28-1 **Typical courtships** of *D. melanogaster* (a) and *D. simulans* (b). The sequence reads from left to right, and the scale shows time units of 1½ seconds. The height of shading indicates the courtship element being performed; the 3 levels are marked on the right (L + A.C. = licking plus attempted copulation. W.D. = wing display, O = orientation). (From A. Manning, *Behavior* **15:**123, 1959, Figure 1.)

each species divergence involved different combinations of genetic change, environmental selection, and random drift. We can, however, identify some common patterns.

## 28.2 Hybridization and Allopolyploidy

Conceptually, the most straightforward mode of speciation is **hybridization:** two distinct species manage to subvert all prezygotic and postzygotic isolation mechanisms and form a fertile hybrid species. This pattern is found primarily in monoecious plants, the obvious reason being that such a hybrid can fertilize itself and thereby avoid the necessity of finding a like-constituted mate. In addition, certain hybrid plants have evolved ways of producing seeds by asexual means.

Natural hybrids usually arise in disturbed environments: when two related species that occupy distinct ecological niches find these niches disrupted by geological or human forces, the reproductive barriers isolating them are often removed and hybrids result. If the hybrid is better adapted to the newly created niches than either parent, its survival is possible. Survival is by no means assured, however, unless the hybrid can maintain its fertility and retain its adaptive features through successive generations of breeding.

A major obstacle to normal fertility in hybrids is usually meiosis, since the two sets of parental genomes are insufficiently similar to pair and assort normally during gametogenesis. A widespread "solution" to this problem, particularly common in plants, is that the hybrid becomes tetraploid: each parental genome doubles itself so that proper pairing of homologues can occur and viable 2C gametes can form (*review Section 5.16*). Such an organism is designated as **allopolyploid** to denote the disparate origin of its two chromosome complements. An allopolyploid usually undergoes a series of back crosses with one or both of its parental diploid ancestors that occupy the same habitat. As a result, its descendents are most accurately called **segmental allopolyploids,** having some of their homologues derived from one parent and some from the other, and it is often difficult to recognize the ancestral species of an established hybrid by examining its karyotype. Each segmental allopolyploid line will, of course, possess unique fitness attributes upon which selection can act, and in the "disturbed environments" in which these events most often take place, the new niches can be rapidly filled by new species.

Speciation by interspecific hybridization and segmental allopolyploidization is believed to have played an important role in the natural history of plants, although its net result is usually the creation of species that are intermediate between the two parents rather than a species with novel traits. Plant breeders have attempted to create allopolyploids that combine the desirable traits of two species, an example being triticale, a hybrid between wheat and rye. These efforts occasionally have amusing consequences: thus the rado-cabbage was bred in hopes of creating a plant with the head of a cabbage and the root of a radish; sure enough, the hybrid proved to combine the root of a cabbage with the leaves of a radish.

## 28.3  Allopatric Speciation

Although hybridization is an important speciation mechanism, it contributes little insight into the question of how two species evolve from a single species. We therefore consider two models that address this question directly.

The first, known as **allopatric** or **geographic speciation,** proposes that a large panmyctic population becomes subdivided into two subpopulations by some geographic barrier; in the case of the well-studied South American species *D. willistonii,* for example, the east coast and west coast flies are separated by the Andes Mountains. With gene flow effectively abolished, each population experiences independent microevolutionary pressures and accumulates sufficient allelic differences that, when individuals are brought together in the laboratory, they copulate readily but exhibit hybrid sterility. They are therefore said to constitute two **subspecies,** which, in the *Drosophila* example, are termed *D. willistoni willistonii* and *D. willistonii quechua.*

The theory of allopatric speciation goes on to propose two alternative courses for the next stage. In some cases, it is suggested, the two subspecies remain separated sufficiently long that they also evolve different prezygotic reproductive isolation mechanisms, just by chance. In other cases, the geographic barrier may be removed or surmounted so that interbreeding is possible. In this case, the occurrence of hybrid sterility imposes a major selective pressure on both populations, and individuals who develop a mechanism to mate exclusively with their own subspecies have a greater fitness than those who squander their gametes in copulations with the other subspecies. In other words, there arises intense selection for alleles that promote prezygotic sexual isolation. If such mechanisms are acquired, then the two subspecies are converted into two **sibling species,** which, although morphologically very similar to one another, are reproductively isolated at both the prezygotic and postzygotic levels. The stage is now set for the two species to coexist in the same environment (if separate niches are found) and to continue to diversify and speciate further.

Support for this model is given by the data in Table 28-2, in which the frequencies of electrophoretic variants, and hence genetic polymorphisms (*Section 26.4*), were analyzed for various groups of *Drosophila.* The data are expressed by the mathematical expression $\bar{D}$, for **Average Genetic Distance,** which estimates the proportion of genes that are different in the groups under comparison. It is seen that individuals sampled from local populations of *Drosophila* are nearly identical in their overall allelic composition ($\bar{D} = 0.03$), whereas they are far more dissimilar ($\bar{D} = 0.23$) when the two subspecies of *D. willistonii* are compared. Somewhat unexpectedly, a similar level of genetic relatedness ($\bar{D} = 0.226$) is found between two

**TABLE 28-2 Average Genetic Distance,** $\bar{D}$, Between Taxa of Various Levels of Evolutionary Divergence in *Drosophila*

| Taxonomic level | $\bar{D}$ |
|---|---|
| Local populations | $0.031 \pm 0.007$ |
| Subspecies | $0.230 \pm 0.016$ |
| Semispecies | $0.226 \pm 0.033$ |
| Sibling species | $0.581 \pm 0.039$ |
| Nonsibling species | $1.056 \pm 0.068$ |

(From F. J. Ayala, *Evol. Biol.* **8**:1, 1975.)

**semispecies** of *D. paulistorum*, which not only display hybrid sterility but also are beginning to develop prezygotic isolation mechanisms. To explain the similar levels of genetic relatedness found in both the subspecies and the semispecies, it is postulated that only a few gene loci need to change in order to affect such behavioral traits as courtship and copulating behavior, whereas a large number of genetic changes are needed to create the genic disharmony that causes hybrid sterility; therefore, **subspecies and semispecies would not be expected to differ significantly in overall polymorphism.** The final entries in Table 28-2 document that further increases in genetic distance are found when sibling (morphologically similar) and nonsibling (morphologically distinct) species are compared.

**Dynamics of Sexual Selection by Female Mating Preference** An interesting phenomenon often accompanies the evolution of prezygotic reproductive isolation mechanisms. Let us suppose that two subspecies of birds are brought together and selection pressures strongly disfavor interbreeding because of the resultant hybrid sterility. Now suppose that a male bird in one subspecies comes to inherit a group of genes that endow him with a longer tail, and a female of his subspecies possesses the ability to discriminate, and prefer, long-tailed males. The lineage descended from this pair will enjoy greater relative fitness because the females will shun the short-tailed males from the other sympatric subspecies. Because of its strong sexual isolation, moreover, the lineage should rapidly acquire the status of a species. There is, however, a potential problem with this system: if a bird of this species acquires alleles for an even *longer* tail, females "programmed" to prefer long tails may choose this male over his competitors. The result may be, in R. A. Fisher's words, a "runaway process": male tail length and female preference for long tails may form a self-reinforcing loop that results in the evolution of male birds with such long tails that they eventually can no longer fly. Obviously, there will at some point be counterselection against male secondary sexual characteristics that are so extreme that they become deleterious for survival, but the phenomenon may account for the extravagant plumage in peacocks and the enormous horns of the elk.

## 28.4 Speciation via the Founder Principle

An alternative to the allopatric mode of speciation is the possibility of speciation by the **founder principle.** In our previous considerations of the founder effect (*Sections 26.8 and 27.10*), we stressed the possibility that small populations derived from a small inbreeding group tend to become homozygous and "bizarre." It is now important to stress another possibility, which is best visualized by a simplified example. Suppose there exists in the original population a multilocus complex concerned with a trait such as mating behavior, and suppose the expression of this complex is strongly influenced by additional loci that are epistatic to it. If the founder population carries with it the multilocus complex but not its epistatic "modifiers," the multilocus complex will experience a saltatory shift in its genetic environment and may, in even one generation, direct a novel pattern of mating behavior that establishes new prezygotic and postzygotic isolation mechanisms. A. Templeton has obtained evidence for such rapid speciation events in *Drosophila* under laboratory conditions, and the mechanism represents an attractive alternative to allopatric speciation, particularly when new niches are opened up to colonization by small groups.

# Molecular Evolution

How do genes and genomes change during the course of evolution? Although this question is the central theme of the remaining sections, it is hardly a new question for readers of this text, for it has been raised repeatedly in preceding chapters. Thus we have devoted major attention to comparing the genetic systems of prokaryotes and eukaryotes; we have pondered the C-value paradox (*Section 18.6*), the maintenance of sequence homogeneity in histone genes (*Section 10.11*), and so on. Therefore, although we will be considering new material, we will also be recapitulating relevant concepts from earlier chapters (*see also* "molecular evolution" entries in the index).

## 28.5 Karyotype Evolution

A karyotype (*Section 2.10*) is diagnostic for a species, and much can often be learned about the evolution of genomes by comparing the karyotypes of related species.

That chromosomes can undergo translocation, inversion, and duplication has been documented in Sections 7.7 to 7.9, and we have already noted the importance to evolution of "supergenes" maintained by inver-

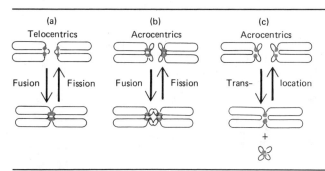

FIGURE 28-2 **Mechanisms for creating new chromosomes.** (a) Reversible Robertsonian fusion of two telocentrics (chromosomes with terminal centromeres) to produce a metacentric. (b) Reversible Robertsonian fusion of two acrocentrics to produce a metadicentric (such chromosomes can become mitotically stable if one centromere becomes inactive or latent). (c) Reciprocal translocation between acrocentrics produces a metacentric and a small centric fragment, which, if subsequently lost, makes this process irreversible. (From G. P. Holmquist and B. Dancis, *Proc. Natl. Acad. Sci. U.S.* **76**:4566, 1979.)

sions (*Section 27.4*) and of gene duplications (*Sections 7.7 and 19.7*) as the source of novel genes. Chromosomes can also undergo what are known as **Robertsonian rearrangements:** as diagrammed in Figure 28-2, a metacentric can undergo fission to generate two telocentrics and, conversely, two telocentrics or acrocentrics can fuse to form a metacentric.

The extent to which chromosomal rearrangements have occurred during evolution can be probed both by direct visualization of banded karyotypes (*Section 2.11*) and by *in situ* hybridization (*Section 4.10*). Figure 28-3 compares the human (H) karyotype with the chromosomes of our primate relatives. The most dramatic difference is that the human chromosome 2 appears to have resulted from the fusion of two small acrocentric chromosomes found in the other hominoids (Figure 28-3, dashed line), reducing the human chromosome number from 24 to 23. In addition, close inspection reveals that at least eight inversions distinguish the human and chimpanzee karyotypes. On the other hand, almost every band present in the human chromosomes is found to have a counterpart in the great-ape chromosomes. Moreover, *in situ* hybridization studies show that there has been strong conservation of gene loci during ape and human evolution: the histone genes are all localized on homologous regions of the counterparts of human chromosome 7, whereas the major histocompatibility genes reside on counterparts of human chromosome 6. The most dramatic example of karyotype conservation is given by the X chromosome; as we noted in Section 6.18, X linkage of a number of loci has been preserved throughout the mammalian lineage and extends as well to the lower vertebrates.

Constitutive heterochromatin is a particularly labile element of the eukaryotic karyotype. This was demonstrated in Figure 3-15, where we showed that the telomeric heterochromatin of two closely related species of

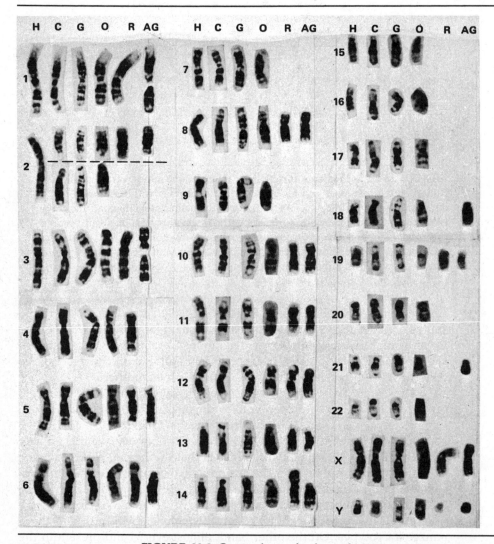

**FIGURE 28-3 Comparison of primate karyotypes.** Giemsa-banded chromosomes of humans and other primates, arranged to show homology with the human karyotype. The human chromosomes are on the left (H) followed by those of chimpanzee (C), gorilla (G), orangutan (O), Rhesus monkey (R), and African Green monkey (AG), respectively. Not all of the monkey chromosomes are displayed. (Courtesy of Dr. J. J. Garver, Instituut voor Anthropogenetica, Leiden.)

the genus *Secale* (rye) had undergone extensive amplification. Since telomeric and/or centromeric DNA sequences are frequently homologous throughout the genome of a species (*recall Figure 4-7*), it appears likely that many chromosome rearrangements are initiated and mediated by homologous pairing between these sequences.

We noted in Section 25.16 that dramatic changes in gene expression often accompany the translocation of chromosomal elements from their "natural" site to a new site, most often the suppression of expression of contiguous gene loci. It is therefore attractive to postulate that chromosomal rearrangements may have produced some of the dramatic saltatory changes in gene expression that are thought to accompany certain speciation events (*Section 28.4*).

## 28.6  Polypeptide Evolution

A more finely tuned analysis of the genetic events that have accompanied speciation can be made by a comparison of gene products. This is most often done for polypeptides, although the ribosomal RNAs have also provided important information.

Numerous comparisons have been made between the amino acid sequences of polypeptides that are shared by diverse taxonomic groups. Most extensively studied have been cytochrome c (present in all eukaryotes) and hemoglobin (present in all vertebrates and related to the more primitive globin polypeptides). If the protein is polymorphic within a species, the most common version is singled out; its amino acid sequence is then compared with the amino acid sequence of the protein in another species, and the minimal number of nucleotide replacements required to shift from one amino acid sequence to the other is calculated.

Figure 28-4 shows a resultant "genealogy" of globin polypeptides, where the lengths in each linking line represent the numbers of nucleotide replacements deduced as having occurred between adjacent ancestor and descendent sequences. The line from a primitive globin to the vertebrate β-α-hemoglobin ancestor is depicted as having given off, along the way, annelid, insect, mollusc, lamprey, and mammalian myoglobin, and the entire genealogy entails 1629 nucleotide replacements. Shown on the ordinate is a time scale, in millions of years, based on paleontological versions of metazoan evolution, and it is seen that the "protein clock" and the "paleontological clock" agree closely.

From such genealogies it becomes possible to estimate nucleotide replacement rates for hemoglobin genes during various evolutionary periods, and here interesting patterns emerge. A very rapid rate of nucleotide replacement is calculated to have occurred during the period when the primitive globin gene, whose polypeptide product functions as a monomer, evolved into a gene whose polypeptides associate as a tetramer capable of cooperative functions. Moreover, the amino acids affected during this "burst" of replacements appear to have been those involved with heme contacts and the interchain cooperativity that facilitates oxygen binding.

A second period of rapid nucleotide replacement is also deduced from the hemoglobin genealogies. This appears to have occurred about 500 million years ago, at the time of the postulated α-β-gene duplication, and

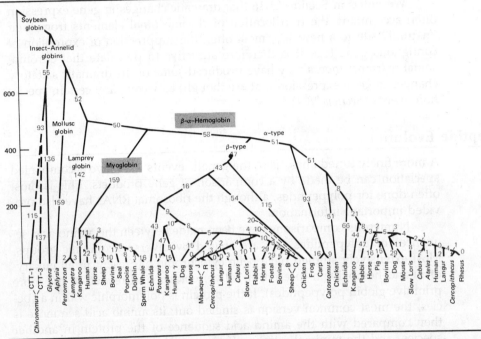

**FIGURE 28-4  Globin genealogical tree.** Link lengths are the numbers of nucleotide replacements between adjacent ancestor and descendent sequences. The ordinate is a time scale in Myr based on paleontological views concerning the ancestral separations of the organisms from which the globins came. (From M. Goodman, G. W. Moore, and G. Matsuda. *Nature* **253**:603, 1975.)

again a functional interpretation can be given. The β₄-tetramer is adaptively inferior to an α₂β₂-tetramer, and it is proposed that following the gene duplication, the newly formed α-gene accumulated a sufficient number of nucleotide substitutions to generate a novel α-polypeptide that could combine with the β-chain.

In summary, therefore, the two peak periods for nucleotide substitutions in hemoglobin genes can be correlated with periods in which natural selection was presumably intensively operating to produce optimal versions of α- and β-chains. During the 300 million years since the α-β divergence, nucleotide replacement rates in the two genes have fallen from a peak of 109 to an average of 15 per 100 codons per $10^8$ years, and these more "conservative" replacements are claimed to have created α- and β-chains that are increasingly better adapted in the transport and release of molecular oxygen. Thus molecular evolutionists at present favor the notion that natural selection has strongly guided the evolution of this particular protein.

The polypeptide studies have two major disadvantages. The first is that amino acid sequencing is time consuming (although improvements are rapidly being made). The second is that they are unable to detect synonymous codon changes (*Section 11.10*) and may therefore underestimate the extent of nucleotide divergence between genes (although for many purposes these mutations can be "averaged out" without compromising the overall significance of the analysis). Therefore, the method of choice has become the analysis of nucleotide sequences. To illustrate the difference in resolution, we can consider the histone protein H4. Histone H4 is the most highly conserved protein known: in calf, pea, pig, and rat, for example, the only amino acid changes are found in the pea, where an isoleucine replaces valine at position 60 and an arginine replaces lysine at position 77. When, however, the H4 DNA from two genera of sea urchin, *Lytechinus* and *Strongylocentrotus,* is compared, a number of nucleotide substitutions are found in the third position of various codons, substitutions that convert one codon into a synonymous codon. Should one now wish to learn whether a third species of sea urchin is more closely related in its histone genes to *Lytechinus* or to *Strongylocentrotus,* it is only necessary to determine its H4 nucleotide sequence and compare it with the published sequence. Such a one-gene comparison, we should stress, is hardly sufficient to establish evolutionary relationships between species, but comparisons of many genes will often provide a sufficiently consistent pattern of sequence relationships to draw strong evolutionary conclusions.

## 28.7 Gene Evolution

The comparison of nucleotide sequences between homologous genes of different species permits evaluation of sequence divergence, not only in the coding regions but also in the introns and in the flanking sequences. It is also possible, as noted previously, to distinguish replacements leading to amino acid substitutions from those leading to synonymous substitutions. Therefore, the evolutionary history of a gene can be studied in exquisite detail.

T. Miyata and his colleagues have recently examined the data on 50 published nucleotide sequences in which two or more homologous genes were compared. They calculated the number of changes per nucleotide site and reached the following conclusions:

1. The rate of nucleotide substitution leading to synonymous codons is in almost all cases considerably greater than the rate leading to amino acid substitutions, as we have already noted for the histone H4 gene in the preceding section. This indicates that most synonymous mutations are virtually immune to selection, although we noted in Section 27.14 that the level of selection experienced by any one synonymous codon may be impossible to evaluate. In other words, nucleotide sequence data support the concept that synonymous mutations are usually "neutral" and that they "drift" through evolutionary time, much as predicted by the Neutral Theory (*Section 27.13*). They are therefore

excellent mutations to follow in evaluating evolutionary distances between closely related species.

2. The rate of nucleotide substitution in noncoding regions of genes (introns and flanking sequences) is also much higher than the amino acid substitution rate, although it is predictably low in the vicinity of such informational domains as splicing sequences, capping sequences, and the like (*recall Table 9-2*). The absolute rate of substitution, in fact, is about $5 \times 10^{-9}$ substitutions per nucleotide site per year, which is about the same as for synonymous substitutions.

3. Whereas the rates of noncoding and synonymous nucleotide replacements are approximately constant over time, rates of amino acid substitutions vary sharply from gene to gene. This observation is fully consistent with the conclusion reached from polypeptide studies (*Section 28.6*), namely, that the coding sectors of genes are modified primarily by the dictates of selection.

## 28.8  Evolution of the Primates

The evolution of our immediate ancestors is, of course, of particular interest. Figure 28-5 shows data on the relatedness of various mammals as detected by immunological analyses, where humans are seen to be extremely close to the gorillas and chimpanzees of Africa. What does this mean in genetic distance?

We have already seen, from Figure 28-3, that the human karyotype is very similar to the gorilla and chimpanzee, differing in the possession of an apparent fusion chromosome and in the presence of several inversions. This level of resolution, however, cannot provide the information we seek.

The most sensitive estimates to date have made use of **restriction-endonuclease cleavage analysis of mitochondrial DNA,** a technique that promises to have widespread use in molecular evolution studies. In such analyses, mitochondrial DNA (*Section 16.1*) is isolated from the organisms to be compared and is subjected to a battery of restriction enzymes (*Section 4.11*), after which the fragments are displayed by gel electrophoresis and ordered into a restriction map (*Section 12.16*). The resultant maps for five primate species are shown in Figure 28-6, where the gorilla is seen to carry a deletion found in no other lineage. Maps are judged most similar when

**FIGURE 28-5 Immunological Distances between selected mammals** are indicated by the separation, as measured along the horizontal axis, between the branches of this "divergence tree." For example, the monotremes (primitive egg-laying mammals) are removed from the marsupials by a distance (in arbitrary units) of only 1.5 but are removed from the chimpanzee by a distance of nearly 17. The distance between humans and the Old World monkeys is a little more than 3, between humans and the Asiatic gibbons 2, and between humans and the gorillas and the chimpanzees of Africa less than 1. (From The Evolution of Man by Sherwood L. Washburn. Copyright © 1978 by Scientific American, Inc. All rights reserved.)

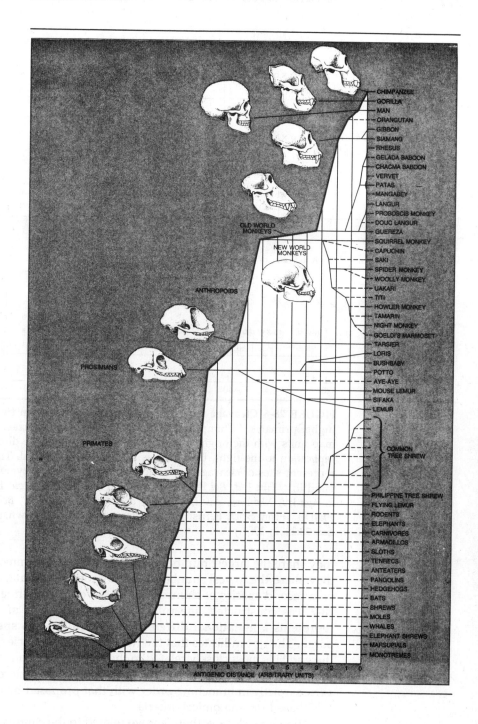

CHIMPANZEE
GORILLA
MAN
ORANGUTAN
GIBBON
SIAMANG
RHESUS
GELADA BABOON
CHACMA BABOON
VERVET
PATAS
MANGABEY
LANGUR
PROBOSCIS MONKEY
DOUC LANGUR
GUEREZA
SQUIRREL MONKEY
CAPUCHIN
SAKI
SPIDER MONKEY
WOOLLY MONKEY
UAKARI
TITI
HOWLER MONKEY
TAMARIN
NIGHT MONKEY
GOELDI'S MARMOSET
TARSIER
LORIS
BUSHBABY
POTTO
AYE-AYE
MOUSE LEMUR
SIFAKA
LEMUR

COMMON
TREE SHREW

PHILIPPINE TREE SHREW
FLYING LEMUR
RODENTS
ELEPHANTS
CARNIVORES
ARMADILLOS
SLOTHS
TENRECS
ANTEATERS
PANGOLINS
HEDGEHOGS
BATS
SHREWS
MOLES
WHALES
ELEPHANT SHREWS
MARSUPIALS
MONOTREMES

OLD WORLD
MONKEYS

NEW WORLD
MONKEYS

ANTHROPOIDS

PROSIMIANS

PRIMATES

17  16  15  14  13  12  11  10  9  8  7  6  5  4  3  2  1  0
ANTIGENIC DISTANCE (ARBITRARY UNITS)

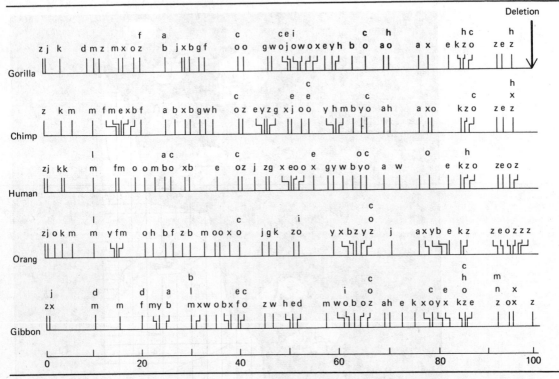

**FIGURE 28-6 Cleavage maps of mitochondrial DNA from five species of higher primates.** "Chimp" refers to the common chimpanzee. The scale is in map units with the origin of replication at 0 units and the direction of replication to the right. Each map is for a single individual. The small letters represent sites of cleavage by the following restriction enzymes: *a*, EcoRI; *b*, HindIII; *c*, Hpa I; *d*, Bgl II; *e*, Xba I; *f*, BamHI; *g*, Pst I; *h*, Pvu II; *i*, Sal I; *j*, Sac I; *k*, Kpn I; *l*, Xho I; *m*, Ava I; *n*, Sma I; *o*, HincII; *w*, BstEII; *x*, Bcl I; *y*, Bgl I, and *z*, FnuDII. (From S. D. Ferris, A. C. Wilson, and W. M. Brown, *Proc. Natl. Acad. Sci.* **78**:2432, 1981.)

they share the largest number of restriction sites and most dissimilar when they share the lowest number of sites. From these data, plus similar data on a second species of chimpanzee, the evolutionary tree drawn in Figure 28-7 was constructed, the rule being that **each loss of a restriction site represents a mutational difference between the two species being compared.** The tree indicates that the two chimpanzee species are most related; the gorilla is most closely related to the chimpanzee pair; and the human is most closely related to the chimpanzee-gorilla trio. Interestingly, this tree agrees exactly in branching order with that proposed by G. G. Simpson in 1963 based on anatomical criteria.

Just how closely, then, are we related to the chimpanzee? M. C. King and A. C. Wilson studied data from enzyme polymorphism and amino acid and nucleotide sequence comparisons and calculated the genetic dis-

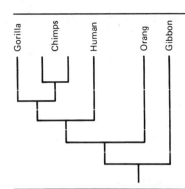

**FIGURE 28-7 Evolutionary tree for mitochondrial DNA of six higher primates** based on the cleavage maps in Figure 28-6. The tree requires a minimum of 67 mutations at 42 positions. (From S. D. Ferris, A. C. Wilson, and W. M. Brown, *Proc. Natl. Acad. Sci. U.S.* **78**:2432, 1981.)

tances between the human races and between humans and chimpanzees, much as genetic distances were estimated for the *Drosophila* groups in Table 28-2. The results are summarized in Figure 28-8. Not surprisingly, the genetic distances between the human races are very small ($D = 0.01$ to $0.02$), comparable to the levels found between local populations of *Drosophila*

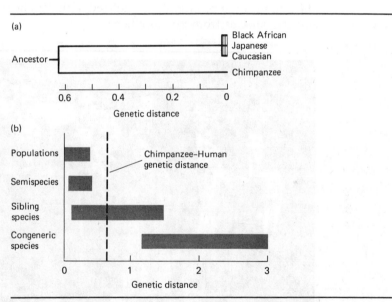

**FIGURE 28-8 Genetic distances among primates.** (a) Phylogenetic relationship between human populations and chimpanzees. The genetic distances are based on electrophoretic comparison of proteins. No human population is significantly closer than another to the chimpanzee lineage. The vertically hatched area between the three human lineages indicates that the populations are not really separate, owing to gene flow. (b) The genetic distance, *D*, between humans and chimpanzees (dashed line) compared with the genetic distances between other taxa. Taxa compared include several species of *Drosophila*, the horseshoe crab *Limulus polyphemus*, salamanders from the genus *Taricha*, lizards from the genus *Anolis*, the teleost fish *Astyanax mexicanus*, bats from the genus *Lasiurus*, and several genera of rodents. The great majority of proteins in these studies are intracellular. (From M. C. King and A. C. Wilson, *Science* **188**:107, 1975.)

(Table 28-2). Surprising, however, was the discovery that humans and chimpanzees, although 25 to 60 times more distant than the human races, are nonetheless separated by $\bar{D} = 0.62$, which is in the range of sibling species in other analogous studies (for example, see Figure 28-8 and Table 28-2).

King and Wilson quickly point out that humans and chimps differ much more than sibling species in their anatomy and their way of life, having very different brain sizes, bone structures, posture, method of procuring food, and so on. On these criteria, they have traditionally been placed in separate families. Why, then, the paradox? Although the answer is of course not known, the best guess is to invoke a saltatory form of evolution (*Section 28.4*). By this hypothesis, key regulatory genes or gene complexes came to differ in prehumans and chimps while the structural loci remained greater than 99 percent identical. The chimp-human example exemplifies in the extreme an important principle of evolutionary genetics: it is not necessarily the *nature* of the molecules produced by the species, but the *way* they are produced and utilized in metabolic pathways, that differentiates species from one another.

(a) A group of population geneticists at a not-so-recent meeting. *Bottom row* (left to right): Daniel Marien (Queens College); Richard Lewontin (Harvard University); Lee Ehrman (State University of New York, Purchase); Theodosius Dobzhansky (Rockefeller University and University of California, Davis); Howard Levene (Columbia University); and Francisco Ayala (University of California, Davis) *Middle row* (left to right): Sergey Polivanov (Catholic University of America); Bruce Wallace (Cornell University); Louis Levine (City College of New York); Irwin Herzkowitz (Hunter College); Marvin Druger (Syracuse University); and Timothy Prout (University of California, Riverside). *Top:* Richard Levins (left) (Harvard University) and Leigh Van Valen (right) (University of Chicago).

(b) A group of population geneticists at a more recent meeting. *Bottom row* (left to right): Sergey Polivanov (Catholic University of America); Theodosius Dobzhansky, Elliot Spiess (University of Illinois, Chicago Circle); Francisco J. Ayala (University of California, Davis); *Middle row* (left to right): Jeffrey R. Powell (Yale University); Wyatt Anderson (University of Georgia); Abd-El Khalek Mourat (University of Cairo, Egypt); Louis Levine (City College of New York); *Top row* (left to right): Kirshna Sankaranarayanan (University of Leiden, Netherlands); Timothy Prout (University of California, Davis); Rollin Richmond (North Carolina State University); George Carmody (Carleton University, Ottawa).

# Questions and Problems

1. How would you distinguish between a species, two subspecies, two semi-species, two sibling species, and two nonsibling species of animals?

2. Would the Neutral Theory (*Section 27.13*) expect the discontinuous rates of nucleotide replacements in hemoglobin genes postulated to have occurred during phylogeny? Explain.

3. Chromosome translocation is thought to be an important speciation mechanism. Explain how translocation could
   (a) Promote postzygotic isolation (*Section 5.15 and 7.9*).
   (b) Alter expression of large blocks of genes (*Section 25.16*).

4. The ability of cow, sheep, and pig genomic DNA to form DNA hybrids was analyzed, and the amino acid sequence of several of their proteins was compared, allowing an estimate of nucleotide substitutions in the genes encoding these proteins. The results are plotted on page 876 (from C. Laird et al., *Nature* **224**:149–154, 1969).

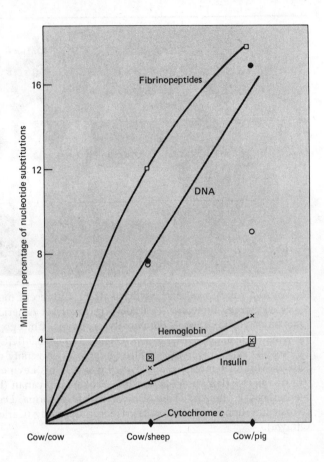

(a) How would you explain the different patterns shown by the different proteins (Note: the fibrinopeptides act as small structural elements in blood clots)?

(b) This study was originally interpreted as showing that DNA evolved much more rapidly than proteins in the artiodactyls (hooved mammals). How would you qualify this conclusion given your understanding of the DNA-DNA hybridization technique?

5. The mitochondrial DNA of pocket gophers (*G. pinetis*) captured in Florida, Alabama, and Georgia was analyzed by restriction enzyme digestion. With *Bam* HI, three patterns were found, labeled N, O, and M on page 878; gophers carrying each pattern were localized geographically as shown on page 878 (from J. C. Avise et al., *Proc. Natl. Acad. Sci. U.S.* **76:** 6694, 1979).

(a) Explain why it is possible to argue that each group represents a clone evolving from a common female parent at some time in the past (*review Sections 16.7 and 16.8*).

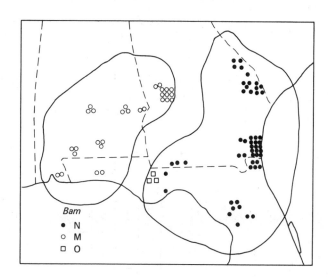

(b) Digestion with five additional restriction enzymes allowed the investigators to subdivide this population into 23 distinct clones. How do you predict they would be distributed on the above map?

(c) Analysis of allozyme polymorphism at 25 loci permitted the investigators to separate the population into the two major East-West regions shown on the map but did not allow the more finely tuned discriminations. Why is the DNA analysis more sensitive?

6. The gophers in the previous study are all of the same species. Members of the eastern population display an average genetic distance $\bar{D}$ of 0.009; members of the western population show $\bar{D} = 0.024$, and the two populations are separated by $\bar{D} = 0.065$ (estimates from the allozyme analysis).

(a) How would you demonstrate that the population contains a single species?

(b) Compared with the data on *Drosophila* (*Table 28-2*) and humans (*Section 28.8*), would you predict that the two groups are approaching subspecies status?

7. (a) Which of the loci listed in Problem 26-8 would you select as particularly "diagnostic" for each of the four species of *Drosophila?* Explain your reasoning.

(b) Might you have been fooled into thinking that two or more of these species were closely related had you considered only one locus? Give an example.

8. Describe how, if at all, each of the following might promote the isolation of subpopulations and the formation of subspecies.

(a) Assortative mating

(b) Inbreeding

(c) Linkage disequilibrium

(d) Allozyme polymorphism

(e) Hybrid dysgenesis

9. Respiration in isolated rat mitochondria is found to occur at comparable rates when rat cytochrome c is washed out and replaced with bovine cytochrome c, whereas the bovine protein is not able to support rat mitochondrial ion transport or to interact properly with rat cytochrome oxidase. What principle is illustrated by these observations?

10. V. Sarich and A. Wilson (*Science* **158**:1200, 1967) constructed the following primate phylogeny based on immunological relatedness of serum albumins.

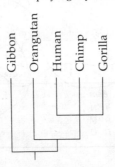

(a) How does this phylogeny compare with that based on mitochondrial DNA?
(b) Which is likely to be the more accurate? Explain.

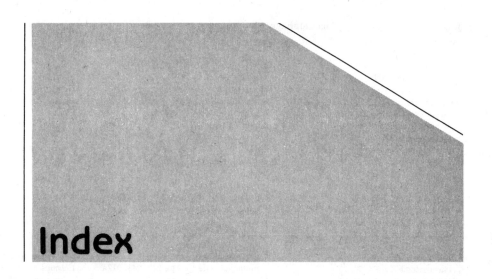

# Index

ABO blood groups, 597–600
  populations and, 793–794, 844–846
  secretor, 615
Abortion, 145, 220, 227, 595, 597
Acentric chromosome, 217
Achtman, M., 427
Acridine, 250
  mutagenesis, 249, 317–318, 326
Acrocentric chromosome, 48
Activator
  corn element, 582–583
  protein, 695, 710, 716
Addition, 202–203, 249–250, 318–320
Adenine, 4
Adhya, S., 586
Aflatoxin B, 246–247
Agouti coat, 659
Albinism, 196–197
Alignment, 539
Alkaline phosphatase gene, 331–332
Alkaptonuria, 592
Alkylating agents
  misrepair and, 246–247
  mutagenesis, 239–241
Allele, 158
  frequency, 804–806, 815
  hetero, 470
  multiple, 470, 793
  neutral, 796, 813
  population patterns, 853–854
Allelic exclusion, 631–633
Allelism. See Polymorphism.
Allolactose, 688
Allopatric, 811
Allopolyploidy, 150, 861
  segmental, 861
Allotype, 623–624
Allozyme, 605, 800–802, 815–816, 838–839, 854
Altman, R., 531

α-amanitin, 263
Amati, P., 559
Amber mutation, 330–332
  suppressor, 332–335
Ames, B., 222, 225–227
Ames Test, 222, 225–226
Amino acids, 63–65
  acidic, 65
  basic, 65
Aminoacyl-tRNA, 310–311
  synthetase, 310–311
Aminopterin, 493–494
Aminopurine, 220, 242
Amitosis, 57
Amniocentesis, 220, 595–596
Ampicillin resistance, 429–430, 579
Amplification of genes, 215, 607
Anaphase
  meiotic, 135
  mitotic, 41–48
Anderson, W., 875
Androgens, 738, 779
Aneuploidy, 145–150
Antennapedia mutation, 774
Antibiotic resistance, 210–212, 425–426, 429–431, 550–553, 578–580
Antibody. See Immunoglobulin.
Anticodon, 286–287, 310
Antigen, 45
Antisense strand, 264
Antiserum, 45
Antiterminator, 715
Aporepressor, 698
Apurinic
  endonuclease, 249
  site, 247
Arabinose operon, 695
Arber, W., 109
Arginine, biosynthesis, 654–655
Artificial selection, 832–835

Ascospore, 142
Ascus, 142
Asellus, 838–839
Ashburner, M., 736
Aspergillus nidulans
  parasexual cycle, 487
  regulation, 733–734
  somatic genetics, 483–490
Aspinwall, N., 854
Assortative mating, 807
Ataxia-telangiectasia (AT), 253–254
Att sequences, 570–573
Attached X chromosomes, 194–195, 566
Attachment point, 27
Attenuator, 699–703
Auerbach, C., 254
Autogamy, 143–144
Autogenous regulation
  5S-RNA, 739
  λ repressor, 722
  SV40, 724
  trp repressor, 699
Autoradiography, 32–33, 69
Autosome, 50
Auxotrophy, 206–208
Avery, O.T., 18
Axel, R., 752
Ayala, F., 815, 854, 874, 875

Bacillus subtilis
  map, 416
  phage infection, 712–713
  sporulation, 727
  synchronization, 417
  transduction, 418
  transformation, 413–417
Bacteriophage. See also specific phage
  strains.

Bacteriophage (Continued)
 female-specific, 427
 genetics, 341–385
 helper, 570
 infection cycles, 19, 341–346
 lysogenic, 342, 345–346
 male-specific, 427
 packaging chromosomes, 345, 367,
  381, 418
 regulation, 709–722
 structure, 20, 341–344
 temperate, 342, 345–346
 virulent, 342
Bacterium. See also specific bacterial
  strains.
 cell cycle, 26–27
 cell structure, 26–27
 chromosome, 27–34, 74, 75
 DNA replication, 29–34
 genetics, 405–421
 regulation, 687–704
 restriction-modification, 412–413
Balanced polymorphism, 812
Balancing selection, 812–813
Baldness, pattern, 789
Banded chromosomes, 51–59, 83, 87–
  88, 89
 C bands, 53, 89
 G bands, 53, 87–88
 mapping and, 450–456
 Q bands, 53
 R bands, 53
Bar eye, 187–189, 617
 position effect, 791
 ultrabar, 617, 791
Barnett, L., 317
Barr, M., 780
Barr body, 780–783
Basal body, 534
Basc X chromosome, 208–209
Base, 4
 analogue mutagenesis, 241–243, 317–
  318, 325–326
 composition, 95–96
 minor, 293–294
 wobble pair, 323–325, 327
Bateson, W., 438
Beckwith, J., 419, 432, 434
Behavioral genetics, 661
Bennett, J., 624
Benzer, S., 350
Benzo [a] pyrene, 247, 499
Bicaudal locus, 759
Biochemical genetics, 654–664
Birds, sex chromosomes, 183–189
Birnsteil, M., 291
Birth defects, 226–227
Bithorax locus, 774–775
Bivalent, 132–134
Blastocyst, 747
Blastoderm, 760–761

Blood typing, 597–600, 615, 793–794,
  844–846
Bloom syndrome, 253–254
Bobbed mutations, 306
Bonner, J. J., 737
Boram, W., 564
Bourgeois, S., 705
Bovari, T., 752
Brachydactyly, 164
Branch, 539
 migration, 540
Breeding, 664–674, 833–838, 861–862
Brenner, S., 317, 419
Bridges, C., 59
Britten, R., 98, 105
5-Bromodeoxyuracil (5-BUdR), 207, 222
 enrichment procedure, 207
 mutagenic effects, 220, 242–243, 317
Brown, D., 296, 299, 306
Buoyant density (DNA), 31, 96
Burst, of phage, 342
Butterfly mimicry, 857

C band, 53, 89
C value, 94, 125
 paradox, 582
cII/cIII protein, 716–717
cDNA, 274
Cairns, J., 32–33, 236
Campbell, A., 571, 589
Cancer. See Carcinogenesis.
Cap, mRNA, 266–267, 310
CAP protein, 687, 695–697, 740
Carbon, J., 502, 506
Carcinogenesis, 224–226
 chromosome changes, 213
 dominance, 499
 environmental induction, 224, 499
 repair and, 253–254
 somatic mutation theory, 499
 tumor-specific antigens and, 790
Carmody, G., 875
Carrier, W., 248
Case, M., 647, 730, 742
Cat
 calico, 780–781
 tortoise-shell, 780–781
Catabolite repression, 695–697
Cell cycle
 bacterial, 27–28
 eukaryotic, 37–38, 73
Central component, 130
Central region, 130
Centriole, 45, 534
Centromere, 41
 behavior in meiosis, 134–135, 466
 heterochromatin, 89, 103, 107–108,
  866

linkage, 466
 suppression of crossing over, 455
Chargaff, E., 8, 22, 96
Chargaff rule, 8, 96
Chase, M., 19
Chassin, L., 492
Chi
 form, 540
 hotspot, 550
Chi-square test, 174–176
Chiasma (chiasmata), 133–134, 439
 crossing over and, 439, 441–444,
  456–457
 terminalization, 134
Chinese hamster ovary (CHO) line,
  491–492
Chironomus tentans, 279, 736
Chlamydomanas reinhardi
 life cycle, 139–140
 map, 467
 mapping, 459–466
 tetrad analysis, 166–167, 179–182,
  459–466
 uniparental inheritance, 536
Chloramphenicol, 513, 515, 527
 resistance, 578–580
Chloroplast, 35–36, 527–531
 DNA, 514, 527
 genetics, 527–531
 map, 528
 ribosomes, 514
CHO (Chinese Hamster Ovary) line,
  491–492
Chromatid
 interference, 448
 sister, 40–41
Chromatin. See also Chromosome.
 loops, 82–83
 10 nm filament, 74–81
 25 nm fiber, 81–82
Chromatography, 67–68
Chromatosome, 77–78
Chromocenter, 58, 89
Chromomere, 88
Chromosomal theory of inheritance,
  194, 438, 457
Chromosome, 25–59, 73–91
 aberration test, 223–224
 bacterial, 27, 74, 75
 bands, 51–55, 57–59, 83–88, 89
 coiling, 41, 81–82, 88–89
 condensation, 81–82, 88–91
 daughter, 41, 45
 elimination, 492–493
 homologous, 50
 lampbrush, 277–279, 754
 mutation, 202–203, 212–220, 588, 776,
  782–783
 number (table), 51
 packaging, 81–82, 88–89
 polytene, 57–59, 83, 88

Chromosome (*Continued*)
  proteins, 73–74
  purification, 55–56, 499
  replication, 29–34. *See also* DNA
    replication.
  scaffold, 86–87
  sex, 50
  staining, 53–54
  structure, 73–91, 741
Chymotryptic peptides, 67
Ciliate. *See Paramecium.*
Circulate permutation, 366–367, 577
*Cis*-dominance, 692, 785
*Cis-trans* test, 350
Cistron, 350–351
Clark, A. J., 427
Clarke, L., 502, 506
Clastogen, 220, 223
Cleavage planes, 755–756, 763
Clever, U., 736
Cline, 844–845
Clonal analysis, 768–769
Clone, 28
Cloning
  DNA, 113–122
  shotgun, 117, 119
  vector, 113–117
Cloverleaf tRNA, 278
Cluster, gene. *See Gene, cluster.*
CMS, 525–527
Code, genetic, 317–335
  degenerate, 317, 320, 322
  evolution, 323, 328–330
  mitochondrial, 328–330, 332
  patterns, 323–325
  table, 322, 323
  triplet, 317–320
  universality, 328–330
Codominant, 592, 599
Codon, 262, 309–310, 321–330
  chain-termination, 314, 322–323
  initiator, 266, 313, 322–323
  nonsense, 314, 322–323
  preference, 327, 704
  terminator, 314, 322–323
Coiling
  chromosome, 88–89
  DNA, 77, 80
Coincidence, coefficient of, 449–450,
  557–561
Col plasmid, 426, 431
Colcemid, 150
Colchicine, 48, 150
Colicin, 431
Colinearity, 327
Colony, 28
Color blindness, 192
Comings, D. E., 91
Compartment, 767–774
  boundary, 768
Competence factor, 413

Competent bacteria, 413
Complementary
  base pairs, 10
  DNA (cDNA), 274
Complementation, 348–352
  *cis-trans* test and, 350
  contrast with recombination, 349
  intragenic, 351–352, 730
  *in vitro*, 396, 661
  negative, 707
  test, 348
Complex locus, 649–650
Conalbumin gene, 301, 304
Concatamer, 345, 367, 499
Conditional mutant, 208–210
Conidium (conidia), 141
Conjugal
  plasmid, 426–431
  transfer, 428–429
Conjugation, bacterial, 400–404
  mapping by, 405–412
  tube, 400
Conjugation, ciliates, 142–144
Consanguinity, 165
Consensus sequence, 266
Constant region of antibody, 621–624
Constitutive
  gene, 251, 682–683
  heterochromatin, 53, 89–91, 108,
    148–149, 782–783, 865–866
  operator, 692, 698
Control
  attenuator, 699–703
  negative, 687
  positive, 687
  transcriptional, 683, 753
  translational, 683, 704, 754
Controlling element
  eukaryotic, 582–584, 740–741, 785
  prokaryotic, 259–260, 683
Conversion, 561–565
*Copia*, 580–581
Corepressor, 686, 698
Corn
  breeding, 664–666, 672–674, 816
  chloroplast, 527–528, 532
  controlling elements, 582–584
  cytoplasmic male sterility, 525–526
  *Ds/Ac*, 582–583
  *Dt*, 583–584
  ear length, 664–666, 672–674
  *iojap*, 529–531
  knobs, 459
  life cycle, 139
  linkage groups, 458–459
  map, 460–461, 528
  plasmid, 527
Correction theory, 607
Correlated response, 833
Correns, C., 438, 528
Cortical inheritance, 533–534

Cot analysis, 100–106
Cotransduction, 419
Cotransformation, 414–415, 502–503
Coupling, 356
Cri-du chat syndrome, 213
Crick, F. H. C., 8, 317, 323, 581
cRNA, 107
Cro protein, 719–721
Crossing over, 130, 439. *See also*
  Recombination.
  absence in male *Drosophila*, 446
  chiasmata and, 439, 441–444
  double, 361–362, 448–450, 557–561
  frequency, 456–457
  meiotic, 130–134
  mitotic, 486–490, 508–509, 768
  single, 361
  suppression, 216–217, 456–457, 776
  unequal, 610, 612, 791
Crow, J. F., 797, 813
CRP protein, 695–697
CsCl gradient centrifugation, 30–31
  GC determination, 96
*Curly* wing, 604
Cyclic AMP (cAMP), 688, 695–697
Cycloheximide, 518
Cytochrome c, 334, 466–472
  evolution, 867
  fine-structure map, 466–472
Cytogenetics, 53
Cytokinesis, 48, 755–756, 763
Cytoplasm
  developmental role, 752–760
  Texas, 525–526
Cytoplasmic
  genetics, 512, 524
  influence in development, 752–760
  male sterility (CMS), 525–527
Cytosine, 4

D loop, 549
Danna, K. J., 378
Darwin, C., 823, 851
*Datura stramonium*, 147
Davenport, C., 668
Davenport, G., 668
Davidson, E. H., 105, 751, 752
Davis, R., 502
DeLucia, P., 236
de Vries, H., 438
Deficiency, 203, 213–215, 242
  heterozygote, 213, 450–451, 583
  mapping, 450–452, 495–496
Degenerate code, 317, 320, 322
Delayed-early genes, 714
Delbrück, M., 211
Deletion, 202–203, 213–215, 249–250,
  318–320, 367–374, 588
  mapping use, 367–374

Deletion (*Continued*)
  overlapping, 368–369
  properties of, 367
Deme, 846
Denaturation map, 375–376
Denaturation-renaturation of DNA, 98
Deoxyribonuclease (DNAse), 19, 75, 79, 786
  I, 786–787
  I hypersensitivity, 786
  I sensitivity, 786–787
Deoxyribonucleic acid. *See* DNA.
Deoxyribose, 5
Derepression, 686
Desynapsis, 132
Determination, 746, 755
Development, 746–787
Diabetes mellitus, 300
Diakinesis, 134
Dicentric chromosome, 217
*Dictyostelium discoideum*, cell cycle, 40
Dictyotene, 134
Diethylnitrosamine, 220
Differentiation
  long-term, 745–787
  short-term, 728–741
Dihybrid ratio, 177, 658, 663–664
  modified, 658–659
*Diplococcus pneumoniae* (pneumococcus), 17
  transformation, 17–19, 413–415, 550–553
Diploid, 50
  partial, 422–423, 689
Diplonema, 132–134
Dihybrid cross, ratio, 177
Dihydrofolate reductase, 215, 301
Directional selection, 823–833
*Disporum sessile*, 441
Dissociation-reassociation, of DNA, 98–106, 395
  kinetics of reassociation, 100–106
Dissociator-activator, 582–583
Disulfide linkage, 72
*dna* mutations, 232–233
DNA, 1, 4–23
  amount per organism, 94–95
  centromeric, 103
  cloning, 113–122
  complementary (cDNA), 274
  composition, 95–96
  denaturation, 98
  dissociation, 98–106
  double fork, 104–105
  double helix, 9
  duplex, 9
  dyad symmetry, 106, 698, 718, 721, 740–741
  grooves, 12, 549–550, 693, 720
  hair pin, 106
  heavy strand, 520

heteroduplex, 546, 550–557
highly repetitive, 102–104
hybrid, 542, 550–552
hybridization, 107
inverted-repeat, 105–106
kinetic classes, 101
library, 121
ligase. *See* Ligase.
light strand, 520
melting, 98
methylatin, 785
middle-repetitive, 104–105, 581
mismatched, 514
palindromic, 105, 109
polymerase. *See* DNA polymerase.
reassociation, 98–106
renaturation, 98
replication. *See* DNA replication.
satellite, 96–98, 102–103, 107–108, 581
selfish, 581–582
sequencing technique, 113–115
single-copy, 102, 251
snap-back, 106
stem-and-loop, 106
sticky-ended, 117
superhelical, 77–80
translocase, 414
unique, 102
DNAse. *See* Deoxyribonuclease.
DNA-dependent RNA polymerase. *See* RNA polymerase.
DNA polymerase, 79, 232–237
  I, 235–236
  III, 232–235
  α, 233
  β, 235
  mitochondrial, 513
  repair role, 248–249
DNA replication, 15, 29–34, 82–83, 231–237
  bidirectional, 32, 38, 409
  elongation, 232–237
  eukaryotic, 38–39, 82–83
  fork, 32
  heterochromatin and, 90–91
  initiation, 232
  integration and, 574–576, 578
  origin, 32, 232, 409–410, 502
  primer sequences, 233, 255
  rate, 32, 38
  repair and, 547–548
  roll-in, 574–575
  rolling circle, 345–346, 401–402, 754
  semiconservatice, 32, 29–34, 38
  terminus, 32
  unwinding, 154–155
Dobzhansky, Th., 831, 858, 874–875
Dominance, 155, 159, 592, 602–604
  *cis*–, 692, 785
  expressivity and, 602–604

hypothesis, 837
  incomplete, 592
  mutations, 602–604
  partial, 592
  penetrance and, 602–604
Doolittle, W. F., 581
*Dorsal* locus, 759–760
Dosage compensation, 781
Double cross method, 835
Double helix, 8–13
Down's syndrome, 148–149
  duplications and, 213
  translocations and, 219
Downstream, 265
Dressler, D., 542
Drift. *See* Genetic drift.
Druger, M., 874
*Drosophila*
  biochemical genetics, 657–658
  chromosome mutations, 214, 830–831
  complex locus, 649–650
  constitutive heterochromatin, 89–90, 123
  courtship, 860
  DNA, 123
  embryogenesis, 749–750, 757–774
  endosymbionts, 531
  eye color, 657–658, 661–663
  gene amplification, 754
  genetic drift, 850–851
  heat shock response, 733–735
  histone genes, 304–305
  hybrid dysgenesis, 584–586
  imaginal discs, 737, 760–774
  independent assortment, 172–179
  inversions, 214–217, 830–831
  larval chromosomes. *See* Polytene.
  life cycle, 137–138, 761
  map, 458
  mapping, 352, 444–459
  maternal effect, 759–760
  nucleolus, 90
  phototaxis, 831–832
  polymorphism, 800–802, 815–816, 839, 854
  polytene chromosomes. *See* Polytene.
  populations, 800–802, 815–816, 827, 839, 850–851, 854, 862–863
  regulation, 734–737, 749–750, 757–774
  salivary gland chromosomes. *See* Polytene.
  satellite DNA, 123
  segregation, 160–164
  selection, 827–828, 831–832, 839, 850–854
  sex chromosomes, 90, 183
  sex determination, 183
  sex-linked inheritance, 184–189
  speciation, 860, 862–864
  transformation, 498
  transposable elements, 580–581, 584–586

*Drosophila* (*Continued*)
  tRNA genes, 290
Drug
  dependency, 210–211
  resistance, 210–211, 425–426, 429–431
Dryer, W., 624
*Dt* (*dotted*) element, 582
Duchenne-type muscular dystrophy,
  477–478
Duffy blood group, 506, 794
Duplex DNA, 9
  patched, 542
  spliced, 542
Duplication, 203, 213–215, 242
  in evolution, 215, 606, 635
  heterozygote, 213
Dyad symmetry, 106, 698, 718, 721,
  740–741


Early genes, 712
East, E., 664, 667
Ecdysone, 736–737, 739, 767
*Eco RI* enzyme, 109–113
Edgar, R., 660
Edman degradation, 67, 70
Effector, 684–685
Egg. *See* Oocyte.
Ehrman, L., 874
Electrophoresis, 69
Electrophoretic variant. *See* Allozyme.
Elgin, S., 786
Elimination of allele, 847, 852
Elongation
  DNA, 232–237
  factors, 312–314, 327
  protein, 313–314
  RNA, 259
Embryogenesis
  amphibian, 752–757
  *Drosophila*, 749–750, 757–774
  mammalian, 747–749, 754–755, 774–
    787
Emerson, R., 478, 664
EMS, 239–240
Endonuclease, 75–78, 109
  V, 548
  integration, 573, 575
  restriction, 109–113
Endoplasmic reticulum, 35
Endopolyploid, 58, 83
Endosperm, 138
Endosymbiosis, 531–532
*Engrailed* locus, 771–772
Enrichment procedure, 207
  5-BUdR, 207
  penicillin, 207
  tritium suicide, 228
Envelope, nuclear, 35, 41, 89, 91
Environmental influence, 603–604, 671–674

Enzyme, 79–80
Ephrussi, B., 515
Ephrussi-Taylor, H., 432
Episome, 425, 427, 570
Epistasis, 660, 822–823, 864
Erythromycin, 513
  resistance, 513
*Escherichia coli*
  biochemical genetics, 654–655
  cell cycle, 26–27
  chromosome, 27, 74, 75, 409–410
  conjugation, 400–404
  DNA replication, 29–34, 231–237,
    409–410
  lysogenic (λ), 345, 570–573
  map, 409–412
  mapping methods, 405–425
  operons, 643–647
  regulation, 687–704
  restriction-modification, 109–113
  rRNA genes, 295
  sexduction, 423–425
  specialized transduction, 289, 421–425
  structure, 26–27
  transduction, 417–421
  transformation, 413
  tRNA genes, 286–290
Esterase, 605–606
Estrogeus, 738
Ethidium bromide, 69
Ethylmethane sulfonate (EMS), 239–240
Ethylnifrosourea, 220, 222
Euchromatin, 89
Eugenics, 826
Eukaryote, 27, 35
Evolution, molecular. *See* Molecular
  evolution.
Evolution, populations, 795–799, 819–
  855. *See also* Molecular evolution.
  Darwin-Wallace·theory, 823
  macro, 858
  micro, 858
  saltatory, 847
Exon, 272, 635
Exonuclease, 235–236
Expressivity, 602–604
Extinction, 847
Extrachromasomal, 512
Extranuclear, 16


*F*[+] bacterium, 400–404
*F*[−] bacterium, 400–404
*F*-duction, 425
$F_1$ generation, 155
$F_2$ generation, 155
F element (agent, factor), 400–404,
  427–428
  immunity, 427

  integration by, 407–408, 428–428
  map, 428
  sexduction by, 424–425
  *tra* operon, 427–428
F' element, 423–425
F pilus, 400
f1 phage, 162
f2 phage, 164
fd phage, 162
Facultative heterochromatin, 779–783
Fanconi's anemia, 253–254
Fate map, 762–764
  surgical, 767
Fecalase, 226
Feedback inhibition, 697
Female-specific phage, 427
Ferritin, 520
Feulgen stain, 53
Fiers, W., 380, 381
Fimbria, sex, 400–401, 427–428
Fink, G., 647
Fitness, 820
  modifiers, 822
  relative, 820–823
Fixation of allele, 847, 852
Fluctuation test, 211–212
Fogel, S., 562
Footprinting, 267, 573
Founder
  cell, 765–767
  effect, 812, 847
  principle, 864
434 phage, 726
Four o'clock, 528
Fox, A., 498
Fox, M., 543–545, 550, 565
Fraenkel-Conrat, H., 22, 23
Frameshift, 250, 318–320, 326
Framework region, 623
Franklin, R., 8, 22
Free radical, 242


G band, 53, 87–88
β-galactosidase, 646. *See also Lac* operon.
Galactose (*gal*)
  gene cluster, 731–732
  operon, 586–587, 643
Galactoside permease, 646. *See also Lac*
  operon.
Galapagos Islands, 851
Gall, J., 122
γ (gamma) irradiation, 220, 222, 242
γ δ insertion sequence, 427, 577, 580
Gamete, 125, 139
Gametophyte, 138
Garcia-Bellido, A., 768, 788
Garen, A., 331, 749
Garrod, A., 592
Gaucher's disease, 594

GC content (percent), 95–96
Gehring, W., 740
Geimsa stain, 53
Gene, 15, 258–281, 285–305, 308–335, 353
 amplification, 607, 753–754
 antisense strand, 264–265
 chromomere relationship, 88
 clusters, 288–289, 297–300, 303–305, 316, 647–649, 729–733
 constitutive, 251, 682–683
 conversion, 561–565
 expression 308–335
 family, 606–614, 619, 635, 651
 flow, 846
 number, 752
 overlapping, 383–385
 pool, 793, 796–797
 regulated, 682–683
 regulatory, 684–685
 sense strand, 263–265
 split, 272. See also Intron.
 structural, 259, 300–305
 super, 822, 831
 transfer, 498–503, 780
Generalized transduction, 417–421
Genetic
 background, 603–604
 counseling, 597
 distance, 862–863
 drift, 796, 813, 846–855
 engineering, 108–122
 load, 757–760, 797–798
Genic disharmony, 860
Genome, 25
 library, 121
 size, 752
Genotype, 158
 frequency, 804–806
Georgopoulos, C., 715
Germ
 line, 137, 756–759
 plasm, 756–759
Gierer, A., 22
Gilbert, W., 113, 277, 581, 705
Giles, N., 647, 730, 742
Glucocorticoids, 738–740
Glucose effect, 695–697
Glycosides, mutagens, 226
Glycosylase, 247, 249
Goldberg, E., 727
Gol lberg-Hogness box, 266, 270
Golgi apparatus, 35–36
Goodman, H., 333
Gorini, L., 335
Grandchildless locus, 760
Grindley, N., 587
Grooves in DNA, 12, 549–550, 693, 720
Groudine, M., 785
GT-AG rule, 275, 631
Guanine, 4

mRNA cap, 266–267, 310
Gurdon, J., 296, 306, 747, 752
Gurney, T., Jr., 550
Gynandromorph, 762–767
Gyrase, 237, 573

H antigen, 597–600
H-2 locus. See Histocompatability.
Haemophilus influenzae, 431
Hairpin loop, 106
Haldane, J. B. S., 813
Half-tetrad analysis, 566
Haploidization, 485, 487
Haploidy, 50, 149
Haplotype, 776, 822
Hardy, G., 802
Hardy-Weinberg principle, 802–811
Harelip, 602
Harris, H., 605, 614
HAT selection, 493–494
Hayes, W., 405
Headful mechanism, 345, 367, 418
Heat, as mutagen, 241
Heat shock response, 733–735, 740–741
Heavy
 chain of antibody, 621
 strand of DNA, 114, 520
Helicase, 237, 548
Helix, double, 8–13
Helix-destabilizing protein, 237
Helper phage, 570
Hemizygous, 186
Hemoglobin
 A, 607
 α, 607–613
 anti-Lepore, 610–612
 β, 335, 607–613
 Buda, 617
 Constant Spring, 338
 δ, 607–613
 embryonic, 607, 784
 ε, 607, 609, 613
 evolution, 608, 867–868
 F, 607
 fetal, 607, 610, 784–785
 γ, 607–613
 gene family, 607–613
 gene structure, 277, 612–613
 HPFH, 610, 784–785
 Kenya, 611–612
 Lepore, 610–612, 616
 mutations, 339, 607–613
 Pest, 617
 processing, 274–276
 pseudogene, 613
 regulation, 783–787
 S, 839–840
 sickle cell, 794–796, 839–840
 structure, 72

thalassemias, 198, 275, 335
 variants, 339, 608–613
 ζ, 607, 613
Hemophilia, A, 190–192
Hemophilus influenzae, 413
Hereditary persistence of fetal
 hemoglobin (HPFH), 610, 784–785
Heritability, 672–674, 833–835
 IQ and, 674
 selection and, 833–835
 within-group, 674
 without-group, 674
Hershey, A. D., 19
Herzkowitz, I., 874
Heterochromatin
 centromeric, 89, 103, 107–108, 866
 constitutive, 89–91, 108, 148–149, 782–783, 865–866
 facultative, 779–783
 regulation and, 779–783
Heteroduplex
 DNA, 546, 550–557
 mapping, 375–378, 587
Heterogametic, 184
Heterogeneous nuclear RNA (hnRNA), 272–274
Heterokaryon, 483, 491
Heterokaryosis, 483
Heterologous base sequence, 546
Heterosis, 837–838
Heterozygote, 158
 advantage, 795–796
 partial, 553
 population frequency, 800–802
 screening, 595
Heterozygous, 158
 DNA, 546
Hfr bacteria, 401–409, 427–428
HGPRT, 493–494, 506, 594–595
High-mobility-group proteins, 74, 787
Histidine
 gene cluster, 647–649
 operon, 643, 645
Histocompatibility locus (MCH), 650–652
 complex, 651
 linkage disequilibrium and, 806, 822
 map, 652
Histones, 73–81
 core, 75–77
 genes, 301–305, 606–607, 869
 nucleosome involvement, 75–81
 phosphorylation, 89
HLA locus. See Histocompatibility.
HMG proteins, 74, 787
Hoechst 33258, 228
Holliday, R., 539, 564, 565
Holliday
 intermediate, 539–543
 model, 539–543
Homoeotic mutation, 771–774

Homogametic, 184
Homokaryon, 483
Homologue, 50, 130, 157–158
  heteromorphic, 441
Homozygous, 158
Hood, L., 625, 637, 640
Hoppe, P., 747
Hormones, 735–738
  ecdysone, 736–737
  polypeptide, 736
  steroid, 738–740, 779
Hot spot
  chi, 550
  F integration, 407–408
  mutational, 371–372, 584
  transposon integration, 580
Housekeeping genes, 750, 752
Housman, D., 496
Howard-Flanders, P., 549
Hubby, J., 800
Huberman, J., 38–39, 59
Human
  albinism, 196–197
  alkaptonuria, 592
  birth defects, 226–227
  blood types, 597–600, 615, 793–794,
    844–846
  brachydactyly, 164
  chromosome aberrations, 145–150,
    213, 218–220
  chromosomes, 49, 52, 865–866
  color blindness, 192
  consanguinity, 165
  crossing-over frequency, 133–134
  Down's syndrome, 148–149, 213, 219
  evolution, 865–866, 870–874, 878
  gametogenesis, 137–138
  Gaucher's disease, 594
  genome size, 60
  harelip, 602
  hemoglobin. *See* Hemoglobin.
  hemophilia, 190–192
  histocompatibility, 650–652, 806, 822
  inborn errors, 592–597
  inbreeding, 165
  insulin gene, 301–303
  IQ, 674
  karyotype, 49, 52, 865–866
  Klinefelter syndrome, 145–146, 183–
    184
  Lesch-Nyhan syndrome, 506, 594–595
  map, 504–505, 524
  mapping, 492–506, 524
  maternal age, 148–149
  mitochondrial genetics, 523–525,
    870–872
  nonsense mutation, 335
  pedigree analysis, 164–166, 190–192,
    523, 870–872
  phenylketonuria, 593–595
  polydactyly, 603

  populations, 793–796, 817–818, 840–
    841, 844–849
  races, 794, 840–841, 844–846, 873–874
  repair deficiency syndromes, 253–254
  Rh blood group, 616, 793–794
  sex determination, 183, 777–779
  sex-linked inheritance, 190–194
  skin color, 668–669, 680
  somatic cell genetics, 492–498
  Tay-Sachs disease, 594–595, 817–818
  testicular feminization (*Tfm*), 779
  thalassemia, 198, 275, 335
  Turner's syndrome, 145–146, 183–184
  xeroderma pigmentosum, 253–254
Huntington's chorea, 615
Hurst, D., 562
H-Y antigen, 777
Hybrid
  clone panel, 498
  crop plants, 835–837
  DNA, 542, 550–552
  dysgenesis, 584–586
  somatic cells, 492–498
  vigor, 837–838
Hybridization
  DNA-DNA, 107, 750
  DNA-RNA, 107, 751
  *in situ*, 107, 290–291, 496
  mapping, 520–521
  somatic cell, 492–498
  Southern blot, 381
  speciation and, 861
Hydroxylamine, 239–241
Hyperploid, 485
Hyperpolyploid, 57
Hypervariable region, 623
Hyphae, 141
Hypoploid, 485
Hypoxanthine, 239–240, 493–494


Idiotype, 623–624
Illmensee, K., 747, 757, 790
Imaginal disc, 737, 760–774
Immediate-early genes, 714
Immunity
  F element, 427
  phage, 721, 726
  protein, 431
Immunoautoradiography, 46
Immunodiffusion test, 46
Immunofluorescence, 46
Immunogenetics, 618–639
Immunoglobulin, 45–46, 619–639
  chain alternation, 636
  chains, 621–622
  classes, 621
  class switching, 636
  domains, 621
  genetic specification, 625–639

  joining reaction, 627–637
  mutation, 637–639
  structure, 620
Inactivation, 220, 241
Inborn errors of metabolism, 592–597
Inbreeding, 165, 807–811, 835–837
  coefficient, 808–811
  depression, 807–808
  selection and, 835–837
Independent assortment, 170–182
Inducer, 686
Inducible
  gene, 251, 729
  operon, 686–687
Induction of prophage, 346, 722
Industrial melanism, 828–829
Initiation
  codons, 266, 313, 322–323, 327–328
  complex, 313
  factors, 312–313
  DNA replication, 232
  transcription, 262
  translation, 266, 310–313, 327–328
Inner cell mass, 747
Inosine, 324, 493–494
Insecticide, 607
Insertion, mutation, 318–320, 584
Insertion sequences (IS), 577–580, 586–
    589
  F element, 427–428
  transposons, 578–580
*in situ* hybridization. *See* Hybridization.
Insulin genes
  cloned, 122
  structure, 300–303
Int protein, 570–573, 716–717
Integration, 407–408, 498–503, 569–589
  base repeats and, 574–581
  F element, 407–408, 427–428
  phage, 570–576
  transformation, 498–503
  transposable element, 577–589
  viral, 576–577
Integrator locus, 733–734
Interband, 58
Intercalation, 249
Interference, 448–450, 556–561
  chiasma, 448–450
  chromatid, 448
  high negative, 556–561
  low negative, 559
  negative, 450, 599
  positive, 450, 599
Interphase, 35
  stages of, 37
Interrupted mating, 406–407
Interstitial region, 441
Intervening sequence. *See* Intron.
Intragenic complementation, 351–352
Intron, 272–277, 300–305
  conalbumin, 301, 304

Intron (*Continued*)
  evolution, 276–277, 581–582, 635
  hemoglobin, 277, 612–613
  insulin, 301–303
  mitochondrial, 522–523, 525, 532
  splicing, 266, 275–276, 293–294, 631
  tRNA, 293–294
Inversion, 202–203, 215–217
  adaptive, 830–831
  crossover suppression, 216–217
  evolution and, 830–831
  heterozygote, 216
  induction, 584, 588
  loops, 216
  paracentric, 216
  pericentric, 216
Inverted repeat (IR), 105–106
  CAP binding site and, 698
  immunoglobulin genes and, 627–628
  integration and, 578–580
  operators and, 698, 718–719, 721
  terminator and, 270–271, 700–703
*lojap* mutation, 529–531
Ionizing radiations, mutagenesis, 242
IPTG, 646, 688
IQ, 674
IS. *See* Insertion sequences.
Iso-1-cytochrome *c*. *See* Cytochrome *c*.
Isoacceptor tRNA, 286, 325
Isoelectric variant. *See* Allozyme.
Isogamous, 139
Isoloci, 604–606
Isotypic exclusion, 632–633
Isozymes, 604–606

Jacob, F., 405, 570, 684
Janssens, F., 439
Jayaraman, R., 727
Joining reaction, 627–637
Judd, B., 452, 475

Kaiser, A. D., 359, 725
Kan, Y. W., 335
Kangaroo, 780
Kao, F., 491, 496
Kappa
  chains of IgG, 623
  particles, 531, 538
Karyotype, 18–59, 83, 87–89, 865–866
  evolution, 864–867
Kettlewell, H., 829
Khorana, G., 321
Kilobase, 83
Kimura, M., 855
Kinetochore, 45
King, M. C., 872
Kleckner, N., 580, 589
Klinefelter's syndrome, 145–146, 183–184, 781
Knob, chromosomal, 169, 459

Kohne, D., 98
Kornberg, A., 236
Kornberg, T., 771

*lac* (lactose) operon, 557–558, 645–646, 687–697, 732
  DNA isolation, 434
  regulation, 687–697
  transduction, 423–424
Lactate dehydrogenase, 653
Lambda chains of IgG, 623
Lambda phage, 342
  chromosome, 395–396, 570–573
  control of gene expression, 713–722
  deletion mapping, 367–375
  excision, 570–573
  helpers, 570
  immunity, 721, 726
  induction, 346, 722
  infection cycle, 345–346, 713–716, 719
  integration, 570–573
  lysogeny, 345–347, 570–573, 716–722
  map, 360
  mapping, 355–364, 367–368, 372–378
  mismatch repair, 554–556
  morphogenesis, 360, 396
  recombination, 355–368, 559–561, 570–573
  repressor, 716–722
  restriction digest, 111
  specialized transduction, 372–375
  sticky ends, 395–396
  virulent, 717–718
Lampbrush chromosomes, 277–279, 754
Langly, C., 800
Larva. *See* Embryogenesis.
Larval chromosomes. *See* Polytene.
Late genes, 712–713
Lateral component (element), 130
Lawn, bacterial, 205
Lawrence, P., 771
Leader sequence, 265, 267, 310
Leaky, 317
Leder, P., 274, 282, 625, 640
Lederberg, E., 405
Lederberg, J., 405
Leigh Brown, A., 800
Leptonema, 126–130
Lerner, I. M., 675
Lesch-Nyhan syndrome, 506, 594
  heterozygote detection, 595
Lethal mutation, 207–208
Leucine operon, 643–644
Levene, H., 874
Levine, L., 874, 875
Levins, R., 874
Lewis, E. B., 773
Lewontin, R., 800, 874
Lex A protein, 251
Library, genomic, 121

Ligase, 117, 119, 236–237
  recombination role, 548
  repair role, 248
Light
  chain g antibody, 621
  strand of DNA, 114, 540
*Limnea*, 755–756
Lin, J. K., 584
Linkage, 341, 492
  disequilibrium, 806, 822, 831, 864
  group, 456–459
  partial, 438
Load, mutational, 797–798
Locus, genetic, 158, 353
Low, B., 408
Luria, S., 211
Luxuriance, 837–838
Lyon, M., 780, 788
Lyonization, 780–783
Lysogeny, 345–346, 570–573, 716–722
  abortive, 367
Lysosomal storage disease, 594
Lysozyme, 326, 342
Lytic infection cycle, 342–345

M protein, 646. *See also Lac* operon.
MacLeod, C. M., 18
Macronucleus, 57, 142–144
Mahowald, A. P., 757
Maize. *See* Corn.
Major groove (DNA), 12
Malaria, 170, 795–796, 840–842
Male-specific phage, 427
McBride, O. W., 498
McCarty, M., 18
McClintock, B., 584, 589
Malignancy. *See* Carcinogenesis.
Map, 341. *See also* Mapping.
  cytological, 450–456
  deletion, 368–372
  denaturation, 375–376
  distance, 358, 450, 489–490
  fate, 762–764
  fine-structure, 368–372, 466–472
  gene product, 470, 509
  heteroduplex, 375–378, 381, 587
  hybridization, 520–521
  peptide, 617
  physical, 375–378
  R-loop, 375, 377
  restriction, 289, 378–381, 520, 523–524
Mapping bacteria
  conjugation, 405–412
  generalized transduction, 417–421
  specialized transduction, 289, 421–425
  transformation, 413–417
Mapping eukaryotes, 444–472
  cytological, 450–456
  human, 492–506, 524
  restriction, 291

Mapping eukaryotes (*Continued*)
  somatic cells, 482–502
  tetrad analysis, 459–472
Mapping organelles
  chloroplast, 527–528
  mitochondria, 520–527
Mapping viruses, 341–389
  deletion, 367–372
  physical approaches, 375–378
  recombination frequency, 352–367
  restriction enzymes, 378–385
  transduction, 372–375
Marien, D., 874
Marker
  effect, 169, 556–557
  genetic, 353
  inside, 545
  outside, 545
  rescue, 397
Maternal
  age, 148–149
  effect mutation, 759–760, 771
  influence, 755–760, 770–771
  inheritance, 523–524, 527–531
  ribosomes, 753
  RNA, 296–297, 753–755
Mating
  assortative, 807
  preference, 863
  random, 807
  type, 139, 141–142
Maturase, 522
Maxam, A., 113
Mealybug, 780
Mean, 670
Megaspore, 138
Megasporophyte, 138
Meiosis, 125–150
  errors during, 145, 151–152, 219–220
  Mendel and, 159–160, 170–172
  mutations affecting, 548
  nondisjunction, 145–146, 151–152
  recombination during, 561–565
  regulation, 752
  segregation bias, 169
Melanin synthesis, 659–660
Mendel, G., 154
Mendelian
  genetics, 154–196
  population, 802–813
Mendel's Laws
  First Law, 159–170
  Second Law, 170–182
Merozygote, 404
Meselson, M., 29–32, 59, 539, 554, 559, 740
Meselson-Stahl experiment, 29–32
Mesosome, 27
Messenger RNA (mRNA), 15, 260–261, 264–281
  artificial, 321–322

hnRNA relationship, 272–274
leader sequences, 265, 267, 310
maternal, 754–755
pre-mRNA, 272–276
processing, 260, 272–277
protein association, 280–281
protein synthesis role, 261
terminator sequence, 260, 264–266, 270–272
trailer sequence, 265
turnover, 273
Metacentric chromosome, 48
Metaphase
  meiotic, 134–135
  mitotic, 41–48
  plate, 45
Metastasis, 224
Methionine synthesis, 656–657
Methotrexate, 215, 607
Methylase, 786
Methylating enzymes, 112–113, 785–786
  gene regulation by, 786
5-Methylcytosine, 785–786
7-Methylguanidine cap, 266–267, 310
Methylmethane sulfonate (MMS), 225, 239–240
MHC. *See* Histocompatibility.
Microcephaly, 165
Micrococcal endonuclease, 75–78
Micronucleus, 57, 142–144
Microsporophyte, 138
Microtubule, spindle, 44
Middle genes, 712
Miescher, F., 17
Migration, 811
  pressure, 842–846
Miller, D. A., 459
Miller, J., 689
Miller, O. J., 459
Minor groove (DNA), 12
Minority area, 766
Mintz, B., 790
*Mirabilis jalapa*, 528
Mirsky, A., 17
Mismatch, 546
  repair, 546, 554–565
Mispairing, 238
Mitochondria, 35–36, 512–517
  code differences, 328–330, 332
  DNA, 513–514, 870–872, 876–877
  genetics, 515–527
  mapping, 520–527
  maps, 521, 524
  protein synthesis, 513, 517, 519
  restriction analysis, 870–872, 876–877
Mitomycin C, 220, 223, 239, 722
Mitosis, 41–48
  chromosome condensation, 88–89, 108
  crossing over during, 486–487, 508–509, 768

Mixed infection, 354
Miyata, T., 869
MMS, 225, 239–240
MNNG, 223, 239–240
Modifier locus, 800, 864
Modification enzymes, 112–113
Modified dihybrid ratio, 658–659
Molecular evolution, 858
  gene, 869–870
  gene clusters, 648–650, 731
  gene families, 305, 606
  hemoglobin, 608–609, 867–868
  introns, 276–277, 581–582, 635
  karyotype, 864–867
  operon, 646–647, 649
  organelles, 531–532
  polypeptide, 867–869
  satellite DNA, 581–582
  selfish DNA, 581–582
  spacer DNA, 304–305, 581–582
  X chromosome, 182–183, 192–194
Monod, J., 684
Monoecious plant, 150
Monomorphic locus, 799
Monosomic, 145–149
Monozygotic twins, 602, 671, 674
Moore, C., 509
Morata, G., 771
Morgan, L. V., 194, 196
Morgan, T. H., 196, 352, 429
Mosaic
  *Drosophila*, 762–774
  mammal, 780–783
  position effect, 782–783
  recombinational, 768–774
  sexual, 762–767
Moth, Peppered, 828–829
Mourat, A. K., 875
Mouse
  coat color, 658–660
  constitutive heterochromatin, 89, 107–108
  embryogenesis, 774–776
  histocompatibility locus, 650–652
  karyotype, 51, 54
  map, 462
  mosaic, 780–783
  nuclear transplantation, 747–749
  satellite DNA, 107–108
  specific locus test, 221–222
  *T* locus, 774–776
  tail length, 774–776
mRNA. *See* Messenger RNA.
MS2 phage, 381–383
  overlapping genes, 383–384
Mu phage, 573–576, 590
Muller, H. J., 227
Müller-Hill, B., 689, 705
Müllerian duct, 777–779
Multiplicity of infection, 355
*Mus. See* Mouse.

Muscular dystrophy, 477–478
Mutagens, 220–227
testing, 221–224
Mutagenesis. *See also* Mutation.
direct, 237–245
*in vitro*, 242–246, 270
misrepair and, 245–254
Mu phage, 590
site-directed, 242–246, 270
topography, 370–372
transposable elements, 584–589
Mutant, 202
Mutation, 3, 202–227, 237–254. *See also*
Mutagenesis.
addition, 202–203, 249–250, 318–320
allelic, 470
*amber*, 330–332
antimutator, 252–253
cancer and, 224–226
chromosomal, 202–203, 212–220,
584, 588, 776, 782–783
cold-sensitive, 209
conditional, 208–210
deficiency. *See* Deficiency.
deletion. *See* Deletion.
duplication. *See* Duplication.
equilibrium, 812, 814–815
forward, 202, 221
frameshift, 250, 318–320, 326
genomic, 202–203
homoeotic, 771–774
hot spot, 371–372
inversion. *See* Inversion.
*in vitro*, 242–246, 270
lethal, 207–208
load, 797–798
missense, 330
mutator, 252, 584, 590
neutral, 796, 813, 853–855, 867–870
nonsense, 330–335
nutritional, 205–207
*ochre*, 330–332
*opal*, 330–332
point, 202–203
polar, 331–332, 586–588
pressure, 798–799
rate, 221–224, 798–799
reverse, 202–203
somatic, 225, 637–639
spontaneous, 220, 231, 372, 586, 797
sterile, 208
substitution, 202–203
suppressor, 203–204, 318–320, 332–
335
synonymous, 323, 853, 869–870
temperature-sensitive, 209
translocation. *See* Translocation.
visible, 204–205
Mutator activity, 252, 584, 590
Myoglobin
evolution, 608, 867–868

structure, 71

N protein, 714–716
Nasmuth, K. A., 502
Nathans, D., 109, 378
Natural selection, 823
Neal, N. P., 816
Negative control, 687
Neocentromere, 169
*Neurospora*
biochemical genetics, 655–656
gene clusters, 647–649
gene regulation, 729–731
heterokaryon, 483
life cycle, 141–142
map, 469
mapping, 466
morphology, 204–205
somatic fusion, 483
tetrad analysis, 168–169
Neutral
allele, 796, 813, 852–855
mutation, 852–855, 869–870
theory, 852–855, 869
Neutral Mutation-Random Drift
Hypothesis, 852–855, 869
Nick translation, 244
*Nicotiana tabacum*, 490, 492, 667–668
Nilsson-Ehle, H., 663
Nirenberg, M., 321
Nitrogen mustard, 220, 722
Nitrosoguanidine (MNNG), 223, 239–
240
Nitrous acid, 220, 239–241
Nomadic element, 580
Nomura, M., 335, 704
Noncoding DNA, 264, 272–277
Nondisjunction
mitotic, 485–486
primary, 145–146
secondary, 151–152
Nonhistone proteins, 73–74, 787
regulation and, 787
Nonleaky, 317, 330
Non-Mendelian genetics, 512, 516
Nonparental ditype, 180
Nonsense mutations, 330–335
Norby, S., 649
Norkin, L., 557
Normal distribution, 670
*Notch* locus, 791
Nuclear
envelope, 35, 41, 89, 91, 130, 135
transplantation, 747–749, 752
Nucleic acid, 7
Nucleohistone, 52
Nucleoid, 27
Nucleolar organizer
amplification, 753–754

chromosome location, 297
Nucleolus, 35, 36, 263, 296–299. *See also*
Ribosomal RNA.
in mitosis, 41
mutation in *Xenopus*, 296–297
oocyte amplification, 753–754
RNA polymerase of, 263
Nucleoside, 5
modification, 293
Nucleosome, 74–81
breathing, 81
core particle, 77–78
phasing, 81, 787
transcription, 270
Nucleotide, 5
complementary, 10
composition, 14
pair, 9
sequence, 14
Nucleotidyl transferase, 293
Nucleus, 35–36
DNA content, 94
envelope, 35, 41, 89, 91, 130, 135
polar, 138
pore, 35
NusA protein, 262–263
Nutrient agar, 205


*Ochre* mutation, 330–332
suppressors, 332–335
Ohno, S., 774
Ohta, T., 855
Okazaki, R., 233
Okazaki fragments, 233–237
Oligonucleotide, 5
Olins, A., 91
Olins, D., 91
Ommochrome synthesis, 657–658, 661
One-factor cross, 355–356
Oocyte, 135, 137
cytoplasm, 752, 755–760, 770–771
differentiation, 752–760, 770–771
meiosis, 133–136, 752
nuclear transplantation, 747–749, 752
RNA, 296–297, 753–755
transcription factor, 753
Oogenesis, 133–138, 149
Oogonium, 137
*Opal* mutation, 330–332
suppressors, 332–335
Operator, 260, 684
constitutive, 692, 698
*lac*, 692–694
λ, 717–719
*trp*, 698–699
Operon, 260, 643–647, 684
inducible, 686–687
repressible, 686, 698
Organelle, 35, 511–532

Organelle (*Continued*)
  evolution of, 531–532
Orgel, L. E., 581
Outcrossing, 807–808
Ovalbumin gene, 301, 738
Overdominance, 837–838
Ovum, 136, 137

*p* arm of chromosome, 55–56
P1 phage, 418–420
P22 phage, 418
Pace, N., 22
Pachynema, 132
Packaging phage, 345, 367, 381, 418
Palindrome, 105, 109
Palmiter, R., 738
Panmyxia, 807
*Paramecium*, 57
  cortical inheritance, 533–534
  doublet, 533
  endosymbionts, 531–532, 535
  killer, 535
  life cycle, 142–144
  nuclei, 57
  preformed structures, 533
Parasexual cycle, 487
Pardue, M. L., 122, 737
Parental ditype, 180
Paris Conference nomenclature, 55–56
Partial
  diploid, 422–423, 689
  heterozygote, 206, 553
  linkage, 438
  trisomy, 220
  zygote, 404
Path analysis, 809–811
Pattern baldness, 789
pBR322 vector, 113–119
  restriction map, 116
PBS1 phage, 418–419
Pea
  genetic analysis, 154–157, 177–179
  sweet, 438–439
Pedigree analysis, 164–166, 190–192,
  523, 870–872
Penetrance, 602–604
Penicillin enrichment, 207
Peptidase A, 600–601
Peptide, 65–71
  bond, 66
  mapping, 617
Peptidyl transferase, 312–313
Permissive growth conditions, 208
Pero, J., 712
*Petite* yeast, 515–518
  neutral, 516
  nuclear, 516
  segregational, 516
  suppressive, 517
*Petunia*, 490

Phage. *See* Bacteriophage.
Phasing, nucleosome, 81, 787
Phenotype, 158
Phenylketonuria (PKU), 593–594
  heterozygote detection, 595
φ80, 345
φX174, 342
  codon utilization, 327, 338
  genome, 384
  mapping, 397
  overlapping genes, 383–385
  reading frame, 383–384
Philadelphia chromosome, 213
Phosphodiester bond, 7
Phosphoglucomutase, 604–605
Photoreactivation, 274
Phototaxis, 831–832
Pilus (pili), 400–401, 427–428
Pingelap population, 847–848
*Pisum sativuum*, 154–157, 177–179
pKM101 plasmid, 222
Plant
  breeding, 149–150, 525–526, 664–666,
    672–674, 833–838
  cells in culture, 490–492
  maternal inheritance, 528–531
Plaque, 205–206, 717
Plasmid, 27, 425–531
  cloning use, 113–119
  Col, 426, 431
  conjugal, 425
  corn, 527
  crossing over in, 542–543
  episomal, 425
  F, 400–404, 427–428
  mobilization of, 426
  nonconjugal, 426
  nontransmissible, 426
  pBR322, 113–119
  pKM101, 222
  pMB9, 543
  R, 425–431
  recombinant, 113–119
  table, 425
  transmissible, 426
  yeast, 502–503
Pleiotropic-negative, 730
Pleiotropy, 594, 653–654
pMB9, 543
Pneumococcus. *See Diplococcus
  pneumoniae.*
Polar
  body, 135–136
  cells, 757–759
  effect, 586–587
  granules, 757
  nuclei, 138
  plasm, 757–759
Polarity, 330–331, 586–587
Polivanov, S., 874, 875
Pollen, 125, 139

gene expression in, 166, 170
Polyadenosine (poly A), 272
Polyclone, 767–774
Polydactyly, 603
Polyethylene glycol, 491
Polygene, 664, 823
Polygenic inheritance, 664–674, 823
Polymerase
  DNA. *See* DNA polymerase.
  RNA. *See* RNA polymerase.
Polymorphism, 793, 799–802, 815–816,
  838–839, 851–855
  balanced, 812
  stable, 812
Polynucleotide, 5
Polyoma, 346
Polypeptide. *See* Protein.
Polyploidy, 149–150
  speciation and, 861
Polyribosome, 309, 314
Polysome, 309, 314
Polytene chromosomes, 57–59, 83, 88,
  91, 123
  mapping, 450–456
  mutations, 214
  puffing, 279, 734–737
Pongs, O., 737
Popp, R., 608
Population, 793–794
  allopatric, 811
  effective size, 846
  genetics, 792–874
  Mendelian, 802–813
  small, 846–851
  splinter, 846–847
  stable, 796
  sympatric, 811
Population cage, 829
Positional information, 770
Position effect, 781–783, 867
Positive control, 687, 695–697, 714–716,
  729–731
Post-translational modifications, 800
Postzygotic isolation, 859–861
Potter, H., 542
Powell, J., 815, 850, 875
Precipitin test, 46
Premutational lesion, 241
Prenatal detection, 595–596
Prezygotic isolation, 859–860
Pribnow box, 266
Primase, 233
Primate evolution, 51, 865–866, 870–
  874, 878
Primer strands for DNA replication, 233,
  235
Probability concepts, 176, 362, 415, 448
Processing RNA, 260
  mRNA, 260, 272–277
  rRNA, 295, 299
  tRNA, 289, 293–294

Proflavin, 249
Prokaryote, 27
Promoter, 260, 264–266, 267–270, 683–684
  internal, 291–292, 300, 739–740, 753
  low-level, 695–696
  mutation, 269
  regulation and, 684, 693–696, 712–717, 723, 739–740, 753
Proofreading enzymes, 236–238
Prophage, 345, 722
Prophase, meiotic, 126, 135
  mitotic, 41–48
Protein
  structure, 67–73
  synthesis, 309–316, 321, 516
Protoperithecium, 141
Prototrophy, 206
Prout, T., 874, 875
Provirus, 576
Pseudodominance, 451
Pseudogene, 613
Pseudo-wild, 318, 326
Ptashne, M., 717, 725
Pterin synthesis, 661–663
Puck, T., 491–506
Puffing, 279
  heat shock, 734–735
  hormones and, 736–737
Pulse-chase experiment, 237
Punnett, R., 173, 438
Punnett Square, 173–174
Purine, 4
  biosynthesis, 493–494
Pyrimidine, 4
  biosynthesis, 649–650
  dimer, 246–247

q arm of chromosome, 55–56
Q band, 53
Q protein, 714–716
qa gene cluster, 729–731
Qβ, 23, 381
Quantitative genetics, 664–674, 823
Quinacrine mustard, 53
Quinic acid, 729–731

R17, 336, 381
rII gene of T4, 205–206, 368–372
  complementation analysis, 348–351
  frameshifts in, 317–320
R band, 53
R-loop mapping, 274–275, 375, 377
R plasmid, 425–531
  multiresistance, 430–431
Race, human, 674, 794, 840–841, 844–846, 873–874
Radding, C. M., 539

Random walk, 852
Reading frame, 319–320, 383–385
  open, 522
Rearrangements, chromosomal, 202–203, 212–220, 588, 776, 782–783
Reassociation kinetics, 98–106
rec genes, 251–252, 548–550
Rec A protein
  independence, 570, 578
  recombination, 548–550
  repair, 251–252
  repressor proteolysis, 722
RecBC protein, 548, 570
Recessive, 155, 159, 592
Reciprocal cross, 185
Recognition, 539, 549–550
Recombinant DNA technology, 108–122
Recombination, 3, 130, 539–565, 569–589. See also Crossing Over
  Integration.
  chiasmata and, 441–444
  contrast with complementation, 349
  double, 361–363, 557–561
  enzymology, 547–550
  frequency, 354, 456–457
  general, 539–565
  heteroallelic, 470
  illegitimate. See Integration.
  immunoblobulin genes and, 625–637
  interallic, 470
  mechanisms, 539–565
  meiotic, 130–131, 561–565
  mitotic, 486–490, 508–509
  nonreciprocal, 543–545
  pairing of chromosomes, 108, 130–131
  percent, 357
  reciprocal, 539
  site-specific, 503, 570–573
  suppression, 216–217, 456–457, 776
Reed, S. I., 502
Regulated gene, 682–683
Regulation
  attenuator, 699–703
  autogenous, 699
  bacterial, 687–704
  eukaryotic, 728–741, 745–787
  general features, 682–686
  phage, 709–722
  transcriptional, 683
  translational, 683
  viral, 723–725
Regulator protein, 684
Regulatory gene, 685, 688–691, 698–699
Relative fitness, 820–823
Relaxed strains, 708
Release factor, 314–315
Renaturation of DNA, 98
Repair, 245–254, 546–558, 554–565
  error-free, 245
  error-prone, 245
  excision, 248–249

genes involved, 251–254
  mismatch, 546, 554–565
  photoreactivation, 247
  recombination and, 546–548, 554–565
  SOS, 251–252, 722
  tract, 554–556
Replica plating, 120, 166–167
Replicase, 232. See DNA polymerase.
Replication. See DNA replication.
Replicon, 38, 82–83
Repressible operon, 686, 698
Repression, translational, 704
Repressor
  lac, 688–694
  λ, 716–722
  Mu, 573
  transposable element, 579–580, 585–586
  trp, 698–699, 702–703
Reproductive isolation, 859–861
Repulsion, 356
Resistance plasmid. See R plasmid.
Restriction endonuclease, 109–113
  applications, 378–381, 520, 523–524, 749, 786, 870–872
  digest, 112
  fragment, 109
  table, 110
Restrictive growth conditions, 208
Retrovirus, 576–577, 627
Reverse transcriptase, 274, 576
Reversion, 202–204
Rh blood group, 616, 793–794
Rho protein, 271, 331
Rho⁻ yeast, 515–518
Rho° yeast, 517
Rhoades, M., 529
Ribonuclease (RNase), 293
  processing role, 293
Ribonucleic acid. See RNA.
Ribonucleoprotein particle (RNP), 280–281
Ribose, 5
Ribosomal RNA (rRNA), 260, 294–300, 607
  chloroplast, 527–528, 532
  complementary to leader, 267
  evolution, 607
  gene location, 297, 300
  genes, 294–300, 607
  mitochondrial, 513, 522
  oocyte, 753–754
  pre-rRNA, 295, 299
  processing, 295, 299
  regulation, 739–740, 753
  synthesis, 263, 295–296
Ribosome, 26, 294, 309–316. See also Ribosomal RNA.
  70S, 294
  80S, 294
  binding sequence, 266–267, 310, 704

Ribosome (*Continued*)
  chloroplast, 527
  control role, 683, 704, 754
  maternal, 753–754
  mitochondrial, 513
  prokaryotic vs. eukaryotic, 294
  protein synthesis role, 309–316
  proteins, 316, 704
  subunits, 294, 310–316
Richmond, R., 850, 875
Rifampicin, 262, 513
Riggs, A., 38–39, 59
Ring X, 763–764
Ris, H., 17
Ritossa, F., 734
RNA, 1
  double helical, 11, 287, 381
  heterogeneous nuclear, 272–274
  maternal, 296–297, 753–755
  messenger. *See* Messenger RNA.
  phage, 381–384, 400
  polymerase. *See* RNA polymerase.
  ribosomal. *See* Ribosomal RNA.
  small nuclear, 276
  transfer. *See* Transfer RNA.
RNA-dependent DNA polymerase. *See*
  Reverse transcriptase.
RNA polymerase, 260–263, 267–272,
  710
  I, 263
  II, 263, 270
  III, 263, 291–293, 300, 739–740, 753
  bacterial, 262–263, 268–271, 715
  core enzyme, 262
  genes encoding, 262–263
  mitochondrial, 513
  phage, 710
  regulation and, 712–713, 715–716
  subunits, 262–263, 712–713
Robash, M., 754
Robertsonian rearrangement, 865
Roeder, R., 739
Rolling-circle replication, 344–345, 401–
  402, 754
Roman, H., 564
Ronen, A., 567
Rosenberg, M., 716
RuBP carboxylase, 527–528
Ruddle, F., 496, 498
Ruderman, J., 754
*Rudimentary* locus, 649–650, 790–791
Russell, L. B., 221
Russell, W. L., 221
Rye, 89–90


Saccharomyces. *See* Yeast.
Sack, G. H., 378
Saedler, H., 586
Sager, R., 534

Salamander, 94, 581
Salivary gland chromosomes. *See*
  Polytene chromosomes.
Salmon, 854
*Salmonella typhimurium*
  mutagen test, 222, 225–226
  operons, 643–647
  transduction, 418
Salts, Y., 567
Sanger, F., 114, 525
Sapienza, C., 581
Satellite DNA, 96–98, 102–103, 107–
  108, 299, 581
Saunders, E., 438
Scaffold, chromosome, 86–87
Schizophrenia, 602
Schramm, G., 22
Screening procedures
  HAT selection, 493–494
  mutants, 204–212
  recombinant clones, 119–122
  SLRL, 208
Sea urchin, 305, 751–752, 754
Sedimentation coefficient, 274
Segregation
  abnormal, 169–170
  bacterial, 27
  distorter, 170
  meiotic, 159–170
  mitotic, 485
  post-meiotic mitotic, 565
  Principle of, 160
  recombination and, 546, 552–553
  second-division, 167–169
Selection
  artificial, 832–838
  balancing, 813
  coefficient, 821, 827–828
  directional, 823–838
  disruptive, 838–839
  diversifying, 838–839
  natural, 823
  stabilizing, 812–813, 839–842
  truncation, 833
Selective value, 796
Selector gene, 770–774
Selfing, 807–808
Semiconservative DNA replication, 32
Semi-dominant, 188
Sendai virus, 491
Sense strand, 263–266
Sequencing
  DNA, 113–115
  protein, 67–71
Setlow, R., 248
Sex chromosomes
  birds, 183
  evolution of, 182–183, 192–194
  mammalian, 50, 145–146, 182–184,
    192–194
Sex determination

birds, 183
  *Drosophila*, 183
  genital-duct, 777–779
  gonadal, 777
  human, 183, 777–779
Sex element. *See* F element.
Sex-limited inheritance, 789, 790
Sex-linked
  inheritance, 182–196
  recessive lethal screen, 208–209
Sexduction, 423–425
Shapiro, J., 434, 586
Sherman, F., 334, 466, 475, 509
*Shigella*, 431
Shimke, R., 215
Shine-Dalgarno sequence, 266–267, 310,
  704
Shoenheimer, R., 17
Shull, G. H., 835
Shuttle vector, 502–503
Sickle-cell, 794–796, 839–842
  anemia, 795, 840
  hemoglobin S, 839–840
  heterozygote, 795, 841–842
  populations and, 745–746, 840–842
  trait, 840
Sidley, C., 738
Sigma
  factor, 262–263, 712–713
  virus, 531
Signal sequence, 301–302, 314
Signer, E., 419
Simmons, M. J., 584
Simpson, G. G., 872
Singer, B., 22, 23
Single-strand binding protein (SSB), 237,
  548–549
Sister-chromatid exchange, 222–223
Site-directed mutagenesis, 243–246
Site-specific recombination, 503
Skin color, 668–670, 680
Small nuclear RNA, 276
Smith, H. O., 109
Smith, M., 472
Snail shell coiling, 755–756
Snell, G., 675
Somatic cell, 124
  fusion, 491
  genetics, 482–506
  hybrid, 490, 492–498, 783
  interspecific, 490, 492–498
  intraspecific, 490–492
  line, 137, 490
  mutation, 225, 637–639
  recombination. *See* Mitosis.
  transformation, 498–503
Somatic pairing, 58, 88
Sonneborn, T., 533–534
Sorting out, 529
SOS repair, 251–252
Southern Blot, 380–381

Southern, E. N., 381
Spacer DNA, 289
  evolution, 304–305, 581–582
  non-transcribed, 299, 304–305
  transcribed, 289
Spassky, B., 831
Specialized transduction, 372–375
Speciation, 859–864
  allopatric, 862–864
  allopolyploid, 861
  founder principle, 864
  geographic, 862
  hybridization, 861–862
  molecular, 862–870
  primate, 865–866, 870–874
Species
  semi, 863
  sibling, 862, 874
  sub, 862
Specific locus test, 221–222
Spermatogenesis, 136, 137
Spermatogonium, 137
Spermatozoa, 137, 170
Spiegelman, S., 22, 107
Spiess, E., 875
Spindle, 44–47
Splicing genes, 275–276, 627–635
Split gene, 272–277
SPO1 phage, 418, 712–713
Spontaneous mutation, 220, 231, 372,
    586, 797–798
  rate, 231, 797–798
Spore, 138
Sporophyte, 138
Sporulation, 142
Spreading effect, 782–783
SSB protein, 237, 548–549
Stable transformant, 498–499, 503
Stadler, L. J., 227
Staggered nicks, 109, 575–577
Stahl, F., 29–34, 59, 550
Standard deviation, 670–671
Starlinger, P., 586
Steitz, J., 282
Stem-and-loop, 106, 270–271, 627–628
Steriles, 208
Stern, C., 150, 443
Steroid hormones, 736–740, 779
Stiff little finger, 602
Strand
  heavy, 520
  light, 520
  switching, 714, 724, 727
Streisinger, G., 249–250, 364
Streptococcus, 413
Stringent control, 707–708
Sturt, 763
Sturtevant, A. H., 352, 440, 763
Substitution, 202–203
Sucrose gradient centrifugation, 273–274
Sueoka, N., 417

Supercoil, 80, 237, 573
Supergene, 822, 831
Superhelical DNA, 77, 80, 237, 573
Superinfection, 721
Suppression, 203–204, 318–320, 332–
    335
  bacterial, 332–334
  frameshift, 318–320
  intergenic, 332
  intragenic, 332
  nonsense, 332–335
  temperature-sensitive, 334
  yeast, 334–335
Suppressor mutation. See Suppression.
Survival value, 832
Sutton, W., 438
SV40
  chromosome, 347
  codon preference, 327
  infection cycle, 346–347
  integration, 576–577
  map, 379, 380, 382
  origin of replication, 723, 787
  overlapping genes, 383–385
  regulation, 723–725
  restriction enzyme analysis, 109–112,
    378–381
  shuttle vector, 503
  transformation, 346–347, 576–577
Svedberg unit (S), 273
Sympatric, 811
Synapsis, 130–133
  aberrant, 213–217
Synaptonymal complex, 130–133
Synkaryon, 483, 491
Synonymous mutation, 853, 869–870
Syntenic, 492, 498


T2 phage, 19–22, 205–206, 343–344
  chromosome, 343, 344
T4 phage
  chromosome, 365–366, 727
  map, 366
  mapping, 364–367
  morphogenesis, 660–662
  r mutant strains, 205–206, 317–320,
    348–351
  rII fine structure, 368–372
  recombination, 364–367
  regulation, 712
T7 phage
  chromosome, 344
  infection cycle, 345
  map, 711
  promoters, 268
  regulation, 710–712
  RNA polymerase, 710–711
T antigen, 347, 382, 385, 723–725
T locus, 774–776

Tandem genes, 606–614, 619, 635, 651
Taster trait, 804
TATA box, 270, 740
Tautomeric shift, 4, 219–220
Taylor, J. H., 222
Tay-Sachs disease,
  heterozygote detection, 595
  population genetics, 817–818
Telomere, 89, 108, 130, 866
Telophase
  meiotic, 135
  mitotic, 48
Temperate phages, 343, 345–346. See
    also Lambda, Mu.
Temperature-sensitive mutant, 209
Template, 15
Templeton, A., 864
Terminal repetition, 365
Terminalization, of chiasma, 134
Termination
  transcription, 260, 264–266, 270–272,
    700–703
  translation, 314–315, 322–323
Terminator
  codon, 314, 322–323
  transcriptional, 260, 264–266, 270–
    272, 700–703
Terzaghi, E., 326
Test cross, 162, 445–446
Testicular feminization syndrome, 779
Testosterone, 779
Tetracycline resistance, 429–430
Tetrad
  meiotic, 135
  ordered, 168
Tetrad analysis, 166–169, 179–182, 459–
    466
  conversion in, 561–565
  mapping by, 459–466
Tetrahymena, 57, 142
Tetraploidy, 149–150
Tetrasomic, 145
Tetratype, 180
Tfm, 779
Thalassemia, 198, 275, 335
Thiamin synthesis, 655–656
Thiogalactoside transacetylase, 646
Thomas, C. A., Jr., 122
Three-factor cross, 419–420, 446–450,
    464–466
Threonine synthesis, 656–657
Thymidine kinase, 493–494
Thymine 4
  dimer, 246–247
Tn. See Transposon.
Tobacco, 490, 492, 667–668
Tobacco mosaic virus, 22
Tompkins, G., 738, 747
Tonegawa, S., 625, 640
Topoisomerase, 237, 573
Trailer sequence, 265

Trait, 158, 591
  composite, 661
  quantitative, 664
Transcription, 258–281
  control, 683, 753
  selective, 750–752
Transducing particle, 418
Transduction
  generalized, 417–421
  high-frequency, 423
  specialized, 289, 372–375, 421–425
Transfection, 354, 499–502
Transfer RNA (tRNA), 260, 286–294
  charging, 310–311
  chloroplast, 527, 532
  genes, 286–294
  initiator, 313, 327–328
  isoaccepting, 286, 325
  mitochondrial, 329, 513, 525
  pre-tRNA, 289, 293–294
  processing, 289, 293–294
  structure, 287, 310
  suppressor, 289, 333–335
Transformation
  bacterial, 17–19, 413–417
  eukaryotic, 57, 498–503, 780
  recombination mechanism, 543–545,
    550–553
  SV40, 346–347, 576–577
Transition, 238
Translation, 261–262, 309–316
  control, 683, 704, 754
  coupling to transcription, 280
  genes involved, 316
  repression, 704
Translocase, 313
Translocation, 202–203, 217–220, 242
  balanced, 218
  Down's syndrome, 219
  heterozygote, 217
  mapping by, 459, 495–496
  overlapping, 459
  position effect and, 782–783, 867
  protein synthesis, 313
  reciprocal, 217
Transplantation
  imaginal discs, 767
  nuclear, 747–749, 752
Transposable element, 569, 627
  eukaryotic, 580–586
  prokaryotic, 577–580, 586–589
Transposase, 578, 580
Transposon (Tn), 429, 578–580, 627
  structure, 429
Transversion, 239
Trihybrid cross ratio, 177, 663
Triplet, 309, 320
Triploidy, 149–150
Trisomy, 145–149
  double, 145
  partial, 220

Tritium suicide, 228
Trophectoderm, 747
Truncation
  point, 833
  selection, 833
Tryptic peptides, 67, 617
Tryptophan (*trp*)
  mutagenesis, 325–326
  operon, 643–644
  regulation, 697–703
  synthesis, 643–644, 653–654
Tsugita, A., 326
Tumor-specific antigens, 790
Tumor virus. *See* Retrovirus, SV40.
Turner's syndrome, 145–146, 183–184
Twins, monozygotic, 602, 671, 674
Two-factor cross, 356–359, 419, 445–446,
  459–464
Tzagaloff, A., 518

U-1, 276
*Ultrabar*, 617, 791
Ultraviolet (uv), mutagenesis, 220, 242,
  722
Undermethylation, 786
Unequal crossing over, 610, 612, 791
Uniparental inheritance, 512, 537
Unscheduled DNA synthesis, 249
Unstable
  element, 582–584
  transformant, 498
Unwinding protein, 237, 548
  recombination role, 548
Upstream, 265
Uracil, 4
*Ustilgo*, 564
*uvr* genes, 251

VanValen, L., 874
Variable region of antibody, 621–624
Variance, 671–673
  environmental, 672
  genetic, 671
  phenotypic, 671
Variegated plant, 529, 582–584
Variegation, position effect, 782–783
Vector
  cloning, 113–117
  shuttle, 502–503
Vegetative cells, 124, 139
Virus, 314–385. *See also* Bacteriophage.
  animal, 109–112, 327, 378–385, 576–
    577, 627, 723–725
  plant, 22
Vitamin C, 853

von Wettstein, D., 150

W chromosome, 183
Wagner, R., 554
Wallace, A., 823
Wallace, B., 874
Watson, J. D., 8
Watts-Tobin, R., 317
Weatherall, D. J., 614
Weinberg, R., 499
Weinberg, W., 802
Weintraub, H., 785
Weissman, S. M., 380
Westergaard, M., 150
Wheat kernel color, 663–665
Wickner, S., 254
Wild type, 158, 202
Wildenberg, J., 554
Wilson, A. C., 872
Wilson, E. B., 752
Witkin, E., 251, 254
Wobble hypothesis, 323–325, 327
Wolffian duct, 777–779
Wollman, E., 405, 570
Wood, W., 660
Wu, C., 786

X chromosome, 50, 182–184, 192–194
  aberrations, 145–146, 183–184, 781
  attached, 194–195, 566
  evolution, 182–183, 192–194
  ring, 763–764
X ray
  mitotic recombination induced, 508–
    509, 768
  mutations induced, 220, 222, 242
Xanthine dehydrogenase, 567
*Xenopus laevis*
  nuclear transplantation, 747, 752
  nucleolar mutant, 296–297, 747
  oocyte, 747, 752–754, 756–757
  rRNA genes, 299
  tRNA genes, 291–292
Xeroderma pigmentosum, 253–254
XYY genotype, humans, 145, 149, 183–
  184

Y chromosome, 50, 182–184, 186, 777
  H-Y antigen, 777
  heterochromatin, 89–90, 103, 108
  XYY, 145, 149, 183–184
Yang, V. W., 276
Yanofsky, C., 325, 335, 698, 701
Yeast
  bud position effect, 537

Yeast (*Continued*)
  conversion, 561–565
  cytochrome *c* gene, 334, 466–472
  *GAL* regulation, 731–732
  gene clusters, 647–649
  *grande*, 515
  *H1S4* locus, 647–649
  life cycle, 142–143, 537
  map, 468
  mapping, 459–472
  mitochondrial mutant, 515–523, 537
  mitotic recombination, 508–509
  *petite*, 515–518
  respiratory deficient, 515–518

  *rho*⁻, 515–518
  shuttle vector, 502–503
  sporulation, 142
  suppressor genes, 334
  transposable element, 581
  tRNA genes, 293
  uniparental inheritance, 537
Yoon, S., 498
Yoshikawa, H., 417


Z chromosome, 183
*Zea mays. See* Corn.

*Zeste-white* locus, 452–454
Zygonema, 130–132
  bouquet, 132
Zygote, 50, 139, 747–748
  partial, 404
Zygotic
  induction, 722
  lethal, 790
Zubay, G., 698